High Performance Computing in Science and Engineering '08

Wolfgang E. Nagel · Dietmar B. Kröner ·
Michael M. Resch
Editors

High Performance Computing in Science and Engineering '08

Transactions of the High Performance Computing Center Stuttgart (HLRS) 2008

Springer

Wolfgang E. Nagel
Zentrum für Informationsdienste
und Hochleistungsrechnen (ZIH)
Technische Universität Dresden
Helmholtzstr. 10
01069 Dresden
wolfgang.nagel@tu-dresden.de

Dietmar B. Kröner
Abteilung für Angewandte Mathematik
Universität Freiburg
Hermann-Herder-Str. 10
79104 Freiburg, Germany
dietmar@mathematik.uni-freiburg.de

Michael M. Resch
Höchstleistungsrechenzentrum
Stuttgart (HLRS)
Universität Stuttgart
Nobelstraße 19
70569 Stuttgart, Germany
resch@hlrs.de

Front cover figure: 3D-flame-modeling for industrial combustion equipment: Computed oxygen concentration in the furnace of a 330 MWe coal fired power plant (RECOM Services GmbH; www.recom-services.de)

ISBN 978-3-642-10000-0 e-ISBN 978-3-540-88303-6

DOI 10.1007/978-3-540-88303-6

Mathematics Subject Classification (2000): 65Cxx, 65C99, 68U20

Cover design: WMXDesign, Heidelberg

Printed on acid-free paper

9 8 7 6 5 4 3 2 1

springer.com

Preface

The discussions and plans on all scientific, advisory, and political levels to realize an even larger "European Supercomputer" in Germany, where the hardware costs alone will be hundreds of millions Euro – much more than in the past – are getting closer to realization. As part of the strategy, the three national supercomputing centres HLRS (Stuttgart), NIC/JSC (Jülich) and LRZ (Munich) have formed the Gauss Centre for Supercomputing (GCS) as a new virtual organization enabled by an agreement between the Federal Ministry of Education and Research (BMBF) and the state ministries for research of Baden-Württemberg, Bayern, and Nordrhein-Westfalen. Already today, the GCS provides the most powerful high-performance computing infrastructure in Europe. Through GCS, HLRS participates in the European project PRACE (Partnership for Advances Computing in Europe) and extends its reach to all European member countries. These activities aligns well with the activities of HLRS in the European HPC infrastructure project DEISA (Distributed European Infrastructure for Supercomputing Applications) and in the European HPC support project HPC-Europa. Beyond that, HLRS and its partners in the GCS have agreed on a common strategy for the installation of the next generation of leading edge HPC hardware over the next five years.

The University of Stuttgart and the University of Karlsruhe have furthermore agreed to bundle their competences and resources. Stuttgart will take a leading role in HPC being responsible for the operation and support for the national HPC facilities. Karlsruhe – with the newly created Karlsruhe Institute of Technology – will integrate its activities in Grid computing to take a lead in the state of Baden-Württemberg in the field of distributed systems and data management. The two centers will collaborate under the umbrella of a common organization.

Moreover, it is expected that in the next few months – following the proposal of the German HPC community, guided by Professor Andreas Reuter (EML) – the reshape of the High Performance Computing in Germany will proceed to form the German HPC "Gauß-Allianz", with the goal to improve and establish competitiveness for the coming years.

Beyond stabilization and strengthening of the existing German infrastructures – including the necessary hardware at a worldwide competitive level – a major software research and support program to enable Computational Science and Engineering on the required level of expertise and performance – which means: running Petascale applications on more than 100.000 processors – has been established by the BMBF. The projects of the first funding round are about to start at the end of 2008. It is expected that – after this first funding round – the next four years another 20 Million Euro will be spend – on a yearly basis – for projects to develop scalable algorithms, methods, and tools to support massively parallel systems. As we all know, we do not only need competitive hardware but also excellent software and methods to approach – and solve – the most demanding problems in science and engineering. The success of this approach is of utmost importance for our community and also will strongly influence the development of new technologies and industrial products; beyond that, this will finally determine if Germany will be an accepted partner among the leading technology and research nations.

Having been awarded funding as part of the German national initiative of excellence in the Cluster of Excellence for Simulation Technology, HLRS has started to further integrate its research with application scientists, computer scientists and mathematicians over the last year. With a focus on workflow management and the programming of large scale systems, HLRS is pursuing a strategy that will improve support for user with the demand to run thousands of jobs in the next years.

In addition, HLRS has significantly strengthened its collaboration within Germany. On March 7th 2008, the Automotive Simulation Center Stuttgart (ASCS) was founded. The focus of the center is on the development for simulation software for research in automotive engineering. Bringing together automotive manufactures and independent software vendors as well as researchers and hardware vendors the center is excellently positioned to not only strongly drive application research for automotive HPC simulation but will also significantly accelerate the know-how transfer from research into industrial development.

Since 1996, HLRS is supporting the scientific community as part of its official mission. Like in the years before, the major results of the last 12 months were reported at the Eleventh Results and Review Workshop on High Performance Computing in Science and Engineering, which was held September 29-30, 2008 at Stuttgart University. This volume contains the written versions of the research work presented. The papers have been selected from all projects running at HLRS and at SSC Karlsruhe during the one year period beginning October 2007. Overall, about 40 papers have been chosen from Physics, Solid State Physics, Computational Fluid Dynamics, Chemistry, and other topics. The largest number of contributions, as in many other years, came from CFD with 18 papers. Although such a small collection cannot represent a large area in total, the selected papers demonstrate the state of the art in high performance computing in Germany. The authors were encouraged to emphasize computational techniques used in solving the problems examined. This often forgotten aspect was the major focus of these proceedings, nevertheless this should not disregard the importance of the newly computed scientific results for the specific disciplines.

We gratefully acknowledge the continued support of the Land Baden-Württemberg in promoting and supporting high performance computing. Grateful acknowledgement is also due to the Deutsche Forschungsgemeinschaft (DFG): many projects processed on the machines of HLRS and SSC could not have been carried out without the support of the DFG. Also, we thank the Springer Verlag for publishing this volume and, thus, helping to position the local activities into an international frame. We hope that this series of publications is contributing to the global promotion of high performance scientific computing.

Stuttgart, October 2008

Wolfgang E. Nagel
Dietmar Kröner
Michael Resch

Contents

Physics

Prof. Dr.-Ing. Michael Resch

Höchstleistungsrechenzentrum Stuttgart (HLRS), Universität Stuttgart, Nobelstr. 19, 70569 Stuttgart, Germany

Three contributions have been included into the review proceedings of HLRS this year. These three papers reflect the main thread of simulation in physics on supercomputers as it has developed over the last years. Two of the papers focus on astrophysical simulations. The third one deals with the investigation of basic physics.

In their contribution "Magnetic Fields in Very Light Extragalactic Jets" Volker Gaibler and Max Camenzind from the Landessternwarte at Heidelberg investigate the propagation of very light astrophysical jets with density contrasts between 10^{-1} and 10^{-4} including magnetic fields into the intra-cluster medium on the galaxy cluster scale, using the magnetohydrodynamic code NIRVANA. The code is run on the NEC SX-8 system very efficiently and enables the authors to perform simulations in both axisymmetry as well as full 3D, using both shared-memory and MPI parallelization. Simulations further supported a stabilization of the contact surface by magnetic fields, as seen in low-frequency radio observations. Efficient storage methods for high time resolution runs as well as visualization methods were implemented and allow in-depth analysis of velocity and magnetic fields.

B. Müller, A. Marek, and H.-Th. Janka from the Max-Planck-Institut für Astrophysik provide an update on their Super Nova project. The paper gives an overview of the problems and the current status of their two-dimensional (core collapse) supernova modeling, and discuss the system of equations and the algorithm for its solution that are employed in the code. In particular they report on recent progress, and focus on the ongoing calculations that are performed on the NEC SX-8 at the HLRS Stuttgart. The paper discusses the case of low-mass progenitors below about ten solar masses, where the authors have obtained robust explosions, as well as the case of more massive progenitors. Several open issues are mentioned, and the need for a larger set of models, evolved to sufficiently late times, is emphasized. The authors stress that a highly parallel code, capable of exploiting the multi-node architecture of the NEC SX-8, is indispensable for addressing these unresolved questions.

J.H. Kühn, P. Marquard, M. Steinhauser and M. Tentyukov from the Institut für Theoretische Teilchenphysik, Universität Karlsruhe give an update on their work on the Standard Model. Their paper "Massless four-loop integrals and the total cross section in $e^+ e^-$ annihilation" is a continuation of their report given last year.

The main motivation behind the calculations performed within this project is the precise determination of input parameters of the theory describing the fundamental interaction of the elementary particles, the so-called Standard Model. In particular, the authors are interested in a precise extraction of one of the three coupling constants, the strong coupling, and three out of the six quark masses, the strange, charm and bottom quark mass, by a comparison of simulation results with experimental data.

Magnetic Fields in Very Light Extragalactic Jets

Volker Gaibler[1] and Max Camenzind[2]

[1] Landessternwarte Königstuhl, 69117 Heidelberg, Germany.
V.Gaibler@lsw.uni-heidelberg.de
[2] Landessternwarte Königstuhl, 69117 Heidelberg, Germany.
M.Camenzind@lsw.uni-heidelberg.de

Summary. We simulate the propagation of very light astrophysical jets with density contrasts between 10^{-1} and 10^{-4} including magnetic fields into the intracluster medium on the galaxy cluster scale, using the magnetohydrodynamic code NIRVANA on the NEC SX-6 and SX-8. The code runs very efficiently and enables us to perform simulations in both axisymmetry as well as full 3D, using both shared-memory and MPI parallelization. Simulations further supported a stabilization of the contact surface by magnetic fields, as seen in low-frequency radio observations. Efficient storage methods for high time resolution runs as well as visualization methods were implemented and allow in-depth analysis of velocity and magnetic fields.

1 Introduction

Extragalactic jets emerge from active nuclei of massive elliptical galaxies and are present over a wide range of distances – from Centaurus A in the local universe at a distance of 4 Mpc[3] to radio galaxies in the early universe at redshifts of $z > 5$. The energy release associated with these objects is enormous and can exceed 10^{40} watts. This power is believed to be released near supermassive black holes with masses up to 10^{10} solar masses. The jets are collimated, bipolar plasma streams with speeds close to the speed of light, which are probably launched by hydromagnetic processes near the accretion disks around these black holes or by electromagnetic extraction of angular momentum from the black hole. They extend over many orders of magnitude in length, observed from several hundred Schwarzschild radii ($< 10^{-3}$ pc) to the 10^6 pc scale, and thus propagate far into the surrounding galaxy cluster medium.

We are especially interested in the interaction of the jet with this intracluster medium on the scale of hundreds of kiloparsecs, which is a length scale that is typically observed for the high-power FR II jets in a wide spectral range

[3] 1 pc = 3.26 light years = 3×10^{16} m

from radio to X-rays. These jets are visible only due to motion of relativistic electrons in magnetic fields as continuum synchrotron radiation in radio frequencies (for other frequency ranges this may also be true, but is still under debate). Thus we try to simulate these jets including magnetic fields to be able to explain several observed properties, such as their emission properties and the bow shock and cocoon shapes (Fig. 1).

Extragalactic jets are assumed to be much lighter than their ambient gas, which makes it very hard for the jet to propagate due to its little momentum. A rough estimate of the propagation speed can be found from momentum balance with a constant working surface, which yields a speed decreasing with the square root of the density contrast between jet and ambient matter. This leads to a very slow motion of the jet head, where the fast jet plasma hits the external medium in the so-called hotspots. To reach the observed sizes and to agree with statistical analyses, the black hole activity has to last for ten of million of years or more. This long time makes it impossible to run the simulations on usual workstations – it would take years – and the use of supercomputers is the only way for numerical studies.

While there are many detailed observations of extragalactic jets, even the most basic parameters of these jets, as density, Mach number and the composition of the plasma, are very hard to constrain. This makes simulations difficult and enforces setups with several different parameters. Most simulations in the literature assumed less extreme density contrasts to reach reasonable jet sizes. We try to use more realistic density contrasts despite the huge computational demands because previous work in this project [4, 5] showed that these very light jets show a different behaviour than their heavier counterparts. Additionally, current observations suggest equipartition magnetic fields in jet cocoons, which makes a magnetohydrodynamical treatment of these objects necessary, even when one assumes somewhat weaker magnetic fields.

2 Numerical Method

The simulations were performed using the NIRVANA code [8], which numerically solves the nonrelativistic magnetohydrodynamic (MHD) equations in three dimensions in cartesian, cylindrical or spherical coordinates. It is based on a finite-differences discretization in an explicit formulation using operator splitting and uses van Leer's interpolation scheme, which is second order accurate. The advection part is solved in a conservative form and the magnetic fields are evolved using the constrained transport method, which conserves $\nabla \cdot \mathbf{B}$ up to machine roundoff errors. The code is vectorized and shared-memory parallelized [3, 6] and now runs very efficiently on the NEC SX machines. We also developed an MPI-parallelized version [3] which is used for 3D simulations and performs well.

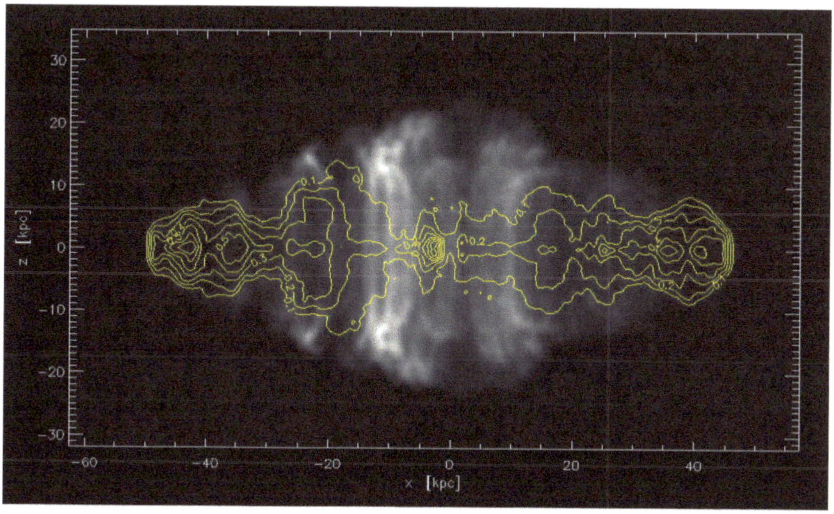

Fig. 1. Emission map synthetically derived from our MHD jet simulations, showing a map of expected cold gas filaments turbulently embedded in the hot cocoon in galaxies at high redshift (several billion years ago) with a synthetic radio map overlaid as contour lines

3 Computational and Scientific Results

We now report on the projects we worked on during the last year. We implemented a method to efficiently store simulations with high time resolution, which could not be done with the previous code version. It now enables us to exactly follow the evolution of shocks in the jet beam. We then worked on automatic computation of the initial magnetic field setup in force-balance, which is more realistic and suitable for very underdense jets. These underdense jets show a pronounced backflow and a very turbulent cocoon. Thus, exploration of the cocoon structure and turbulence is another goal which only can be achieved using special visualization techniques we implemented. We found, that due to the helical magnetic field in the jet beam, kinetic energy can be transferred to magnetic energy, generating stronger magnetic fields in the cocoon than expected. More details about these projects are given in the following text.

For the simulations, we used both the SX-6 and the SX-8. For single-node axisymmetric simulations, we achieved 19.2 Gflops on the SX-6 with our optimized code version [3, 6] (8 CPUs, vector operation ratio 99.5%). A single simulation run takes $\sim$ 3 000 CPU hours for a very light jet with density contrast 10^{-4}. Multi-node runs are used for the 3D simulations on the SX-8. On 32 nodes, we reached 550–600 Gflops so far (256 CPUs, vector operation ratio 99.0%), still working on improvement of the MPI communication to increase this further before starting very long runs (we estimate that a factor of nearly 2 might be possible). Unfortunately, this MPI-parallelized version is not suitable for axisymmetric simulations, as splitting the computational domain further for different nodes makes these regions too small to reach as good vectorization and shared-memory parallelization as we get for single-node runs. The main computational demand is the large number of time steps we have to compute (several millions), which cannot be parallelized due to causality. Hence, we use both the single-node and multi-node code versions, depending on where they are appropriate.

3.1 Jet Simulations with High Time Resolution

We simulate the propagation of extragalactic jets into the ambient cluster gas on the length scale of 200 kpc and assume a jet radius of 1 kpc. The jets are very light with respect to the ambient density – down to 10^{-4} for our study. A reasonable resolution of the jets thus yield a grid resolution of $4\,000 \times 1\,600$ cells. To reach typical lifetimes of jets as inferred from observations, we need several million timesteps, which corresponds to tens of millions of years simulated time. Analyzing the evolution with high temporal resolution is desirable due to various reasons, mainly: the ability to trace the evolution of shocks and plasma flow and to have a larger sample, how corresponding observations could look like.

Saving all our variables (≈ 400 MB) in short time intervals would require a large amount of disk space and much computational time for analysis and would pretty much slow down our simulation due to I/O activity. Data compression is no solution, as the compression rates for light jets with highly turbulent cocoons are too small (typically 50%). Thus we decided to save a reduced dataset in shorter time intervals. One third of the spatial resolution is used for this and only scalar variables (density, pressure and magnetic pressure) are saved. Since the range of values is high (for densities more than four orders of magnitude, for pressure even more), the values are scaled logarithmically into byte range. This still gives accuracy better than 5%. The full dataset is still available at coarser time intervals for more detailed analysis.

This method naturally gives a strong (but lossy) compression by a factor of ≈ 200 and allows saving 10 000 snapshots per run without storage difficulties. It proved very practical and allowed high-resolution animations to be made and analyzed. Hence, we are now able to follow the rise and decline of individual shocks, which are the best visible regions in jet systems due to their enhanced magnetic fields.

3.2 Magnetic Field Configuration

For very light jets, a careful treatment of the jet inlet is necessary. To get realistic morphologies for the jets, these have to be simulated bipolarly, meaning the jet nozzle is a boundary condition *inside* the computational domain. This can already yield difficulties in pure hydrodynamics, as in the midplane the

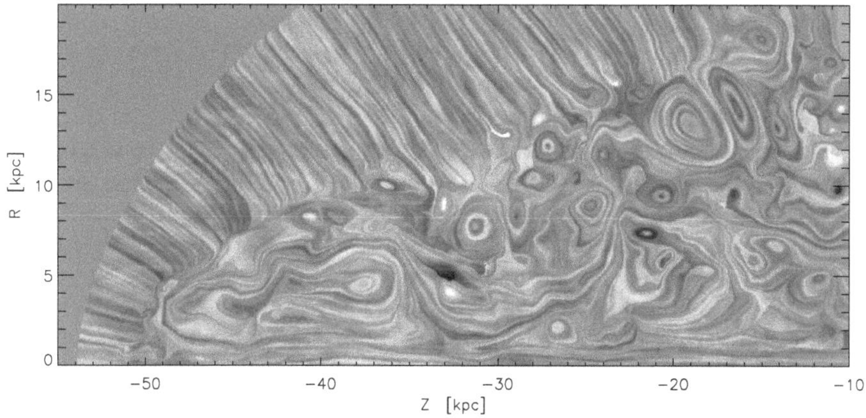

Fig. 2. LIC image of the jet head region for a simulation with density contrast 10^{-3} after 15 million years. Even smallest vortices in the cocoon are visible which would be missed with other visualization techniques

backflows converge and dense ambient gas can enclose the nozzle and create undesirable gas accumulation there. In magnetohydrodynamics, however, these difficulties are more serious, especially for strong fields. The magnetic field has to be kept divergence-free (as imposed by Maxwell's equations) – which generally is insured by divergence-free initial conditions and the constrained transport scheme employed in NIRVANA.

Setting fixed boundary conditions inside the computational domain inevitably violates this constraint. Thus we let the poloidal magnetic field component evolve without boundary conditions other than zero-gradient but try to prescribe initial conditions, which are in force balance and give a stable magnetic field at the jet nozzle. For this we evaluate hydromagnetic force balance in the midplane, prescribing only the nozzle-averaged values for pressure, poloidal and toroidal magnetic field, and use this as initial conditions. The setup was then used with strong magnetic fields ($\beta = 8\pi p/B^2 = 3$) and high temporal resolution.

3.3 Visualization of Turbulent Vector Fields

Analyzing the impact of magnetic fields on the propagation of very light jets is one of the main goals for our project. When a jet of very low density supersonically hits the ambient gas, it is reflected as if it were hitting a solid wall and streams sideways and backwards, inflating a pronounced cavity ("cocoon"), which is highly turbulent. Both the velocity field and the magnetic field thus show very fine structures. These vector fields are difficult to analyze and visualize with arrows without missing the numerous small-scale structures. Flow directions are hard to see since the array directions appear to be random (due to insufficient resolution).

A visualization technique suitable for large and fine-structured vector fields is Line Integral Convolution (LIC) [1]. Working on a white noise image of the same size as the vector field, every pixel of the resulting image is assigned the value of the line integral of the noise image along field lines of fixed length through its position (Fig. 2). This has the effect of blurring the noise in direction of the vector field and makes even small vortices easily visible. The disadvantage of this method is its rather high computational cost – but it is perfectly parallelizable due to the independence of all resulting pixels.

Unfortunately, the field strength cannot be represented by this, but if it changes by large factors, the LIC representation may give a misleading impression of the field strength. Regions of significant field strength simply cannot be recognized in the image.

As this visualization method is not available in any analysis software we know of and to overcome this limitation, we implemented this method ourselves and work in HLS color space. We encode the LIC streamlines in lightness and the magnitude in color (hue) and saturation (Fig. 3). While many other combinations of the data are possible, HLS encoding clearly separates both contributions, although a compromise between clearly visible streamlines and color values has to be made. This method also proved useful when other scalar fields were used instead of the field magnitude, such as density or plasma composition. It enables us to perform in-depth analysis of cocoon turbulence.

3.4 The Role of Magnetic Fields in Extragalactic Jets

Jet plasma is only visible due to the relativistic electrons, which spiral around magnetic field lines and emit sychrotron radiation. Therefore the presence of magnetic fields in jet beams and cocoons is undisputed. However, it is unclear, whether and how magnetic fields influence the dynamics in the system.

Our simulations of magnetized jets show pronounced jet heads which are more stable than their hydrodynamic counterparts [2]. Kelvin–Helmholtz instabilities at the contact surface between jet and ambient matter are damped by the magnetic tension, leading to less entrainment of dense gas, which could destabilize the flow. Observations of jets show smooth contact surfaces (eg. [7] for Hercules A), which could be explained by stabilizing magnetic fields.

Assuming the general case of helical magnetic fields in the jet beam, we found that these trigger rotation of the jet plasma. When it then hits the terminal shock and streams away from the axis, the magnetic field is sheared and amplified (Fig. 4), transferring kinetic energy into magnetic energy. In our axisymmetric simulations, this shearing and amplification could be measured. Though, further away from the jet head region, the toroidal field component artificially dominates, but it can not damp the instabilities. Our work on 3D simulations will allow us to overcome this limitation and a balance between

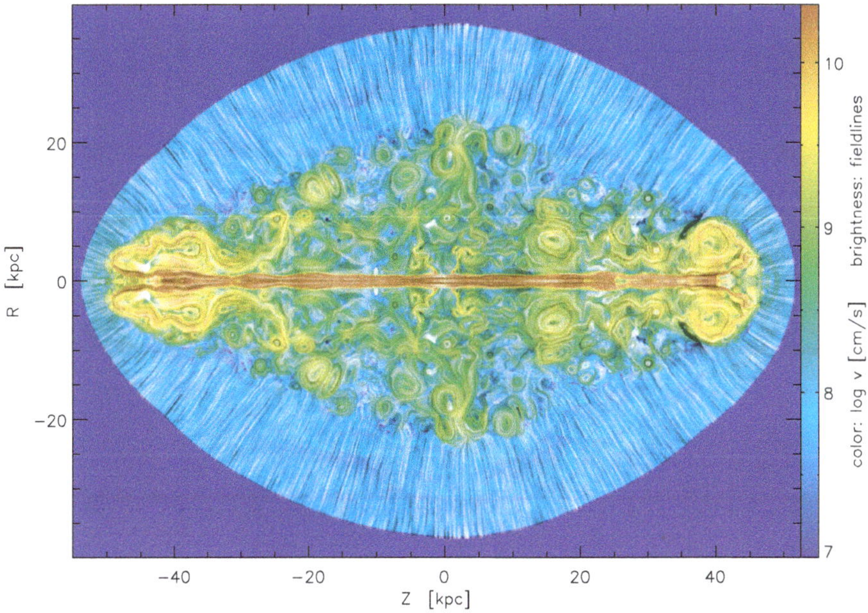

Fig. 3. HLS-encoded LIC image of an axisymmetric jet simulation with density contrast 10^{-3} after 15 million years – showing velocity stream lines in brightness information and field magnitude in color (logarithmically, in cm/s). Not only the small-scale vortices are visible – velocities easily show the jet beam region (red, $\sim 10^{10}$ cm/s), the backflow (yellow, several 10^9 cm/s) and the continuation to the highly turbulent cocoon (green)

toroidal and poloidal field is expected in the cocoon (due to the strong turbulence). In 3D, poloidal and toroidal field components then could damp the entrainment of dense gas, as we were able to show for the jet head in axisymmetric runs. The amplified magnetic field provides further support against instabilities at the contact surface.

Considering all magnetic field components in the simulations, we could compute some emission maps (as in Fig. 1), where projection of magnetic fields was described correctly and allows direct comparison with observations.

3.5 Ongoing Work

As shown before, there is still plenty of work left. Here we are persueing two goals: improvement and in-depth analysis of axisymmetric simulations as well as complementary fully 3D simulations. It is still necessary to follow both directions, as 3D simulations are not yet possible with the good resolution as we reach in axisymmetry, but they remove artificial symmetry restrictions with a still reasonable resolution and are necessary for a realistic description of the magnetic field in the turbulent cocoon. In addition, propagation of jets

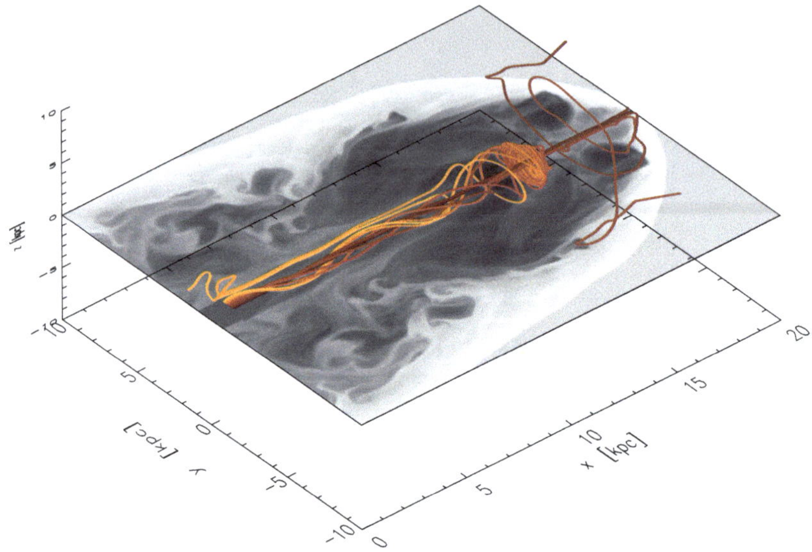

Fig. 4. Selected field lines from within the jet beam. The magnetic field in the jet beam is not only compressed in the terminal shock, but also sheared due to differential rotation and is thus found mostly toroidal in the head region and amplified in the cocoon

through clumpy media, as expected for high redshift sources, is reasonable only in 3D.

Much progress has been made with axisymmetric simulations by now. We reached good resolution, a high number of timesteps and have a stable magnetic field setup, which allows exploration of near equipartition magnetized jets. Still, impact of different magnetic field topologies and strenghts as well as beam stability are to be explored. Furthermore, analysis of high-performance simulations is an important part of supercomputing and also needs large computing power and available memory. Emission map computation of synchrotron, bremsstrahlung and inverse-Compton radiation and subsequent structural analysis will be the most important elements thereof, which we just began.

4 Summary

Our code is running very efficiently on the vector machines at the HLRS and enables us to perform large-scale simulations of the jet–cluster gas interaction, which would be impossible otherwise. But even with the SX-8 we are mostly restricted by the high number of necessary timesteps, which can not be resolved by parallelization.

We were able to follow the evolution of very light magnetized jets with high temporal resolution using an efficient storage method especially tailored to that problem, and used initial conditions in hydromagnetic force balance. The resulting velocity and magnetic fields were analyzed using Line Integral Convolution images and showed fine-grained structure of the cocoon plasma. Shearing of magnetic field due to differential rotation was found to amplify the magnetic field in the cocoon and further encouraged us to examine the magnetic field evolution in three dimensions, as is possible with our NIRVANA version on the SX-8.

Acknowledgments

This work was also supported by the Deutsche Forschungsgemeinschaft (Sonderforschungsbereich 439).

References

1. Cabral, B., Leedom, L. 1993, Imaging Vector Fields Using Line Integral Convolution. SIGGRAPH, Computer Graphics Proceedings, 263–270
2. Gaibler, V., Krause, M., Camenzind, M., 2008, in preparation
3. Gaibler V., Vigelius, M., Krause, M., and Camenzind, M. MHD Code Optimizations and Jets in Dense Gaseous Halos. In *High Performance Computing in Science and Engineering '06, eds.: Nagel, W.E., Jäger, W., and Resch, M., Springer*, 2006
4. Krause, M., 2003, A&A, 398, 113
5. Krause, M., 2005, A&A, 431, 45
6. Krause, M., Gaibler, V., and Camenzind, M. Simulations of Astrophysical Jets in Dense Environments. In *High Performance Computing in Science and Engineering '05, eds.: Nagel, W.E., Jäger, W., and Resch, M., Springer*, 2005
7. Nulsen, P.E.J., Hambrick, D.C., McNamara, B.R., Rafferty, D., Birzan, L., Wise, M.W., David, L.P., 2005, ApJ, 625, L9
8. Ziegler, U., Yorke, H.W., 1997, Computer Physics Communications, 101, 54

The SuperN-Project: Status and Outlook

B. Müller, A. Marek, and H.-Th. Janka

Max-Planck-Institut für Astrophysik, Karl-Schwarzschild-Strasse 1, Postfach 1317, D-85741 Garching bei München, Germany
bjmuellr@mpa-garching.mpg.de

Summary. We give an overview of the problems and the current status of our two-dimensional (core collapse) supernova modelling, and discuss the system of equations and the algorithm for its solution that are employed in our code. In particular we report on our recent progress, and focus on the ongoing calculations that are performed on the NEC SX-8 at the HLRS Stuttgart. We discuss the case of low-mass progenitors below about ten solar masses, where we have obtained robust explosions, as well as the case of more massive progenitors, exemplified by a 15 $M_\odot$ mass star, for which we have also observed a developing explosion in one simulation. Several open issues are mentioned, and the need for a larger set of models, evolved to sufficiently late times, is emphasized. We stress that a highly parallel code, capable of exploiting the multi-node architecture of the NEC SX-8, is indispensable for addressing these unresolved questions.

1 Introduction

A star more massive than about 8 solar masses ends its live in a cataclysmic explosion, a supernova. Its quiescent evolution comes to an end, when the pressure in its inner layers is no longer able to balance the inward pull of gravity. Throughout its life, the star sustained this balance by generating energy through a sequence of nuclear fusion reactions, forming increasingly heavier elements in its core. However, when the core consists mainly of iron-group nuclei, central energy generation ceases. The fusion reactions producing iron-group nuclei relocate to the core's surface, and their "ashes" continuously increase the core's mass. Similar to a white dwarf, such a core is stabilised against gravity by the pressure of its degenerate gas of electrons. However, to remain stable, its mass must stay smaller than the Chandrasekhar limit. When the core grows larger than this limit, it collapses to a neutron star, and a huge amount ($\sim 10^{53}$ erg) of gravitational binding energy is set free. Most ($\sim 99\%$) of this energy is radiated away in neutrinos, but a small fraction is transferred to the outer stellar layers and drives the violent mass ejection which disrupts the star in a supernova.

Despite 40 years of research, the details of how this energy transfer happens and how the explosion is initiated are still not well understood. Observational evidence about the physical processes deep inside the collapsing star is sparse and almost exclusively indirect. The only direct observational access is via measurements of neutrinos or gravitational waves. To obtain insight into the events in the core, one must therefore heavily rely on sophisticated numerical simulations. The enormous amount of computer power required for this purpose has led to the use of several, often questionable, approximations and numerous ambiguous results in the past. Fortunately, however, the development of numerical tools and computational resources has meanwhile advanced to a point, where it is becoming possible to perform multi-dimensional simulations with unprecedented accuracy. Therefore there is hope that the physical processes which are essential for the explosion can finally be unravelled.

An understanding of the explosion mechanism is required to answer many important questions of nuclear, gravitational, and astro-physics like the following:

- How do the explosion energy, the explosion timescale, and the mass of the compact remnant depend on the progenitor's mass? Is the explosion mechanism the same for all progenitors? For which stars are black holes left behind as compact remnants instead of neutron stars?
- What is the role of the – poorly known – equation of state (EoS) for the proto neutron star? Do softer or stiffer EoSs favour the explosion of a core collapse supernova?
- What is the role of rotation during the explosion? How rapidly do newly formed neutron stars rotate?
- How do neutron stars receive their natal kicks? Are they accelerated by asymmetric mass ejection and/or anisotropic neutrino emission?
- What are the generic properties of the neutrino emission and of the gravitational wave signal that are produced during stellar core collapse and explosion? Up to which distances could these signals be measured with operating or planned detectors on earth and in space? And what can one learn about supernova dynamics from a future measurement of such signals in case of a Galactic supernova?

2 Numerical Models

2.1 History and Constraints

According to theory, a shock wave is launched at the moment of "core bounce" when the neutron star begins to emerge from the collapsing stellar iron core. There is general agreement, supported by all "modern" numerical simulations, that this shock is unable to propagate directly into the stellar mantle and envelope, because it looses too much energy in dissociating iron into free nucleons while it moves through the outer core. The "prompt" shock ultimately stalls.

Thus the currently favoured theoretical paradigm needs to exploit the fact that a huge energy reservoir is present in the form of neutrinos, which are abundantly emitted from the hot, nascent neutron star. The absorption of electron neutrinos and antineutrinos by free nucleons in the post shock layer is thought to reenergize the shock, and lead to the supernova explosion.

Detailed *spherically symmetric* hydrodynamic models, which recently include a very accurate treatment of the time-dependent, multi-flavour, multi-frequency neutrino transport based on a numerical solution of the Boltzmann transport equation [1, 2, 3], reveal that this "delayed, neutrino-driven mechanism" does not work as simply as originally envisioned. Although in principle able to trigger the explosion (e.g., [4], [5], [6]), neutrino energy transfer to the postshock matter turned out to be too weak. For inverting the infall of the stellar core and initiating powerful mass ejection, an increase of the efficiency of neutrino energy deposition is needed.

A number of physical phenomena have been pointed out that can enhance neutrino energy deposition behind the stalled supernova shock. They are all linked to the fact that the real world is multi-dimensional instead of spherically symmetric (or one-dimensional; 1D) as assumed in the work cited above:

(1) Convective instabilities in the neutrino-heated layer between the neutron star and the supernova shock develop to violent convective overturn [7]. This convective overturn is helpful for the explosion, mainly because (a) neutrino-heated matter rises and increases the pressure behind the shock, thus pushing the shock further out, and (b) cool matter is able to penetrate closer to the neutron star where it can absorb neutrino energy more efficiently. Both effects allow multi-dimensional models to explode easier than spherically symmetric ones [8, 9, 10].
(2) Recent work [11, 12, 13, 14] has demonstrated that the stalled supernova shock is also subject to a second non-radial low-mode instability, called SASI, which can grow to a dipolar, global deformation of the shock [14, 15].
(3) Convective energy transport inside the nascent neutron star [16, 17, 18, 19] might enhance the energy transport to the neutrinosphere and could thus boost the neutrino luminosities. This would in turn increase the neutrino-heating behind the shock.

This list of multi-dimensional phenomena awaits more detailed exploration in multi-dimensional simulations. Until recently, such simulations have been performed with only a grossly simplified treatment of the involved microphysics, in particular of the neutrino transport and neutrino-matter interactions. At best, grey (i.e., single energy) flux-limited diffusion schemes were employed. All published successful simulations of supernova explosions by the convectivly aided neutrino-heating mechanism in two [8, 9, 20] and three dimensions [21, 22] used such a radical approximation of the neutrino transport.

Since, however, the role of the neutrinos is crucial for the problem, and because previous experience shows that the outcome of simulations is indeed very sensitive to the employed transport approximations, studies of the explosion mechanism require the best available description of the neutrino physics. This implies that one has to solve the Boltzmann transport equation for neutrinos.

2.2 Recent Calculations and the Need for TFlop Simulations

We have recently advanced to a new level of accuracy for supernova simulations by generalising the VERTEX code, a Boltzmann solver for neutrino transport, from spherical symmetry [23] to multi-dimensional applications [24, 25]. The corresponding mathematical model, and in particular our method for tackling the integro-differential transport problem in multi-dimensions, will be summarised in Sect. 3.

Results of a set of simulations with our code in 1D and 2D for progenitor stars with different masses have recently been published by [25, 26], and with respect to the expected gravitational-wave signals from rotating and convective supernova cores by [27]. The recent progress in supernova modelling was summarised and set in perspective in a conference article by [24].

Our collection of simulations has helped us to identify a number of effects which have brought our two-dimensional models close to the threshold of explosion. This makes us optimistic that the solution of the long-standing problem of how massive stars explode may be in reach. In particular, we have recognised the following aspects as advantageous:

- The details of the stellar progenitor (i.e. the mass of the iron core and its radius–density relation) have substantial influence on the supernova evolution. Especially, we found explosions of stellar models with low-mass (i.e. small) iron cores [28, 26], whereas more massive stars resist the explosion more persistent [25]. Thus detailed studies with different progenitor models are necessary.
- Stellar rotation, even at a moderate level, supports the expansion of the stalled shock by centrifugal forces and instigates overturn motion in the neutrino-heated postshock matter by meridional circulation flows in addition to convective instabilities.

All these effects are potentially important, and some (or even all of them) may represent crucial ingredients for a successful supernova simulation. So far no multi-dimensional calculations have been performed, in which two or more of these items have been taken into account simultaneously, and thus their mutual interaction awaits to be investigated. It should also be kept in mind that our knowledge of supernova microphysics, and especially the EoS of neutron star matter, is still incomplete, which implies major uncertainties for supernova modelling. Unfortunately, the impact of different descriptions for this input physics has so far not been satisfactorily explored with respect to the neutrino-heating mechanism and the long-time behaviour of the supernova shock, in particular in multi-dimensional models. However, first multi-dimensional simulations of core collapse supernovae with different nuclear EoSs [29, 19] show a strong dependence of the supernova evolution on the EoS.

From this it is clear that rather extensive parameter studies using multi-dimensional simulations are required to identify the physical processes which are essential for the explosion. Since on a dedicated machine performing at a sustained speed of about 30 GFlops already a single 2D simulation has a turn-around time of more than half a year, these parameter studies are not possible without TFlop simulations.

3 The Mathematical Model

The non-linear system of partial differential equations which is solved in our code consists of the following components:

- The Euler equations of hydrodynamics, supplemented by advection equations for the electron fraction and the chemical composition of the fluid, and formulated in spherical coordinates;
- the Poisson equation for calculating the gravitational source terms which enter the Euler equations, including corrections for general relativistic effects;
- the Boltzmann transport equation which determines the (non-equilibrium) distribution function of the neutrinos;
- the emission, absorption, and scattering rates of neutrinos, which are required for the solution of the Boltzmann equation;
- the equation of state of the stellar fluid, which provides the closure relation between the variables entering the Euler equations, i.e. density, momentum, energy, electron fraction, composition, and pressure.

In what follows we will briefly summarise the neutrino transport algorithms. For a more complete description of the entire code we refer the reader to [25], and the references therein.

3.1 "Ray-by-Ray Plus" Variable Eddington Factor Solution of the Neutrino Transport Problem

The crucial quantity required to determine the source terms for the energy, momentum, and electron fraction of the fluid owing to its interaction with the neutrinos is the neutrino distribution function in phase space, $f(r, \vartheta, \phi, \epsilon, \Theta, \Phi, t)$. Equivalently, the neutrino intensity $I = c/(2\pi\hbar c)^3 \cdot \epsilon^3 f$ may be used. Both are seven-dimensional functions, as they describe, at every point in space (r, ϑ, ϕ), the distribution of neutrinos propagating with energy ϵ into the direction (Θ, Φ) at time t (Fig. 1).

The evolution of I (or f) in time is governed by the Boltzmann equation, and solving this equation is, in general, a six-dimensional problem (as time is usually not counted as a separate dimension). A solution of this equation by direct discretisation (using an S_N scheme) would require computational resources in the PetaFlop range. Although there are attempts by at least one group in the United States to follow such an approach, we feel that, with the currently available computational resources, it is mandatory to reduce the dimensionality of the problem.

Actually this should be possible, since the source terms entering the hydrodynamic equations are *integrals* of I over momentum space (i.e. over ϵ, Θ, and Φ), and thus only a fraction of the information contained in I is truly required to compute the dynamics of the flow. It makes therefore sense to consider angular moments of I, and to solve evolution equations for these moments, instead of dealing with the Boltzmann equation directly. The 0th to 3rd order moments are defined as

$$J, \boldsymbol{H}, \mathbf{K}, \mathbf{L}, \ldots (r, \vartheta, \phi, \epsilon, t) = \frac{1}{4\pi} \int I(r, \vartheta, \phi, \epsilon, \Theta, \Phi, t)\, \boldsymbol{n}^{0,1,2,3,\ldots}\, \mathrm{d}\Omega \qquad (1)$$

where $\mathrm{d}\Omega = \sin\Theta\, \mathrm{d}\Theta\, \mathrm{d}\Phi$, $\boldsymbol{n} = (\cos\Theta, \sin\Theta\cos\Phi, \sin\Theta\sin\Phi)$, and exponentiation represents repeated application of the dyadic product. Note that the moments are *tensors* of the required rank.

This leaves us with a four-dimensional problem. So far no approximations have been made. In order to reduce the size of the problem even further, one needs to resort to assumptions on its symmetry. At this point, one usually employs azimuthal symmetry for the stellar matter distribution, i.e. any dependence on the azimuth angle ϕ is ignored, which implies that the hydrodynamics of the problem can be treated in two dimensions. It also implies $I(r, \vartheta, \epsilon, \Theta, \Phi) = I(r, \vartheta, \epsilon, \Theta, -\Phi)$. If, in addition, it is assumed that I is even independent of Φ, then each of the angular moments of I becomes a *scalar*, which depends on two spatial dimensions, and one dimension in momentum space: $J, H, K, L = J, H, K, L(r, \vartheta, \epsilon, t)$. Thus we have reduced the problem to three dimensions in total.

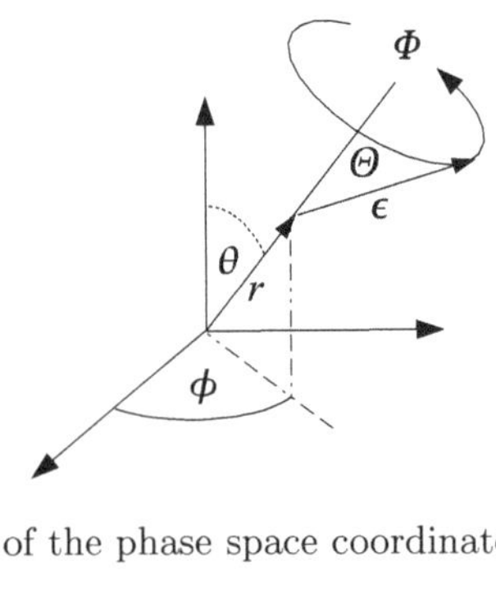

Fig. 1. Illustration of the phase space coordinates (see the main text)

The System of Equations

With the aforementioned assumptions it can be shown [25], that in order to compute the source terms for the energy and electron fraction of the fluid, the following two transport equations need to be solved:

$$
\begin{aligned}
&\left(\frac{1}{c}\frac{\partial}{\partial t}+\beta_r\frac{\partial}{\partial r}+\frac{\boldsymbol{\beta_\vartheta}}{\boldsymbol{r}}\frac{\boldsymbol{\partial}}{\boldsymbol{\partial\vartheta}}\right)J+J\left(\frac{1}{r^2}\frac{\partial(r^2\beta_r)}{\partial r}+\frac{1}{\boldsymbol{r\sin\vartheta}}\frac{\boldsymbol{\partial(\sin\vartheta\beta_\vartheta)}}{\boldsymbol{\partial\vartheta}}\right)\\
&+\frac{1}{r^2}\frac{\partial(r^2H)}{\partial r}+\frac{\beta_r}{c}\frac{\partial H}{\partial t}-\frac{\partial}{\partial\epsilon}\left\{\frac{\epsilon}{c}\frac{\partial\beta_r}{\partial t}H\right\}-\frac{\partial}{\partial\epsilon}\left\{\epsilon J\left(\frac{\beta_r}{r}+\frac{1}{\boldsymbol{2r\sin\vartheta}}\frac{\boldsymbol{\partial(\sin\vartheta\beta_\vartheta)}}{\boldsymbol{\partial\vartheta}}\right)\right\}\\
&\qquad-\frac{\partial}{\partial\epsilon}\left\{\epsilon K\left(\frac{\partial\beta_r}{\partial r}-\frac{\beta_r}{r}-\frac{1}{\boldsymbol{2r\sin\vartheta}}\frac{\boldsymbol{\partial(\sin\vartheta\beta_\vartheta)}}{\boldsymbol{\partial\vartheta}}\right)\right\}\\
&\qquad+J\left(\frac{\beta_r}{r}+\frac{1}{\boldsymbol{2r\sin\vartheta}}\frac{\boldsymbol{\partial(\sin\vartheta\beta_\vartheta)}}{\boldsymbol{\partial\vartheta}}\right)\\
&\qquad+K\left(\frac{\partial\beta_r}{\partial r}-\frac{\beta_r}{r}-\frac{1}{\boldsymbol{2r\sin\vartheta}}\frac{\boldsymbol{\partial(\sin\vartheta\beta_\vartheta)}}{\boldsymbol{\partial\vartheta}}\right)+\frac{2}{c}\frac{\partial\beta_r}{\partial t}H=C^{(0)}, \qquad (2)
\end{aligned}
$$

$$
\begin{aligned}
&\left(\frac{1}{c}\frac{\partial}{\partial t}+\beta_r\frac{\partial}{\partial r}+\frac{\boldsymbol{\beta_\vartheta}}{\boldsymbol{r}}\frac{\boldsymbol{\partial}}{\boldsymbol{\partial\vartheta}}\right)H+H\left(\frac{1}{r^2}\frac{\partial(r^2\beta_r)}{\partial r}+\frac{1}{\boldsymbol{r\sin\vartheta}}\frac{\boldsymbol{\partial(\sin\vartheta\beta_\vartheta)}}{\boldsymbol{\partial\vartheta}}\right)\\
&\qquad+\frac{\partial K}{\partial r}+\frac{3K-J}{r}+H\left(\frac{\partial\beta_r}{\partial r}\right)+\frac{\beta_r}{c}\frac{\partial K}{\partial t}-\frac{\partial}{\partial\epsilon}\left\{\frac{\epsilon}{c}\frac{\partial\beta_r}{\partial t}K\right\}\\
&\qquad-\frac{\partial}{\partial\epsilon}\left\{\epsilon L\left(\frac{\partial\beta_r}{\partial r}-\frac{\beta_r}{r}-\frac{1}{\boldsymbol{2r\sin\vartheta}}\frac{\boldsymbol{\partial(\sin\vartheta\beta_\vartheta)}}{\boldsymbol{\partial\vartheta}}\right)\right\}\\
&-\frac{\partial}{\partial\epsilon}\left\{\epsilon H\left(\frac{\beta_r}{r}+\frac{1}{\boldsymbol{2r\sin\vartheta}}\frac{\boldsymbol{\partial(\sin\vartheta\beta_\vartheta)}}{\boldsymbol{\partial\vartheta}}\right)\right\}+\frac{1}{c}\frac{\partial\beta_r}{\partial t}(J+K)=C^{(1)}. \qquad (3)
\end{aligned}
$$

These are evolution equations for the neutrino energy density, J, and the neutrino flux, H, and follow from the zeroth and first moment equations of the comoving frame (Boltzmann) transport equation in the Newtonian, $\mathcal{O}(v/c)$ approximation. The quantities $C^{(0)}$ and $C^{(1)}$ are source terms that result from the collision term of the Boltzmann equation, while $\beta_r = v_r/c$ and $\beta_\vartheta = v_\vartheta/c$, where v_r and v_ϑ are the components of the hydrodynamic velocity, and c is the speed of light. The functional dependences $\beta_r = \beta_r(r, \vartheta, t)$, $J = J(r, \vartheta, \epsilon, t)$, etc. are suppressed in the notation. This system includes four unknown moments (J, H, K, L) but only two equations, and thus needs to be supplemented by two more relations. This is done by substituting $K = f_K \cdot J$ and $L = f_L \cdot J$, where f_K and f_L are the variable Eddington factors, which for the moment may be regarded as being known, but in our case is indeed determined from a separate simplified ("model") Boltzmann equation.

A finite volume discretisation of Eqs. (2–3) is sufficient to guarantee exact conservation of the total neutrino energy. However, and as described in detail in [23], it is not sufficient to guarantee also exact conservation of the neutrino number. To achieve this, we discretise and solve a set of two additional equations. With $\mathcal{J} = J/\epsilon$, $\mathcal{H} = H/\epsilon$, $\mathcal{K} = K/\epsilon$, and $\mathcal{L} = L/\epsilon$, this set of equations reads

$$\left(\frac{1}{c}\frac{\partial}{\partial t}+\beta_r\frac{\partial}{\partial r}+\boldsymbol{\frac{\beta_\vartheta}{r}\frac{\partial}{\partial\vartheta}}\right)\mathcal{J}+\mathcal{J}\left(\frac{1}{r^2}\frac{\partial(r^2\beta_r)}{\partial r}+\boldsymbol{\frac{1}{r\sin\vartheta}\frac{\partial(\sin\vartheta\beta_\vartheta)}{\partial\vartheta}}\right)$$
$$+\frac{1}{r^2}\frac{\partial(r^2\mathcal{H})}{\partial r}+\frac{\beta_r}{c}\frac{\partial\mathcal{H}}{\partial t}-\frac{\partial}{\partial\epsilon}\left\{\frac{\epsilon}{c}\frac{\partial\beta_r}{\partial t}\mathcal{H}\right\}-\frac{\partial}{\partial\epsilon}\left\{\epsilon\mathcal{J}\left(\frac{\beta_r}{r}+\boldsymbol{\frac{1}{2r\sin\vartheta}\frac{\partial(\sin\vartheta\beta_\vartheta)}{\partial\vartheta}}\right)\right\}$$
$$-\frac{\partial}{\partial\epsilon}\left\{\epsilon\mathcal{K}\left(\frac{\partial\beta_r}{\partial r}-\frac{\beta_r}{r}-\boldsymbol{\frac{1}{2r\sin\vartheta}\frac{\partial(\sin\vartheta\beta_\vartheta)}{\partial\vartheta}}\right)\right\}+\frac{1}{c}\frac{\partial\beta_r}{\partial t}\mathcal{H}=\mathcal{C}^{(0)}, \quad (4)$$

$$\left(\frac{1}{c}\frac{\partial}{\partial t}+\beta_r\frac{\partial}{\partial r}+\boldsymbol{\frac{\beta_\vartheta}{r}\frac{\partial}{\partial\vartheta}}\right)\mathcal{H}+\mathcal{H}\left(\frac{1}{r^2}\frac{\partial(r^2\beta_r)}{\partial r}+\boldsymbol{\frac{1}{r\sin\vartheta}\frac{\partial(\sin\vartheta\beta_\vartheta)}{\partial\vartheta}}\right)$$
$$+\frac{\partial\mathcal{K}}{\partial r}+\frac{3\mathcal{K}-\mathcal{J}}{r}+\mathcal{H}\left(\frac{\partial\beta_r}{\partial r}\right)+\frac{\beta_r}{c}\frac{\partial\mathcal{K}}{\partial t}-\frac{\partial}{\partial\epsilon}\left\{\frac{\epsilon}{c}\frac{\partial\beta_r}{\partial t}\mathcal{K}\right\}$$
$$-\frac{\partial}{\partial\epsilon}\left\{\epsilon\mathcal{L}\left(\frac{\partial\beta_r}{\partial r}-\frac{\beta_r}{r}-\boldsymbol{\frac{1}{2r\sin\vartheta}\frac{\partial(\sin\vartheta\beta_\vartheta)}{\partial\vartheta}}\right)\right\}$$
$$-\frac{\partial}{\partial\epsilon}\left\{\epsilon\mathcal{H}\left(\frac{\beta_r}{r}+\boldsymbol{\frac{1}{2r\sin\vartheta}\frac{\partial(\sin\vartheta\beta_\vartheta)}{\partial\vartheta}}\right)\right\}$$
$$-\mathcal{L}\left(\frac{\partial\beta_r}{\partial r}-\frac{\beta_r}{r}-\boldsymbol{\frac{1}{2r\sin\vartheta}\frac{\partial(\sin\vartheta\beta_\vartheta)}{\partial\vartheta}}\right)$$
$$-\mathcal{H}\left(\frac{\beta_r}{r}+\boldsymbol{\frac{1}{2r\sin\vartheta}\frac{\partial(\sin\vartheta\beta_\vartheta)}{\partial\vartheta}}\right)+\frac{1}{c}\frac{\partial\beta_r}{\partial t}\mathcal{J}=\mathcal{C}^{(1)}. \quad (5)$$

The moment equations (2–5) are very similar to the $\mathcal{O}(v/c)$ equations in spherical symmetry which were solved in the 1D simulations of [23] (see Eqs. 7,8,30, and 31 of the latter work). This similarity has allowed us to reuse a good fraction of the one-dimensional version of VERTEX, for coding the multi-dimensional algorithm. The additional terms necessary for this purpose have been set in boldface above.

Finally, the changes of the energy, e, and electron fraction, Y_e, required for the hydrodynamics are given by the following two equations

$$\frac{\mathrm{d}e}{\mathrm{d}t}=-\frac{4\pi}{\rho}\int_0^\infty\mathrm{d}\epsilon\sum_{\nu\in(\nu_\mathrm{e},\bar\nu_\mathrm{e},\dots)}C_\nu^{(0)}(\epsilon), \quad (6)$$

$$\frac{\mathrm{d}Y_\mathrm{e}}{\mathrm{d}t}=-\frac{4\pi\, m_\mathrm{B}}{\rho}\int_0^\infty\mathrm{d}\epsilon\left(\mathcal{C}_{\nu_\mathrm{e}}^{(0)}(\epsilon)-\mathcal{C}_{\bar\nu_\mathrm{e}}^{(0)}(\epsilon)\right) \quad (7)$$

(for the momentum source terms due to neutrinos see [25]). Here m_B is the baryon mass, and the sum in Eq. (6) runs over all neutrino types. The full system consisting of Eqs. (2–7) is stiff, and thus requires an appropriate discretisation scheme for its stable solution.

Method of Solution

In order to discretise Eqs. (2–7), the spatial domain $[0,r_\mathrm{max}]\times[\vartheta_\mathrm{min},\vartheta_\mathrm{max}]$ is covered by N_r radial, and N_ϑ angular zones, where $\vartheta_\mathrm{min}=0$ and $\vartheta_\mathrm{max}=\pi$

correspond to the north and south poles, respectively, of the spherical grid. (In general, we allow for grids with different radial resolutions in the neutrino transport and hydrodynamic parts of the code. The number of radial zones for the hydrodynamics will be denoted by N_r^{hyd}.) The number of bins used in energy space is N_ϵ and the number of neutrino types taken into account is N_ν.

The equations are solved in two operator-split steps corresponding to a lateral and a radial sweep.

In the first step, we treat the boldface terms in the respectively first lines of Eqs. (2–5), which describe the lateral advection of the neutrinos with the stellar fluid, and thus couple the angular moments of the neutrino distribution of neighbouring angular zones. For this purpose we consider the equation

$$\frac{1}{c}\frac{\partial \Xi}{\partial t} + \frac{1}{r \sin\vartheta}\frac{\partial(\sin\vartheta\, \beta_\vartheta\, \Xi)}{\partial \vartheta} = 0\,, \tag{8}$$

where Ξ represents one of the moments J, H, $\mathcal{J}$, or $\mathcal{H}$. Although it has been suppressed in the above notation, an equation of this form has to be solved for each radius, for each energy bin, and for each type of neutrino. An explicit upwind scheme is used for this purpose.

In the second step, the radial sweep is performed. Several points need to be noted here:

- terms in boldface not yet taken into account in the lateral sweep, need to be included into the discretisation scheme of the radial sweep. This can be done in a straightforward way since these remaining terms do not include derivatives of the transport variables (J, H) or $(\mathcal{J}, \mathcal{H})$. They only depend on the hydrodynamic velocity v_ϑ, which is a *constant* scalar field for the transport problem.
- the right hand sides (source terms) of the equations and the coupling in energy space have to be accounted for. The coupling in energy is non-local, since the source terms of Eqs. (2–5) stem from the Boltzmann equation, which is an integro-differential equation and couples all the energy bins
- the discretisation scheme for the radial sweep is *implicit* in time. Explicit schemes would require very small time steps to cope with the stiffness of the source terms in the optically thick regime, and the small CFL time step dictated by neutrino propagation with the speed of light in the optically thin regime. Still, even with an implicit scheme $\gtrsim 10^5$ time steps are required per simulation. This makes the calculations expensive.

Once the equations for the radial sweep have been discretized in radius and energy, the resulting solver is applied ray-by-ray for each angle ϑ and for each type of neutrino, i.e. for constant ϑ, N_ν two-dimensional problems need to be solved.

The discretisation itself is done using a second order accurate scheme with backward differencing in time according to [23]. This leads to a non-linear system of algebraic equations, which is solved by Newton-Raphson iteration

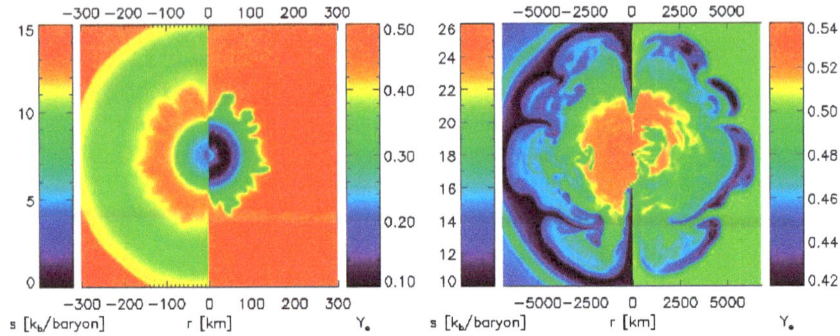

Fig. 2. Snapshot showing the gas entropy s (left half of panels) and the electron-to-nucleon ratio Y_e (right half of panels) for the explosion of the O-Ne-Mg core for two different times. Convection develops soon after the onset of the explosion (left panel, 97 ms after core bounce), and results in the ejection of pockets of rather neutron-rich material (with a low $Y_e \approx 0.43$), clearly visible in the right panel (363 ms after bounce)

with explicit construction and inversion of the corresponding Jacobian matrix with the Block-Thomas algorithm.

4 Recent Results and Ongoing Work

We make use of the computer resources available to us at the HLRS to address some of the important questions in SN theory (see Sect. 1) with 2D-simulations. At the HLRS, we typically run our code on one node of the NEC SX-8 (8 processors, OpenMP-parallelised) with 98.3% of vector operations and up to 30000 MFLOPS per second. The performance of the code has improved significantly (by 25% to 30%) due to recent optimizations. In particular, we now employ highly optimized routines for the Thomas algorithm, written by K. Benkert [30, 31].

In the following we present some of our results from these simulations that are currently conducted at the HLRS. For the neutrino interaction rates we use the full set as described in [32], and general relativistic effects are taken into account according to [33].

4.1 Successful Multi-Dimensional Explosion Models of Oxygen-Neon-Magnesium Cores

In recent simulations at HLRS, we focussed on low-mass progenitors with an O-Ne-Mg core instead of an iron core. Stars in this mass range develop cores composed of oxygen, neon, and magesium instead of iron, with an extremely sharp density gradient at the surface shock front to expand continuosly as it propagates into rapidly diluting infalling material. As the shock moves steadily

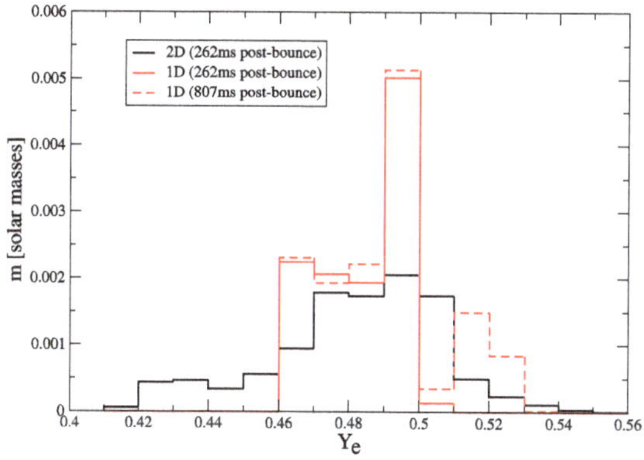

Fig. 3. Histogram for the electron fraction Y_e in the ejected material for the one-dimensional (red lines) and two-dimensional models (black lines) of the O-Ne-Mg core. For the one-dimensional simulation, we also show the distribution at a very late stage (807 ms after shock formation). Note that the two-dimensional model gives a significantly broader distribution with more neutron-rich (low Y_e) and and proton-rich (high Y_e) ejecta

outward, the accreted material stays in the heating region for a longer time, and due to these favourable conditions a neutrino-driven explosion sets in within less than 100 ms after shock formation.

While multi-dimensional hydrodynamical instabilities like the standing accretion shock instability (SASI) are not crucial for the explosions of oxygen-neon-magnesium (O-Ne-Mg) cores, multi-dimensional effects are nonetheless of tremendous importance in these cases: For example, convection behind the shock (see Fig. 2) leads to a noticeable increase in the explosion energy compared to the 1D case, and produces inhomogeneities in the ejecta (see also below). We also find a hemispherical asymmetry in the early neutrino driven wind, which tends to be more proton-rich (high Y_e) along the polar axis in one hemisphere. Moreover, 2D simulations are also necessary to determine the initial kick of the newly-born neutron star. Since we have been able to extend our simulations to more than 360 ms after core bounce in 2D (and ≈ 700 ms in 1D), i. e. well beyond the onset of the explosion (≈ 100 ms after bounce), we can now address several important questions conerning the post-explosion phase, e. g. the production of heavy elements in the supernova event, or the observational signatures in earth-bound neutrino detectors.

The nucleosynthesis conditions for our one-dimensional models have been discussed in [34] with the help of detailed nuclear reaction network calculations. We found a signficant overprodution of closed neutron-shell nuclei in the ($A \approx 90, N = 50$) region, like ^{90}Zr. Although the high yields are possibly in conflict with constraints from Galactic chemical evolution, these results represent a significant step towards linking our numerical models to astronomical observations, and suggest that uncertainties in the progenitor evolution, or

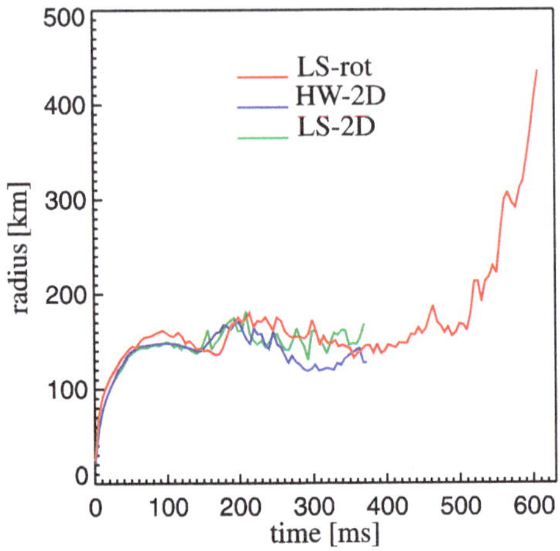

Fig. 4. The average shock radius for the rotating $15\,M_\odot$ star with rotation and a soft equation of state (red, LS-rot). For the non-rotating case, results for a soft (LS-2D, green) and a stiff equation of state (HW-2D, blue) are displayed

multi-dimensional effects (convection, rotation, magnetic fields) may significantly affect the production of heavy elements in O-Ne-Mg supernovae. As mentioned before, our 2D calculations already show that the composition of the ejected material is considerably different from the spherically symmetric case (see Figs. 2 and 3) due to convective overturn behind the outgoing shock front. Our simulations also shed light on the recent suggestion by [35], according to which the rapid neutron capture process (r-process) may operate in O-Ne-Mg supernovae, producing many of the heaviest elements in nature, such as *Au* and *Pb*. Judging from our calculations, this r-process scenario does not seem viable [36].

Data from our models have been used to analyse certain properties of the neutrino signals from these low-mass progenitors by Lunardini et al. [37], where the effects of neutrino-matter oscillations and collective $\nu\bar{\nu}$ flavour transfornations in the exploding star are taken into account. Their investigation revealed that the steep density gradient and the fast propagation of the shock lead to a detection signature clearly distinct from that of more massive stars with iron cores, implying that several features of models are amenable to direct observational tests in the event of a nearby Galactic O-Ne-Mg supernova.

4.2 Long-Time Simulations of Massive Progenitor Models

We have also continued our long-time simulations of several massive progenitor models at HLRS (see [29, 38] for earlier status reports). Simulations partly performed on the SX-8 at HLRS (using different equations of state, and covering both the rotating and the non-rotating case) provided the basis for a

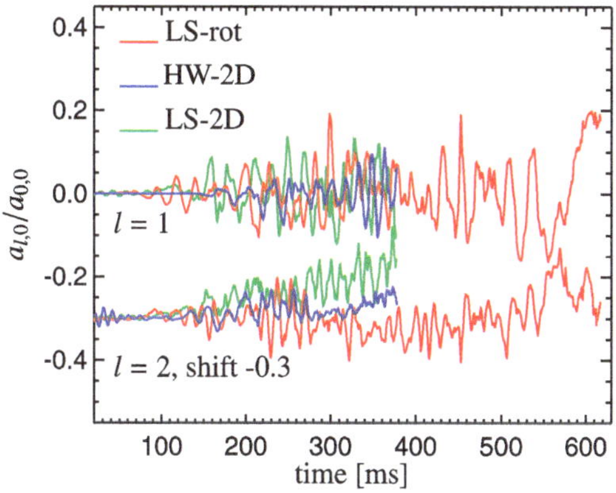

Fig. 5. Time-dependent coefficients $a_{l,0}$ of the decomposition of the shock position into spherical harmonics (cf. Eqn. 9). The first two expansion coefficients, corresponding to a dipolar (l=1) or quadrupolar (l=2) shock deformation are shown. Note that the $l = 2$ case is displayed with an offset of -0.3

detailed analysis of the evolution of a $15\,\mathrm{M}_{\odot}$ star [39]. An incipient explosion, aided by strong SASI activity, was observed for the rotating model LS-rot, which shows rapid shock expansion starting 500 ms after core bounce (see Fig. 4). Since none of our non-rotating models has been calculated to such late times, no final conclusion can be reached as to whether rotation is crucial for this explosion. However, for the corresponding non-rotating model LS-2D the neutrino heating conditions, the average shock radius (Fig. 4) are quite similar until 400 ms, as is the amplitude of the SASI oscillation shown in Fig. 5. There we display the first two coefficients of an expansion of the angle-dependent shock position into spherical harmonics,

$$r_{\mathrm{shock}}(\theta, t) = \sum_{l=0}^{\infty} a_{l,0}(t) Y_{l,0}(\theta). \tag{9}$$

The similar evolution of the models LS-rot and LS-2D suggests that rotation may not be the decisive factor. On the other hand, the non-rotating model HW-2D, computed with a stiffer equation of state, does not develop strong SASI activity comparable to model LS-2D until 400 ms after bounce, indicating that a soft equation may be more favourable for an explosion.

5 Conclusions and Outlook

We continued to simulate well-resolved 2D models of core collapse supernovae with detailed neutrino transport at the HLRS. Explosions have been obtained for an O-Ne-Mg core, and we are presently working on analysing the post-explosion phase in more detail, with a particular focus on nucleosynthesis aspects. Future work will have to address the harmful overproduction of certain neutron-rich isotopes, and clarify the influence of multi-dimensional effects and variations in the progenitor structure.

In addition, we confirmed that non-radial hydrodynamic instabilities support the onset of supernova explosions, and for a $15\,M_\odot$ progenitor model we obtained an explosion at a time of roughly 600 ms after shock formation. While our set of models already allows us to assess the importance of rotation and the equation of state for the explosion to a certain extent, simulations beyond 500 ms after bounce for a sufficiently broad selection of models are necessary before any definite conclusions can be reachead.

However, such parameter studies require tremendous computational resources, and can only be carried out with a very efficient and highly parallel code. While recent optimizations have improved the single-node performance of VERTEX significantly, a multi-node version with good scaling is clearly desirable. For this reason, a MPI version of VERTEX is currently being developed within the framework of the Teraflop Workbench.

Acknowledgements

Support from the SFB 375 "Astroparticle Physics", SFB/TR7 "Gravitationswellenastronomie", the SFB/TR27 "Neutrinos and Beyond" of the Deutsche Forschungsgemeinschaft, and the Cluster of Excellence EXC 153 "Origin and Structure of the Universe" (http://www.universe-cluster.de) and computer time at the HLRS and the Rechenzentrum Garching are acknowledged. We also thank M. Galle and R. Fischer for performing the benchmarks on the NEC machines. We thank especially K. Benkert for further optimising our code for the SX-8 architecture.

References

1. Rampp, M., Janka, H.T.: Spherically Symmetric Simulation with Boltzmann Neutrino Transport of Core Collapse and Postbounce Evolution of a $15 M_\odot$ Star. Astrophys. J. **539** (2000) L33–L36
2. Mezzacappa, A., Liebendörfer, M., Messer, O.E., Hix, W.R., Thielemann, F., Bruenn, S.W.: Simulation of the Spherically Symmetric Stellar Core Collapse, Bounce, and Postbounce Evolution of a Star of 13 Solar Masses with Boltzmann Neutrino Transport, and Its Implications for the Supernova Mechanism. Phys. Rev. Letters **86** (2001) 1935–1938
3. Liebendörfer, M., Mezzacappa, A., Thielemann, F., Messer, O.E., Hix, W.R., Bruenn, S.W.: Probing the gravitational well: No supernova explosion in spherical symmetry with general relativistic Boltzmann neutrino transport. Phys. Rev. D **63** (2001) 103004–+
4. Bethe, H.A.: Supernova mechanisms. Reviews of Modern Physics **62** (1990) 801–866

5. Burrows, A., Goshy, J.: A Theory of Supernova Explosions. Astrophys. J. **416** (1993) L75
6. Janka, H.T.: Conditions for shock revival by neutrino heating in core-collapse supernovae. Astron. Astrophys. **368** (2001) 527–560
7. Herant, M., Benz, W., Colgate, S.: Postcollapse hydrodynamics of SN 1987A - Two-dimensional simulations of the early evolution. Astrophys. J. **395** (1992) 642–653
8. Herant, M., Benz, W., Hix, W.R., Fryer, C.L., Colgate, S.A.: Inside the supernova: A powerful convective engine. Astrophys. J. **435** (1994) 339
9. Burrows, A., Hayes, J., Fryxell, B.A.: On the nature of core-collapse supernova explosions. Astrophys. J. **450** (1995) 830
10. Janka, H.T., Müller, E.: Neutrino heating, convection, and the mechanism of Type-II supernova explosions. Astron. Astrophys. **306** (1996) 167–+
11. Thompson, C.: Accretional Heating of Asymmetric Supernova Cores. Astrophys. J. **534** (2000) 915–933
12. Foglizzo, T.: Non-radial instabilities of isothermal Bondi accretion with a shock: Vortical-acoustic cycle vs. post-shock acceleration. Astron. Astrophys. **392** (2002) 353–368
13. Blondin, J.M., Mezzacappa, A., DeMarino, C.: Stability of Standing Accretion Shocks, with an Eye toward Core-Collapse Supernovae. Astrophys. J. **584** (2003) 971–980
14. Scheck, L., Plewa, T., Janka, H.T., Kifonidis, K., Müller, E.: Pulsar Recoil by Large-Scale Anisotropies in Supernova Explosions. Phys. Rev. Letters **92** (2004) 011103–+
15. Scheck, L.: Multidimensional simulations of core collapse supernovae. PhD thesis, Technische Universität München (2006)
16. Keil, W., Janka, H.T., Mueller, E.: Ledoux Convection in Protoneutron Stars—A Clue to Supernova Nucleosynthesis? Astrophys. J. **473** (1996) L111
17. Burrows, A., Lattimer, J.M.: The birth of neutron stars. Astrophys. J. **307** (1986) 178–196
18. Pons, J.A., Reddy, S., Prakash, M., Lattimer, J.M., Miralles, J.A.: Evolution of Proto-Neutron Stars. Astrophys. J. **513** (1999) 780–804
19. Marek, A.: Multi-dimensional simulations of core collapse supernovae with different equations of state for hot proto-neutron stars. PhD thesis, Technische Universität München (2007)
20. Fryer, C.L., Heger, A.: Core-Collapse Simulations of Rotating Stars. Astrophys. J. **541** (2000) 1033–1050
21. Fryer, C.L., Warren, M.S.: Modeling Core-Collapse Supernovae in Three Dimensions. Astrophys. J. **574** (2002) L65–L68
22. Fryer, C.L., Warren, M.S.: The Collapse of Rotating Massive Stars in Three Dimensions. Astrophys. J. **601** (2004) 391–404
23. Rampp, M., Janka, H.T.: Radiation hydrodynamics with neutrinos. Variable Eddington factor method for core-collapse supernova simulations. Astron. Astrophys. **396** (2002) 361–392
24. Janka, H.T., Buras, R., Kifonidis, K., Marek, A., Rampp, M.: Core-Collapse Supernovae at the Threshold. In Marcaide, J.M., Weiler, K.W., eds.: Supernovae, Procs. of the IAU Coll. 192, Berlin, Springer (2004).
25. Buras, R., Rampp, M., Janka, H.T., Kifonidis, K.: Two-dimensional hydrodynamic core-collapse supernova simulations with spectral neutrino transport. I.

Numerical method and results for a 15 Mȯ star. Astron. Astrophys. **447** (2006) 1049–1092
26. Buras, R., Janka, H.T., Rampp, M., Kifonidis, K.: Two–dimensional hydrodynamic core–collapse supernova simulations with spectral neutrino transport. II. Models for different progenitor stars. Astron. Astrophys. **457** (2006) 281–308
27. Müller, E., Rampp, M., Buras, R., Janka, H.T., Shoemaker, D.H.: Toward Gravitational Wave Signals from Realistic Core-Collapse Supernova Models. Astrophys. J. **603** (2004) 221–230
28. Kitaura, F.S., Janka, H.T., Hillebrandt, W.: Explosions of O–Ne–Mg cores, the Crab supernova, and subluminous type II–P supernovae. Astron. Astrophys. **450** (2006) 345–350
29. Marek, A., Kifonidis, K., Janka, H.T., Müller, B.: The SUPERN-project: Understanding core collapse supernovae. In Nagel, W.E., Jäger, W., Resch, M., eds.: High Performance Computing in Science and Engineering 06, Berlin, Springer (2006)
30. Benkert, K., Fischer, R.: An efficient implementation of the Thomas-algorithm for block penta-diagonal systems on vector computers. In Shi, Y., van Albada, G.D., Dongarra, J., Sloot, P.M., eds.: Computational Science – ICCS 2007. Volume 4487 of LNCS., Springer (2007) 144–151
31. Müller, B., Marek, A., Benkert, K., Kifonidis, K., Janka, H.T.: Supernova simulations with the radiation hydrodynamics code prometheus/vertex. In Resch, M., Roller, S., Lammers, P., Furui, T., Galle, M., Bez., W., eds.: High Performance Computing on Vector Systems 2007, Berlin, Springer (2007)
32. Marek, A., Janka, H.T., Buras, R., Liebendörfer, M., Rampp, M.: On ion-ion correlation effects during stellar core collapse. Astron. Astrophys. **443** (2005) 201–210
33. Marek, A., Dimmelmeier, H., Janka, H.T., Müller, E., Buras, R.: Exploring the relativistic regime with Newtonian hydrodynamics: an improved effective gravitational potential for supernova simulations. Astron. Astrophys. **445** (2006) 273–289
34. Hoffman, R.D., Müller, B., Janka, H.T.: Nucleosynthesis in O-Ne-Mg Supernovae. Astrophys. J. **676** (2008) L127–L130
35. Ning, H., Qian, Y.Z., Meyer, B.S.: r-Process Nucleosynthesis in Shocked Surface Layers of O-Ne-Mg Cores. Astrophys. J. **667** (2007) L159–L162
36. Janka, H.T., Müller, B., Kitaura, F.S., Buras, R.: Dynamics of shock propagation and nucleosynthesis conditions in O-Ne-Mg core supernovae. ArXiv e-prints **712** (2007), submitted to Astron. Astrophys.
37. Lunardini, C., Müller, B., Janka, H.T.: Neutrino oscillation signatures of oxygen-neon-magnesium supernovae. ArXiv e-prints **712** (2007), submitted to Phys. Rev. D.
38. Marek, A., Kifonidis, K., Janka, H.T., Müller, B.: The SUPERN-project: Current progress in modelling core collapse supernovae. In Nagel, W.E., Kröner, D., Resch, M., eds.: High Performance Computing in Science and Engineering 07, Berlin, Springer (2007)
39. Marek, A., Janka, H..: Delayed neutrino-driven supernova explosions aided by the standing accretion-shock instability. ArXiv e-prints **708** (2007), submitted to Astron. Astrophys.

Massless Four-Loop Integrals and the Total Cross Section in $e^+ e^-$ Annihilation

J.H. Kühn, P. Marquard, M. Steinhauser, and M. Tentyukov

Institut für Theoretische Teilchenphysik, Universität Karlsruhe, 76128 Karlsruhe, Germany

This is the report for the project ParFORM for the period June 2007 to June 2008.

1 Technical and Physical Framework

A detailed motivation for the current project has been given in the previous reports. Thus, let us only provide a brief account of the physical framework of our calculations.

The main motivation behind the calculations performed within this project is the precise determination of input parameters of the theory describing the fundamental interaction of the elementary particles, the so-called Standard Model. In particular, we are interested in a precise extraction of one of the three coupling constants, the strong coupling, and three out of the six quark masses, the strange, charm and bottom quark mass, by a comparison of our results with experimental data.

The quantity which has to be computed to high precision is the vacuum polarization of the photon, the force particle of Quantum Electrodynamics. This quantity depends on the momentum of the photon, q, and is usually denoted by $\Pi(q^2)$.

An elegant framework to perform calculations within particle physics is based on Feynman diagrams which represent complicated mathematical expressions in simple and intuitive graphical form. The main difficulty in practical applications is the occurrence of closed loops at higher orders in perturbation theory. This project deals at the forefront of what is currently possible and considers Feynman diagrams up to five loops. Some sample Feynman diagrams for $\Pi(q^2)$ up to this order are shown in Fig. 1.

Apart from the external momentum $\Pi(q^2)$ also depends on the mass of the quark, m_q, which is virtually generated for a short amount of time. Beyond two loops an analytical evaluation of $\Pi(q^2)$ is not available. However, for many

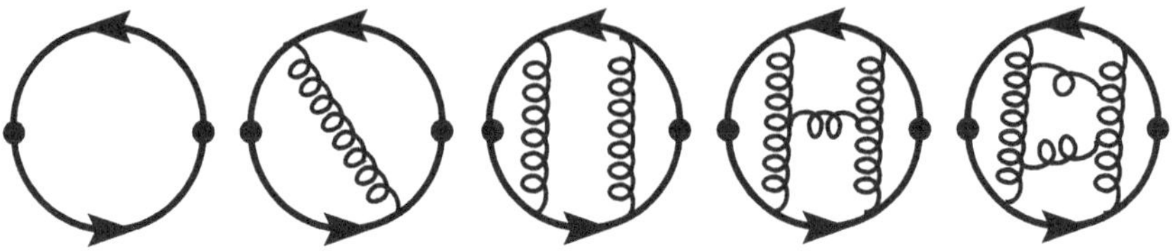

Fig. 1. Sample diagrams contributing to $\Pi(q^2)$

physical applications it is sufficient to have approximations for $m_q^2 \ll q^2$ or $m_q^2 \gg q^2$.

In this project we are interested in the imaginary part of the five-loop contribution for $m_q = 0$ which leads to the physical cross section $R(s)$ defined through

$$12\pi \,\mathrm{Im}\left[\Pi(q^2 = s + i\epsilon)\right] = R(s) = \frac{\sigma(e^+e^- \to \text{hadrons})}{\sigma(e^+e^- \to \mu^+\mu^-)} \,. \tag{1}$$

$R(s)$ can be expanded in the coupling α_s

$$R(s) = \sum_{i \geq 0} \left(\frac{\alpha_s}{\pi}\right)^i \delta R^{(i)} + \text{quark mass effects}\,, \tag{2}$$

and thus provides the possibility to extract α_s by comparing the theory prediction with experimental measurements. Applications related to the τ lepton decay and the strange quark mass have been discussed in the previous reports.

The inverse limit where $m_q^2 \gg q^2$ leads to theoretical input necessary for the determination of the charm and bottom quark mass. In Ref. [1] the first non-trivial four-loop result has been incorporated in the analysis and very precise values for the quark masses has been extracted. The computation of the next expansion term in q^2/m_q^2 is part of this project. Its evaluation constitutes one of the most complicated calculations which are currently undertaken in particle physics.

The workhorse for the complicated calculations involving huge intermediate expressions is the computer algebra program `FORM` [2] and its parallel versions `ParFORM` [3] and `TFORM` [4]. In particular in the context of the evaluation of the four-loop corrections to $\Pi(q^2)$ a `C++` program, `crusher`, has been developed in our group [5]. Both `FORM` and `crusher` are described in the next Section. The physical results obtain within the last year are discussed in Sections 3 and 4.

2 Parallel Computer Algebra

As already mentioned, most of the computations performed in this project are performed by means of `ParFORM` [3], which, in contrast to `Mathematica` or Maple, is designed for the manipulation of huge expressions ranging up to several tera bytes. The effective manipulation of large expressions requires that all algebraic instructions are applied to a big sequence of terms. The sheer size of the intermediate results prevents the storage of more than a single version of an expression.

The internal specifications allow FORM to deal with expression which are much large than the available memory (RAM). The only restriction for the size of an expression is the disk space which nowadays is rather cheap. As a consequence, the complexity of a problem solvable by FORM is practically restricted only by time. In this context an improvement of the efficiency is very important.

The efficiency could basically be improved in three ways:

1. faster hardware;
2. better algorithms;
3. parallelization.

As far as the first point is concerned one has to mention that fast processors and fast disks are available at reasonable prices and thus available for many researchers. Concerning the algorithms we want to mention that in the meantime there is a constant improvement of FORM for more than 15 years and this will be continued also in the coming years. Thus, parallelization is most promising to increase the performance and, in fact, is the only way to reduce the wall-clock time , i.e., the elapsed time between start and end of a compute job.

There are essentially two implementations of ParFORM: one is based on MPI ("Message Passing Interface") which is quite good for systems that have processors with separated memory. The other one is specially adopted to Symmetric-Multi-Processor (SMP) architectures with shared memory implemented using the "Non-Uniform-Memory-Access" (NUMA) technology. Most calculations of the project were done on the machines XC4000 and XC6000. Both computers are clusters with distributed memory, and thus only the MPI-based version of ParFORM can be used. The cluster XC6000 has Itanium2 processors and consists of 108 2-way nodes and twelve 8-way nodes with Quadrics QsNet2 interconnection. XC4000 is based on Opteron processors with 750 compute nodes each with four cores, interconnected by 4X DDR InfiniBand.

Let us mention that recently a different concept for the parallelization of FORM has been developed. The basic idea is the use of POSIX[1] threads in order to realize the communication between the various processors of a shared memory machine. The main application is thus centered around multi-core machines with two, four or eight cores. First tests of TFORM [4] were quite successful and comparable speedup to ParFORM could be achieved. In future we will continue to further development of TFORM. Due to the hardware

[1] "a Portable Operating System Interface for uniX"

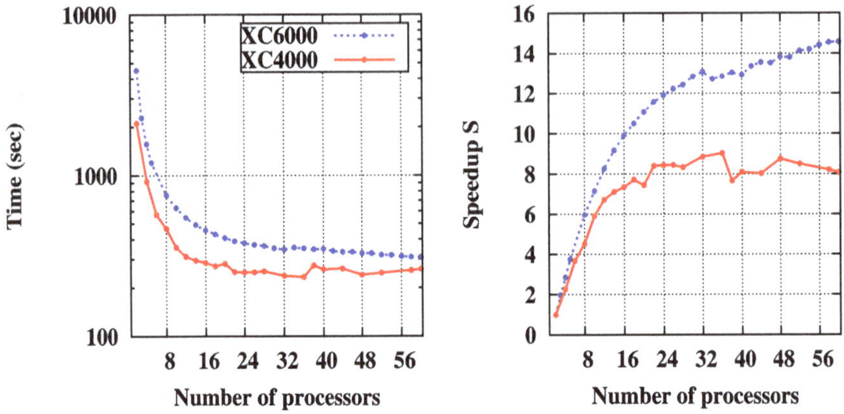

Fig. 2. CPU-time and speedup curve for a typical job on the XC6000 and XC4000

structure of the Landeshöchstleistungsrechner this is very promising. It is particularly tempting to combine `TFORM` and `ParFORM` in order to reach an optimal speedup.

The run-time of our problems varies from a few days or weeks up to about two months. Due to the very structure of `FORM` and due to the organization of the calculation it is not possible to set check points. Since the maximum CPU time at the Landeshöchstleistungsrechner is limited to about seven days it is only possible to submit small and medium-sized jobs.

In Fig. 2 the performance of `ParFORM` is shown for a typical job where up to 60 processors have been used. The most important characteristic of scalability is a speedup on p parallel workers, $S(p) = T_1/T_p$, the ratio of the time spent by one worker for solving the problem to the time spent by p workers. For the XC6000 a good scaling behaviour is observed up to about 16 processors. Above approximately 24 processors the saturation region starts and only a marginal gain is observed once 60 processors are employed. The cluster XC6000 has only about 300 processor cores, and the communication media, QsNet2, has dynamical balancing while the XC4000 cluster is much larger and the communication media, InfiniBand, does not have dynamical balancing. This is probably the reason that for the XC4000 the situation is much worse. Beyond about ten processors the system is very unstable and is thus less attractive for our applications. It seems that the interconnection of the individual nodes is much worse than for the XC6000 cluster.

In the following we describe a simple model which provides a quantitative description of the communication overhead. Let α be the essentially non-parallelizable part of the job. Then the time spent by one worker is $T_1 = \alpha T_1 + (1 - \alpha)T_1$, and p workers need the time

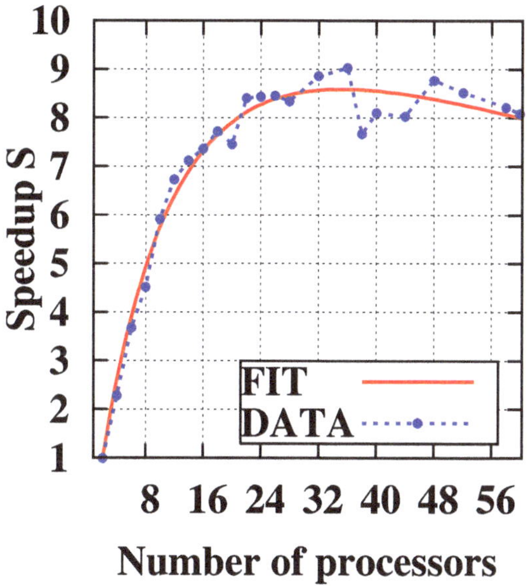

Fig. 3. Speedup curve for XC4000 and fit result of our model (see text)

$$T_p = \alpha T_1 + \frac{(1-\alpha)T_1}{p} + t_d\,, \tag{3}$$

where t_d is the communication overhead. The speedup

$$S(p) = \frac{T_1}{T_p} \;=\; \frac{T_1}{\left(\frac{1-\alpha}{p} + \alpha\right) T_1 + t_d}\,, \tag{4}$$

with $t_d = 0$ corresponds to the Amdahl's law [6].

Let us assume that the cluster XC6000 has a "good" communication medium, i.e., t_d = const = b, while the cluster XC4000 has a "bad" one, $t_d = b_1 + b_2 \times p$. We would like to fit numerically Amdal's constant α from the the XC6000 data and use it to estimate b_1 and b_2 for the XC4000 data. The constant t_d can be absorbed by the α. Indeed, let $\beta = \alpha + \frac{b}{T_1}$, which leads to

$$S(p) = \frac{1}{\frac{1-\beta}{p} + \beta + \frac{b}{pT_1}}\,. \tag{5}$$

Assuming that $b \ll T_1$, the term $\frac{b}{pT_1}$ can be neglected and one obtains from a fit to the XC6000 data the value $\alpha = 0.042 (\approx 4\%)$ (Note, that the communication overhead for the XC6000 machine is included in α). The subsequent fit to the XC4000 date leads to b_1 = 38.0 s and b_2 = 1.7 s. In Fig. 3 we compare the speedup curve and the fit result from our model which seems to work quite good. From the above results for b_1 and b_2 one can see that for our benchmark job the XC4000 spends about 40 s for communications (in addition to the time spent by the XC6000 machine). Furthermore there are about 2 s per worker due to latency. Despite this bad behaviour we currently compute most of our tasks on the XC4000 since the individual processors are significantly faster than the one of the XC6000 cluster.

A further program, `Crusher`, which has been developed in our group, implements the so-called Laporta-algorithm [7, 8]. In high-energy physics the Laporta-algorithm is a widespread tool which is used to reduce the huge number of Feynman integrals occuring in the calculation of physical quantity to a small set of basis integrals. The reduction is essentially based on a particular implementation of the Gauss elimination algorithm which is applied to a system of equation obtained from the original Feynman diagrams. The main problem in the practical implementation of the Laporta-algorithm is that the number of equations ranges typically up to several millions and thus an effective method is mandatory. `Crusher` is written in `C++` and uses the computer algebra program `Fermat` [9] for the manipulation of the coefficients which are rational functions in the space-time, masses and momenta involved in the the problem. The thread-based version of `Crusher`, `TCrusher`, has been further improved to include an adaptive solving algorithm which reduces the number of equations considerably. We can reach a speedup of 3 on a computer with four cores.

3 Massless Four-Loop Integrals: $\sigma(e^+e^- \to \text{hadrons})$

For most of the compute-jobs connected to the massless four-loop propagators integrals we have used 12 processors which leads to the total amount of 120 processors assuming 10 jobs in the batch queue.

$R(s)$ is currently known to order α_s^3 and the corresponding theoretical uncertainty in the value of $\alpha_s(M_Z)$ is around 3% which is the same as the experimental one. In future the experimental errors will be decreased and thus it is necessary to compute the $\mathcal{O}(\alpha_s^4)$ term in order to reduce also the uncertainties in the theoretical prediction. The calculation is highly non-trivial and requires a lot of preparation work to be done which is necessary to fine tune the programs and to accumulate experience.

The preparation work has been under way during last years. A few related projects have been successfully completed and their results have been published [10, 11, 12, 13, 14, 15]. Thus, the theoretical possibility of the complete calculation has been demonstrated.

As mentioned in Section 1 the order α_s^4 contribution to $R(s)$ is related to the absorptive part of the five-loop vector current correlator, whose calculation can be reduced to the evaluation of the four-loop propagator-type integrals (p-integrals).

In order to cope with the problem the special package, `BAICER`, has been created. This is a `FORM` package capable to analytically compute p-integrals up to four loops, based on the approach [16, 17, 18].

The package computes coefficients in decomposition of a given p-integral into the fixed basis of known ones. The coefficients are known to be rational functions of the space-time dimension D and are computed as expansion over $1/D$ as $D \to \infty$. From the knowledge of sufficiently many terms in the expansion one can reconstruct their exact form. The terms in the $1/D$ expansion are expressed in terms of simple Gaussian integrals. The number of the latter for typical 4-loop physical problem is large (of order 10^{10}), but their calculation can be efficiently parallelized.

During 2007 and 2008 the following problems have been computed with `BAICER` on the basis of our local SGI multi-processor computer and the XC4000 in the Rechenzentrum.

1. The evaluation of the β-function for the so-called Quenched[2] QED (QQED) in the five loop approximation. In fact, the quantity can be considered as an important gauge independent contribution to R(s) at order α_s^4 (namely the one proportional to the colour structure C_F^4). The QQED β-function is also a fascinating theoretical object by its own right:
   (i) It is scheme independent in all orders.
   (ii) Its coefficients are simple rational numbers at one, two, three and four loops: $(4/3, 4, -2, -46)$.
   (iii) There is a belief that this characteristic reflects some deep not yet understood property of the quantum field theory which should be valid in all orders (see, e.g., Refs. [19, 20]).

   The result of our calculation for the five-loop term of the QQED β-function unexpectedly contains an irrational contribution proportional to $\zeta(3)$. At the moment we have completed the check of our calculations and submitted the results for publication [21].
2. The corrections of order α_s^4 for the cross section of electron-positron annihilation into hadrons and for the decay rates of the Z boson and the τ lepton into hadrons are evaluated [22]. This allows for an extraction of α_s both for $q^2 = m_\tau^2$ and $q^2 = M_Z^2$. The new terms lead to a significant stabilization of the perturbative series, to a reduction of the theory uncertainly in the strong coupling constant α_s, as extracted from these measurements, and to a small shift of the central value, moving two central values closer together. The agreement between two values of α_s measured at vastly different energies constitutes a striking test of asymptotic freedom. Combining the results from Z and tau decays we find $\alpha_s(M_Z) = 0.1198 \pm 0.0015$ as one of the most precise and presently only next-to-next-to-next-to-leading order result for the strong coupling constant.

[2] i.e., diagrams with closed fermion loops are not considered

4 Massive Vacuum Integrals: $\Pi(q^2)$ to Four Loops

The main software tool in the part of the project dealing with the massive polarization function to four-loop approximation is the program `TCrusher`.

The status of this sub-project is as follows: `TCrusher` has successfully been applied to the three-loop diagrams. The reduction to master integrals took about 48 CPU hours where 4 processors have been used. Once the result is expressed in terms of basis integrals the latter have to be computed. A promising approach is based on an expansion for small external momentum. Using this approach we calculated the low energy expansion of four different currents, namely the vector, axial-vector, scalar and pseudo-scalar current [23].

Concerning the four-loop order, we completed very recently the reduction of the integrals needed for the contribution involving two closed fermion loops. Work on the full result is still ongoing.

Acknowledgements

The computations presented in this contribution were performed on the Landeshöchstleistungsrechner XC4000 and XC6000.

References

1. J.H. Kuhn, M. Steinhauser and C. Sturm, "Heavy quark masses from sum rules in four-loop approximation," arXiv:hep-ph/0702103.
2. `FORM` version 3.0 is described in: J.A.M. Vermaseren, "New features of FORM," arXiv:math-ph/0010025;
   for recent developments, see also: M. Tentyukov and J.A.M. Vermaseren, "Extension of the functionality of the symbolic program FORM by external software," arXiv:cs.sc/0604052;
   `FORM` can be obtained from the distribution site at http://www.nikhef.nl/~form.
3. M. Tentyukov, D. Fliegner, M. Frank, A. Onischenko, A. Retey, H.M. Staudenmaier and J.A.M. Vermaseren, "ParFORM: Parallel Version of the Symbolic Manipulation Program FORM," arXiv:cs.sc/0407066;
   M. Tentyukov, H.M. Staudenmaier and J.A.M. Vermaseren, "ParFORM: Recent development," Nucl. Instrum. Meth. A **559** (2006) 224.
   H.M. Staudenmaier, M. Steinhauser, M. Tentyukov, J.A.M. Vermaseren, "'ParFORM,"', Computeralgebra Rundbriefe **39** (2006) 19.
   See also http://www-ttp.physik.uni-karlsruhe.de/~parform.
4. M. Tentyukov and J.A.M. Vermaseren, "The multithreaded version of FORM," arXiv:hep-ph/0702279.
5. P. Marquard, D. Seidel, unpublished.
6. Gene Amdahl, "Validity of the Single Processor Approach to Achieving Large-Scale Computing Capabilities", AFIPS Conference Proceedings, (30), pp. 483-485, 1967.

7. S. Laporta and E. Remiddi, "The analytical value of the electron (g-2) at order alpha 3 in QED," Phys. Lett. B **379** (1996) 283 [arXiv:hep-ph/9602417].
8. S. Laporta, "High-precision calculation of multi-loop Feynman integrals by difference equations," Int. J. Mod. Phys. A **15**, 5087 (2000) [arXiv:hep-ph/0102033].
9. R.H. Lewis, Fermat's User Guide, http://www.bway.net/~lewis.
10. P.A. Baikov, K.G. Chetyrkin and J.H. Kuhn, "Five-loop vacuum polarization in pQCD: O(alpha(s)**4 N(f)**2) results," Nucl. Phys. Proc. Suppl. **116** (2003) 78.
11. P.A. Baikov, K.G. Chetyrkin and J.H. Kuhn, "Vacuum polarization in pQCD: First complete O(alpha(s)**4) result," Nucl. Phys. Proc. Suppl. **135** (2004) 243.
12. P.A. Baikov, K.G. Chetyrkin and J.H. Kuhn, "Strange quark mass from tau lepton decays with O(alpha(s**3)) accuracy," Phys. Rev. Lett. **95** (2005) 012003 [arXiv:hep-ph/0412350].
13. P.A. Baikov, K.G. Chetyrkin and J.H. Kuhn, "Scalar correlator at O(alpha(s)**4), Higgs decay into b-quarks and bounds on the light quark masses," Phys. Rev. Lett. **96** (2006) 012003 [arXiv:hep-ph/0511063].
14. P.A. Baikov and K.G. Chetyrkin, "New four loop results in QCD," Nucl. Phys. Proc. Suppl. **160** (2006) 76.
15. P.A. Baikov and K.G. Chetyrkin, "Higgs decay into hadrons to order alpha(s)**5," Phys. Rev. Lett. **97** (2006) 061803 [arXiv:hep-ph/0604194].
16. P.A. Baikov, "Explicit solutions of the multi-loop integral recurrence relations and its application," Nucl. Instrum. Meth. A **389** (1997) 347 [arXiv:hep-ph/9611449].
17. P.A. Baikov, "Explicit solutions of the three loop vacuum integral recurrence relations," Phys. Lett. B **385** (1996) 404 [arXiv:hep-ph/9603267].
18. P.A. Baikov, "The criterion of irreducibility of multi-loop Feynman integrals," Phys. Lett. B **474** (2000) 385 [arXiv:hep-ph/9912421].
19. D.J. Broadhurst, "Four-loop Dyson-Schwinger-Johnson anatomy," Phys. Lett. B **466** (1999) 319 [arXiv:hep-ph/9909336].
20. A. Connes and D. Kreimer, "Renormalization in quantum field theory and the Riemann-Hilbert problem," JHEP **9909** (1999) 024 [arXiv:hep-th/9909126].
21. P.A. Baikov, K.G. Chetyrkin and J.H. Kuhn, "Massless propagators: applications in QCD and QED", To be published in the procceedings of the 8th International Symposium on Radiative Corrections: (RADCOR 2007): Application of Quantum Field Theory to Phenomenology, Florence, Italy, October 1-5, 2007"
22. P.A. Baikov, K.G. Chetyrkin and J.H. Kuhn, "Hadronic Z- and tau-Decays in Order $alpha_s^4$", arXiv:0801.1821 [hep-ph]
23. A. Maier, P. Maierhofer and P. Marquard, at Nucl. Phys. B **797** (2008) 218 [arXiv:0711.2636 [hep-ph]].

Solid State Physics

Prof. Dr. Werner Hanke

Institut für Theoretische Physik und Astrophysik, Universität Würzburg, Am Hubland, 97074 Würzburg, Germany

P. Nielaba from the University of Constance is an internationally recognized expert in dealing with nanostructures in reduced geometry. In this field, computer simulations have become an indispensable tool, due to the fact that these nanosystems contain about up to 104 particles. Many important results for these nanosystems in reduced geometry, which are of significant technological relevance, have already been obtained by earlier support of HPC centers as overviewed and discussed already in the proceedings of the last years. The present report by P. Nielaba contains several new insights dealing with quantum effects in nanowires, phase transitions of model colloids in external fields, dynamics of micro-channels and the electronic properties of clusters. The main numerical tool is the path integral Monte Carlo simulation, which, in particular, makes very efficiently use of an efficient scaling concerning the numbers of parallel processors. The Monte Carlo procedure employed requires the computation of statistical averages, which can be done if averages of system replicas with different initial conditions are computed in parallel on several processors. Especially this topic is discussed in detail in Nielabas article.

The group F. Bechstedt with R. Leitsmann and F. Ortmann from the University of Jena has investigated structural and electronic properties of PbTe nanocrystals embedded in solid state matrices (for example, CdTe), using a density-functional method supplemented with repeated supercell approximations. This project has led to new insights from the electronic structure point of view and the corresponding optical properties. The question of light sources in the mid-inference spectral region, is crucial for many applications, for example for medical diagnostics and gas-sensor systems. To describe such real-life systems like, e.g. bulk materials, their interfaces and surfaces or nanowires and nanodots, one has, in principle, to solve the stationary Schrödinger question for the corresponding ensemble of interacting electrons and ions. The so-called adiabatic approximation allows for a decoupling of ionic and electronic degrees of freedom but still one faces, in principle, an extremely complicated many-body problem with 1023 interacting particles. The main idea also followed in the Bechstedt et. al. work, is to employ the density functional theory (DFT) in the so-called local-density approximation (LDA). Here, one solves an effective single-particle equation self-consistently to obtain the ground state density and from it all physical properties that are a functional of this density. Despite this enormous simplification, where the electron is treated as a single particle embedded in an "effective mean-field", the numerical solution is still extremely demanding by the presence of the complicated lattice structure of the PbTe nanocrystals embedded in a CdTE matrix. The difficulties related to this are discussed convincingly in the article on the "Quantum confined Stark-effect in embedded PbTE nanocrystal" by the above group of authors.

T. Ulbricht and P. Schmitteckert from the University of Karlsruhe discuss the signal transport and the conductance of correlated nanostructures in their article. They give a insight into the fact why the transport properties of strongly interacting quantum many-body systems are a major challenge in todays condensed matter theory. The breakthrough in this project is that the authors managed to obtain the finite bias conductances from real-time dynamics, working consistently from small to large voltage regimes. The calculations are, in particular, based on the density matrix renormalization group method using algorithmic developments to parallelize the code as described in detail in the present and earlier articles.

An interesting paper by the group around A. Muramatsu from the University of Stuttgart, concerning the possibility of supersolid fermions inside a harmonic trap, concludes the interesting papers of the solid state area. This latter paper is in the stream of a variety of new works following a renewed interest in the possible coexistence of superfluidity and crystalline order, a phase named supersolid. The present project uses Quantum Monte Carlo simulations to study the properties of ultra cold fermionic atoms on optical lattices when confined in a one-dimensional harmonic trapping potential. For this system, the high-performance computer simulation established a supersolid phase, also inside a confined geometry.

Computer Simulations of Complex Many-Body Systems

C. Schieback, F. Bürzle, K. Franzrahe, J. Neder, M. Dreher, P. Henseler, D. Mutter, N. Schwierz, and P. Nielaba

Physics Department, University of Konstanz,
78457 Konstanz, Germany
peter.nielaba@uni-konstanz.de

Summary. The static and dynamic properties of model magnetic systems have been studied by the Landau-Lifshitz-Gilbert equation. Soft matter systems have been investigated by Monte Carlo and Brownian Dynamics simulations. In particular the behaviour of two dimensional binary hard disk mixtures in external periodic potentials has been studied as well as the transport of colloids in micro-channels and the features of lipid bilayers under tension. Certain aspects of star cluster formation processes have been computed using smoothed particle hydrodynamics. The conductance of ferromagnetic atomic-sized contacts has been analyzed by Molecular Dynamics simulations with respect to their conductance and structural properties under stretching. In the next sections we give an overview on our recent results.

1 Simulations of Spin Structures and Spin Dynamics in Nano-Structures

Nano-structured magnetic materials form the basic building blocks for several devices of the next generation [1, 2, 3]. Magnetic nano-structures reveal new properties and phenomena, when the system sizes are comparable to characteristic lengths as the spin diffusion length, the mean free path length or the domain wall (DW) size.

By computer simulation methods using a Heisenberg model [4] we investigated the effects of finite system sizes and confining geometry on the spin structure and stability of magnetic DW types as well as the effect of spin transfer torque on the DW motion including temperature effects.

The spin structure of DWs in constrictions down to 20 nm was investigated with numerical simulations using a Heisenberg model. The simulation results are compared to experiments [5]. As shown in Fig. 1 symmetric and asymmetric transverse DWs for different constriction sizes w_c are observed, consistent with experiments. The symmetric transverse DW is obtained in small constrictions, with $w_c = 20 - 80$nm, and exhibits an elliptical shape. For wider constrictions, $w_c \geq 160$nm, an asymmetric spin structure is favored. For intermediate $w_c = 120$nm, we find both wall types depending on the element width w_e. The experimentally observed asymmetric transverse walls can be further divided into tilted and buckled walls, the latter being an intermediate state just before the vortex nucleation.

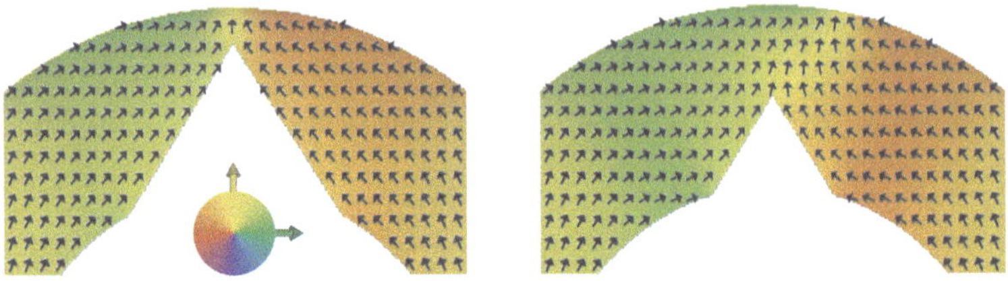

Fig. 1. Magnetization configurations of symmetric (right hand side) and asymmetric (left hand side) transverse DWs in constrictions [5]; shown are parts of simulated permalloy structure with thickness 4nm and element width $w_e = 400$nm; $w_c = 80$nm for the symmetric transverse DW; $w_c = 200$nm for the asymmetric DW. The color code and the arrows indicate the magnetization direction

The key energy contributions to the DWs are the exchange energy, which favors large wall widths, and the stray field energy (shape anisotropy), which favors alignment of the spins parallel to the element edges. The increasing influence of the stray field energy results in smaller DW widths w_{DW} for smaller constrictions as shown in Fig. 2. As the constriction width decreases, the DW width decreases faster than linearly, which leads to very narrow DWs for narrow constrictions.

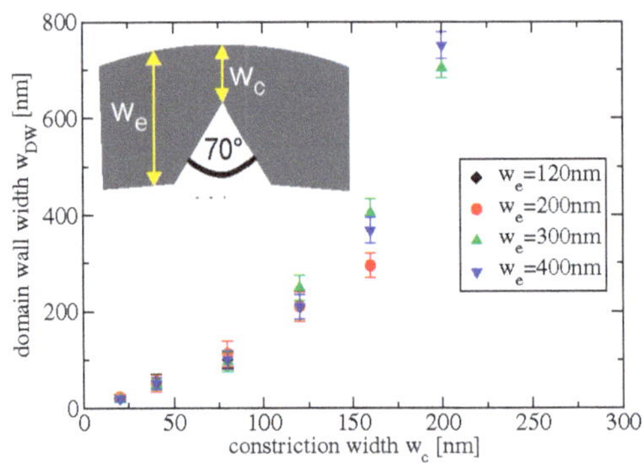

Fig. 2. Dependence of the DW width w_{DW} on the constrictions width w_c [5]. The inset shows a schematic of the permalloy element geometry, with element width w_e, constriction width w_c and notch angle 70°

Interesting effects of electric current on the dynamics of DWs have been studied theoretically recently [6, 7, 8]. Analytical and numerical results for the

DW velocity, the reversible displacement x_{DW} and the deformation of the DW have been obtained for the case of a low current $u_x < u_c$ below the Walker breakdown [6] and $\beta = 0$. In [7] an approximate analytical prediction for the long term DW velocity v as a function of effective velocity u_x has been derived: $v = \sqrt{u_x^2 - u_c^2}/(1+\alpha^2)$ for effective velocities exceeding a critical effective velocity u_c. These predictions have been reproduced by our simulations [9].

So far the analytical predictions and numerical calculations do not include temperature effects. In Fig. 3 simulation results of the DW displacement x_{DW} are shown including temperature effects $k_B T/J \neq 0$. For currents $u_x < u_c$ below the Walker breakdown at $T = 0K$ the DW does not move any further in contrast to simulations performed at finite temperatures.

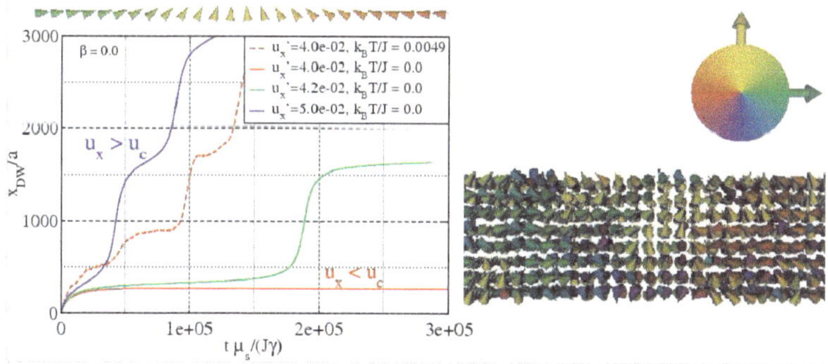

Fig. 3. Left: Time dependence of the DW displacement x_{DW}. Simulation results for a finite temperature ($k_B T/J \neq 0$) and 0K are shown for a spin chain (d=1). Right: Snapshot of a transverse DW in a wire (d=3) at finite temperature under the influence of a current

A related system of magnetic moments confined to the caps of colloids has been studied in our group with computer simulations [10], motivated by the behavior of ferromagnetic Co/Pd caps under the influence of external magnetic fields. These caps can be regarded as multilayers on the top of non-magnetic spheres with diameters in the nanometer range. As Albrecht *et al.* pointed out, such systems are a promising approach for a further reduction of the magnetic storage density in hard-disks [11].
The magnetic volume was modelled by 3-dimensional spins on a cubic lattice which interact with each other by exchange and dipolar-coupling [4]. The interlayer anisotropy between Co- and Pd-layers points always perpendicular to the surface of the sphere.

The dynamics of the system was analyzed by numerical solving of the Landau-Lifshitz-Gilbert equation. Hysteresis curves were observed and the corresponding coercivities as a function of the applied field angle are shown in Figure 4 for different cap-sizes. The curves differ from the theoretical behavior of a classical Stoner-Wohlfahrt particle, a uniformly magnetized prolate ellipsoid with uniaxial anisotropy. The origin of this difference is the underlying reversal mechanism at the critical field strength, which, in case of the caps, takes place in form of a domain-wall propagation ("*DWP*") instead of coherent rotation. By analyzing caps with uniaxial anisotropy, the Stoner-Wohlfahrt behavior is much better reproduced (Figure 4). This shows, that the special feature of the interlayer anisotropy causes the reversal by DWP.

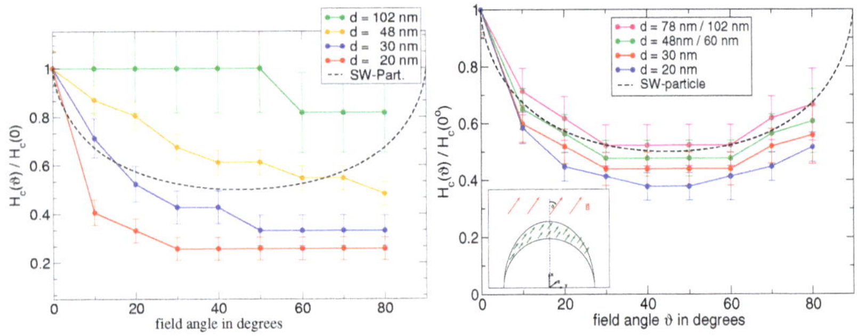

Fig. 4. Coercivities as a function of field-angles. Left: spherical anisotropy, Right: uniaxial anisotropy

2 Two-Dimensional Colloidal Systems in Periodic External Fields

Minimization trends in physics and technology have caused a lot of interest in monolayers and their interactions with a substrate lately. The assembly of nano-particles into spatially extended regular structures is the first step towards a new generation of materials and devices. Colloidal dispersions in external fields or under confinement are valuable model systems for the systematic study of such settings, as they are experimentally directly accessible via laser scanning microscopy. The interactions within the monolayer can be altered by changing the interaction potential of the colloids, while the shape and strength of the substrate potential can be modeled by external light fields. The advantage of the model system is, that via laser scanning microscopy direct access to the particle configurations is given. In this way it is possible to gain insight in the relative importance of the various possible physical processes that occur. From the theoretical point of view, even the relatively simple combination of a monodisperse system in a one-dimensional, spatially periodic light field shows a highly non-trivial phase behavior as the amplitude of the external field is raised: Laser Induced Freezing (LIF) and Laser Induced Melting (LIM). In particular we explored [13] a hard disk system with commensurability ratio $p = \sqrt{3}a_s/(2\lambda) = 2$, where a_s is the mean distance between the disks and λ the period of the external potential. Three phases, the modulated liquid (ML), the locked smectic (LSm) and the locked floating solid (LFS) have been observed, in agreement with other experimental [14] and analytical [15] studies. Various statistical quantities like order parameters, their cumulants and response functions, have been used to obtain a phase diagram for the transitions between these three phases [13].

We address the question of how the addition of another length scale into such two-dimensional systems influences the intricate competition between adsorbate-adsorbate interaction and adsorbate-substrate interaction by studying a binary 50% mixture under the influence of a one dimensional (1d) spatially periodic substrate potential. We show, that adding another competing length scale generates novel, interesting phenomena [16, 17, 18, 19].

Defining the binary mixture, we set the mixing ratio to $x_A = x_B = 50\%$ and the diameter ratio to $\sigma_B/\sigma_A = 0.414$. The external potential is spatially periodic in the x- direction with

$$V(\mathbf{r}) = V_0^* \sin(\mathbf{K} \cdot \mathbf{r})$$

$V_0^* = V_0/k_BT$ is the reduced potential strength, where k_B is the Boltzmann constant and T the temperature. The wave vector is $\mathbf{K} = \frac{4\pi}{a}(1, 0)$, where a is the lattice constant of the $S_1(AB)$ square lattice. The corresponding wave-

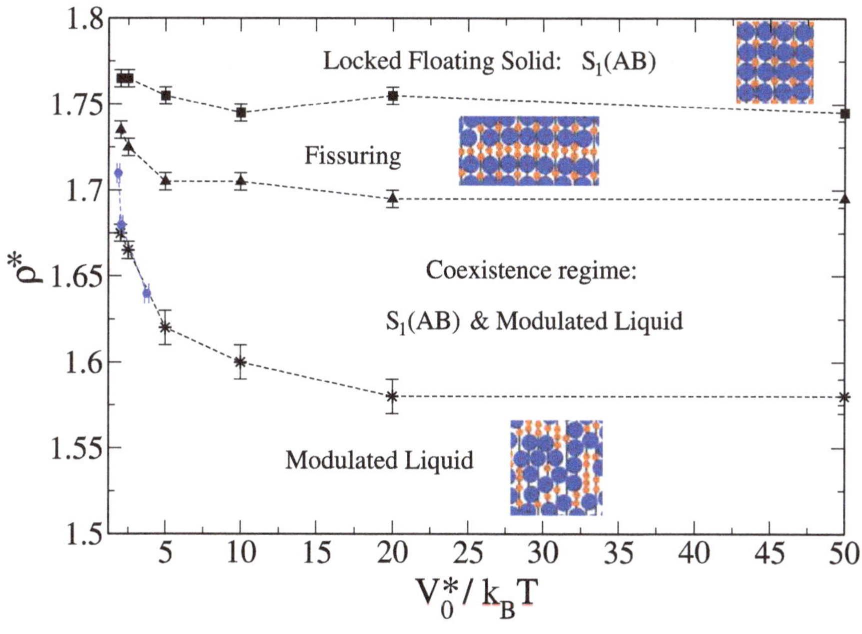

Fig. 5. The $\varrho^* - V_0^*$ plane of the phase diagram of an equimolar binary mixture ($\sigma_B/\sigma_A = 0.414$) for a system in which only the smaller component of the mixture interacts with the external field [16]

length of the external field was chosen to be commensurate to the square lattice, which yields the highest packing fraction for the given mixture. The commensurability ratio, i.e. the ratio of the wave vector to the corresponding parallel reciprocal lattice vector is therefore $p = 2$. Simulations in the NVT-Ensemble were carried out. In order to facilitate equilibration a cluster move by L. Lue and L.V. Woodcock [12] was used. In addition special large moves, which attempt to translate the particle over integer multiples of the wavelength of the external field, were employed, besides the standard Metropolis algorithm. Periodic boundary conditions were employed in all simulations. Figure 5 shows the phase diagram as calculated from the simulations for the case, that only the smaller component interacts with the external field [16]. It was taken by lowering the dimensionless number density $\varrho^* = \varrho a^2$. We took a commensurate path through phase space, i.e. the wavelength of the external field $\lambda = 1/(\sqrt{2\varrho^*})$ was adjusted to the corresponding ϱ^*. Part of the phase diagram was obtained by raising the potential strength V_0^* at constant ϱ^*. Simulations carried out in an incommensurate setting, i.e. λ is kept constant independent of ϱ^*, intersect the phase diagram consistently. At low external field strengths ($V_0^* \leq 1.5$) we observe a laser induced demixing [17, 19]. A small component enriched binary fluid coexists with a monodisperse triangular lattice formed by the larger component. The mixture does not phase separate in the field free case. This laser induced phase separation is driven by the attempts of the smaller components to form chains along the minima of the external field. At higher external field strengths a laser induced freezing (LIF) into the commensurate $S_1(AB)$ square lattice structure takes place. Depending on the overall number density ϱ^* the $S_1(AB)$ locked floating solid is either in the one phase regime or coexists with an equimolar binary fluid. A fissuring regime separates these two regimes from each other. Here the square lattice of the larger component grows fissures perpendicular to the minima of the external field. The smaller component accumulates in these fissures. A laser induced melting transition as it is observed in monodisperse systems does not occur in the binary case. Such a transition is geometrically blocked in the analyzed binary mixture exposed to a commensurate external field with $\lambda = 1/(\sqrt{2\varrho^*})$.

3 Transport of Colloids in Micro-Channels

We studied the motion of superparamagnetic particles through narrow channels in a two-dimensional (2D) planar system of size $L_x \times L_y$ under influence of a constant driving force by Brownian Dynamics (BD) simulations [20, 21]. These systems are the dynamic generalizations of the static system studied in [22, 23].

The pair interaction $V(r) = (\mu_0/4\pi)M^2/r^3$ (M is the dipole moment) is purely repulsive and can be characterized by the dimensionless interaction strength $\Gamma = \mu_0 M^2 \rho^{2/3}/(4\pi k_B T)$.

The interactions between the particles are adjusted to a range where the unrestricted system is hexagonally ordered. Therefore, in presence of parallel confining walls, layers of particles can be seen in the motion of the particles. Additionally, a density gradient forms along the channel due to the motion of the particles and the interparticle interactions. This density gradient forces a rearrangement in the layering of the system. The particles are driven by a constant external force from left to right (positive x-direction). As a result of the longitudinal density gradient more layers of particles form near the channel entrance compared to the channel exit [21, 20, 25]. Consequently, there exist multiple regions where the particles rearrange. The positions of these layer reductions do not move together with the particles under stationary non-equilibrium conditions. We therefore do not observe plug flow of a crystal, but rather a dynamic behavior of particles moving in layers and adapting to the external potential. Studies of the local particle separation in x- (along the channel) and y- (perpendicular to the motion of the particles) directions show that the system transforms from a situation, in which it is stretched in flow direction, to a situation, in which it is compressed in flow direction. This fact can be explained by estimating the energies for a crystal in the channel with different numbers of layers [21].

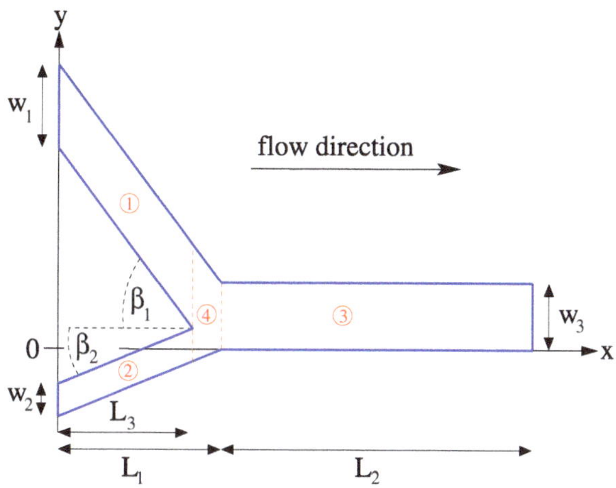

Fig. 6. Sketch of the general channel geometry of a junction as being realized in the simulation code. This geometry has four different regions: the three arms with numbers 1, 2, and 3 plus the mixing region 4

Several interesting aspects of this system have been analyzed [20], ranging from the effect of *single file diffusion* in very narrow channels, which can be filled with one layer of particles only, to the effect of optical traps on the structure formation, and to the effects of *counterflow* on the formation of lane structures.

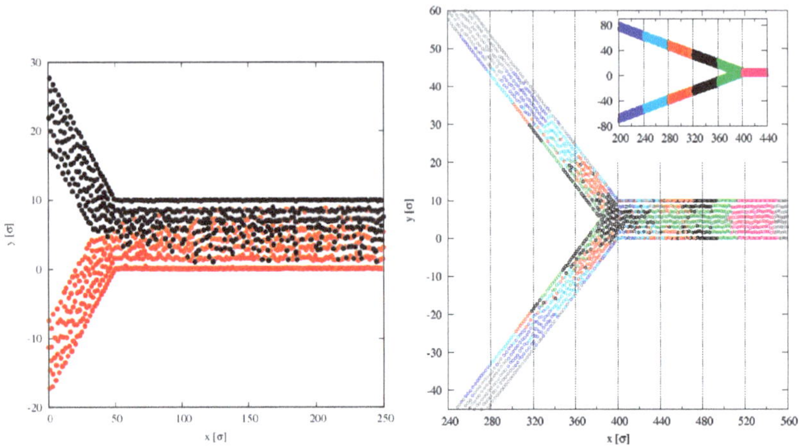

Fig. 7. (a) Snapshot of a symmetric junction with $\beta_1 = \beta_2 = 20^0$ and $\rho = 0.324$, $L_1 = L_2 = 400$, $w_1 = w_2 = w_3 = 10$. The magnetic field strength $B = 0.14mT$ is very weak and the system is in the fluid state. The slope is chosen as $\alpha = 0.2^0$. Under these parameters the particles originating from the two inputs slightly mix. (b) Mixing behavior of a symmetric junction for the same parameters as in (a) at a magnetic field strength $B = 0.24mT$. The particles are all identical and color coded according to the time origin (see inset). The time origin has been chosen after the system has been driven in positive x-direction for $2.5 \cdot 10^6$ BD time steps. All particles of the system which are not shown in the inset have gray color. Shown is a snapshot after the system has evolved over $5 \cdot 10^5$ BD time steps, i.e. $\Delta t = 39.5\tau_B$

The effect of the dimensionality on the layer reduction scenario and on the counterflow has been analyzed by corresponding simulations of systems in three dimensional channels [26, 20]. In three dimensional channels with square shaped cross sections, a reduction of planes was found [26]. Similar effects can be found in systems confined between two parallel ideal hard walls [20].

In microfluidic applications and for lab-on-a-chip technologies the mixing behavior of different chemical reactants is of high importance and subject of intense research. On microscopic length scales the fluid flow is always laminar (low Reynolds numbers), which makes it much more difficult compared to turbulent mixing.

To study the mixing behavior occurring by diffusion alone, simulations have been done [20] with different junction geometries with two inputs and a single outflow channel or vica versa for opposite driving field. The general setup is sketched in figure 6. Variable parameters of this "Y-junction" are the lengths L_1 and L_2, the widths w_1, w_2, and w_3 of each arm, and the opening angles β_1 and β_2. The driving force still acts in x-direction. Modification of the opening angles β_1 and β_2 thus allows for different particle flow velocities in the arms 1 and 2.

An analysis of the system time evolution, as shown in figure 7, reveals that the flow is mixing (a) and that particles at the outer parts of the channel move much faster than the particles close to the symmetry axis (b), and particles close to the inner corner may even move "backwards", opposite to the overall flow direction, apparently due to formation of vortices.

In future studies we plan to extend our studies to explore the flow behavior in dependence of the particle interaction range, the characteristics of the channel walls, and the channel geometry (bottlenecks, barriers). Experiments on the above mentioned geometries are planned as well, the optical masks, which are required to produce such structures, have been already fabricated [27].

4 Lipid Bilayers Under Tension

Lipid bilayers forming biological membranes display a great variety of different physical phenomena. Using a generic model recently developed by O. Lenz and F. Schmid [29, 30] we are investigating the effect of an applied tension to a bilayer.

In our lipid model molecules are represented by chains made of one head bead and six tail beads. Neighboring beads at a distance r within the molecule interact via a FENE potential and the angle θ between subsequent bonds in the lipid gives rise to a stiffness potential. Tail beads, which are not direct next neighbors, interact via a truncated and shifted Lennard-Jones potential. Head-head, head-tail, head-solvent and tail-solvent interactions are modelled by the repulsive part of the Lennard-Jones potential.

The tension is put into effect by an additional energy term $-\gamma A$ to the Hamiltonian of the system, where A is the projected area of the bilayer onto the xy-plane. Therefore, the effective Hamiltonian for our Monte Carlo simulation reads

$$H_{eff} = H + PV - \gamma A - Nk_BT \ln(V/V_0) \quad . \tag{1}$$

H is the interaction energy, V the volume of the simulation box, V_0 an arbitrary reference volume and N the total number of beads [28, 31].

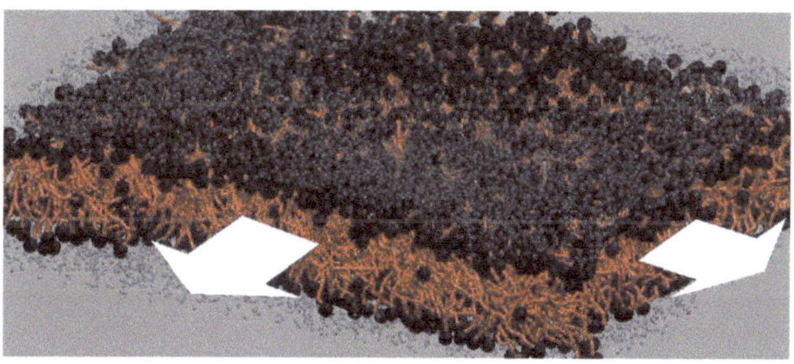

Fig. 8. A model bilayer under lateral tension

At lower temperatures the model system forms a well ordered bilayer phase, which can be identified with the experimentally known tilted gel phase $L_{\beta'}$. At higher temperatures the system forms a less ordered phase, which corresponds to the fluid phase L_α. We measured the area per lipid and the bilayer thickness for different temperatures at increasing lateral tension. In the gel phase the system is much less extensible compared to the fluid phase, which can be seen in figure 9.

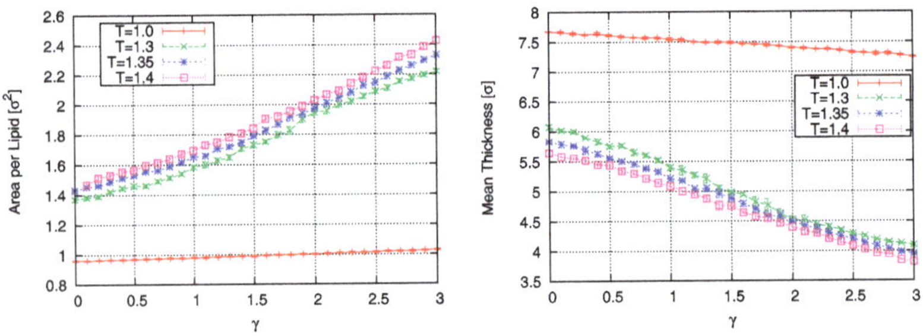

Fig. 9. Area per lipid (left) and bilayer thickness (right) at different temperatures for increasing tension γ

When looking at the fluctuation spectra of systems under increasing lateral tension one finds a decrease of the fluctuation height $\langle |h(q)^2| L^2 \rangle$ (figure 10). This means that the additional tension is damping the fluctuations of the bilayer around its average midplane. Moreover, fluctuations with shorter wavelength (corresponding to higher values of $\langle |q^2| \rangle$) are also less likely when increasing the tension on the bilayer.

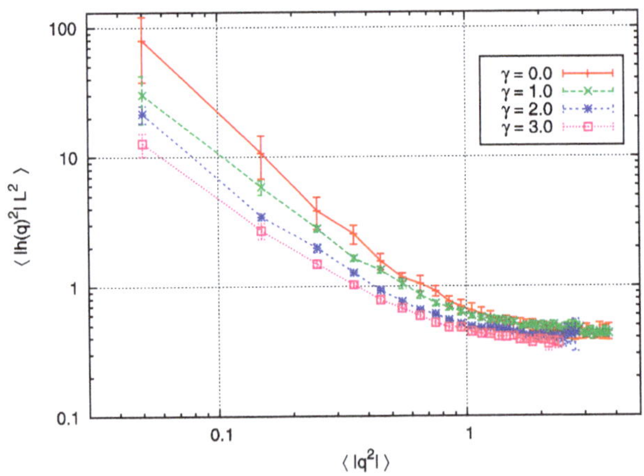

Fig. 10. Lateral tension reduces fluctuations in the bilayer

5 Magnetic Fields in Star Cluster Formation

A new research project is aimed at star cluster formation, specifically at the role magnetic fields play in molecular clouds prior to as well as during the gravitational collapse.

In general, the interest in star formation has grown significantly within the last decade, essentially due to a new understanding of the star formation process, based on the interplay between supersonic turbulence and self-gravity of the interstellar gas (see, e. g., [33]).

Although the importance of magnetic fields in the star formation process was recognized early, the details remain still unclear. From a theoretical point of view this is due to the complexity of the involved physical processes, imposing severe restrictions on any computer simulation. Appropriate simulations should treat a highly compressible, turbulent, self-gravitating fluid governed by the equations of (non-ideal) magnetohydrodynamics (MHD) including chemical reactions and radiation transport. Contemporary simulations include only parts of these processes, and especially the treatment of MHD is known as being notoriously difficult.

Basically, there exist two numerical schemes to attack these problems. One is the well know grid CFD, based on finite differences, and the second is the particle method SPH (Smoothed Particle Hydrodynamics) [34]. In this project, the latter approach is utilized using the N-body/SPH code Gadget-2 [38] which was developed for use on massively parallel computers with distributed memory. Our present version of the code includes additional, state of the art modifications, making it more suitable for star formation studies (courtesy of Ralf Klessen's star formation group, University of Heidelberg). However, due to numerical difficulties, merging MHD with SPH has not been successful so far. But recently, new algorithms have been developed to overcome this issue (see [37] and references therein). The utilization of these new algorithms will be addressed within this project.

Several preliminary studies have already been performed in this project. One of those, which we are going to present here, concentrates on the collapse and fragmentation of molecular cloud cores with initial conditions leading to binary star formation. This is a well know problem and important test case for our future work, since [36] used a variation of this problem to test their MHD implementation. For a more detailed description of the parameters used, see [32] and references therein.

Within the simulation, the initial cloud was chosen to be perfectly spherical with mass $M = 1\mathrm{M}_\odot$, radius $R = 4.99 \times 10^{16}\mathrm{cm}$, density $\varrho_0 = 3.86 \times 10^{-18}\mathrm{gcm}^{-3}$ and uniform angular velocity $\omega_0 = 7.2 \times 10^{-13}\mathrm{s}^{-1}$. The thermodynamics of this model is identical to an ideal gas with mean molecular weight $\mu \approx 3$. The whole setup thus corresponds to ratios of thermal and rotational energies to the absolute value of gravitational potential energy of $\alpha \approx 0.26$ and $\beta \approx 0.16$, respectively. Additionally, a small perturbation of the form $\varrho = \varrho_0[1 + a\cos(m\varphi)]$ (with $m = 2, a = 0.1$) is applied on the underlying

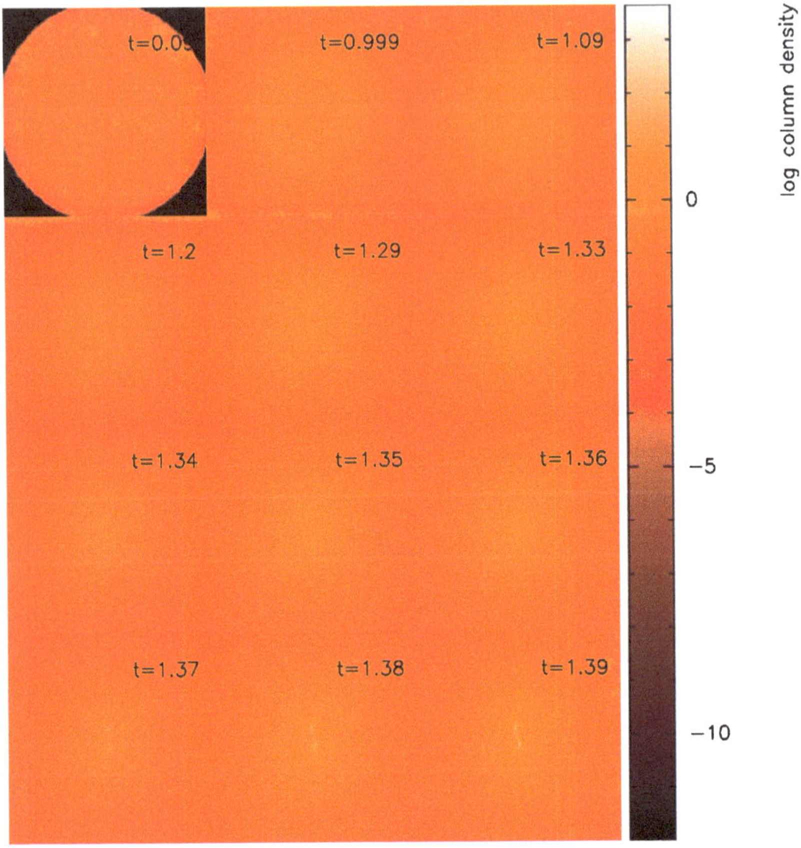

Fig. 11. The figure shows the column density distribution of the whole evolution from the initial cloud to a binary star (time in code units). The image was processed from raw data using the special purpose rendering software SPLASH by D. J. Price, University of Exeter [35]

uniform density distribution, φ being the azimuthal angle with respect to the z-axis.

The energy equation is not solved within the simulation, but the change in temperature during the collapse is taken into account by the barotropic equation of state $p = c_{\text{iso}}^2 \varrho + K\varrho^\gamma$ with the adiabatic exponent $\gamma = 5/3$ (rotational and vibrational degrees of freedom are frozen out), the isothermal sound speed c_{iso} and $K = c_{\text{iso}}^2 \varrho_{\text{crit}}^{1-\gamma}$. At $\varrho = \varrho_{\text{crit}}$ the isothermal and adiabatic parts of the barotropic equation are equal. For the free parameter ϱ_{crit} we chose the value $\varrho_{\text{crit}} = 5.0 \times 10^{12}\text{gcm}^{-3}$.

The simulation presented here was done with resolution of 1.29×10^6 particles, thus implying a particle mass $m_p \approx 7.75 \times 10^{-7}\text{M}_\odot$, so that the Jeans condition at $\varrho = \varrho_{\text{crit}}$

$$m_p < \frac{6c_{\mathrm{iso}}^3}{N_{\mathrm{neigh}}\varrho_{\mathrm{crit}}^{1/2}}\left(\frac{2}{G}\right)^{3/2} \tag{2}$$

is fulfilled. A visualization of results from the performed simulations is supplied in Figure 11, showing the evolution of a molecular cloud with initial conditions as stated above. In this figure the projection of the density distribution of the cloud is shown.

6 The Conductance of Ferromagnetic Atomic-Sized Contacts

Metallic nanowires fabricated by means of scanning-tunneling microscope and break-junction techniques have turned out to be a unique playground to test basic concepts of electronic transport at the atomic scale [39]. Usually the conductance of these contacts is described by the Landauer formula $G = G_0 \sum_n T_n$, where the sum runs over all the available conduction channels, T_n is the transmission for the nth channel, and $G_0 = 2e^2/h$ is the quantum of conductance. In the case of broken spin symmetry one should include a sum over spin and replace G_0 by e^2/h in the previous formula. It has been shown that the number of channels in a one-atom contact is mainly determined by the number of valence orbitals of the central atom, and the transmission of each channel is fixed by the local atomic environment [40, 41, 42].

During the last years a lot of attention has been devoted to the analysis of contacts of magnetic materials [43, 44, 45, 46, 47, 48, 50]. In these nanowires the spin degeneracy is lifted, which can potentially lead to interesting spin-related phenomena in the transport properties. For instance, different groups have reported the observation of half-integer conductance quantization either induced by a small magnetic field [45] or even in the absence of a field [47, 48]. These observations are quite striking since such quantization requires simultaneously the existence of a fully spin-polarized current and perfectly open conduction channels [49]. With our present understanding of the conduction in these metallic junctions, it is hard to believe that these criteria can be met, in particular, in the ferromagnetic transition metals (Ni, Co, and Fe). As a matter of fact, in a more recent study by Untiedt *et al.*,[50] carried out at low temperatures and under cryogenic vacuum conditions, the complete absence of quantization in atomic contacts of Ni, Co and Fe has been reported, even in the presence of a magnetic field as high as 5 T. Several recent model calculations [51, 52, 53, 54, 55, 56, 57, 58, 59, 60] have convincingly shown that neither conductance quantization nor full spin polarization are to be expected in contacts of this kind.

In spite of the coherent picture that is emerging, one still misses in the literature a comparative analysis of the ferromagnetic 3*d* materials (Fe, Co, and Ni) that clarifies basic issues like which orbitals are relevant for the transport, the role of atomic disorder, or the dependence of the spin polarization

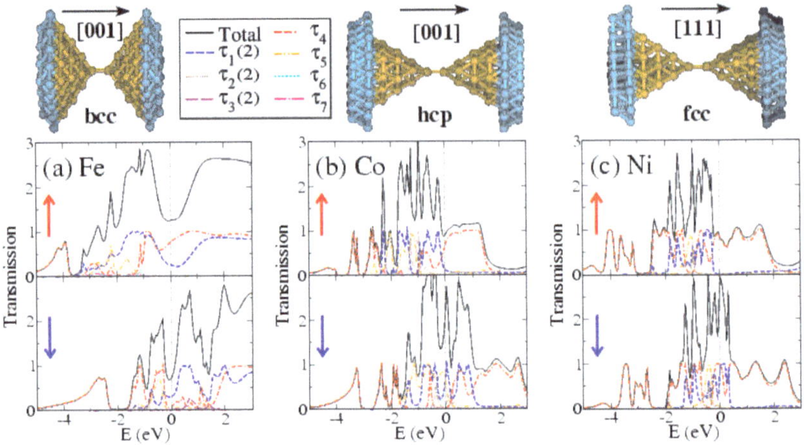

Fig. 12. Transmission as a function of energy for the three dimer structures of (a) Fe, (b) Co, and (c) Ni, which are shown in the upper panels. We present the total transmission (black solid line) for both majority spins and minority spins as well as the transmission of individual conduction channels that give the most important contribution at Fermi energy, which is indicated by a vertical dotted line. The blue, brown and violet dash-dotted lines of τ_1, τ_2, and τ_3 refer to twofold degenerate conduction channels. The legends in the upper graphs indicate, in which direction the contacts are grown. These contacts contain in the central region 59 atoms for Fe, 45 for Co, and 39 for Ni. The blue atoms represent a part of the atoms of the leads (semi-infinite surfaces) that are coupled to the central atoms in our model

of the current on the thickness of the contact. To fill this gap, we calculated the conductance of ferromagnetic atomic contacts of Fe, Co, and Ni [61], s. Figure 12. In our present calculations we follow our previous work [62, 63, 64] and study ideal contact geometries using a realistic tight-binding model. We only consider situations where the contact region is a single magnetic domain. Our results indicate that for a few-atom contact of the three materials one can draw the following general conclusions: (i) there is no conductance quantization, mainly due to the partially open conduction channels of the minority spin electrons, (ii) the last plateau has typically a conductance above $G_0 = 2e^2/h$, (iii) the current is not fully spin-polarized and both spin species contribute to the transport, and (iv) both the value of the conductance and the current polarization are very sensitive to the contact geometry and disorder. The origin of all these findings can be traced back to the fact that the d bands of these transition metals play a very important role in the electrical conduction. This is in contrast with other physical situations such as tunnel junctions or bulk systems. Finally, we find that in the tunnel regime, which is reached when the contacts are broken, the nature of the conduction changes qualitatively and almost fully spin-polarized currents are indeed possible.

The computation of structural and electronic properties of Si_n clusters has been done [65, 66, 67, 19] by density functional methods [68].

Acknowledgments

We gratefully acknowledge useful discussions with K. Binder, C. Cuevas, A. Erbe, J. Klapp, M. Kläui, R. Klessen, P. Leiderer, U. Nowak, F. Pauly, U. Rüdiger, E. Scheer, F. Schmid, S. Sengupta, B. West and financial support by the SFB 513, SFB 767, SFB TR6 and the Landesstiftung Baden-Württemberg.

References

1. M.M. Miller *et al.*, Appl. Phys. Lett. **81**, 2211 (2002).
2. G.A. Prinz, J. Magn. Magn. Mat., **200**, 57 (1999).
3. B.D. Terris and T. Thomson, J. Phys. D: Appl. Phys. **38**, R199 (2005).
4. U. Nowak, Ann. Rev. of Comp. Phys. **9**, 105 (2001).
5. D. Backes, C. Schieback, M. Kläui, F. Junginger, H. Ehrke, P. Nielaba *et al.*, Appl. Phys. Lett. **91**, (2007).
6. Z. Li *et al.*, Phys. Rev. **B70**, 024417 (2004).
7. A. Thiaville, Y. Nakatani, J. Miltat, N. Vernier, J. Appl. Phys. **95**, 7049 (2004).
8. A. Thiaville, Y. Nakatani, J. Miltat, Y. Suzuki, Europhys. Lett. **69**, 990 (2005).
9. C. Schieback *et al.*, Eur. J. Phys. B **58**, 429 (2007).
10. D. Mutter: Diplomarbeit, Univ. Konstanz (2007).
11. M. Albrecht et al.: *Nat. Mater.* **4**, 203 (2005).
12. Lue, L. and Woodcock, L.V. (1999) *Mol. Phys.* **96** 1435.
13. F. Bürzle, P. Nielaba, Phys. Rev. **E76**, 051112 (2007).
14. J. Baumgartl, M. Brunner, C. Bechinger, Phys. Rev. Lett. **93**, 168301 (2004).
15. E. Frey, D.R. Nelson, L. Radzihovsky, Phys. Rev. Lett. **83**, 2977 (1999); L. Radzihovsky, E. Frey, D.R. Nelson, Phys. Rev. **E63**, 031503 (2001).
16. Franzrahe K. and Nielaba P. (2007) *Phys. Rev. E* **76** 061503.
17. Franzrahe K., Nielaba P., Ricci A., Binder K., Sengupta S., Keim P. and Maret G., *J. Phys.: Condens. Mat.* in press (2008).
18. Franzrahe K. *et al.*, (2005) *Comp. Phys. Commun.* **169** 197.
19. Franzrahe K. *et al.*, in *"High Performance Computing in Science and Engineering'07"*, ed. by Nagel W.E., Kröner D., Resch M., Springer Verlag (2007), 83.
20. P. Henseler, Dissertation, U. Konstanz (2008).
21. M. Köppl, P. Henseler, A. Erbe, P. Nielaba, and P. Leiderer. Phys. Rev. Lett. **97**, 208302 (2006).
22. R. Haghgooie, C. Li, P. Doyle, Langmuir **22**, 3601 (2006).
23. R. Haghgooie and P.S. Doyle. Phys. Rev. E **72**, 11405 (2005).
24. M.P. Allen and D.J. Tildesley. *Computer Simulation of Liquids* (Oxford Science Publications, 1987).
25. P. Henseler *et al.*, in preparation (2008).
26. N. Schwierz, Diplomarbeit, U. Konstanz (2008).
27. A. Erbe, *private communication.*

28. F. Schmid, D. Düchs, O. Lenz, B. West, Comp. Phys. Commun. **177**, 168 (2007).
29. O. Lenz and F. Schmid, J. Mol. Liq. **117**, 147 (2005).
30. O. Lenz and F. Schmid, Phys. Rev. Lett. **98**, 058104 (2007).
31. O. Lenz, Dissertation, Univ. Bielefeld (2007).
32. G. Arreaga-García, J. Klapp, L. Di, G. Sigalotti, R. Gabbasov, ApJ, 666, 290 (2007).
33. J. Ballesteros-Paredes, R.S. Klessen, M.-M. Mac Low, E. Vazquez-Semadeni, in *Protostars & Planets V*, eds. B. Reipurth, D. Jewitt, K. Keil, University of Arizona Press, 63 (2007).
34. J.J. Monaghan, Rep. Prog. Phys. 68, 1703 (2005).
35. D.J. Price, Publications of the Astronomical Society of Australia, 24, 159 (2007).
36. D.J. Price, M.R. Bate, MNRAS, 377, 77 (2007).
37. S. Rosswog, D.J. Price, MNRAS, 379, 915 (2007).
38. V. Springel, MNRAS, 364, 1105 (2005).
39. N. Agraït, A. Levy Yeyati, and J.M. van Ruitenbeek, Phys. Rep. **377**, 81 (2003).
40. J.C. Cuevas, A. Levy Yeyati, A. Martín-Rodero, Phys. Rev. Lett. **80**, 1066 (1998).
41. E. Scheer, N. Agraït, J.C. Cuevas, A. Levy Yeyati, B. Ludoph, A. Martín-Rodero, G. Rubio, J.M. van Ruitenbeek and C. Urbina, Nature **394**, 154 (1998).
42. J.C. Cuevas, A. Levy Yeyati, A. Martín-Rodero, G. Rubio Bollinger, C. Untiedt, and N. Agraït, Phys. Rev. Lett. **81**, 2990 (1998).
43. J.L. Costa-Krämer, Phys. Rev. B **55**, R4875, (1997).
44. F. Ott *et al.*, Phys. Rev. B **58**, 4656 (1998).
45. T. Ono, Y. Ooka, H. Miyajima, Y. Otani, Appl. Phys. Lett. **75**, 1622 (1999).
46. M. Viret *et al.*, Phys. Rev. B **66**, 220401(R) (2002).
47. F. Elhoussine *et al.*, Appl. Phys. Lett. **81**, 1681 (2002).
48. V. Rodrigues, J. Bettini, P.C. Silva, D. Ugarte, Phys. Rev. Lett. **91**, 96801 (2003).
49. More generally, the half-integer conductance quantization could also arise from a perfectly polarized current, where the channel transmissions of the transmitted spin-component add up to 1.
50. C. Untiedt *et al.*, Phys. Rev. B **69**, 081401(R) (2004).
51. A. Martín-Rodero, A. Levy Yeyati, J.C. Cuevas, Physica C **352**, 67 (2001).
52. A. Smogunov, A. Dal Corso, E. Tossati, Surf. Sci. **507**, 609 (2002); **532**, 549 (2003).
53. A. Delin and E. Tosatti, Phys. Rev. B **68**, 144434 (2003).
54. J. Velev and W.H. Butler, Phys. Rev. B **69**, 094425 (2004).
55. A. Bagrets, N. Papanikolaou, and I. Mertig, Phys. Rev. B **70**, 064410 (2004).
56. A.R. Rocha and S. Sanvito, Phys. Rev. B **70**, 094406 (2004).
57. D. Jacob, J. Fernández-Rossier, J.J. Palacios, Phys. Rev. B**71**, 220403(R) (2005).
58. M. Wierzbowska, A. Delin, and E. Tosatti, Phys. Rev. B **72**, 035439 (2005).
59. H. Dalgleish and G. Kirczenow, Phys. Rev. B **72**, 155429 (2005).
60. D. Jacob and J.J. Palacios, Phys. Rev. B **73**, 075429 (2006).
61. M. Häfner, J.K. Viljas, D. Frustaglia, F. Pauly, M. Dreher, P. Nielaba, and J.C. Cuevas, Phys. Rev. B **77**, 104409 (2008).
62. M. Dreher, Dissertation, U. Konstanz (2008).
63. M. Dreher, F. Pauly, J. Heurich, J.C. Cuevas, E. Scheer, and P. Nielaba, Phys. Rev. B **72**, 075435 (2005).

64. F. Pauly, M. Dreher, J.K. Viljas, M. Häfner, J.C. Cuevas, and P. Nielaba, Phys. Rev. B **74**, 235106 (2006).
65. W. Quester, Dissertation, U. Konstanz (2008).
66. F. von Gynz-Rekowski, W. Quester, R. Dietsche, Dong Chan Lim, N. Bertram, T. Fischer, G. Ganteför, M. Schach, P. Nielaba, Young Dok Kim, Eur. Phys. J. **D45**, 409 (2007).
67. M. Schach, Diplomarbeit, U. Konstanz (2007).
68. CPMD. Copyright IBM Corp 1990–2001, Copyright MPI für Festkörperforschung Stuttgart 1997–2004. http://www.cpmd.org/.

Quantum Confined Stark Effect in Embedded PbTe Nanocrystals

R. Leitsmann, F. Ortmann, and F. Bechstedt

European Theoretical Spectroscopy Facility (ETSF) and
Institut für Festkörpertheorie und -optik
Friedrich-Schiller-Universität Jena
Max-Wien-Platz 1, 07743 Jena, Germany
roman@ifto.physik.uni-jena.de

Summary. We investigate structural and electronic properties of PbTe nanocrystals (NCs) embedded in a CdTe matrix using an *ab initio* pseudopotential method and a repeated super cell approximation. The dot-matrix interface structure was found to be in good agreement with observations at planar PbTe/CdTe interfaces. In particular the so called rumpling effect at Te-terminated {100}PbTe/CdTe interfaces is of the same order of magnitude. Calculations concerning the stability of the embedded PbTe NCs confirm the assumption of a rhombo-cubo-octahedron equilibrium crystal shape (ECS). The occurrence of internal electrostatic fields as a consequence of the polar dot-matrix interfaces is demonstrated. The resulting internal quantum confined Stark effect (IQCSE) leads to a spatial separation of electron and hole wavefunctions in the NCs and, hence, to a reduction of the optical oscillator strength at low temperatures.

1 Introduction

Light sources in the mid-infrared spectral region are crucial for many applications, e.g. for gas sensor systems or medical diagnostics. Nanostructuring of semiconductor materials is one promising way to generate desirable properties, by changing the system dimension and shape. Recently, huge efforts towards the development of quantum dot (QD) laser systems in this spectral region has been made using PbTe QDs embedded in a CdTe matrix. [1] The highly symmetric QDs with rhombo-cubo-octahedron shape exhibit a high mid-infrared luminescence yield, [2, 3] which is a consequence of the peak-like density of states in such zero-dimensional systems. Nevertheless there are still many open questions concerning, e.g. the influence of internal electrostatic fields [1] on the electronic states and optical properties.

[1] Due to the almost ionic character of PbTe we expect hugh internal electrostatic fields created by interface charges at the polar dot-matrix interface facets.

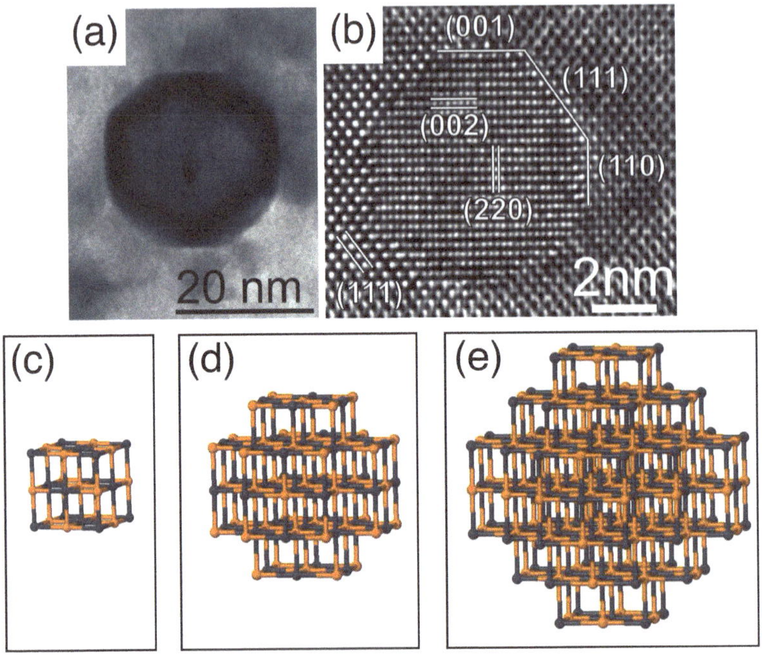

Fig. 1. (a) TEM image of a typical PbTe QD in a CdTe matrix. [6] (b) HRXTEM image of a similar but smaller QD with indications of the symmetry-allowed lattice planes and interface orientations. [6, 8] (c-e) Schematic stick and ball representation of three different PbTe NC-shapes and NC-sizes: (c) 6.41 Å cube, (d) 12.82 Å oct(1) and (d) 19.23 Å oct(2). Pb atoms are shown in black, while Te atoms are shown in yellow

Typically those questions can be addressed using state of the art density functional theory (DFT). However, the *ab intio* description of Pb- and Cd-salts is a challenging task for several reasons. First of all Cd exhibits very shallow *d*-states. [4] Therefore, we have to treat (beside the s, p states) the outermost Pb- and Cd-*d* states as valence electrons. On the other hand for a precise description of heavy elements like Pb the inclusion of relativistic effects (e.g. spin-orbit coupling) is necessary. [5] This increases the computational coast of those calculations considerable. In addition we have to use hugh supercells ($\approx$ 1000 atoms) to be able to compare our results to experimental observations. In former publications [3, 6, 7] we have already analyzed the geometry and electronic properties of PbTe/CdTe interfaces and there influence on the equilibrium crystal shape (ECS) of embedded PbTe QDs.

In the present study we focus on the electronic and optical properties of PbTe QDs embedded in a CdTe matrix. In particular the influence of internal electrostatic fields on the spatial localization of the electron and hole wavefunctions, i.e. the quantum confined Stark effect will be discussed in detail.

2 Computational Method

2.1 Kohn-Sham Energy Functional and PAW Method

To describe real-life systems like, e.g. bulk materials, there interfaces and surfaces or nanowires and nanodots, one has in principle to solve the stationary Schrödinger equation for the corresponding ensemble of ions and electrons. However, the fact that the mean kinetic energy of the nuclei is of the order $m/M \approx 10^3$ smaller than the mean kinetic energy of the electrons justifies the use of the Born-Oppenheimer approximation, [9] i.e a full decoupling of ionic and electronic degrees of freedom. Therefore the total energy of a system of interest can be written as the sum of the electron energy for a given atomic configuration and an ionic contribution, which can (in the most cases) be treated classically.

To solve the remaining Schrödinger equation for the electronic wavefunction we are using the DFT based on theorems of Hohenberg and Kohn. [10] In this theory the ground state energy E_0 of N interacting electrons in an external potential $V_{\mathbf{ion}}$ is a unique functional of the electron density $n(\mathbf{r})$. Kohn and Sham [11] devised a scheme to exploit the Hohenberg-Kohn theorems for actual calculations. The energy functional is written

$$E[n] = T_{\mathbf{s}}[n] + \int d\mathbf{r} n(\mathbf{r}) V_{\mathbf{ion}}(\mathbf{r}) + \frac{1}{2} \int d\mathbf{r} d\mathbf{r}' \frac{n(\mathbf{r}) n(\mathbf{r}')}{|\mathbf{r} - \mathbf{r}'|} + E_{\mathbf{XC}}[n], \quad (1)$$

where $T_{\mathbf{s}}$ is the functional of the kinetic energy of non-interacting electrons end $E_{\mathbf{XC}}$ is the energy functional containing the exchange and correlation contributions as well as corrections to the kinetic energy arising from the electron-electron interactions. Minimization of this energy functional with respect to the electron density under the condition of particle conservation leads to the so-called Kohn-Sham equation

$$\left\{ -\frac{\hbar^2}{2m} \nabla^2 + V_{\mathbf{ion}}(\mathbf{r}) + V_{\mathbf{H}}[n](\mathbf{r}) + V_{\mathbf{XC}}[n](\mathbf{r}) \right\} \Psi_{n\mathbf{k}}(\mathbf{r}) = \varepsilon_{n\mathbf{k}} \Psi_{n\mathbf{k}}(\mathbf{r}), \quad (2)$$

where $V_{\mathbf{H}}$ is the classical Hartree potential. The exchange-correlation potential is defined as the density variation of the XC energy functional $\delta E_{\mathbf{XC}} / \delta n(\mathbf{r}) = V_{\mathbf{XC}}[n](\mathbf{r})$ and will be calculated using the local-density approximation (LDA). Solving equations (2) self-consistently one obtains the exact ground-state density and thus all physical properties that are functionals of this density.

For the numerical solution of equation 2 we expand the electron wavefunctions Ψ in the space regions between the cores into a plane-wave basis set, which is very efficient for systems with periodic boundary conditions like, e.g. bulk crystalline structures or repeated supercells. In this way an efficient evaluation of the action of the Hamiltonian using Fast Fourier Transforms (FFT) can be achieved. To further reduce computational coasts we are using the Projector Augmented Wave (PAW) method [12], which gives rise to wavefunctions

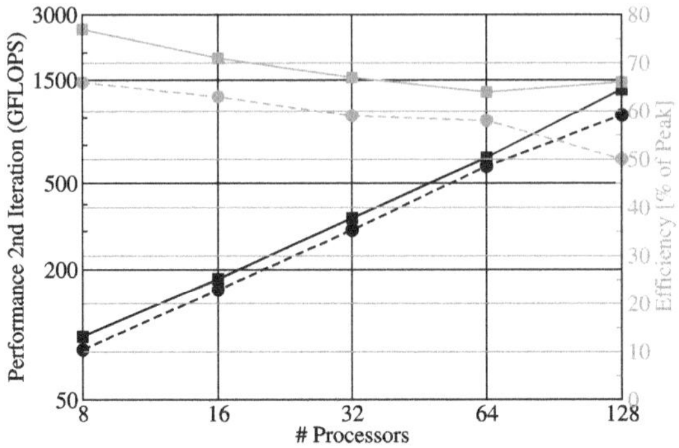

Fig. 2. Performance on the NEC SX-8 for two different PbTe nanodot systems containing 1000 (squares) and 512 (circles) atoms. Additionally we show the efficiency with respect to the peak performance in percent for both systems

of the valence electrons with the correct nodal structure. To diagonalize the Kohn-Sham matrix we employ the Residual Minimization Method with Direct Inversion in Iterative Subspace (RMM-DIIS) [13] as implemented in the Vienna Ab-initio Simulation Package (VASP) [14, 15]. Parallelization is done using the Message Passing Interface (MPI).

2.2 Computational Cost

Since the Kohn-Sham matrix is diagonal in the index n of the eigenstate ("inter-band-distribution") the diagonalization can be efficiently parallelized. If there are enough nodes available, the diagonalization for the n-th state may be parallelized as well ("intra-band-distribution"). A limiting factor is, however the communication overhead required for the redistribution of the wavefunctions between all nodes, which is necessary during the orthogonalization procedure of the eigenstates. Therefore, the scalability of the VASP-code depends strongly on the size of the actual problem as can be seen in Fig. 2. In this graph we compare two systems containing 512 (circles) and 1000 (squares) valence electrons, which corresponds to Kohn-Sham matrices of the size $(2810 \text{ x } 108383)^2$ and $(5445 \text{ x } 211441)^2$, respectively. The calculations carried out on the NEC SX-8 system demonstrate the very good scaling behavior of the code, which is in few cases biased by different numbers of electronic iteration cycles per ionic step. The computation is dominated by complex matrix-matrix multiplications (CGEMM). The sustained iteration performance for both cases exceeds 1 TFLOPS already on 16 nodes NEC SX-8 (Fig. 2). The sustained efficiency is between 79 and 50 % [16]. Compared to other high-computation facilities and local machines the implementation on the NEC SX-8 system is the most efficient one (see Tab. 1). This fact is a consequence of the vector architecture of the NEC SX-8, which can handle CGEMM operations in a very efficient way.

Table 1. Wall-time in seconds of a system with a Kohn-Sham matris of the size $(168 \times 15399)^2$ computed on 8 CPUs at three different machines

machine	NEC SX-8	SGI Altix 4700	Cray XD1 (local)
wall-time [sec]	1908.65	3148.33	8893.56

3 Nanocrystal Construction Using Supercells

The PbTe NCs embedded in CdTe matrix are modeled using the so-called supercell method, [17] i.e., periodic boundary conditions are applied. Therefore, a three-dimensional arrangement of nanocrystals embedded in a matrix material is assumed. We use a simple-cubic arrangement of supercells. Each of them contains a PbTe QD and a certain amount of the matrix material (CdTe). The size of the simple-cubic supercells vary with the diameter of the PbTe QDs and the size of the CdTe matrix.

The constructed supercell systems with different dot and matrix sizes are denoted as shown in Table 2. The denotation nm(a/c) characterizes the number n of nearest-neighbor atom shells counted from the NC center, the edge length of the supercell ma_0 in units of the lattice constant a_0, and the central atom either a Te anion (a) or a Pb cation (c). According to the construction three different QD shapes occur - cubes, rhombo-cubo-octahedrons, and spheres. They differ with respect to the relative contributions from the {100}, {110}, and {111} facets. In the case of the cubes only {100} facets occur. The rhombo-cubo-octahedrons are regular octahedrons with {111} facets truncated at each apex by {100} planes perpendicular to the cube axis. They are distorted by additional {110} facets between the other ones. The areas of the six {100} facets, the twelve {110} facets, and the eight {111} facets occur in

Table 2. Structural parameter of the constructed supercells: O-shell - outermost nanocrystallite shell, C-atom - atom in nanocrystallite center, SC-size - supercell size in a_0, NC-size - nanocrystallite diameter in Å, NC-sh - nanocrystallite shape

notation	33a	33c	43a	43c	44a	44c	45a	45c	55a	55c	55sph
O-shell	3	3	4	4	4	4	4	4	5	5	5
C-atom	Te	Pb	Te	Pb	Te	Pb	Te	Pb	Te	Pb	Te
SC-size	3	3	3	3	4	4	5	5	5	5	5
NC-size	6.41	6.41	12.82	12.82	12.82	12.82	12.82	12.82	19.23	19.23	19.23
NC-sh	cube	cube	oct(1)	oct(1)	oct(1)	oct(1)	oct(1)	oct(1)	oct(2)	oct(2)	sphere

two different ratios, $1 : \sqrt{2} : \frac{1}{3}$ [oct(1)] and $1 : 2\sqrt{2} : \frac{4}{3}$ [oct(2)]. In the case of the largest NCs under consideration we also study a more or less spherical shape denoted by "sph".

Independent of the shape and the facets all generated structures exhibit a C_{3v} symmetry with the symmetry axis along the [111] direction. This symmetry reduction of the original O_h (PbTe) point group is related to the inequality of oppositely terminated but for reasons of electrostatic neutrality pairwise occurring dot-matrix interfaces. [18] Interestingly, those facet orientations lead to a dominance of the cation-terminated facets at the $[\bar{1}\bar{1}\bar{1}]$ halves of the NCs and vice versa of the anion-terminated facets at the [111] halves. Therefore, one expects strong electrostatic fields along the [111] direction, which will have a considerable influence on the geometry and the electronic states of the embedded QDs.

4 Results and Discussion

4.1 Geometric Characterization

The geometry optimization with respect to the Hellmann-Feynman forces yields slightly disturbed lattice structures. A special measure of such a lattice distortion are the nearest-neighbor distances in the PbTe NCs. They deviate from the bulk value $a_0/2 = 3.205$ Å as can be seen in Figs. 3, which compares the Pb-Te bond lengths obtained for PbTe NCs with the same inter-dot distances, but three different dot sizes and different central atoms. The results are in contrast to those obtained for hydrogenated Si and Ge NCs, which show the general tendency to increase the bond lengths (tensile strain) in the NC core region, but to decrease the bond lengths in the outermost regions (compressive strain). [19] Thereby, these effects are most pronounced for dots with small radii. Here, we find no clear radial dependence of the bond length distribution for the embedded PbTe NCs. Enlargement and shortening of the Te-Pb distances occur for all radial distances from the QD center. Nevertheless, two main effects can be identified.

First of all we observe a tendency to form bilayers along the symmetry axis [111]. This results in two different bond lengths: an intra-bilayer bond length and an inter-bilayer bond length. The averages of both are indicated in Fig. 3 by solid and dashed lines, respectively. The deviations from the bulk PbTe bond length are of about 9% for 33c, 6% for 33a, 44a, 44c, 55a, and 3% for 55c. Hence, no clear size dependence can be stated. However, the formation of these polar bilayers is a consequence of the occurrence of the electrostatic field along the [111] symmetry direction. The second effect visible in Figs. 3 is the so called rumpling effect. [20] As we have shown in Ref. [3] for flat Cd-terminated PbTe/CdTe(100) interfaces one observes an inward shift of the Pb atoms of the order of 0.4 Å, resulting in an enhanced Pb-Te interface bond length. This is in excellent agreement with the 0.4-0.5 Å larger Pb-Te bond length observed

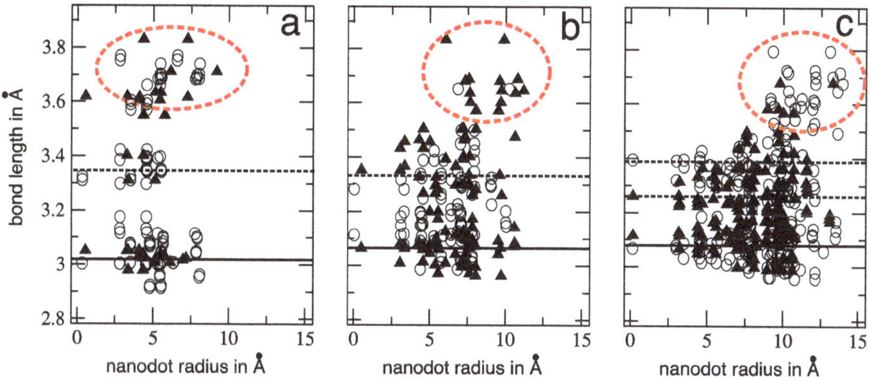

Fig. 3. Pb-Te bond length for NCs with different diameter 6.41 (a), 12.82 (b), and 19.23 Å (c). Open circles represent NCs with Te atom in the center, while filled triangles represent NCs with Pb atom in the center. The dashed horizontal lines represent the averaged inter bilayer bond length, while the continuos horizontal lines represent the inner bilayer bond length. The increased bond length (indicated by red dashed ellipses) at the $(\bar{1}00)$, $(0\bar{1}0)$, and $(00\bar{1})$ dot matrix interfaces are related to a rumpling effect. [18]

at the $(\bar{1}00)$, $(0\bar{1}0)$, and $(00\bar{1})$ dot matrix interface facets. In Fig. 3 we have indicated those values by red dashed ellipses. In contrast, the (100), (010), and (001) dot matrix interface facets exhibit a much smaller rumpling effect, which is again in excellent agreement with our results for flat Te-terminated PbTe/CdTe(100) interfaces. Due to the combination of facets to an entire dot-matrix interface, i.e., due to symmetry reasons the lateral offset parallel to PbTe/CdTe(110) interfaces discussed in Ref. [3] cannot occur at interfaces of embedded PbTe NCs. Experimentally, the main aspects of the theoretical predictions for the {100} interface geometries (at both planar and dot-matrix interfaces) are confirmed. [1, 21]

4.2 Nanocrystallite Stability

Assuming that the system is formed by atoms from reservoirs with the chemical potentials μ_{Te}, μ_{Pb}, and μ_{Cd} one can calculate the so-called dot formation energy

$$E_f = E_{\mathrm{tot}} - \left[(N_{\mathrm{Te}}^{\mathrm{d}} + N_{\mathrm{Te}}^{\mathrm{m}})\mu_{\mathrm{Te}} + N_{\mathrm{Pb}}\mu_{\mathrm{Pb}} + N_{\mathrm{Cd}}\mu_{\mathrm{Cd}} \right]. \quad (3)$$

Thereby the total number of Te atoms [2] in a supercell is given by $N_{\mathrm{Te}} = N_{\mathrm{Te}}^{\mathrm{d}} + N_{\mathrm{Te}}^{\mathrm{m}}$. Due to the applied stoichiometry and the electrostatic neutrality $N_{\mathrm{Te}} = N_{\mathrm{Pb}} + N_{\mathrm{Cd}}$. The experimental preparation of the NCs starts from

[2] The indizes d and m denote the quantum dot and matrix region, respectively.

relatively thick, [001] oriented CdTe and PbTe layers. [8] These bulk layers form the reservoir for the atoms contributing to the dot-matrix system. Then a rapid thermal annealing (RTA) happens. We assume that the resulting PbTe NCs and the matrix material are near the thermal equilibrium, i.e.,

$$\mu_{\mathrm{CdTe}}^{\mathrm{bulk}} = \mu_{\mathrm{Cd}} + \mu_{\mathrm{Te}} \text{ and } \mu_{\mathrm{PbTe}}^{\mathrm{bulk}} = \mu_{\mathrm{Pb}} + \mu_{\mathrm{Te}}. \tag{4}$$

With the definition (3) this leads to

$$E_f = E_{\mathrm{tot}} - N_{\mathrm{Pb}}\mu_{\mathrm{PbTe}}^{\mathrm{bulk}} - N_{\mathrm{Cd}}\mu_{\mathrm{CdTe}}^{\mathrm{bulk}}. \tag{5}$$

The formation energy (5) can also be interpreted as an interface energy between the dot and the matrix. Since it depends on the actual size of the NC[3] or more precisely on the size of the dot-matrix interface A_{O} we introduce a normalized energy

$$\gamma = \frac{E_f}{A_{\mathrm{O}}}, \tag{6}$$

which is essentially the NC interface energy per unit area. The stability of a certain dot-matrix system depends very much on the interface shape and the central atom. Table 3 indicates that faceted interfaces with the simultaneous appearance of {100}, {110}, and {111} facets give rise to the most stable dot-matrix systems. The two extremum shapes, cube with {100} facets and sphere, are energetically less favorable. The ratio of the facet areas on the distorted rhombo-cubo-octahedrons depends on the chemical nature of the central atom. In the case of an anion in the center the lowest energy appears for a ratio of the {100}, {110}, and {111} areas as in oct(1) NCs, while for a cation in the center the oct(2) NCs are preferred. The strong dependence of the stability on the central atom is the result of differences in the atomic displacements at the dot-matrix interfaces. As can be seen in Figure 3 the distortion of the Te-Pb interface bond length is larger for 44c and 55a compared to 44a and 55c. Such an increase of the bond length will of course lead to an increased interface energy γ and hence a decreased NC stability.

4.3 Internal Electrostatic Fields and Quantum Confined Stark Effect

The considered NCs exhibit dot-matrix interface facets with different polarities. Therefore, one expects a residual electrostatic field created by interface

Table 3. Normalized NC interface energy γ for different NC-shapes. [18]

NC-shape	cube	oct(1)	oct(2)	sphere
C-atom: Te	1.91	1.00	1.26	1.30
C-atom: Pb	2.07	1.36	0.86	-

[3] In the limits N_{Cd} or $N_{\mathrm{Pb}} \to 0$ it vanishes.

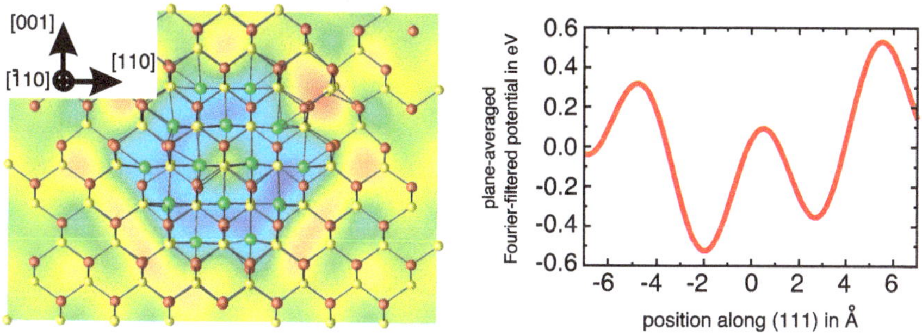

Fig. 4. Left hand side - Fourier-filtered electrostatic potential shown in the ($\bar{1}10$) plane. Blue colors corresponds to negative and red colors to positive values. The atomic positions of PbTe NC and matrix are indicated by a stick and ball model. Cd-, Pb-, and Te-atoms are shown by red, green, and yellow balls, respectively. Right hand side - plane average of the Fourier-filtered electrostatic potential along the [111] direction. [22]

charges at the polar {111} and {100} interface facets. To visualize this effect we have calculated the electrostatic potential acting on a Kohn-Sham electron. We are interested only in macroscopic or mesoscopic fields, i.e., in fields with a periodicity of the order of the NC size. Variations as a consequence of the electron distributions around individual atoms should not be considered. For that reason we average over characteristic distances of the order of the bond lengths or even smaller using a Fourier filter.

In Fig. 4 (left hand side) the results for the Fourier-filtered electrostatic potential using a cutoff wavelength of 6.41 Å are shown. Red colors correspond to high potential values, while blue colors represent small potential values. The offset between the averaged electrostatic potential in the dot and the matrix region is clearly visible. It is of the same order ($\sim$ 3 eV) as observed at flat PbTe/CdTe interfaces.[3] It indicates the electronic confinement within the PbTe QD. The asymmetry of the potential along the [111] axis is a consequence of the distribution of Cd- and Te-terminated interface facets. Similarly to what one knows from layered structures (superlattices) where the atomic oscillations of the electrostatic potential are superimposed by a mesoscopic or macroscopic saw-tooth potential in one direction the dot-matrix system exhibits such an effect in three dimensions. This can be seen at the right hand side of Figure 4, where we show the plane average of the Fourier-filtered electrostatic potential along the [111] symmetry axis of the system. Superimposed on the long-periodic dot-matrix oscillations one finds a triangular potential. In reality this additional mesoscopic/macroscopic potential represents a rounded saw-tooth potential in [111] direction with a characteristic increase/decrease in the dot and matrix region. It is the result of the interface polarization charges. Its consequences on the structural properties have already been discussed in the section 4.1.

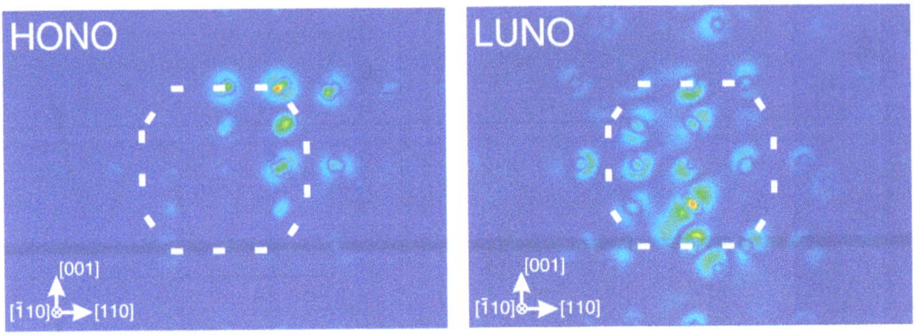

Fig. 5. Probability density of the HONO (left hand side) and LUNO (right hand side) in a [$\bar{1}$10] plane at the center of QD 44a. Blue (red) colors correspond to low (high) values. The location of the PbTe QD is indicated by withe dashed lines

4.4 Electronic and Optical Properties Versus Experimental Observations

In addition to the impact on the atomic structure, such an electrostatic field significantly influences the electronic properties of the NC-matrix system. In order to demonstrate the field influence on the probability distribution of the highest occupied NC orbital (HONO) and the lowest unoccupied NC orbital (LUNO) both of them are plotted in Fig. 5. The different spatial distributions along the [111] axis is clearly visible. The corresponding maxima of the probability to find a hole or an electron are spatially separated. This IQCSE leads to a localization of the HONO and LUNO in different space regions and decreases the energy separation of the HONO and LUNO levels.

While the former has a drastic influence on the optical oscillator strengths, the later gives rise to a small red shift of transition energies. With increasing size of the NCs the oscillator strength quenching should result in disappearing photoluminescence (PL) at low temperatures. Such a reduction of PL-intentsity can be indeed observed at embedded PbTe QDs with diameters of about 10-30 nm. [23, 22]

5 Summary and Outlook

We have investigated nanocrystals consisting of heavily ionic materials (PbTe) embedded in a matrix of a polar semiconductor (CdTe) by means of ab initio methods. Quantum dots with three types of interface shapes - cube, rhombo-cubo-octrahedron with varying ratios of facet areas and sphere - have been studied. A large influence of the electrostatic fields created by the interface charges on the structural properties have been demonstrated. In particular, quantum dots with polar interfaces show the tendency to form bilayers along the [111] symmetry axis as a consequence of the dominance of anion-terminated interface facets around the [111] corner and of cation-terminated interface facets around the $[\bar{1}\bar{1}\bar{1}]$ corner of the nanocrystal.

The accompanying electrostatic potential within the dot-matrix system is strongly influenced by the interface charges and exhibits a C_{3v} symmetry. In direction parallel to [111] it is superimposed by a potential variation over the characteristic extents of the dot and matrix with similarities to the saw-tooth potential appearing in superlattices. The accompanying quantum confined Stark effect shifts the maxima of the electron and hole distributions in different directions along the [111] axis. This affects the optical oscillator strengths. Due to the decreasing overlap of the wave functions for larger nanocrystals the emission strength decreases rapidly as confirmed by low temperature photoluminescence experiments.

Acknowledgement

We acknowledge valuable discussions with colleagues of our group J. Furthmüller and F. Fuchs. In addition we thank Prof. Heiss and Prof. Schäffler from the University of Linz for providing us the experimental HRXTEM- and PL-data. The work was financially supported through the Fonds zur Förderung der Wissenschaftlichen Forschung (Austria) in the framework of SFB25, Nanostrukturen für Infrarot-Photonik (IR-ON) and the EU NANOQUANTA network of excellence (NMP4-CT-2004-500198). Grants of computer time from the Höchstleistungsrechenzentrum Stuttgart are gratefully acknowledged.

References

1. H. Groiss, E. Kaufmann, G. Springholz, T. Schwarzl, G. Hesser, F. Schäffler, W. Heiss, K. Koike, T. Itakura, T. Hotei, M. Yano, and T. Wojtowicz, Appl. Phys. Lett. **91**, 222106 (2007).
2. W. Heiss, H. Groiss, E. Kaufmann, M. Böberl, G. Springholz, F. Schäffler, K. Koike, H. Harada, and M. Yano, Appl. Phys. Lett **88**, 192109 (2006).
3. R. Leitsmann, L.E. Ramos, and F. Bechstedt, Phys. Rev. B **74**, 085309 (2006).
4. S.-H. Wei and A. Zunger, Phys. Rev. B **37**, 8958 (1988).
5. E.A. Albanesi, C.M.I. Okoye, C.O. Rodriguez, E.L.P. y Blanca, and A.G. Petukhov, Phys. Rev. B **61**, 16589 (2000).
6. R. Leitsmann, L.E. Ramos, F. Bechstedt, H. Groiss, F. Schäffler, W. Heiss, K. Koike, H. Harada, and M. Yano, New J. Phys. **8**, 317 (2006).
7. R. Leitsmann and F. Bechstedt, Phys. Rev. B **76**, 125315 (2007).
8. W. Heiss, H. Groiss, E. Kaufmann, M. Böberl, G. Springholz, F. Schäffler, K. Koike, H. Harada, and M. Yano, Appl. Phys. Lett. **88**, 192109 (2006).
9. M. Born and J.R. Oppenheimer, Ann. Phys. **84**, 457 (1927).
10. P. Hohenberg and W. Kohn, Phys. Rev. **136**, B864 (1964).
11. W. Kohn and L.J. Sham, Phys. Rev. **140**, A1133 (1965).
12. P.E. Blöchl, Phys. Rev. B **50**, 17953 (1994).

13. P. Pulay, Chem. Phys. Lett. **73**, 393 (1980).
14. G. Kresse and J. Furthmüller, Comp. Mat. Sci. **6**, 15 (1996).
15. G. Kresse and J. Furthmüller, Phys. Rev. B **54**, 11169 (1996).
16. S. Haberhauer, NEC - High Performance Computing Europe GmbH **SHaberhauer@hpce.nec.com** (2006).
17. M.C. Payne, M.P. Teter, D.C. Allan, T.A. Arias, and J.D. Joannopoulos, Rev. Mod. Phys. **64**, 1045 (1992).
18. R. Leitsmann, F. Ortmann, and F. Bechstedt, Phys. Rev. B **to be published** (2008).
19. H.C. Weissker, J. Furthmüller, and F. Bechstedt, Phys. Rev. B **67**, 245304 (2003).
20. A.A. Lazarides, C.B. Duke, A. Paton, and A. Kahn, Phys. Rev. B **52**, 14895 (1995).
21. H. Groiss, Diplomarbeit (2006).
22. R. Leitsmann, F. Ortmann, and F. Bechstedt, Nature Phys. **submitted** (2008).
23. R.E. de Lamaestre, H. Bernas, D. Pacifici, G. Franzo, and F. Priolo, Appl. Phys. Lett. **88**, 181115 (2006).

Signal Transport in and Conductance of Correlated Nanostructures

Tobias Ulbricht[1,2] and Peter Schmitteckert[2]

[1] Institut für Theorie der Kondensierten Materie
Wolfgang Gaede Straße 1
Universität Karlsruhe
D-76128 Karlsruhe

[2] Institut für Nanotechnologie
Research Center Karlsruhe
D-76021 Karlsruhe

Transport properties of strongly interacting quantum systems are a major challenge in todays condensed matter theory. In our project we apply the density matrix renormalization group method [2, 3, 4, 8, 9] (DMRG) method to study non-equilibrium transport properties of quantum devices attached to metallic leads. After struggling a long time with the details within this project we finally managed to get the approach of obtaining finite bias conductances from real-time dynamic working consistently from small to large voltage regime and to obtain exciting results as explained in section 1.

In our last report [9] we have demonstrated that one has to be careful in calculating the transport properties of a spinful Fermi system and showed that by taking a large number of states per block one can actually get reliable results, which is in contrast a previous work [11] where the promised accuracy was not achieved. We have now extended our code to include an explicit creation of wave packets using creation and annihilation operators. We have applied this approach to study spin-charge separation in a transport setup, see section 2.

We have also developed a technique to extract exact functionals for density functional theory [21] which allows us to establish a connection between our work and the main tool in molecular electronics, see section 3.

In a first step towards including quantum chemistry in our code we looked at the fractional quantum hall effect (FQH). This problem is similar to a quantum chemistry type calculation and allows us to check the implementation of the code and get more experience with these kind of systems, see section 4.

In our DMRG code we use Posix threads to parallelize the code which is described in detail in [8]. We spent a large part of the time frame of this report on optimizing our serial code and to improve the used linear algebra algorithms. This is displayed in Fig. 1 where we plot the remaining CPU hours within the time frame of this report. It reflects that until Dezember 2007 we have mainly been occupied with coding and improving the algorithms. The strong increase of the CPU usage was governed by the finite bias calculations for the interacting resonant level model, see section 1.

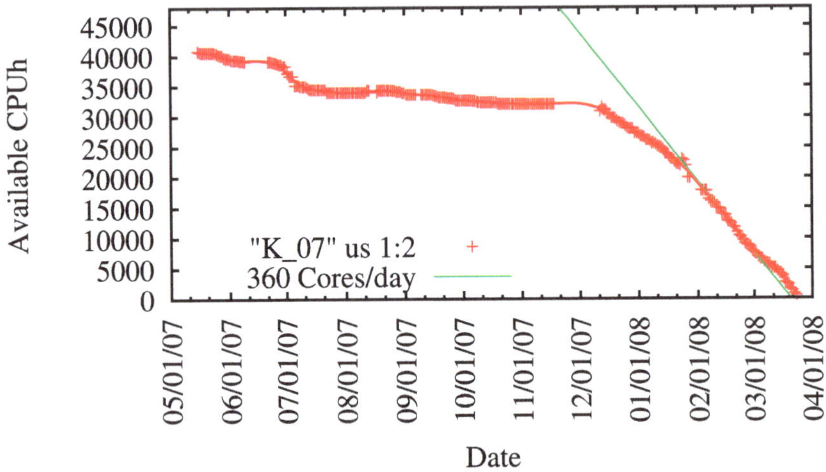

Fig. 1. Remaining CPUh at the XC2 vs. date

1 Differential Conductance in the Interacting Resonant Level Model

Formally the conductance of a quantum device attached to leads is given by the Meir Wingreen formula.[1] Besides the special case of proportional coupling, the Meir Wingreen formula can usually only be treated within perturbative approaches. The major problem in non-equilibrium dynamics consists in the fact that the stationary Schrödinger equation is now replaced by the time-dependent Schrödinger equation. Therefore an *eigenvalue problem* is replaced by a *boundary problem* and one has to take care of the initial state. Therefore one has to be very careful by sending all difficulties to time equal minus infinity since at some time t_0 one hits the initial state. In our approach the answer to this problem is to actually start with an initial state and to perform the full time integration of the time dependent Schrödinger equation via a time evolution operator given by the matrix exponential $U(t) = \mathrm{e}^{-i(\mathcal{H}-E_0)t}$.[7] The method is described in detail in [5, 8].

In this project we concentrated on the interacting resonant level model. In this model a single level is attached to non-interacting leads via a hybridization of t' and an interaction U on the contact links. Mehta und Andrei [15] claim to have solved the non-equilibrium transport properties via a scattering Bethe ansatz. In our previous work using the Kubo approach within DMRG [6, 9, 14] we showed that repulsive interaction on the contact link leads to an increase of the resonance width of the linear conductance vs. gate voltage up to an interaction strength of the order of the Fermi velocity of the leads. Here we are looking at the finite bias conductance where the level is on resonance, i.e. we use a particle hole symmetric interaction $U(\hat{n}_{x_0-1} + \hat{n}_{x_0+1} - 1)(\hat{n}_{x_0} - 1/2)$ without an additional gate voltage, where x_0 denotes the position of the impurity.

In pushing the calculations to the strong bias voltage regime we realized that one has to be careful with the initial state as one can get stuck with an excited state which leads to a voltage drop which is slightly smaller than the correct applied value. In addition we realized that the adaptive time evolution scheme is not reliable in the regime of strong bias voltages. Therefore we developed the following scheme. First we perform a ground state DMRG calculation without time evolution to find the correct initial state. This step typically takes only a few hours on a quad core XC2 node and up to a day for the 3000 states per block calculation for 120 lattice sites. We then perform a full time evolution using the matrix exponential evaluated using a Krylov space approximation.[7] We choose the time frame to be large enough to cover most of the transient regime. This is step takes typically a week of on a quad core node and up to three weeks for the 3000 states per block calculation on a 120 site system. Finally we continue with an adaptive time evolution to cover a larger time frame within the quasi stationary regime.

In Fig. 2 we plot the current $I(t)$ divided by the applied voltage V_{SD} for a 96 site system with a non-interacting coupling using $t' = 0.5$. Here we used a full time DMRG only up to $t = 5$. The small symbols correspond to 1000 states per block, while the larger symbols correspond to 2000 states per block. The lines are given by the analytical result for the noninteracting system for infinite tight binding leads

$$I(V_{\mathrm{SD}}, t') = 32\left(t'^2 - 1\right)^2 t'^4 \frac{\arctan \frac{(4t'^2-2)V_{\mathrm{SD}}}{8t'^2\sqrt{1-2t'^2}}}{8t'^2 * \sqrt{1-2t'^2}(2t'^2-1)} + \frac{t'^4 V_{\mathrm{SD}}}{2t'^2-1} . \qquad (1)$$

The drop of the current around time $t = 48$ is related to the back reflection of the wave packets at the end of the leads and corresponds to the transit time. For longer time scales we would have to use longer leads. The deviation of the current from a straight line after the settling time is related to truncation errors of the adaptive time evolution scheme. The sharp drops in this regime are related to a missing renormalization of the wave function at the restart of the adaptive time evolution scheme. Actually, due to unitarity of the time evolution operator the time dependent wave function should always be normalized. However, due to the projective nature of the adaptive time evolution scheme one looses weight at each DMRG step. We are now normalizing the wave function in the adaption scheme at each step and at the restart since it gives results which agree for a longer time frame with the analytical results in the non-interacting limit. While the results are fine in the not too

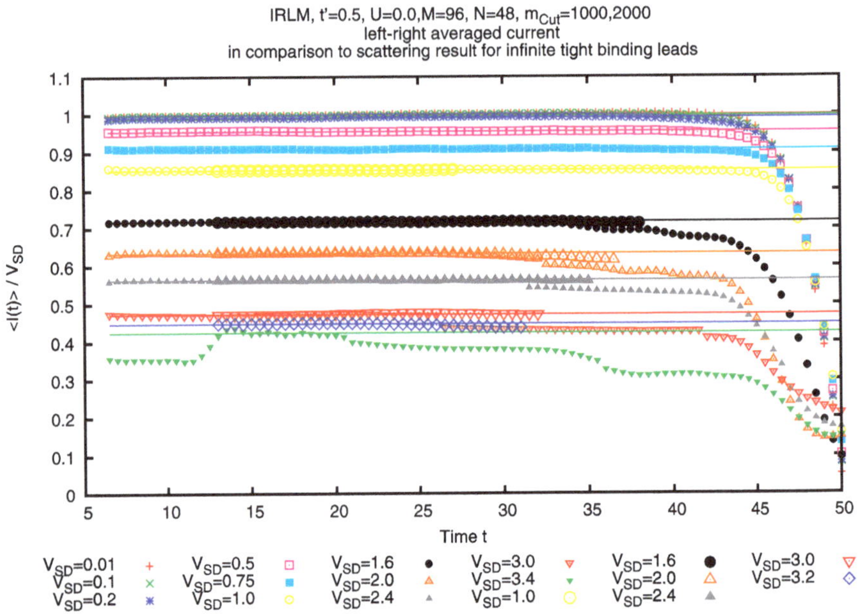

Fig. 2. Current I(t) divided by the the applied voltage vs. time for a set of applied source drain voltages. the smaller symbols correspond to 1000 states per block, the larger symbols to 2000 states per block. Results are from adaptive time sweeps after an initial full time DMRG up to t=5

strong voltage regime they also show that one has to be very careful in the large voltage regime. Since we are especially interested in the latter regime we employed the approach described above and double checked our results by taking up to 3000 states per block. Finally we increased the full time DMRG regime to $T = 10$. By carefully fitting the current left and right the impurity vs. time T and checking for the $I + a\cos(V_{\rm SD}T + \eta)$ oscillations, compare [8], in the quasi stationary regime we obtain the results displayed in Fig. 3 for a hybridization of $t' = 0.5$. The red line shows the analytical result of Eq. 1. The red plusses (crosses) are obtained by fitting the current left (right) of the impurity. The results show that calculations using 2000 states per block reproduce the analytical result even in the large bias regime. Note that the units are given by the hopping elements of the tight binding leads leading to a band width of 4. The results are done for $M = 96$ sites and 48 fermions using 2000 states per block if not stated otherwise in the legend.

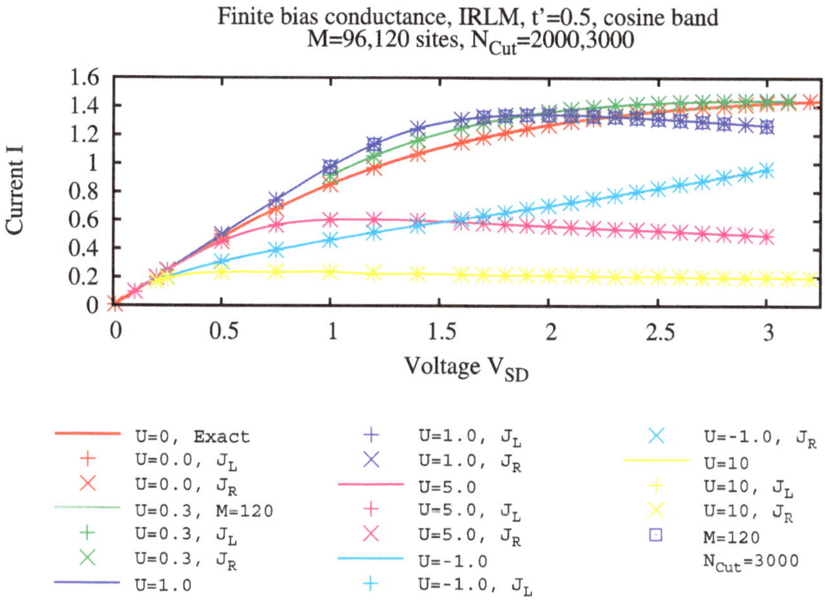

Fig. 3. Current I vs. applied source drain voltage V_{SD}

By switching on interaction one sees that the system still displays a conductance of $1e^2/h$ for small voltage in agreement with our Kubo calculations. For $U = 0.3$ there is a slight enhancement in the current as compared to the noninteracting case, which is even stronger for $U = 1.0$. For larger interaction, $U = 5, 10$ we do not see a corresponding enhancement, i.e. the resonance width is now reduced. Most strikingly the $U \geq 1.0$ results show a clear negative differential conductance, i.e. the current gets reduced by increasing the voltage. In order to be sure that this effect is really given by the system and not by truncation errors of the numerics we double checked the results using 3000 states per block for the same system size. In addition we increased the system size to 120 sites to check for finite size effects. For the $V_{\mathrm{SD}} = 1.0$ results (blue symbols, the line is a guide to the eyes) the comparison shows perfect agreement. The negative differential conductance is also very pronounced for strong interaction of $U = 5.0$, while for $U = 10$ only a very small differential conductance survives. However, for attractive interaction $U = -1$ the negative differential conductance is not present. We would like to remark that this effect can currently not be deduced from the scattering Bethe ansatz of Mehta and Andrei.

2 Spin Charge Separation in a Transport Setup

As already described in [9] wave packets in a repulsive Hubbard model undergo a spin charge separation. The question we ask in this section is whether one should be able to observe the spin charge separation in a transport setup. We have therefore attached an interacting region to noninteracting leads. We then created a single, left moving hole excitation in the right lead and let it evolve. At the end of this scattering process of a single electronic excitation, one may ask what we end up with. The main question that arises is, whether the outcome will be well defined spin-charge separated wave packets, or if a hole will be reconstructed, since we took out one electron of the system in the beginning, or if there will emerge an incoherent superposition of many excitations.

The transport setup consists of 100 sites, divided into 41 left lead sites, 29 interacting sites and 30 right lead sites. Upon the ground state we created a single hole excitation in the right lead using a Gaussian distribution of annihilation operators with momenta k, around k_0 and of width σ and N is the normalization to one.

$$\frac{1}{N}\sum_k e^{-\frac{(k-k_0)^2}{2\sigma^2}} c_k|\Psi_0\rangle \tag{2}$$

With 48 down and 48 up electrons the non-interacting system was at half-filling and the interacting system with an onsite Hubbard interaction was kept at a filling of ~ 0.43 and the injected hole had an average momentum of $k_0 = 0.43\pi - 2\sigma$, where $\sigma = 0.03$. This ensured the ability of the hole to tunnel into and out of the interacting region and keeping the transmission amplitude maximal.

In Fig. 4 we display the time evolution of the hole excitation for several time steps, where we have subtracted the background of the system (n_0) without an excitation. Additionally, we averaged over Friedel oscillations of $2k_F$. The system was calculated using 800 states per block using adaptive time dependent DMRG only. The data nicely shows that the wave packet undergoes a spin-charge separation and finally one ends up with a charge and a spin excitation travelling separately but with equal speed in the left lead. To this end, there is no reason for a recombination of spin and charge degrees of freedom to a single hole excitation. Nevertheless the outgoing wave packets are well defined and, in principle, the charge density and the spin density should be measurable in a time-resolved measurement of a spin-polarized charge density.

3 Exact Functionals for Density Functional Theory

Density functional theory (DFT) is currently the most applied numerical tool to study the electronic transport through molecules. The Hohenberg-Kohn

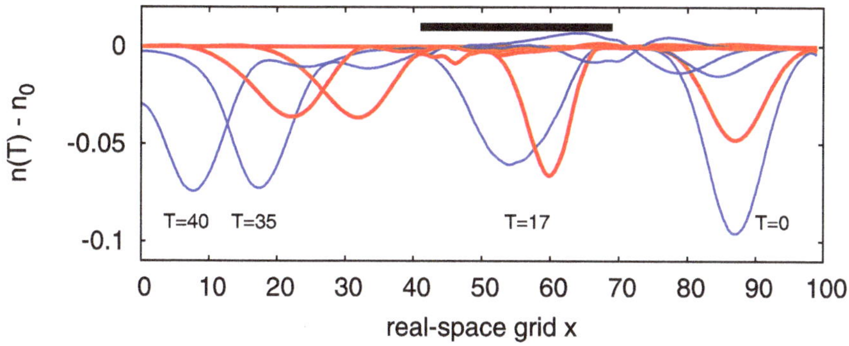

Fig. 4. Spin charge separation in a transport experiment: Spin (thick,red) and charge (thin, blue) densities were subtracted from the background of the ground state system without an additional excitation. A hole created in the right lead (T=0) passes the interacting nanostructure undergoing SCS (T=17). In the left lead spin and charge densities travel independently with equal velocity (T=35,40)

theorem [17] for density functional theory guarantees that all ground state observables can be obtained from a density functional evaluated with the ground state density and that the ground state density minimizes the energy functional. In addition the theorem by Kohn and Sham [18] provides a route to perform DFT calculation practically as they showed that there is a one to one correspondence between an interacting electron system and an auxiliary free noninteracting Fermi system where one replaces the interaction by density potentials v_j, keeping the kinetic term and the local potentials of the interacting system. The theorem now states that the ground state densities of the fully interacting system and that of the Kohn-Sham auxiliary system are identical. In addition the Kohn-Sham potentials are unique if the ground state is non-degenerate. While the energy functional is therefore known to exist, the explicit form of the functional for interacting systems is not known and one has to resort to approximations. In DFT calculations one typically resorts to a local density approximation which may be improved by gradient expansions. One then obtains the eigen functions of the Kohn-Sham system and uses those levels to calculate transport properties.

Allthough this is a widely used approach there a two fundamental problems. First it is not clear whether a local approximation to the functional describes interaction effects correctly and second it is unclear whether the Kohn-Sham levels are the correct objects to be used for transport calculations. In this project we extended earlier ideas of calculating exact density functionals from DMRG by Gunnarsson and Schönhammer [19, 20] to inhomogenous systems. In short, since we can calculate the local densities from DMRG and due to the uniqueness of the Kohn-Sham potentials, we can start from the exact density and can calculate the corresponding Kohn-Sham po-

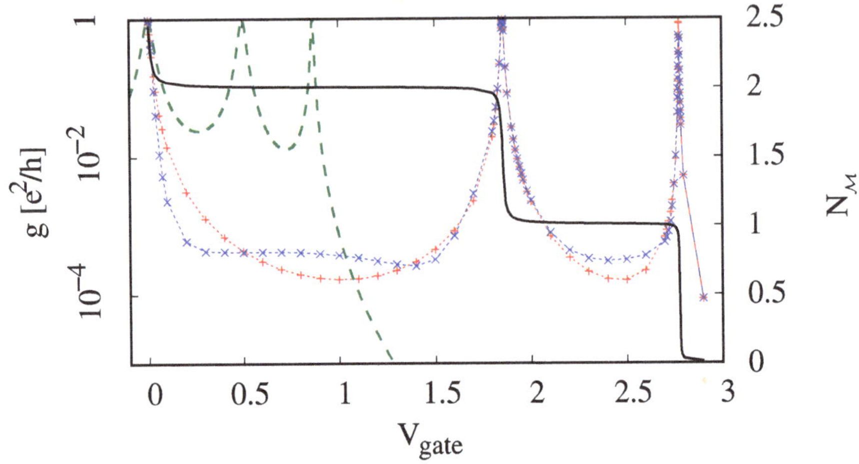

Fig. 5. Comparison of the exact conductance (+, dotted line as a guide) and the ground state DFT approximation (×, dashed line) for a five site system (t'=0.2, $t_{\mathrm{dot}} = 0.5$, $U = 2.0$). For comparison the conductance of the noninteracting system (U=0) is shown as well (long dashed line). The solid line indicates the particle number $N_{\mathcal{M}}(V_{\mathrm{gate}})$ of the molecule. The resonances of g are situated at $V_{\mathrm{gate}} = 0$, 1.854, and 2.779 with resonance widths of $\Gamma = 0.026$, 0.015, and 0.0033

tentials by a multidimensional steepest descent method. Details are explained in [21]. We then have the exact DFT Kohn-Sham potentials corresponding to our model. The question we now asked is whether by applying the standard procedure of using a Kubo formula for the non-interacting Kohn-Sham levels will actually give the same conductance as the full Kubo calculation within DMRG, compare Bohr and Schmitteckert.[14] In this work we coupled a five site nano structure with a nearest neighbor interaction of $U_{\mathrm{Dot}} = 2.0$ and a hopping element $t_{\mathrm{Dot}} = 0.5$ with a hopping of $t' = 0.2$ to 5 real space lead sites which are then coupled to leads described in energy space, for a detailed description see [14, 9, 21].

In Fig. 5 we compare the linear conductance obtained from the DMRG calculation with the one obtained from the Kohn-Sham levels. The results show a surprising agreement close to the resonance capturing the shift in position (Coulomb blockade) and the change in the resonance width. Only in the tails of the resonances there is a significant deviation. However, a careful analysis of the data reveals that the exact Kohn-Sham functional is non-local. For instance it is not single valued if the local potentials are plotted against the local densities. In addition, a plot of the Kohn-Sham potential for the real-space sites vs. gate voltage shows that the exact functional contains jumps in the potentials at the resonances. as displayed in Fig. 6. There we plot the Kohn-Sham potential vs. the applied gate voltage for the ten real space sites. The increase of the Kohn-Sham potential for the $x = 4, 6, 8$ sites after resonances is responsible for shifting the resonances to large gate voltages (Coulomb blockade), while the sharp drop at the resonances is responsible for the actual conductance peak. In conclusion we have demonstrated that

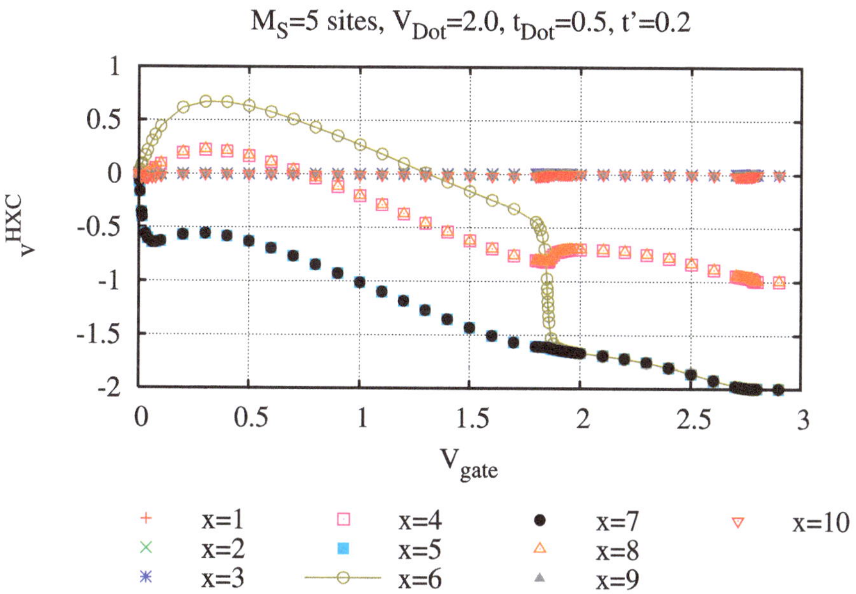

Fig. 6. The Kohn-Sham, or Hartree-Exchange-Correlation potential for the nano structure an the attached real space sites

the Kohn-Sham level may indeed be sufficient to describe linear transport, at least for well isolated resonance. However one should go beyond local density approximations and one has to include discontinuities in the functionals describing the molecules.

4 Fractional Quantum Hall States: Towards Quantum Chemistry

In the projects discussed so far the calculations have been restricted to model Hamiltonians. In a first step to simulate quantum chemistry type models described by

$$\mathcal{H} = \sum_{p,q,l,m} V_{p,q,\ell,m}\ \hat{c}_p^+ \hat{c}_q^+ \hat{c}_\ell \hat{c}_m \tag{3}$$

we implemented the above Hamiltonian with $V_{p,q,\ell,m}$ given by the problem of a two dimensional electron gas in a strong magnetic field where the angular degrees of freedom are integrated out. For this one has to carefully implement the minus sign appearing from the fermionic nature of the electrons. Due to momentum conservation the number of summands of $\mathcal{H}$ is $\sim M^3$. By implementing an implicit partial sum of the terms we could reduce the number of summands to be $\sim M^2$. A comparison with an exact diagonalization programme of Xin Wan showed that we successfully implemented this Hamiltonian, see Hu, Wan and Schmitteckert.[22]. In order to test the feasibility of this kind of Hamiltonian in a larger calculation we looked at the distribution of the eigenvalues of the reduced density matrices of the left and the right block in the symmetric calculation as displayed in Fig. 7. It shows that there is

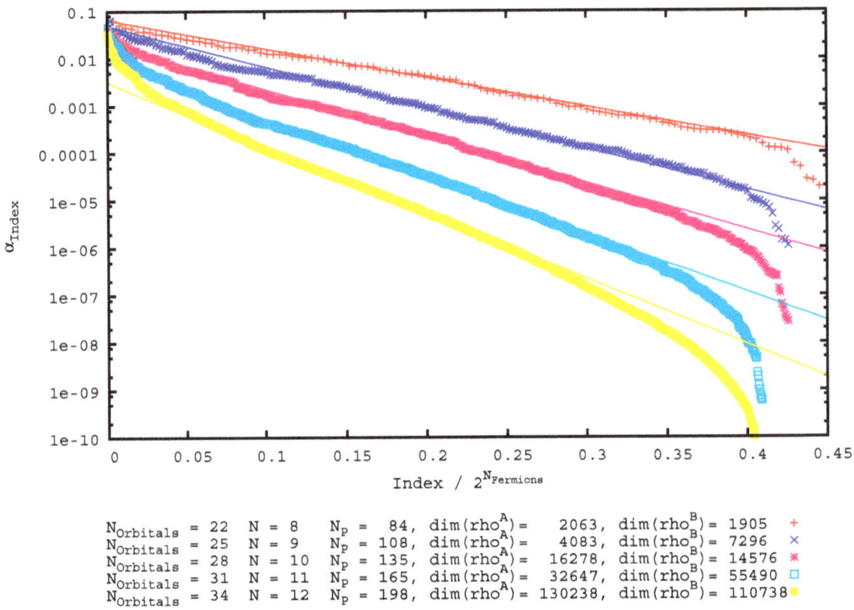

Fig. 7. Distribution of reduced density matrix eigen values in the symmetric blocking for Laughlin pseudo potentials obtained by taking the complete Hilbert space into account

indeed an exponential decay of the eigenvalues of the reduced density matrix, therefore a DMRG truncation procedure in this problem should be possible. However, the decay rate appears to grow exponentially with the number of electrons, so there will be a limitation on system sizes that will be achievable. Nevertheless this test reveals the possibility to look into the problem of topological quantum computation with fractional quantum hall states and we will check whether we will be able to follow this route.

5 Further Projects

We have continued our project on applying the Kubo formalism within DMRG. Currently Dan Bohr is investigating interacting ring structures and studies the interplay of interaction and interference in quantum transport phenomena. During the continuation of this HPC project we found excellent agreement beween our numerical results for the differential conductance for the interacting resonant level model and an analytical approach based on the (usual, non-scattering) Bethe ansatz.[23]

Acknowledgments

The work on the fractional quantum hall effect was performed together with Xin Wan. The Density Functional Theory calculations have been performed together with Ferdinand Evers. Most of the calculations have been performed at the XC2 of the SCC Karlsruhe under the grant number RT-DMRG.

References

1. Y. Meir and N.S. Wingreen, Phys. Rev. Lett. **68**, 2512 (1992).
2. S.R. White, Phys. Rev. Lett. **69**, 2863 (1992).
3. S.R. White, Phys. Rev. B **48**, 10345 (1993).
4. Density Matrix Renormalization – A New Numerical Method in Physics, edited by I. Peschel, X. Wang, M. Kaulke, and K. Hallberg (Springer, Berlin, 1999); Reinhard M. Noack and Salvatore R. Manmana, Diagonalization- and Numerical Renormalization-Group-Based Methods for Interacting Quantum Systems, AIP Conf. Proc. 789, 93-163 (2005).
5. Günter Schneider and Peter Schmitteckert: *Conductance in strongly correlated 1D systems: Real-Time Dynamics in DMRG*, condmat-0601389.
6. Dan Bohr, Peter Schmitteckert, Peter Wölfle: *DMRG evaluation of the Kubo formula – Conductance of strongly interacting quantum systems*, Europhys. Lett., **73** (2), 246 (2006).
7. Peter Schmitteckert: *Nonequilibrium electron transport using the density matrix renormalization group*, Phys. Rev. B **70**, 121302 (2004).
8. Günter Schneider and Peter Schmitteckert: *Signal transport and finite bias conductance in and through correlated nanostructures* p. 113–126 in W.E. Nagel, W. Jäger, M. Resch (Eds.), "High Performance computing in Science and Engineering '06", Springer Verlag Berlin Heidelberg 2007, ISBN 978-3-540-36165-7.
9. Günter Schneider, Peter Schmitteckert: *Signal transport in and conductance of correlated nanostructures* in W.E. Nagel, W. Jäger, M. Resch (Eds.), "High Performance computing in Science and Engineering '06", Springer Verlag Berlin Heidelberg 2008.
10. E.A. Jagla, K. Hallberg, C.A. Balseiro, Phys. Rev. B 47, 5849 (1993).
11. C. Kollath, U. Schollwoeck, W. Zwerger, Phys. Rev. Lett. 95, 176401 (2005).
12. S.R. White and A.E. Feiguin, Phys. Rev. Lett. 93, 076401 (2004).

13. A.J. Daley, C. Kollath, U. Schollwöck, and G. Vidal, J. Stat. Mech.: Theor. Exp. P04005 (2004).
14. Dan Bohr and Peter Schmitteckert: *Strong enhancement of transport by interaction on contact links*, Phys. Rev. B **75**, 241103(R) (2007).
15. Pankaj Mehta and Natan Andrei, Phys. Rev. Lett. 96, 216802 (2006); Erratum: Phys. Rev. Lett. 100, 086804 (2008).
16. Rafael A. Molina, Jorge Dukelsky, and Peter Schmitteckert, *Commensurability effects for fermionic atoms trapped in 1D optical lattices*, arXiv:0707.3209, accepted by Phys. Rev. Lett. (2007).
17. P. Hohenberg and W. Kohn, Phys. Rev. **136**, B864 (1964)
18. W. Kohn and L.J. Sham, Phys. Rev. **140**, A1133 (1965)
19. O. Gunnarsson and K. Schönhammer, Phys. Rev. Lett. **56**, 1968 (1986).
20. K. Schönhammer, O. Gunnarsson and R.M. Noack, Phys. Rev. B **52**, 2504 (1995).
21. Peter Schmiteckert and Ferdinand Evers, *Exact ground state density functional theory for impurity models coupled to external reservoirs and transport calculations*, Phys. Rev. Lett. **100**, 086401 (2008).
22. Zi-Xiang Hu, Xin Wan, and Peter Schmitteckert,
23. Edouard Boulat, Hubert Saleur, and Peter Schmitteckert *Out of equilibrium transport in interacting nanostructures: a double step forward*, arXiv:0806.3731.

Supersolid Fermions Inside Harmonic Traps

F. Karim Pour[1], M. Rigol[2], S. Wessel[1], and A. Muramatsu[1]

[1] Institut für Theoretische Physik III, Universität Stuttgart, Pfaffenwaldring 57, D-70550 Stuttgart, Germany.
[2] Department of Physics and Astronomy, University of Southern California Los Angeles, CA 90089, USA

Summary. We use quantum Monte Carlo simulations to study the properties of ultra-cold fermionic atoms on optical lattices when confined in one-dimensional harmonic trapping potentials. In particular, we analyze the case of attractive interactions by considering the attractive fermionic Hubbard model. We find that in the trapped case, the density-density and pairing correlations are characterized by the anomalous dimension of a corresponding periodic system. The fluctuations corresponding to these quantum critical behavior render the SU(2) symmetry breaking by the trapping potential irrelevant, and the system can establish a supersolid phase also inside a confined geometry.

1 Introduction

Recently, there has been renewed interest in the possible coexistence of superfluidity and crystalline order, a phase named supersolid [1]. This novel interest drives from the experimental observation of non-classical rotational inertia in solid ^{4}He [2]. However, early arguments against a supersolid state [3] now seem to be validated by quantum Monte Carlo simulations [4] and theoretical work [5], showing that a defect free crystal of ^{4}He does not show off-diagonal long-range order. Further numerical simulations suggest that off-diagonal long-range order may arise from domain walls [6] or in a meta-stable state [7]. Hence, the existence of a pure supersolid phase in ^{4}He is still under debate [8].

It was suggested that supersolid phases of matter could however be realized for quantum gases on optical lattices, e.g. based on bosons with dipolar interactions [9], Bose-Fermi mixtures [10], frustrated lattice geometries [11] or extended interactions [12] by loading the atoms in higher bands [13]. Some of these proposals however are based on mean-field approximations that tend to become unstable with respect to phase separation as shown by quantum Monte Carlo simulations [12]. Furthermore, with respect to experimental realizations, the presence of a confining potential must be accounted for, which is however generally neglected in the above mentioned studies. In fact, it is known that trapping potentials lead to the coexistence of phases [14, 15]. This could eventually obscure the experimental determination of supersolidity [16]. It is thus important to study the relevance of a trapping potential for the emergence of a supersolid state.

Another system that is known to exhibit a supersolid phase is the fermionic Hubbard model with an attractive contact interaction. In this model, the simultaneous presence of diagonal and off-diagonal long range order results from an inherent SU(2) symmetry, that relates density and pairing amplitudes at half-filling (density $n = 1$) [17]. In fact, systems of fermions with attractive contact interaction were recently achieved with quantum gases by means of Feshbach resonances, leading to paring and superfluidity [18, 19, 20]. Motivated by the observation of superfluidity with fermions in an optical lattice for a dense system ($n \sim 1$) [21], we analyzed in [22] the possible realization of supersolidity in the trapped attractive fermion gas. In fact, we found, that even though the SU(2) symmetry is explicitly broken by the presence of a trapping potential, conditions can be reached, where the structure form factors of both correlation functions diverge with the same power as the system tends towards the thermodynamic limit. Hence, a proper treatment of long-ranged quantum fluctuations shows that the SU(2) symmetry-breaking becomes irrelevant, thus giving rise to a supersolid state.

The rest of this report is organized as follows: In the following section, we define the model for our study of trapped attractive fermions. Then, we present the results from our detailed analysis of the relevant correlations in the trapped system. This allows us to make an interesting connection to the uniform system, i.e. in the absence of a trap. In the final section, we review our evidence that the symmetry-breaking by the trapping potential is rendered irrelevant by the quantum fluctuations, in that a supersolid phase can indeed be realized in this system, by establishing an appropriate filling of the optical lattice.

Apart from the work detailed below, we considered in the last grant period the following topics within the project CorrSys: (i) In [23], we analyzed the properties of ultra-cold bosonic atoms on the honeycomb lattice, where we found a bosonic supersolid phase in the quantum phase diagram. For this study we employed the stochastic series expansion Quantum Monte Carlo method with an improved directed-loop update[24]. (ii) We also analyzed the various supersolid phases that emerge for hard-core bosons on the square lattice with next-nearest neighbor interactions, which in particular can stabilize a supersolid phase near quarter filling[25]. (iii) We studied the properties of quantum antiferromagnets on the quasiperiodic Penrose tiling, exhibiting a rich variety of magnetic environments, with self-similar structures in the local order parameter distribution[26]. (iv) In [27], we performed a Monte Carlo study of the magnetic phase diagram of anisotropic two-dimensional antiferromagnets, contrasting the classical and the quantum case. We exhibited the relevance of biconical structures in such magnetic systems at low temperatures. (v) We analyzed the time-evolution of strongly correlated fermions after a sudden increase of the interaction strength from a metallic region to the insulating regime. For this study, which is motivated by recent experimental progress in the field of ultra-cold atoms out of equilibrium, we employed the time-dependent density matrix renormalization group (t-DMRG) method, to trace the time evolution of such a strongly correlated quantum system[28]. (vi) In [29], we improved our previously presented extended ensemble quantum Monte Carlo method[30], by employing an iterative ensemble optimization strategy, that improves the stochastic sampling near first-order phase transitions, or in rough energy landscapes. (vii) On the basis of t-DMRG we could show [31] that the correlations of an expanding cloud of strongly interacting fermions can be described by a reference system*in equilibrium* in spite of the fact that the system is in a transient regime *out of equilibrium.*

2 Attractive Hubbard Model

We consider in this report the Hamiltonian of the attractive Hubbard model,

$$H = -t\sum_{j,\sigma}\left(c^{\dagger}_{j\sigma}c_{j+1\sigma} + h.c.\right) - U\sum_{j} n_{j\uparrow}n_{j\downarrow} + V\sum_{j=0}^{N}\left(x_j - \frac{Na}{2}\right)^2 n_j \,, \qquad (1)$$

where $c^{\dagger}_{j\sigma}$ and $c_{j\sigma}$ are creation and annihilation operators, respectively, for fermions on site j (position $x_j = ja$, where a is the lattice constant) with spin $\sigma =\uparrow, \downarrow$. The local density is $n_j = n_{j\uparrow} + n_{j\downarrow}$, where $n_{j\sigma} = c^{\dagger}_{j\sigma}c_{j\sigma}$. $U > 0$ denotes the attractive contact interaction, and the last term represents the confining potential. The quantum Monte Carlo simulation were performed using a $T = 0$ projector algorithm [32, 33, 34].

Applying a particle-hole transformation $d^{\dagger}_{j\uparrow} = (-1)^j\, c_{j\uparrow}$, $d^{\dagger}_{j\downarrow} = c^{\dagger}_{j\downarrow}$, to the above Hamiltonian leads to a *repulsive* Hubbard model in the presence of an inhomogeneous magnetic field $h_j = 2V\,(x_j - Na/2)^2$. In case the original system is spin-neutral, $\langle n_{j\uparrow}\rangle = \langle n_{j\downarrow}\rangle$, we get $n^d_j \equiv \langle d^{\dagger}_{j\downarrow}d_{j\downarrow}\rangle + \langle d^{\dagger}_{j\uparrow}d_{j\uparrow}\rangle = 1$, so the transformed system is exactly at half-filling. For the uniform system, i.e. for $V = 0$, we get $h_j = 0$, and a Mott-insulator with dominating antiferromagnetic correlations results for the repulsive system [35]. For the original attractive model, this implies that for $V = 0$ and $n = 1$ the density-density correlation function function $N_{j\ell} \equiv \langle n_j n_\ell\rangle - \langle n_j\rangle\langle n_\ell\rangle$ and the pairing correlation function $P_{j\ell} \equiv \langle \Delta_j \Delta^{\dagger}_{\ell}\rangle$, with $\Delta_j \equiv c_{j\uparrow}c_{j\downarrow}$, obey the following identity:

$$N_{j\ell} = 2(-1)^{|j-\ell|}P_{j\ell} \,. \qquad (2)$$

They will be the relevant correlation functions, and exhibit the same power-law decay, and thus the system forms a supersolid state. In the presence of a confining potential, with $V \neq 0$, this SU(2) symmetry is broken, so that for $V \neq 0$, the supersolid might be expected to be destroyed. However, we found, that quantum critical fluctuations allow to recover the supersolid even for $V \neq 0$[22], i.e. in the presence of a finite trapping potential, as will be the case in experimental realizations of our system.

3 Correlation Function Analysis

We now proceed to study the density-density and pairing correlations, using our quantum Monte Carlo simulations. Here, we show results for a trapped system of $N_f = 74$ fermions inside a harmonic trap with $Va^2/t = 1.826 \times 10^{-4}$, $U/t = 4$, at a characteristic density $\tilde{\rho} = 1$, for which the extent of the fermion cloud covers a range of approximately 150 lattice sites. Here, the characteristic density [36] is defined as $\tilde{\rho} = N_f a(V/t)^{1/2}$, and relates systems with different sizes, number of particles, and confining harmonic potentials in the same way as does the particle density for periodic systems of different sizes.

In the period case, the presence of quasi-long range order is indicated by a divergence at a particular wavevector k. To identify a similar indication for long-range order in the case of a trapped system, we examine the eigenvalue equation for density-density correlations

$$\sum_{\ell} N_{j\ell}\, \phi^{\mu}_{\ell} = N^{\mu} \phi^{\mu}_{j} \; . \qquad (3)$$

This approach reduces to a Fourier transformation for a periodic system, and the eigenvectors ϕ^{μ}_{j} would be plane waves, each mode being characterized by a particular k-vector. In our study, we found, that even in the trapped case, the modes ϕ^{μ}_{j} can still be assigned each to a characteristic wavevector. In order to illustrate this, we consider the moduli of the Fourier transform $[\sim \sum_j \phi^{\mu}_{j} \exp(ikx_j)]$ of each eigenmode ϕ^{μ}_{j}. This is seen in Fig. 3 (a), that shows the Fourier spectra of all the modes, labeled by μ in ascending order of N^{μ}. We find that most of the modes exhibit dominant Fourier components that can be related each to a reflexion symmetric pair of wavevectors $\pm k$. Regions, where such an identification is not possible, we found to correspond to the lowest eigenvalues, which arise due to the vanishing density at the edges of the fermionic cloud. Results of a similar analysis for the eigenvectors φ^{μ}_{ℓ} of the pairing correlations $P_{j\ell}$

$$\sum_{\ell} P_{j\ell}\, \varphi^{\mu}_{\ell} = P^{\mu} \varphi^{\mu}_{j} \; , \qquad (4)$$

are shown in Fig. 3 (b). It is interesting to note, that from such an analysis, we find that the highest eigenvalues of the pairing correlation function are

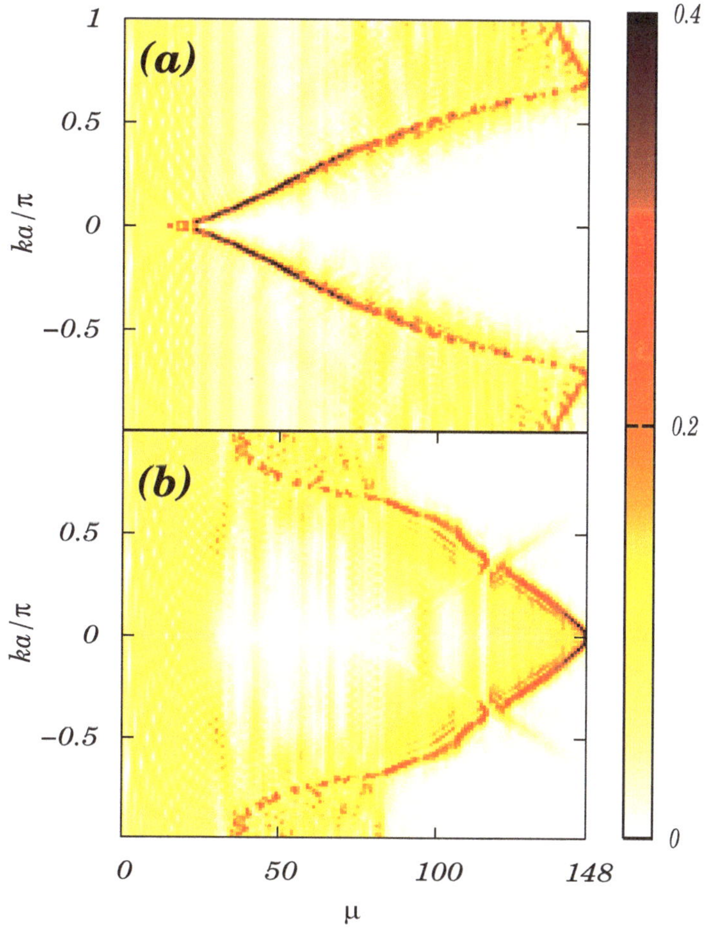

Fig. 1. Level plot of the moduli of the Fourier spectrum of the eigenmodes of (a) the density-density correlation function and (b) the pairing correlation function of attractive fermions on an optical lattice inside a harmonic trapping potential. In both cases, the eigenmodes are labeled by an index μ in increasing order of the corresponding eigenvalue N^{μ}.

dominated at k-vectors around $k = 0$. This indicates that pairs are formed by fermions with opposite momenta. Thus in trapped systems, where translation invariance is broken, and momentum is not a "good" quantum number, pairing correlations are dominantly observed between atoms with opposite momenta [37].

As we found that each mode μ can be assigned rather clearly to a wavevector k, we can also label each eigenvalue N^{μ} by the corresponding wavevector

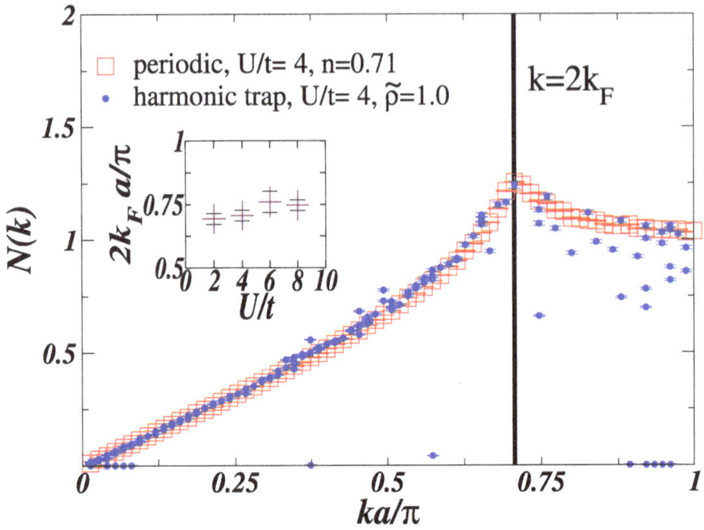

Fig. 2. $N(k)$ for confined (•) attractive fermions on an optical lattice inside a harmonic trap. Superimposed are the results for a periodic system (□) with $2k_F$ given by the black vertical line. The resulting density is $n = 0.71$. Inset: position of the $2k_F$ peak as a function of U in the harmonic trap for $\tilde{\rho} = 1$.

as $N(k)$, similar as one does in a periodic system. Figure 2 shows the result of such an assignment (•), based on the data shown in Fig. 3 (a). Also shown in Fig. 2 are the same results for a specific periodic system (□): namely, for a period system for which we adjusted the particle number such that the wavevector corresponding to the maximum in $N(k)$ of the confined system equals $2k_F$ of the periodic system (black vertical line). Here, k_F denotes the Fermi momentum. This correspondence is motivated by the very fact, that in a 1D fermionic system with attractive interactions, the $2k_F$ oscillations in the charge channel dominate. As Fig. 2 clearly shows, this fact allows us to establish a link between a trapped system with a characteristic density $\tilde{\rho}$ and a periodic system of density $n = 2k_F a/\pi$.

In order to test, if this correspondence holds also for other parameter values, we show in the inset the position of the peak as a function of U. For the range $2 \leq U/t \leq 8$, this does not change appreciably. Hence, the above mapping from a trapped system to a period system is possible over a wide range of parameters and does not depend on any particular fine-tuning.

The identification of $N(k)$ in the periodic and the confined system is remarkably good close to $k = 0$. For a periodic system the density-density correlation function obeys $N(k) \to K_\rho |k| a/\pi$ for $k \to 0$ [35]. Hence, the anomalous dimension K_ρ that determines the power-law decay of the $2k_F$ oscillations that dominate at large distances, can be directly obtained from the slope of $N(k)$ for $k \to 0$. From Figure 2 we find that the mapping proposed here leads to the identification of K_ρ also for the confined system. We find [22], that such an identification is indeed possible over a broad parameter regime, by directly comparing the values of K_ρ obtained for a confined system with those obtained by the Bethe-Ansatz solution [35].

4 Conditions for Supersolidity

We can now discuss the conditions that are necessary for realizing a fermionic supersolid state inside the harmonic trap. In the period case, the supersolid phase emerges for the half-filled system, i.e. for a density $n = 1$, corresponding to a Fermi momentum $2k_F a = \pi$. In a similar fashion, in order to obtain supersolidity inside the harmonic trap, we thus increase $\tilde{\rho}$ until the maximum of $N(k)$ in the confined system similarly shifts towards $k = \pi/a$.

As shown in Fig. 3, the mapping described in the previous section for the eigenvalues of the density-density correlations $[N(k)]$ and the pairing correlations $[P(k)]$ also holds in this case. In the period system, where the SU(2) symmetry is not broken, the relationship in Eq. (2) holds between the correlation functions. For a period system, this implies that $P(k) = N(\pi/a - k)/2$, a relation that is indeed obeyed in Fig. 3 (□). Furthermore, we find that to a very good approximation, this relation is seen to be obtained also for the trapped system (•). Remarkably, we thus find that in spite of the explicit breaking of the SU(2) symmetry by the trapping potential, this symmetry is recovered in the dominant correlation functions, that display quasi-long range order. We thus find that once the appropriate characteristic density is obtained, the SU(2) symmetry is restored due to the long-ranged fluctuations with a high accuracy.

The scaling behavior of the largest eigenvalues $N(k = \pi)$ and $P(k = 0)$ for the harmonically confined system as a function of the number of particles for $\tilde{\rho} = 2.34$ and $U/t = 4$ is shown in the inset of Fig. 3. We find that both quantities diverge with the very same power-law, and differ in magnitude within statistical errors by a factor of 2, as expected when the SU(2) symmetry is present. This is explicit evidence, that quasi-long range order is present in both the charge density and pairing channel with a unique exponent, giving rise to a supersolid fermionic state also inside the harmonic trap.

5 Summary

In conclusion, based on extensive quantum Monte Carlo simulations, we have found that the dominant correlations in a trapped system of attractive fermions are characterized by an unique exponent K_ρ of a related periodic system. We constructed a prescription on how to relate both systems. Based on this explicit mapping, we were able to determine the conditions, under which diagonal and off-diagonal quasi-long range order arise with the same

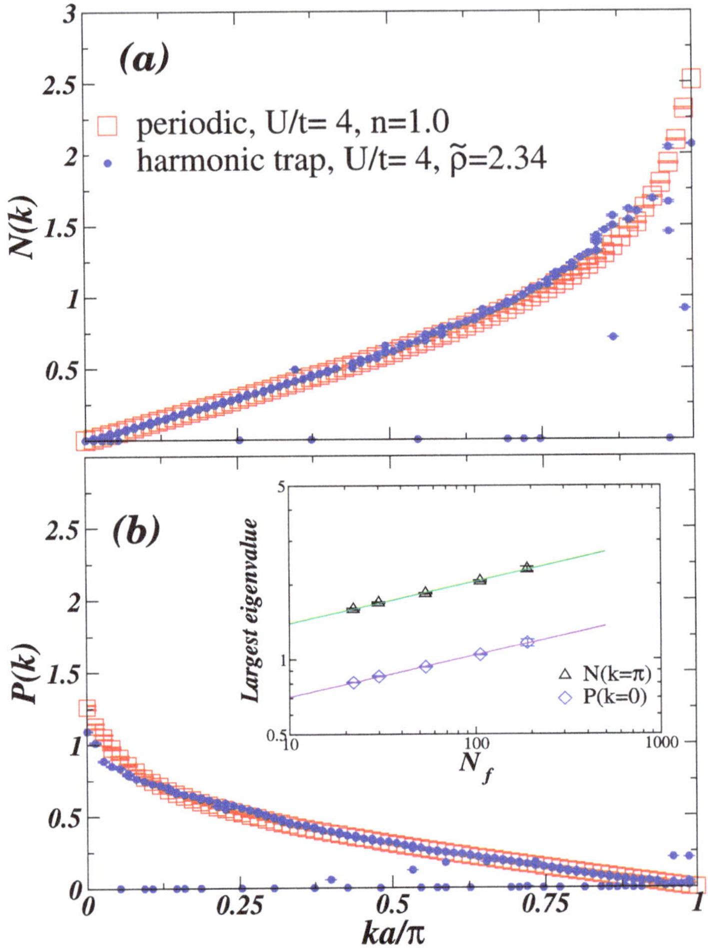

Fig. 3. (a) $N(k)$ for the confined case for $\tilde{\rho} = 2.34$ (•). Superimposed are the results for a periodic system (□) with density $n = 1.0$ (b) $P(k)$ for the confined (•) and periodic systems (□) with the same parameters as in (a). Inset: Scaling of the largest eigenvalues for the density-density correlations ($N(k = \pi)$) and pairing correlations ($P(k = 0)$) *vs.* the number of fermions N_f in a double logarithmic scale.

scaling exponent inside the trap. This gave rise to a restoration of the SU(2) symmetry in the correlation functions, even though it is explicitly broken by the trapping potential. The state emerging under such conditions corresponds to a supersolid in this one-dimensional, inhomogeneous fermion system.

Acknowledgments

We wish to thank HLRS-Stuttgart (Project CorrSys) and NIC-Jülich for the allocation of computer time. We acknowledge financial support by the DFG programs SFB/TRR 21.

References

1. A.F. Andreev and I.M. Lifshitz, Sov. Phys. JETP **29**, 1107 (1969).
2. E. Kim and M.H.W. Chan, Nature **427**, 225 (2004).
3. O. Penrose and L. Onsager, Phys. Rev. **104**, 576 (1956).
4. D.M. Ceperley and B. Bernu, Phys. Rev. Lett. **93**, 155303 (2004).
5. N.V. Prokof'ev and B.V. Svistunov, Phys. Rev. Lett. **94**, 155302 (2005).
6. E. Burovski *et al.*, Phys. Rev. Lett. **94**, 165301 (2005).
7. M. Boninsegni, N. Prokof'ev, and B. Svistunov, Phys. Rev. Lett. **96**, 105301 (2006).
8. E. Kim and M.H.W. Chan, Phys. Rev. Lett. **97**, 115302 (2006).
9. K. Góral, L. Santos, and M. Lewenstein, Phys. Rev. Lett. **88**, 170406 (2002).
10. H.P. Büchler and G. Blatter, Phys. Rev. Lett. **91**, 130404 (2003).
11. S. Wessel and M. Troyer, Phys. Rev. Lett. **95**, 127205 (2005).
12. P. Sengupta *et al.*, Phys. Rev. Lett. **94**, 207202 (2005).
13. V.W. Scarola and S.D. Sarma, Phys. Rev. Lett. **95**, 033003 (2005).
14. G. Batrouni, V. Rousseau, R.T. Scalettar, M. Rigol, A. Muramatsu, P.J.H. Denteneer, and M. Troyer, Phy. Rev. Lett. **89**, 117203 (2002).
15. M. Rigol, A. Muramatsu, G. Batrouni, and R. Scalettar, Phy. Rev. Lett. **91**, 130403 (2003).
16. V.W. Scarola, E. Demler, and S.D. Sarma, Phys. Rev. A **73**, 051601(R) (2006).
17. R. Micnas, J. Ranninger, and S. Robaszkiewicz, Rev. Mod. Phys. **61**, 113 (1990).
18. C.A. Regal, M. Greiner, and D.S. Jin, Phys. Rev. Lett. **92**, 040403 (2004).
19. M. Bartenstein *et al.*, Phys. Rev. Lett. **92**, 120401 (2004).
20. M.W. Zwierlein *et al.*, Phys. Rev. Lett. **92**, 120403 (2004).
21. J.K. Chin *et al.*, Nature **443**, 961 (2006).
22. F.K. Pour, M. Rigol, S. Wessel, and A. Muramatsu, Phys. Rev. B **75**, 161104(R) (2007).
23. S. Wessel, Phys. Rev. B **75**, 174301 (2007).
24. F. Alet, S. Wessel, and M. Troyer, Phys. Rev. E **71**, 036706 (2005).
25. Y.-C. Chen, R. Melko, S. Wessel, and Y.-J. Kao, Phys. Rev. B **77**, 014524 (2008).
26. A. Jagannathan, A. Szallas, S. Wessel, and M. Duneau, Phys. Rev. B **75**, 212407 (2007).
27. M. Holtschneider, S. Wessel, and W. Selke, Phys. Rev. B **75**, 224417 (2007).
28. S. Manmana, S. Wessel, R. Noack, and A. Muramatsu, Phys. Rev. Lett. **98**, 210405 (2007).
29. S. Wessel *et al.*, J. Stat. Mech. P12005 (2007).
30. M. Troyer, F. Alet, and S. Wessel, Phys. Rev. Lett. **90**, 120201 (2003).
31. F. Heidrich-Meisner, M. Rigol, A. Muramatsu, A.E. Feiguin, and E. Dagotto, arXiv:0801.4454.
32. G. Sugiyama and S.E. Koonin, Annals of Phys. **168**, 1 (1986).

33. S. Sorella, S. Baroni, R. Car, and M. Parinello, Europhys. Lett. **8**, 663 (1989).
34. E.Y. Loh and J.E. Gubernatis, in *Modern Problems of Condensed Matter Physics*, edited by W. Hanke and Y. Kopaev (North Holland, Amsterdam, 1992).
35. T. Giamarchi, *Quantum Physiscs in One Dimension* (Clarendon Press, Oxford, 2004).
36. M. Rigol and A. Muramatsu, Phy. Rev. A **69**, 053612 (2004).
37. M. Greiner *et al.*, Phys. Rev. Lett. **94**, 110401 (2005).

Chemistry

Prof. Dr. Christoph van Wüllen

Fachbereich Chemie, Technische Universität Kaiserslautern,
Erwin-Schrödinger-Straße, D-67663 Kaiserslautern,
Germany

For a long time, the use of supercomputers was considered somewhat esoteric, at least in the community of application-oriented quantum chemists. Meanwhile we see these facilities are more and more used to tackle problems that arise, for example in the field of materials science and chemical engineering. These calculations are often intimately related to what is being done in the chemical laboratory or even in the chemical reactor, and the results are indispensable to make application-oriented development faster, more rational and target-oriented.

There is much interest in molecules which have a specific function, "nanodevices" is just one of the many buzz-words one encounters in this field. Konopka et al. report calculations on azobenzene derivatives. Azobenzene is an interesting molecule because it exists in two stable configurations which are quite different in diameter. One can use light to switch from one configuration to the other (this is not too different from the vision process in our eyes where a retinal molecule is involved). However, in the absence of light, also mechanical stress can induce a configuration change. In an actual device, the azobenzene unit must be "welded" to external tips, most popular are bonds between sulfur atoms (that are part of the azobenzene derivative) and gold surfaces. Because it is not clear from the outset that this connection is "innocent", that is, does not affect the functionality of the azobenzene, the influence of the gold tips on the electronic structure of azobenzene has also been investigated.

Electrochemical devices such as batteries or fuel cells basically consist of two volumes (which are connected to the electrodes) which should be kept as separate as possible (to prevent shortcuts) while allowing a flow of charge (to keep charge neutrality) between them. High (mechanical and thermal) stability can be expected if membranes separating the electrolytes or even the electrolytes themselves are solids in which however certain anions can move more or less freely. In most cases these materials are oxides or nitrides doped such that they contain anion vacancies where moving anions can hop to. This process has been simulated, on an atomic level, in the contribution by Wolff et al. Important questions here are the relative mobility of oxide and nitride anions in quaternary M/Ta/O/N (for various metals M) materials, as well as the energy profile that a moving anion encounters while moving from its current lattice position to a neighbouring vacancy.

The huge amount of substance present in a reactor in a chemical plant make precise knowledge of thermophysical data highly important, otherwise efficiency and safety could not be guaranteed. Macroscopic quantities like these can be derived from interactions at the atomic level through the machinery of statistical mechanics. *"Molecular modeling"* is the name of various computational techniques to do so. While for pure substances such data is probably more easily accessible from experiment, calculations become attractive if properties of *mixtures* are of interest. The equilibrium between vapour and the liquid phase, structural properties of the liquid state, and the nucleation process (droplet formation) when a saturated vapour starts condensing are investigated in the contribution of Eckl et al. They describe both the parametrization of the interaction potential as well as molecular simulations using the force field generated this way.

Although the systems of interest were of very different sizes, ranging from a nano-switch with few nano-meter diamater, over a fuel cell that one person can carry, to a chemical plant in which dozens of people could be working, the common feature of all these investigations is that the system is actually looked at *at the atomic level*: let it be the shape change mostly induced by rotation about a nitrogen-nitrogen chemical bond, the migration of an oxygen atom from one lattice position to a neighbouring one, or the interaction between different sites of two interacting oxirane (ethylene oxide) molecules. Since some time, the software used to do such calculations is well adapted to parallel computers, such that these programs can unleash the power of present-day supercomputers and turn the investment required for setting up and maintaining supercomputer centers into scientific output ready for the improvement of products and production processes.

Azobenzene–Metal Junction as a Mechanically and Opto–Mechanically Driven Switch

Martin Konôpka[1], Robert Turanský[1], Nikos L. Doltsinis[2,3], Dominik Marx[2], and Ivan Štich[1,4]

[1] Center for Computational Materials Science, Department of Physics, Slovak University of Technology (FEI STU), 81219 Bratislava, Slovakia
{martin.konopka,robert.turansky}@stuba.sk
[2] Lehrstuhl für Theoretische Chemie, Ruhr–Universität Bochum, 44780 Bochum, Germany
dominik.marx@theochem.ruhr-uni-bochum.de
[3] Department of Physics, King's College London, London WC2R 2LS, United Kingdom
nikos.doltsinis@kcl.ac.uk
[4] Institute of Physics, Slovak Academy of Sciences, 84511 Bratislava, Slovakia
ivan.stich@savba.sk

Summary. Mechanically and opto–mechanically controlled azobenzene (AB) switch based on AB–metal break–junction have been studied using *ab–initio* simulations. It was found that both *cis*→*trans* and *trans*→*cis* mechanically driven switchings in the lowest singlet state are possible. Bidirectional optical switching of mechanically strained AB through first excited singlet state was also predicted provided that the length of the molecule is adjusted towards the target isomer equilibrium length.

1 Introduction

Rapidly developing field of molecular electronics is directed towards single–molecule devices which are based on specific responses of certain molecules. The aim is to build electrically, optically and mechanically driven molecular switches, sensors or data storage media [1, 2, 3, 4]. A lot of attention has been focused on azobenzene (AB) which is a molecule capable of optical switching [5]. Its two different conformations — *trans* (TAB) and *cis* (CAB) — can be optically switched using laser light of appropriate wavelength [6]. The *trans*→*cis* isomerisation is accompanied by a significant decrease in the length of the molecule (by ≈ 2.4 Å). AB can be used as a molecular engine [7] driven by optical pulses.

In molecular–electronics applications AB is a subsystem of a larger device. For this purpose the molecule is functionalised by sulphur atoms which serve as very convenient bridges between carbons and metals like gold (also copper at low temperatures, see recent computational studies [8, 9]). Hence the anchoring is done by functionalising the AB molecule by the *thiolate* bonds between sulphurs and phenyl rings, see Figs. 1 and 2. Embedding AB into junctions introduces new complexities such as the modification of mechanical, electronic and chemical properties of AB with possible modification of the isomerisation mechanisms. Indeed, experiments on anchored AB [32] and related photochromic molecules [10] indicate that two–way photo–switching of these systems may not be straightforward.

The key issues raised by embedding AB in a mechanically controllable break–junction (MCB) [11] are as follows. (1) In which way does the anchoring to the gold tips alter the electronic structure of AB and hence its physical properties? (2) In which way can the optical switching probabilities be manipulated by applying mechanical strain? (3) Can AB be switched by purely mechanical means? (4) Can this switching be bidirectional, i.e. both *cis*→*trans* and *trans*→*cis*? (5) What is the interplay between optical [12, 13] and mechanical [14] switching forces?

2 Models and Methods

In this report we present *ab initio* modelling of both mechanical and opto–mechanical isomerisation of a 4,4′–dithioazobenzene (DAB) bridge connected to two gold electrodes [15]. Mechanical *cis*→*trans* switching has been modelled using two differently shaped tips each composed of $40 \div 60$ Au atoms. These tips, interconnected through periodic boundary conditions, are large enough to allow for elastic/plastic deformations arising during mechanical manipulation. A similar model with smaller (≈ 20 Au atoms) tips was used for mechanical *trans*→*cis* simulations. We performed ground state DFT calculations employing the Perdew–Burke–Ernzerhof (PBE) functional [16], a 19–electron (semi–core) pseudopotential for Au, and an all–electron representation for the other species. The calculations were carried out using the DMol3 code [17] with a DNP (double numerical plus polarisation) basis set, and a 0.005 Ha electronic smearing. To obtain a starting point for the DFT calculations, the gold tips were first prepared using empirical interatomic potentials [18] by a MCB [11] type of procedure. A gold rod coupled to two gold (110) plates was generated. The system was then heated and let to evolve under Langevin dynamics followed by a relatively fast pulling process up to the point of breakage of the Au nanojunction. The tips so formed were then allowed to evolve at low temperatures (5 K) for a few picoseconds. All remaining modelling was performed using DFT techniques: the tips resulting from the dynamics were relaxed, their separation adjusted so as to accommodate a relaxed CDAB or TDAB (stripping the hydrogen atoms off the terminating SH groups). The junction was then relaxed by DFT geometry optimisation at different constrained inter–tip distances, typically in steps of 0.2 Å. Straining

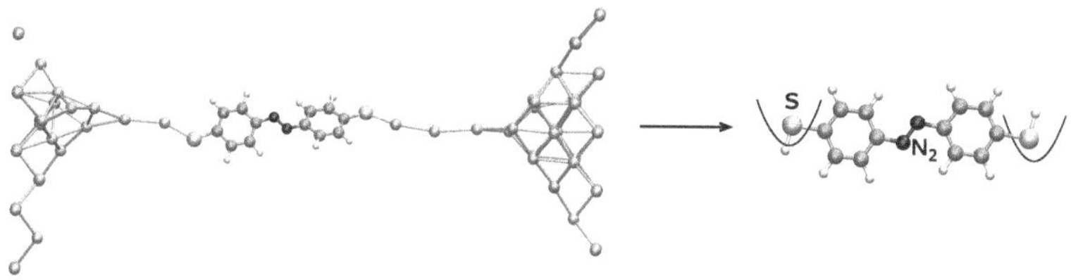

Fig. 1. A model of azobenzene junction between gold tips used in mechanical switching simulations and the schematics of the model with harmonic restraints applied on the sulphur atoms and used for the opto–mechanical simulations. The chosen configuration with the tips corresponds to extension $D = 15.2$ Å of the system described up to 8.0 Å by Fig. 2

(both pulling and compressing) of the junction in static manner was continued until rupture. The pulling simulations have been applied mainly to *cis* isomer (which is the shorter one in equilibrium) while the compressing strain was exerted in most cases on *trans* isomers. Several realisations of the junction have been modelled, differing by the model tips used and/or sulphur bonding site on the tip. Three of such realisations have been used to model pulling of the *cis* (also *trans* after the pulling–induced conversion) isomer and two realisations to study compression of the *trans* isomer.

To study opto–mechanical switching in the S_1 excited state we employed the generalised restricted open–shell Kohn–Sham (gROKS) method [19, 20], which has been successfully applied to the AB chromophore previously [13, 21]. In the present scenario, the electronic structure of AB is considerably complicated by the presence of the gold electrodes. In order to disentangle mechanical and optical/chemical effects, we adopted a simplified model to study opto–mechanical switching, in which the mechanical action of the tips were modelled by harmonic restraining potentials in which the hydrogen–terminated sulphur atoms of DAB are allowed to move during dynamical evolution in the S_1 excited state, see Fig. 1. One choice of the force constant was 0.167 a.u., corresponding to the value of the Au–S molecule. The other value was twice of this, capable to model harder metal–sulphur of semiconductor–sulphur bondings. Since the model neglects chemical effects of gold on AB electronic structure we analyse and discuss them below in subsection 3.3. The model was found qualitatively robust also against the choice of the force constants. Varying the distance between the two restraining centres mimics variations of a tip distance of an STM/AFM apparatus. Straining of the system is continued until rupture of the molecule and, in case of the *trans* conformation, we studied also its compression.

The excited state simulations were performed with the CPMD code [22, 23] using a plane–wave basis set truncated at 70 Ry, Goedecker pseudopotentials [24, 25], and the PBE functional [16]. Comparative ground state calculations were carried in the same setup. We do not perform a non–adiabatic simulation but rather generate a complete set of ground–state energies for

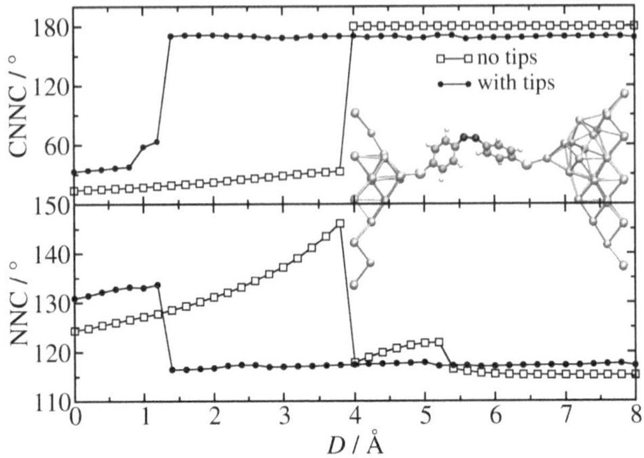

Fig. 2. Simulation of the mechanical *cis*→*trans* switching. The system with tips (see Figs. 1 and 2) was simulated using 130 Au atoms in the supercell. *Upper panel:* CNNC dihedral angle. *Lower panel:* Average NNC bond angle. The extension D is defined as the displacement of the fixed Au atoms plane from the equilibrium value. For the system without tips the displacement of the pulled sulphur atom is considered as D. The inset shows the starting configuration which is the *cis*–equilibrium one. See also the left–hand side of Fig. 1 where the stretched configuration (*trans*–conformation) at $D = 15.2$ Å is drawn in the same zoom as the inset. All the results have been calculated using numerical localised basis sets and the DMol3 code [17]

configurations visited in the S_1 state, see subsection 3.2. Based on a comparison with experimental results [26], we have added a uniform shift of 0.7 eV to all gROKS (S_1) energies, see the values in Tab. 1 to be discussed below.

3 Results

3.1 Mechanical Switch

We have found that both *cis*→*trans* and *trans*→*cis* conformation changes in the S_0 electronic state are possible if appropriate mechanical manipulation is applied. *Cis*→*trans* isomerisation occurred under an external pulling action while for the opposite isomerisation direction a compressing force was used. Basically this behaviour is consistent with the different equilibrium lengths of the two isomers which in terms of sulphur–sulphur distances are about 9 and 12.6 Å for *cis* and *trans* DAB, respectively.

Simulation results of mechanical *cis*→*trans* switching induced by the external pulling force are shown in Fig. 2. The quantities are plotted against the extension D which is defined as the displacement of the fixed Au atoms plane from its equilibrium position. In the case without tips D is just the displacement of the pulled sulphur atom. The results indicate that mechanical *cis*→*trans* switching in the lowest singlet is easily feasible requiring fairly modest forces (~1 nN) irrespective of the details of the tips used. Isomerisation proceeds primarily via torsion about the N=N bond as can be seen from the changes in the CNNC dihedral angle. The dihedral angle is a soft degree–of–freedom, and therefore the switching proceeds on a flat part of the potential energy surface (PES) and changes only weakly with the model tips used. However, after the *cis*→*trans* isomerisation has occurred, further force application to the *trans* isomer proceeds on much steeper part of the PES with a number of elastic/plastic tip(s) deformations. The *trans* isomer presents a relatively stiff structure without any significant structural changes during the pulling. Fragmentation of the junction occurs by breaking an Au–Au bond in the monatomic gold nanowire(s) pulled out of the tip(s). See Fig. 1 with the configuration one simulation step ($\Delta D = 0.2$ Å) before the gold nanowire at the right–hand sulphur atom breaks. Starting the mechanical treatment from the *trans* isomer (not shown) reproduces the behaviour observed after the *cis*→*trans* switching corroborating the fact, that the *trans* isomer is rather stiff and the gold tips rather ductile.

The mechanical effect of gold tips on both molecular isomerisation and breaking of the junction can be appreciated by comparison to the same processes but generated using hard distance constraints on the two S atoms instead of gold tips, see "no tips" plots on Fig. 2. Clearly, the system without the tips is much stiffer due to the missing elastic degrees of freedom in the tips and thus the PES is much steeper and the force required for isomerisation much larger (~3 nN). Qualitatively, the strain–induced processes with and without tips have several common features, e.g. *cis*→*trans* isomerisation at fairly small extensions (at $\approx 1.5 \div 5$ Å). There are, however, some important differences between the scenarios with and without tips. First, without tips breakage occurs at much smaller extensions, ≈ 6 Å compared to $9 \div 16$ Å with tips. This is an obvious consequence of the tips deformations accounted to the extension D. Moreover, without tips the isomerisation takes place primarily via the *inversion mechanism*, i.e. by change of NNC bond angle, and not via the (mostly) rotation mechanism observed with the tips. In both cases however the isomerisation mechanisms are neither pure rotation nor pure inversion but mixtures of both.

Our simulations have shown that, contrary to intuition, the reverse mechanical *trans*→*cis* switching is also possible [27]. I.e. the process of TDAB compression in the S_0 singlet state can turn the molecule into the *cis* conformation. Again, we have examined the process on the two levels of modelling, with just dithioazobenzene alone and with the molecule attached to small gold tips of 40 atoms together (18 of them fixed). In both cases we observed the switching. The switching without tips occured at a compression corresponding to sulphur–sulphur distance $d^{\mathrm{sw}}_{\mathrm{S–S}} = 5.6$ Å. On the other hand, for azobenzene attached to the gold tips, the isomer was changed at $d^{\mathrm{sw}}_{\mathrm{S-S}} = 7.4$ Å. The minimum of energy for the structure with the small tips has an inter–

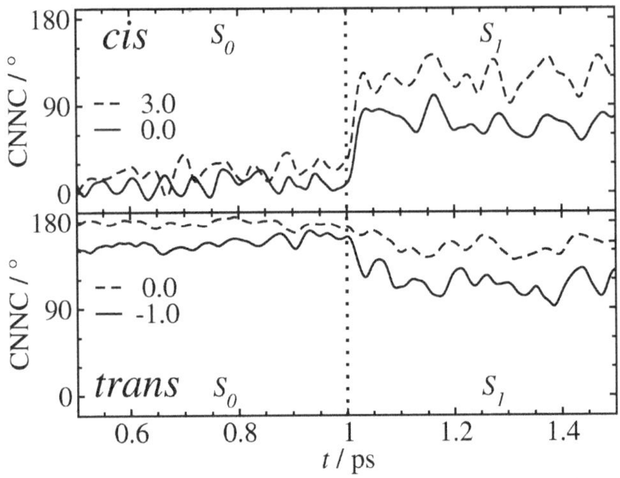

Fig. 3. Simulation of the opto–mechanical switching. Time evolution of the CNNC dihedral angles. Interval [0, 1] ps (second half shown) presents the equilibration stage in the S_0 electronic state. Interval [1, 1.5] ps corresponds to the excited S_1 state dynamics. The dotted vertical line represents the instant of the vertical $S_0 \rightarrow S_1$ excitation. *Upper graph:* evolution starting from the *cis* conformation. *Lower graph:* evolution from the *trans* conformation. Extension D (different values distinguished by different lines) measures for each of the isomers the distance difference between the harmonic restraining potential centres and the equilibrium isomer length (8.6 Å for *cis* and 12.6 Å for *trans* structure). The restraining parameter was $\kappa = 0.167$ a.u.

sulphur distance corresponding to a partially compressed AB molecule which leads to difficulties when comparing the results in terms of the quantity D. The minimum–energy structure with the tips is a result of the tip preparation procedure and accompanying structural changes of Au–Au bonds in the tips. The compression regime of AB attached between gold tips is still under investigation.

3.2 Opto–Mechanical Switch

The protocol of the dynamical simulations is as follows. (1) First dynamics of DAB in its S_0 ground–state at a given stretch for given isomer is simulated at 300 K for 1 ps. (2) A laser pulse applied in an experiment is modelled by a vertical $S_0 \rightarrow S_1$ electronic excitation. (3) System dynamics in the S_1 state is followed for 0.5 ps. The S_0 energies for the excited state trajectories are simultaneously generated and the $S_1 - S_0$ energy separation, crucial for non–adiabatic relaxation, calculated. Steps (1)–(3) are repeated for several stretch values for both *cis* and *trans* isomers. If not otherwise said the results shown corresponds to the restraining parameter $\kappa = 0.167$ a.u.

Fig. 3 shows the time evolution of the CNNC dihedral angle leading up to and after photo–excitation of the *cis* and *trans* isomers at different tip

Table 1. Average de–excitation energies $\langle E_{\text{deexc}} \rangle$ in eV's for the set of simulations shown in Fig. 3. The legends *cis*/*trans* denote the *initial* conformations of the isomers. For the definition of the extension D see Fig. 3 and its caption

D/Å	-1.0	0.0	3.0
cis	—	1.00	1.24
trans	1.19	1.75	—

separations. (The tips are modelled by the harmonic restraints in this case.) Extensions $0.0 \div 4.5$ Å for the *cis* and $-2.0 \div 1.6$ Å for the *trans* isomer have been studied, several representative cases are shown in the figure. Comparisons of "no tips" vs. "tips" plots on Fig. 2 indicate that the model with restraints qualitatively reproduces mechanical effects of the model with the tips only for small extensions D. In Fig. 3 an abrupt response to the vertical $S_0 \rightarrow S_1$ excitation at time $t = 1$ ps can be seen, in particular for the *cis* isomer which undergoes changes in the CNNC dihedral by about $90 \div 150°$ with 10 fs. These CNNC dihedral changes are the larger the larger is the tip separation (Fig. 3). The amplitude of the rotational relaxation of the *trans* isomer is much less pronounced although still very fast. An extremely fast partial ($15 \div 20°$) opening of NNC angles takes effect especially at larger stretches (not shown in figures) for the *trans* isomer. For negative stretches the CNNC variations upon photo–excitation are considerably larger since the DAB molecule is allowed sufficient freedom to relax to the excited state global minimum rotamer structure [29]).

So far we have discussed the dynamical evolution in the excited state. However, photoisomerisation is only complete after the chromophore has relaxed back to the electronic ground state and its excess kinetic energy has dispersed into the environment. An important criterion controlling non–radiative decay is the $S_1 - S_0$ "vertical" energy gap, E_{deexc}. In order to extract trends for the efficiency of non–radiative decay pathways at the different elongations, we computed the time averages $\langle E_{\text{deexc}} \rangle$ over the 0.5 ps excited–state trajectories for simulations starting from both isomers, see Tab. 1. For the *cis* isomer the gap size is increased when the extension of 3 Å is applied. The pronounced feature of the *trans*-DAB is the fall of $\langle E_{\text{deexc}} \rangle$ in the compression regime. The effect of extension/compression on $\langle E_{\text{deexc}} \rangle$ can qualitatively be understood as to follow from profiles of the S_1 and S_0 PESs [30, 29]; the average de–excitation gap is a function of the molecule length and has its minimum in between equilibrium *cis* and *trans* lengths. These trends depend only weakly on the force constant of the restraining potentials [27].

Analysis of time evolution of E_{deexc} shows that the gap is smallest for CNNC dihedrals around 100°. This value is in agreement with recent quantum chemical calculations [29]. The CNNC dihedral is thus the major structural parameter governing the de–excitation gap and consequently the probability of $S_1 \leftrightarrow S_0$ switch.

The compressive regime in *trans*→*cis* switching is important also in view of counteracting the large length mismatch between the *trans* and *cis* isomers which is in equilibrium about 4 Å measured in terms of sulphur–sulphur separations. In other words, if the *trans* isomer is kept by the restraints or sulphur–gold bonds and its length is 3 ÷ 4 Å longer than the equilibrium *cis* length then the probability of *trans*→*cis* conformation change is quite small due to the mechanical (length) restriction. The restriction highly suppresses the internal degrees of freedom of the molecule which are important for the isomer change. In addition, as can be seen from Tab. 1, the average de–excitation gap is relatively large for equilibrium *trans* length thus making $S_1 \rightarrow S_0$ transition less likely to occur. Hence both mechanical effects and magnitude of the de–excitation gap decrease the probability of the *trans*→*cis* switching for *trans* lengths close to or larger than the equilibrium *trans* length. Contrary, in the compressive regime both $\langle E_{\mathrm{deexc}} \rangle$ drops hand–in–hand with reduction of the mechanical hindrance. Hence, the *trans*→*cis* switching may only be possible in compressive regime. This finding may explain the lack of experimental observation of a two–way switching of photochromic molecules anchored to tips [31, 32].

3.3 Chemical Effects of Gold on Azobenzene Electronic Structure

In addition to mechanical hindrance to photo–switching of anchored photochromic molecules mentioned above, the other factor limiting the efficiency of the optical isomerisation in these systems may be chemical quenching of the $n \rightarrow \pi^*$ (HOMO → LUMO) excitation on the nitrogen atoms due bonding to the tips. To examine chemical effects of the tips on the AB electronic structure we have prepared isolated complexes of the pattern Au_{12}–S–*AB*–S–Au_{12} for both *trans* and *cis* isomers. The structures have been roughly relaxed and Kohn–Sham eigenstates $\psi_i(\mathbf{r})$ with energies E_i, including also sufficient numbers of empty states, have been calculated. These sets of states have then been used to calculate the local densities of states (LDOSs)

$$\rho(E, \mathbf{r}) = \sum_i |\psi_i(\mathbf{r})|^2 \delta(E - E_i) \ .$$

For plotting purposes the delta functions have been replaced by gaussians of a finite width. As we are now primarily interested how the optical switching of AB through the S_1 state is modified by the gold tips we focus our attention on the central azo region. The two most important electron orbitals for the $S_0 \leftrightarrow S_1$ switching, the n and π^* orbitals, are primarily localised in this part of the system. Hence the LDOS was integrated over a spherical region of 1.8 Å radius and centred in the midst of the N=N bond. The resulting function of energy for both isomers are shown on Fig. 4. The full lines show (integrated) LDOSs for pure free azobenzene. These are reference systems for which it is also experimentally known that it is possible to switch them

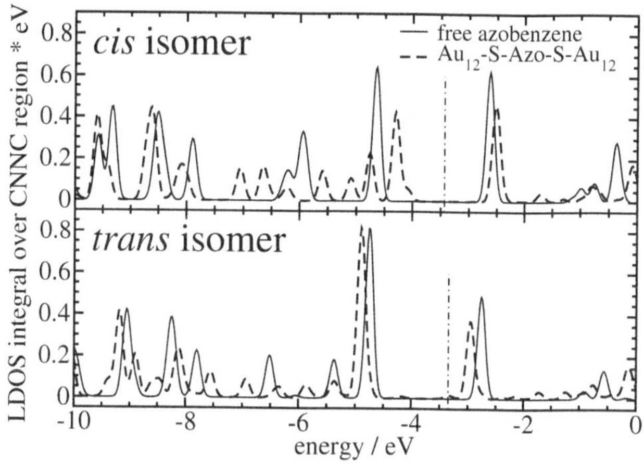

Fig. 4. Local density of states integrated over the CNNC regions (spheres of radii of 1.8 Å centred in between the N atoms) for Au_{12}–S–azobenzene–S–Au_{12} complexes and for the ordinary azobenzene isomers for reference. The dashed–dotted vertical lines represent the chemical potential values and distinguish occupied and unoccupied states. See also text

optically in both isomerisation directions. We can see perfectly distinguished peaks corresponding to the HOMO and LUMO. In order the switching be possible also upon attachment of AB to the gold peaks via the sulphur atoms it is necessary that the distinguished peaks survive (although they need not to be highest occupied and lowest unoccupied states of the whole metal–molecule system). The actual situation is shown in dashed plots. We clearly observe that the peaks are still present although there is significant red shift of $E_{\mathrm{LUMO}} - E_{\mathrm{HOMO}}$ in the *cis* case. The peaks are spread out by the interaction with gold, more significantly for larger gold clusters [32]. Hence we conclude this paragraph by stating that the probability of optical switching is reduced by the interaction with gold and the optical frequencies used to induced the transitions may have changed but the most important features of the electronic structure, the n and π^* orbitals, are qualitatively preserved.

4 Computer Resources and Code

As described in details in previous subsections, the whole project consists of the two closely related parts, first of which deals with mechanical straining of the azobenzene–gold system in the *ground state* and *zero temperature* while the other part presents a model of the azobenzene opto–mechanical switch including the *excited state* and using *dynamical simulations*. Computer resources of the Scientific Supercomputing Center in Karlsruhe, namely the Itanium2–based HP-XC machine, have been used for this second part (about half of it) of the project.

Table 2. Summary of the tasks performed on the HP-XC Itanium2–based supercomputer, all done using the CPMD code [22]. GOpt stands for geometry optimisation and MD for molecular dynamics. Number of tasks of each kind was approximately 30 in the whole project. Roughly half of them used the HP-XC machine at SSCK. There were also several test tasks in additions and tens of less time–consuming single–point energy and forces calculations. The total usage was about 40000 CPU hours

task type	# of CPU hours per task
ground state GOpt	20–100
excited state GOpt	45–300
ground state MD	300
excited state MD	1000–1300

In subsection 3.2 we have sketched the basic schedule of the Born–Oppenheimer molecular dynamic (MD) simulations in the S_0 (ground) and S_1 (excited) states; see Fig. 3 and Tab. 1. In addition to the MD simulations it was necessary to first prepare optimised ground–state structures (used as starting points for the simulations) at each constrained sulphur–sulphur distance using a geometry optimisation method (typically LBFGS, see [22] and references therein). Further, we have calculated optimised excited–state structure at each constraint to get static (zero–temperature) reference results. Another large part of calculations have been done to determine vertical de–excitation gaps along the excited–state trajectories. In other words, ground–state energy for each of the structures occurring in the excited–state MD simulations was computed. This task thus presented a large number ($\propto 10^4$) of short ($1 \div 2$ minutes of real time) single–point energy calculations which have been done on our local resources.

Table 2 shows typical computing times for the computational stages realised (at least to some extent) on the HP-XC machine. We estimate the total CPU time spent of the SSCK resources to be about 40000 CPU hours of the HP-XC machine.

All these simulations employed the CPMD code [22], in most cases version 3.11.1 (with a few of our own modifications). As written also in section 2 this code implements the density functional theory with the plane–wave basis set and pseudopotentials. A great advantage of the plane–wave basis set is the spatially uniform (unbiased) accuracy of the description of the given system. The accuracy can be very high for sufficiently large plane–wave cutoff. The cost for the basis set reliability, compared to localised basis set codes, is higher computational demands. In our simulations a plane–wave energy cutoff of 70 Ry (a necessary minimum for the employed Goedecker norm–conserving pseudopotentials) was used which resulted in number of plane waves typically about 120000.

CPMD as any plane–wave electronic structure code heavily uses three–dimensional discrete Fourier transforms in its efficient implementation – the fast Fourier transform algorithm (FFT). A typical FFT mesh was $240 \times 128 \times 128$. Roughly half of total elapsed time in a typical MD run was spent in FFT–related subroutines and about 20% of the time was spent in subroutines dealing with gradient corrections to the density functional. Details of the implementation of the code can be found in [23] and references therein.

Code performance largely depends also on mathematical libraries performing linear algebra operations ranging from simple ones like vector products to more complex ones like multidimensional arrays multiplications. We used the Intel Mathematical Kernel Library (MKL) available on the HP-XC nodes. We also used available version of the Intel Fortran compiler. However since most time–expensive parts of the code are executed in specialised libraries like MKL the choice of the compiler or compilation options is not dramatically important for overall performance. The most important factor is usage of highly optimised mathematical libraries. Overall performance of the CPMD code for our simulations was $1.5 \div 2$ times larger than on an analogous parallel Opteron244–based (1.8 GHz) machine, also with two processors per node and with a high–speed Myrinet network which did not cause (similarly as the network on the HP-XC machine) any significant performance drops.

Large data arrays such as wavefunctions or electron density (represented either in reciprocal or in real space) used in the simulations have been distributed among a number of processors. The calculations typically run on 8, 16 or 24 processors. Memory requirements were not substantial: about 3.2 GB for whole task and memory usage per node was approximately number–of–nodes times smaller. The parallelisation of CPMD is realised through the MPI library calls. A parallel task is divided between number of copies each one working with its own portion of data. Each process of a parallel run uses serial implementation of specialised mathematical libraries (like MKL) for linear algebra and discrete Fourier transforms. I/O operations are realised by only one of the in–parallel running processes. The parallelisation efficiency in our tests, measured in terms of (1/elapsed time), have been found to scale about linearly or even slightly better that linearly with number of processors in the tested range up to 16 processors of the HP-XC machine. The super–linear scaling is a results, apart from perfect parallelisation of the code and a high–speed network, of better cache utilisation when a larger number of processors is used; a task using larger number of processor has larger portion of its data stored in CPU caches. An interesting finding is that the super–linear scaling was obtained only for jobs running on the 2–processor 1.5 GHz Itanium2 nodes. The parallelisation scaling was noticeably (although not significantly) sub–linear on the 16–processor 1.6 GHz Itanium2 nodes (with whole job running on single physical node, which was however divided into two logical nodes because of operating system restrictions). Finally, tests on other machines, IBM SP2 and SP5 for example, (not all of them at SSCK) has also shown noticeably super–linear scaling on 2–processor nodes and sub–linear one on a

16–processor node. Detailed report on CPMD parallelisation (also vectorisation) efficiency especially for much higher numbers of processors than we had available can be found in [34].

5 Conclusions

We have carried out *ab initio* molecular dynamics simulations, structural optimisations and electronic structure analysis of mechanically and opto–mechanically controlled molecular junctions based on a dithioazobenzene bridge between gold tips. Purely mechanical switching from *cis* isomer to *trans* one in the lowest singlet state S_0 was found possible for the tips externally pulled away of each other. The isomerisation in the presence of the gold tips proceeds primarily via the rotation mechanism. Reverse process of mechanically induced *trans*→*cis* switching was also observed.

Optical switching via the first excited singlet state S_1 was explored, subject to the mechanical effect of the electrodes modelled by harmonic restraints. The vertical de–excitation gap crucial for non–radiative relaxation was found to significantly vary with the extension/compression applied to the molecule. In line with experiments [32, 10], efficient switching has been shown to be mechanically hindered for stiff junctions under certain conditions due to significantly different isomer lengths, especially in the *trans* → *cis* direction. To avoid the hindrance, the length of the molecule must be mechanically adjusted or the contacts to the tips must not be stiff.

Chemical effects of gold tips on azobenzene orbitals have been found to decrease the efficiency of the optical switching and to red–shift the $S_0 \rightarrow S_1$ excitation relative to an isolated molecule. Quantitative measure of these effects depends on tips sizes and shapes but significant evidence of the optical switching ability have been observed for each of the analysed systems.

Acknowledgements

Computer resources from Scientific Supercomputing Center Karlsruhe (the HP-XC supercomputer), BOVILAB@RUB, Recherverbund–NRW, and CCMS as well as financial support from Volkswagen–Stiftung (STRESSMOL), APVT (20-019202), DFG, and FCI are gratefully acknowledged.

References

1. Ellenbogen, J.C., Love, J.C.: Architectures for Molecular Electronic Computers. IEEE, New York, (2000)
2. Joachim, C., Gimzewski, J.K., Aviram, A.: Electronics using hybrid–molecular and mono–molecular devices. *Nature*, **408**, 541–548 (2000)

3. Park, J., Pasupathy, A.N., Goldsmith, J.I., Chang, C., Yaish, Y., Petta, J.R., Rinkoski, M., Sethna, J.P., Abruña, H.D., McEuen, P.L., Ralph, D.C.: Coulomb blockade and the Kondo effect in single–atom transistors. *Nature*, **417**, 722–725 (2002)
4. Liang, W., Shores, M.P., Bockrath, M., Long, J.R., Park, H.: Kondo resonance in a single–molecule transistor. *Nature*, **417**, 725–729 (2002)
5. Dürr, H., Bouas-Laurent, H. (ed.): *Photochromism. Molecules and Systems.* Elsevier, Amsterdam (1990).
6. Hartley, G.S.: The Cis–form of Azobenzene. *Nature*, **140**, 281–281 (1937)
7. Hugel, T., Holland, N.B., Cattani, A., Moroder, L., Seitz, M., Gaub, H.E.: Single–Molecule Optomechanical Cycle. *Science*, **296**, 1103–1106 (2002)
8. Konôpka, M., Rousseau, R., Štich, I., Marx, D.: Detaching Thiolates from Copper and Gold Clusters: Which Bonds to Break? *J. Am. Chem. Soc.*, **126**, 12103–12111 (2004)
9. Konôpka, M., Rousseau, R., Štich, I., Marx, D.: Electronic Origin of Disorder and Diffusion at a Molecule–Metal Interface: Self–Assembled Monolayers of CH_3–S on Cu(111). *Phys. Rev. Lett.*, **95**, 096102–1–4 (2005)
10. Dulić, D., van der Molen, S.J., Kudernac, T., Jonkman, H.T., de Jong, J.J.D., Bowden, T.N., van Esch, J., Feringa, B.L., van Wees, B.J.: One–Way Optoelectronic Switching of Photochromic Molecules on Gold. *Phys. Rev. Lett.*, **91**, 207402–1–4 (2003)
11. Smit R.H.M., Noat, Y., Untiedt, C., Lang, N.D., van Hemert, M.C., van Ruitenbeek, J.M.: Measurement of the conductance of a hydrogen molecule. *Nature*, **419**, 906–909 (2002)
12. Cimelli, C., Granucci, G., Persio, M.: Are azobenzenophanes rotation–restricted? *J. Chem. Phys.*, **123**, 174317–1–10 (2005)
13. Nonnenberg, C., Gaub, H., Frank, I.: First–Principles Simulation of the Photoreaction of a Capped Azobenzene: The Rotational Pathway is Feasible. *ChemPhysChem*, **7**, 1455–1461 (2006)
14. Konôpka, M., Turanský, R., Reichert, J., Fuchs, H., Marx, D., Štich, I.: Mechanochemistry and Thermochemistry are Different: Stress–Induced Strengthening of Chemical Bonds. *Phys. Rev. Lett.*, **100**, 1155031–1–4 (2008)
15. Turanský, R., Konôpka, M., Reichert, J., Fuchs, H., Marx, D., Štich, I.: Mechanical and Opto–Mechanical Switching of Azobenzene Metal–Organic Junctions. (In preparation)
16. Perdew, J.P., Burke, K., Ernzerhof, M.: Generalized Gradient Approximation Made Simple. *Phys. Rev. Lett.*, **77**, 3865–3868 (1996); *Phys. Rev. Lett.*, **78**, 1396–1396 (1997)
17. see `http://www.accelrys.com`
18. Jacobsen, K.W., Norskov, J.K., Puska, M.J.: Interatomic interactions in the effective–medium theory. *Phys. Rev. B*, **35**, 7423–7442 (1987)
19. Frank, I., Hutter, J., Marx, D., Parrinello, M.: Molecular dynamics in low–spin excited states. *J. Chem. Phys.*, **108**, 4060–4069 (1998)
20. Grimm, S., Nonnenberg, C., Frank, I.: Restricted open–shell Kohn–Sham theory for $\pi - \pi^*$ transitions. I. Polyenes, cyanines, and protonated imines. *J. Chem. Phys.*, **119**, 11574–11584 (2003)
21. Böckmann, M., Doltsinis, N.L., Marx, D.: Submitted (2007)
22. CPMD, Copyright IBM Corp 1990–2006, Copyright MPI für Festkörperforschung Stuttgart 1997–2001

23. Marx, D., Hutter, J.: Ab initio molecular dynamics: Theory and Implementation. In: Grotendorst, J. (ed) *Modern Methods and Algorithms of Quantum Chemistry.* NIC, FZ Jülich (2000), pp. 301-449; for downloads see: www.theochem.ruhr-uni-bochum.de/go/cprev.html
24. Goedecker, S., Teter, M., Hutter, J.: Separable dual–space Gaussian pseudopotentials. *Phys. Rev. B*, **54**, 1703–1710 (1996)
25. Hartwigsen, C., Goedecker, S., Hutter, J.: Relativistic separable dual–space Gaussian pseudopotentials from H to Rn. *Phys. Rev. B*, **58**, 3641–3662 (1998)
26. The gROKS $n \rightarrow \pi^*$ excitation energies are calculated to be 2.21 eV for *trans* AB (2.13 eV for *cis* AB) compared to 2.82 eV (2.91 eV) from gas–phase experiments [28] and 2.84 eV (3.0 eV) from correlated Coupled Cluster calculations [33]. Similar errors are expected for DAB isomers.
27. Turanský, R., Konôpka, M., Reichert, J., Fuchs, H., Marx, D., Štich, I.: (In preparation)
28. Andersson, J.-Å., Petterson, R., Tegnér, L.: Flash photolysis experiments in the vapour phase at elevated temperatures I: spectra of azobenzene and the kinetics of its thermal cis–trans isomerization. *J. Photochem.*, **20**, 17–32 (1982)
29. Cembran, A., Bernardi, F., Garavelli, M., Gagliardi, L., Orlandi, G.: On the Mechanism of the cis–trans Isomerization in the Lowest Electronic States of Azobenzene: S_0, S_1, and T_1. *J. Am. Chem. Soc.*, **126**, 3234–3243 (2004)
30. Gagliardi, L., Orlandi G., Bernardi, F., Cembran, A., Garavelli, M.: A theoretical study of the lowest electronic states of azobenzene: the role of torsion coordinate in the cis–trans photoisomerization. *Theor. Chem. Acc*, **111**, 363–372 (2004)
31. Choi, B.-Y., Kahng, S.-J., Kim, S., Kim, H., Kim, H.W., Song, Y.J., Ihm, J., Kuk, Y.: Conformational Molecular Switch of the Azobenzene Molecule: A Scanning Tunneling Microscopy Study. *Phys. Rev. Lett.*, **96**, 156106–1–4 (2006)
32. Reichert, J., Klein, S., Konôpka, M., Turanský, R., Marx, D., Štich, I., Fuchs, H.: Conductance of an Illuminated Metal–Molecule–Metal Junction Utilizing a Near–Field Probe as Counterelectrode. [Submitted to *Rev. Sci. Inst.* (2008)]
33. Fliegl, H., Köhn, A., Hättig, C., Ahlrichs, R.: Ab Initio Calculations of the Vibrational and Electronic Spectra of *trans*– and *cis*–Azobenzene. *J. Am. Chem. Soc.*, **125**, 9821–9827 (2003)
34. Hutter, J., Curioni, A.: Car–Parrinello Molecular Dynamics on Massively Parallel Computers. *ChemPhysChem*, **6**, 1788–1793 (2005)

A Density-functional Study of Nitrogen and Oxygen Mobility in Fluorite-type Tantalum Oxynitrides

Holger Wolff, Bernhard Eck, and Richard Dronskowski

Institut für Anorganische Chemie
RWTH Aachen University
Landoltweg 1, 52056 Aachen, Germany
drons@HAL9000.ac.rwth-aachen.de

Summary. In this contribution we present results of our theoretical studies of nitrogen mobility in solid M–Ta–O–N systems. Periodic supercell calculations at density-functional level have been performed to investigate the local structure of N-doped oxynitrides and the anion-diffusion mechanisms. The migration pathways and activation barriers were calculated using the nudged elastic band method with the climbing-image enhancement. We show that the defect migration is mainly caused by the diffusion of oxygen anions. The activation energy can be lowered by increasing the defect concentration, and it is, to a large extent, depending on the dopant size.

1 Introduction

Solid electrolytes are desirable in cells or batteries where liquid electrolytes may be undesirable, such as in implantable medical devices. Preferred solid electrolytes include materials that are solid at room temperature, electronically insulating and ionically conducting. The ionic conductivity is strongly affected by the anion-vacancy concentration which is enhanced by doping with aliovalent cations or with nitrogen [Ler96, LLM06]. Anion diffusion in ternary Zr–O–N and quaternary Y–Zr–O–N materials has been investigated experimentally by single-crystal neutron diffraction [KBS05], tracer diffusion [KAB03, KTA04] and impedance spectroscopy [LLM06]. Only a small number of theoretical studies has been performed so far on M–O–N systems, mainly focusing on the description of well-defined bulk structures. It is well-known that lattice distortions determine the ionic conductivity to a large extent [MJS95, HJ00], and it has been argued that conductivity activation energies depend on vacancy ordering in the anion sublattice [Wac04].

Recently, yttrium-doped tantalum oxynitrides have been synthesized for the first time, and these materials crystallize with defect fluorite-type structures [SWD06, WSL06]. The energy barriers for ionic conductivity in such compounds have been calculated from first principles, and we present a comparative view touching upon dopant size, the amount of dopants, lattice distortions and vacancy ordering.

2 Computational Method

Modelling

A twofold unit cell containing eight formula units of $M_x\text{Ta}_{1-x}(\text{O,N},\square)_2$ was used in this study. The migration pathway of the vacancy ($\square$) and, also, the energy barrier to migration were determined by finding the minimum energy path from one lattice site to an adjacent site using the nudged elastic-band method [MJS95] in its climbing image variant [HJ00, HUJ00, HMJ98] as implemented in the density-functional Vienna *ab initio* simulation package (VASP) [KH93, KH94, KF96, KF96a].

First, the starting and final configurations were determined by optimizing the structure with the vacancy at each of the two adjacent sites. A set of seven intermediate configurations were then optimized such as to converge to points lying on the minimum-energy path. All nine configurations are shown in Figure 1. Harmonic-spring interactions were used to connect the different images of the system along the path. Calculations were performed using the GGA functional of Perdew and Wang [PW92] and ultrasoft pseudopotentials of Vanderbilt type [Van90]. The kinetic cutoff-radius was chosen to be 490 eV, calculating at the Gamma point only with about 70,000 plane-waves in a total of about 70 bands included, the exact numbers depending on the system beheld.

System Requirements

To perform the nudged elastic-band method with VASP, the number of CPUs must be divisible by the number of intermediate configuration images. VASP divides the processors into groups with each group calculating one image. Here, the total number of iterations until convergence is reached can be up to 50,000.

Earlier performance tests on a Sun Fire E25K-Cluster concerning the parallelization performance of VASP have shown that an efficiency of $t/t_{\text{opt}} = 99\%$ can be reached up to 16 CPUs per image, so that a maximum of $7 \times 16 = 112$ CPUs makes economic sense. On the IBM p690 Cluster JUMP, it is reasonable to calculate each image on one complete node and thereby minimizing ethernet traffic, giving a request for $7 \times 4 = 28$ CPUs. A performance test with a total of only 7 processors has shown that, on the IBM p690

Cluster, calculations undergo a speedup of 40 % compared with a Sun Fire E25K-Cluster. Since VASP was in earlier years optimized for vector machines, the next logical step was to use the SX8 at HLRS. After initial compiling problems, which could be solved with the help from NEC personnel and the staff at Stuttgart, a typical speedup of 2–10 times compared to the Sun Fire cluster was achieved, in some special cases even more. VASP heavily relies on FFT and is therefore a good candidate for vector machines with optimized libraries. Tests on the SunFire and the IBM machines showed that the use of those libraries alone led to a speedup of about 20 % compared to the standard lapack routines. On all machines the best performance can only be achieved if the job is computed within one node. A slowdown can be observed on all platforms if the job is distributed over different nodes, but the slowdown is smaller and less important for larger calcalutions.

Concerning memory requirement, VASP relies on dynamic memory allocation. From our experience, the charge density symmetrization usually takes the highest amount of memory. Additionally, the executable and all tables have to be held on all processors as well as a complex array of the size $N_{\text{bands}} \times N_{\text{bands}}$ leading to an average memory allocation of about 4 GBytes per CPU and peak memory allocations of up to 9 GBytes per CPU. The latter memory requirement could only be fulfilled satisfactorily at the SX8, thus enabling us to compute the large cell at high precision.

An average job at the SX8 uses 7 CPUs and 3–8 GBytes of memory, and the time from submission to running was exceptionally short.

3 Results and Discussion

In the system Ta–O–N, the fluorite-type structure is stabilized upon doping with medium-sized trivalent cations such as rare-earth elements [MTO05]. For reasons of electroneutrality, a number of vacancies is introduced into the anionic sublattice, making these compounds promising candidates for ionic (nitrogen and oxygen) conductors. It is often argued that lattice distortions due to ionic radii mismatch lead to an increasing activation barrier for ionic conductivity [MLB04]. In order to study this influence of dopant size for the case of fluorite-type tantalum oxynitrides, we modelled several—partly hypothetical—fluorite-related compounds in the system M–Ta–O–N with four different cations M, namely Fe^{3+}, Sc^{3+}, In^{3+}, and Y^{3+}. The ionic radii [Sha76] of Fe^{3+} and Ta^{5+} are almost identical and they are slightly increasing towards Y^{3+} (see Table 1).

First, structural relaxation was allowed to ensure that the chosen compounds make sense in terms of structure and stability. The resulting structures were stable but heavily distorted; nonetheless, the classification of these compounds as being of a defect fluorite-type is sustainable [WSL06]. The energy–volume functions for $M_{0.125}Ta_{0.875}O_{0.875}N\square_{0.125}$ are shown in Figure 2.

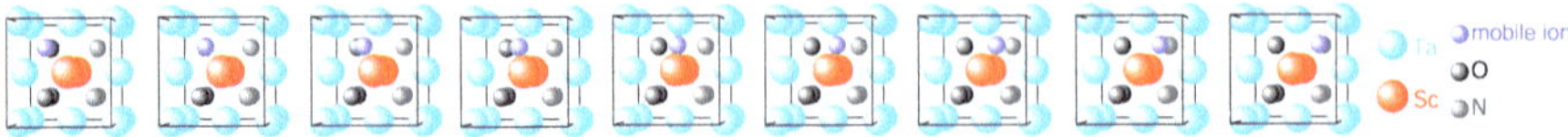

Fig. 1. Initial guess for the calculation of a diffusion path in $M_{0.125}Ta_{0.875}O_{0.875}N\square_{0.125}$

Table 1. Effective ionic radii as taken from reference [Sha76]

M	r_{eff} (Å)
Fe^{3+}	0.78
Sc^{3+}	0.87
In^{3+}	0.92
Y^{3+}	1.02
Ta^{5+}	0.74

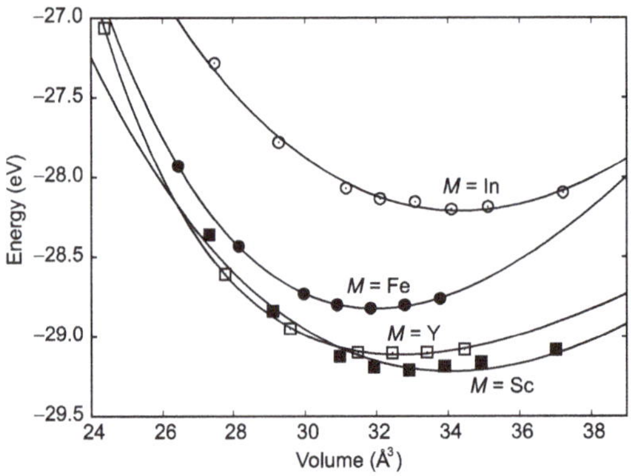

Fig. 2. Total energies for differently doped tantalum oxynitrides $M_{0.125}Ta_{0.875}O_{0.875}N\square_{0.125}$

The migration pathways and activation barriers were then calculated for both cases of oxygen and nitrogen movement. In the fluorite structure, the anions are located in the tetrahedral sites within the face-centered cubic cation framework. Judging from chemical intuition, one would expect that the minimum-energy path for the diffusion of an anion to an adjacent vacant site touches the empty octahedral sites, resulting in *two* saddle points on the energy hypersurface according to the two sharing faces of the octahedron and the two tetrahedra. On the other hand, experimental studies on conductivity of cubic ZrO_2 hint at a diffusion pathway along the edge which is connecting the adjacent tetrahedra [KHB07].

The calculated diffusion path, which is shown in Figure 3, using the example of $Sc_{0.25}Ta_{0.75}O_{0.75}N\square_{0.25}$, shows an unusual square profile at first sight,

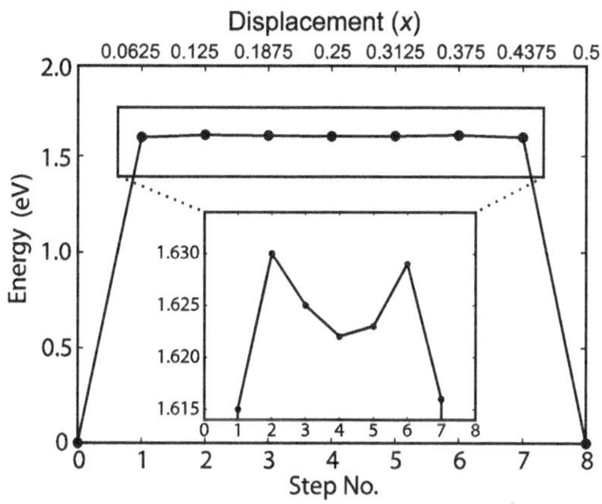

Fig. 3. Calculated energies along the minimum-energy path for oxygen migration in $Sc_{0.25}Ta_{0.75}O_{0.75}N\square_{0.25}$. The central parts are enlarged in the middle

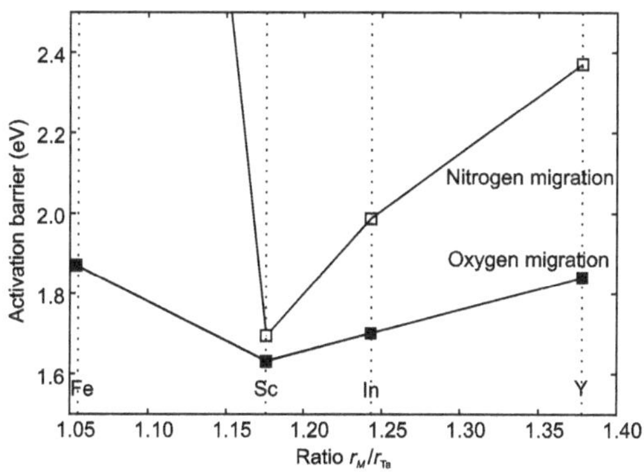

Fig. 4. Calculated energy barriers for nitrogen migration (empty boxes) and oxygen migration (filled boxes) in $M_{0.25}Ta_{0.75}O_{0.75}N\square_{0.25}$, as a function of the ratio of the ionic radii $r_{M^{3+}}/r_{Ta^{5+}}$

without any distinct saddle point. Only the presentation of the energy at the intermediate steps—shown in the center of Figure 3—reveals *two* energetic maxima, as one would expect for a diffusion path along the octahedral site. Nonetheless, the energy differences within this intermediate range are very small (about 10^{-2} eV) if compared with the actual activation barrier. Obviously, the complete activation energy is needed even for a small dislocation of the diffusing atom. This abnormal diffusion property is caused by the low symmetry of the local structure. Due to the high defect concentration, the atoms are not located at the ideal lattice sites of the fluorite structure, making it impossible to distinguish between octahedral and tetrahedral sites. Figure 6

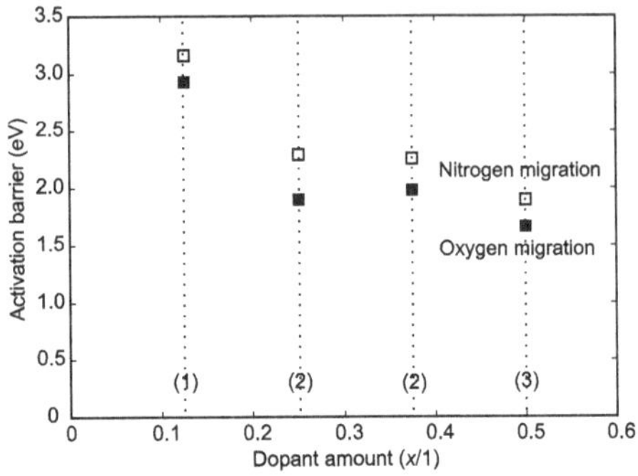

Fig. 5. Energy barriers for nitrogen migration (empty boxes) and oxygen migration (filled boxes) in $Y_xTa_{1-x}(O,N,\square)_2$, as a function of the dopant amount and the number of vacancies per unit cell (in parentheses)

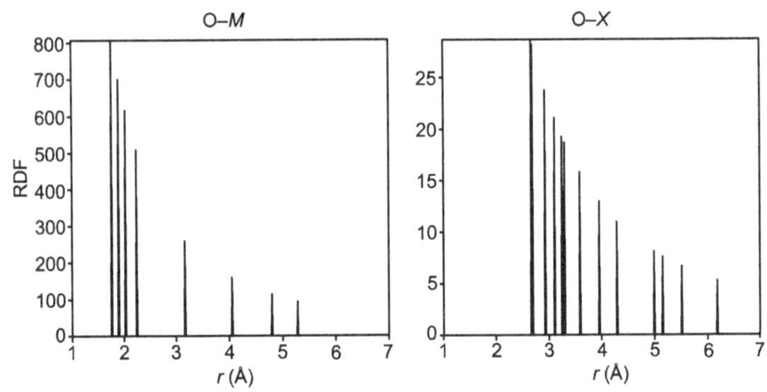

Fig. 6. Radial distribution of cations M and anions X around a moving oxygen atom in $Sc_{0.25}Ta_{0.75}O_{0.75}N\square_{0.25}$

shows the radial distribution function of a moving oxygen atom, giving insight to its local vicinity. In an ideal crystal lattice, the local vicinity of a moving atom can be described as a relaxation around a point defect and the ions move only little from their ideal lattice sites. In a highly defective structure, however, structural relaxation means a *complete* rearrangement of the lattice. In Figure 7 dislocation of the nearest neighbors around a diffusing oxygen atom is shown using the example of $Sc_{0.25}Ta_{0.75}O_{0.75}N\square_{0.25}$. The dislocation is shown in relation to the movement of the oxygen atom by means of a vector **u**. Obviously, the length of **u** is similar for both the diffusing and the surrounding atoms. At the end of the diffusion path, all atoms have moved to new equilibrium sites. Alternatively speaking, *all* atoms move as soon as *one* atom moves, and this is observed as a high activation barrier. Comparing the activation barriers for oxygen and nitrogen movement (see Table 1), it turns out that oxygen is generally more mobile, which can be traced back

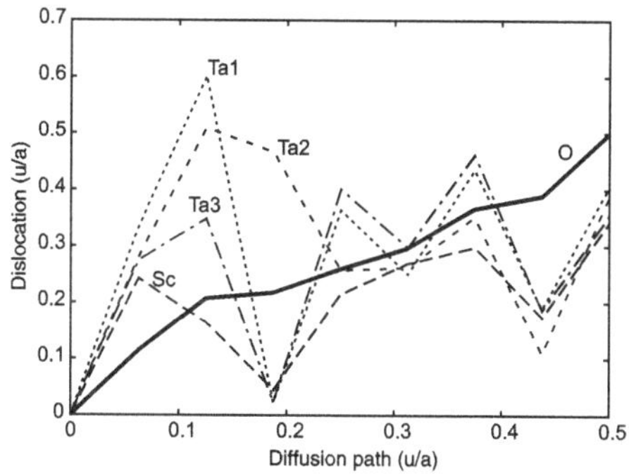

Fig. 7. Ionic movement during the diffusion of an oxygen atom in fluorite-type $Sc_{0.25}Ta_{0.75}O_{0.75}N\square_{0.25}$. The length of the dislocation vector is given as a rate of the lattice constant of the ideal cubic unit cell and is related to the original sites

to the higher charge of the N^{3-} anion. Decreasing the dopant size and thus approximating a ratio $r_{M^{3+}}/r_{Ta^{5+}} \approx 1$, the energy threshold lowers, as can be seen in Figure 4. While this is true for Sc, In and Y, the anionic movement in the Fe-doped compounds is dramatically hindered. Compared with the other three dopant cations, which achieve noble-gas configurations in their trivalent state, Fe^{3+} has a $3d^5$ configuration. Thus, the lowered anionic mobility probably goes back to the more covalent bonding between Fe and N.

Increasing the dopant amount leads to a growing number of vacancies, which causes a higher ion mobility and a decrease of the activation barrier. In Figure 5 the energy barriers are shown in relation to both dopant amount and vacancy concentration. It is obvious that the energy threshold is independent from the dopant concentration but, instead, determined by the vacancy concentration: Turning from $M_{0.25}Ta_{0.75}O_{0.75}N\square_{0.25}$ to $M_{0.375}Ta_{0.625}ON_{0.75}\square_{0.25}$ the activation energy does not significantly change, since the number of vacancies is two in both unit cells.

In $M_{0.5}Ta_{0.5}O_{0.875}N_{0.75}\square_{0.375}$, on the other hand, one additional defect is brought into the unit cell, and the barrier is again lowered.

4 Conclusions

Density-functional calculations show that the energy barrier for anion diffusion in fluorite-type tantalum oxynitrides is closely related to both dopant size and dopant amount. Increasing the dopant amount causes a higher vacancy concentration, making it easier for the ions to move inside the material. The ratio of the cationic radii determines the activation threshold to a very large extent, making it decrease while $r_{M^{3+}}/r_{Ta^{5+}}$ approaches unity. The defect migration, however, is mainly caused by the diffusion of oxide anions, which are slightly more mobile than the nitride anions.

Acknowledgments

The authors wish to thank the computing centers at RWTH Aachen University, Jülich Research Center and HLRS Stuttgart for providing CPU time. This project is supported by Deutsche Forschungsgemeinschaft in the priority program "Substitutional effects in ionic solids".

References

[Ler96] Lerch, M.: Nitridation of Zirconia. J. Am. Ceram. Soc., **79**, 2641–2644 (1996)

[LLM06] Lee, J.-S., Lerch, M., Maier, J.: Nitrogen-doped Zirconia: A comparison with cation stabilized zirconia. J. Solid State Chem., **179**, 270–277 (2006)

[KBS05] Kaiser-Bischoff, I., Boysen, H., Scherf, C., Hansen, T.: Anion diffusion in Y- and N-doped ZrO_2. Phys. Chem. Chem. Phys., **7**, 2061–2067 (2005)

[KAB03] Kilo, M., Argirusis, C., Borchardt, G., Jackson, R.A.: Oxygen diffusion in yttria stabilised zirconia—experimental results and molecular dynamics calculations. Phys. Chem. Chem. Phys., **5**, 2219–2224 (2003)

[KTA04] Kilo, M., Taylor, M.A., Argirusis, C., Borchardt, G., Lerch, M., Kaitasov, O., Lesage, B.: Nitrogen diffusion in nitrogen-doped yttria stabilised zirconia. Phys. Chem. Chem. Phys., **6**, 3645–3649 (2004)

[MJS95] Mills, G., Jónsson, H., Schenter, G.K.: Reversible work transition state theory: Application to dissociative adsorption of hydrogen. Surf. Sci, **324**, 305–337 (1995)

[HJ00] Henkelman, G., Jónsson, H.: Improved tangent estimate in the nudged elastic band method for finding minimum energy paths and saddle points. J. Chem. Phys., **113**, 9978–9985 (2000)

[Wac04] Wachsmann, E.D.: Effect of oxygen sublattice order on conductivity in highly defective fluorite oxides. J. Europ. Ceram. Soc., **24**, 1281–1285 (2004)

[SWD06] Schilling, H., Wolff, H., Dronskowski, R., Lerch, M.: Fluorite-type Solid Solutions in the System Y–Ta–O–N: A Nitrogen-rich Analogue to Yttria-stabilized Zirconia (YSZ). Z. Naturforsch. B, **61**, 660–664 (2006)

[WSL06] Wolff, H., Schilling, H., Lerch, M., Dronskowski, R.: A Density-Functional and Molecular Dynamics Study on the Properties of Yttrium-doped Tantalum Oxynitride. J. Solid State Chem., **179**, 2265–2270 (2006)

[HUJ00] Henkelman, G., Uberuaga, B.P., Jónsson, H.: A climbing image nudged elastic band method for finding saddle points and minimum energy paths. J. Chem. Phys., **113**, 9901–9904 (2000)

[HMJ98] Jónsson, H., Mills, G., Jacobsen, K.W.: Nudged Elastic Band Method for Finding Minimum Energy Paths of Transitions. In: Berne, B.J., Cicotti, G., Coker, D.F. (eds) Classical and Quantum Dynamics in Condensed Phase Simulations. World Scientific, 1998.

[KH93] Kresse, G., Hafner, J.: Ab initio molecular dynamics for liquid metals. Phys. Rev. B, **47**, 558–561 (1993)

[KH94] Kresse, G., Hafner, J.: Ab initio molecular-dynamics simulation of the liquid-metal–amorphous-semiconductor transition in germanium. Phys. Rev. B, **49**, 14251–14269 (1994)

[KF96] Kresse, G., Furthmüller, J.: Efficiency of ab initio total energy calculations for metals and semiconductors using a plane-wave basis set. Comput. Mat. Sci., **6**, 15–50 (1996)

[KF96a] Kresse, G., Furthmüller, J.: Efficient iterative schemes for ab initio total-energy calculations using a plane-wave basis set. Phys. Rev. B, **54**, 11169–11186 (1996)

[PW92] Perdew, J.P., Wang, Y.: Accurate and simple analytic representation of the electron-gas correlation energy. Phys. Rev. B, **45**, 13244–13249 (1992)

[Van90] Vanderbilt, D.: Soft self-consistent pseudopotentials in a generalized eigenvalue formalism. Phys. Rev. B, **41**, 7892–7895 (1990)

[MTO05] Maillard, P., Tessier, F., Orhan, E., Chevire, F., Marchand, R.: Thermal Ammonolysis Study of the Rare-Earth Tantalates $RTaO_4$. Chem. Mater., **17**, 152–156 (2005)

[MLB04] Mogensen, M., Lybye, D., Bonanos, N., Hendriksen, P.V., Poulsen, F.W.: Factors controlling the oxide ion conductivity of fluorite and perovskite structured oxides. Solid State Ionics, **174**, 279–286 (2004)

[Sha76] Shannon, R.D.: Revised Effective Ionic Radii and Systematic Studies of Interatomic Distances in Halides and Chalkogenides. Acta Cryst. A, **51**, 751–767 (1976)

[KHB07] Kilo, M., Homann, T., Bredow, T.: Molecular dynamics calculations of anion diffusion in nitrogen-doped yttria-stabilized zirconia. Philos. Mag., **87**, 843–852 (2007)

Molecular Modeling and Simulation of Thermophysical Properties: Application to Pure Substances and Mixtures

Bernhard Eckl, Martin Horsch, Jadran Vrabec, and Hans Hasse

Institut für Technische Thermodynamik und Thermische Verfahrenstechnik, Universität Stuttgart, D-70550 Stuttgart, Germany
vrabec@itt.uni-stuttgart.de

1 Introduction

For development of new processes and optimization of existing plants, the knowledge of reliable thermophysical data is crucial. Classical methods, next to the often expensive and time consuming experimental determination, are based on models of Gibbs excess enthalpy (G^E-models) or equations of state. These phenomenological approaches usually include a large number of adjustable parameters to describe the real behavior quantitatively correct. As in most cases these parameters have no physical meaning, their methodic deduction is rarely possible and a fit to extensive experimental data is unavoidable. Furthermore, extrapolations may be too unreliable for engineering applictions.

Molecular modeling and simulation offers another approach to thermophysical properties on a much better physical basis. First simulations of this type were done by Metropolis *et al.* in 1953 [1]. By solely defining the interactions between molecules, it is possible to calculate the relevant thermophysical properties with statistical methods. There are two basic techniques which are fundamentally different: molecular dynamics (MD) and Monte Carlo (MC). The goal of both approaches is to sample the phase space, i.e. all available positions and orientations of the molecules under the given boundaries, and an estimation of derivatives of the partition function. Macroscopic properties like pressure or enthalpy are calculated with statistical thermodynamic methods. Detailed descriptions are given, e.g. by Allen and Tildesley [2].

In the present project, the molecular interactions are described by effective pair potentials. These potentials distinguish dispersive, repulsive, and electrostatic interactions as well as interactions resulting from hydrogen bonding. The dispersive and repulsive part is approximated by Lennard-Jones sites due to their computational efficiency. Electrostatics is modeled by partial charges, ideal point dipoles and ideal point quadrupoles. For hydrogen bonding good results were obtained using eccentric partial charges.

Due to a consistent modeling of the pure substances, an application to mixtures is straightforward. Mixed electrostatic potentials are given directly by the underlying laws of electrostatics. The parameters of the unlike Lennard-Jones interactions are taken from combining rules, where good results are obtained with the Lorentz-Berthelot rule [3, 4]. If needed, one adjustable state-independent parameter can be introduced to fit the binary simulation results to experimental data. No further parameters were to be used for the description of ternary or higher mixtures.

In the present work, different applications of molecular modeling and simulation are presented. Within the project MMSTP, two new molecular models for ethylene oxide and ammonia were developed and subsequently used for prediction of different pure substance properties. Next to this, molecular modeling and simulation was applied on the prediction of nucleation processes. Furthermore, within the present project a comprehensive study on the determination of vapor-liquid equilibria of binary and ternary mixtures is performed. This is a still ongoing task and thus is not included in the present report.

Results of this work are consistently published in peer-reviewed international journals. The following publications contribute to the present project:

- B. Eckl, J. Vrabec, M. Wendland, and H. Hasse: Thermophysical properties of dry and humid air by molecular simulation - dew point calculations in a new ensemble. In preparation.
- B. Eckl, J. Vrabec, and H. Hasse: A set of new molecular models based on quantum mechanical *ab initio* calculations. *J. Phys. Chem. B*, in press (2008).
- M. Horsch, J. Vrabec, and H. Hasse: Molecular dynamics based analysis of nucleation and surface energy of droplets in supersaturated vapors of methane and ethane. *ASME J. Heat Transfer*, in press (2008).
- M. Horsch, J. Vrabec, and H. Hasse: Modification of the classical nucleation theory based on molecular simulation data for critical nucleus size and nucleation rate. *Phys. Rev. E*, 78: 011603 (2008).
- M. Horsch, J, Vrabec, M. Bernreuther, S. Grottel, G. Reina, A. Wix, K. Schaber, and H. Hasse: Homogeneous nucleation in supersaturated vapors of methane, ethane, and carbon dioxide predicted by brute force molecular dynamics. *J. Chem. Phys.*, 128: 164510 (2008).
- B. Eckl, J. Vrabec, and H. Hasse: An optimized molecular model for ammonia. *Mol. Phys.*, 106: 1039 (2008).
- B. Eckl, J. Vrabec, and H. Hasse: On the Application of Force Fields for Predicting a Wide Variety of Properties: Ethylene Oxide as an Example. *Fluid Phase Equilib.*, in press (2008).

The present status report briefly shows results of the pure substances ethylene oxide and ammonia, demonstrating the outstanding predictive power of the applied methods. Secondly, the application to nucleation processes is presented. Next to a pure description, the authors were able to further study the underlying effects on the molecular level and to deduce a modification to classical nucleation theory. Futher details are given in the references mentioned above.

2 Prediction of a Wide Variety of Properties for Ethylene Oxide

The predictive and extrapolative power of molecular modeling and simulation is of particular interest for industrial applications, where a broad variety of properties is needed but often not available. To discuss this issue, the Industrial Simulation Collective [5] has organized the Fourth Industrial Fluid Properties Simulation Challenge as an international contest. The task in 2007 was to calculate for ethylene oxide on the basis of a single molecular model a total of 17 different properties from three categories. It should be noted that our contribution was awarded with the first price [5].

Ethylene oxide (C_2H_4O) is a widely used intermediate in the chemical industry. In 2006, 18 million metric tons were produced mostly by direct oxidation of ethylene, over 75 % of which were used for ethylene glycols production. Despite its technical and economical importance, experimental data on thermophysical properties of ethylene oxide are rare, apart from basic properties at standard conditions [6]. This lack of data is mainly due to the hazardous nature of ethylene oxide. It is highly flammable, reactive, explosive at elevated temperatures, toxic, carcinogenic, and mutagenic. Therefore, it is an excellent example to show that molecular modeling and simulation can serve as a reliable route for obtaining thermophysical data in cases, where avoiding experiments is highly desirable.

In the present work, a new molecular model for ethylene oxide was developed. This model is based on prior work at our institute [7] and was further optimized to experimental vapor-liquid equilibria (VLE), i.e. saturated liquid density, vapor pressure, and enthalpy of vaporization. Using this model, phase equilibria, thermal, caloric, transport properties, and surface tension were predicted and compared to experimental results, where possible.

2.1 Optimization of the Molecular Model for Ethylene Oxide

The molecular model for ethylene oxide consists of three Lennard-Jones (LJ) sites (one for each methylene group and one for the oxygen atom) plus one static point dipole. Thus, the united-atom approach was used. Due to the fact that ethylene oxide is a small molecule, the internal degrees of freedom may be neglected and the model was assumed to be rigid. Experimental data on the molecular structure was used to specify the geometric location of the LJ sites, cf. [7]. The dipole was located in the center of mass.

A set of five adjustable parameters, i.e., the four LJ parameters and the dipole moment, was optimized. This was done by a Newton scheme using correlations to experimental bubble density, vapor pressure, and enthalpy of vaporization data over the full range of the VLE between triple point and critical point.

Following the optimization procedure of Stoll [7], an optimized parameter set was obtained.The model parameters are listed in Table 1. In Figure 1, saturated densities, vapor pressure, and enthalpy of vaporization are shown. The present model describes the vapor pressure p_σ, the saturated liquid density ρ', and the enthalpy of vaporization Δh_v with mean relative deviations in the complete VLE range from triple point to critical point of $\delta p_\sigma = 1.5$ %, $\delta\rho' = 0.4$ %, and $\delta\Delta h_\mathrm{v} = 1.8$ %, respectively.

The model, as specified in Table 1, was used for calculating other properties discussed in the subsequent section. No further parameter adjustments were made and thus all results for properties except p_σ, ρ', and Δh_v are fully predictive.

Table 1. Coordinates and parameters of the present molecular model for ethylene oxide. The three Lennard-Jones sites are denoted by the molecular group which they represent, while the single dipolar site is denoted "Dipole". All coordinates are in principal axes with respect to the center of mass. The orientation of the electrostatic site is definded in standard Euler angles, where φ is the azimuthal angle with respect to the x-y-plane and θ is the inclination angle with respect to the z-axis

Interaction Site	x Å	y Å	z Å	σ Å	ε/k_B K	θ deg	φ deg	μ D
$CH_2(1)$	0.78000	0	-0.48431	3.5266	84.739	—	—	—
$CH_2(2)$	-0.78000	0	-0.48431	3.5266	84.739	—	—	—
O(3)	0	0	0.73569	3.0929	62.126	—	—	—
Dipole	0	0	0	—	—	0	0	2.459

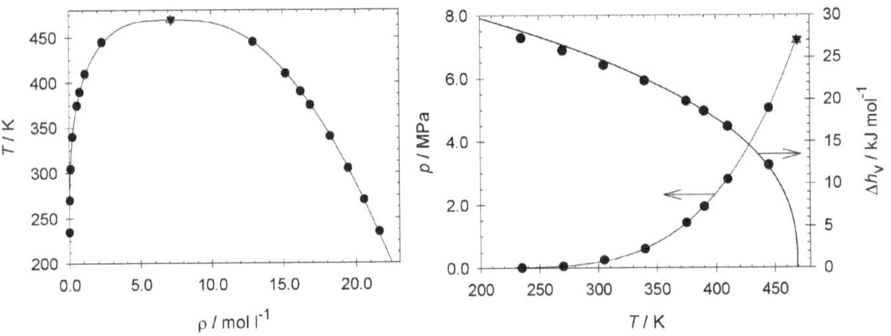

Fig. 1. Vapor-liquid equilibrium of ethylene oxide. Saturated densities (left) and vapor pressure and enthalpy of vaporization (right): ● simulation results, — experimental data [6]

2.2 Results and Comparison to Experimental Data

Individual VLE simulation results from the present model between 235 and 445 K deviate by less than 3 % in vapor pressure and heat of vaporization, and less than 0.5 % in saturated liquid density.

Saturated densities, vapor pressure, and enthalpy of vaporization of ethylene oxide at 375 K are all reproduced by the present model with deviations below 1 % and therefore within the experimental error given in [6].

Critical properties are in excellent agreement with experimental data, being within 1.3 %. The normal boiling temperature deviates by less than 0.5 % from experimental data for the present model.

The second virial coefficient was calculated by direct evaluation of the intermolecular potential. It is underpredicted from simulation by 3.4 %.

Experimental data on isobaric heat capacity and isothermal compressibility for the saturated fluid states at 375 K have not been available to us prior to the deadline of the Simulation Challenge. Later, the IFPSC published reference values, which are used for comparison here. In the saturated liquid, the simulation overpredicts isobaric heat capacity by 13 %, while in the saturated vapor, it is overpredicted by 18 %. Reference values for isothermal compressibility are obtained from the Brelvi-O'Connell correlation [8] for the saturated liquid and from the virial equation for the saturated vapor. Simulations for the present model yield a 39 % higher result for the liquid state and a 5.6 % higher result for the vapor state.

While the deviation from the reference value for the isothermal compressibility in the saturated vapor state is within the assumed uncertainty of 9.2 %, the one in saturated liquid state is significantly above the assumed uncertainty of 23.1 %. This is astonishing as the densities predicted by the present model along the bubble line are in good agreement with experimental ethylene oxide data. An additional check on our value after the deadline of the Simulation Challenge, evaluating simulation runs at different pressures for 375 K, yielded a value within the statistical uncertainty of the previous simulation result. On the other side, the Brelvi-O'Connell correlation is known to give systematic deviations to too low compressibilities for polar substances similar to ethylene oxide, e.g. sulfur dioxide or acetone, when using the critical density for reduction [8]. Therefore, the authors believe that molecular modeling and simulation yields more reliable results here than the standard method does.

Experimental surface tension data are available between 200 and 296 K and may be extrapolated to higher temperatures using a correlation taken from [6]. The simulation results give deviations of −17 %.

Experimental transport properties for ethylene oxide are scarce. In fact, for viscosity and thermal conductivity at 375 K they are only available in the saturated vapor state. Here, the results from the present model agree with the experimental values within their statistical uncertainty. After the deadline of the Simulation Challenge, the IFPSC has given reference values for viscosity in the saturated liquid state from an extrapolation of experimental data from 223 to 282 K and for thermal conductivity from the Missenard method [9]. Simulation results agree with the reference values within their statistical uncertainties and the uncertainties of the reference values proposed by IFPSC.

It can be seen that the molecular model for ethylene oxide is capable to reasonably predict the wide variety of properties. This underlines that molecular modeling and simulation in combination with the chosen molecular modeling route, using experimental bubble density, vapor pressure, and enthalpy of vaporization only, can be followed for predicting properties where experimental data is insufficient.

3 An Optimized Molecular Model for Ammonia

Ammonia is a well-known chemical intermediate, mostly used in fertilizer industries; another important application is its use as a refrigerant. Due to its simple symmetric structure and its strong intermolecular interactions it is also of high academic interest both experimentally and theoretically.

In the present work a new molecular model for ammonia is proposed, which is based on the work of Kristóf *et al.* [10] and improved by including data on geometry and electrostatics from *ab initio* quantum mechanical calculations.

3.1 Selection of Model Type and Parameterization

For the present molecular model for ammonia, a single Lennard-Jones potential was assumed to describe the dispersive and repulsive interactions. The electrostatic interactions as well as hydrogen bonding were modeled by a total of four partial charges. This modeling approach was also followed by Kristóf *et al.* [10] for ammonia.

The Lennard-Jones site and the partial charges were placed according to the nucleus positions obtained from a quantum mechanical geometry optimization. For an initial model, the magnitude of the charges were taken directly from quantum mechanics. This initial model was subsequently optimized to experimental VLE data. The optimized parameters are given in Table 2.

To optimize the molecular model, the two Lennard-Jones parameters were adjusted to experimental saturated liquid density, vapor pressure, and enthalpy of vaporization using a Newton scheme as proposed by Stoll [7]. These properties were chosen for the adjustment as they all represent major characteristics of the fluid region. Furthermore, they are relatively easy to be measured and are available for many components of technical interest.

Table 2. Parameters of the molecular model for ammonia ($e = 1.6021 \cdot 10^{-19}$ C is the elementary charge)

Interaction Site	x Å	y Å	z Å	σ Å	ε/k_B K	q e
N	0	0	0.0757	3.376	182.9	-0.9993
H(1)	0.9347	0	-0.3164	—	—	0.3331
H(2)	-0.4673	0.8095	-0.3164	—	—	0.3331
H(3)	-0.4673	-0.8095	-0.3164	—	—	0.3331

3.2 Results and Discussion

Vapor-Liquid Equilibria

VLE results for the new model are compared to data obtained from a reference quality equation of state (EOS) [11] in Figure 2. This figure also includes the results that we calculated using the model from Kristóf *et al.* [10].

The model of Kristóf *et al.* shows noticeable deviations from experimental data. The mean unsigned errors over the range of VLE are 1.9 % in saturated liquid density, 13 % in vapor pressure and 5.1 % in enthalpy of vaporization. With the new model, a significant improvement was achieved compared to the model from Kristóf *et al.* The description of the experimental VLE is very good, the mean unsigned deviations in saturated liquid density, vapor pressure and enthalpy of vaporization are 0.7, 1.6, and 2.7 %, respectively.

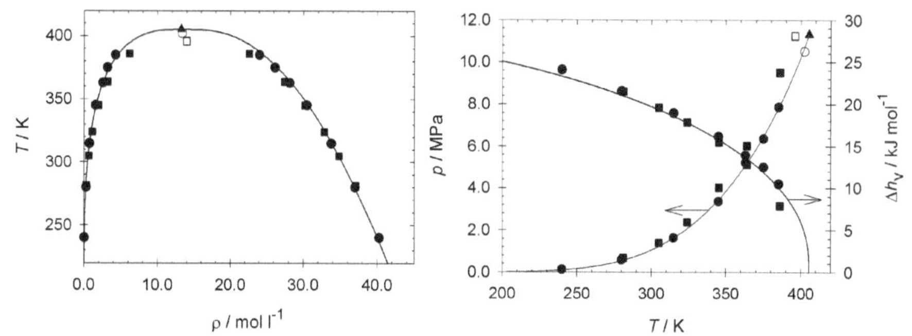

Fig. 2. Vapor-liquid equilibrium of ammonia. Saturated densities (left) and vapor pressure and enthalpy of vaporization (left): ● simulation results from the new molecular model, ■ simulation results from the model by Kristóf *et al.* [10], — experimental data [6]

Mathews [12] gives experimental critical values of temperature, density and pressure for ammonia: T_c =405.65 K, ρ_c =13.8 mol/l, and p_c =11.28 MPa. Following the procedure suggested by Lotfi *et al.* [13] the critical properties T_c =395.82 K, ρ_c =14.0 mol/l, and p_c =11.26 MPa for the model of Kristóf *et al.* were calculated, where the critical temperature is underestimated by 2.4 %. For the new model T_c =402.21 K, ρ_c =13.4 mol/l, and p_c =10.52 MPa were obtained. The new model gives very good results for the critical temperature, while it underpredicts the critical pressure slightly.

Homogeneous Region

In many technical applications thermodynamic properties in the homogeneous fluid region are needed. Thus, the new molecular model for ammonia was tested on its predictive capabilities for such states.

Thermal and caloric properties were predicted with the new model in the homogenous liquid, vapor and supercritical fluid region. In total, 70 state points were regarded, covering a large range of states with temperatures of up to 700 K and pressures of up to 700 MPa. In Figure 3, relative deviations between simulation and reference EOS [11] in terms of density and enthalpy are shown. The deviations are typically below 3 % for density with the exception of the extended critical region, where a maximum deviation of 6.8 % is found, and below 5 % for enthalpy.

These results confirm the modeling procedure. By adjustment to VLE data only, quantitatively correct predictions in most of the technically important fluid region can be obtained.

Structural Quantities

Due to its scientific and technical importance, experimental data on the microscopic structure of liquid ammonia are available. Ricci *et al.* [14] applied neutron diffraction and published all three types of atom-atom pair correlations, namely nitrogen-nitrogen (N-N), nitrogen-hydrogen (N-H), and hydrogen-hydrogen (H-H). In Figure 4, these experimental radial distribution functions for liquid ammonia at 273.15 K and 0.483 MPa are compared to present predictive simulation data based on the new ammonia model.

It was found that the structural properties are in very good agreement, although no adjustment was done regarding these properties. The atom-atom distance of the first three layers is predicted correctly, while only minor overshootings in the first peak are found. Please note that the first experimental peak in $g_{\mathrm{N-H}}$ and $g_{\mathrm{H-H}}$ show intramolecular pair correlations, which are not included in the present model.

In the experimental radial distribution function $g_{\mathrm{N-H}}$ the hydrogen bonding of ammonia can be seen at 2–2.5 Å. Due to the simplified approximation by eccentric partial charges, the molecular model is not capable to describe this effect completely. But even with this simple model a small shoulder at 2.5 Å is obtained.

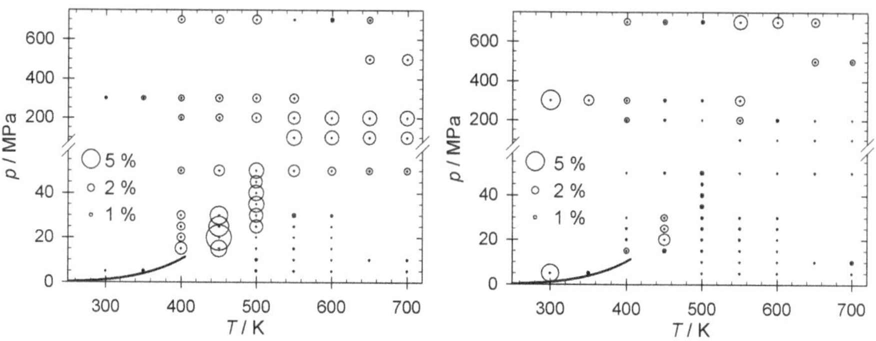

Fig. 3. Relative deviations for the density (left) and the enthalpy (right) between simulation and reference EOS [11] ($\delta z = (z_{\mathrm{sim}} - z_{\mathrm{eos}})/z_{\mathrm{eos}}$) in the homogeneous region: O simulation data of new model, — vapor pressure curve. The size of the bubbles denotes the relative deviation as indicated in the plot

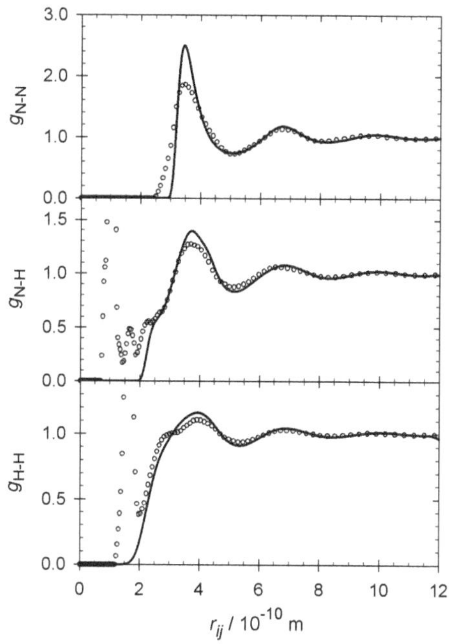

Fig. 4. Pair correlation functions of ammonia: — simulation data with the new model, ∘ experimental data [14]

4 Nucleation Processes During Condensation

Understanding homogeneous nucleation is required to develop an accurate theoretical approach to nucleation that extends to more complex and technically more relevant heterogeneous systems. MD simulations can be used to investigate the condensation of homogeneous vapors at high supersaturations.

The nucleation rate J is influenced to a large extent by the surface energy of droplets, which also determines how many droplets are formed and from which size on they become stable. The accuracy of different theoretical expressions for the surface energy can be assessed by comparison to results from MD simulations.

4.1 Nucleation Theory

Classical nucleation theory (CNT) was developed by Volmer and Weber [15] in the 1920s and further extended through many contributions over the following decades [16]. It is founded on the capillarity approximation: droplets emerging during nucleation are assumed to have the same thermodynamic properties as the bulk saturated liquid. In particular, the specific surface energy of the emerging nano-scaled droplets is assumed to be the surface tension of the planar phase boundary in equilibrium.

Laaksonen, Ford, and Kulmala [17] (LFK) proposed a surface energy coefficient that depends on the number of molecules in the droplet, i.e. the size of the droplet. Tanaka *et al.* [18] found that this expression leads to nucleation rates which agree with their simulation results. It was shown both theoretically [19, 20] and by simulation [21, 22] that the surface tension acting in the curved interface of nano-scaled droplets is actually much lower than in a planar interface. This finding underlines the necessity of a correction to CNT.

4.2 Simulation Method

Both the critical droplet size and the nucleation rate can be determined by molecular simulation. After an initial period of equilibration, the steady state distribution of droplets is established. The critical droplet size, i.e. the size from which on droplets more probably grow than decay, can either be calculated from the equilibrium distribution [23] or by the "nucleation theorem" [24]. Both methods require data on the nucleation rate J, i.e. the number of macroscopic droplets emerging per volume and time in a steady state at constant supersaturation. From an MD simulation of a supersaturated vapor, this rate can straightforwardly be extracted by counting the droplets that exceed a certain threshold size. This method was proposed by Yasuoka and Matsumoto [23], who found that as long as this threshold is larger than the critical droplet size, its precise choice hardly affects the observed value of J.

The nucleation rate J is the number of nuclei formed per volume and time in a supersaturated vapor

$$J(\iota) = \sum_{n>\iota} \frac{\partial \rho_n}{\partial t} , \tag{1}$$

for a threshold size ι larger than the critical size, measured after an initial temporal delay required to form sufficiently large nuclei. For nucleation rates obtained from molecular simulations, a very low size threshold is used, roughly of the same order of magnitude as the critical size. Because the choice of the threshold may influence the result, it must be indicated explicitly, i.e. as $J(\iota)$ instead of J.

For simulations in the canonical ensemble, it has to be taken into account that as the condensation proceeds, the density of the remaining vapor decreases and the pressure in the vapor is reduced significantly. This causes larger nuclei to be formed at a lower rate. Thus, the nucleation rates are given together with pressure values which were taken in the middle of the interval where the value of $J(\iota)$ was obtained by linear approximation.

For the present study, methane was modeled as a LJ fluid and ethane as a rigid two-center LJ fluid with an embedded point quadrupole [25]. Furthermore, the truncated and shifted LJ (LJTS) fluid with a cutoff radius of 2.5 times the size parameter σ was studied. The LJTS fluid is an accurate model for fluid noble gases and methane [21].

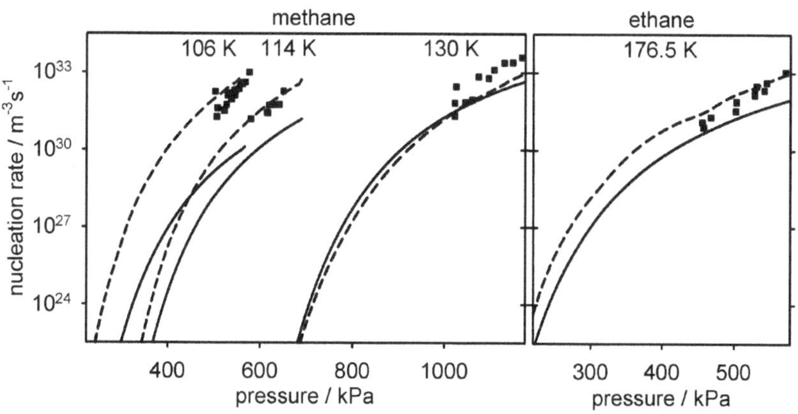

Fig. 5. Nucleation rates of methane and ethane at low temperatures from simulation according to the Yasuoka-Matsumoto method and different threshold sizes (■) as well as CNT (—) and LFK (- - -)

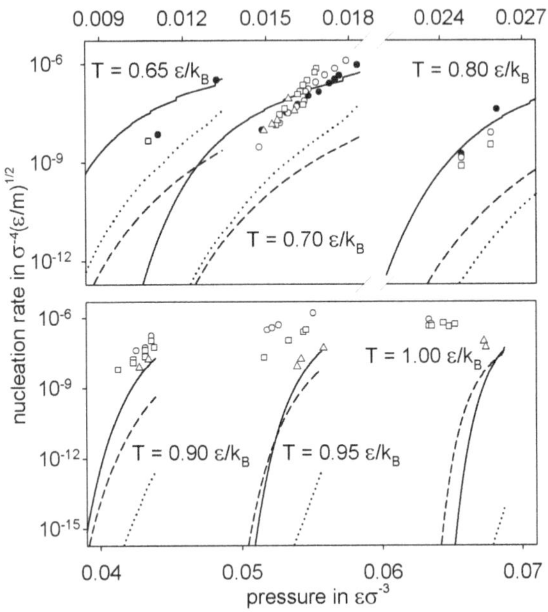

Fig. 6. Nucleation rate of the LJTS fluid over supersaturated pressure from the present direct simulations for different threshold values (● $i = 25$, ○ $i = 50$, ■ $i \in \{75, 100\}$, △ $i \geq 150$) and following the standard CNT (- - -), the LFK modification of CNT (···), and the new SPC modification of CNT (—)

Within a typical computation time of about 24 hours, a time interval of a few nanoseconds can be simulated for systems with a volume of 10^{-21} m^3. Given that the nucleation rate is determined as units per volume and time, only values that exceed 10^{30} m^{-3}s^{-1} can be obtained from molecular simulation. However, only nucleation rates below 10^{23} m^{-3}s^{-1} can be measured in experiments at present. Therefore, computational power is crucial to keep the gap between simulation and experiment as small as possible.

In Figure 5 the nucleation rates from MD simulation are shown in comparison to predictions from CNT and LFK theories. At the given state points, simulation results confirm both CNT and LFK, deviations are throughout lower than three orders of magnitude.

4.3 Simulation Results

However, from a series of simulations it was seen that standard CNT only predicts the nucleation rate J with an acceptable accuracy, but leads to deviations for the critical droplet size; the LFK modification provides excellent predictions for the critical size but not for the temperature dependence of J. LFK underpredicts the nucleation rate at high temperatures by several orders of magnitude. Both theories assume an inappropriate curvature dependence of the surface tension, although for this essential property of inhomogeneous systems a qualitatively correct expression is known since the 1940s [26]. With the collected simulation data on critical droplet size and nucleation rate over a broad range of temperatures, enough quantitative information is available to formulate a more adequate modification of CNT. For the LJTS fluid, a surface property corrected (SPC) modification of CNT was developed which postulates a non-spherical surface and is based on surface tension values obtained from molecular simulation. As shown in Figure 6, the SPC modification accurately describes the nucleation rate of the LJTS fluid.

5 Computing Performance

All simulations presented in Sections 2.1 and 3 were carried out with the MPI based molecular simulation program *ms*2 developed in our group. The parallelization of the molecular dynamics part of *ms*2 is based on Plimptons force decomposition algorithm [27]. The nucletion simulations were carried out with the massivly parallel program *ls*1. Here, the parallelization relies on spatial domain decomposition for parallel simulations [28] due to the large size of the simulated systems (up to 2 million particles).

With *ms*2 typical simulation runs to determine the vapor-liquid equilibrium employ 4–8 CPUs running for 4–6 hours. For model optimization or the comprehensive study on mixtures, a large number of independent simulations are necessary and can be performed in parallel. For the prediction of other thermophysical properties, additional simulations were performed using up to 32 CPUs and running up to 72 hours depending on the system size and desired property.

For the simulation of nucleation processes, very large systems must be considered. Here, the massively parallel program *ls*1 was used. Table 3 demonstrates the good scaling of the program on the HP XC4000 cluster at the Steinbuch Center for Computing, Karlsruhe.

Table 3. Scaling of the massively parallel program *ls*1 on HP XC4000. Particle numbers N are 0.5–2 million

CPUs	CPU time (per time step and molecule)
24	0.022 ms
32	0.021 ms
40	0.021 ms
100	0.025 ms
125	0.023 ms

References

1. N. Metropolis, A.W. Rosenbluth, M.N. Rosenbluth, A.N. Teller, and E. Teller. Equation of state calculations by fast computing machines. *J. Chem. Phys.*, 21:1087, 1953.
2. M.P. Allen and D.J. Tildesley. *Computer Simulation of Liquids*. Clarendon, Oxford, 1987.
3. H.A. Lorentz. Über die Anwendung des Satzes vom Virial in der kinetischen Theorie der Gase. *Annalen der Physik*, 12:127, 1881.
4. D. Berthelot. Sur le mélange des gaz. *Comptes rendus hebdomadaires des séances de l'Académie des Sciences*, 126:1703, 1898.
5. Industrial Fluid Properties Simulation Collective, http://www.ifpsc.org.
6. Design Institute for Physical Property Data / AIChE. DIPPR project 801 (full version), 2005.
7. J. Stoll. *Molecular Models for the Prediction of Thermophysical Properties of Pure Fluids and Mixtures*. Number 836 in VDI Fortschritt-Berichte (Reihe 3). VDI, Düsseldorf, 2005.
8. S.W. Brelvi and J.P. O'Connell. Corresponding states correlations for liquid compressibility and partial molal volumes of gases at infinite dilution in liquids. *AIChE J.*, 18:1239, 1972.
9. F.-A. Missenard. Méthode additive pour la détermination de la chaleur molaire des liquides. *Comptes rendus hebdomadaires des séances de l'Académie des Sciences*, 260:5521, 1965.
10. T. Kristóf, J. Vorholz, J. Lisze, B. Rumpf, and G. Maurer. A simple effective pair potential for the molecular simulation of the thermodynamic properties of ammonia. *Mol. Phys.*, 97:1129, 1999.
11. R. Tillner-Roth, F. Harms-Watzenberg, and H.D. Baehr. Eine neue Fundamentalgleichung für Ammoniak. In *Tagungsband zur 20. Kälte-Klima-Tagung*, page 167. Deutscher Kälte- und Klimatechnischer Verein, 1993.
12. J.F. Mathews. The critical constants of inorganic substances. *Chem. Rev.*, 72:71, 1972.
13. A. Lotfi, J. Vrabec, and J. Fischer. Vapour liquid equilibria of the Lennard-Jones fluid from the NpT plus test particle method. *Mol. Phys.*, 76:1319, 1992.
14. M. Ricci, M. Nardona, F. Ricci, C. Andreani, and A. Soper. Microscopic structure of low temperature liquid ammonia: A neutron diffraction experiment. *J. Chem. Phys.*, 102:7650, 1995.
15. M. Volmer and A. Weber. Keimbildung in übersättigten Gebilden. *Z. phys. Chem. (Leipzig)*, 119:277, 1926.
16. J. Feder, K.C. Russell, J. Lothe, and G.M. Pound. Homogeneous nucleation and growth of droplets in vapours. *Advances in Physics*, 15:111, 1966.
17. A. Laaksonen, I.J. Ford, and M. Kulmala. Revised parametrization of the Dillmann-Meier theory of homogeneous nucleation. *Phys. Rev. E*, 49:5517, 1994.
18. K.K. Tanaka, K. Kawamura, H. Tanaka, and K. Nakazawa. Tests of the homogeneous nucleation theory with molecular-dynamics simulations. I. Lennard-Jones fluids. *J. Chem. Phys.*, 122:184514, 2005.
19. V.G. Baidakov and G.Sh. Boltachev. Curvature dependence of the surface tension of liquid and vapor nuclei. *Phys. Rev. E*, 59:469, 1999.
20. J.G. Kirkwood and F.P. Buff. The statistical mechanical theory of surface tension. *J. Chem. Phys.*, 17:338, 1949.

21. J. Vrabec, G.K. Kedia, G. Fuchs, and H. Hasse. Comprehensive study on vapour-liquid coexistence of the truncated and shifted Lennard-Jones fluid including planar and spherical interface properties. *Mol. Phys.*, 104:1509, 2006.
22. I. Napari and A. Laaksonen. Surface tension and scaling of critical nuclei in diatomic and triatomic fluids. *J. Chem. Phys.*, 126:134503, 2007.
23. K. Yasuoka and M. Matsumoto. Molecular dynamics of homogeneous nucleation in the vapor phase. *J. Chem. Phys.*, 109:8451, 1998.
24. D.W. Oxtoby and D. Kashchiev. A general relation between the nucleation work and the size of the nucleus in multicomponent nucleation. *J. Chem. Phys.*, 100:7665, 1994.
25. J. Vrabec, J. Stoll, and H. Hasse. A set of molecular models for symmetric quadrupolar fluids. *J. Phys. Chem. B*, 105:12126, 2001.
26. R.C. Tolman. The effect of droplet size on surface tension. *J. Chem. Phys.*, 17:333, 1949.
27. S. Plimpton. Fast parallel algorithms for short-range molecular dynamics. *J. Comp. Phys.*, 117:1, 1994.
28. M. Bernreuther and J. Vrabec. Molecular simulation of fluids with short range potentials. In M. Resch, Th. Bönisch, K. Benkert, T. Furui, and W. Bez, editors, *High Performance Computing on Vector Systems*, page 187. Springer, Berlin – Heidelberg – New York, 2006.

Flow with Chemical Reactions

Prof. Dr. Dietmar Kröner

Abteilung für Angewandte Mathematik, Universität Freiburg,
Hermann-Herder-Str. 10, 79104 Freiburg, Germany

In this section the contributions concerning reactive flows are collected. In the article "A hybrid finite-volume/transported PDF model for simulations of turbulent flames on vector machines" by S. Lipp, U. Maas, P. Lammers the turbulence-chemistry interaction is investigated. The special subject of this contribution is the application of the "intrinsing low-dimensional manifold" method. It means that the chemical kinetics are taken into account with a reduced chemical reaction mechanism. Furthermore the turbulence-chemistry interaction is described by solving the joint probability density function of velocity and scalars. The obtained results in this project show that the qualitative behaviour of the flame could be predicted at least for the stationary case.

In the paper about "Numerical investigations of model scramjet combusters" by M. Kindler, T. Blacha, M. Lempke, P. Gerlinger, M. Aigner different ways of injecting fuel into supersonic combustion ramjet combusters are compared. In the first case the injection of fuel is realised by a lobed strut that is located in the middle of the combustion chamber. While in the second case the fuel is injected by several holes in the wall of the combuster. The numerical code has been used to investigate different configurations of combusters.

A Hybrid Finite-Volume/Transported PDF Model for Simulations of Turbulent Flames on Vector Machines

Stefan Lipp[1], Ulrich Maas[1], and Peter Lammers[2]

[1] University Karlsruhe, Institute for Technical Thermodynamics,
lipp,mass@itt.mach.uni-karlsruhe.de
[2] University Stuttgart, High Performance Computing Center Stuttgart
lammers@hlrs.de

Summary. The mathematical modeling of turbulent flames is a difficult task due to the intense coupling between turbulent transport processes and chemical kinetics. The mathematical model presented in this paper is focused on the turbulence-chemistry interaction. The method consists of two parts. Chemical kinetics are taken into account with a reduced chemical reaction mechanism, which has been developed using the ILDM-Method ("Intrinsic Low-Dimensional Manifold"). The turbulence-chemistry interaction is described by solving the joint probability density function (JPDF) of velocity and scalars. Simulations of test cases with simple geometries verify the developed model.

1 Introduction

Reliable predictive models of turbulent flames are important in many industrial applications, e.g. the design and improvement of industrial burners used in gas turbines [1]. In particular new emission goals and new demands concerning fuel efficiency require detailed models which can treat the combustion chemistry without using oversimplified models. The standard methods for non-reacting flows (RANS, LES) cannot satisfactorily tackle the problem of the strong non-linearity of the chemical source term and often suffer from a poor modeling of the turbulence-chemistry interaction. However, a detailed modeling of this effect is possible for instance by applying probability density function methods (PDF). They show a high capability for modeling turbulent flames because these methods treat convection and finite rate non-linear chemistry exactly [2, 3]. Only the effect of molecular mixing has to be modeled [4, 5, 6].

Two basically different PDF approaches exist in the literature. Some authors use stand-alone PDF models in which all hydrodynamic and thermokinetic properties of the flow are computed from a joint probability density function [7, 8, 9, 10]. Other authors use the PDF transport equation only to calculate a certain number of the hydrodynamic and thermokinetic properties and calculate the remainder of the properties with an ordinary CFD solver [11, 12, 13].

In order to obtain the propability density function the repective transport equation must be solved. It can be derived from the Navier-Stokes equations [2]. As previously mentioned the chemistry term together with the body forces and the mean pressure gradient appear already in closed form here, only the terms describing the fluctuating pressure gradient and molecular mixing need to be modeled. Numerically the treatment of the PDF transport equation is quite different from the Navier Stokes equations. In contrast to the system of partial differential equations which is formed by the Navier-Stokes equations the transport equation for the PDF is a high dimensional scalar transport equation. In general it has $7 + n_S$ dimensions which consist of three dimensions in space, three dimensions in velocity space, the time and the number of species n_S used for the description of the thermokinetic state. Because of its high dimensionality it is not feasible to solve the equation using finite-difference or finite-volume methods. For that reason Monte Carlo methods have been employed, which are widely used in computational physics to solve problems of high dimensionality, because the numerical effort increases only linearly with the number of dimensions.

This solution method takes advantage of the fact that the PDF can be represented as an ensemble of stochastic particles [14]. The transport equation for the PDF is transformed to a system of stochastic ordinary differential equations. This system is constructed in such a way that the particle properties, e.g. velocity, scalars, and turbulent frequency, represent the same PDF as in the turbulent flow. In order to fulfill consistency of the modeled PDF, the mean velocity field derived from an ensemble of particles needs to satisfy the mass conservation equation [2]. This requires the pressure gradient to be calculated from a Poission equation. The available Monte Carlo methods cause strong bias solving a Poisson equation. This leads to stability problems calculating the pressure gradient. Different methods to calculate the mean pressure gradient where used in order to avoid these instabilities. One possibility is to couple the particle method with an ordinary finite-volume or finite-difference solver to obtain the mean pressure field from the Navier-Stokes equations. These so called hybrid PDF/CFD methods are widely used by different authors for many types of flames [15, 16, 17, 18, 19, 20].

In the presented paper a hybrid scheme is presented and used as well. The fields for mean pressure gradient and a turbulence charactaristic, e.g. the turbulent time scale, are derived solving the Favre averaged conservation equations for momentum and mass for the flow field using a finite-volume method. The effect of turbulent fluctuations is modeled using a k-τ model [21]. Chemical kinetics is taken into account by using the ILDM method to get reduced chemical mechanisms [22, 23]. In the presented case the reduced mechanism describes the reaction with two parameters which on the one hand are few enough to limit the simulation time to an acceptable extent and on the other hand sufficient to get a detailed description of the chemical reaction.

The test case for the developed model is a model combustion chamber investigated by serveral authors [24, 25, 26, 27]. With their data the results of the presented simulations are validated.

2 Numerical Model

The numerical model used is this paper is a hybrid CFD/PDF model and consists of two parts. A finite-volume solver for the Navier-Stokes equations (CFD) which provides the hydrodynamic quantities and a Monte Carlo solver for the probability density function (PDF) which gives the thermokinetic state of the flow. The principles of the solution procedure shall be briefly overviewed before going into details and discussing consistency and numerical matters.

A sketch of the method is shown in Fig. 1. The calculation starts with a CFD step in which the Navier-Stokes equations for the flow flield are solved. As an intermediate result the mean pressure gradient together with the mean velocities and the turbulence characteristics is handed over to the PDF part. Here the transport equation for the joint probability density function of the scalars describing the thermokinetic state and the velocities is solved. The result of the previous CFD step is considered at this point. The change of the thermokinetic state due to chemical reactions is calculated from a lookup table (ILDM). This table is based on a detailed chemical reaction mechanism which was in a preprocessing step reduced using the ILDM method. As a result of the PDF step the new mean molar mass and the mean temperature are returned to the CFD part. These internal iteration steps are performed until global convergence is achieved.

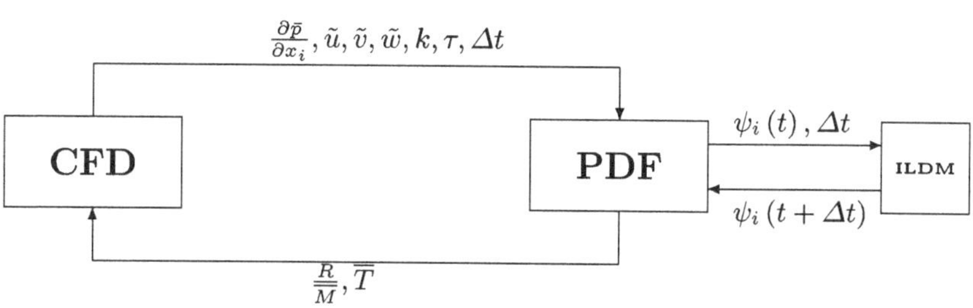

Fig. 1. Scheme of the coupling of CFD and PDF

2.1 CFD Model

The utilized CFD code SPARC is an inhouse development of the Department of Fluid Machinery at Karlsruhe University [28]. It is a finite volume based solver for the compressible Navier-Stokes equations on block structured domains. Several turbulence closure models are implemented in this code. In the presented work a two equation model is applied which solves a transport equation for the turbulent kinetic energy and a turbulent time scale [21]. This model was selected due to being a good compromise in modeling accuracy and numerical stability.

Navier-Stokes Equations

The system of partial differential equations which are solved are the mass conservation equation and the momentum conservation equation. All quanities appear in Favre-averaged manner. The Favre-average of any quantity ζ is calculated according to

$$\tilde{\zeta} = \frac{\overline{\rho \cdot \zeta}}{\bar{\rho}} \quad . \tag{1}$$

In detail the equations read

$$\frac{\partial \bar{\rho}}{\partial t} + \frac{\partial (\bar{\rho}\tilde{u}_i)}{\partial x_i} = 0 \tag{2}$$

$$\frac{\partial (\bar{\rho}\tilde{u}_i)}{\partial t} + \frac{\partial}{\partial x_j}\left(\bar{\rho}\tilde{u}_i\tilde{u}_j + \overline{\rho u_i^{''} u_j^{''}} + \bar{p}\delta_{ij} - \bar{\tau}_{ij}\right) = 0 \tag{3}$$

which are the conservation equations for mass and momentum in Favre average notation, respectively. In this model the solution of the energy equation is not necessary since the mean temperature is calculated in the PDF part from the variables desribing the thermokinetic state.

Turbulence Model

The unclosed cross correlation term $\overline{\rho u_i^{''} u_j^{''}}$ in the momentum conservation equation is modeled using the Boussinesq approximation

$$\overline{\rho u_i^{''} u_j^{''}} = \bar{\rho}\mu_T \left(\frac{\partial \tilde{u}_i}{\partial x_j} + \frac{\partial \tilde{u}_j}{\partial x_i}\right) \quad . \tag{4}$$

The turbulent viscosity μ_T is given by a two equation model of Speziale et al. [21] with

$$\mu_T = C_\mu f_\mu k \tau \quad . \tag{5}$$

The parameter C_μ is an empirical constant with a value of $C_\mu = 0.09$ and f_μ accounts for the influence of walls. The turbulent kinetic energy k and the turbulent time scale τ are calculated from their transport equation which are [21]

$$\bar{\rho}\frac{\partial k}{\partial t} + \bar{\rho}\tilde{u}_j\frac{\partial k}{\partial x_j} = \tau_{ij}\frac{\partial \tilde{u}_i}{\partial x_j} - \bar{\rho}\frac{k}{\tau} + \frac{\partial}{\partial x_i}\left[\left(\mu + \frac{\mu_T}{\sigma_k}\right)\frac{\partial k}{\partial x_j}\right] \tag{6}$$

$$\begin{aligned}\bar{\rho}\frac{\partial \tau}{\partial t} + \bar{\rho}\tilde{u}_j\frac{\partial \tau}{\partial x_j} &= (1 - C_{\epsilon 1})\frac{\tau}{k}\tau_{ij}\frac{\partial \tilde{u}_i}{\partial x_j} + (C_{\epsilon 2} - 1)\,\bar{\rho} + \\ &\frac{\partial}{\partial x_j}\left[\left(\mu + \frac{\mu_T}{\sigma_{\tau 2}}\right)\frac{\partial k}{\partial x_j}\right] + \frac{2}{k}\left(\mu + \frac{\mu_T}{\sigma_{\tau 1}}\right)\frac{\partial k}{\partial x_k}\frac{\partial \tau}{\partial x_k} - \\ &\frac{2}{\tau}\left(\mu + \frac{\mu_T}{\sigma_{\tau 2}}\right)\frac{\partial \tau}{\partial x_k}\frac{\partial \tau}{\partial x_k} \quad .\end{aligned} \tag{7}$$

Here $C_{\epsilon 1} = 1.44$ and $\sigma_{\tau 1} = \sigma_{\tau 2} = 1.36$ are empirical model constants. The parameter $C_{\epsilon 2}$ is calculated from the turbulent Reynolds number $\mathrm{Re_t}$.

$$\mathrm{Re_t} = \frac{k\tau}{\mu} \tag{8}$$

$$C_{\epsilon 2} = 1.82\left[1 - \frac{2}{9}\exp\left((-\mathrm{Re_t}/6)^2\right)\right] \tag{9}$$

2.2 Joint PDF Model

In the literature many different joint PDF models can be found, for example models for the joint PDF of velocity and composition [29, 30] or for the joint PDF of velocity, composition and turbulent frequency [31]. A good overview of the different models can be found in [17].

In the work presented here a joint PDF of velocity and composition vector is employed. This is a one-time, one-point joint probability density function which has the main advantage to treat chemical reactions exactly without any modeling assumptions [2]. However, the effect of molecular mixing has to be modeled.

Joint Probability Density Function

The state of a reacting fluid flow at one point in space and time can be fully described by the velocity vector $\mathbf{V} = (V_1, V_2, V_3)^T$ and the composition vector $\mathbf{\Psi}$ containing the mass fractions of $n_S - 1$ species and the enthalpy h $\left(\mathbf{\Psi} = (\Psi_1, \Psi_2, \ldots, \Psi_{n_s-1}, h)^T\right)$. The joint probability density function is defined as that function which gives, when integrated by the hole state space, the probability that at one point in space and time one realization of the flow falls within the interval

$$\mathbf{V} \le \mathbf{U} \le \mathbf{V} + \mathrm{d}\mathbf{V} \tag{10}$$

for its velocity vector and

$$\mathbf{\Psi} \leq \mathbf{\Phi} \leq \mathbf{\Psi} + \mathrm{d}\mathbf{\Psi} \tag{11}$$

for its composition vector. Thus the the joint PDF reads

$$f_{\mathbf{U}\phi}\left(\mathbf{V},\mathbf{\Psi};\mathbf{x},t\right)\mathrm{d}\mathbf{V}\mathrm{d}\mathbf{\Psi} = \mathrm{Prob}\left(\mathbf{V} \leq \mathbf{U} \leq \mathbf{V} + \mathrm{d}\mathbf{V}, \mathbf{\Psi} \leq \mathbf{\Phi} \leq \mathbf{\Psi} + \mathrm{d}\mathbf{\Psi}\right) \tag{12}$$

PDF Transport Equation

According to [2] a transport equation can be derived for the joint PDF $\tilde{f}$ of velocity and composition. Under the assumption that the effect of pressure fluctuations on the fluid density is negligible the transport equation writes

$$\underbrace{\rho(\mathbf{\Psi})\frac{\partial \tilde{f}}{\partial t}}_{I} + \underbrace{\rho(\mathbf{\Psi})U_j\frac{\partial \tilde{f}}{\partial x_j}}_{II} + \underbrace{\left[\left(\rho(\mathbf{\Psi})g_j - \frac{\partial \langle p\rangle}{\partial x_j}\right)\right]\frac{\partial \tilde{f}}{\partial U_j}}_{III} + \underbrace{\frac{\partial}{\partial \Psi_\alpha}\left[\rho(\mathbf{\Psi})S_\alpha(\mathbf{\Psi})\tilde{f}\right]}_{IV}$$
$$= \underbrace{\frac{\partial}{\partial U_j}\left[\left\langle -\frac{\partial \tau_{ij}}{\partial x_i} + \frac{\partial p'}{\partial x_i}|\mathbf{U},\mathbf{\Psi}\right\rangle \tilde{f}\right]}_{V} + \underbrace{\frac{\partial}{\partial \Psi_\alpha}\left[\left\langle -\frac{\partial J_i}{\partial x_i}|\mathbf{U},\mathbf{\Psi}\right\rangle \tilde{f}\right]}_{VI}. \tag{13}$$

Term I describes the instationary change of the PDF with $\rho(\mathbf{\Psi})$ beeing the fluid density as a function of the composition vector $\mathbf{\Psi}$, Term II its change due to convection in physical space with U_j beeing the component of the velocity vector in the spacial direction x_j and Term III takes into account the influence of gravity (g_j) and the mean pressure gradient ($\frac{\partial\langle p\rangle}{\partial x_i}$) on the PDF. Term IV includes the chemical source term which describes the change of the PDF in composition space due to chemical reactions. All terms on the left hand side of the equation appear in closed form, e.g. the chemical source term. In contrast, the terms on the right hand side are unclosed and need further modeling. Many closing assumptions for these two terms exist. In the following only the ones that are used in the present work shall be explained further.

Term V describes the influence of pressure fluctuations p' and viscous stresses τ_{ij} on the PDF. Commonly a Langevin approach [32, 33] is used to close this term. In the presented case the SLM (Simplified Langevin Model) is used [2]. More sophisticated approaches that take into account the effect of non-isotropic turbulence or wall effects exist as well [32, 34]. But in the presented case of a free stream flame the closure of the term by the SLM is assumed to be adequate and was chosen because of its simplicity and numerical robustness.

Term VI regards the effect of molecular diffusion within the fluid. This diffusion flattens the steep composition gradients which are created by the strong vortices in a turbulent flow. The model uses in this work has originally been proposed by Curl [35], then been modified by Janika et al. [36] and Pope [37] and is used here in this modified form.

Solution of the PDF Transport Equation

As previously mentioned it is numerically unfeasable to solve the PDF transport equation with finite-volume or finite-difference methods because of its high dimensionality. Therefore a Monte Carlo method is used to solve the transport equation making use of the fact that the PDF of a fluid flow can be represented as a sum of δ-functions.

$$f^*_{\mathbf{U},\phi}\left(\mathbf{U},\boldsymbol{\Psi};\mathbf{x},t\right)=\sum_{i=1}^{N(t)}\delta\left(\mathbf{v}-\mathbf{u}^i\right)\delta\left(\phi-\boldsymbol{\Psi}^i\right)\delta\left(\mathbf{x}-\mathbf{x}^i\right) \tag{14}$$

Utilizing this fact the PDF of the fluid flow can be disreticed by a set of stochastic particles. In all following equations a star indicates quatities which are particle quantities[3]. For these particles equations of motion in space, velocity space and compostion space can be derived. Now, instead of the high dimensional PDF transport equation a set of (stochastic) ordinary differential equations are solved. The evolution of the particle position $\mathbf{X}^*_\mathrm{i}$ reads

$$\frac{\mathrm{d}\mathbf{X}^*_\mathrm{i}}{\mathrm{d}t}=\mathbf{U}^*_\mathrm{i}\left(t\right) \tag{15}$$

in which $\mathbf{U}^*_\mathrm{i}$ is the velocity vector for each particle.

The evolution of the particles in the velocity space can be calculated according to the Simplified Langevin Model [2] by

$$\mathrm{d}\,U^*_i=-\frac{\partial\bar{p}}{\partial x_i}\mathrm{d}t-\left(\frac{1}{2}+\frac{3}{4}C_0\right)\left[U^*_i-\langle U_i\rangle\right]\frac{\mathrm{d}t}{\tau}+\sqrt{\frac{C_0\,k}{\tau}}\mathrm{d}\,W_i \quad . \tag{16}$$

Here the equation is only written for the U component of the velocity vector $\mathbf{U}=(U,V,W)^T$ belonging to the spacial coordinate x ($\mathbf{x}=(x,y,z)^T$) for reasons of simplicity. The equations of the other components V,W look accordingly. In eq. 16 $\partial\bar{p}/\partial x_i$ denotes the mean pressure gradient, $\langle U_i\rangle$ the mean particle velocity, t the time, $\mathrm{d}\,W_i$ a differential Wiener increment, C_0 a model constant, k and τ the turbulent kinetic energy and the turbulent time scale, respectively.

Finally the evolution of the composition vector can be calculated as

$$\frac{\mathrm{d}\boldsymbol{\Psi}}{\mathrm{d}t}=S+M \tag{17}$$

in which S is the chemical source term (appearing in closed form) and M denotes the effect of molecular mixing. This term is unclosed and as mentioned previously is modeled by a modified Curl model [36].

---

[3] e.g. $\mathbf{x}$ denotes a space vector in general, $\mathbf{x}^*$ the space vector of one stochastic particle

Chemical Kinetics

The source term appearing in eq. 17 is calculated from a lookup table which is created using automatically reduced chemical mechanisms. The deployed technique to create these tables is the ILDM method ("Intrinsic Low-Dimensional Manifold") by Maas and Pope [22, 23].

The basic idea of this method is the identification and separation of fast and slow time scales. In typical turbulent flames the time scales governing the chemical kinetics range from 10^{-9}s to 10^{2}s. This is a much larger spectrum than that of the physical processes (e.g. molecular transport) which only vary from 10^{-1}s to 10^{-5}s. Reactions that occur in the very fast chemical time scales are in partial equilibrium and the species are in steady state. These are usually responsible for equilibrium processes. Making use of this fact it is possible to decouple the fast time scales. The main advantage of decoupling the fast time scales is that the chemical system can be described with a much smaller number of variables (degrees of freedom).

In our test case the chemical kinetics are described with only two parameters namely the mixure fraction and the specific mole number of CO_2 instead of the 34 species (degrees of freedom) appearing in the detailed methane reaction mechanism. Further details of the method and its implementation can be found in [22, 23].

2.3 Coupling of the Solution Procedure

The coupling of the CFD and the PDF part of the model is done with the calculation of the pressure via the equation of state in the CFD part which closes the equation system.

$$\overline{p} = \overline{\rho} \cdot \frac{R}{\overline{M}} \cdot \overline{T} \tag{18}$$

Here the mean temperature $\overline{T}$ is calculated (together with the mean molar mass $\overline{M}$) from the PDF solver instead of from the energy equation as done in standard compressible flow solvers. The density $\overline{\rho}$ is calculated in the CFD part from the mass conservation equation.

3 Performance Optimization for SX-8

In the Teraflop-Workbench project which is a collaboration between NEC and the HLRS we are optimizing the presented model, especially the PDF part, for the NEC SX architecture.

We started with significant changes in the data structure of the PDF solver. For instance, we now use modern Fortran language constructs to increase the solvers flexibility in handling the block structured grids of the CFD solver. This results in changes of the arrays holding the information of the stochastic particles as well as of the arrays for exchanging quantities between CFD and PDF solver etc..

At first some general remarks may help to understand the expected performance on the SX. As the code implements a hybrid method in the sense that some quantities are calculated by the CFD code and others by a PDF Monte-Carlo method one has to continuously exchange information between the quantities defined on the grid nodes of the CFD solver and the particles of the PDF part and vice versa. This requires the maintenance of list vectors in each time step which hold the relation between particle position and grid nodes/cells. Consequently, this results further in two kinds of loop structures appearing in the solver. For the first kind of loops in which the inner loop goes over all particles within the domain the vector length is always sufficient. On the other hand, there are a lot of cases in which the inner loop has to run over the particles located in one cell of the CFD solver grid. We currently use a number of particles per cell which lies around 100 and is therefore not sufficient for the SX-8. But, we will increase this number in the future as the quality of the statistics will improve naturally with a larger number of particles per cell. Then, the PDF part of the solver is likely to benefit from the SX architecture.

Beside changing the data structures we have already started to work on performance improvements of the algorithm for solving the joint PDF transport equation itself. It is done with a so called fractional step algorithm which consists of five steps:

$$\frac{\partial \mathcal{F}}{\partial t} = P_1 + P_2 + P_3 + P_4 + P_5 \tag{19}$$

P_1 and P_4 deal with the convection of the particles in each time step. Each part applies eq. 15 for $\Delta t/2$. With the discretization chosen so far we talk algorithmically about a daxpy loop for the three particle positions. In a two dimensional model the particles moving outside the plan have to be projected back into it. Now, everything is done in one collapsed loop over all particles. Additionally the routine consist of two debugging loops, which have been vectorized. Altogether the performance of P_1 and P_4 has been improved by a factor of more than 20. The next step P_2 deals with the changes in the particles velocity by applying the simplified Langevin model eq. 16. The corresponding loop is now vectorized over all particles. But, beforehand it is necessary to get several quantities from the flow solver and interpolate them to the particles position. This is an example for loops where the vector length is restricted to the number of particles inside a cell. So far, we achieved an improvement of a factor three in comparison to the original version. What remains to be done for P_2 is the vectorization of the Wiener increment determination (see below).

We conclude this section with some bullet points regarding our plans for future changes and improvements:

- P_2: The currently used random number generators are not suitable for large scale computations. Therefore, they have to be substituted anyway. We

will implement ASL (Advance Scientific Library) routines with a very long period. If necessary, the algorithms will be changed to enable vectorization (Wiener increment).
- P_3: Vectorization of the mixing model.
- P_5 The layout of the tables containing the chemical reactions will be changed. This should offer huge possibilities to improve the solvers performance.
- Improving the administrative part needed for coupling the PDF and the CFD parts.
- Parallel version of the PDF part.

4 Results and Discussion

Results of simulations of a simple swirling premixed confined turbulent flame are shown. Turbulent swirling flames which are aerodynamically stabilized by an internal recirculation zone were used in many industrial applications such as jet engines and gas turbines. However, some major physical effects appearing in these flames are still not entirely understood. That includes also effects like the combustion induced vortex breakdown or presizing vortex cores which might be critical with respect to a save operation. Parameter studies on test rigs with a simplified geometry should clarify the influence of global operation parameters like the equivalence ratio and the swirl number on these effects. Details of the test rig simulated in this work and the experimental data can be found in [25, 26, 27]. A validation of the flow field for the non-reacting case was subject of previous work. The results can be found in [38]. The results presented in this paper will therefore focus mainly on the reacting case. Due to a lack of experimental data only qualitative statements can be made. Fig. 2 shows a sketch of the test rig investigated. It consists of three parts: a plenum,

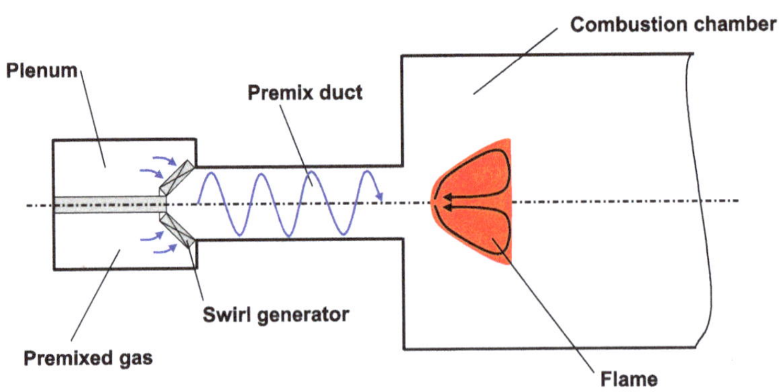

Fig. 2. Sketch of the investigated combustion chamber

the so-called premix duct and the combustion chamber. In the plenum a perfectly premixed methane-air mixture is aided through the swirl generator into the premixing duct. The swirling flow continues through the premix duct into the combustion chamber. Here a vortex breakdown occurs which creates the internal recirculation zone. There the flame is aerodynamically stabilized by hot exhaust gas which enables a save and stable operation of the hole burner. Fig. 3 shows the position of the computational domain with respect to the

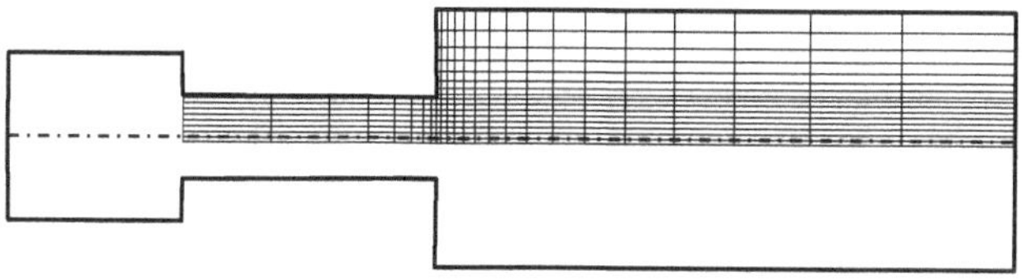

Fig. 3. Position of the computational domain

physical domain. Only the premixing duct and the combustion chamber are simulated. The spatial model is reduced to a 2D axisymmetric solution domain which makes parameter studies faster and more convenient. To account for the effect of the swirl generator profiles of all hydrodynamic properties are used as boundary conditions. These profiles are taken from the work of Kiesewetter et al. [24] and are derived from detailed simulations of the entire burner including the plenum and the swirl generator. The same author also shows that this mapping from 3D to 2D axisymmetric solution domain leads to reasonable results.

The global operation parameters are an equivalence ratio of $\phi = 0.556$, a total inlet mass flux of $70\,{}^{\mathrm{g}}/_{\mathrm{s}}$ and a swirl number of $S = 0.5$. The preheat temparture of the mixture is $T = 298\mathrm{K}$.

All the results shown are for the reactive case. Fig. 4 depicts the axial component of the velocity vector. Shown is a contour plot spanning over the streamwise (x) and radial (y) coordinate in space. Even though the simulations are undertaken making use of the axial symmetry the results are mirrored at the symmetry line in the shown figures for reasons of clarity. The contour plot in fig. 4 shows two areas with negative axial velocity: One in the corner of the combustion chamber caused by the backward facing step and another on the symmetry line aerodynamically caused by the swirl in the main flow. This internal recirculation zone is used to stabilize the flame by holding hot exhaust gases.

This can be observed in fig. 5 as well. This figure shows the temperature field of the flame. The temperature as explained above is calculated from a lookup table. One can see that the reaction of the premixed fresh gas starts on the symmetry line where the hot exhaust gas is transported by the internal

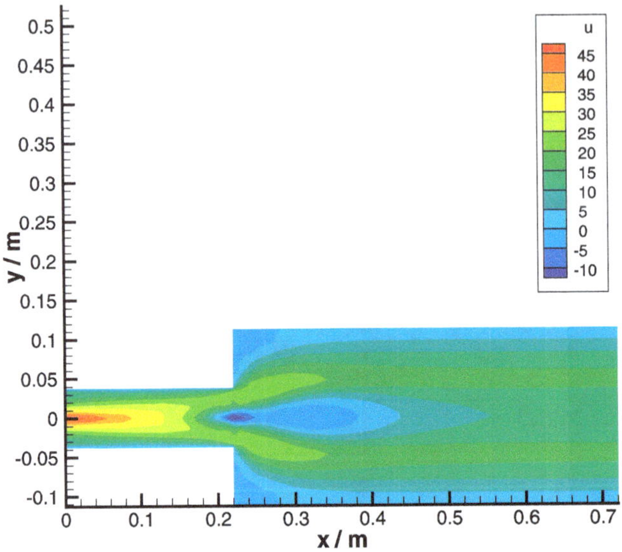

Fig. 4. Contourplot of the axial velocity component ($u/\frac{\mathrm{m}}{\mathrm{s}}$)

recirculation zone. A steep temperature gradient can be observed and the temperature after the flame front is resonably well predicted by the presented model.

An aspect of main interest in this studies is the stability of the flame under different global operation condition (e.g. different equivalence ratios). Therefore fig. 6 shows a closeup of the combustion chamber at the position of the backward facing step. The figures depicts again a contour plot. In the upper half the temperature field is shown while in the lower half the axial velocity can be seen. Additionally the $u = 0\,\mathrm{m/s}$ isoline is plotted. The isoline marks the boundary of the internal recirculation zone. One can see that the flame is located within the recirculation bubble and a slight increase of temperature due to prereactions can be observed close to the symmetry line in front of the bubble in regions of slow axial valocity. A stable operation of the flame is possible as long as the recirculation bubble is stationary in space.

5 Conclusion

Simulations of premixed swirling methane-air flames are presented. The model consists of two parts: a finite-volume solver for the hydrodynamic quantities and a Monte Carlo solver for the transport equation of the JPDF of velocity and scalars. This provides a detailed modeling of the turbulence chemistry interaction. Chemical kinetics are described by automatically reduced mechanisms created with the ILDM method.

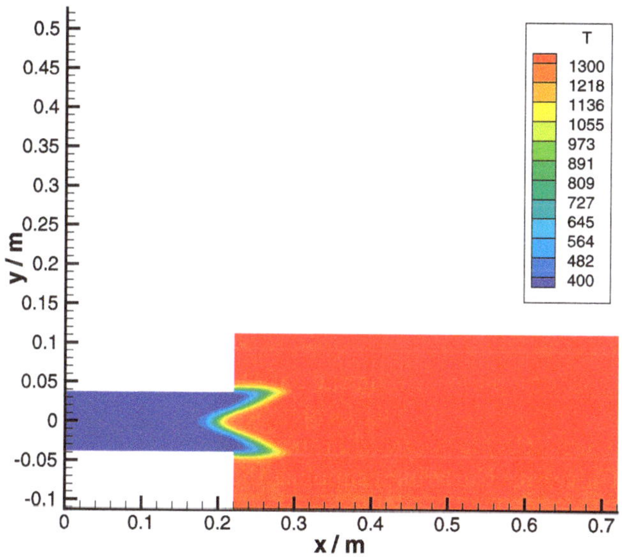

Fig. 5. Contourplot of the temperature (T/K)

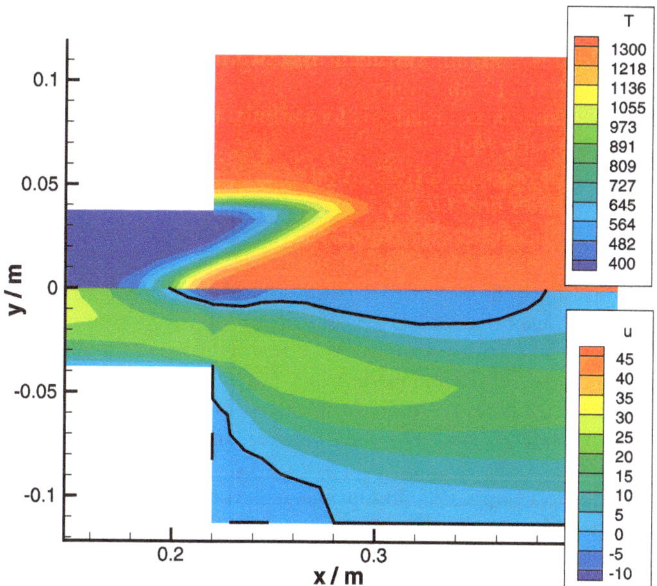

Fig. 6. Temperature and axial velocity

The presented results show that a at least the qualitative behaviour of the flame could be predicted for the stationary case. Although quantitative comparison is not possible due to a lack of detailed experimental data. The potential of the model to predict turbulent flames quantitatively is currently investigated using for some test cases where large experimental data sets are available. An analysis of instationary effects like the combustion induced vortex breakdown which shall be trigged by a variation of the equivalence ratio is subject of future research work.

Acknowledgment

The simulations were performed on the national super computer NEC SX-8 at the High Performance Computing Center Stuttgart (HLRS) under the grant number FLASPARC. The work of Franco Magagnato (Department of Fluid Machinery, University Karlsruhe) suppling the CFD code SPARC and assisting with its use is gratefully acknowledged.

References

1. C. Duwig, D. Stankovic, L. Fuchs, G. Li, and E. Gutmark. Experimental and numerical study of flameless combustion in a model gas turbine combustor. *Combustion Science and Technology*, 180(2):279–295, 2008.
2. S.B. Pope. Pdf methods for turbulent reactive flows. *Progress in Energy Combustion Science*, 11:119–192, 1985.
3. S.B. Pope. Lagrangian pdf methods for turbulent flows. *Annual Review of Fluid Mechanics*, 26:23–63, 1994.
4. Z. Ren and S.B. Pope. An investigation of the performence of turbulent mixing models. *Combustion and Flame*, 136:208–216, 2004.
5. B. Merci, D. Roekaerts, B. Naud, and S.B. Pope. Comparative study of micromixing models in transported scalar pdf simulations of turbulent nonpremixed bluff body flames. *Combustion and Flame*, 145:109–130, 2006.
6. B. Merci, D. Roekaerts, and B. Naud. Study of the perfomance of three micromixing models in transported scalar pdf simulations of a piloted jet diffusion flame ("delft flame iii"). *Combustion and Flame*, 144:476–493, 2006.
7. P.R. Van Slooten and S.B Pope. Application of pdf modeling to swirling and nonswirling turbulent jets. *Flow Turbulence and Combustion*, 62(4):295–334, 1999.
8. V. Saxena and S.B Pope. Pdf simulations of turbulent combustion incorporating detailed chemistry. *Combustion and Flame*, 117(1-2):340–350, 1999.
9. S. Repp, A. Sadiki, C. Schneider, A. Hinz, T. Landenfeld, and J. Janicka. Prediction of swirling confined diffusion flame with a monte carlo and a presumed-pdf-model. *International Journal of Heat and Mass Transfer*, 45:1271–1285, 2002.
10. K. Liu, S.B. Pope, and D.A. Caughey. Calculations of bluff-body stabilized flames using a joint probability density function model with detailed chemistry. *Combustion and Flame*, 141:89–117, 2005.

11. M. Nau. *Berechnung turbulenter Diffusionsflammen mit Hilfe eines Verfahrens zur Bestimmung der Wahrscheinlichkeitsdichtefunktion und automatisch reduzierter Reaktionsmechanismen.* PhD thesis, Universität Stuttgart, Fakultät für Energietechnik, 1997.
12. R. Bender, U. Maas, S. Böckle, J. Kazenwadel, C. Schulz, and J. Wolfrum. Monte-carlo-pdf-simulation and raman/rayleigh-measurement of a turbulent premixed flame. In *Twenty-Eighth Symposium (International) on Combustion.* The Combustion Institute, Pittsburgh, PA, 2000.
13. Y.Z. Zhang and D.C. Haworth. A general mass consistency algorithm for hybrid particle/finite-volume pdf methods. *Journal of Computational Physics*, 194:156–193, 2004.
14. S.B Pope. A monte carlo method for pdf equations of turbulent reactive flow. *Combustion, Science and Technology*, 25:159–174, 1981.
15. P. Jenny, M. Muradoglu, K. Liu, S.B. Pope, and D.A. Caughey. Pdf simulations of a bluff-body stabilized flow. *Journal of Computational Physics*, 169:1–23, 2000.
16. A.K. Tolpadi, I.Z. Hu, S.M. Correa, and D.L. Burrus. Coupled lagrangian monte carlo pdf-cfd computation of gas turbine combustor flowfields with finite-rate chemistry. *Journal of Engineering for Gas Turbines and Power*, 119:519–526, 1997.
17. M. Muradoglu, P. Jenny, S.B Pope, and D.A. Caughey. A consistent hybrid finite-volume/particle method for the pdf equations of turbulent reactive flows. *Journal of Computational Physics*, 154:342–370, 1999.
18. M. Muradoglu, S.B. Pope, and D.A. Caughey. The hybid method for the pdf equations of turbulent reactive flows: Consistency conditions and correction algorithms. *Journal of Computational Physics*, 172:841–878, 2001.
19. G. Li and M.F. Modest. An effective particle tracing scheme on structured/unstructured grids in hybrid finite volume/pdf monte carlo methods. *Journal of Computational Physics*, 173:187–207, 2001.
20. V. Raman, R.O. Fox, and A.D. Harvey. Hybrid finite-volume/transported pdf simulations of a partially premixed methane-air flame. *Combustion and Flame*, 136:327–350, 2004.
21. H.S. Zhang, R.M.C. So, C.G. Speziale, and Y.G. Lai. A near-wall two-equation model for compressible turbulent flows. In *Aerospace Siences Meeting and Exhibit, 30th, Reno, NV*, page 23. AIAA, 1992.
22. U. Maas and S.B. Pope. Simplifying chemical kinetics: Intrinsic low-dimensional manifolds in composition space. *Combustion and Flame*, 88:239–264, 1992.
23. U. Maas and S.B. Pope. Implementation of simplified chemical kinetics based on intrinsic low-dimensional manifolds. In *Twenty-Fourth Symposium (International) on Combustion*, pages 103–112. The Combustion Institute, 1992.
24. F. Kiesewetter, C. Hirsch, J. Fritz, M. Kröner, and T. Sattelmayer. Two-dimensional flashback simulation in strongly swirling flows. In *Proceedings of ASME Turbo Expo 2003*, 2003.
25. M. Kröner. *Einfluss lokaler Löschvorgänge auf den Flammenrückschlag durch verbrennungsinduziertes Wirbelaufplatzen.* PhD thesis, Technische Universität München, Fakultät für Maschinenwesen, 2003.
26. J. Fritz. *Flammenrückschlag durch verbrennungsinduziertes Wirbelaufplatzen.* PhD thesis, Technische Universität München, Fakultät für Maschinenwesen, 2003.

27. F. Kiesewetter. *Modellierung des verbrennungsinduzierten Wirbelaufplatzens in Vormischbrennern.* PhD thesis, Technische Universität München, Fakultät für Maschinenwesen, 2005.
28. F. Magagnato. SPARC *Structured PArallel Research Code.* Department of Fluid Machinery, Karlsruhe University, Kaiserstrasse 12, 76131 Karlsruhe, Germany, 1998.
29. D.C. Haworth and S.H. El Tahry. Propbability density function approach for multidimensional turbulent flow calculations with application to in-cylinder flows in reciproating engines. *AIAA Journal*, 29:208, 1991.
30. S.M. Correa and S.B. Pope. Comparison of a monte carlo pdf finite-volume mean flow model with bluff-body raman data. In *Twenty-Fourth Symposium (International) on Combustion*, page 279. The Combustion Institute, 1992.
31. W.C. Welton and S.B. Pope. Pdf model calculations of compressible turbulent flows using smoothed particle hydrodynamics. *Journal of Computational Physics*, 134:150, 1997.
32. D.C. Haworth and S.B. Pope. A generalized langevin model for turbulent flows. *Physics of Fluids*, 29:387–405, 1986.
33. H.A. Wouters, T.W. Peeters, and D. Roekaerts. On the existence of a generalized langevin model representation for second-moment closures. *Physics of Fluids*, 8(7):1702–1704, 1996.
34. T.D. Dreeben and S.B. Pope. Pdf/monte carlo simulation of near-wall turbulent flows. *Journal of Fluid Mechanics*, 357:141–166, 1997.
35. R.L. Curl. Dispersed phase mixing: 1. theory and effects in simple reactors. *A.I.Ch.E. Journal*, 9:175, 181, 1963.
36. J. Janicka, W. Kolbe, and W. Kollmann. Closure of the transport equation of the probability density function of turbulent scalar flieds. *Journal of Non-Equilibrium Thermodynamics*, 4:47–66, 1979.
37. S.B Pope. An improved turbulent mixing model. *Combustion, Science and Technology*, 28:131–135, 1982.
38. S. Lipp and U. Maas. Simulations of premixed swirling flames using a hybrid finite-volume/transported pdf approach. In *High Performance Computing on Vector Systems*, pages 181–193, 2007.

Numerical Investigations of Model Scramjet Combustors

Markus Kindler, Thomas Blacha, Markus Lempke, Peter Gerlinger, and Manfred Aigner

Institut für Verbrennungstechnik der Luft- und Raumfahrt, Universität Stuttgart Pfaffenwaldring 38-40, 70569 Stuttgart, Germany

Summary. In the present paper different types of scramjet (supersonic combustion ramjet) combustors are investigated. Thereby the main difference between the combustors is the way of injecting the fuel into the combustion chamber. The first investigated concept of fuel injection is the injection by strut injectors. Here the injection of fuel is realized by a lobed strut that is located in the middle of the combustion chamber. The second concept for fuel supply is the wall injection of hydrogen. Here the fuel is injected by several holes in the wall of the combustor. Both concepts of fuel injection have different advantages and disadvantages which are explained in detail. Although different performance parameters for both scramjet combustors are introduced this paper will not compare the different techniques among each other. Because of the high Reynolds numbers in scramjet combustors, the need to resolve the boundary layers and the necessity of detailed chemistry, the simulation of scramjets is extremely CPU time demanding.

1 Introduction

Due to the high velocities in a scramjet combustor the residence time of air and fuel in the combustion chamber is extremely short. Additionaly the mixing rates at high flow Mach numbers are inherently low. Hence techniques are required to achieve a rapid and efficient mixing of fuel and air and thus a stable combustion and a complete burnout of the fuel. There are mainly two concepts of fuel injection: 1. wall injectors [1, 2, 3] and 2. strut injectors [4, 5, 6, 7]. In case of a wall injection the fuel is injected through the wall into the air flow. Wall injectors are easy to manufacture and easy to cool, have a good near field mixing and cause no pressure loss when they are turned off. On the other hand in real size combustors there might be problems of the penetration depth of the fuel into the air flow and a strong blockage might occur in the combustion chamber. Strut injectors, where the fuel is injected through a strut directly into the core of the air flow, cannot be removed from the flow and hence cause pressure losses even when they are switched off.

Additionally they have to be cooled and usually need mixing enhancement techniques. The advantages of strut injectors are the absence of strong shock waves due to a blockage caused by the fuel jet. In the present paper both concepts are investigated numerically.

2 Governing Equations and Numerical Scheme

The scientific inhouse code TASCOM3D (Turbulent All Speed Combustion Multigrid) describes reacting flows by solving the full compressible Navier-Stokes, species and turbulence transport equations. Additionally an assumed PDF (probability density function) approach is used to take turbulence chemistry interaction into consideration. Therefore two additional equations (for the variance of the temperature and the variance of the sum of species mass fractions) have to be solved. Thus the described set of averaged equations in three-dimensional conservative form is given by

$$\frac{\partial \mathbf{Q}}{\partial t} + \frac{\partial(\mathbf{F} - \mathbf{F}_\nu)}{\partial x} + \frac{\partial(\mathbf{G} - \mathbf{G}_\nu)}{\partial y} + \frac{\partial(\mathbf{H} - \mathbf{H}_\nu)}{\partial z} = \mathbf{S}\,, \tag{1}$$

where

$$\mathbf{Q} = \left[\bar{\rho}, \bar{\rho}\tilde{u}, \bar{\rho}\tilde{v}, \bar{\rho}\tilde{w}, \bar{\rho}\tilde{E}, \bar{\rho}q, \bar{\rho}\omega, \bar{\rho}\sigma_T, \bar{\rho}\sigma_Y, \bar{\rho}\tilde{Y}_i\right]^T, \quad i = 1, 2, \ldots, N_k - 1\,. \tag{2}$$

The variables in the conservative variable vector $\mathbf{Q}$ are the density $\bar{\rho}$ (averaged), the velocity components (Favre averaged) $\tilde{u}$, $\tilde{v}$ and $\tilde{w}$, the total specific energy $\tilde{E}$, the turbulence variables $q = \sqrt{k}$ and $\omega = \epsilon/k$ (where k is the kinetic energy and ϵ the dissipation rate of k), the variance of the temperature σ_T and the variance of the sum of the species mass fractions σ_Y and finally the species mass fractions Y_i ($i = 1, 2, \ldots, N_k - 1$). Thereby N_k describes the total number of species that are used for the description of the gas composition. The vectors $\mathbf{F}$, $\mathbf{G}$ and $\mathbf{H}$ specify the inviscid fluxes in x-, y- and z- direction, $\mathbf{F}_\nu$, $\mathbf{G}_\nu$ and $\mathbf{H}_\nu$ the viscous fluxes, respectively. The source vector $\mathbf{S}$ in Eq. (1) results from turbulence and chemistry and is given by

$$\mathbf{S} = \left[0, 0, 0, 0, 0, \bar{S}_q, \bar{S}_\omega, \bar{S}_{\sigma_T}, \bar{S}_{\sigma_Y}, \bar{S}_{Y_i}\right]^T, \quad i = 1, 2, \ldots, N_k - 1\,, \tag{3}$$

where $\bar{S}_q$ and $\bar{S}_\omega$ are the averaged source terms of the turbulence variables, $\bar{S}_{\sigma_T}$ and $\bar{S}_{\sigma_Y}$ the source terms of the variance variables (σ_T and σ_Y) and $\bar{S}_{Y_i}$ the source terms of the species mass fractions. For turbulence closure a two-equation low-Reynolds-number q-ω turbulence-model is applied [8]. The momentary chemical production rate of species i in Eq. (1) is defined by

$$S_{Y_i} = M_i \sum_{r=1}^{N_r} \left[\left(\nu''_{i,r} - \nu'_{i,r} \right) \left(k_{fr} \prod_{l=1}^{N_k} c_l^{\nu'_{l,r}} - k_{br} \prod_{l=1}^{N_k} c_l^{\nu''_{l,r}} \right) \right], \tag{4}$$

where k_{f_r} and k_{b_r} are the forward and backward rate constants of reaction r (defined by the Arrhenius function), the molecular weight of a species M_i, the species concentration $c_i = \rho Y_i/M_i$ and the stoichiometric coefficients $\nu'_{i,r}$ and $\nu''_{i,r}$ of species i in reaction r. The averaged chemical production rate for a species i due to the use of the assumed PDF approach and the assumption of statistical independence of temperature, gas composition and density is given by

$$\bar{S}_{Y_i} = \int S_i\left(\hat{T}, \hat{c}_1, \cdots, \hat{c}_{N_k}\right) P\left(\hat{T}, \hat{c}_1, \cdots, \hat{c}_{N_k}\right) d\hat{T} d\hat{c}_1 \cdots d\hat{c}_{N_k} , \qquad (5)$$

where

$$P\left(\hat{T}, \hat{c}_1, \cdots, \hat{c}_{N_k}\right) = P_T\left(\hat{T}\right) P_Y\left(\hat{Y}_1, \cdots, \hat{Y}_{N_k}\right) \delta\left(\hat{\rho} - \bar{\rho}\right) . \qquad (6)$$

P_T defines the temperature PDF desribed by a Gaussian distribution

$$P_T\left(\hat{T}\right) = \frac{1}{\sqrt{2\pi\sigma_T}} \exp\left[-\frac{\left(\hat{T} - \tilde{T}\right)^2}{2\sigma_T}\right] , \quad \sigma_T = \widetilde{T''^2} \qquad (7)$$

that is clipped at lower and upper temperature limits due to the limitations of the Arrhenius equation [9]. The PDF of the gas composition P_Y is described by the multi-variate β-pdf proposed by Girimaji [10]

$$P_Y\left(\hat{Y}_1, \cdots, \hat{Y}_{N_k}\right) = \frac{\Gamma\left(\sum_{m=1}^{N_k} \beta_m\right)}{\prod_{m=1}^{N_k} \Gamma\left(\beta_m\right)} \left[\delta\left(1 - \sum_{m=1}^{N_k} \hat{Y}_m\right) \prod_{m=1}^{N_k} \hat{Y}_m^{\beta_m - 1}\right] \qquad (8)$$

where

$$\beta_m = \tilde{Y}_m B, \quad B = \left[\frac{\sum_{m=1}^{N_k} \tilde{Y}_m \left(1 - \tilde{Y}_m\right)}{\sigma_Y} - 1\right] , \quad \sigma_Y = \sum_{m=1}^{N_k} \widetilde{Y''^2_m} . \qquad (9)$$

δ in Eq. (6) and (8) denotes the δ-function. The unsteady set of equations (1) is solved using an implicit Lower-Upper-Symmetric Gauss-Seidel (LU-SGS) [11, 12, 13, 14] finite-volume algorithm, where the finite-rate chemistry is treated fully coupled with the fluid motion. More details concerning TASCOM3D may be found in [13, 14, 15, 16, 17, 18]

3 Investigations of Lobed Strut Injectors

In this section a scramjet combustion chamber with a strut injector is investigated. Main focus is the effect of different amounts of circulation produced by the lobed strut on the mixing and hence on the combustion process. Thereby hydrogen is injected in axial flow direction by a lobed strut injector (see Fig. 1) which creates streamwise vortices to enhance the mixing (compared to planar struts) [19]. In order to produce different amounts of circulation the length of the lobed structure of the strut is varied. Fig. 1 shows the struts investigated which all have 60 mm in width and 8 mm in height, while the length of the lobed parts differ. The area for hydrogen injection is the same in all three cases.

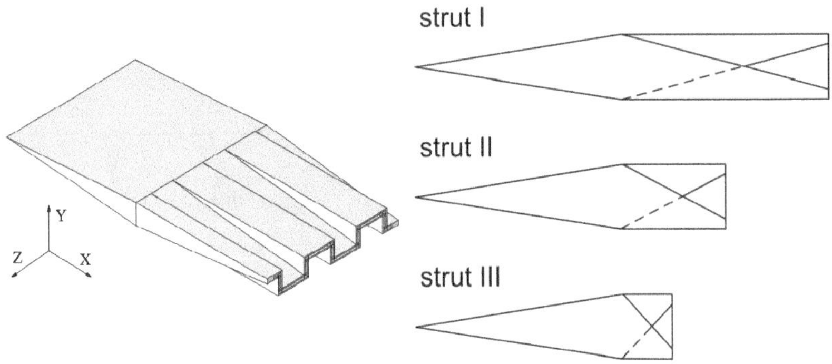

Fig. 1. Investigated struts with different lengths and amount of vorticity production

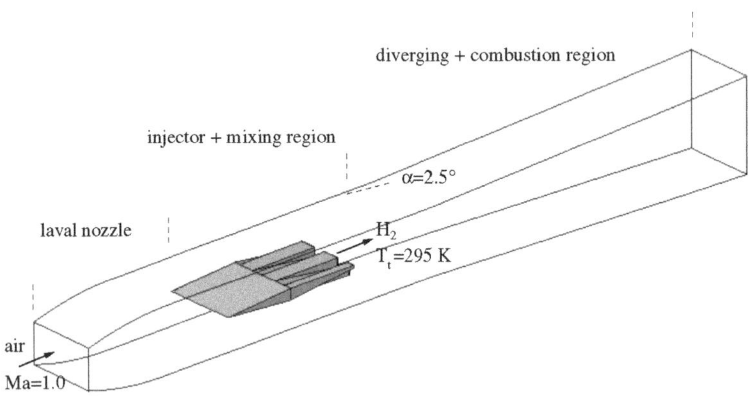

Fig. 2. Sketch of the combustion chamber with a lobed strut injector

In case of strut I the length is 80 mm. For strut II the lobed part is shortened by a factor of 2 (60 mm total length) and for strut III by a factor of 4 in comparison to strut I (50 mm total length). A reduction of the strut length increases the angles for changes in flow direction and thus the strength of the expansion fans at the middle of the strut and the pressure ratio in cross stream direction. Thus the shorter struts create stronger vortices. Fig. 2 demonstrates a sketch of the combustion chamber with a lobed strut injector. The dimensions of the combustor are 536 mm x 60 mm x 38 mm. Thereby the computional domain includes only a section of the combustion chamber using periodic and symmetry boundary comditions. Furthermore the simulations have been performed in two steps: a two-dimensional simulation up to the middle of the strut and three-dimensional simulations of the remaining combustion chamber. The inflow conditions for the simulations are summarized in Tab. 1. More details concerning the model scramjet combustor geometry and the lobed strut injectors may be found in [20]. The grid of the computational domain has about 800.000 volumes (220 x 70 x 54) and is strongly refined at all near wall regions as well as in the main combustion zone. Although the quantity of the volumes is moderate the computational costs are high due to the additional transport equations for the variances and especially the species as well as the numerical stiffness caused by combustion. Fig. 3 shows representatively for all configurations the H_2O, H_2 and OH distributions of the model scramjet combustor using strut I, respectively. The hydrogen distribution nicely shows the production of vortices due to the lobed strut injector right after the injection. The H_2O and OH distributions represent the combustion process and show that in all cases (strut I - strut III) lifted flames are obtained. As observed in previous investigations [19], ignition takes place near the point where the diverging channel part begins. The positions for ignition vary between 85 mm (strut I), 105 mm (strut II) and 110 mm (strut III), respectively. Further downstream the flame becomes wider and the separated structures grow together till the flame has a W-shape at the outlet of the combustion chamber. In order to describe the effect of the struts on the combustion process calculated H_2O distributions are plotted in Fig. 4 for different y-z slices at channel positions ranging from 100-400 mm. 100 mm downstream of the hydrogen injection the shortest strut has a very small amount of H_2O, because ignition has just occured. In case of the longest

Table 1. Inflow conditions for the simulation of the combustion chamber with the lobed strut injector

	air	strut
pressure p [Pa]	211000	50000
temperature T [K]	1300	196
velocity u [$\frac{m}{s}$]	723	2281
Mach-number Ma [-]	1	2
mass fraction Y_{H_2}	0	1
mass fraction Y_{O_2}	0.23	0
mass fraction Y_{N_2}	0.77	0
$\dot{m}$ [$\frac{g}{s}$]	628	4.18

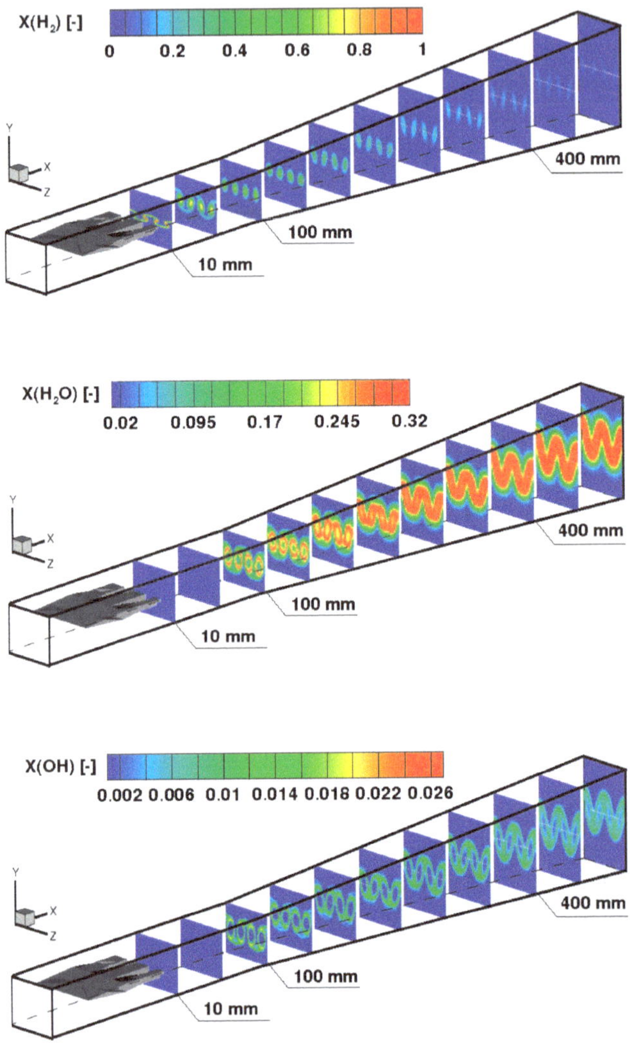

Fig. 3. Calculated H_2, H_2O and OH distributions (from top to bottom) of the model scramjet combustor using strut I, respectively

strut the combustion process is already in progress and the amount of H_2O is much higher compared to the other struts. The main combustion zone in all three cases forms two circles which are turning to a more elliptical shape with an increased strut length. With increasing distance from the fuel injector the flame spreads in vertical direction and the circle-like shapes grow together. In case of the shortest strut the flame shows the most homogeneous H_2O distribution, whereas the differences are quite small at the end of the combustion

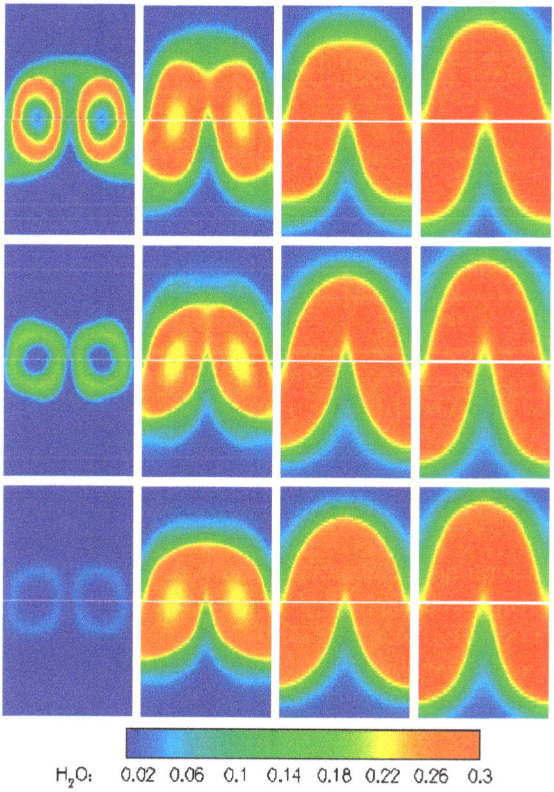

Fig. 4. Calculated H_2O molar fraction in y-z cross sections at x = 100, 200, 300, and 400 mm for the struts I to III (from top to bottom)

chamber. Finally in Fig. 5 a comparison of the three struts by means of performance parameters (hydrogen mass flux, total pressure) is given. Strut I has the best mixing in the near field of the injector which causes an early ignition and burning of the hydrogen (indicated by the rapid decrease of the hydrogen mass flux at the normalized channel length of 0.15). However, with increasing distance from the strut, vorticity seems to be too weak to transport air from the upper and lower channel walls into the main combustion zone. This results in a higher amount of unburned hydrogen at the end of the combustor compared with strut II and III. Strut III has the fewest amount of unburned hydrogen at the outlet. The differences in loss in total pressure between the three struts are relatively small.

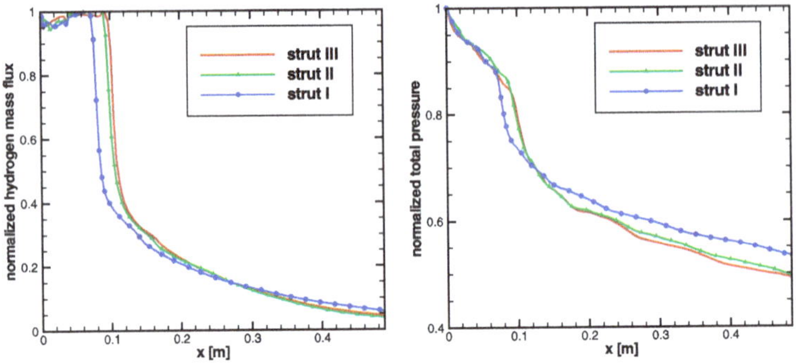

Fig. 5. Normalized hydrogen mass flux and total pressure over the normalized channel length for strut I to III, respectively

4 Investigations of Wall Injectors

In this section a scramjet combustion chamber with wall injection is investigated. Thereby the geometry and flow conditions are based on the HyShot flight experiments performed since 2001 by the University of Queensland to investigate supersonic combustion in flight. Since the beginning of the program four launches have been realized (HyShot I-IV) in which supersonic combustion has been achieved at HyShot II (2002) and HyShot III (2006). Additionally in the year 2003 and again in 2007 an identical model of the HyShot combustor has been investigated in the High Enthalpy Shock Tunnel in Göttingen (HEG) [21]. Thereby several freestream conditions, equivalence ratios and angles of attack have been investigated. Fig. 6 demonstrates a sketch of the combustion chamber. The flow approaches an intake ramp (363 x 100 mm) with a Mach-number of about 7.8. Due to the induced shocks

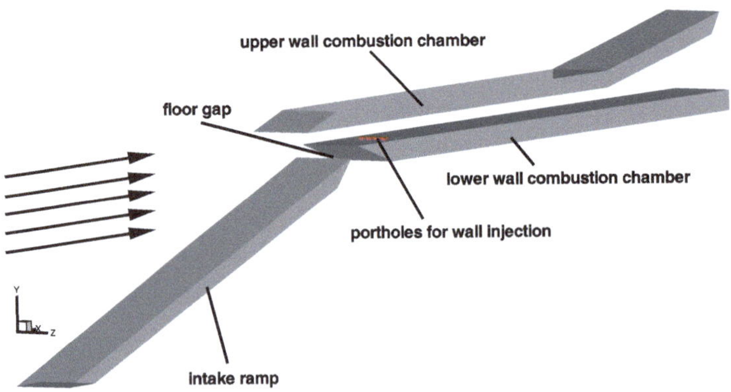

Fig. 6. Sketch of the combustion chamber with wall injectors

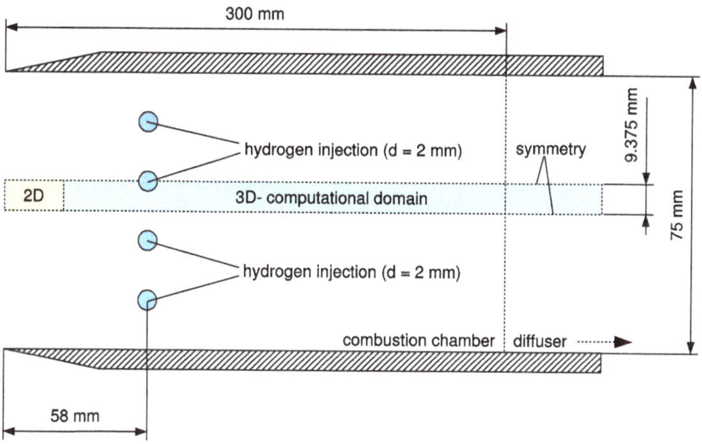

Fig. 7. Computational domain for the numerical simulation of the combustion chamber with wall injectors

at the leading edges of the intake ramp and the combustion chamber (300 x 75 x 9.8 mm) the flow reaches a Mach-number of 2.65 in front of the wall injectors. To ensure a relatively homogenous flow field in the combustion chamber the boundary layer is bled and shocks are by-passed through the floor gap indicated in Fig. 6. Again the numerical simulation has been performed in two steps: a two-dimensional simulation for the intake ramp and a three-dimensional simulation for the combustion chamber and the diffuser. The computational domain (Fig. 7) covers a slice from the middle of a wall injector to the symmetry line between two portholes using symmetry boundary conditions Thereby the grid of the computational domain has about 1.500.000 cell volumes. Tab. 2 summarizes the inflow conditions for the numerical sim-

Table 2. Inflow conditions for the simulation of the combustion chamber with the lobed strut injector

	air	wall injector
pressure p [bar]	1725	258135
temperature T [K]	221	250
velocity u [$\frac{m}{s}$]	2328	1250
Mach-number Ma [-]	7.8	1
mass fraction Y_{H_2}	0	1
mass fraction Y_{O_2}	0.23	0
mass fraction Y_{N_2}	0.77	0
$\dot{m}$ [$\frac{g}{s}$]	422.78	4.24

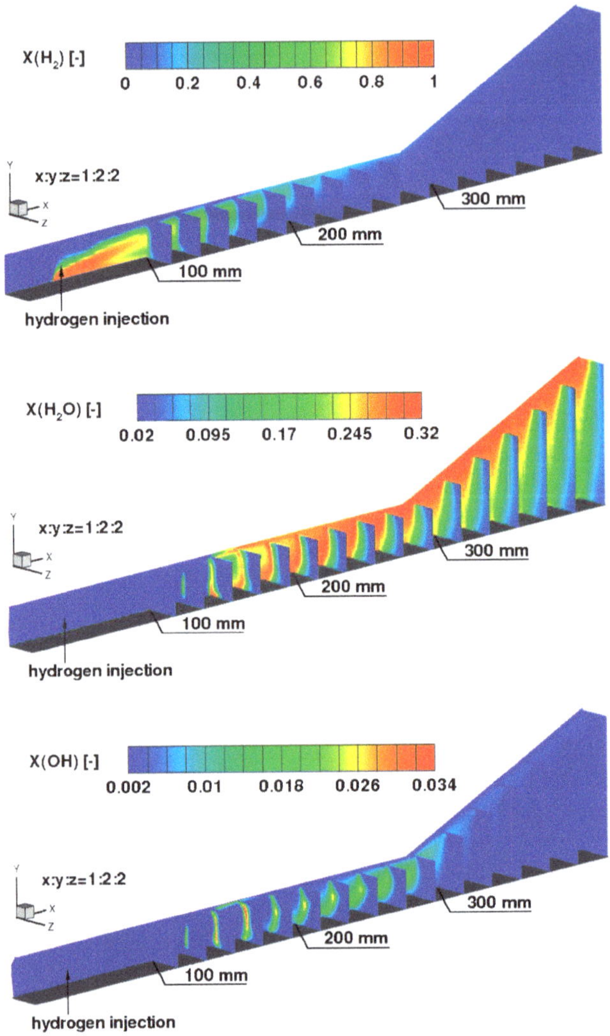

Fig. 8. Calculated H_2,H_2O and OH distribution (from top to bottom) of the HyShot scramjet combustor, respectively

ulation of the combustion chamber. Fig. 8 shows the calculated H_2, H_2O and OH distribution of the HyShot combustion chamber with hydrogen injection into air. A lifted flame is obtained and combustion starts about 60 mm downstream of the hydrogen injection. Fig. 8 shows in the beginning a relatively homogeneous hydrogen distribution in vertical direction. The majority of the unburned hydrogen at the end of the combustor (about 200-300 mm distance downstream of the combustor entrance) is located at the upper wall. The

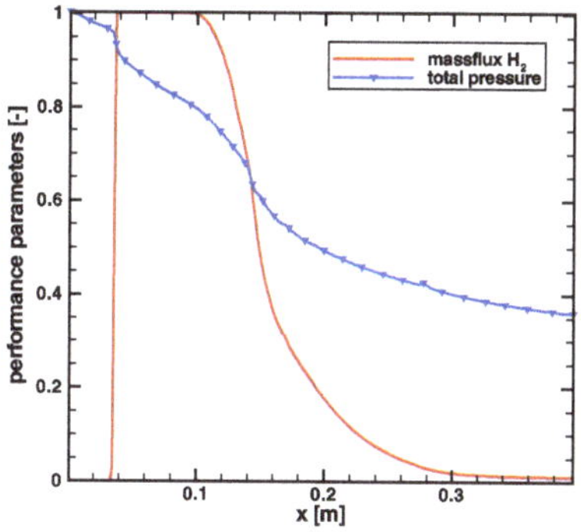

Fig. 9. Normalized hydrogen mass flux and total pressure over the channel length for the HyShot scramjet combustor, respectively

H_2O-distribution demonstrates that combustion starts at the outer regions of the area covered by hydrogen. After ignition the flame has a circle-like shape and turns to a relatively homogeneous distribution at a distance of about 200 mm from the combustor entrance. Further downstream the main combustion zone is located at the upper wall. Similar observation are obtained from the OH-distribution. Fig. 9 describes the normalized hydrogen mass flux and normalized total pressure along the channel length, respectively. Again the ignition delay can be identified by the rapid decrease of the hydrogen mass flux taking place at a distance of about 60 mm downstream of the fuel injection. At the combustor exit nearly no unburned fuel is left. The total pressure shows two dips: the first due to the injection of the hydrogen at about $x = 0.04$ m and the second due to the combustion process at about $x = 0.11$ m. The total pressure loss at the end of the combustion chamber is about 64% compared to the value at the combustor entrance.

5 Performance

Previous investigations of the performance of TASCOM3D on the NEC SX-8 [22] have shown that the performance in case of combustion simulations is quite good. Thereby a huge part of the whole computational time is consumed by a subroutine calculating PDF averaged chemical production rates and source term Jacobians. Due to the fact that chemistry is a local phenomena, a good vector lengths, quantity of MFLOPS and vector operation ratio are reached in the corresponding routine. Another huge part is consumed by the implicit solver used which is required because of the high numerical stiffness caused by combustion. The implicit solver is vectorized along diagonals (planes in 3D) to prevent data dependencies. The length of these diagonals depends on the size of the block and hence is dependent on the geometry which could yield to relatively short vector lengths in blocks with few cells. In this paper additionally the MPI - parallelization of TASCOM3D has been reviewed and tested in order to achieve good performance in parallel runs. Therefore a testcase (i.e. two dimensional channel flow of an inert gas) with about 2.000.000 cell volumes has been calculated by a one block serial run and parallel runs with two, four and eight blocks, respectively. Tab. 3 shows the number of CPUs, the total CPU time consumed by the simulation, the averaged vector length and vector operation ratio as well as the MFLOPs reached for 1000 iterations. For the serial run TASCOM shows an excellent

Table 3. Perfomance parameter for simulation with 1, 2, 4 and 8 CPUs

CPUs [-]	total CPU-time [s]	aver. v.len [-]	v.op. ratio [-]	MFLOPS [-]
1	1008	243.9	99.2	6133.8
2	1079	242.7	99.09	5731.8
4	1165	234.8	99.01	5307.7
8	1473	205.8	98.80	4200.4

performance. The averaged vector length and vector operation ratio reaches nearly the maximum values and the total CPU-time is 1008 s. With increasing number of CPUs the total CPU-time raises up to 1473 s for 8 CPUs (which is a factor of 1.46). The main reason can be found in the averaged vector length which becomes shorter with increasing number of CPUs. The total number of cell volumes per CPU becomes smaller due to the fragmentation of the grid. Hence the diagonals for the implicit solver are shorter and thus the vector length. Therefore the number of cell volumes per CPU should be sufficient high in every simulation in order to reach a good performance. The problem of the shortening of vector lengths would be less dramatic in three-dimensional test-cases (which have not been investigated in this section) because the vector length significantly increases while the implicit solver vectorizes along planes compared to diagonals. Nevertheless the performance of the parallelization has to be investigated more detailed and further improved.

6 Conclusion

The scientific inhouse code TASCOM3D has been used to investigate different configurations of scramjet combustors. One investigated combustor configuration uses lobed strut injectors while the other configuration uses wall injection. In case of the lobed strut injection the effect of different amounts of circulation on the combustion process is investigated additionally. The code has been tested on the NEC SX-8 and has shown to have a good vector performance. Also the implemented MPI - parallelization has been tested on the NEC SX-8.

It has been shown that the number of cell volumes per CPU should be as high as possible in order to reach a good performance.

Acknowledgements

This work was performed within the 'Long-Term Advanced Propulsion Concepts and Technologies' project investigating high-speed airbreathing propulsion. LAPCAT, coordinated by ESA-ESTEC, is supported by the EU within the 6th Framework Programme Priority 1.4, Aeronautic and Space, Contract no.: AST4-CT-2005-012282. Further information on LAPCAT can be found on http://www.esa.int/techresources/lapcat The simulations were performed on the national super computer NEC SX-8 at the High Performance Computing Center Stuttgart (HLRS) under the grant number scrcomb.

References

1. Belanger, J., Hornung, H.G., Transverse Jet Mixing and Combustion Experiments in Hypervelocity Flows, Journal of Propulsion and Power, **12**, pp. 186-192, 1996.
2. Riggins, D.W., McClinton, C.R., Rogers, R.C., Bittner, R.D.: Investigation of Scramjet Injection Strategies for High Mach Number Flows, Journal of Propulsion and Power, 11, pp. 409-418, 1995.
3. Baurle, R.A., Fuller, R.P., White, J.A., Chen, T.H., Gruber, M.R., Nejad, A.S: An Investigation of Advanced Fuel Injection Schemes for Scramjet Combustion, AIAA paper 98-0937, 1998.
4. Glawe, D.D., Samimiy, M., Nejad, A.S., Cheng, T.H.: Effects of Nozzle Geometry on Parallel Injection from Base of an Extended Strut into a Supersonic Flow, AIAA paper 95-0522, 1995.
5. Strickland, J.H., Selerland, T., Karagozian, A.R: Numerical Simulation of a Lobed Fuel Injector, Physics of Fluids, 10, pp. 2950-2964, 1998.
6. Charyulu, B.V.N., Kurian, J., Venugopalan, P., Sriamulu, V.: Experimental Study on Mixing Enhancement in Two Dimensional Supersonic, Experiments in Fluids, 24, pp. 340-346, 1998.
7. Sunami, T., Wendt, M., Nishioka, M.: Supersonic Mixing and Combustion Control Using Streamwise Vorticity, AIAA paper 98-3271, 1998.
8. Coakley, T.J., Huang, P.G., Turbulence Modeling for High Speed Flows, AIAA paper 92-0436, 1992.
9. Gerlinger, P., Möbus, H., Brüggemann, D.: An Implicit Numerical Scheme for Turbulent Combustion Using an Assumed PDF Approach, AIAA paper 99-3775, 1999.
10. Girimaji, S.S.: A Simple Recipe for Modeling Reacting-Rates in Flows with Turbulent Combustions, AIAA paper 91-1792, 1991.
11. Jameson, A., Yoon, S.: Lower-Upper Implicit Scheme with Multiple Grids for the Euler Equations, AIAA Journal, **25**, pp. 929-937, 1987.
12. Shuen, J.S.: Upwind Differencing and LU Factorization for Chemical Non-Equilibrium Navier-Stokes Equations, Journal of Computational Physics, **99**, pp. 233-250, 1992.

13. Gerlinger, P., Brüggemann, D.: An Implicit Multigrid Scheme for the Compressible Navier-Stokes Equations with Low-Reynolds-Number Turbulence Closure, Journal of Fluids Engineering, **120**, pp. 257-262, 1998.
14. Gerlinger, P., Möbus, H., Brüggemann, D.: An Implicit Multigrid Method for Turbulent Combustion, Journal of Computational Physics, **167**, pp. 247-276, 2001.
15. Gerlinger, P., Brüggemann, D.: Numerical Investigation of Hydrogen Strut Injections into Supersonic Air Flows, Journal of Propulsion and Power, **16**, pp. 22-28, 2000.
16. Gerlinger, P.: Investigations of an Assumed PDF Approach for Finite-Rate-Chemistry, Combustion Science and Technology, **175**, pp. 841-872, 2003.
17. Stoll, P., Gerlinger, P., Brüggemann, D.: Domain Decomposition for an Impicit LU-SGS Scheme Using Overlapping Grids, AIAA-paper 97-1896, 1997.
18. Stoll, P., Gerlinger, P., Brüggemann, D.: Implicit Preconditioning Method for Turbulent Reacting Flows, Proceedings of the 4th ECCOMAS Conference, **1**, pp. 205-212, John Wiley & Sons, 1998.
19. Gerlinger, P., Stoll, P., Kindler, M., Schneider, F. and Aigner, M.: Numerical Investigation of Mixing and Combustion Enhancement in Supersonic Combustors by Strut Induced Ttreamwise Vorticity, Aerospace Science and Technology, **12**, pp. 159-168, 2008.
20. Kindler, M., Gerlinger, P. and Aigner, M.: Numerical Investigations of Mixing Enhancement by Lobed Strut Injectors in Turbulent Reactive Supersonic Flows, ISABE-2007-1314, 2007.
21. Gardner, A.: HyShot Scramjet Testing in the HEG,FB 2007-14,University of Queensland, Australia, 2007.
22. Kindler, M., Gerlinger, P. and Aigner, M.: Assumed PDF Modeling of Turbulence Chemistry Interaction in Scramjet Combustors, High Performance Computing in Science and Engineering '07, pp. 203-213, 2008.

Computational Fluid Dynamics

Prof. Dr.-Ing. Siegfried Wagner

Institut für Aerodynamik und Gasdynamik, Universität Stuttgart, Pfaffenwaldring 21, 70550 Stuttgart

The impact of computer simulation in engineering has been significant and continues to grow. Simulation allows the development of highly optimised designs, the investigation of hazards too dangerous to test and reduced development costs. In parallel new scientific investigations are developing understanding in areas such as turbulence and flow control necessary for future engineering concepts.

However, since the flow phenomena are usually very complex highly sophisticated numerical procedures and algorithms as well as high performance computers (HPCs) had to be developed. Because the flow processes in daily life are turbulent and may even include chemical reactions, phase changes, heat transfer and interference with structural movement the high performance computers that are available at present are still by far too small in order to simulate for instance the turbulent flow around a complete aircraft. Despite this fact the following chapter on results in the field of computational fluid dynamics (CFD) that were obtained at HLR Stuttgart and SSC Karlsruhe will demonstrate the usefulness of numerical simulation and the progress in gaining more insight into complex flow phenomena when the computer capacity is increased and the corresponding numerical methods are improved.

The highly sophisticated numerical methods necessary for simulations on HPCs include Direct Numerical Simulation (DNS), Large Eddy Simulation (LES), numerical solutions of the Reynolds Averaged Navier-Stokes (RANS) equations, Detached Eddy Simulation (DES) that is a combination of LES and RANS, Lattice Boltzmann Method (LBM), combinations of DES and LES and finally Finte Element Methods (FEM) to investigate both flow and structural problems.

The high performance computers that were applied in the present studies were mostly the vector computer NEC SX-8 and the CRAY Opteron Cluster of HLRS as well as the massively parallel computers HP XC-4000 and HP-XC-2 of SSC Karlsruhe. However, some of the authors also compared the performance of these computers with CRAY XT and IBM BLUE GENE (e.g. paper of Zeiser et al.) and with JUMP (e.g. paper of Buijssen et al.).

It is not surprising that most of the papers use highly sophisticated numerical methods that undoubtedly require HPCs. More specifically, the distribution of the papers shows that six are devoted to DNS, five use LES, two apply the Lattice Boltzmann Method, two papers investigate flow-structure interactions, one uses RANS and optimization methods, one applies Eddy Dissipation/Finte Rate Combustion Model (EDM/FRC) and finally one paper investigates the performance of FEM.

Memory requirements did not seem to be a problem so far. At NEC SX-8 memory used ranged between 15 and 1086 GB whereas at HP XC-4000 only one information was available, namely 1.04 GB. The maximum number of processors used was 128 on the NEC SX-8 and 1024 on the HP XC-2. The maximum achieved performance approached 166.4 GFLOP/s on the NEC SX-8. Despite this big performance wall clock time was 6 hours in this case whereas in other cases wall clock time approached 85.7 hours. For the HP XC-4000 no performance numbers were presented. The jobs were big. For instance one run on the HP XC-2 with 1024 processors required 80,000 CPU hours. Turn around times went up to 60 days. These numbers show that turn around times would have been unacceptable without the access to HPC. Thus, the performance seems to be the bottle neck and should be increased by the next generation of high performance computers.

At present investigations in the field of DNS and LES have to be restricted to small Reynolds numbers and simple geometries i.e. Reynolds numbers far below those of technical applications. The present situation will be demonstrated by the following example that stems from flow-structure interactions in helicopter aeromechanics.

The CPU requirements for future simulations of helicopter aeromechanics are not yet known exactly but are bound to be substantial. Since little experience is available for full-scale/full-helicopter simulations researchers are still in the process of discovering the required grid resolution. An accurate simulation of an isolated rotor on today's supercomputer systems (3 GHz processor) would roughly require 5 million grid points per blade, 2000 time steps for each turn of the rotor, around 5 rotor revolutions until a trimmed state is reached and around 30000 CPU hours. This may well increase as the fidelity of the data required by the researchers and engineers increases. Accurate drag and sectional pitching-moment predictions require viscous flow simulations and this can easily triple the above estimate. Adding a fuselage and a tail rotor will increase this requirement to around 150000 CPU hours. This is due to the bluff body aerodynamics of the fuselage with shed vortices and massive flow separation but also due to the complex interaction between the fuselage and the rotors. Several calculations that are an order of magnitude larger would allow an assessment of issues such as grid dependency and establish confidence levels for smaller simulations on local facilities. The incorporation of flow control devices will potentially boost the computational requirements significantly. Finally, the helicopter system requires a delicate balance between the aeromechanics of the aircraft, the flight control system and the pilot. Simulation of manoeuvres, in contrast to design conditions, becomes an inherently multi-disciplinary problem with contributing modules requiring different treatments when it comes to parallel computing. The aerodynamic analysis is to have the lion's share but also a large effort will be required to simulate the effect of the pilot's action and of the control systems on the flight mechanics. This creates a challenging project where different solvers and modelling techniques must come together in a single parallel environment and require a PFlop/s sustained performance.

The presentation shows that a vector computer like the NEC SX-8 would be advantageous for many applications. The next generation, the NEC SX-9, will provide both several times higher performance and new cache-like memory concepts. These new concepts would probably require much more effort to optimally adapt existing vectorized codes to this new design than by the movement from NEC SX-4 to NEC SX-8. On the other hand increasing the performance of parallel systems by increasing the number of CPUs up to 100,000 and more will also be a big task in software development. The maximum number of CPUs in the papers presented was only 1024.

Direct Numerical Simulation of Film Cooling in Hypersonic Boundary-Layer Flow

J. Linn and M.J. Kloker

Institut für Aerodynamik und Gasdynamik, Universität Stuttgart, Pfaffenwaldring 21, D-70550 Stuttgart, Germany, email: lastname@iag.uni-stuttgart.de

Summary. Effusion cooling by discrete slits and holes in various laminar zero-pressure gradient super- and hypersonic boundary layers is investigated using direct numerical simulation. For an adiabatic Mach-6 boundary layer it was found that slits are better than holes due to the lower blowing velocity. Slit blowing causes a destabilisation of 2nd-mode disturbances, and a complete stabilisation of 1st modes despite the generated maxima of the spanwise vorticity inside the boundary layer. Hole blowing gives rise to counter-rotating streamwise vortices, with a noticeable laminar-flow destabilisation only for large spanwise hole spacings. Computational aspects such as the performance of the numerical code on the supercomputer NEC SX-8 and optimization of the code are also discussed.

1 Introduction

For aerospace or hypersonic cruise vehicles the state of the boundary layer is of major importance because for turbulent boundary layers, the thermal loads and skin friction are higher than in laminar boundary layers. Therefore, knowledge of cooling features and laminar-turbulent transition is necessary for the design of the thermal protection system (TPS). Different strategies are used to reduce the thermal loads of hypervelocity vehicles, e.g. radiation, ablation, transpiration or effusion cooling. Direct numerical simulations (DNS) are carried out to investigate the effect of effusion cooling by blowing through spanwise slits and discrete holes in a laminar flat-plate boundary layer at Mach 6. The numerical method and boundary conditions are described in section 2. The performance of the numerical code NS3D is shown in section 4. Furthermore we discuss the influence on numerical performance by replacing the spanwise Fourier spectral approach with compact finite differences. The results are summarized in section 5.

2 Numerical Method

2.1 Governing Equations

The numerical method is based on the complete 3-d unsteady compressible Navier-Stokes equations, the continuity equation and the energy equation. These equations can be written in dimensionless form as:

$$\frac{\partial \rho}{\partial t} + \nabla \cdot (\rho \vec{u}) = 0 \tag{1}$$

$$\frac{\partial (\rho t)}{\partial t} + \nabla \cdot (\rho \vec{u}\,\vec{u}) + \nabla p = \frac{1}{Re} \nabla \cdot (\vec{\sigma}\,\vec{u}) \tag{2}$$

$$\frac{\partial (\rho e)}{\partial t} + \nabla \cdot (p + \rho e) \vec{u}) = \frac{1}{(\kappa - 1)\, Re\, Pr\, Ma^2} \nabla \cdot (\vartheta \cdot \nabla T) + \frac{1}{Re} \nabla \cdot (\vec{\sigma}\,\vec{u}) \quad , \tag{3}$$

where

$$\vec{\sigma} = \mu \cdot \left((\nabla \vec{u} + \nabla \vec{u}^T) - \frac{2}{3} (\nabla \cdot \vec{u}) I \right) \tag{4}$$

is the viscous stress tensor and

$$e = c_v \cdot T + \frac{1}{2}(u^2 + v^2 + w^2) \tag{5}$$

is the internal energy per mass unit. The air is considered as a non-reacting calorically perfect gas [4, 15]

$$p = \frac{1}{\kappa\, Ma^2} \cdot \rho T \tag{6}$$

with a constant Prandtl number ($Pr = 0.71$) and specific heat ratio of $\kappa = c_p / c_v = 1.4$. The viscosity is calculated using Sutherland's law [16]. All length scales are dimensionless with respect to a reference length

$$L^\star = \frac{\nu^\star_\infty \cdot Re}{u^\star_\infty} \quad . \tag{7}$$

Reference values for velocity, density temperature, viscosity and conductivity are their freestream values at the inflow (indicated by subscript ∞). The pressure is normalised by $\rho^\star_\infty {u^\star_\infty}^2$, where the superscript $\star$ denotes dimensional quantities. With these definitions, the global and running-length Reynolds numbers are respectively defined as

$$Re = \frac{u^\star_\infty \cdot L^\star}{\nu^\star_\infty} = 10^5 \tag{8}$$

and

$$Re_x = \frac{u^\star_\infty \cdot x^\star}{\nu^\star_\infty} = x \cdot 10^5 \quad . \tag{9}$$

2.2 Spatial and Time Discretisation

The Navier-Stokes equations are solved in a rectangular integration domain (Fig. 1) on the flat plate, well below the shock wave induced by the leading edge. In streamwise (x-) and wall-normal (y-) direction, the discretisation is realized by splitted compact finite differences of 6th order [9]. In the spanwise (z-) direction, the flow is assumed to be periodic, thus a Fourier spectral representation is employed. The time integration is done with a classical 4th-order Runge-Kutta method. A detailed description of the discretisation and algorithm is reported in [1].

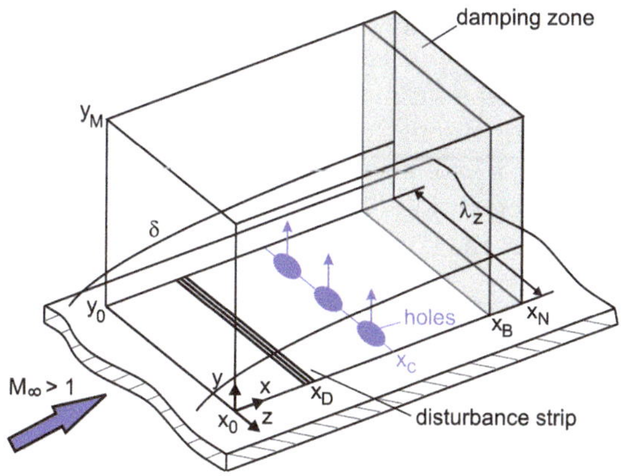

Fig. 1. Integration domain

2.3 Initial and Boundary Conditions

The numerical simulation is performed in two steps. First, the steady base flow is calculated by solving the Navier-Stokes equations using a pseudo time stepping for integrating the time-dependent equations to a steady state. For real unsteady simulation this base flow is used as initial state ($t = 0$). Disturbance waves are introduced for $t > 0$ by localized periodic blowing and suction in a disturbance strip, and the spatial downstream development of the disturbance waves is calculated from the full equations. We use a disturbance flow formulation, meaning that all flow quantities are splitted in their base-flow and disturbance part ($\phi = \phi_{BF} + \phi'$), to ease the formulation of specific boundary conditions. The full equations are used and a non-linear generated time mean is contained in the disturbance flow ($\langle\phi'\rangle \neq 0$). At the inflow boundary ($x = x_0$), profiles from boundary-layer theory are fixed for all variables, and the disturbances are zero. For the base-flow boundary condition at the outflow ($x = x_N$), all equations are solved neglecting the second x-derivative terms and for the disturbance flow, all disturbances are damped to zero in a damping zone shortly upstream the outflow boundary. At the

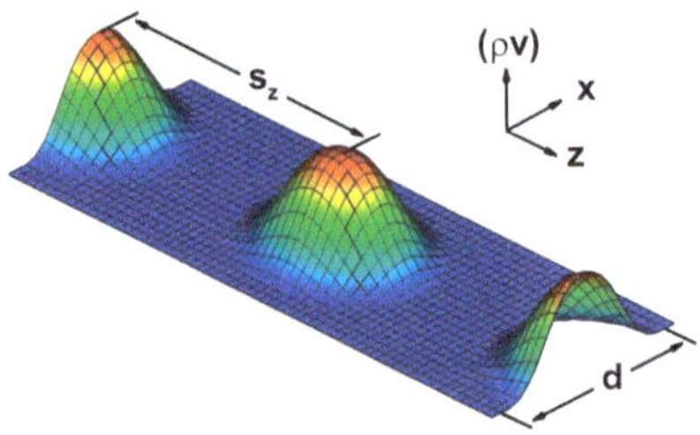

Fig. 2. (ρv)-distruibution at the wall for one row of holes

freestream boundary ($y = y_M$) for the base flow, the gradient of the flow variables is set to zero along spatial characteristics [4]. An exponential decay condition is used for the disturbance flow [15]. At the wall, all velocity components are zero, except within the slits, holes, and disturbance strip. The steady blowing of cold air through holes at the wall with a radius r_c (Fig. 2) is modelled by prescribing a wall-normal mass-flux distribution where $(\rho v)_{c,max}$ is the maximum blowing ratio. The wall temperature distribution over the blowing is prescribed for the steady base and disturbance flow by

$$T_c = T_w \cdot (1 - c(r)) + T_{c,core} \cdot c(r) \quad , \tag{10}$$

where $T_{c,core}$ is the core temperature of the cold air and T_w is the local wall temperature at the edge of the hole. The distribution function $c(r)$ is a polynomial of 5^{th}order, which has already been used in [14] for suction and blowing to generate disturbances at the wall. Both the gradient and curvature are zero at $r = 0$ and $r = r_c$:

$$c(r) = 1 - 6 \cdot \left(\frac{r}{r_c}\right)^5 + 15 \cdot \left(\frac{r}{r_c}\right)^4 - 10 \cdot \left(\frac{r}{r_c}\right)^3 \tag{11}$$

$$r = \sqrt{(x - x_c)^2 + (z - z_c)^2} \quad , \quad 0 \leq r \leq r_c \quad . \tag{12}$$

x_c and z_c stand for the center coordinates of the hole. Outside the hole, (ρv) is zero at the wall. T_w is outside the blowing region either the local adiabatic wall temperature defined by

$$\left.\frac{\partial(T_{BF} + T')}{\partial y}\right|_w = 0 \tag{13}$$

or has a constant value (isothermal wall; $T_{BF} = const.$; $T' = 0$). The pressure gradient in wall-normal direction at the wall, holes, slits and disturbance strip is zero. For steady blowing through a spanwise slit of width $b_c = 2 \cdot r_c$, the distribution function $c(r)$ in Eq. 11 is independent from the z-coordinate, i.e. $z = z_c$. Disturbance waves are introduced within a disturbance strip by time-periodic simultaneous blowing and suction, which is modelled by a distribution of the wall-normal mass flux $(\rho v)'$ over the strip [4].

Grid-refinement studies varying Δx, Δy, Δz and Δt for exemplary cases were performed and convergence was always found. A comparison by experiments with effusion cooling shows good agreement, see [10].

3 Results

3.1 Comparison with Experiments at Mach 2.67

In this section, we compare our simulation results with the measurements by Heufer and Olivier [5, 6] at the Shock Wave Laboratory Aachen (SWL). They investigated an isothermal laminar boundary layer on a wedge with a deflection angle of 30° and a given post-shock Mach number 2.67, with $M_\infty = 7$ and $T^\star_\infty = 1368K$. Cold air is blown through one spanwise slit. The given post-shock freestream temperature is $T^\star_\infty = 564K$ ($\approx 0.45 \cdot T^\star_{rec}$), the pressure bar $p^\star_\infty = 0.1489 bar$ ($\rightarrow L^\star = 24.57mm$, $Re^\star_{unit} = 4.07 \cdot 10^6\ 1/m$), and the wall temperature is $T^\star_\infty - 293K = const.$ ($\approx \frac{1}{4} T^\star_{rec}$). This wall temperature means that the wall itself is already strongly cooled. Investigations based on the Linear Stability Theory (LST) have shown (see, e.g. [12]), that wall cooling stabilises 1st mode (vorticity) disturbances and destabilises the 2nd-mode (acoustic) disturbances that do not exist at Mach numbers lower than approximately 3.5. In addition, the basic boundary layer investigated is subcritical ($R_{x,crit} \gg R_x(x_N)$) due to the strong wall cooling. The core temperature of the effusion air is $T^\star_{c,core} = 293K$ ($= T^\star_w$) and the slit width is $b^\star_c = 0.5mm$ corresponding to $0.57 \cdot \delta^\star_c$, where δ_c is the boundary-layer thickness without blowing at the slit position. The cooling effectiveness is defined by $\circ q$

$$\eta_c = 1 - \frac{\dot{q}}{\dot{q}_{ref}} \tag{14}$$

where $\dot{q}_c = \lambda \cdot \partial T/\partial y|_w$ is the heat flux into the wall with effusion cooling and $\dot{q}_{ref}$ without effusion cooling. The cooling effectiveness behind the slit is shown over the blowing rate in Fig. 3. It increases linearly with the blowing rate in the simulation and is slightly lower than in the experiment. Note that no experimental data of the boundary-layer evolution are available and thus the local thickness parameters and Reynolds numbers may differ. Lower cooling effectiveness was also found by tentative numerical simulations at SWL. A longitudinal cut of the temperature field with streamlines is shown in Fig. 4 for the blowing rate $(\rho v)^\star_c/(\rho u)^\star_\infty = 0.065$. In front of the slit is a reversed-flow region with a clockwise rotating vortex with its center marked by the dot. For both blowing rates, no instability regions were found using spatial LST despite a separation region exists in front of the slit. The basic cooling by the cool wall is so strong that it stabilizes even the blowing. Here effusion cooling is applied in a case where the flow is already strongly cooled by a cool wall, thus this case is unrealistic. A simple transfer of the results to cases with significantly different wall temperature gradients is not possible as DNS not presented here have shown.

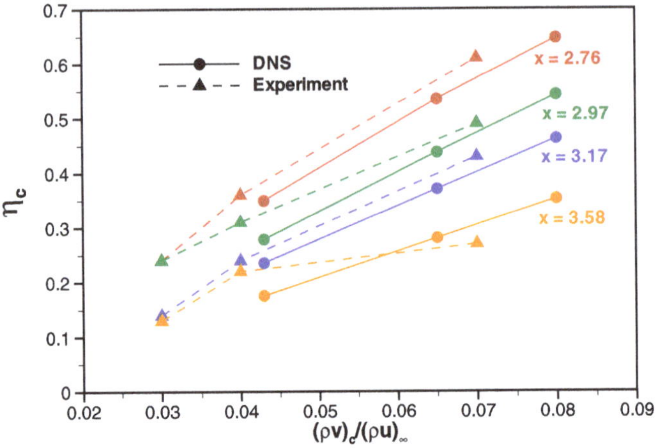

Fig. 3. Cooling effectiveness η_c from simulation and experiment as function of the blowing rate at three downstream positions for an effusion-cooled boundary layer at Mach 2.67 (line with dots – simulation; lines with triangle - experiment)

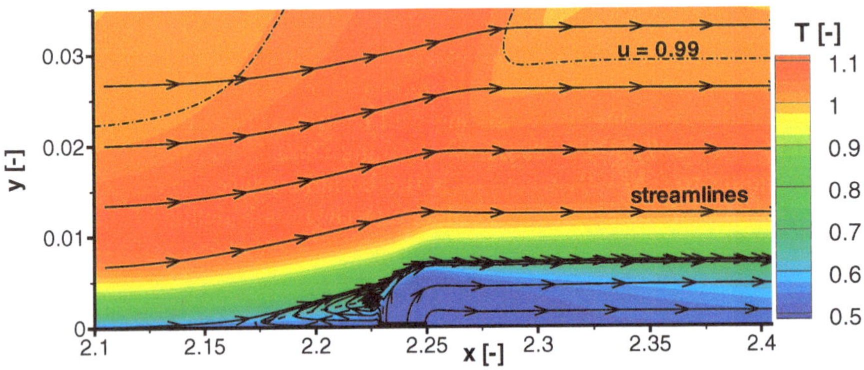

Fig. 4. Visualisation of the temperature field with streamlines in a longitudinal cut for the effusion-cooled boundary layer at Mach 2.67 ($(\rho v)_{c,max} = 0.065$). Isolines of the u-velocity for $u = 0$ (dashed line) and $u = 0.99$ (dashed dotted line). $\Delta x = 0.25\ 10^{-2}$ and $\Delta y = 0.6\ 10^{-3}$

3.2 Comparison of Effusion-Cooling Configurations at Mach 6

In this section we investigate an adiabatic boundary layer at an edge Mach number 6 in which cold air is blown through spanwise slits and rows of holes. The freestream temperature is $T^\star_\infty = 89K$ ($\approx \frac{1}{7}T^\star_{rec}$) and the pressure is $p^\star_\infty = 0.0038bar$ ($\rightarrow L^\star = 36.28mm$, $Re^\star_{unit} = 2.8 \cdot 10^6 1/m$), matching the flow parameters of experiments in the hypersonic wind tunnel H2K of DLR-Köln [2]. Table 1 summarizes the parameters. Two successive slits were used in case $\mathcal{A}$, piecewise homogeneous blowing (one wide slit) in case $\mathcal{B}$, and holes in cases $\mathcal{C}$, $\mathcal{D}$. The integrally injected mass flow and the cooling gas temperature

Table 1. Parameters of the slits and holes configurations for cases at $M_e = 6$

case	slits	row of holes	$(\rho v)_{c,max}$	hole diameter or slit width d	streamwise spacing s_x	spanwise spacing s_z	rows z-offset
$\mathcal{A}$	2	-	0.0284	$0.058 \approx 0.6\ \delta$	$0.1378 \approx 1.4\ \delta$	-	-
$\mathcal{B}$	1	-	0.0284	0.087	0.1378	-	-
$\mathcal{C}$	-	2	0.15	0.058	0.1378	0.1378	-
$\mathcal{D}$	-	2	0.15	0.058	0.1378	0.1378	$s_z/2$
$x_0 = 0.225$, $x_N = 7.33$, $y_M = 0.54 \approx 4\ \delta$ at $x = x_N$, blowing starts at $x_c = 2.205$ ($x_c^\star = 80mm$)							

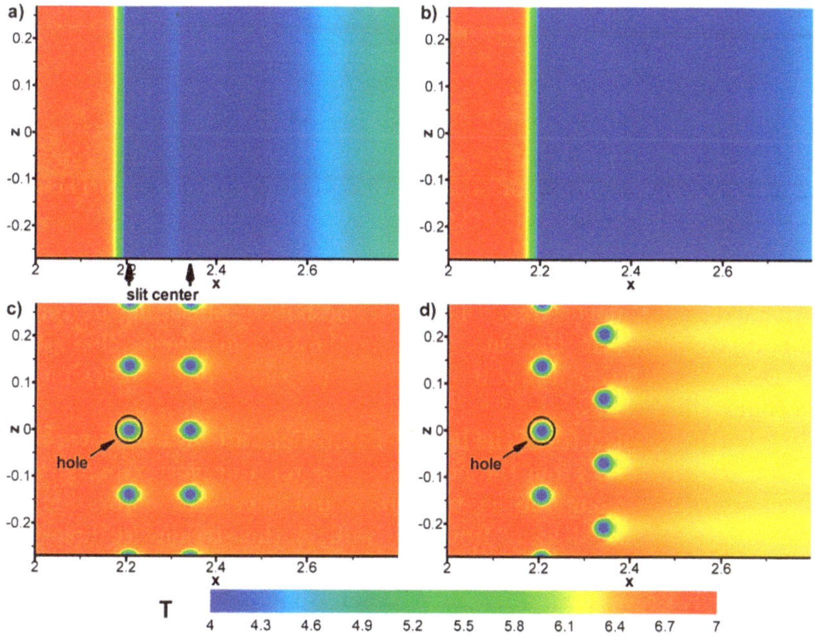

Fig. 5. Wall temperature for steady blowing into an adiabatic flat-plate boundary layer at Mach 6 through two spanwise slits ((a), case $\mathcal{A}$), piecewise homogeneous blowing (one wide slit, (b), case $\mathcal{B}$), two aligned rows of holes ((c), case $\mathcal{C}$), and two $s_z/2$-staggered rows of holes ((d), case $\mathcal{D}$)

$T^\star_{c,core} = 293K$ ($\approx \frac{1}{2}T^\star_\infty$) are in all cases identical. In case $\mathcal{C}$, the two rows of holes are aligned in contrast to case $\mathcal{D}$ where the rows are $s_z/2$ staggered. The resulting wall temperature is shown in Fig. 5. Cases $\mathcal{A}$ and $\mathcal{B}$ show a significant lower T_w than the other two cases. The "homogeneous" blowing model with its low wall-normal velocity (case $\mathcal{B}$) has the lowest T_w. (We remark that, due to the used model, T_w would keep low even if the blowing vanished.) In the "aligned" case $\mathcal{C}$, T_w is only slightly reduced and strongly varies in the z-direction. In the "staggered" case $\mathcal{D}$, T_w is lower than in case $\mathcal{C}$, and does not vary as strong in the z-direction. The aligned rows blow more cold gas from the wall into the boundary layer and show stronger $\partial u/\partial z$-gradients

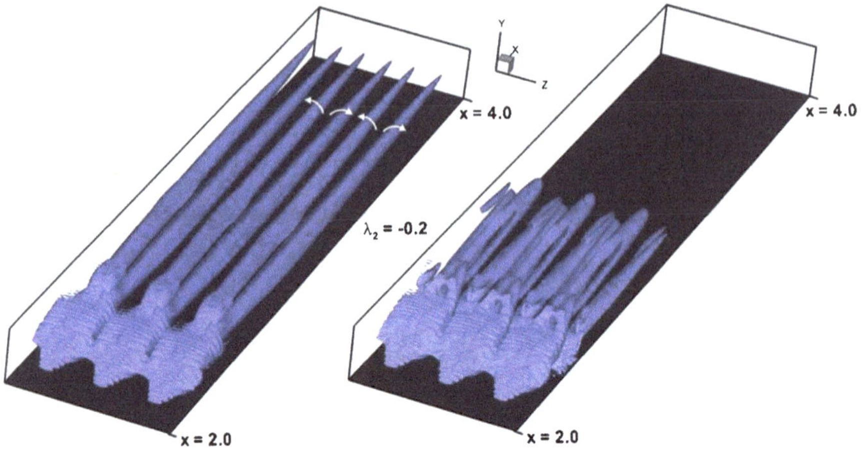

Fig. 6. Visualisation of vortical structures via λ_2-isosurface ($\lambda_2 = -0.2$) for aligned rows of holes ($\mathcal{A}$ - left) and for staggered rows of holes ($\mathcal{B}$ - right). The arrows indicate the rotation sense

than the staggered rows. The reason why the slits are more efficient is that the blowing surface is much larger than with the holes, translating into a lower wall-normal velocity in the slits. Thereby the cold gas keeps closer to the wall. Decreasing the spanwise and streamwise spacing of the holes increases the cooling effectiveness.

The vortical structures of the hole configurations are visualised via the λ_2-criterion [7] in Fig. 6. From the holes, counter-rotating vortex pairs (CVPs) emerge which are along the jet trajectory and have such a rotation sense that gas is transported away from the wall in the streamwise hole center line. Furthermore exists a toroidal neck vortex at each hole edge. It has a counter-clockwise rotation sense in the center-line plane upstream the hole, in contrast to the considered slit case. A horseshoe vortex is not observed in the simulations due to the low blowing ratio. In studies of jets in crossflow (JICF), where typically a horseshoe vortex (with a rotation sense opposite to the CVP and neck vortex) is found, $(\rho v)_{c,max} = \mathcal{O}(1)$ and $d > \delta$. In the aligned case (Fig. 6a), the second row enhances the CVPs from the first row and the vortices lay wall-parallel. In contrast, the CVPs from the second row are pushed downwards, keeping the cold gas at the wall. Moreover, the CVPs of both cases decay downstream.

For a Mach-6 boundary layer, the strongest amplified disturbance mode is the 2nd mode as a 2-d wave (spanwise wave number / $\gamma = 0$) after Linear Stability Theory (LST) [11]. We found some other almost neutral eigenvalues in the region of the slits, however the eigenvalue of the 2nd-mode disturbance is the by far most amplified in this region. Figure 7 shows the N-factors for case $\mathcal{A}$, where

$$N = -\int_{x_0}^{x} \alpha_i(x)\, dx = ln\left(\frac{A}{A_0}\right) \quad . \tag{15}$$

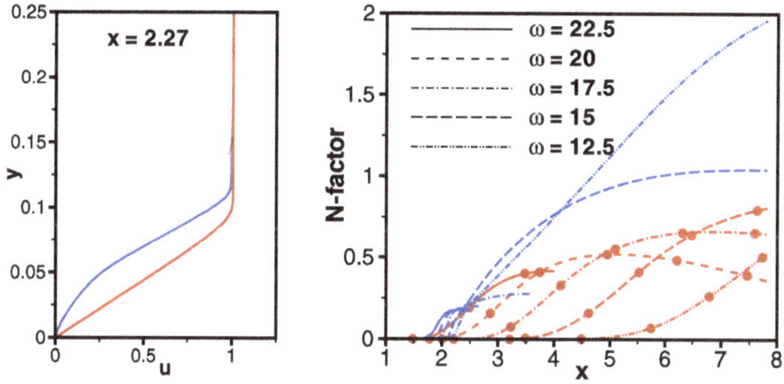

Fig. 7. u-velocity profile (left) and N-factors (LST) for 2-d disturbances from case $\mathcal{A}$ with blowing (red) and without blowing (blue) for various frequencies, $\omega = 15$ is $f^\star = 74.67kHz$

For 2nd-mode 2-d disturbances the N-factor of the angular frequency $\omega = 12.5$ $(= 2\pi \frac{L^\star}{u_\infty^\star} f^\star)$is approximately four times higher at the end of the considered streamwise domain than without blowing. For 1st-mode 3-d disturbances a stabilisation by effusion cooling can be observed. This is non-trivial since not only the wall but also the boundary layer itself is cooled, and, at the same time, the u-velocity profile has an inflection point (Fig. 7). Primary LST uses the assumption that the spanwise base-flow gradients are zero. Thus it can not be used to predict the instability of cases $\mathcal{C}$, $\mathcal{D}$. Recall that enhanced laminar instability can compromise the cooling effect.

3.3 Instability Investigations of an Effusion-Cooled Adiabatic Mach 6 Boundary Layer

Here we investigate the same adiabatic Mach-6 boundary layer as in the section before, but now we blow cold air through four spanwise rows of holes and add unsteady 2-d background disturbances prescribed in front of the holes at the wall. Four rows of aligned holes are used because of the stronger and persistent flow deformation of the boundary layer. Two cases are investigated: case $\mathcal{E}$ with a small spanwise spacing ($s_{z,a} = 0.1378 = 3d$), and case $\mathcal{F}$ with four times larger spacing ($s_{z,b} = 4 \cdot s_{z,a}$), see Fig. 8. The hole diameter d, the cooling gas temperature $T_{c,core}$ and the blowing ratio $(\rho v)_{c,max}$ are in both cases equal, corresponding to case $\mathcal{B}$ of section 3.2, table 1. Thus the mass-flow through the holes in case $\mathcal{F}$ is only one quarter of case $\mathcal{E}$ per spanwise unit. The hole region lies within $2.205\ 10^5 \leq Re_x \leq 2.756\ 10^5$. A crosscut of the u-velocity field downstream the rows is shown in Fig. 9. In case $\mathcal{F}$, right, the boundary layer is deformed stronger than in case $\mathcal{E}$, left, both showing mushroom-like structures by the action of the CVPs.

Upstream of the holes, a packet of unsteady 2-d disturbance waves is generated by timewise periodic suction and blowing within a disturbance strip

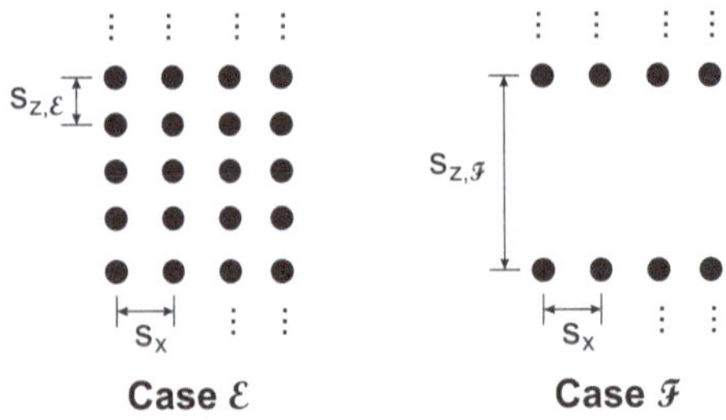

Fig. 8. Sketch of the hole configuration of case $\mathcal{E}$ and $\mathcal{F}$

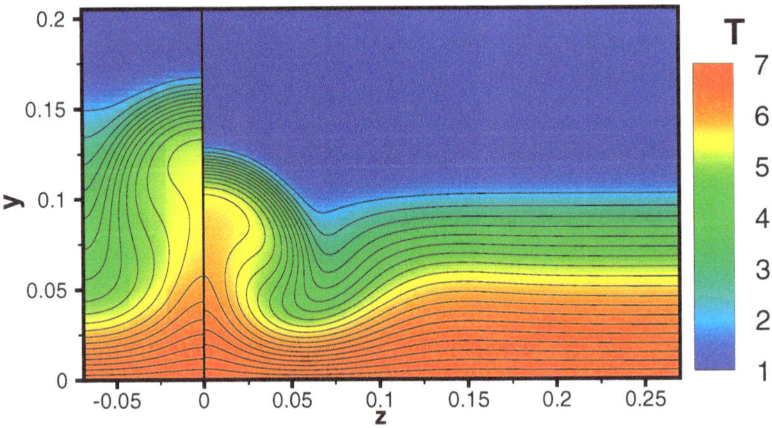

Fig. 9. Temperature field and u-velocity isolines in a crosscut at $Re_x = 3.128\ 10^5$ downstream of the holes for case $\mathcal{E}$ (left) and case $\mathcal{F}$ (right). Half the respective spanwise domain width is shown

at the wall ($Re_x = 1.78\ 10^5$) for a bunch of frequencies to check for laminar instability. Note that due to the large steady vortices 3-d unsteady disturbances are nonlinearly generated with the 2-d packet, and that due to the physically fixed streamwise extent of the strip, matched approximately to the streamwise wavelength of the ($\omega = 10$)-disturbance, the receptivity is lower for other frequencies. Figure 10 shows the downstream development of the u'-disturbance amplitudes (u'_h – maximum over y and z) from a timewise Fourier analysis for both cases. The curve (0,0) represents the timewise and spanwise mean deformation of the 2-d boundary layer and the other curves represent the maximum over y and z of the u-disturbances (2-d and 3-d together) for a specific angular frequency $\omega = 2\pi \cdot f^\star \cdot L^\star / u^\star_\infty$ ($\omega = 10$ is $f^\star = 49.78kHz$). In case $\mathcal{E}$ the mean flow deformation (0,0) is stronger than in case $\mathcal{F}$ due to the higher injected mass flow, but all frequencies are damped or neutral for $Re_x > 5 \cdot 10^5$, except frequencies near $\omega = 10$, being also amplified in the pure 2-d base flow as a 2nd mode. Here, however, it is a disturbance localized in the y-z plane as a localized secondary mode (Fig. 11). The maximum of the

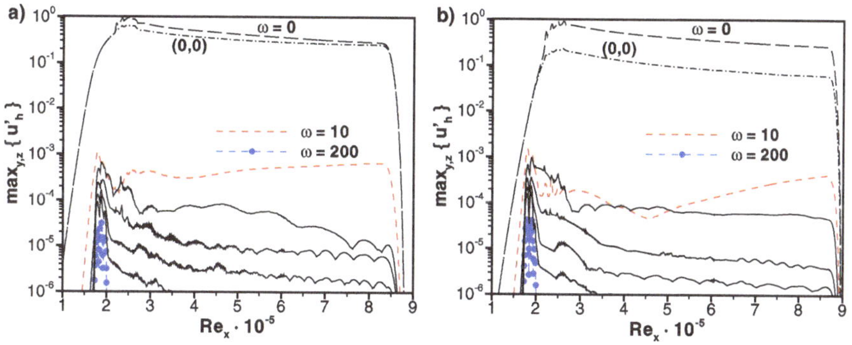

Fig. 10. Downstream t-modal amplitude development (u'_h – maximum over y and z) for cases $\mathcal{E}$ (a) and $\mathcal{F}$ (b), $\omega = 10$ is $f^\star = \omega\ u^\star_\infty/(2\pi\ L^\star) = 49.78kHz$

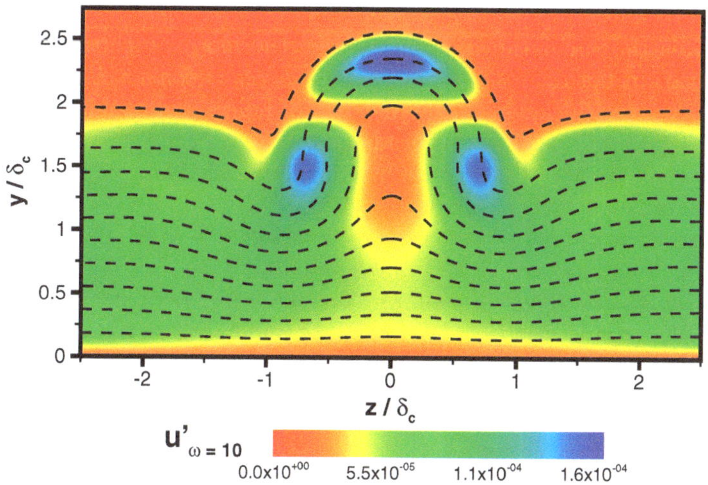

Fig. 11. Amplitude distribution (coloured) and mean flow (isolines) at $Re_x = 6.5 \cdot 10^5$ for case $\mathcal{F}$; $0.1 \leq u \leq 1$, $\Delta u = 0.1$

disturbance amplitude is related to the strongest $\frac{\partial u}{\partial y}$ and $\frac{\partial u}{\partial z}$ gradient in the crosscut. Low frequencies are neutral or damped like in case $\mathcal{E}$ (we checked down to $\omega = 1$). Thus the steady 3-d flow deformation by blowing does not invoke sudden transition in the young boundary layer in the front part of the plate. A small spanwise hole spacing is preferable due to larger cooling effectiveness and lower amplification of unsteady disturbances. "Isolated" holes give rise to enhanced instability.

From investigations of incompressible boundary layers it is known that blowing often strongly destabilizes the boundary layer. Thus a Mach-0.6 case ($\mathcal{G}$) is simulated and compared with case $\mathcal{F}$ to clarify the effects. We chose an equal $R_x = 469.57 = \sqrt{Re_x}$ at the first row of holes in both cases. Table 2

Table 2. Parameters of the Mach-6 ($\mathcal{F}$) and Mach-0.6 case ($\mathcal{G}$)

	Case $\mathcal{F}$	Low Mach case $\mathcal{G}$
M_∞	6.0	0.6
$T^\star_\infty$	89 K	290 K
$u^\star_\infty$	1135.8 m/s	204.8 m/s
$p^\star_\infty$	0.0038 bar	0.203 bar
$L^\star$	36.28 mm	36.28 mm
Re_{unit}	$2.8 \cdot 10^5$ $1/m$	$2.8 \cdot 10^5$ $1/m$
Wall condition	adiabatic	adiabatic
$T^\star_{rec}$	629 K	307.6 K
$T^\star_{c,core}$	293 K	290 K
$v^\star_c/u^\star_\infty$	0.45	0.15
d_c/λ_z	0.105	0.105
$\delta^\star_c/L^\star$	0.095	0.023

summarizes the simulation parameters. Furthermore we chose equal length scales ratios in both cases ($s_x/\delta_c = const.$; $s_z/\delta_c = const.$; $d/\delta_c = const.$; δ_c - boundary-layer thickness at the first row of holes) and equal blowing ratios $(\rho v)^\star_c/(\rho v)^\star_\infty = 0.15$. The wall gradient $\partial u/\partial y|_w \cdot \delta_c$ in the Mach-0.6 case is larger than in the Mach-6 case. With increasing Mach number, the skin friction and thus $\partial u/\partial y$ at the wall decreases.

In case $\mathcal{G}$, we did not obtain a steady solution, due to self-excited unsteadiness of the boundary layer at the fourth hole row. Therefore in Fig. 12 the time mean of $\partial u/\partial y$ and $\partial u/\partial z$ are shown. The deformation of the boundary layer in case $\mathcal{G}$ is stronger than in the other case and closer to the wall due to the different blowing velocity. It is in the Mach-6 case $\mathcal{F}$ three times larger than in the Mach-0.6 case $\mathcal{G}$, due to the different effusion densities ($\rho_{Mach-0.6} \approx 1$; $\rho_{Mach-6} \approx 1/3$).

In Fig. 12 the gradients $\partial u/\partial y$ and $\partial u/\partial z$ are shown in the same crosscut as in the previous figure. The gradients in the low Mach number case ($\mathcal{G}$) are much larger than in case $\mathcal{F}$ due to the blowing. Furthermore the maximum of the gradients lay closer to the wall in case $\mathcal{G}$. We note that, for a free shear layer, the dimensional disturbance growth rate is proportional to $\partial u/\partial y \cdot (u^\star_\infty/\delta^\star_c)$, the expression in brackets being almost equal for the two cases.

The vortical structures (λ_2-criterion, [7]) are shown in Fig. 13. In the Mach-0.6 case (snapshot) downstream travelling Λ-vortices are directly generated downstream at the fourth hole row. This is in contrast to the Mach-6 case, where only a pair of steady longitudinal vortices is generated at each hole that decays downstream.

The spanwise vorticity ω_z is shown in longitudinal cuts through the holes ($z = 0$) for both cases in Fig. 14. In the Mach-0.6 case ($\mathcal{G}$) four time instances are plotted in Fig. 14a-d in which the shear-layer formation and dynamics is seen. In case $\mathcal{G}$ the instability of the shear layer causes its breakup, and structures known from the disturbance peak plane in the K-Breakdown scenario [9]

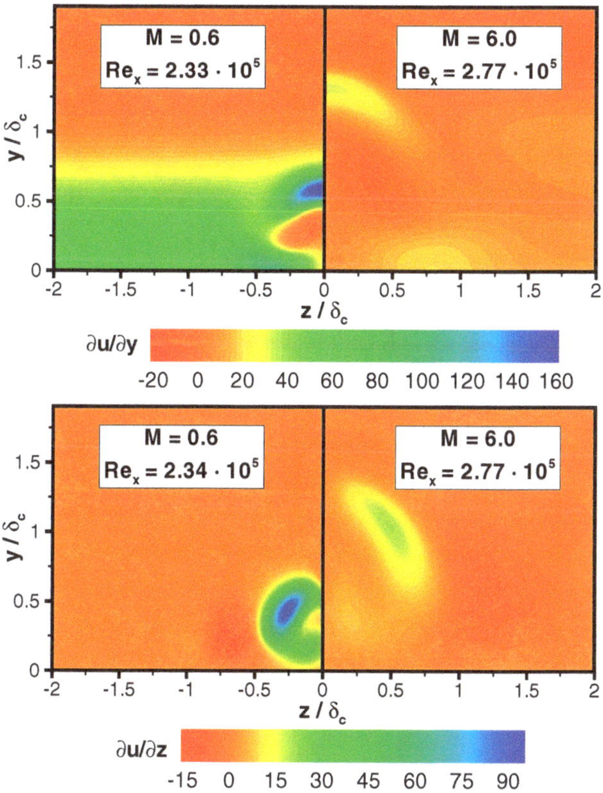

Fig. 12. Contour plot of $\partial u/\partial y$ (upper figure) and $\partial u/\partial z$ (lower figure) in a crosscut at three boundary-layer thicknesses downstream the last row of holes for case $\mathcal{G}$ (left) and $\mathcal{F}$ (right)

are formed. The dominant wavelength is however by a factor 5 smaller than caused by the most unstable 2-d Tollmien-Schlichting wave.

4 Computational Aspects

All DNS simulations were performed on the NEC SX-8 supercomputer at the High Performance Computing Center Stuttgart, using 3 nodes for case $\mathcal{F}$ in section 3.3, corresponding to 24 processors. On each node, one MPI process was executed, each with shared-memory parallelization having eight tasks. The shared-memory parallelization is done in spanwise ($z-$) direction. For a detailed description of the algorithm and the parallelization see [1]. The computation of 60000 time steps required 42 hours wall-clock time. This leads to a total CPU-time of nearly 1008 hours and a specific computational time of $1.69\mu s$ per gridpoint and time step (including four Runge-Kutta subcycles).

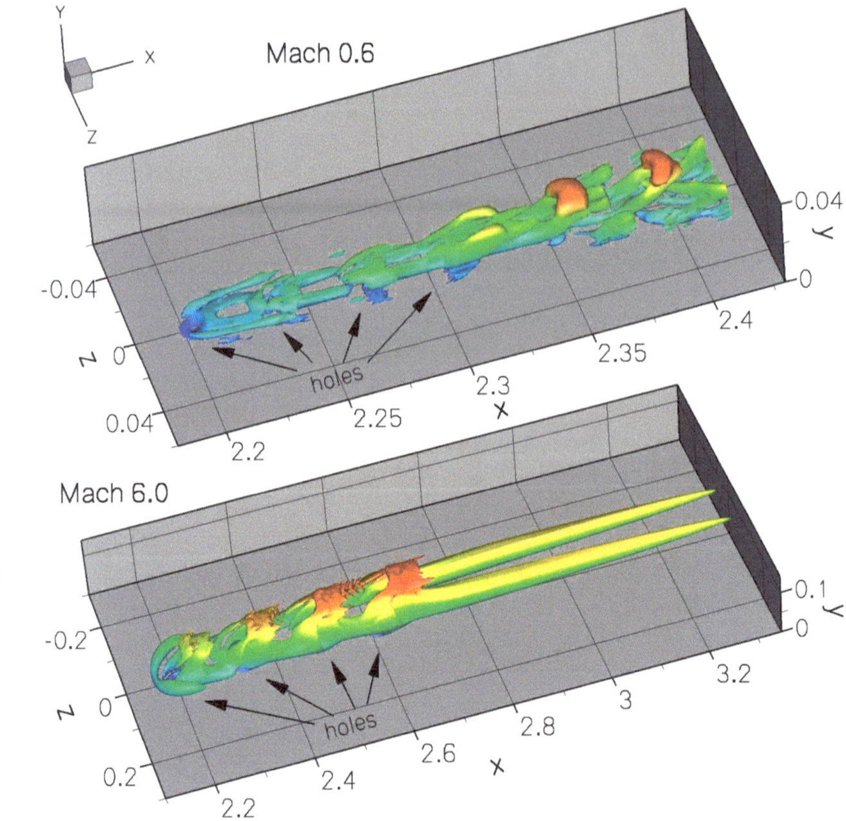

Fig. 13. Visualisation of the vortical structures via λ_2-isosurfaces for the Mach-0.6 (case $\mathcal{G}$, snapshot) and Mach-6 case (case $\mathcal{F}$)

With a sustained performance of 138.1 GFLOP/s, 36% of the theoretical peak performance per CPU (16 GFLOPS theoretical peak performance per CPU) are reached. The code shows a vector operation ratio of 99.59% with an average vector length of 211 and a total memory size of 38.4 GB. As the array sizes of each domain are equal, only slight performance differences between the MPI processes exist. Detailed profiling of the simulations shows that the main computational time (40%) is spent in the Fourier-transformation.

Due to the large computational time for the Fourier spectral approach in z-direction, a compact finite-difference (KFD) discretisation was tested to compute the z-derivatives. As a test case a fundamental breakdown simulation of an incompressible boundary layer by Kloker [8] is employed for comparison. Here a 2-d wave $(1, 0)$ and a steady 3-d wave $(0, 1)$ are generated in a disturbance strip and interact leading to fundamental, Klebanoff-type breakdown. In our simulation a Mach number 0.3 is used. It is a symmetric flow-field computation and the spanwise resolution is not a multiple of 8 with the Fourier

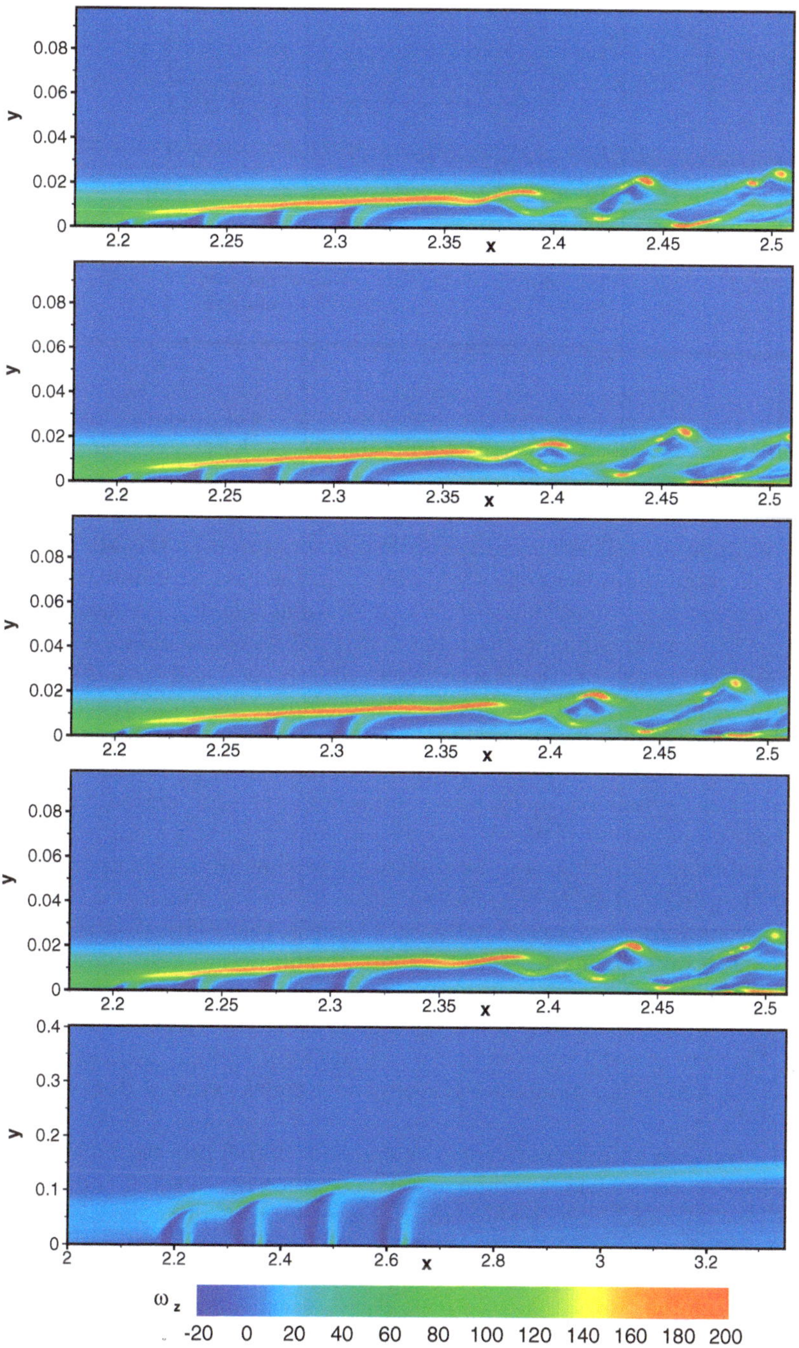

Fig. 14. Contour plot of ω_z $(= \frac{\partial u}{\partial y} - \frac{\partial v}{\partial x})$ in the $z = 0$ plane (through the hole centers) for case $\mathcal{G}$ (a-d; 4 different moments; $Ma = 0.6$) and case $\mathcal{F}$ (e; $Ma = 6$)

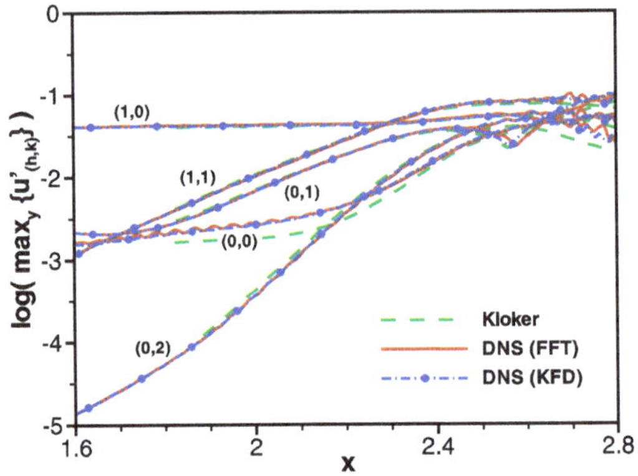

Fig. 15. Downstream t-z-modal amplitude development of u-velocity for the fundamental breakdown case with compact finite differences (KFD) and Fourier spectral approach (FFT) in z-direction

spectral approach. This means that a few processors pass the Microtasking-loop even more times than other processors. In the case with finite differences in z-direction a gridpoint number, which is a multiple of 8, is chosen. In our test case the z-direction is discretised with 33 physical points (32 modes, 21 de-aliased modes) in the Fourier spectral approach and with 40 points in the finite difference case. Identical splitted compact finite differences are now used in x-, y- and z-direction. The forward/backward alternating finite differences in z-direction produce in the symmetric case problems because of the employed symmetric boundary conditons. In the full periodic case these problems do not occur. Thus we also tested central compact finite differences in the z-direction. In this case the simulation becomes unstable due to missing dissipation, in contrast to the splitted finite differences which add dissipation of high-wavenumber modes. Therefore a filter in the z-direction is necessary with central differencing to secure stability, resulting in larger computational time.

Both simulations show good agreement with the results from Kloker (Fig. 15). The growth rates of the waves are very similar. We use a coarser mesh in x and y-direction than Kloker and wiggles occur at the end of the domain ($x = 2.6$).

The simulations were performed on one node with 8 processors. Peak/average performance was 43.8 and 56.08 GFLOP/s for Fourier and compact finte difference approach, respectivly thus compact finite differences performed about 28% better. Comparison of the computional time per gridpoint and time step shows a decrease of 30% from 1.73 μs in the Fourier case to 1.21 μs in the finite difference case. Beyond, this the total computational time is reduced by 15.7 %, but this may not be the case in all simulations and depends on spanwise resolution.

These results look promising but further testing is required. Additionally a replacement of our FFT routines [3] for spectral discretisation with the routines of the SX-8 internal library may also achieve improvements. In case of the incompressible N3D code an increase of 20-30% in speed was obtained [13] by use of the internal libraries.

5 Conclusion

Effusion cooling by slits and holes in hypersonic boundary layers has been investigated using direct numerical simulations (DNS). Focus of the study using air as flow and cooling gas is not only the investigation of the cooling effectiveness of various cooling configurations for a supposedly laminar flow but also the alteration of the laminar stability properties. Enhanced laminar instability by inducing shear layers with blowing and cooling the hypersonic boundary layer can compromise the cooling if the flow transitions to turbulence due to the cooling.

The presented results for effusion-cooling configurations with successive slits and holes of an adiabatic Mach 6 boundary layer at wind-tunnel conditions show that slits are better than (a few) holes. The slit-blowing velocity is, at same injected massflow, smaller than that of (a few) holes, where we had blowing ratios $(\rho v)_{c,max}/(\rho u)_\infty$ of 3% for the slits and 15% for the holes. At lower blowing ratios the coolant gas keeps closer to the wall.

The analysis of the two-dimensional flow with slit blowing with primary linear stability theory shows that the maximum amplification of a single frequency rises by a factor of 4 and the amplified frequency band is shifted to lower frequencies for the adiabatic case. The 1st mode is completely stabilised despite the prononunced inflection point in the decelerated u-velocity profile.

Aligned rows of holes induce a strong spanwise variation of the wall temperature and less cooling effectiveness compared to staggered rows of holes. A counter-rotating longitudinal vortex pair (CVP) is generated at each hole, decaying downstream, that pushes the coolant gas of the successive, staggered row down to the wall.

For the instability analysis of the real 3-d flow field, unsteady background disturbances have been added upstream of the holes at the wall and their timewise and downstream evolution computed by DNS. In a case with four rows of aligned holes it turned out that a small spanwise spacing of the holes is preferable over a large spacing, i.e. that the spanwise spacing should roughly be less than 2.5 boundary layer thicknesses (δ) for hole diameters less than $0.5 \cdot \delta$ ($s_z < 5\, d$). The steady 3-d deformation is than less detrimental.

It appears that effusion cooling at low blowing rates of $(\rho v)_{c,max}/(\rho u)_\infty < 5-10\%$ does not significantly increase laminar instability, at least as for modal growth of disturbances. This is especially true for slits and narrow-spaced holes. If the blowing is more localized and stronger, longitudinal vortices are generated that lead to strong 3-d deformations of the mean flow. But also here, the laminar instability is much weaker than at subsonic Mach numbers.

The performance of our code for the present case was found satisfactory with 27,6% of the teoretical peak performance per CPU on the NEC SX-8 for a typical run. By changing the spanwise discretisation from a Fourier spectral approach to compact finite differences we see a decrease in the computational time per grid point and time step from 1.73 μs to 1.21 μs and of total computational time by 15.7%. More testing is necessary to verify these results. In addition a replacement of our FFT routines with the SX-8 internal library may also achieve the same improvements like in the incompressible N3d code [13].

Acknowledgements

The partial financial support of this work by the Helmholtz-Gemeinschaft HGF within project A8 of the RESPACE group is gratefully acknowledged. We thank the Höchstleistungsrechenzentrum Stuttgart (HLRS) for provision of supercomputing time and technical support within the project "LAMTUR".

References

1. A. Babucke, J. Linn, M. Kloker, and U. Rist. Direct numerical simulation of shear flow phenomena on parallel vector computers. In *High performance computing on vector systems: Proceedings of the High Performance Computing Center Stuttgart 2005*, pages 229–247. Springer Verlag Berlin, 2006.
2. M. Bierbach. Untersuchung zur aktiven kühlung der grenzschichtströmung an einem plattenmodell. Master's thesis, Univerität Darmstadt, 2005.
3. EAS3 project. http://sourceforge.net/projects/eas3.
4. W. Eissler. *Numerische Untersuchungen zum laminar-turbulenten Strömungsumschlag in Überschallgrenzschichten.* PhD thesis, Universität Stuttgart, 2004.
5. K. Heufer and H. Olivier. Film cooling for hypersonic flow conditions. In *Proc. 5th European Workshop on Thermal Protection Systems and Hot Structures 2006*, 2006.
6. K. Heufer and H. Olivier. Film cooling of inclined flate plate in hypersonic flow. In *AIAA-Paper 2006-8067*, 2006.
7. J. Jeong and F. Hussain. On the identification of a vortex. *J. Fluid Mech.*, 285:69–94, 1995.
8. M. Kloker. *Direkte numerische Simulation des laminar-turbulenten Strömungsumschlages in einer stark verzögerten Grenzschicht.* PhD thesis, Universität Stuttgart, 1993.
9. M.J. Kloker. A robust high-resolution split-type compact FD scheme for spatial DNS of boundary-layer transition. *Appl. Sci. Res.*, 59:353–377, 1998.
10. J. Linn and M.J. Kloker. Numerical investigations of film cooling and its influence on the hypersonic boundary-layer flow. In *New Results in Numerical and Experimental Fluid Mechanics VI*, volume 98 of *NNFM: RESPACE – Key Technologies for Reusable Space Systems*, 2008.

11. L. Mack. Boundary-layer linear stability theory. In *AGARD Spec. Course on Stability and Transition of Laminar Flow*, volume R-709, 1984.
12. M. Malik. Prediction and control of transition in supersonic and hypersonic boundary layers. *AIAA Journal*, 27:1487–1493, 1989.
13. R. Messing, U. Rist, and F. Svenson. Control of turbulent boundary-layer flow using slot actuators. In *High performance computing on vector systems: Proceedings of the High Performance Computing Center Stuttgart 2006*, Springer Verlag Berlin, 2007.
14. C. Stemmer and M. Kloker. Interference of wave trains with varying phase relations in a decelerated 2-d boundary layer. In *Recent Results in Laminar-Turbulent Transition (ed. S. Wagner, M. Kloker, U. Rist)*, volume 86 of *NNFM*.
15. A. Thumm. *Numerische Untersuchungen zum laminar-turbulenten Strömungsumschlag in transsonischen Grenzschichtenströmungen.* PhD thesis, Universität Stuttgart, 1991.
16. F. White. *Viscous Fluid Flow.* McGraw-Hill, 1991.

Two-Point Correlations of a Round Jet into a Crossflow – Results from a Direct Numerical Simulation

J.A. Denev[1], J. Fröhlich[2], and H. Bockhorn[1]

1 Institute for Technical Chemistry and Polymer Chemistry, University of Karlsruhe (TH), Kaiserstraße 12, D-76128 Karlsruhe, Germany
{denev,bockhorn}@ict.uni-karlsruhe.de

2 Institute for Fluid Mechanics, Technical University of Dresden
George-Bähr Straße 3c, D-01062 Dresden, Germany
{jochen.froehlich}@tu-dresden.de

Summary. The paper presents results for the two-point correlation coefficient of two velocity components and a passive scalar for jets in crossflow. The data were obtained from two Direct Numerical Simulations carried out at two different Reynolds numbers (650 and 325) with a jet-to-crossflow velocity ratio of 3.3 in both cases. Results along the trajectory of the jet show the larger size of the turbulent eddies for the smaller Reynolds number. The integral scale of the turbulent eddies increases downstream. In the studied region, this scale appears to be larger along the crossflow than along the direction of the jet. The required resources for the investigation are described.

1 Introduction

The jet in crossflow (JICF) is a typical flow configuration used in many technical devices like gas turbines and mixers for the chemical industry. The main advantage of the JICF lies in its increased mixing capabilities when compared to straight jets as e.g. addressed by [1]. While the literature for the jet in crossflow is abundant [12], up to now most of the publications focus on the flow structures identifying different interacting vortex systems [7]. Statistical data on this issue like two-point correlations etc., have so far not been reported. Furthermore, the concentration field has so far not been investigated up to the same level of detail. In order to understand the processes behind the mixing capabilities of this flow, integrated studies for both scalar fields and flow fields are required. And while state-of-the-art in experiments allows the simultaneous study of separate planes in the flow (e.g. [9], [2], [15]), DNS allows the

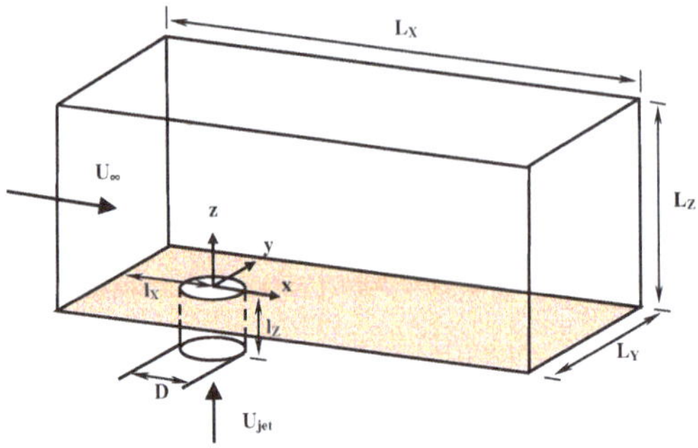

Fig. 1. Geometry of the computational domain and the axes of the coordinate system. lx=3D, lz=2D, Lx=20D, Ly=Lz=13.5D

three-dimensional simultaneous study of both scalars and velocity field in the complete physical domain ([5], [13], [14]). Understanding of turbulent mixing requires knowledge of both flow and mixing scales. To meet this requirement, the present paper deals with the evaluation of turbulent quantities describing this issue, like the correlation coefficient for velocity components and for a passive scalar introduced with the jet. In the present flow configuration both the jet and the crossflow enter the domain as laminar flows and a laminar-to-turbulent transition occurs within the domain. As the correlation coefficient is influenced by the location of this transition, the latter is also presented in the paper.

2 Cases Simulated

Two direct numerical simulations were performed out: one at $Re = 325$ and one at $Re = 650$, where the Reynolds number is defined with the velocity in the middle of the crossflow channel (U_∞ in Fig. 1) and the diameter D of the jet. A sketch of the flow is presented in Fig. 1. The jet-to-crossflow velocity ratio is $U_{jet}/U_\infty = 3.3$ for both cases, with U_{jet} being the bulk velocity of the jet. The inflow conditions are laminar. The pipe flow has a parabolic profile while a Blasius boundary layer is used for the crossflow. Its thickness is $\delta_{99} = 1.03D$. A passive scalar is introduced with the jet imposing a mass fraction (equal to the molar concentration) equal to $c = 1$ at the pipe inlet. The Schmidt number of the scalar is equal to $Sc = 1$. Further details of the configuration are given in [5].

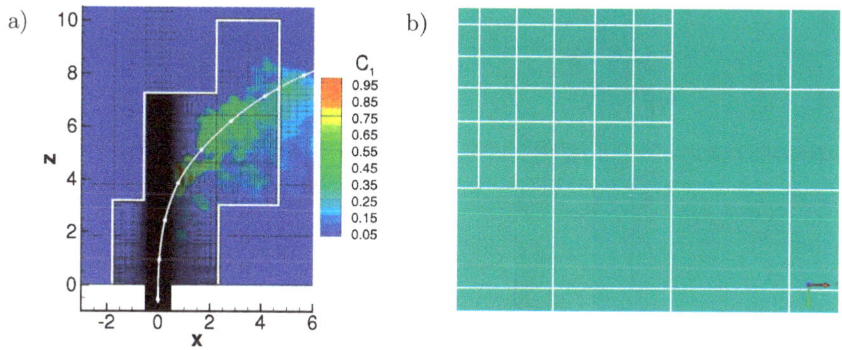

Fig. 2. The locally refined grid near the jet exit. a) Computational cells in the centerplane near the jet outlet in the symmetry plane $y/D = 0$. b) Zoom to illustrate the tangential refinement at block-boundaries with a factor 3 : 1

3 Numerical Method

The finite-volume method for solving the Navier-Stokes equations on block-structured grids is used. The numerical code LESOCC2 written in FORTRAN90 uses explicit Runge-Kutta time stepping and a pressure-correction equation. The method is second-order accurate both in time and space. Further details of the numerical method and the code LESOCC2 were presented by [10]. Advantage has been taken of the possibility in the code of changing the spatial resolution between adjacent blocks. This was used to refine the grid near the outlet of the jet as depicted in Fig. 2. The refinement factor was 3 in all directions, illustrated by the right part of the figure. More details of this feature together with a study of the accuracy obtained by this technique are available in [8].

4 Organization of Simulations and Required Resources

Two separate series of computations were performed, one for each Reynolds number, i.e. for $Re = 650$ and $Re = 325$. In each series, a first part served to accumulate one-point statistics for all variables, averaging over a dimensionless time interval of ≈ 100 (final statistics obtained for each run are 258.9 and 227.7 time units long, respectively). The dimensionless time units are based on the diameter of the pipe and the speed of the crossflow in the middle of the channel here. This first part allowed to determine the jet trajectory, together with the location of the laminar-turbulent transition. The purpose of the second part was to obtain time series at specific points which would constitute the basis of computing two-point statistics. For this to do, specific points were selected on the jet trajectory and close to it as depicted in Fig. 3. Actually, the

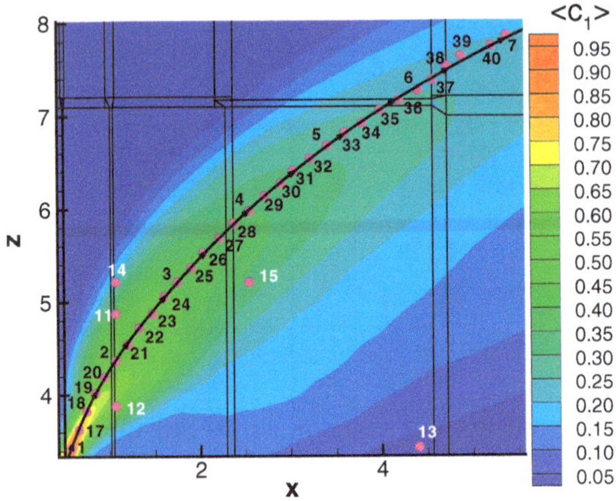

Fig. 3. Part of the trajectory just behind the transition and the averaged scalar concentration for $Re = 650$. Points No 1 to 7 are at a distance $\Delta s = D$ apart from each other along the trajectory, those with the higher numbers inserted to yield $\Delta s = 0.2D$

trajectories obtained for the two Reynolds numbers are merely the same (see fig. 7 below) so that the same points were used for both Reynolds numbers. The second part of the simulations was then continued over 72 696 time steps equal to 59.6 dimensionless time units for $Re = 325$ and 57 361 or 60.2 units for $Re = 650$, respectively. Signals of velocity components and concentration were stored for the points in the domain just mentioned. In total 118 timesignals were recorded, 111 of which are located on the trajectory with a spacing of $\Delta s/D = 0.05$, where s is the curvilinear coordinate along the trajectory. Some of the points, with a spacing of $0.2D$, are shown in Fig. 3, the ones with finer spacing being removed for clarity. The two-point correlations, determined using the whole set of time series, are reported below for a selected set of points, the coordinates of which are given in Tab. 1. The strategy applied in the present investigation was to define the locations where two-point correlations are determined in a physically meaningful manner. For this purpose the jet trajectory was chosen as a reference since it is the feature which is best defined in such a configuration. As a consequence, the desired locations can not be fixed prior to staring the computation but require a first part of the simulation to determine the mean flow. Finally, it should be mentioned that all points selected for storage of the signals and subsequent computation of two-point correlations are located in the region of local grid refinement, where the simulation indeed is of DNS-quality as assessed in previous studies. The time intervals given above over which the simulations were pursued to accumulate the two-point statistics were selected by performing analyses at intermediate instances as reported in Fig. 4. The figure shows the same quantity, calculated with different averaging times. In such tests it was found

Table 1. Coordinates of those points for which the correlation coefficient is reported in the figures below. All points are in the symmetry plane $y/D = 0$

Point No.	x/D	z/D	s/D
1	0.57	3.45	3.50
2	1.06	4.38	4.50
3	1.74	5.21	5.50
4	2.54	5.98	6.50
5	3.41	6.68	7.50
6	4.38	7.29	8.50
7	5.39	7.83	9.50

necessary to average over a longer laps of time for the lower Reynolds number since in this case the vortex structures are larger and hence less frequent, as illustrated by Figures 5 and 6 below. The time step used in the simulations was adjusted instantaneously inside the code by means of a stability criterion. This yielded a 14% larger time step for the lower Reynolds number. All simulations were carried out on the HP XC 4000 of the SCC Karlsruhe. A number of 31 processors were used which was found optimal for the present block-structured grid of 22.3 Mio points in 219 numerical blocks and resulted in a parallel efficiency of 91% [4]. Under these conditions the computation of one time unit required around 10 hours wall clock time. The binary restart files containing the solution at one instant in time and all information for its seamless continuation require 17 GB memory on disk and the 118 time signals stored as zipped ascii files have a size of 1.4 GB.

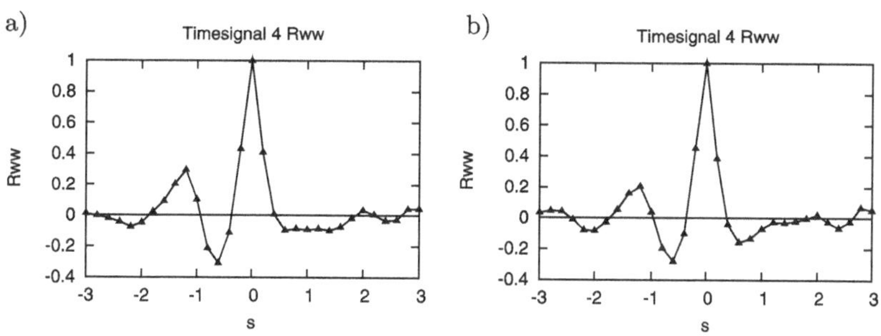

Fig. 4. The correlation coefficient for the vertical velocity component w at point 4 for $Re = 325$. a) Averaging over 32 768 time steps, b) averaging over 42 444 time steps

5 Results and Discussion

5.1 Instantaneous Solution

We shall now briefly address the instantaneous solution in the two cases, $Re = 325$ and $Re = 650$ to lay the ground for the subsequent discussion. Further details concerning average data and analyses of coherent structures can be found in earlier papers of the authors ([5], [3]). Figures 5 and 6 display two cuts through the scalar field for each Reynolds number. The higher turbulence level with $Re = 650$ results in substantially finer coherent structures and an earlier transition to turbulence. Particularly interesting in Fig. 5 are the longish accumulations of jet scalar on the axis just behind the jet. They have been described in detail by [3] and result from the inception of the counter-rotating vortex pair transporting jet fluid away from the center plane and then back to it. Fig. 6 illustrates this mechanism which in the average leads to a kidney shape of the scalar concentration [11].

The unambiguous determination of a transition point along the jet is a difficult matter. Animations with views like those reported in Fig. 5 show that the instantaneous point of transition fluctuates in time. In [3] the present authors therefore suggested to employ centerplane plots of the turbulent kinetic energy to objectively locate the transition point in an average sense. This procedure works very well although a bit of visual inspection is still needed. In the present cases the resulting transition points are $s_{trans} = 4.4D$ for $Re = 325$ and $s_{trans} = 3.5D$ for $Re = 650$, respectively, where s is the curvilinear coordinate along the trajectory. The transition point for $Re = 650$ coincides with Point 1 in Fig.3, while for $Re = 325$ transition is very close to Point 2 in this figure.

5.2 Jet Trajectory and Transition

Figure 7 shows the jet trajectory obtained from the average velocity field for both cases. The trajectory is defined here as the average streamline originating from the center of the jet, which in the present setting coincides with the origin of the coordinate system. The impact of the Reynolds number on this curve is very small, in particular for small distances from the outlet. This motivated the choice of the same ensemble of points for the determination of the two-point correlations.

5.3 Definition of Correlation Coefficients and Organization of Results

The correlation coefficient R_{ij} of two signals ϕ_i and ϕ_j is defined as

$$R_{ij} = \frac{\langle \phi_i \phi_j \rangle}{\sqrt{\langle \phi_i^2 \rangle \langle \phi_j^2 \rangle}} \tag{1}$$

a)

b)

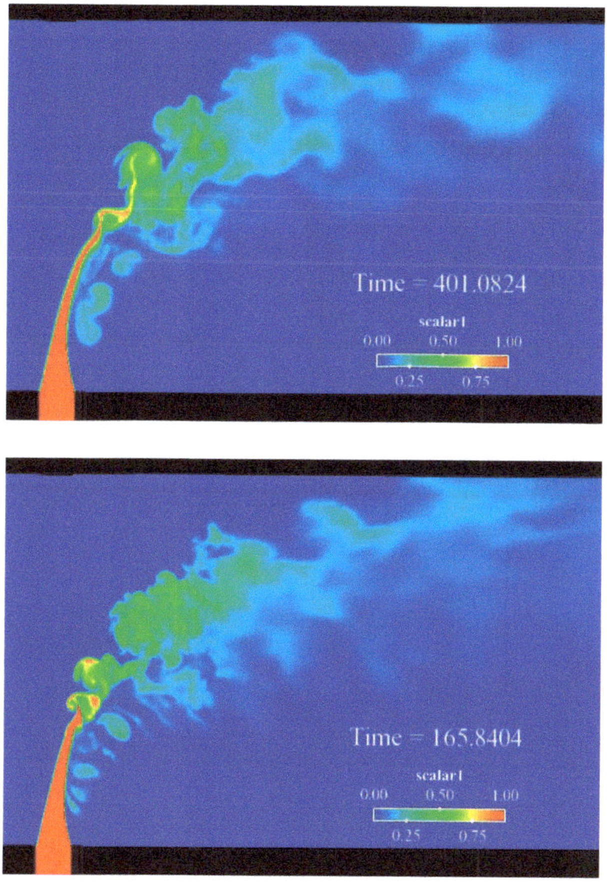

Fig. 5. Instantaneous distribution of the scalar in the symmetry plane $y/D = 0.0$. a) $Re = 325$, b) $Re = 650$. Only part of the computational domain is shown

Here, the variable ϕ represents fluctuations of a velocity component (u or w) or scalar concentration c. In the sequel results will be reported for two-point correlations along the jet trajectory determined by the time signals detailed above. The index in (1) hence relates to the location where the signal was taken. In figures and text below, however, it is more convenient to use $Ruu(s)$, etc. in order to designate the correlation along the trajectory. Results will be presented for the cases individually and furthermore will be compared between the two Reynolds numbers considered. We decided not to compare the correlation data for the same points since in physical terms these are positioned at different locations, e.g. with respect to the transition point. As the transition point moves upstream by about $0.9D$ when increasing the Reynolds number, it was decided for $Re = 650$ to use the corresponding points

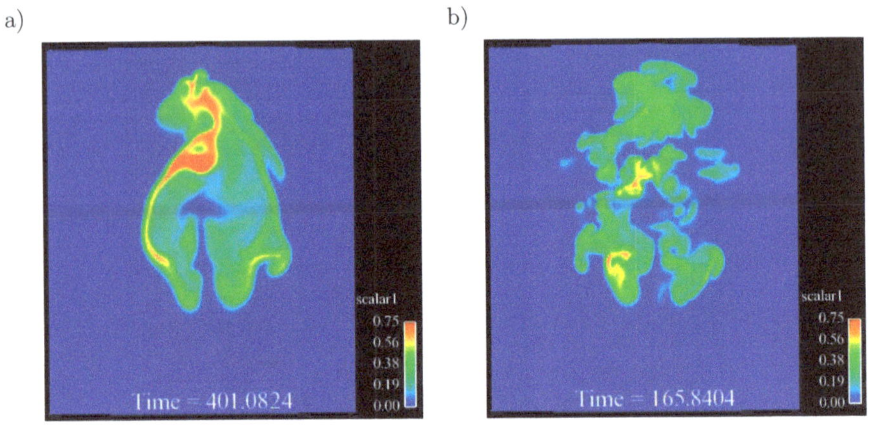

Fig. 6. Instantaneous scalar distribution downstream of the outlet at $x/D = 2.0$. a) $Re = 325$, b) $Re = 650$

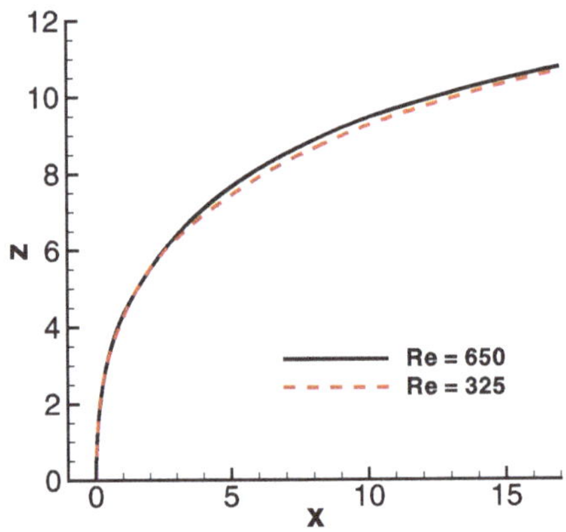

Fig. 7. Jet trajectories for the two Reynolds numbers considered

$1D$ further upstream when discussing the correlation data. Consequently, the result for $Re = 325$ at Point 4 in Tab. 1 will be compared to the data at Point 3 for $Re = 650$, for example. Assessing the sampling error of the two-point correlations is a delicate task. Computing results with time signals of different length as described above was performed to address this issue. Still, as the convergence for the samples is low for mathematical reasons, longer sampling times might be advantageous. The experience with the present data suggests that the two-point correlations are reliable over a distance of $2D$ on both sides. Data beyond this distance should be taken with caution.

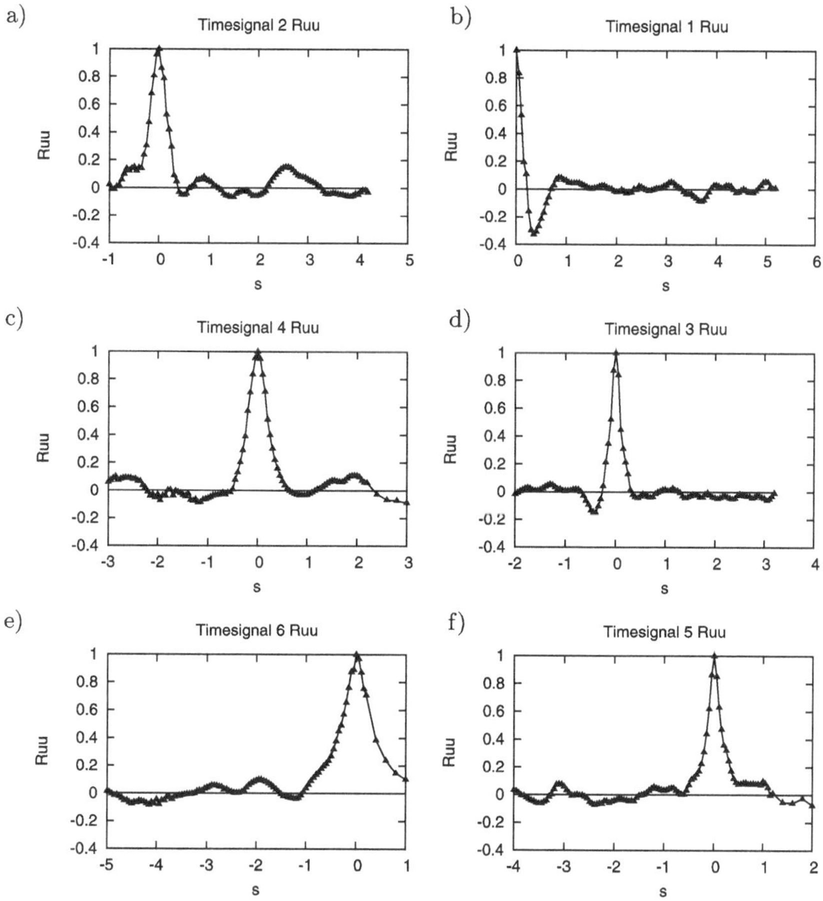

Fig. 8. Two-point correlation coefficient of the u-velocity component calculated for points with similar distance from the transition point. a), c), e): $Re = 325$ and Points 2, 4, 6; b), d), f): $Re = 650$ and Points 1, 3, 5 (according to Tab. 1)

5.4 Results for Two-Point Correlations

Figures 8 to 10 compile the data computed according to the procedure described above. By definition, the correlation coefficient is unity if the signals are identical, i.e. if the displacement between the signals involved is zero. The strong peak around the origin of the correlation plots monitors the size of the coherent structures displayed by the considered quantity. Its width was evaluated by means of the zero crossings, if these follow immediately, if not, the width has to be assessed by its value at a certain level of R. Furthermore, pronounced negative values close to the central peak reflect systematic changes in sign and hence the presence of pronounced and well-defined coherent structures in the quantity investigated. The level of organization is lower if these are not present. The integral length scale of turbulence is defined as the integral of the two-point correlation [16]. If the correlation has zero-crossings, however, the definition is not unproblematic and often the range of integration is restricted to the interval between the zeros. A quantitative result would hence require an extended definition of a macro-length. Using the zero crossing for this purpose, however, is delicate as these are very sensitive even to small sampling errors. For the time being we hence refrain from a definition of this type and restrict ourselves to a qualitative assessment.

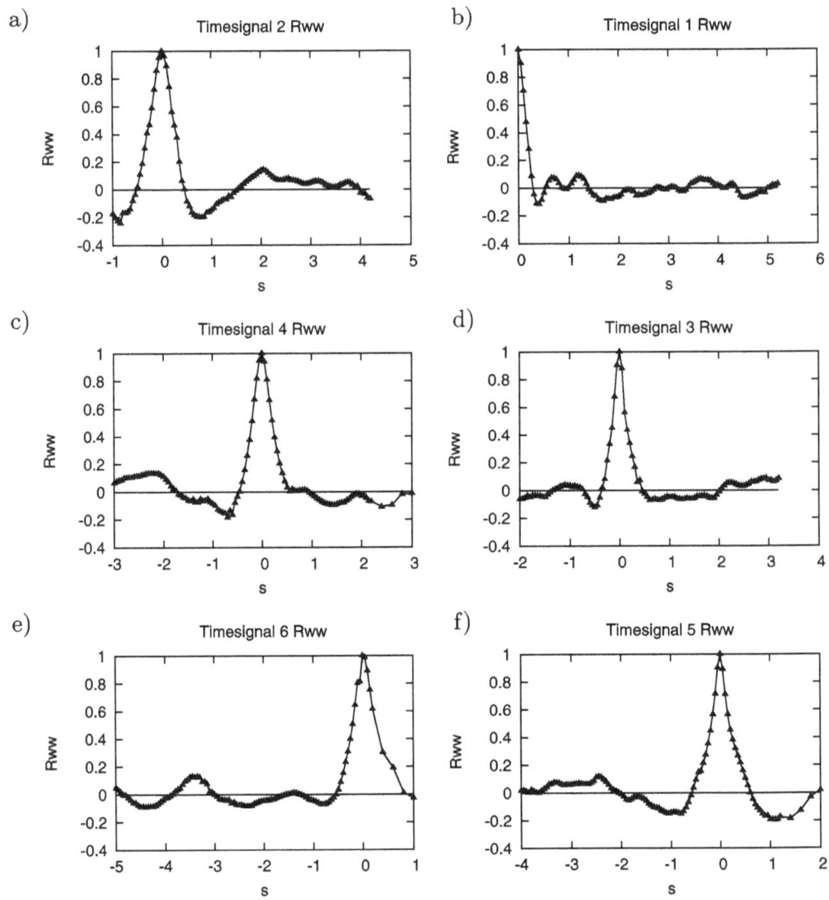

Fig. 9. Two-point correlation coefficient of the w-velocity component calculated for points with similar distance from the transition point. a), c), e): $Re = 325$ and Points 2, 4, 6; b), d), f): $Re = 650$ and Points 1, 3, 5 (according to Tab. 1)

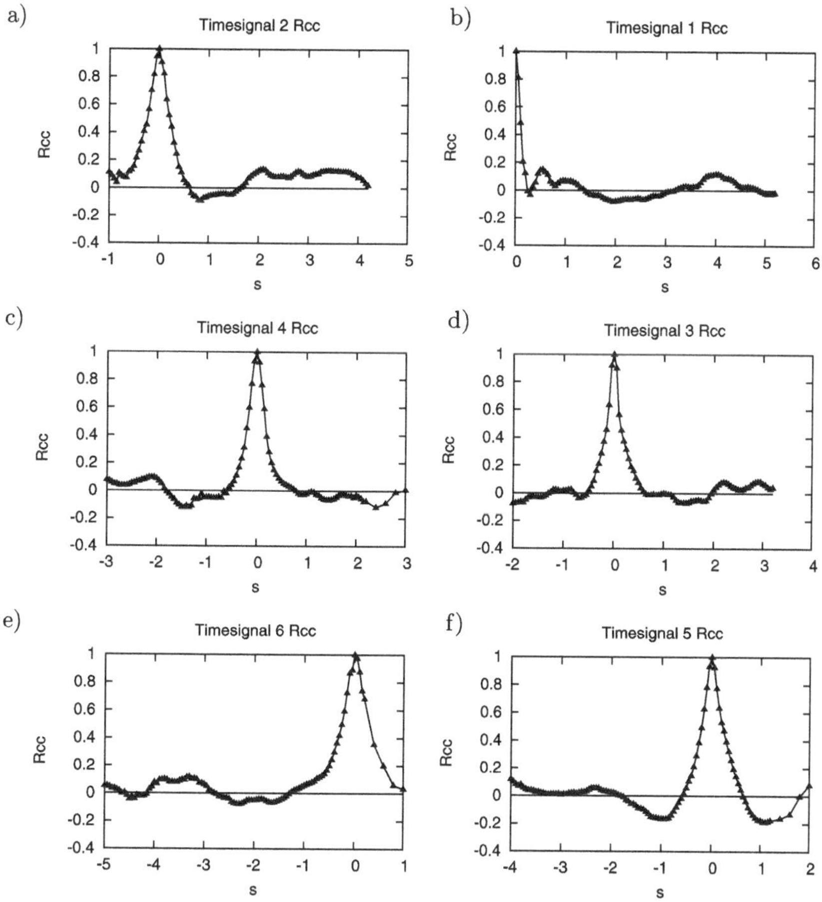

Fig. 10. Two-point correlation coefficient of the scalar calculated for points with similar distance from the transition point. a), c), e): $Re = 325$ and Points 2, 4, 6; b), d), f): $Re = 650$ and Points 1, 3, 5 (according to Tab. 1)

Systematically, the width of the peaks increases with downstream distance along the jet. This reflects the growth of the turbulent structures in this direction, visible in the instantaneous plots of Figure 5. The behavior is expected since the size of the largest structures in a jet correlates with the width of the jet which in turn increases along the trajectory. Similar observations were made by [6]. Furthermore, the correlation peaks are wider for $Re = 325$ than for $Re = 650$, reflecting a smaller size of the dominating structures with higher Reynolds number. This results from the velocity gradients being steeper with higher Reynolds number, as can be appreciated in Figure 5, and from the earlier transition at a distance where the jet has widened to a lower extent. The width of the peaks of R_{uu} is smaller than observed for R_{ww}. With the orientation of the jet more in z- than in x-direction at the locations considered slow oscillations of the jet position would contribute stronger to the w-fluctuations than to the u-fluctuations but the origin could also reside in a different structure of the oscillations. The correlations for the scalar fluctuations, on the other hand, are very similar to the ones for w, for both Reynolds numbers. In particular, far downstream well-defined large scale structures are observed with a period length of about $2D$. In general, no symmetry of the correlation curves with respect to positive or negatives distances is observed. Rather, there is a trend from exhibiting pronounced asymmetry closer to the transition point towards a more symmetrical behavior further downstream. This is understandable due to the decay of the gradients of all mean quantities in downstream direction and the natural tendency towards statistical uniformization of turbulent flows.

6 Conclusions

Extensive time series data were generated from two DNS of a transitional jet into laminar crossflow, at $Re = 325$ and $Re = 650$. These data allowed to determine the two-point correlations along the jet trajectory with high spatial resolution. It was found that the size of the structures can be determined in a qualitative way. A quantitative result would require an extended definition of a macro-length which was not undertaken. The convergence of the correlation data was observed to be slow for the low Reynolds numbers considered which constitutes a severe challenge, so that further sampling might be suitable. On the other hand, the two-point correlation data constitute an objective measure of the size of the coherent structures which, for a jet in crossflow has not jet been addressed in this manner. They may serve for validation of simulations, assessing assumptions in statistical turbulence models and physical interpretation.

Acknowledgements

The present project is funded through the DFG priority program SPP 1141. The simulations were performed on the national super computer HP XC4000 at the High Performance Computing Center Stuttgart (HLRS) under the grant with acronym "DNS-jet". The authors like to thank the staff of HLRS and the Super Computing Center (SCC) Karlsruhe for their cooperation and their efficient support. Carlos Falconi helped with preparing the figures.

References

1. J.E. Broadwell and R.E. Breidenthal. Structure and mixing of a transverse jet in incompressible flow. *J. Fluid Mech.*, 148:405–412, 1984.

2. C. Cárdenas, R. Suntz, J.A. Denev, and H. Bockhorn. Two-dimensional estimation of Reynolds-fluxes and -stresses in a Jet-in-Crossflow arrangement by simultaneous 2D-LIF and PIV. *Applied Physics B - Lasers and Optics*, 88(4):581–591, 2007.
3. J.A. Denev, C. Falconi, J. Fröhlich, and H. Bockhorn. DNS and LES of a jet in crossflow – Evaluation of turbulence quantities and modelling issues. In: *Procs. of 7th Int. ERCOFTAC Symposium on Engineering Turbulence Modelling and Measurements, ETMM7*, volume 2, pages 587–592, Limassol, Cyprus, 4-6 June 2008.
4. J.A. Denev, J. Fröhlich, and H. Bockhorn. Direct numerical simulation of a round jet into a crossflow - analysis and required resources. In W.E. Nagel, D. Kroener, and M. Resch, editors, *High Performance Computing in Science and Engineering 07, Transactions of the High Performance Computing Center Stuttgart*, pages 339–350. Springer - Berlin - Heidelberg, 2008.
5. J.A. Denev, J. Fröhlich, and H. Bockhorn. Direct numerical simulation of a transitional jet in crossflow with mixing and chemical reactions. In R. Friedrich, N.A. Adams, J.K. Eaton, J.A.C. Humphrey, N. Kasagi, and M.A. Leschziner, editors, *Proc. 5th Int. Symp. on Turbulence and Shear Flow Phenomena, TSFP-5*, volume 3, pages 1243–1248, Garching, Germany, August 27–29 2007.
6. M. Freitag, M. Klein, M. Gregor, A. Nauert, D. Geyer, Ch. Schneider, A. Dreizler, and J. Janicka. Mixing analysis of a swirling recirculating flow using DNS and experimental data. In R. Friedrich, N.A. Adams, J.K. Eaton, J.A.C. Humphrey, N. Kasagi, and M.A. Leischziner, editors, *Proc. 4th Int. Symp. on Turbulence and Shear Flow Phenomena, TSFP-4*, volume 2, pages 491–496, Williamsburg, Virginia, June 27–29 2005.
7. T.F. Fric and A. Roshko. Vortical structure in the wake of a transverse jet. *J. Fluid Mech.*, 279:1–47, 1994.
8. J. Fröhlich, J.A. Denev, C. Hinterberger, and H. Bockhorn. On the impact of tangential grid refinement on subgrid-scale modelling in large eddy simulations. In T. Boyanov et al., editor, *Lecture Notes in Computer Science, LNCS 4310*, pages 550–557. Springer Verlag Berlin Heildelberg, 2007.
9. E.F. Hasselbrink and M.G. Mungal. Transverse jets and jet flames. Part 2. Velocity and OH field imaging. *J. Fluid Mech.*, 443:27–68, 2001.
10. C. Hinterberger. *Three-dimensional and depth-averaged Large-Eddy-Simulation of flat water flows.* PhD thesis, Inst. Hydromechanics, Univ. of Karlsruhe.
11. T.T. Lim, T.H. New, and S.C. Luo. On the development of large-scale structure of a jet normal to a cross flow. *Phys. Fluids*, 13(3):770–775, 2001.
12. R.J. Margason. 50 years of jet in cross flow research. *Computational and experimental assessment of jets in crossflow*, pages 1.1–1.41, 1993. AGARD-CP-534.
13. S. Muppidi and K. Mahesh. Passive scalar mixing in jets in crossflow. In *44th AIAA Aerospace Sciences Meeting and Exhibit*, Reno, Nevada, Jan 9-12 2006. AIAA Paper 2006-1098.
14. S. Muppidi and K. Mahesh. Direct numerical simulation of passive scalar transport in transverse jets. *J. Fluid Mech.*, 598:355–360, 2008.
15. O. Özcan, K.E. Meyer, P.S. Larsen, and C.H. Westergaard. Simultaneous measurement of velocity and concentration in a jet in channel-crossflow. In *Proceedings of ASME FEDSM'01, New Orleans, Louisiana, May 29 - 31, 2001*, 2001.
16. H. Tennekes and J.L. Lumley. *A first course in turbulence.* MIT Press, 1972.

The Influence of Periodically Incoming Wakes on the Separating Flow in a Compressor Cascade

Jan G. Wissink and Wolfgang Rodi

Institute for Hydromechanics, University of Karlsruhe
Kaiserstr. 12, D-76128 Karlsruhe
rodi@uka.de

Summary. Two Direct Numerical Simulations (DNS) of flow in the V103 Low-Pressure (LP) compressor cascade with incoming wakes have been performed employing different intensities for the incoming wakes. The computational geometry was chosen in accordance with the setup of the experiments performed by Hilgenfeld and Pfitzner at the University of the Armed Forces in Munich. The computations were carried out on the NEC SX-8 using 64 processors and 85 million grid points. The incoming wakes stem from a separate DNS of flow around a circular cylinder. While in the simulation with the weaker wakes the flow along the suction side was found to remain separated for all phases, in the simulation with the stronger wakes separation was periodically completely suppressed. As the boundary layer separated, it was found to roll up due to a Kelvin-Helmholtz instability triggered by the periodically passing wakes. Inside the rolls further transition to turbulence was observed.

1 Introduction

Because of the increase in pressure from the inlet to the outlet, flows around compressor blades are prone to separation. Separation drastically alters the aerodynamical behaviour of the blades and may even lead to mechanical failure. It is relied upon the periodic unsteadiness – induced by impinging wakes stemming from the upstream row of blades – to limit the adverse effects of separation. There is need for good quality data to elucidate the underlying physical mechanisms and to serve as reference data with which to improve models for transition. By employing such improved models in industrial codes, more efficient blades can be designed.

Recently, periodic unsteady flow in LP turbines has reached much attention. By employing powerful high-performance computers, nowadays it has become feasible to perform Large-Eddy Simulations (LES) (see for instance Michelassi et al. [M03] and Raverdy et al. [R03]) and DNS (such as the ones performed by Wu and Durbin [W01], Kalitzin et al. [K03], Wissink [W03] and Wissink et al. [W06]) of flow around a section at midspan of a linear turbine cascade. Wu and Durbin [W01] simulated flow in a T106 cascade with periodically incoming wakes. They discovered that wakes – which were passively convected by the free stream – induced the formation of longitudinal vortical structures along the pressure side of the turbine blade. Kalitzin et al. [K03] also performed a DNS of flow in the T106 cascade using incoming free-stream turbulence instead of wakes. With a free-stream turbulence level of $Tu = 5\%$ at the inlet, the boundary layer flow in the adverse pressure-gradient region along the suction side was found to undergo by-pass transition. In Wissink [W03] and Wissink et al. [W06] results were reported of simulations of flow in a T106 cascade at a lower Reynolds number than the one employed in two simulations discussed above. Because of the relatively low Reynolds number and the large inflow angle, in the absence of wakes the flow along the downstream half of the suction side was found to separate. In the presence of realistic wakes, this separation was found to be periodically completely suppressed. By removing the small-scale fluctuations from the wakes, it was found that the time-periodic large-scale movement of the wake as a negative jet was sufficient to trigger a Kelvin-Helmholtz instability of the separated boundary layer. Further transition to turbulence, which in the presence of realistic wakes was found to take place inside the rolls of recirculating flow, was not observed. Seeding by small-scale fluctuations was found to be required in order for kinetic energy to be produced at the apex of the deformed wakes in the passage between blades.

Typical Reynolds numbers of flows in LP compressors are comparable with the Reynolds numbers for flows in LP turbines. As discussed at the beginning of this section, the boundary layer around a compressor blade is prone to separation. The presence of free-stream fluctuations is usually sufficient to completely suppress boundary-layer separation around a turbine blade, while for a compressor blade it can only be relied upon the impinging free-stream fluctuations to suppress separation to such a degree that its adverse effects are limited. To study how impinging wakes interact with the separated boundary layer along a compressor blade, we were given the opportunity to perform a direct numerical simulation of this computational expensive flow problem on the NEC SX-8.

2 Computational Details

The three-dimensional, incompressible Navier-Stokes equations were discretised using a finite-volume method with a collocated variable arrangement. A second-order central method was used for the discretisation in space, which was combined with a three-stage Runge-Kutta method for the time-integration. To avoid the decoupling of the velocity field and the pressure field, the momentum-interpolation procedure of Rhie and Chow [R83] was employed. The Poisson equation for the pressure was solved using the SIP solver of Stone [S68]. A more complete description of the basic computational code can be found in Breuer and Rodi [B96].

The code was parallelised using the standard Message Passing Interface (MPI) protocol. The computational mesh was subdivided into 64, partially overlapping, submeshes (blocks) of equal size. Each block was assigned to a unique processor to ensure a near-optimal load balancing. Because of the usage of an implicit Poison solver for the pressure, a frequent exchange of data between neighbouring blocks needed to be carried out. Hence, the simulation speed benefits greatly when running on a computational platform with powerful processors and very fast intra and internodal communication between processors. Both simulations were performed on the NEC SX-8 using 64 processors on 8 nodes. The vector operation ratio was 98% with an average vector length of 117. The simulations reached an average speed of 2.6 *GFlop/s* per processor. In both simulations about 294 400 time steps were performed using chain jobs with a duration of about 6 clock hours. To finish both runs in total 1224 clock hours were required corresponding to 1224*64=78336 CPU hours.

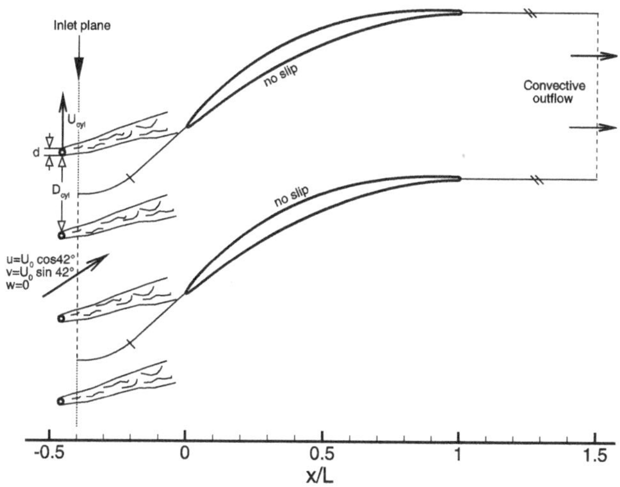

Fig. 1. Spanwise cross-section through the computational domain

Figure 1 shows a spanwise cross-section through the computational domain. The Reynolds number, $Re = 138\,500$, was based on the mean inflow velocity U_0 and the axial chord-length L. At the surface of the blades a no-slip boundary condition was prescribed, at the outlet a convective outflow boundary condition was used and in the spanwise direction as well as in the vertical direction – both upstream and downstream of the blades – periodic boundary conditions were used. The spanwise size of the computational domain was $l_z = 0.15L$ and the pitch, P, between blades was $P = 0.5953L$. At the inlet plane, located at $x/L = -0.4$, wakes were introduced which were generated using a separate DNS of flow around a circular cylinder at $Re_d = 3300$, where Re_d was based on the diameter of the cylinder d and the free-stream velocity of the oncoming, uniform flow. In the simulation, the wakes introduced at the inlet plane corresponded to wakes generated by a row of cylinders located in the plane $x/L = -0.05625$ that moved in the positive y-direction with velocity $U_{cyl} = 0.30U_0$. The diameter of the cylinders for the simulation with weak wakes was $d = 0.009375L$, while for the strong wake simulation this diameter was $d = 0.01875L$. The period of the flow was $T = 0.9922L/U_0$. The wake data introduced at the inlet was interpolated from a series of 1057 snapshots of the instantaneous flow field in a vertical plane downstream of the cylinder. The plane was located at a distance of $6d$ from the axis of the cylinder. The series of snapshots covered 12 vortex-shedding cycles and was chosen such that the phase of the first snapshot was the same as the phase of the $1057{+}1^{th}$ snapshot. Filtering was subsequently applied to smoothen the transition from the last snapshot to the first snapshot of the series, hence enforcing a periodicity More details are given in Wissink and Rodi [W08].

Figure 2 shows every eighth grid-line in a cross-section through the computational domain at midspan. The distribution of grid points in the $1030 \times 640 \times 128$ point mesh was based on experience gained previously in performing various DNS of flow in LP turbine cascades, (see Michelassi et al. [M02] and Wissink [W03]) and produced an adequate resolution of both the suction-side and the pressure-side boundary layers. The mesh was generated using the elliptic mesh generator of Hsu and Lee [H91] and is near-orthogonal. Table 1 shows an overview of the two simulations performed. The

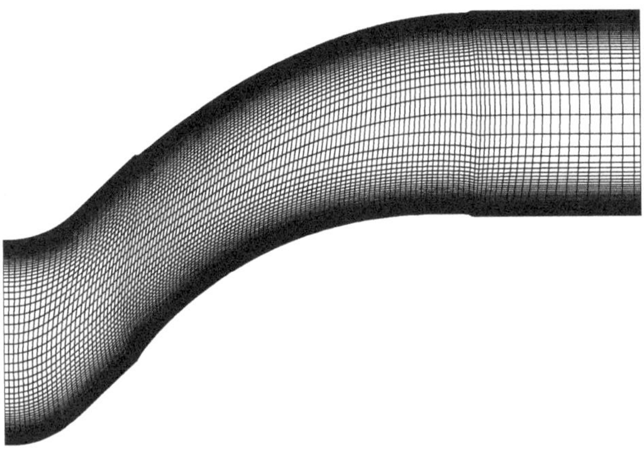

Fig. 2. Mesh at midspan showing every eighth grid-line

Table 1. Overview of the simulations performed

Simulation	Mean wake width	Mean velocity deficit
A	$0.05L$	16%
B	$0.10L$	16%

only difference between the two simulations is the width of the wake which is twice as large in Simulation B than in Simulation A.

3 Results

Figure 3 shows the wall static-pressure distribution $C_p = \frac{\bar{p}-\bar{p}_{ref}}{\frac{1}{2}U_{ref}^2}$ along the compressor blade. In both simulations the streamwise pressure-gradient along the pressure side, upstream of $x/L = 0.8$, is found to be slightly adverse. Downstream of this location the pressure gradient becomes favourable causing the boundary layer flow to accelerate. Along the suction side, in both simulations the pressure gradient is favourable for $x/L < 0.2$ and becomes adverse farther downstream. In the adverse pressure-gradient region of Simulation A a clear kink – identified by the arrow – can be seen in the C_p curve,

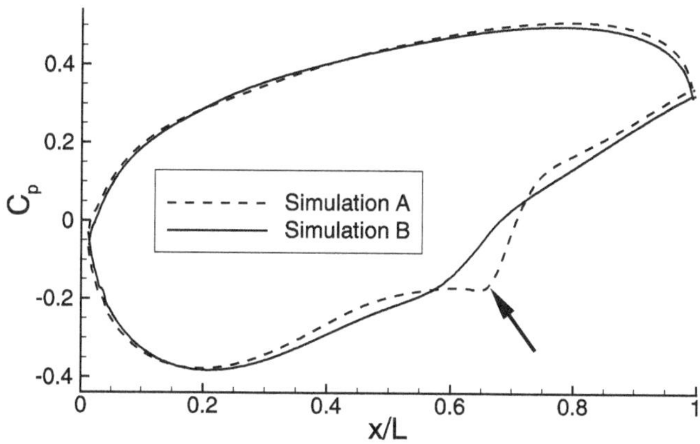

Fig. 3. Mean wall static-pressure distribution

which is an indication for boundary-layer separation. In Simulation B this kink is much less pronounced. That the boundary layer along the suction side in Simulation A indeed separates, is confirmed by the negative values of the mean friction coefficient C_f, observed in Figure 4. The locations of separation and re-attachment are identified by the two red arrows labelled “S” and “R”. The highest values of C_f in both Simulations A and B are found near the leading edge, where the flow is accelerating and the boundary layer is very

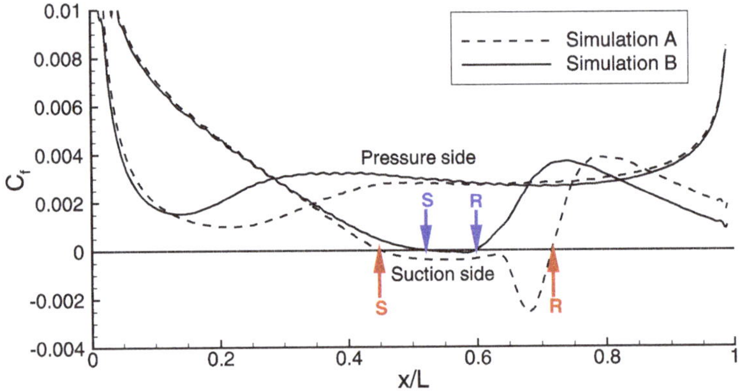

Fig. 4. Simulations A and B: Mean friction velocity along the suction and pressure side

thin. Downstream of the location of re-attachment, the moderate values of the friction coefficient obtained in both simulations indicate that the boundary layer does not reach a fully turbulent state. Compared to Simulation A, in Simulation B the separation bubble along the suction side has significantly reduced in size. The area of separation is identified by the two blue arrows labelled “S” for separation and “R” for reattachment.

Even though in both simulations the pressure gradient is adverse along a significant portion of the pressure side, the mean friction-coefficient is strictly positive which implies that the time-averaged boundary layer does not separate. In both simulations, the highest values for the friction coefficient are observed near the leading edge. Immediately upstream of the trailing edge, C_f is found to rise sharply, which usually indicates that the boundary layer flow undergoes transition but may also be partly caused by the streamwise pressure-gradient turning favourable. The latter may reduce the thickness of the boundary layer and increase the friction velocity.

The snapshot of the instantaneous spanwise vorticity at midspan from Simulation A, shown in Figure 5, clearly illustrates the boundary-layer separation on the suction side. Downstream of the separation bubble, a wake-like flow adjacent to the surface of the blade is observed. The periodically passing wakes are also partially visible in this figure. Inside the passage between blades, the path of the wakes is seen to alter slightly. Along the pressure side, in the adverse pressure gradient region the wakes impinge at a non-zero angle of attack, farther downstream – in the favourable pressure-gradient region located immediately upstream of the trailing edge – the wakes are observed to

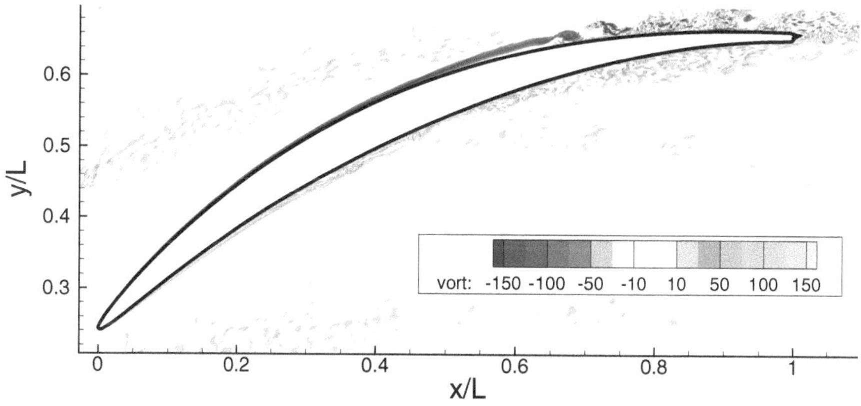

Fig. 5. Simulation A: Contours of the instantaneous spanwise vorticity at midspan for $t/T = 14.333$

align with the direction of flow and impinge at a very small angle of attack. As the wakes are passively convected by the mean flow, they only slightly alter their paths. Compared to the simulations of flow in the T106 turbine cascade, where the wakes were subjected to severe straining and stretching by the mean flow, in the compressor cascade the stretching and straining action of the mean flow on the wakes is almost negligible. Consequently, no production of kinetic energy in the free-stream part of the passage between blades is observed. Along the suction side, the separated boundary layer is found to roll-up due to a Kelvin-Helmholtz (KH) instability triggered by the passing wakes. As observed in earlier simulations [W03], production of kinetic energy is subsequently found to take place inside the KH rolls. The patches of fluctuations, introduced into the pressure side boundary-layer by the impinging wakes, manage to completely suppress separation along this side of the blade. Along the pressure side, the sudden rise in the friction velocity in this region immediately upstream of the trailing edge indicates that the boundary layer flow is undergoing transition, though the thinning of the boundary layer owing to the flow acceleration in this region is also likely to contribute to this phenomenon.

Figure 6 shows two snapshots with contours of the streamwise velocity fluctuations along the suction surface of the blade from Simulation B. The snapshots clearly illustrate the short-time evolution of a turbulent spot. The spot is triggered by the periodically passing wakes. As it is convected downstream it can be seen to expand. Shortly after $t/T = 5.475$ it finally merges with the fully turbulent flow downstream. A closer examination (not shown here) learns that the periodically appearing turbulent spots manage to locally completely suppress separation. Once the spots have passed, the boundary

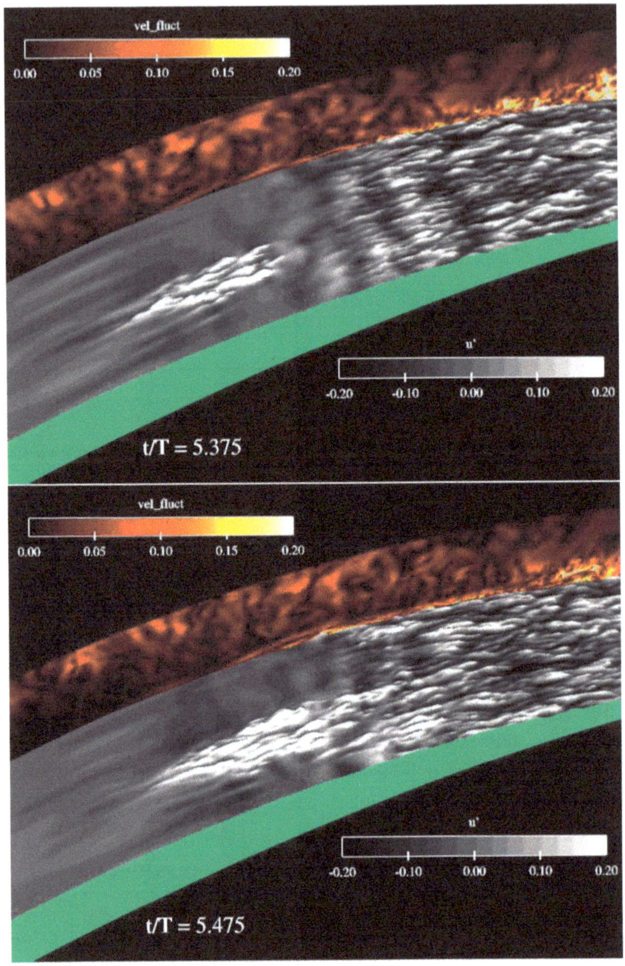

Fig. 6. Simulation B: Elevated contours of the streamwise velocity fluctuations adjacent to the suction surface illustrating the evolution of a turbulent spot. The coloured contours of the magnitude of the velocity fluctuations are shown at the back to identify the location of the wakes.

layer relaminarises and – after some transient time – again starts to separate.

Figure 7, finally shows contours of the phase-averaged kinetic energy of the fluctuations which identify the path of the wake. It can be seen clearly from this figure that the path of the wakes is only very marginally distorted as they are convected by the mean flow through the passage between blades. Compared to the simulations of flow in the passage between two turbine blades (see [W01, W03]), the stretching and straining action that the wakes undergo

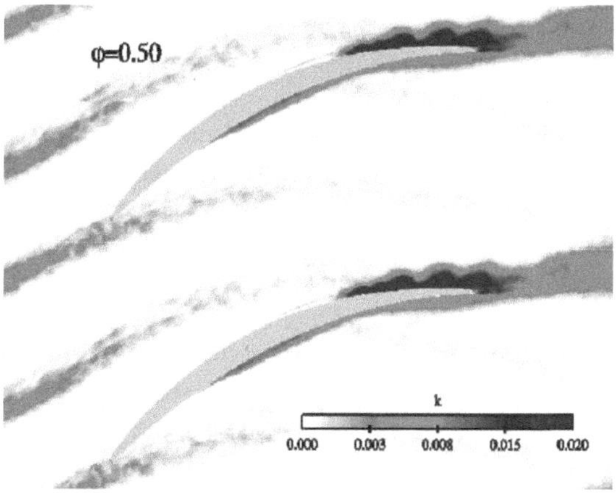

Fig. 7. Simulation A: Contours of the phase-averaged kinetic energy of the fluctuations at $\phi = 0.50$.

in the passage is almost negligible and, as a consequence, no production of kinetic energy is observed. The small regions with high concentrations of kinetic energy that are found side by side along the downstream half of the suction side, is a reflection of the production of phase-averaged kinetic energy in the Kelvin-Helmholtz rolls (see Figure 5).

4 Conclusions

The possibility to perform three-dimensional Direct Numerical Simulations (DNS) of flow in a linear, low-pressure compressor cascade with periodically passing wakes of varying strength on the NEC SX-8 supercomputer in Stuttgart, allowed a comprehensive and accurate investigation of the effects of impinging wakes on the separating boundary layer along a modern low pressure compressor blade.

For the simulations, we reach the following conclusions:

- The wakes, travelling through the passage between the compressor blades, were observed to only slightly alter their trajectory due to the straining and stretching action by the mean flow.
- In both simulations, along the suction side the boundary layer flow was observed to separate. Compared to Simulation A, the boundary layer separation observed in Simulation B was almost negligible. In Simulation A, in the first stage of the transition the separated boundary layer was observed to roll up due to a Kelvin-Helmholtz instability. Inside the rolls, kinetic energy was found to be produced, resulting in a turbulent, wake-like flow adjacent to the surface of the blade downstream of the separation bubble.In Simulation B the wake was observed to periodically trigger turbulent spots

upstream of the location of separation. As the turbulent spots were convected downstream, they were observed to grow and to locally completely suppress separation.

- Along the pressure side, in both simulations the boundary layer was observed to remain attached for all phases. As the wakes impinged on the adverse-pressure gradient portion, fluctuations were introduced into the boundary-layer flow resulting in a patch of mildly turbulent flow. This patch was subsequently observed to be convected downstream. Once entering the favourable pressure-gradient region the fluctuations in the patch were found to be damped. The presence of mildly turbulent patches along the pressure side was found to be sufficient to completely suppress separation for all phases.

Acknowledgements

The authors would like to thank the German Research Foundation (DFG) for funding this project and the Steering Committee for the Supercomputing Facilities in Stuttgart for granting computing time on the NEC SX-8.

References

[B96] Breuer, M. and Rodi, W.: Large eddy simulation for complex flow of practical interest. In: Flow Simulation with High-Performance Computers II, Notes on Num. Fluid Mech., **52**, Vieweg Verlag, Braunschweig (1996)

[H91] Hsu, K. and Lee, L.: A numerical technique for two-dimensional grid generation with grid control at the boundaries. J. Comp. Phys., **96**, 451–469 (1991)

[K03] Kalitzin, G., Wu, X., Durbin, P.A.: DNS of fully turbulent flow in an LPT passage. Int. J. Heat and Fluid Flow, **24**, 636–644 (2003)

[M02] Michelassi, V., Wissink, J.G., Rodi, W.: Analysis of DNS and LES of flow in a low-pressure turbine cascade with incoming wakes and comparison with experiments. Flow, Turbulence and Combustion, **69**, 295–329 (2002)

[M03] Michelassi, V., Wissink, J.G., Fröhlich, J., Rodi, W.: Large-eddy simulation of flow around a low-pressure turbine blade with incoming wakes. AIAA J., **41**, 2143–2156 (2003)

[R03] Raverdy, B., Mary, I., Sagaut, P., Liamis, N.: High-resolution large-eddy simulation of flow around low-pressure turbine blade. AIAA J., **41**, 390–398 (2003)

[R83] Rhie, C.M. and Chow, W.L.: Numerical study of the turbulent flow past an airfoil with trailing edge separation. AIAA J., **21**, 1525–1532 (1983)

[S68] Stone, H.L.: Iterative Solution of Implicit Approximations of Multidimensional Partial Differential Equations. SIAM J. Numerical Analysis, **5**, 530–558 (1968)

[W03] Wissink, J.G.: DNS of separating, low-Reynolds number flow in a turbine cascade with incoming wakes. Int. J. Heat and Fluid Flow, **24**, 626–635 (2003)

[W06] Wissink, J.G., Rodi, W., Hodson, H.P.: The influence of disturbances carried by periodically incoming wakes on the separating flow around a turbine blade. Int. J. Heat and Fluid Flow, **27**, 721–729 (2006)

[W08] Wissink, J.G. and Rodi, W.: Numerical study of the near wake of a circular cylinder. Accepted for publication in: Int. J. Heat and Fluid Flow, (2008)

[W01] Wu, X. and Durbin, P.A.: Evidence of longitudinal vortices evolved from distorted wakes in a turbine passage. J. Fluid Mech., **446**, 199–228 (2001)

Turbulence and Internal Waves in a Stably-Stratified Channel Flow

Manuel García-Villalba[1] and Juan C. del Álamo[2]

[1] Institut für Hydromechanik, Universität Karlsruhe
[2] Department of Mechanical and Aerospace Engineering, University of California San Diego
villalba@ifh.uka.de, jalamo@ucsd.edu

Summary. Direct numerical simulations (DNS) of stably-stratified, turbulent channel flow at moderate Reynolds number are currently being performed on the XC-4000. A wide range of stratification levels is being considered and large computational boxes are being employed to study carefully the effects of stratification on the wall turbulence. First and second order statistics are discussed. The characteristics of the flow are elucidated using instantaneous snapshots of velocity and density fluctuations. It is shown that, for high stratification levels, the flow remains turbulent close to the walls while internal gravity waves dominate the core of the channel.

1 Introduction

Stratified turbulent shear flows are relevant to many applications in environmental engineering and geophysics. These flows are characterized by a variation of fluid density in the vertical direction that often results in alterations of the flow patterns by buoyancy. Stable stratification suppresses vertical mixing and momentum. However, a stably-stratified fluid can support internal waves and turbulence that play an important role in transport and mixing.

Few computational studies of stratified channel flows have been reported to date. Large Eddy Simulations (LES) have been reported in [1, 2] and direct numerical simulations (DNS) in [3, 4, 5]. These simulations are limited by low friction Reynolds number ($Re_\tau < 200$) and small computational boxes. Moreover, most of these works cover only narrow ranges of stratification. For these reasons, even basic aspects of stratified turbulent channel flow, such as the maximum stratification level for which turbulence can be sustained, are still subject of controversy.

The purpose of the present project is to generate a high-quality database of stably-stratified turbulent channel flow. For this purpose, we 1) consider higher Reynolds numbers than previous works, namely $Re_\tau = 550$, 2) use very large computational domains to ensure that the internal gravity waves, which can be very long and wide, are not constrained and 3) consider a wide range of stratification levels. The resulting database will allow us to improve our understanding of stratified wall-bounded flows and, in particular, will be used to test turbulence models (RANS and LES) employed in engineering and geophysics simulations.

2 The Numerical Experiments

In this study, stably-stratified turbulent plane channel flow is simulated directly. A constant pressure gradient drives the flow and a stable stratification is maintained by imposing a constant upper-wall density which is smaller than the constant bottom-wall density. We consider thermally-stratified air and therefore the Prandtl number is $Pr = 0.7$. The relevant non-dimensional parameters of this flow are the friction Reynolds number Re_τ based on the channel half-height h and the friction velocity u_τ and the Richardson number $Ri = \Delta\rho g h/\rho_0 u_r^2$, where ρ_0 is a reference density, $\Delta\rho$ is the difference in density between both walls, g is the gravitational acceleration and u_r is a reference velocity. If the reference velocity is the friction velocity u_τ, Ri_τ is the friction Richardson number, and if the reference velocity is the bulk velocity u_b, Ri_b is the bulk Richardson number. In an analogous way, Re_b is the bulk Reynolds number. An important non-dimensional number is the Nusselt number, Nu, which measures the increase of wall mass transport due to turbulence with respect to its laminar value.

The governing equations are the Navier-Stokes equations under the Boussinesq approximation,

$$\frac{\partial u_i}{\partial x_i} = 0, \tag{1}$$

$$\frac{\partial u_i}{\partial t} + u_j \frac{\partial u_i}{\partial x_j} = -\frac{\partial p}{\partial x_i} + \frac{1}{Re_\tau} \frac{\partial^2 u_i}{\partial x_j \partial x_j} - Ri_\tau \rho' \delta_{i2}, \tag{2}$$

$$\frac{\partial \rho}{\partial t} + u_j \frac{\partial \rho}{\partial x_j} = \frac{1}{Re_\tau Pr} \frac{\partial^2 \rho}{\partial x_j \partial x_j}, \tag{3}$$

where u_i is the velocity field, ρ the density field, ρ' the density fluctuations and p the pressure field that remains after removing the component that is in hydrostatic balance with the mean density field.

The numerical code integrates the governing equations in the form of evolution problems for the wall-normal vorticity ω_y, the Laplacian of the wall-normal velocity $\nabla^2 v$ and the density ρ, following the formulation of Kim, Moin and Moser (1987) [6]. After some algebra, equations (1-2) can be reduced to yield

$$\frac{\partial}{\partial t} \nabla^2 v = h_v + \frac{1}{Re_\tau} \nabla^4 v - Ri_\tau \left(\frac{\partial^2 \rho}{\partial x^2} + \frac{\partial^2 \rho}{\partial z^2} \right), \tag{4}$$

$$\frac{\partial}{\partial t}\omega_y = h_g + \frac{1}{Re_\tau}\nabla^2\omega_y, \tag{5}$$

where h_v and h_g are the nonlinear terms (see ref. [6] for explicit expressions). Equation (3) is not modified.

The spatial discretization uses dealiased Fourier expansions in the wall-parallel planes and Chebyshev Polynomials in y. The streamwise and spanwise coordinates and velocity components are respectively x, z and u,w. The temporal discretization is third-order, semi-implicit Runge-Kutta [7]. The numerical code has been extensively validated for the unstratified case [8, 9]. Validations for the stratified case have been performed at a lower Reynolds number, $Re_\tau = 180$, [10]. The computational domain is $L_x = 8\pi h$ long in the streamwise direction, $L_y = 2h$ high in vertical direction and $L_z = 3\pi h$ wide in spanwise direction. The number of grid points used in each direction is $N_x = 1536$, $N_y = 257$ and $N_z = 1280$, that results in a grid resolution in wall units of $\Delta x^+ = 8.9$, $\Delta y^+_{max} = 6.7$ and $\Delta z^+ = 4$ after dealiasing with the 3/2 rule.

3 Results

To investigate the effect of increasing the stratification, several simulations have been performed varying Ri_τ. An overview of these numerical experiments is given in Table 1. In the laboratory, an increase in stratification leads to a decrease in the friction at the wall [11]. In the present simulations, however, the pressure gradient that drives the flow and the wall friction are kept constant ($Re_\tau = 550$), and increasing the level of stratification accelerates the flow. In other words, the bulk velocity increases with Ri_τ, and so does the Re_b, as can be seen in Table 1. The Nusselt number quantifies the increase of wall mass transport due to turbulence (in a laminar flow $Nu = 1$). The simulation with the highest stratification level, S4, has a $Nu = 3.36$, indicating that the flow is still significantly turbulent. For comparison, Armenio & Sarkar [2] at $Re_\tau = 180$ and $Ri_\tau = 480$ report $Nu = 1.28$.

Table 1. Overview of the simulations and some important quantities obtained from them. The color column refers to the color lines used in the figures to identify the simulations

Case	Ri_τ	Ri_b	Re_b	Nu	Color
S1	0	0.	10 140	16.45	Black
S2	60	0.072	11 220	6.95	Red
S3	120	0.129	11 882	5.40	Green
S4	480	0.333	14 759	3.36	Blue

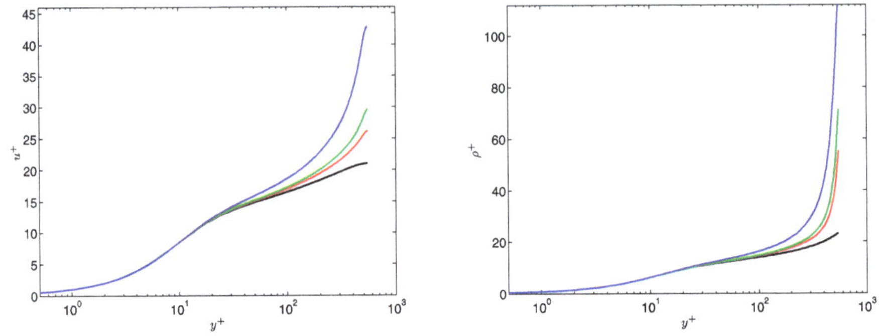

Fig. 1. Left, mean streamwise velocity profile scaled in wall units. Right, mean density profile scaled in wall units. Line styles, see Table 1

3.1 Turbulence Statistics

The increase in the bulk velocity can be clearly seen in the mean streamwise velocity profile displayed in Fig. 1*a*. When using scaling in wall units, all simulations collapse near the wall below $y^+ \sim 20$. Beyond there, the simulation without stratification, S1, presents the typical logarithmic distribution while the others progressively deviate from the logarithmic profile. The mean density profile behaves in a similar way, as shown in Fig. 1*b*.

The turbulence intensities for the three velocity components are displayed in Fig. 2 and the rms density in Fig. 3. They are shown both in wall scaling (u_τ or ρ_τ) and in outer scaling (u_b or $\Delta\rho$). In wall units, all the rms velocity profiles collapse close to the wall and differ around the centre of the channel, similar to the mean profiles. In the central region, the streamwise u' and spanwise w' velocity fluctuations are significantly damped with increasing Ri_τ. The vertical velocity fluctuations v' keep roughly their magnitude but change their y-dependence. The rms density Fig. 3 also collapses in wall units close $y = 0$ but in the centre the density fluctuations appear to increase significantly for $0 < Ri_\tau < 120$, and decay slowly for higher stratifications. As expected for stable stratification, both the velocity and density profiles decrease in outer scaling with increasing Ri_τ, since stratification suppresses vertical mixing and momentum exchange. The figures of density fluctuations display an exception to this behavior at relatively low Ri_τ where the fluctuations are larger at the core of the channel than for the unstratified case. This suggests a possible instability that deserves attention.

3.2 Flow Visualizations

This section compares instantaneous realizations of the flows at lowest and highest Richardson numbers, cases S1 and S4. The inspection of mean and rms profiles suggests that the flow characteristics close to the wall should be

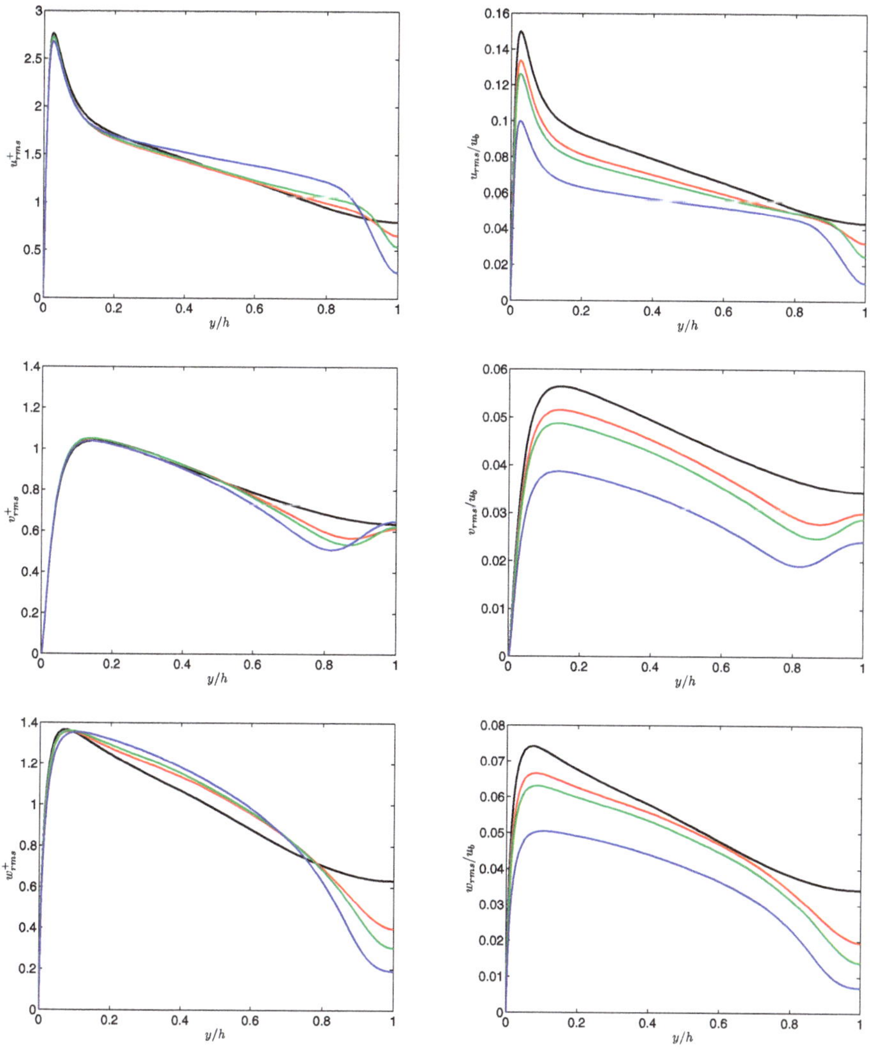

Fig. 2. Top, RMS streamwise velocity profiles. Middle, RMS vertical velocity profiles. Bottom, RMS spanwise velocity profiles. Left, scaling with u_τ. Right, scaling with u_b. Line styles, see Table 1

independent of the level of stratification. This idea is tested in Fig. 4 by looking at the streamwise velocity fluctuations u' in a horizontal plane close to the bottom wall ($y^+ \approx 15$). Low and high speed streaks with similar characteristics are observed in both cases. However, the $Ri_\tau = 0$ case shows a large-scale modulation of the intensity of the streaks that is not observed for high stratification. This modulation is caused by the large-scale structures of

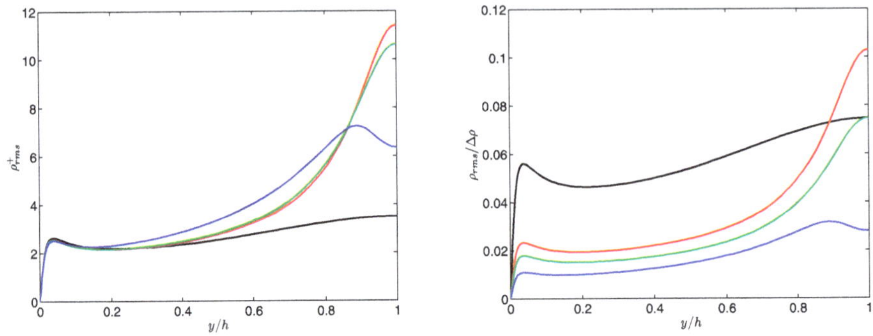

Fig. 3. RMS density profile. Left, scaling with ρ_τ. Right, scaling with $\Delta\rho$. Line styles, see Table 1

the outer region, which penetrate into the buffer layer [12, 8, 13]. The implication is that either these large structures have disappeared due to stratification or they do not penetrate into the wall region. The behavior of the u' profile in the outer region (see Fig. 2 a) favors the first of these options but further analysis needs to be carried out to confirm this idea.

Fig. 5 shows the wall-normal velocity fluctuations v' in a vertical plane parallel to the mean flow. Fig. 6 displays the streamwise velocity fluctuations u' in the same plane. The most striking differences are observed near the centre of the channel. Simulation S4 presents alternating regions of positive and negative wall-normal velocity in this zone, revealing the appearance of internal waves. The analysis of the streamwise velocity fluctuations shows that in the highly stratified case S4 a laminar buffer zone appears in the centre of the channel, which is not crossed by streamwise velocity fluctuations. The existence of this zone is expected to restrict exchange of momentum and vorticity between each half of the channel.

The internal waves can also be appreciated using visualizations of vertical velocity fluctuations v' and density fluctuations ρ' in the horizontal centre-plane, Figs. 7 and 8, respectively. The fluctuations are organized in rather homogeneous spanwise stripes, indicating that the waves propagate in streamwise direction. It is clearly seen that for S4, v' and ρ' have almost opposite phases, which is a typical characteristic of internal gravity waves. For comparison the v' and ρ' fields of S1 show typical characteristics of turbulent flows. Recall that S1 and S4 are relatively similar close to the wall, Fig. 4 but differ substantially in the core region.

4 Computational Details

The original Navier-Stokes code for channel flow solves equations (1-2) without the buoyancy term. In order to consider stratification, an additional equation

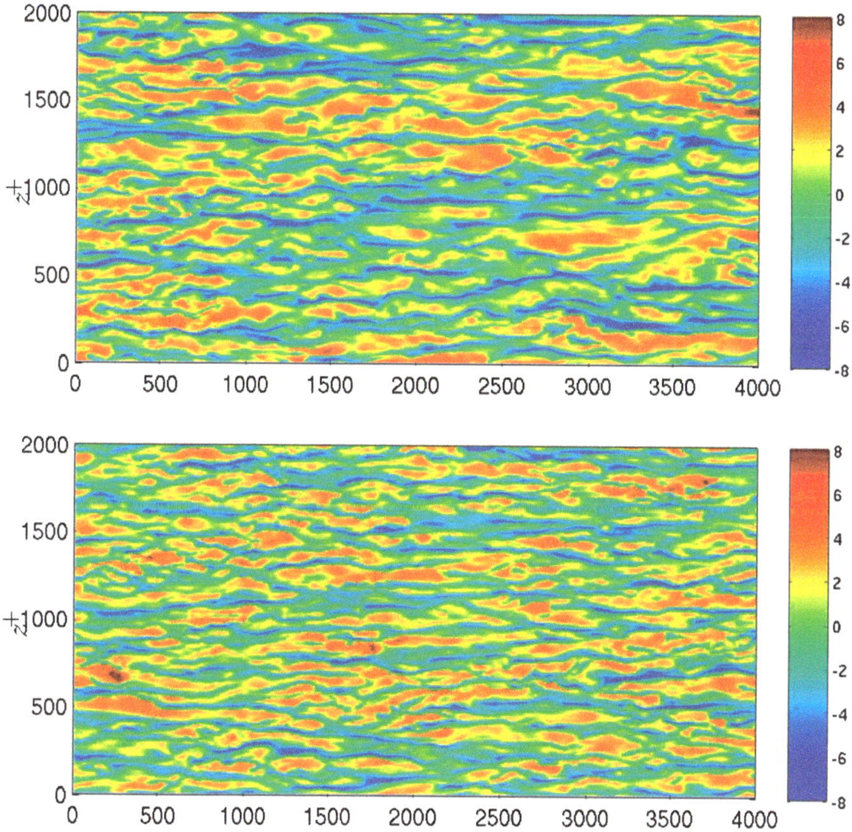

Fig. 4. Instantaneous streamwise velocity fluctuation u'^+ in a horizontal plane at $y^+ \sim 15$. Top, $Ri_\tau = 0$. Bottom, $Ri_\tau = 480$

(3) has been implemented in the code. The computational cost has been seen to increase 30% with respect to the original code due to this addition. To study the performance of the code short test runs of 10 time steps have been performed for various number of processors. Fig. 9 displays the speed up of parallelization up to 128 processors. A typical run uses 128 processors and it consists of two phases. First, for a given Ri_τ, say 120, the flow is initialized with the flow field from the previous Ri_τ, say 60. The flow evolves for a long transient period until it reaches the new steady state. Then, statistics can be collected. To reduce the computational effort, the transient period is run with a lower resolution. The second phase of a run consist of at least 15 time units (h/u_τ) or approximately 240 000 time steps. This corresponds to a clock time of approximately 50 days and a total number of CPU hours of about 150 000.

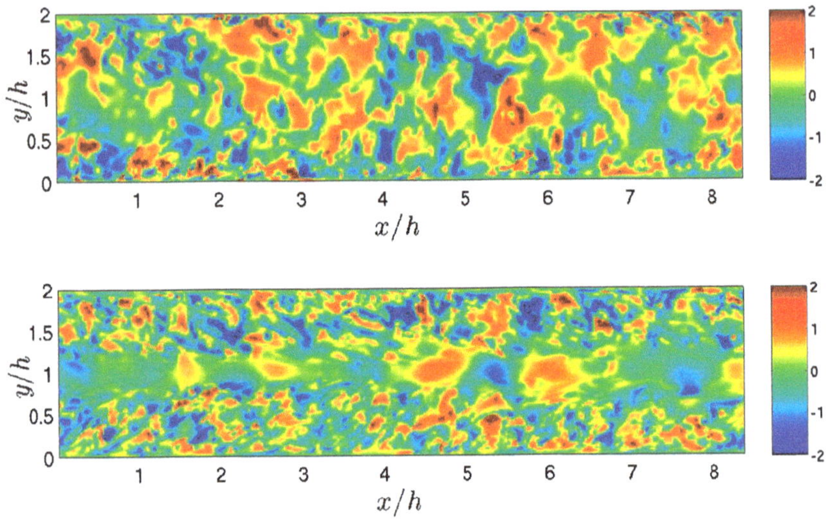

Fig. 5. Instantaneous vertical velocity fluctuation v'^+ in a vertical plane. Top, $Ri_\tau = 0$. Bottom, $Ri_\tau = 480$

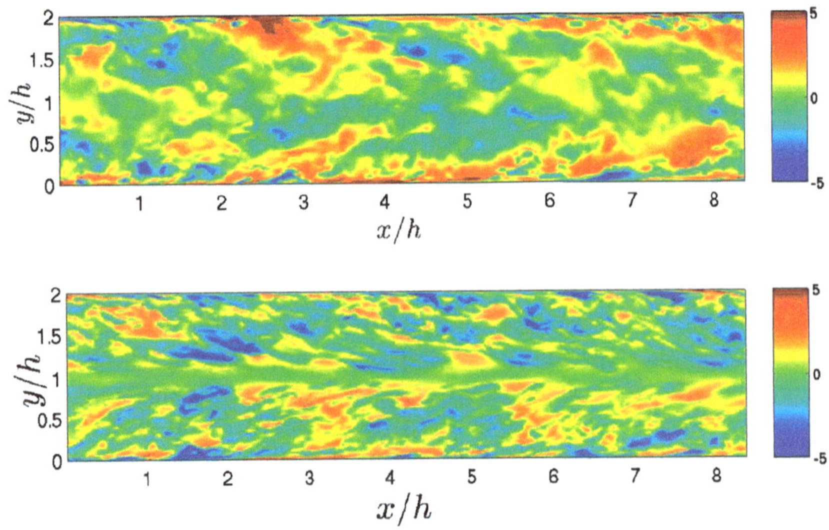

Fig. 6. Instantaneous streamwise velocity fluctuation u'^+ in a vertical plane. Top, $Ri_\tau = 0$. Bottom, $Ri_\tau = 480$

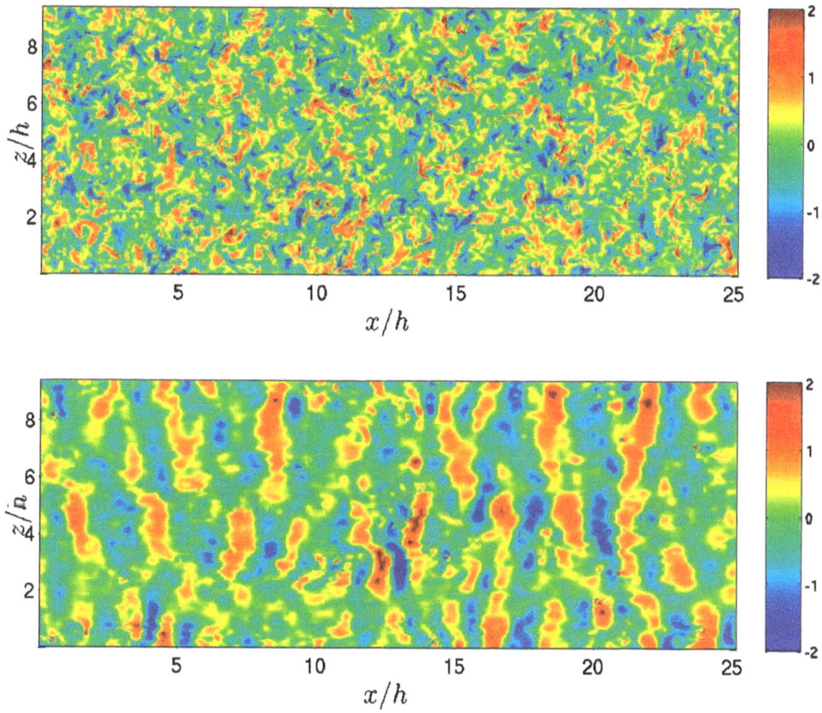

Fig. 7. Instantaneous vertical velocity fluctuation v'^{+}. Horizontal plane in the middle of the channel, $y/h = 1$. Top, $Ri_\tau = 0$. Bottom, $Ri_\tau = 480$

5 Conclusions

In this paper, preliminary results from direct numerical simulations of stably-stratified turbulent channel flow at $Re_\tau = 550$ and various Ri_τ have been reported. It has been shown that the mean and rms distributions, collapse close to the wall if scaled in viscous units. If scaled in outer units, the rms distributions show that increasing stratification reduces the level of fluctuations. Stably-stratified turbulent channel flow, combines wall-turbulence, in the classical way, and internal waves in the central region, for high levels of stratification. The central region, where the internal waves are present, restricts the exchange of vorticity and momentum. A database has been generated that can be used for the validation of turbulence models.

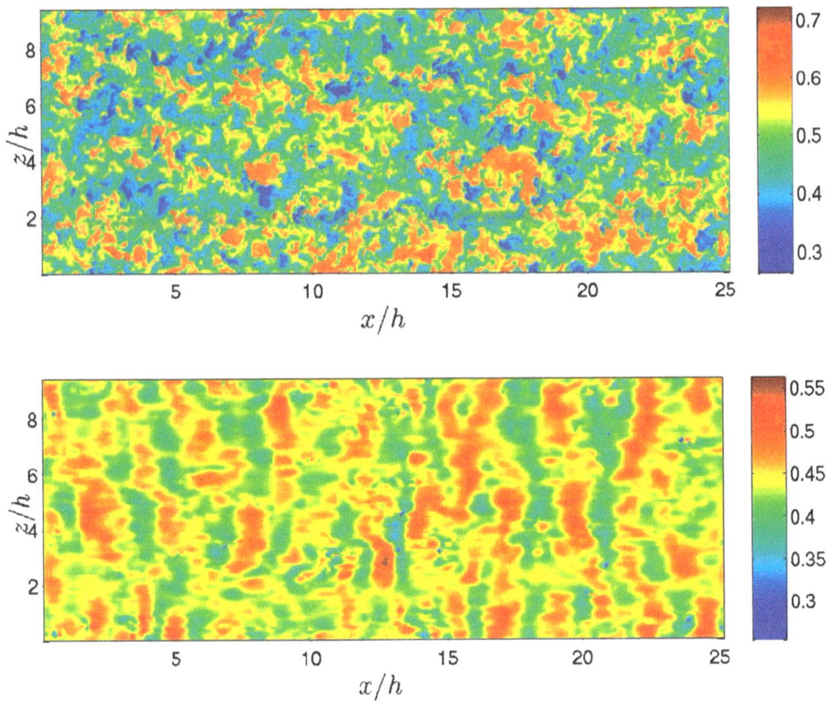

Fig. 8. Instantaneous density fluctuation ρ'^{+}. Horizontal plane in the middle of the channel, $y/h = 1$. Top, $Ri_\tau = 0$. Bottom, $Ri_\tau = 480$

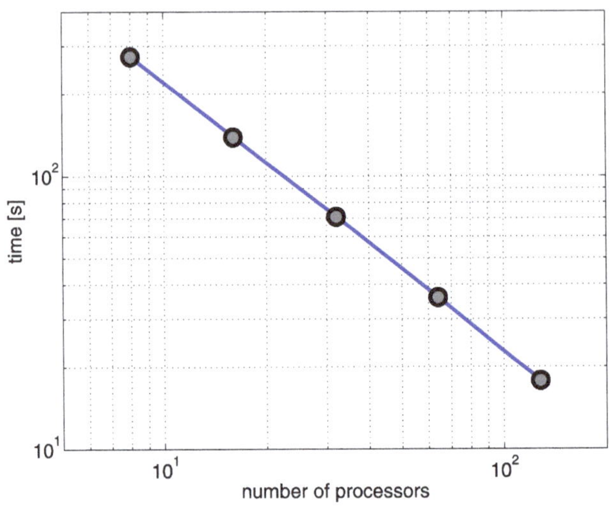

Fig. 9. Time versus number of processors for a single time step

Acknowledgments

The authors are grateful to the steering committee of the supercomputing facilities in Karlsruhe for granting computing time on the XC-4000.

References

1. R.P. Garg, J.H. Ferziger, S.G. Monismith, and J.R. Koseff. Stably stratified turbulent channel flows. I. Stratification regimes and turbulence suppression mechanism. *Phys. Fluids*, 12:2569–2594, 2000.
2. V. Armenio and S. Sarkar. An investigation of stably stratified turbulent channel flow using large-eddy simulation. *J. Fluid Mech.*, 459:1–42, 2002.
3. O. Iida, N. Kasagi, and Y. Nagano. Direct numerical simulation of turbulent channel flow under stable density stratification. *Int. J. Heat Mass Transfer*, 45:1693–1703, 2002.
4. Y.H. Dong and X.Y. Lu. Direct numerical simulation of stably and unstably stratified turbulent open channel flows. *Acta Mechanica*, 177:115–136, 2005.
5. F.T.M. Nieuwstadt. Direct numerical simulation of stable channel flow at large stability. *Boundary Layer Meteorology*, 116:277–299, 2005.
6. J. Kim, P. Moin, and R. Moser. Turbulence statistics in fully developed channel flow at low Reynolds number. *J. Fluid Mech.*, 177:133–166, 1987.
7. P.R. Spalart, R.D. Moser, and M.M. Rogers. Spectral methods for the Navier-Stokes equations with one infinite and two periodic directions. *J. Comp. Phys.*, 96:297–324, 1991.
8. J.C. del Álamo and J. Jiménez. Spectra of the very large anisotropic scales in turbulent channels. *Phys. Fluids*, 15:L41–L44, 2003.
9. J.C. del Álamo, J. Jiménez, P. Zandonade, and R.D. Moser. Scaling of the energy spectra in turbulent channels. *J. Fluid Mech.*, 500:135–144, 2004.
10. M. García-Villalba and J.C. del Álamo. Direct numerical simulation of stably-stratified turbulent channel flow. In preparation, 2008.
11. J.S. Turner. *Buoyancy effects in fluids.* Cambridge Univ. Press, 1973.
12. K.J. Bullock, R.E. Cooper, and F.H. Abernathy. Structural similarity in radial correlations and spectra of longitudinal velocity fluctuations in pipe flow. *J. Fluid Mech.*, 88:585–608, 1978.
13. S. Hoyas and J. Jiménez. Scaling of the velocity fluctuations in turbulent channels up to $Re_\tau = 2003$. *Phys. Fluids*, 18:011702, 2006.

High Resolution Direct Numerical Simulation of Homogeneous Shear Turbulence

Lipo Wang

Institut für Technische Verbrennung
RWTH-Aachen, Aachen
Templergraben 64,52056, Aachen, Germany wang@itv.rwth-aachen.de

Summary. Benefitting from the modern computer technology, Direct numerical simulation (DNS) plays a more and more important role in studying turbulence. To obtain some fine topological structures of turbulent flows, a finer resolution is needed. Because of the better physical properties, larger grid number simulations are more appealing, which, however, become more time-consuming and require more hardware resources. The only method to achieve this is MPI parallelization. In this report presents first some post-processed results for an overall understanding of the physical problem. Then mainly the relevant optimization techniques on NEC SX-8 have been discussed and the reported performance testing showed that the code could run at a reasonable level.

1 Background

Because of the apparent frustrations from the mixture of chaos and order and the multiscale behavior, exact theoretical solution of turbulence does not exist, even for the simplest cases. Experimentation, as a commonly used method for data collecting, has many remarkable advantages. However, due to some inevitable pitfalls in experimentation, such as the response lag, perturbation and poorly changeability of turning parameters etc, direct numerical simulation (DNS) plays a very important and indispensable role in the research of turbulence. Even for theoretical research, DNS becomes possible only in recently years, benefitting from the modern computer technology.

Different from other numerical works, DNS, on the one hand, pursues the Reynolds number as high as possible; one the other hand, must be constringed by the resolution requirements, which make DNS quite hard to reach the applicable level for engineering application. Particularly for our concerns of the fine topological structure in turbulent flows, the requirement of DNS is even harsher. In this report, after a brief introduction of some calculated results for a better understanding of the physical background, mainly we will concentrate on the performance checking on the vector machine (NEC SX-8).

1.1 Direct Numerical Simulation (DNS)

For different application purposes, various efforts have been made for different turbulent flow configurations numerically. In this HRC-DNS (High Resolution Calculation of DNS) project, we have mainly been interested in homogenous shear turbulence, whose configuration can schematically be illustrated in fig. 1.

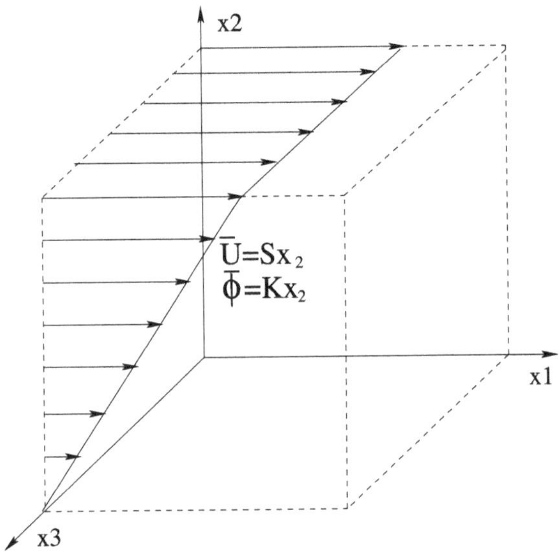

Fig. 1. The physical configuration of homogenous shear turbulence

The time-evolving flow field and passive scalar field are calculated simultaneously. Turbulent motion can be sustained because of the presence of the external shear forcing, and similarly the random fluctuation of passive scalar is maintained by the mean scalar gradient. The mean velocity gradient S and passive scalar gradient K, two free tuning parameters for different simulations, are exerted in the same direction.

To simulate turbulent flows, the most strict requirement is how to ensure the accuracy requirement of the motion of different scales and, at the same, to control the pollution from Boundary Conditions(B.C.). On the one hand, as much as possible the motrion from large scales should be contained to weaken the influence from B.C.; on the other hand, the structures of smallest events should also be resolved to lessen the effect from numerical diffusion. The largest and smallest scales in turbulence are regarded as the integral scalar L and the Kolmogorov dissipative scale η, respectively. However, for the homonomous shear flow, L will increase exponentially with time until the largest eddies hit the boundaries and then it may fluctuate within the domain box. Therefore no much concern is needed for the scale of L for our simulation. A special consideration must be paid to resolving the smallest scales η. Roughly speaking, the grid number N in each direction need to be

$$N \approx Re^{\frac{3}{4}} \approx Re_{\lambda}^{\frac{3}{2}}, \tag{1}$$

where Re and Re_{λ} are the Reynolds number based on the integral scale and the Taylor scale, respectively.

Spectrum method is specially advantageous in DNS because of the high numerical fidelity to solve the structure of small scales. FFT (Fast Fourier Transform) is one of the most favorite spectrum methods. It is numerically perfect to avoid the discretization error, because derivative in physical space can be simply expressed as multiplication with the wavenumber in Fourier space and therefore no finite difference scheme is needed at all. Once the B.C.s are periodic in certain spacial directions, FFT can fit very well to simulate homogenous turbulent flows. In the regard of periodic boundary condition, a moving coordinate attached with the mean flow must be adopted and, correspondingly remeshing (regriding) is needed to adjust the deformation of total calculation domain.

1.2 Dissipation Element Analysis

Because of some prohibitive difficulties of solving turbulence mathematically from the Navier-Stokes equations, alternatively a better understanding of turbulence may rely more on physics than mathematics. There are many efforts for the analysis of turbulence from geometrical aspects. The advantage and importance of the decomposition of flow field into geometrical elements in physical space has been realized and attacked for quite a long time. However due to the difficulty of clearly defining shapes and scales in turbulence, there was no success to get a determined and complete partition of the total turbulent field into many units. One needs to construct a suitable method which can identify specific geometrical elements in the turbulent flow. Which method one should choose is by no means evident.

The application of dissipation element analysis in turbulence is a new method to decompose the entire flow fields into relative simple small geometrical units according to the distribution of some pre-defined scalars [6]. In a turbulent flow field where certain scalar distribution is fully determined, most part of the flow field could be treated as locally monotonous, except these regions around extremal-points. Starting from any material point, its trajectory could be determined by tracing along descending and ascending directions, which is normal to the iso-surfaces of scalar, till extremal points are reached. The ensemble of material points whose trajectories share the same pair of minimum and maximum points defines a spatial region which will be called a dissipation element. From the above definition, we can see that the

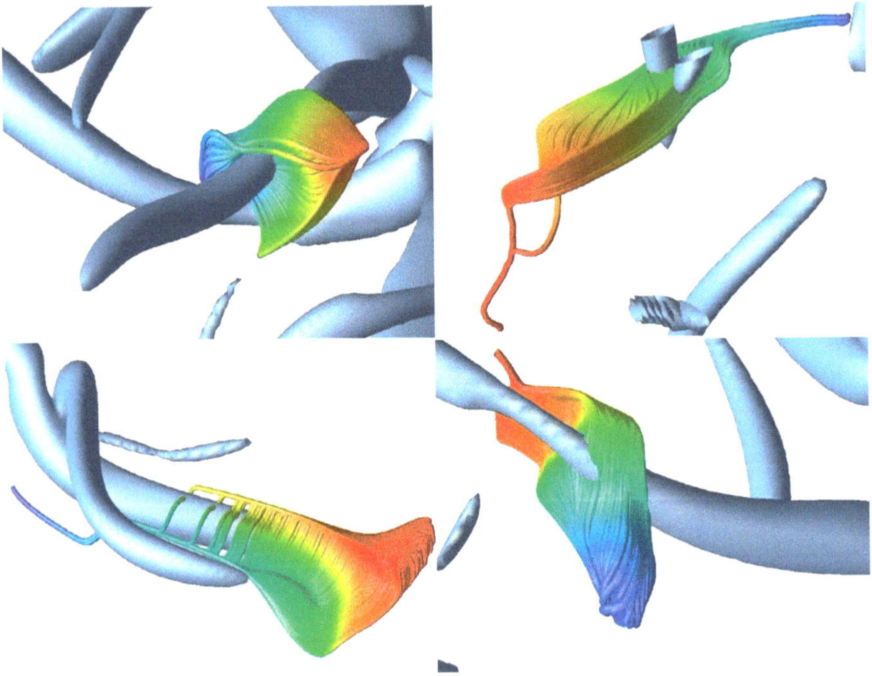

Fig. 2. Examples of the interaction of dissipation elements with vortex tubes

trajectory and DE are totally deterministic objects without any geometrical arbitrariness.

Different with other geometrical conceptions, dissipation elements are space-filling, which means that the total flow field can exactly be partitioned. From DNS data, some examples of dissipation elements, interacting together with vortex tubes, are shown in Fig. 2, where the color shows the value of scalar along trajectories. It can be shown that generally the overall orientations of dissipation elements are perpendicular to vortex tubes, which can be explained by the compressive strain exerted by vortex tubes on the passive scalar. The further important physical results from dissipation element analysis have been introduced in the work by L.Wang and N.Peters [6].

2 Numerical Implementation

2.1 Specialities of HRC-DNS

The spacial resolution criterion (1) is only a rough estimation of the order of magnitude of grid number N. The smallest lengthscale to be resolved is required to be of $O(\eta)$, of the same magnitude order of the Kolmogorov scale

η. Practically depending on different investagations, this criterion may vary to some extent. For example, if DNS is only used for combustion problems, Δx, the grid size in the calculation domain, can be 3 or 4 times η. For some overall flowing properties, for instance, the mean kinetic energy and energy dissipation, 2 times η is a adequate choice for Δx to obtain the first and second order statistics, from the consideration of energy conservation [3]. Therefore it is claimed that requiring η to be resolved as the smallest scale is probably too stringent [2].

But these criteria above become loose when higher order derivatives or special quantities need to be investigated. More strictly, Sreenivasan [5] pointed out that because of intermittency, scales which are much smaller than η can be established locally in turbulence. To resolve also these local small scales or spots of violent events in a strict sense, the grid size must be determined in a more stringent way. Numerical tests [4] [1] have proved this statement. If we want to analyze the behaviors of dissipation elements in turbulence, a finer resolution becomes necessary.

As shown from the examples in Fig. 2, usually dissipation elements assume quite complicated structures in 3D space. Therefore a relative higher numerical resolution is required to discern the fine structure of elements as much as possible without much loss of accuracy. The number of dissipation elements in a flow field is determined by the number of extreme points. Fig. 3 shows a typical result form DNS of the distribution of extremal points and the connections, each of which represents one dissipation element. It may be seen clearly that quite frequently extremal points locate in cluster, with separation distances $\sim \eta$.

To well resolve extremal points in cluster, Δx should be smaller than η to ensure a smooth scalar variation. A benchmark checking goes as follows. For a DNS case with $Re_\lambda = 127$ and $\Delta x/\eta = 1.88$, there exist totally about 60000 extremal points and about 220000 dissipation elements. Once the grid number is doubled in each direction and the resolution becomes as high as $\Delta x/\eta = 0.94$, the number of extremal points and elements drop to about 20000 and 70000, respectively, because higher resolution can smooth the scalar field and many irrelevant kinks and burrs for additional inaccuracy, might be excluded.

For another DNS case with $Re_\lambda = 75$, also two different resolutions are compared. It is found that results form $\Delta x/\eta = 0.44$ and $\Delta x/\eta = 0.85$ remain almost same: the number of extremal points is about 2500. This suggests that $\Delta x/\eta < 1.0$, a more stringent criterion than that for conventional DNS, may approximately work as a resolution cutoff to investigate structures of dissipation elements in turbulence.

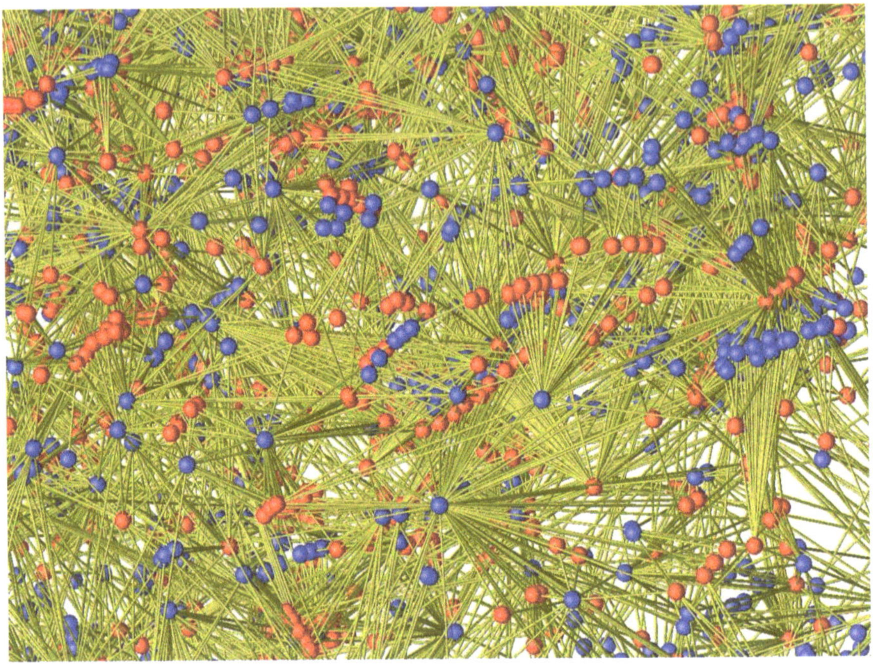

Fig. 3. The clustering structure of extremal points and the corresponding connections there among

2.2 MPI Parallelization

To solve ever larger, more memory intensive problems, or get relative simple problems faster, parallelization is the only possible method. The advantages of approaching more hardware resources makes parallelization very popular for scientific and engineering calculations. The performance of serial computers is becoming saturate, but the combination of multiple processors in a parallel computing architecture may keep the maximum capacity ever increasing steadily.

MPI(Message Passing Interface) is a library of functions (in C) or subroutines (in Fortran) used together with the source code to perform data communication between processes on a distributed memory system. MPI is very effective and necessary for the huge jobs with special demand of memory. Because in this spectrum DNS code, most CPU time will be consumed in the frequently invoked FFT subroutines, to fit the code better with the architecture of the NEC vector machine, an effective MPI-parallelized FFT library is very important. Our great acknowledgment is given to Prof. D.Takahashi (University of Tsukuba, Japan) for the necessary instruction of a parallelized 3D FFT [7] library, which is specially developed for vector machines. Therefore the hardware resource can be effectively utilized. Under the control of

MPI, the shared and private data on different processors communicate with each other for different calculations.

The schematics diagram of the inner structure of the parallelized DNS code is shown in Fig. 4. The original huge 3D arrays $X(1024, 1024, 1024)$ are decomposed in the third direction z into 128 slabs and each of the them is assigned to different processors, under the control of MPI. The communication among the subarrays with shared variables and arrays is done via BUS.

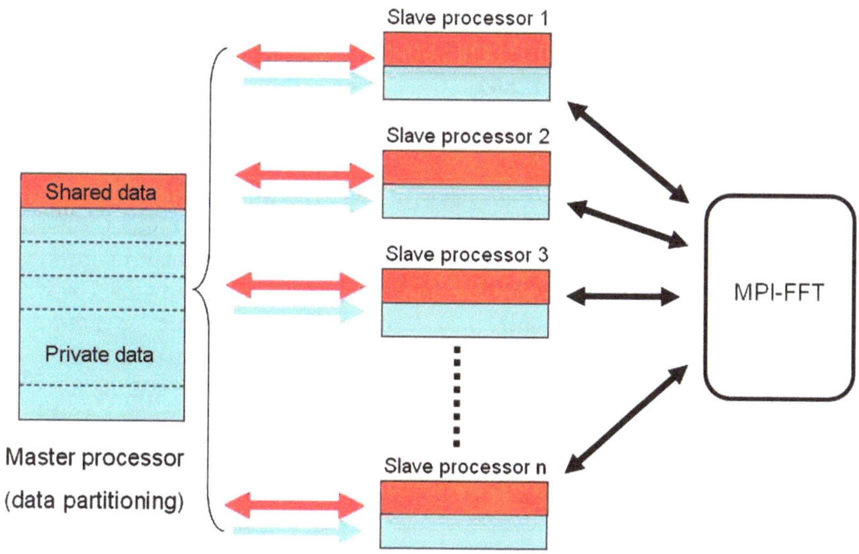

Fig. 4. The structure of MPI parallelized data

3 Performance Testing

Given certain hardware specification, the improvement of computation efficiency is very important to make better use of the available resources. According to different job structures, optimization needs to be considered differently.

3.1 Optimization Measures

For a same code, at different optimization levels, there could be huge performance difference. When jobs become larger, the gains from optimization become larger as well. For the project of HRC-DNS, besides these common treatments of MPI parallelization, such as partition of main large arrays and the using of coding structures fitting vector machines, some further improvements have been adopted as follows.

(1) *I/O*
Because of the execution time limit for each submitted job($< 24hours$), much attention need to be paid to slash the total portion of the CPU time on I/O. For the case of 1024^3 grid points, each time the output needed for next restart read-in is about 60G bytes. In the subroutine **output**, the following statement can ask each thread produce their partitioned outputs parallelly.

```
call tremain(cputime)
if(cputime.le.time) then
call output(filename); here each thread will do the output
call MPI_BARRIER(MPI_COMM_WORLD,IERR)
call MPI_FINALIZE(IERR)
stop
endif
```

The advantage of this parallel I/O lies in that the output can fully be controlled by the function *tremain*, which tells the system how much CPU time is still left. By selecting an appropriate variable of *time*, the output subroutine can get ready just shortly before the job's ending time. Meanwhile, to write huge data blocks out, it is useful to set a reasonable buffer size for these data blocks. It has been tested that the following setting is almost optimal.

```
export F_SETBUF57=1048576 export MPIEXPORT="F_UFMTENDIAN
F_SETBUF00 F_SETBUF06 F_SETBUF57 MPISEPSELECT"
```

where 57 is the ID of the I/O files. For each processor, $1G$ is about the optimal setting.
(2) Parallelized FFT library
For DNS codes using FFT for discrete difference, subroutines related to FFT library belong to the most time-consuming parts. The FFT library we are using is a parallelized 3D FFT, specially designed for vector machines by Dr. D.Takahashi in Japan. For a small (256^3 grid numbers) testing case of our HRC-DNS code, the most intensively invoked routines rank as follows.

Time	Seconds	Cumsecs	#Calls	msec/call	Name
29.2	1026.78	1026.78	1002614784	0.0010	fft4b_
14.6	513.45	1540.22	501307392	0.0010	fft8a_
10.3	362.47	1902.69	40739	8.8974	pzfft3df_
9.9	348.52	2251.21	40854	8.5309	pzfft3db_
6.6	233.88	2485.08	501307392	0.0005	fft235_
5.9	206.94	2692.02	40854	5.0654	ft3dcr_
4.1	143.47	2835.49			ex_qipwr
2.3	82.19	2917.69	1002614784	0.0001	fft4_
1.9	66.47	2984.17	501307392	0.0001	fft8_
1.8	64.76	3048.93	309414	0.2093	hwrite
1.6	57.10	3106.03	75949	0.7518	setzero_
1.4	49.36	3155.39	511	96.59	powerspec1d_

The ratio above remains almost the same for larger cases with different array sizes. Therefore it can be seen clearly that the final performance improvement will depend strongly on the efficiency of the subroutines in the FFT library. It has been tried that using **inline** technique does not help much. However the subroutine *ztranb*() was found to be the most expensive one. From a small test run, it shows that *ztransb*() takes about (33%) of the total CPU time, while the intensively used routines, such as $fft4b()$ and $fft8b()$, take only 8% and 6% CPU time, respectively. Actually *ztransb*() is a copy routine and does not perform FLOPS. For larger cases it may become even more expensive. If *ztransb*() takes a higher share of the run-time, the average performance is reduced. Fortunately it has been tested that the performance can be improved drastically once two compiler directives are inserted into *ztransb*() as follows.

```
!cdir NOLOOPCHG
        do loop1
        enddo
          .
          .
          .
!cdir NOLOOPCHG
        do loop2
        enddo
          .
          .
          .
      RETURN
```

3.2 Performance Report

For the DNS case with total grid numbers of 1024^3, 16 nodes (128 processors) of *NEC SX-8* are needed. The final performance checking is shown in table 1.

Here $[U, R]$ specifies the Universe and the Process Rank in the Universe and the data above is measured from MPI_{Init} till $MPI_{Finalize}$.

The overall data are listed in table 2.

An average performance of 1.5 GFLOPS per process on 128 processes and an Average Vector Length of 146 are reported. The total memory used is 1086 *GB* or 8.6 *GB* per process. This performance is reasonable for DNS codes based on FFT library, which usually is poorly vectorized to some extent, compared with finite difference schemes, because of the frequent invoking of small primary decomposition subroutines.

Typically if the initial input is well given, for example an interpolation from some fully developed field, then usually after 1 integral time the new calculations can provide reasonable results. With the above performance, it has been checked that the running speed is about 0.04 integral time per day

Table 1. MPI performance checking list

Global Data of 128 processes	Min [U,R]	Max [U,R]	Average
Real Time(sec)	85188.687[0,79]	85215.645[0,10]	85202.114
User Time(sec)	84722.290[0,0]	84984.510[0,54]	84909.579
System Time(sec)	42.854[0,113]	128.673 [0,0]	53.058
Vector Time(sec)	55638.568[0,15]	56724.445 [0,120]	55753.603
Instruction Count	14349842109815[0,15]	14514705074132[0,127]	14397468213577
Vector Instruction Count	2492855241312[0,15]	2524846761191[0,120]	2502589365614
Vector Element Count	364238907618599[0,15]	374844704146882[0,120]	368088834670924
FLOP Count	127163910426668[0,123]	127171796964727[0,64]	127167067567418
MOPS	4429.860 [0,15]	4554.280 [0,120]	4475.159
MFLOPS	1496.411 [0,54]	1501.026 [0,0]	1497.677
Average Vector Length	146.113 [0,15]	148.462 [0,120]	147.083
Vector Operation Ratio(%)	96.847 [0,15]	96.914 [0,120]	96.870
Memory size used(MB)	8691.033 [0,0]	8955.643[0,127]	8693.101
Global Memory size(MB)	16.000 [0,1]	32.000[0,0]	18.000
MIPS	169.015 [0,14]	171.004 [0,127]	169.562
Instruction Cache miss (sec)	78.181 [0,0]	82.531 [0,40]	80.749
Operand Cache miss (sec)	4492.328 [0,36]	4517.017 [0,24]	4500.858
Bank Conflict Time (sec)	4590.677 [0,56]	4618.210 [0,79]	4608.131

Table 2. Overall parameters

Real Time (sec)	85215.645
User Time (sec)	10868426.104
System Time (sec)	6791.422
Vector Time (sec)	7136461.222
GOPS (rel. to User Time)	572.820
GFLOPS (rel. to User Time)	191.703
Memory size used (GB)	1086.638
Global Memory size used (GB)	2.250

and approximately 25 days are needed to reach a resolved turbulent field, from which the physical properties can be fully analyzed.

Acknowledgement

The project of HRC-DNS has been continuously sponsored by HLRS. We greatly appreciate the necessary technical and financial support from HLRS. Our special acknowledgement will be given to Mr. Stefan Haberhauer (NEC). Because of the many patient and enlightening discussions with him, our work could proceed smoothly, else the results we have achieved today would be impossible. The necessary instruction from Dr. D.Takahashi in Japan in accessing a parallelized 3D FFT library [7], which is specially developed for vector machines, plays a central role in the calculation.

References

1. Donzis, D.A., Yeung, P.K. et al.: Dissipation and enstrophy in isotropic turbulence: Resolution effects and scaling in direct numerical simulations. Physics of Fluids, **20**, 045108 (2008)
2. Moin, P., Mahesh, K.: Direct Numerical Simulation: A Tool in Turbulence Research. Annual Review of Fluid Mechanics, **30**, 539-578 (1998)
3. Pope, S.B.: Turbulent flows, S.B. Pope. Cambridge University Press (2000)
4. Schumacher, J., Sresnivasan, K.R., Yeung, P.K.: Very fine srtuctures in scalar mixing. J. Fluid Mech., **531**, 113-122 (2005)
5. Sreenivasan, K.R., Schumacher, J., Yakhot, V.: Intermittency and Direct Numerical Simulations. 58th Annual Meeting of the Division of Fluid Dynamics, APS, Chicago, USA. (2005)
6. Wang, L., Peters, N.: The length-scale distribution function of the distance between extremal points in passive scalar turbulence. J. Fluid Mech., **554**, 457-475 (2006)
7. http://www.ffte.jp

Direct Numerical Simulation (DNS) on the Influence of Grid Refinement for the Process of Splashing

Hassan Gomaa[1], Bernhard Weigand[1], Mark Haas[2], and Claus Dieter Munz[2]

[1] Institut für Thermodynamik der Luft- und Raumfahrt, Universität Stuttgart, Pfaffenwaldring 31, 70569 Stuttgart, Germany,
hassan.gomaa@itlr.uni-stuttgart.de
[2] Institut für Aerodynamik und Gasdynamik, Universität Stuttgart, Pfaffenwaldring 31, 70569 Stuttgart, Germany

Summary. The physical process of a drop impact on a thin liquid film is studied numerically with a VOF based code for a Weber number close to the critical Weber number. At this, the influence of the grid resolution is examined. Therefore, the process was simulated on four different three-dimensional grids with a refinement factor of two in each dimension. Maximum grid sizes of about 135 million cells were reached. The investigation covers comparisons of local interface topologies, integral energy balances and secondary droplet properties. Moreover, the numerical phenomenon of flotsam inherent to VOF is examined. Furthermore, the influence of surface tension is analyzed. The simulations were performed on the NEC SX-8 platform of the HLRS.

1 Introduction

Water is injected into the compressor of stationary gas turbines to increase the overall efficiency. So called High Fogging systems inject such a big amount of water into the compressor so that the droplets do not completely evaporate, therefore enter the compressor cascade area and influence the flow topology around the vanes which are usually optimized for dry flow. Accordingly, the droplets in the flow lead to problems such as corrosion or change of the flow angle at the vanes' trailing edge, entailing an immediate deterioration of the compressors efficiency. Hence, it is a major issue of research to understand the processes resulting from the droplets in the flow. Within this scope, the interaction of droplets and wetted walls is examined. The considered process is highly unsteady and its appearance is dominated by occurring instabilities that have a wide range of different scales. Thus, it is a demanding challenge to achieve an accurate numerical simulation of this process.

This is achieved by direct numerical simulations (DNS) of that particular process using the inhouse 3D CFD code free surface 3D (FS3D), developed at the Institue of Aerospace Thermodynamics (ITLR). More specifically, the results of a particular case were examined with a specific focus on the influence of different grid resolutions. Four different resolutions were studied and analyzed.

Firstly, a short description of the physical phenomenon is given. Subsequently, the numerical method of FS3D is presented to be followed by a description of the numerical setup. The presentation of the results is started with a comparison of the flow topologies gained for the different grid resolutions. Afterwards, the occurring energies are balanced to evaluate the numerical performance of the code for the applied grids. Furthermore, the influence of artificial perturbations is studied for the investigated resolutions. Moreover, the integral properties of secondary droplets and flotsam particles are presented. Finally, the influence of the surface tension within the performed simulations is examined.

2 Physical Phenomenon

In general a drop impact on a wetted surface may result into two phenomena. If the momentum of the droplet is sufficiently large, secondary drops are ejected and the process is referred to as *splashing*, as depicted on the right side of Fig. 1. The term *deposition* indicates an impact without production of secondary droplets.

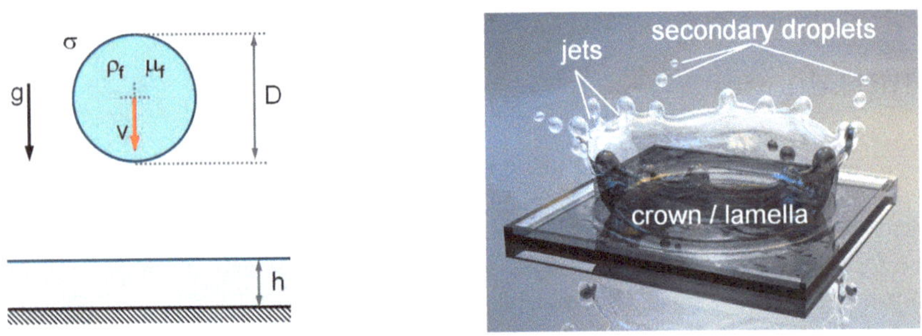

Fig. 1. Schematic of the characteristic variables governing the interaction of a drop and a wetted wall (left) and calculated topology of a splash [6] (right).

The characteristics of the process are mainly determined by the properties of the impacting drop and the film (density, viscosity, surface tension, droplet diameter and velocity, impact angle) which are depicted on the left side of Fig. 1. These properties are represented by four dimensionless numbers, namely the Weber number ($We = \rho v^2 D/\sigma$), the Ohnesorge number ($Oh = \mu/\sqrt{D \rho \sigma}$), the non-dimensional film thickness ($\delta = h/D$) and the dimensionless time ($T = t D/v$), where T is set to zero at the moment of direct contact between drop and film. As shown later, gravity plays a negligible role so that the Bond number need not to be considered for the present investigation.The critical Weber number determines a limit that allows an approximate prediction whether splashing or deposition is expected for the considered case. For configurations with Weber numbers above the critical value splashing and beneath it deposition is expected. An accordant critical Weber number limit is for instance given by Maichle [1], who determined the limit numerically. It depends on the Ohnesorge number and the non-dimensional film thickness.

Referring to Cossali et al. [2] the splash can be subdivided into four temporal phases:

1. **The drop impact with lamella formation:** The impact results in a strong pressure increase within the impact area, leading to a propagation of a pressure wave in radial direction. This causes a radial flux of the fluid, resulting in the formation of the lamella.
2. **The crown formation:** In this phase the lamella continues to propagate in radial direction and grows in height, forming the crown. Along the crown surface two perturbation families can be observed; the longitudinal and the azimuthal waves. The longitudinal waves propagate vertically along the crown surface forming the crown, while the azimuthal waves yield to periodical nodes along the crowns' rim.
3. **The jet formation and break up:** The mentioned azimuthal waves appear to be in correspondence of the roots of the jets that are formed within this phase. It shall be mentioned that a complete theory about the relation of perturbations in the rim and the raising of the jets has not yet been developed. However, due to surface tension the jets constrict and break up through a Rayleigh process which leads to separation of secondary droplets.
4. **The crown collapse:** Resulting from the opposing action of surface and inertial forces the crown collapses within this last phase. During the crown collapse some larger secondary drops are ejected by the surviving jets.

3 Numerical Method

As mentioned before, the numerical results presented here were generated by the inhouse 3D CFD program FS3D, which was specially developed to compute the Navier Stokes Equations for incompressible flows with free surfaces using Direct Numerical Simulation (DNS). The flow field is computed by solving the governing equations for mass conservation

$$\frac{\partial}{\partial t}\int\limits_{\Omega} \rho dV + \int\limits_{\partial\Omega} (\rho\mathbf{u})\cdot\mathbf{n}dS = 0 \tag{1}$$

and momentum conservation

$$\frac{\partial}{\partial t}\int\limits_{\Omega} \rho\mathbf{u}dV + \int\limits_{\partial\Omega} ((\rho\mathbf{u})\circ\mathbf{u})\cdot\mathbf{n}dS = \int\limits_{\partial\Omega} (\mathbf{S}-\mathbf{I}p)\cdot\mathbf{n}dS + \int\limits_{\Omega} (\rho\mathbf{g}+\mathbf{f}_\gamma)dV \tag{2}$$

for two immiscible fluids. Here Ω denotes the volume and $\partial\Omega$ the boundary surfaces of the considered control volume. For Newtonian fluids the shear stress tensor $\mathbf{S}$ is given by

$$\mathbf{S} = \mu\left[\nabla\circ\mathbf{u} + (\nabla\circ\mathbf{u})^T\right]. \tag{3}$$

The surface tension, which is stated by the capillary stress tensor $\mathbf{f}_\gamma$, is handled as a volume force that acts on every cell containing an interface. It is computed by the conservative continuous surface stress (CSS) model suggested by Lafurie et al. [3]. Due to the mentioned incompressibility and the assumed constant fluid properties in each phase the energy equation is decoupled from the momentum equation and not solved within this study.

To describe the interface the Volume of Fluid (VOF) method by Hirt and Nichols [4] is used, where an additional variable f, specifying the liquid volume fraction, is introduced as follows:

$$f(\mathbf{x},t) = \begin{cases} 0 & \text{whithin the gaseous phase} \\ [0;1] & \text{along the interface cells} \\ 1 & \text{whithin the liquid phase} \end{cases} \tag{4}$$

From this, by simple scaling with the properties of the gas and the liquid, the density and viscosity field, might be calculated to

$$\rho(\mathbf{x},t) = \rho_g + (\rho_f - \rho_g)f(\mathbf{x},t) \tag{5}$$
$$\mu(\mathbf{x},t) = \mu_g + (\mu_f - \mu_g)f(\mathbf{x},t) \tag{6}$$

To describe the temporal and spatial evolution of the volume fraction f an additional transport equation, analogously to Eq. (1), is defined by

$$\frac{\partial}{\partial t}\int\limits_{\Omega} f dV + \int\limits_{\partial\Omega} (f\mathbf{u})\cdot\mathbf{n}dS = 0\,. \tag{7}$$

The presented equations are discretized by finite volume schemes with an accuracy of second order in space on a staggered grid and first order in time. The calculation of the fluxes in Eq. (7) is based on a piecewise linear reconstruction of the interface in each cell, the so called piecewise linear interface reconstruction (PLIC) method proposed by Rider and Kothe [5], which ensures a sharp interface. The mass and momentum convection, as well as the momentum diffusion and the occurring body forces are treated explicitly.

The Poisson equation for the pressure resulting from the assumption of incompressibility has to be solved implicitly. In FS3D this is carried out by a multigrid solver that uses Red-Black Gauss Seidel as smoothing algorithm. A more detailed description of the used numerical method is given in Rieber [6].

4 Numerical Setup

Within the presented study the physical characteristics of the examined process were chosen to resemble the conditions in actual compressors. Accordingly, the liquid was chosen to have common properties as water, and the gas was set to air. The droplet at initialization is idealized as a sphere with the diameter $D = 40\mu m$ and falls from a height $H = 2\,D$ on a film with the dimensionless thickness of $\delta = 0.2$. For the considered case the Weber and Ohnesorge number were chosen to $We = 404$ and $Oh = 1.5 \cdot 10^{-2}$. Applying the correlation given in Maichle [1] the critical Weber number is about $We_{crit} = 294$, therefore, lies underneath the Weber number for the considered case. Hence, splashing is expected for the simulated configuration.

Since the droplet falling direction is chosen to be normal to the surface, symmetry can be utilized twice. Hence, the two corresponding boundary conditions are chosen to be symmetrical and only a quarter of the physical domain is computed. The floor is modeled as a wall with no slip condition, and the remaining three sides are open, thus, continuous boundaries. This is necessary to allow separated secondary droplets to leave the computational domain without distorting the results. Of course, taking symmetry into account introduces additional stiffness into the simulation, but the possible increase of resolution is considered to be more significant in this case.

To analyze the grid convergence, calculations are performed on equidistant Cartesian grids with four different resolutions, that are generated by a refinement factor of two in each dimension. The discretization of the computational domain varies from 64 to 512 cells in each dimension. Thus, the total amount of cell numbers results in a range of about $250 \cdot 10^3$ to $130 \cdot 10^6$. Table 1 gives a survey of the mentioned grid properties. Moreover, the absolute resolutions and the number of cells discretizing one droplet diameter are presented for the different grid configurations.

To ensure the existence of disturbances that initiate the break up process in phase three, the velocity field is initially disturbed by a random undirected noise field that is characterized by its standard deviation. For the considered case the standard deviation is set to 5% of the drop's falling velocity.

Table 1. Overview of the calculated grids

Grid	No. of Cells	Resolution	Cells per D
64^3	$0.26 \cdot 10^6$	$3.125\,\mu m$	≈ 13
128^3	$2.10 \cdot 10^6$	$1.563\,\mu m$	≈ 26
256^3	$16.78 \cdot 10^6$	$0.781\,\mu m$	≈ 51
512^3	$134.22 \cdot 10^6$	$0.391\,\mu m$	≈ 102

5 Results

5.1 Flow Topology

As a start, a comparison of the gained morphology is made. In Fig. 2 the results for the selected dimensionless time $T = 7.9$ are presented for the different grids. To illustrate the interface, isosurfaces of the f-variable with the value $f = 0.1$ are plotted in the graphs.

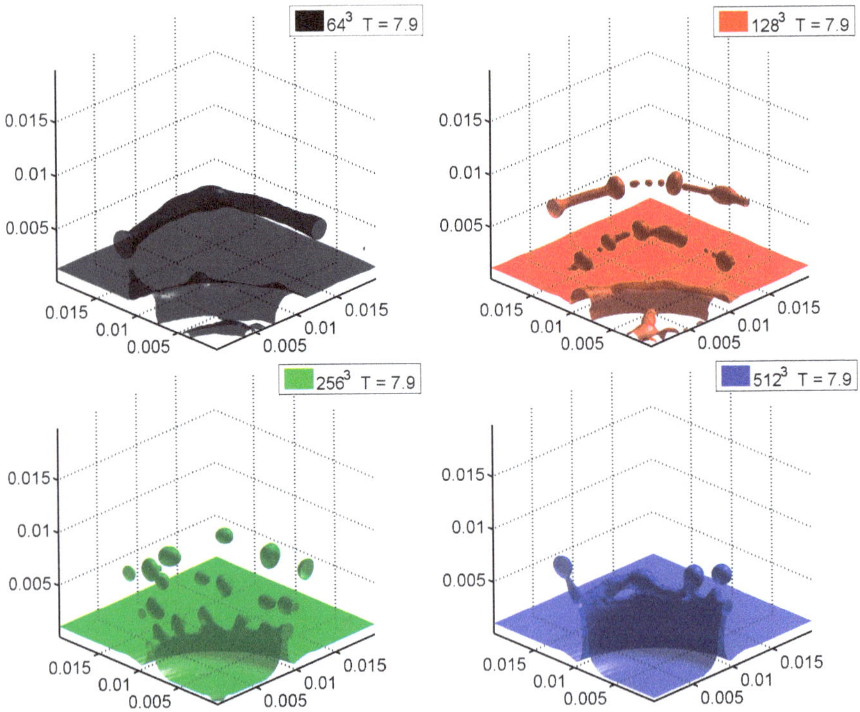

Fig. 2. Interface topologies for the different grids at $T = 7.9$

The first thing to notice is the distinct difference of the interface topology for the different grids. For the two coarser grids ring shaped structures are recognizable, which partly disintegrate into secondary droplets. Thus, it appears that these grids are simply too coarse to be able to gather the subtle structures dominating the separation process and therefore the flow topologies.

By contrast, the reproduction of the physical process by the two finer grids is basically reasonable, since the droplets are in fact generated through a break up process of the jets. The visible deviations of the interface topology occurring for the two finer grids are primarily influenced by the moment of lamella break up. Generally, it was found that increased grid resolutions lead to longer conservation of the lamella. This causes a general temporal offset of the physical phases within the whole process for different grid resolutions. It is therefore not reasonable to compare the interface topologies for similar time steps, since deviations appear larger than they actually are. Instead the comparisons have to be drawn with regard to the appearance of the physical events as done in Fig. 3 for the two finer grids. It is obvious that an improved agreement is achieved.

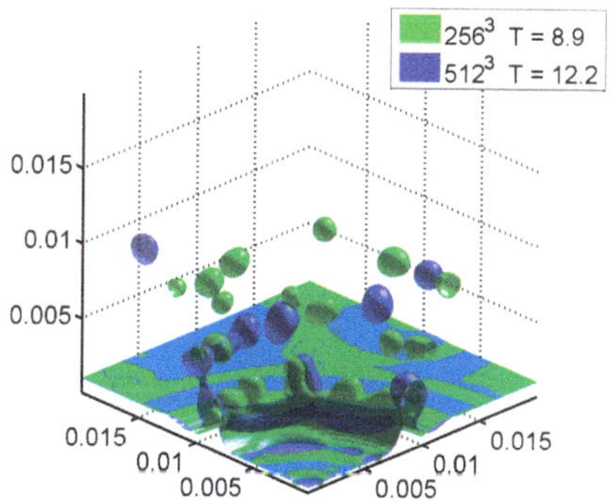

Fig. 3. Interface topology for 256^3 and 512^3 grid at similar process phases

However, the temporal offset of lamella break up causes an extension of the crown growth phase resulting in rising surface energies and decreasing kinetic energies and therefore also has physical consequences. Thus, it is expected that with increasing grid refinement separated secondary droplets have lower kinetic energies.

In fact, regarding the temporal evolution of the two finer grids, the separated droplets within the 256^3 grid case shoot out of the computational domain while the generated droplets of the 512^3 case mainly remain at their position. In this connection it shall become clear that the moment of disintegration is the decisive factor for the whole further appearance of the interface topology and the residual deviations in Fig. 3 are explicable.

It is also interesting to observe that in addition to the main separation cycle in the case of the 512^3 grid, a prompt splash in the initial phase immediately after impact is visible, which does not occur for the coarser 256^3 grid. Overall, there are some visible differences of the local interface topologies depending on the used grid.

Nevertheless, the arising topologies are quite adequate with regard to the high complexity of the simulated process.

5.2 Energy Balance

In order to evaluate the numerical performance of the applied code the mechanical energy is balanced in the following. Due to the assumed incompressibility and constant fluid properties, the energy equation is decoupled from the momentum equation presented above in Eq. (2). Moreover, constant temperature was assumed throughout the whole field; thus, the energy equation given in Eq. (8) is automatically satisfied.

$$\frac{\partial}{\partial t}\int\limits_{\Omega}\left[\frac{1}{2}\rho\mathbf{u}^2-\rho\mathbf{g}\mathbf{x}+\sigma a_\gamma\right]dV+\frac{\partial}{\partial t}\int\limits_{\partial\Omega_\Gamma}\sigma cos\theta_0\,dS = \tag{8}$$

$$\int\limits_{\partial\Omega}-\mathbf{u}\cdot\left[\frac{1}{2}\rho\mathbf{u}^2-\rho\mathbf{g}\mathbf{x}+\sigma a_\gamma\right]\cdot\mathbf{n}\,dS+\int\limits_{\partial\Omega}\mathbf{u}\cdot(\mathbf{S}-\mathbf{I}p)\cdot\mathbf{n}\,dS+\int\limits_{C_\gamma}\mathbf{u}\cdot\sigma\mathbf{n}_\tau\,dC\,.$$

Here $a_\gamma = d\Gamma/dV$ indicates the so called interface density representing the interface area within a cell. The second integral on the left side of Eq. (8) representing the temporal evolution of the surface energy along the wetted wall $\partial\Omega_\Gamma$ in dependence on the contact angle θ_0 is negligibly small for our case and is therefore not considered. Moreover, the last integral on the right side representing the accomplished work due to surface tension along the boundaries of the computational domain is neglected. Based on the assumed incompressibility, pressure waves leave the computational domain infinitely fast. Hence, it is not possible to evaluate the work induced by the pressure wave flux at the boundaries in the second integral on the right side.

The temporal evolution of the remaining terms of Eq. (8) for the two finer grids are depicted in Fig. 4. Here E_{kin} denotes the balanced kinetic energy, E_σ the surface energy, E_μ the viscously dissipated energy, E_{pot} the potential energy and ΣE the effective sum of the considered energies. Moreover the initial value of the energy sum is depicted for comparison. An ideal, compressible and nondissipative numerical algorithm would lead to a temporal evolution of the energy sum that is identical to its initial value. All energies are scaled by $\rho_f\, D^3\, v^2$, where ρ_f is the density of the fluid.

Examining Fig. 4, the expected insignificance of the potential energy due to the marginal spatial scales for both depicted cases is noticeable. Furthermore, the interaction between kinetic and surface energy is vividly visible. Immediately after the impact at $T = 0$ the fluid is strongly decelerated, so the kinetic energy decreases. At the same time the lamella is built up leading to an increase of the surface energy. At the subsequent interval of jet break up and droplet separation around $T \approx 9$, the descent of surface energy and rise of kinetic energy is evident.

Within the initial phase before impact it is interesting to notice, that the viscously dissipated energy quickly jumps to a finite value, while the kinetic energy diminishes by approximately the same value. As shown later, this is linked to the artificially added velocity distortions. E_μ also increases signifi-

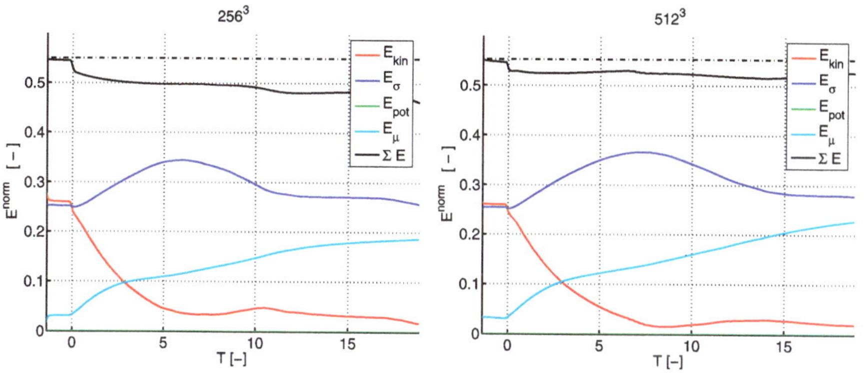

Fig. 4. Energy balance for the 256^3 and 512^3 grid

cantly straight after impact due to the large velocity gradients appearing in that phase.

Before the impact E_μ corresponds to E_{kin}, hence the energy sum almost stays constant. During the impact a considerable drop of ΣE is visible, which declines with increasing grid resolution. Most likely this is due to pressure waves that run infinitely fast through the computational domain and could therefore not be captured in the balance. In the further course, the energy sum ΣE only slightly decreases due to numerical dissipation, while the major offset is generated at the impact time.

Despite the visible deviations between the interface topologies, there is a widely good agreement of the energy evolutions for the two grids presented. In particular, the plots of kinetic and surface energy correspond well for the two finer grids. The increasing grid resolutions lead to an ascending slope of E_μ. The deviations between the energy sum and its initial value reduce with higher refinements.

5.3 Disturbances

In the preceding section, the connection between the imposed velocity field disturbances and the strong descent of the kinetic energy in the phase before impact was mentioned. To clarify the relevance of the added disturbances additionally to the 5% standard deviation, a 50% deviation case and a non-disturbed case were simulated. The results are shown in Fig. 5. Immediately after initialization, the kinetic energy of both disturbed cases approach towards the plateau of the non-disturbed one. Here, the random disturbance of the velocity field causes nondirectional velocity gradients that are damped due to viscous shear. Accordingly the jump of E_μ occurs before impact, as shown in Fig. 4.

As illustrated in Fig. 5 the kinetic energy of the disturbances are damped faster with increasing grid resolutions. An analysis of the appearing frequen-

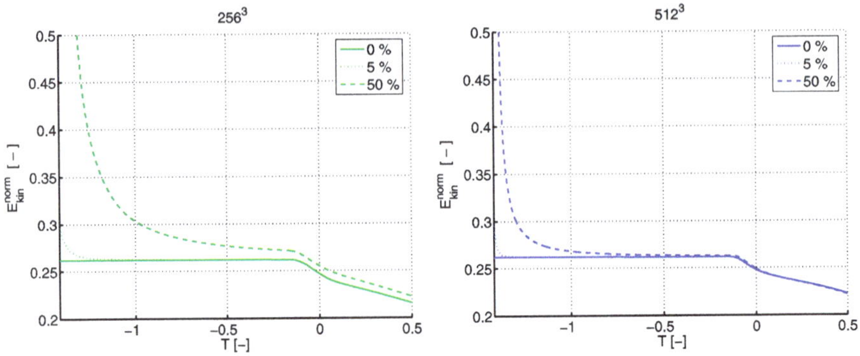

Fig. 5. Temporal evolution of the kinetic Energy, due to articicial velocity disturbances for the 256^3 and 512^3 grids

cies for moderate initial disturbances showed that the imposed perturbations are widely vanished in the moment of the impact. However, perturbations that dominate in regard to their magnitude are generated along the droplets' surface during its downward movement. This effect is amplified with increasing grid resolutions, as well.

5.4 Secondary Droplets

After having studied the properties of the whole system, it is dealt with the properties of the separated droplets, which surely are of major importance with regards to the introduced technical application. Connected droplets and ligaments are identified by a region growing algorithm applied on the f field. Within this section the spuriously, numerically generated flotsam particles are filtered out. They are studied in the next section.

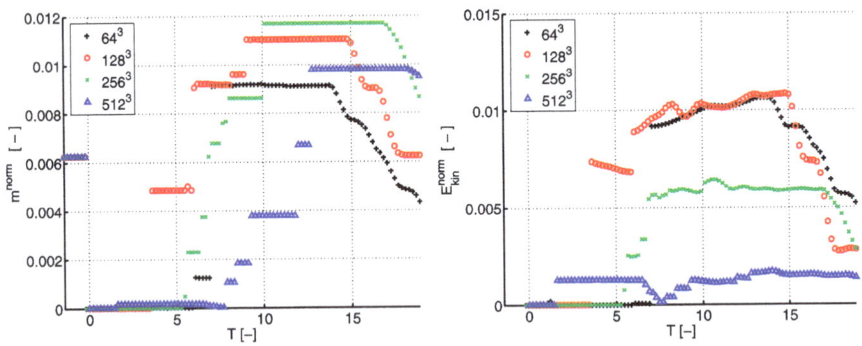

Fig. 6. Temporal evolution of secondary droplets' masses and kinetic energies for the different grids

First, it shall be mentioned that around the moment $T \approx 13$ in all four cases separation is completed and the number of secondary droplets are within the same range of about 10 droplets. Examining the temporal evolution of the droplets' mass shown on the left side of Fig. 6, which is scaled by the initial fluid mass in the system the grid dependent moments of break up can be identified. For the finest grid very small droplets that leave the computational domain comparatively fast are foremost generated in a first separation cycle at $T = 2$. Altogether, it is observable that after separation different levels of the droplets' masses arise for the different resolutions. Nevertheless, it is incidental that for the two finest resolutions nearly the same secondary droplet masses are settled considering both separation cycles for the finest grid. However, the size spectrum is different for these two grids, which is peripheral for our purposes. Assuming that the droplets can be idealized as spheres it is directly deducible that also the surface energies lie within the same range.

The evolution of the kinetic energy is depicted on the right side of Fig. 6, scaled by $\rho_f D^3 v^2$. As expected, the kinetic energy level of the droplets is increasing for coarser grids due to the earlier moment of break up. Including the energies of the first break up cycle of the 512^3 grid, similar levels for the kinetic energies of the two finer grids follow which is an essential result for the process modeling.

To sum up generally good agreement for the integral properties of the whole system and the secondary droplets was found, even though some deviations appeared for the values of the studied local fields.

5.5 Flotsam

This section deals with the so called flotsam particles. The appearance of flotsam is a numerical phenomenon and a result of applying the PLIC method to reconstruct the interface in connection with the used VOF method. During the linear reconstruction, minor f-values are detached from the actual interface and lead to midget droplets moving through the computational domain. By virtue of their emergence, these flotsam particles have volumes smaller than a cell in the computational area. Indeed, flotsam is mentioned frequently in connection with the VOF method; however, an analysis of their relevance was not found in open literature. Therefore, their properties are studied in the following.

On the left side of Fig. 7 the amount of occurring flotsam particles is plotted in a logarithmic diagram. Apparently the amount of flotsam strongly increases with finer resolutions. A spectrum from about 200 particles for the coarsest grid up to nearly 6000 particles for the finest grid is noticed.

Examining the mass of all flotsam particles shown on the right side of Fig. 7, scaled in the same way as the secondary droplets above, a clear tendency is recognizable. With increasing grid resolution the sum of all particle masses strongly decreases, which is an essential result regarding the consistency of the applied numerical algorithm. At the same time it is visible that

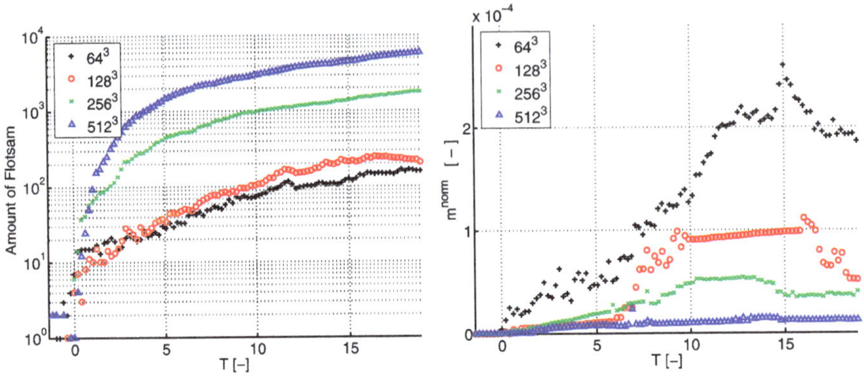

Fig. 7. Temporal evolution of flotsam particle amount and masses for the different grids

the scale of the particle masses is lower by a factor 100 than the masses of the secondary droplets, and therefore completely insignificant for the results of the simulations. Similar results are gained for the flotsams' surface and kinetic energy.

5.6 Influence of Surface Tension

To investigate the cause of the temporal topological grid dependency that was detected before, amongst others the surface tension was set to zero. The corresponding interface pictures are depicted on the right side of Fig. 8. On the left side the standard case is presented for comparison. $T = 6.5$ was chosen as a moment of advanced progress, where all essential effects are visible.

To begin with the coarser 64^3 and 128^3 grids, the characteristic ring structures are visible again, indicating the unsuitability of these grids to reproduce the studied process. The difference of the two finer grids for the standard case on the left side is clearly visible along the crown rims. It was mentioned before that this deviation is based on the different temporal moments of break up. In case of deactivated surface tension, the interfaces of the 256^3 and 512^3 grids agree to a large extent. For this case, it is particularly interesting to observe the disintegration process of the rim. The only action opposing the initial forces is caused by viscosity; hence, droplets can physically not be formed due to the missing surface tension. Instead, the rim constricts to a thin film which disintegrates by reason of perturbations along the interface and the application of PLIC. In case of the 128^3 grid the droplets appearing on the right side of Fig. 8 are generated through disintegration of the preceding ring structures by PLIC. For the two finer grids, the lamellas are extremely thin and lie in the range of single cell sizes, which on the one hand leads to detachment by PLIC. On the other hand the appearance of droplets also result as a visualization problem using isosurfaces with a fixed f-value.

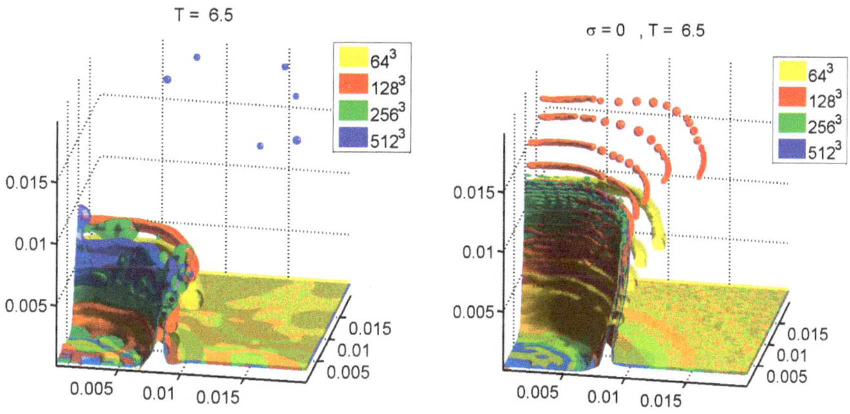

Fig. 8. Interface topology for the standard case (left) and for $\sigma = 0$ (right) at $T = 6.5$

These results are also valuable for the physical perception of assuming a negligible surface tension, as done by Yarin and Weiss [7], who derived analytical models for the process.

6 Performance

All presented results were computed on the NEC SX-8 platform at HLRS. As a result of the explicit calculation of the diffusive terms and the corresponding stability limits for the allowed time steps ($\Delta t \propto 1/n^2$) a massive increase of the computing time is caused by the performed grid refinement. Thus, the computation time grows from 0.5 CPUh for the coarsest 64^3, to 4 CPUh for the 128^3 grid, 104 CPUh for the 256^3 grid, up to 2700 CPUh for the finest 512^3 grid.

The percental division of the computing time of the particular subprocesses is depicted in Fig. 9. For reasons of comparability, all presented results were computed with a parallel version of the program on one node with 8 CPUs. As common for incompressible cases the main part of CPU-time is required for solving the pressure Poisson equation using the multigrid solver. Two phase flows with high density ratios automatically lead to high multigrid iterations cycles until convergence. In combination with the accordant computer architecture multigrid solvers become extremely efficient on very large grids and the solver applied in FS3D is specially optimized for these purposes. The number of pre- and post smoothing steps is adapted during runtime and overrelaxation of the Gauss-Seidel scheme is introduced depending on the convergence rate of the scheme.

As shown in Fig. 9 the optimization is worth the effort, since the percental CPU-time requirement decreases with increasing resolutions. For the finest 512^3 grid the multigrid solver reaches high performances up to 5 GFLOPS

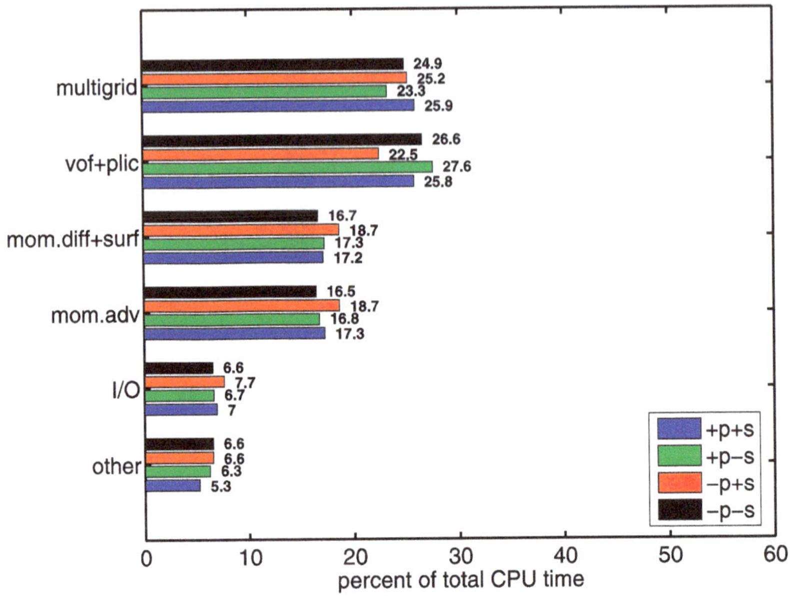

Fig. 9. Overall CPU-time requirement categorized by subprocesses for all grids

and V. op. Ratios ≈ 99%. The momentum diffusion and advection schemes, as well as the volume fraction advection schemes reach performances up to 10 GFLOPS and V. op. Ratios > 99%. Altogether a performance of about 4.5 GFLOPS, V. op. Ratios ≈ 98% and vector lengths of 215 are achieved, assuring a high computational performance.

7 Conclusion

Four different grids with a refinement factor of two were applied to simulate the process of splashing in order to study the influence of the grid resolution on the computed process. Comparisons of the local interface topologies at similar time steps showed noticeable deviations for the different grids that were widely reduced by choosing time steps with regard to the physical events. Good agreement was found for the integral properties of the whole system and the secondary droplets. The detected accordance of the droplets' properties is particularly important for technical applications to model the whole splashing process. For the simulated cases, investigations on the numerically generated flotsam particles showed an insignificance that increases with higher grid resolutions. Moreover, results of cases with deactivated surface tension were presented. It was found that for this case also temporal topological agreement is achieved for the two finest resolutions.

Acknowledgments

The authors greatly appreciate the *High Performance Computing Center Stuttgart* (HLRS) for support and supply of computational time on the NEC SX-8 platform under the Grant No. FS3D/11142.

References

1. F. Maichle. *Numerische Untersuchung und Modellierung von Wandinteraktionen in Zweiphasenströmungen.* PhD thesis, Universität Stuttgart, 2006.
2. G.E. Cossali, A. Cohge, and M. Marengo. The impact of a single drop on a wetted solid surface. *Experiments in Fluids*, Vol. 22, 1997.
3. B. Lafaurie, C. Nardone, R. Scardovelli, S. Zaleski, and G. Zanetti. Modelling merge and fragmentation in multiphase flows with SURFER. *Journal of Computational Physics*, Vol. 113, pp. 134–147, 1994.
4. C.W. Hirt and B.D. Nichols. Volume of fluid (VOF) method for the dynamics of free boundaries. *Journal of Computational Physics*, Vol. 39, pp. 201–225, 1981.
5. W.J. Rider and D.B. Kothe. Reconstructing volume tracking. *Journal of Computational Physics*, Vol. 141, pp. 112–152, 1998.
6. M. Rieber. *Numerische Modellierung der Dynamik freier Grenzflächen in Zweiphasenströmungen.* PhD thesis, Universität Stuttgart, 2004.
7. A.L. Yarin and D.A. Weiss. Impact of drops on solid surfaces: Self-similar capillary waves, and splashing as a new type of kinematic discontiuity. *Journal of Fluid Mechanics*, Vol. 283, 1995.

Implicit LES of Passive-Scalar Mixing in a Confined Rectangular-Jet Reactor

A. Devesa, S. Hickel, and N.A. Adams

Technische Universität München, Institute of Aerodynamics, 85747 Garching, Germany. E-mail: Antoine.Devesa@aer.mw.tum.de

Summary. Recently, the implicit SGS modeling environment provided by the Adaptive Local Deconvolution Method (ALDM) has been extended to Large-Eddy Simulations (LES) of passive-scalar transport. The resulting adaptive advection algorithm has been described and discussed with respect to its numerical and turbulence-theoretical background by Hickel *et al.*, 2007. Results demonstrate that this method allows for reliable predictions of the turbulent transport of passive-scalars in isotropic turbulence and in turbulent channel flow for a wide range of Schmidt numbers. We now intend to use this new method to perform LES of a confined rectangular-jet reactor and compare obtained results to experimental data available in the literature.

1 Introduction: Experimental Configuration

The mixing of passive-scalars at very high Schmidt numbers was recently studied under the conditions of laboratory experiments by the group of Rodney Fox at Iowa State University [1].

In this experiment, simultaneous velocity and scalar concentrations were carried out in confined rectangular jet and wake flows. For the purpose of our study, we will first focus on the confined planar jet. A bidimensional sketch of this test rig is displayed in Figure 1. The flow system was designed to provide a shear flow with a Reynolds number based on the channel hydraulic diameter between $5,000$ and $100,000$. Computations corresponding to the Reynolds number $50,000$ are envisaged in our work. The Schmidt number of the flow is $1,250$.

The measurements are carried out in a Plexiglas test section with a rectangular cross-section measuring 60 mm (height) by 100 mm (wide) and with an overall length of 1 m. The width of each of the inlet channels, separated by two splitters, is 20 mm and the aspect ratio of the rectangular jet is 5. Volumetric flow rates are respectively 1.0, 2.0 and 1.0 L/s, corresponding to free stream velocities of respectively 0.5, 1.0, and 0.5 m/s in the top, center and bottom inlet channels.

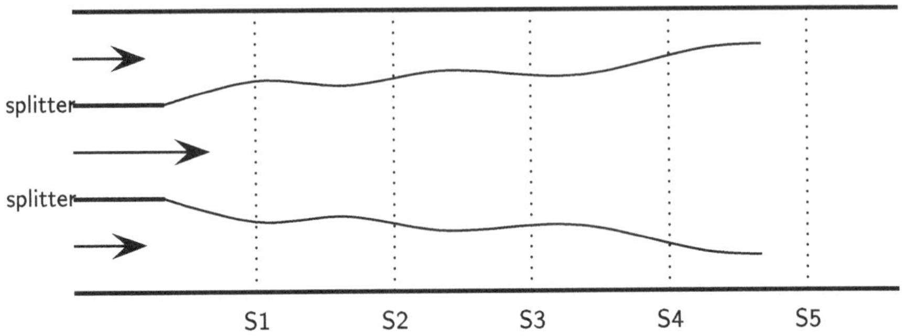

Fig. 1. Schematic of the confined rectangular jet experimental setup

Particle Image Velocimetry (PIV) was used to measure the instantaneous velocity field in 5 planar cross sections of the observed flow, corresponding to the 5 stations shown on Figure 1, S1 to S5. Scalar concentration measurements were carried out simultaneously at the same locations using Planar Laser-Induced Fluorescence (PLIF), so that velocity-scalar correlations were computed. Therefore, an experimental database was generated, including first- and second-order moments for velocity and scalar concentration, as well as cross-correlations.

For further details of the experimental configuration, please refer to the latest publications of the group on this topic [1, 2].

In close collaboration with the experimentalists and in the framework of the validation and assessment of implicit LES, which basic concept is presented in Section 2, we would like to perform LES of the mixing and transport of a passive-scalar in the confined rectangular-jet reactor configuration. An adequate grid size for the computation of such flow requires approximately 15×10^6 points. Our implicit LES code is based on the Adaptive Local Deconvolution Method (ALDM), presented in Section 3. In Section 4, some details of the extension of our subgrid-scale modelling approach to passive-scalar transport are given, before dealing with the performance of our code on the HLRS Supercomputers, Section 6, and presenting the first results of the computations, Section 7.

2 Implicit LES

We consider LES of turbulent flows which are governed by the Navier-Stokes equations and by the incompressible continuity equation. A finite-volume discretization is obtained by convolution with the top-hat filter $\boldsymbol{G}$:

$$\frac{\partial \bar{\boldsymbol{u}}_N}{\partial t} + \boldsymbol{\nabla} \cdot \bar{\boldsymbol{N}}_N(\boldsymbol{u}_N) - \nu \boldsymbol{\nabla} \cdot \boldsymbol{\nabla} \bar{\boldsymbol{u}}_N = -\boldsymbol{\nabla} \cdot \bar{\boldsymbol{\tau}}_{SGS} \tag{1a}$$

$$\boldsymbol{\nabla} \cdot \bar{\boldsymbol{u}}_N = 0 \tag{1b}$$

where an overbar denotes the filtering $\bar{\boldsymbol{u}} = \boldsymbol{G} * \boldsymbol{u}$. The nonlinear term is abbreviated as $\boldsymbol{\nabla} \cdot \boldsymbol{N}(\boldsymbol{u}) = \boldsymbol{\nabla} \cdot \boldsymbol{u}\boldsymbol{u} + \boldsymbol{\nabla}\, p$, where $\boldsymbol{u}$ is the velocity field and p is the pressure. The employed filter approach [3] implies a subsequent discretization of the filtered equations. The subscript N indicates the resulting grid functions obtained by projecting continuous functions on the numerical grid. This projection corresponds to an additional filtering in Fourier space with a sharp cut-off at the Nyquist wavenumber $\xi_C = \pi/h$, where h is a constant grid spacing. The subgrid-stress tensor:

$$\boldsymbol{\tau}_{SGS} = \boldsymbol{N}(\boldsymbol{u}) - \boldsymbol{N}_N(\boldsymbol{u}_N) \tag{2}$$

originates from the discretization of the non-linear terms and has to be approximated by a model for closing Eq. (1). To certain extents common explicit models are based on sound physical theories. Solved numerically, however, the discrete approximation of the explicit SGS model competes against the truncation error of the underlying numerical scheme. A theoretical analysis performed by Ghosal [4] comes to the conclusion that even a fourth-order central difference discretization has a numerical error which can have the same order of magnitude as the SGS model. This fact is exploited for implicit large-eddy simulation where no SGS model terms are computed explicitly. Rather the truncation error of the numerical scheme is used to model the effects of unresolved scales. A recent review on previous implicit LES approaches is provided, e.g. by Grinstein & Fureby [5].

The Modified Differential Equation (MDE) for an implicit LES scheme is given by:

$$\frac{\partial \bar{\boldsymbol{u}}_N}{\partial t} + \widetilde{\boldsymbol{G}} * \widetilde{\boldsymbol{\nabla}} \cdot \widetilde{\boldsymbol{N}}_N(\widetilde{\boldsymbol{u}}_N) - \nu \boldsymbol{\nabla} \cdot \boldsymbol{\nabla} \bar{\boldsymbol{u}}_N = \boldsymbol{0} \tag{3a}$$

$$\boldsymbol{\nabla} \cdot \bar{\boldsymbol{u}}_N = 0 \tag{3b}$$

where $\widetilde{\boldsymbol{u}}_N$ denotes an approximant of the velocity $\boldsymbol{u}_N$. The local Riemann problem is solved by a consistent numerical flux function $\widetilde{\boldsymbol{N}}_N$. The symbols $\widetilde{\boldsymbol{G}}$ and $\widetilde{\boldsymbol{\nabla}}$ indicate that $\boldsymbol{G}$ and $\boldsymbol{\nabla}$ are replaced by their respective numerical approximations. In fact $\widetilde{\boldsymbol{G}} * \widetilde{\boldsymbol{\nabla}}$ can be a nonlinear operator. The truncation error is accordingly:

$$\boldsymbol{\mathcal{G}}_N = \boldsymbol{G} * \boldsymbol{\nabla} \cdot \boldsymbol{N}_N(\boldsymbol{u}_N) - \widetilde{\boldsymbol{G}} * \widetilde{\boldsymbol{\nabla}} \cdot \widetilde{\boldsymbol{N}}_N(\widetilde{\boldsymbol{u}}_N) \tag{4}$$

For implicit SGS modeling the discretization scheme is specifically designed so that the truncation error $\boldsymbol{\mathcal{G}}_N$ has physical significance, i.e.:

$$\boldsymbol{\mathcal{G}}_N \approx -\boldsymbol{G} * \boldsymbol{\nabla} \cdot \boldsymbol{\tau}_{SGS} \tag{5}$$

3 The ALDM Approach

With the adaptive local deconvolution method (ALDM) the local approximation $\widetilde{\boldsymbol{u}}_N$ is obtained from a solution-adaptive combination of deconvolution polynomials. Numerical discretization and SGS modeling are merged entirely. This is possible by exploiting the formal equivalence between cell-averaging and reconstruction in finite-volume discretizations and top-hat filtering and deconvolution in SGS-modeling. Instead of maximizing the order of accuracy, deconvolution is regularized by limiting the degree of local interpolation polynomials and by permitting lower-order polynomials to contribute to the truncation error. Adaptivity of the deconvolution operator is achieved by weighting the respective contributions by an adaptation of WENO smoothness measures [6]. The approximately deconvolved field is inserted into a consistent numerical flux function. Flux function and nonlinear weights introduce free parameters. These allow for controlling the truncation error which provides the implicit SGS model.

4 Implicit Subgrid-Scale Modeling for Passive-Scalar Transport

We consider the turbulent transport of passive-scalars, which do not measurably affect the velocity field. This case represents a one-way coupling of the scalar to the fluid. Hence, the closure problem is restricted to the scalar transport equation. Turbulence modeling and discretization for the momentum equations remain unchanged.

The transport of a passive-scalar c in an incompressible fluid is governed by:

$$\partial_t c + \boldsymbol{\nabla} \cdot \boldsymbol{F}(\boldsymbol{u}, c) = 0 \; , \tag{6}$$

supplemented with appropriate initial and boundary conditions. The scalar flux function is:

$$\boldsymbol{F}(\boldsymbol{u}, c) = \boldsymbol{u} c - \frac{1}{\mathrm{ScRe}} \boldsymbol{\nabla} c \; , \tag{7}$$

where $\boldsymbol{u}$ and Re are velocity vector and Reynolds number of the transporting flowfield. Sc is the Schmidt number. $\mathrm{Sc} = \nu/\kappa$ is defined as the ratio of kinematic viscosity ν and the diffusivity κ associated with the scalar quantity c. Depending on the application, c can be concentration, temperature, or any kind of passive marker.

Following Leonard [3], the discretized equation is obtained by convolution with a homogeneous filter kernel G:

$$\partial_t \bar{c} + G * \boldsymbol{\nabla} \cdot \boldsymbol{F}(\boldsymbol{u}, c) = 0 \tag{8}$$

and subsequent discretization of the filtered equations:

$$\partial_t \bar{c}_N + G * \boldsymbol{\nabla} \cdot \boldsymbol{F}_N(\boldsymbol{u}_N, c_N) = -G * \boldsymbol{\nabla} \cdot \boldsymbol{\tau}_{SGS} \; . \tag{9}$$

The overbar denotes the filtering $\bar{c} = G * c$ and the subscript N indicates grid functions obtained by projecting continuous functions onto a numerical grid with finite resolution.

The flux (7) is formally linear in c. However, the evolution of a non-uniform scalar field is subject to the velocity dynamics. Small-scale fluctuations of velocity and scalar are correlated in the presence of a scalar-concentration gradient. The subgrid tensor:

$$\boldsymbol{\tau}_{SGS} = \boldsymbol{F}(\boldsymbol{u}, c) - \boldsymbol{F}_N(\boldsymbol{u}_N, c_N) \tag{10}$$

originates from the grid projection of advective terms. It represents the effect of the action of subgrid scales and has to be approximated by a SGS model for closure of Eq. (9). This modeling task is far from trivial. One reason is that the various regimes that exist for the passive-scalar variance spectrum have to be recovered by the SGS model. These regimes originate in the difference between typical length scales characterizing the viscous cutoff of the velocity field and the diffusive range of the scalar field [7]. In the following, two cases that are of particular interest in LES are discussed for homogeneous, isotropic turbulence.

The scalar fluctuations are driven by the stirring induced by the velocity field. Different scalar-transport regimes are associated with certain ranges of Reynolds and Schmidt numbers. The first regime is associated with small Schmidt numbers $\mathrm{Sc} \leq 1$.

With respect to LES, this regime is most relevant for large Reynolds numbers, where the kinetic-energy spectrum develops a broad inertial range:

$$E(\xi) = C_K \varepsilon^{2/3} \xi^{-5/3} \tag{11}$$

at wavenumbers $\xi \ll \xi_K = (\varepsilon/\nu^3)^{1/4}$. In this regime, the scalar-variance behaves similar to the kinetic-energy. Obukhov and Corrsin applied Kolmogorov's equilibrium theory to the scalar variance and derived a diffusive cutoff at $\xi_D = (\varepsilon/\kappa^3)^{1/4}$. The scalar-variance spectrum depends on the kinetic-energy transfer ε and the scalar diffusion χ. From simple dimensional arguments [8] follows that it exhibits an inertial-convective range:

$$E_c(\xi) = C_{OC} \varepsilon^{-1/3} \chi \xi^{-5/3} \tag{12}$$

with scaling $E_c(\xi) \sim \xi^{-5/3}$ at wavenumbers $\xi \ll \xi_D$. C_K and C_{OC} are Kolmogorov constant and Obukhov-Corrsin constant, respectively. An analysis of the shape of $E_c(\xi)$ in the diffusive-range $\xi_D \ll \xi$ is presented by [9].

In LES, the filter width is typically chosen in such a way that the numerical cutoff wavenumber ξ_c is within the inertial range (11). For a coarse representation of the scalar dynamics, we also assume that $\xi_c \ll \xi_D$ such that the SGS energy transfer does not directly depend on Re and Sc. A representative example for this regime is LES of isotropic turbulence at $\mathrm{Sc} \approx 1$ and $\mathrm{Re} \gg 1$.

A more complex situation is encountered if the Schmidt number is much larger than unity $\mathrm{Sc} \gg 1$. In this regime two distinct inertial ranges exist for the scalar-variance spectrum. An inertial-convective range (12) is observed for scales within the Kolmogorov inertial range $\xi \ll \xi_K$ of the kinetic-energy spectrum. At smaller scales, the energy spectrum already decays exponentially whereas the scalar fluctuations are not yet affected by diffusion. A second inertial range, the viscous-convective range, is observed for the scalar-variance spectrum at $\xi_K \ll \xi \ll \xi_B$, ξ_B being the Batchelor wavenumber $(\varepsilon/\nu\kappa^2)^{1/4}$. Based on an analytical model for the distortion of small scalar blobs, [7] derived:

$$E_c(\xi) = C_B \nu^{1/2} \varepsilon^{-1/2} \chi \xi^{-1} \tag{13}$$

for the viscous-convective range and an exponential decay in the viscous-diffusive range $\xi \gg \xi_B$. Employing a more sophisticated statistical method, the Lagrangian-history direct-interaction (LHDI) approximation, [10] found further evidence for the ξ^{-1} scaling in the viscous-convective range. However, the LHDI approximation leads to a less rapid decay in the viscous-diffusive range. Recent numerical studies [11] tend to favour the results of Kraichnan. The numerical parameter C_B is the Batchelor constant.

Various experimental, analytical, and numerical determinations of the numerical constants C_B, C_K, and C_{OC} have been proposed. Unfortunately, the reported values scatter a lot.

If the grid cutoff ξ_C lies within the inertial-convective range the same SGS modeling can be used as for the Sc ≤ 1 case. A different approach is required if the numerical cutoff is chosen within the viscous-convective range. This velocity-resolving case is typically associated with low Reynolds numbers in LES. Generic sketches of corresponding kinetic energy spectra, scalar variance spectra, and numerical cutoff for the two regimes Sc ≤ 1 and Sc $\gg 1$ are shown in Fig. 2.

Another difficulty in solving scalar transport equations is associated with a numerical problem. At high Schmidt numbers, even an incompressible and smooth flowfield will generate a filtered scalar field with steep concentration gradients that can only be captured and not resolved by the numerical discretization. Standard centered differencing schemes tend to non-physical oscillations unless they are supplemented with a numerical regularization.

This analysis also revealed that different parameters are required for the two Schmidt-number regimes. Then, two sets of parameters were determined representing two SGS models: one for the low-Schmidt-number regime and one for the high-Schmidt-number regime.

The ALDM approach for passive-scalar mixing has been validated for several canonical flows [12]. For further details on the subgrid-scale model and its implementation, please refer to Hickel *et al.* [12]. The computation of the experimental setup presented in Section 1 will then assess these models in a more complex configuration.

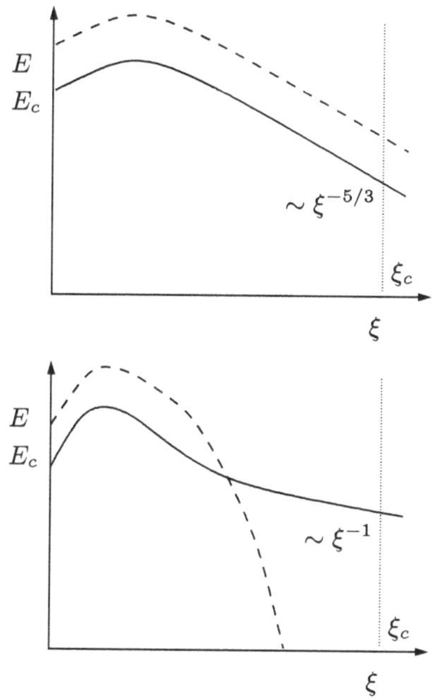

Fig. 2. Critical test cases for predicting the proper subgrid diffusion in large-eddy simulations of scalar mixing. *Top:* Low Schmidt number regime. *Bottom:* High Schmidt number regime at moderate Reynolds number. —— scalar variance; - - - - kinetic energy; · · · · · numerical cutoff wavenumber

5 Numerical Method and Computational Domain

The flow is described by the incompressible Navier-Stokes equations. The equations are discretized on a staggered Cartesian mesh. For time advancement the explicit third-order Runge-Kutta scheme of [13] is used. The time-step is dynamically adapted to satisfy a Courant-Friedrichs-Lewy condition with CFL = 1.0. The pressure-Poisson equation and diffusive terms are discretized by second-order centered differences. The Poisson solver is based on the stabilized bi-conjugate gradient (BiCGstab) method [14]. The Poisson equation is solved at every Runge-Kutta substep. All results presented in this document are obtained by the Simplified Adaptive Local Deconvolution (SALD) method [15] that represents an implementation of the Adaptive Local Deconvolution Method (ALDM) with improved computational efficiency. The validity of this numerical methodology has been established for plane channel flow [16], for turbulent boundary layer separation [17], and for passive scalar mixing [12]. The undergoing work represents namely the further step in the validation of the code in this latter field.

The computational domain used for reproducing the experimental configuration described in Section 1 is divided in two parts. The inlet part, composed by 3 inlet channels, are located upstream from the measurements domain. Each of the inlet channels is periodic in streamwise and spanwise directions, while bounded by solid walls in the transverse direction, and contains $16 \times 90 \times 160$ points. In the wall-normal direction a hyperbolic stretching is used to increase resolution near the wall, while the distribution is equidistant in the two other directions. The measurement section of the domain is confined by side walls in spanwise and transverse directions, while inlet is given by the flow conditions at the exit of the inlet channels, and oulet condition is a Neumann boundary condition for the pressure. This section contains $316 \times 270 \times 160$ finite volumes.

In total, the adequate grid size for the computation of such flow requires approximately 14.4×10^6 points, based on a resolution of about $\Delta y^+ = 1$ close to the walls.

6 Performance

The flow solver used, INCA [15], is optimized for parallel high-performance vector computers. The present simulations are performed on a NEC SX-8 cluster at the Stuttgart high-performance computing center (HLRS). The CPU-time consumption is distributed as follows: 99 percents are used by the iterative solver for the pressure Poisson equation, 1 percent is dedicated to output and statistics purposes, as well as the rest of the computational routines, such as the calculation of the right-hand side of the Navier-Stokes equation.

The performance of the Poisson solver, that has then the most significant influence on the computation, is about 6 GFlops, which is approximately the same value as the overall performance of the computation. The computations use approximately 15 GBytes RAM and the calculation of one flow-through time requires 4600 CPU hours (cf. extent of the computational domain in Section 5 and the first paragraph in Section 7).

7 First Results and Current Status

The current status of the numerical computation we are carrying out does not allow us to present converged statistics of the flow field. Statistics are accumulating at the moment. As the computation is fully 3D because of the presence of lateral solid walls, i.e. there is no homogeneous direction, the statistics converge very slowly. In order to reach converged statistics, between 15 to 20 flow-trough times are required. One flow-through time requires approximately 6,000 iterations, corresponding to 4600 CPU hours (MPI parallelized on 24 processors). Some instantaneous snapshots in the symmetry plane (cut in the spanwise direction) of the computation are shown for the velocity in Fig. 3 and for the scalar concentration fields in Fig. 4.

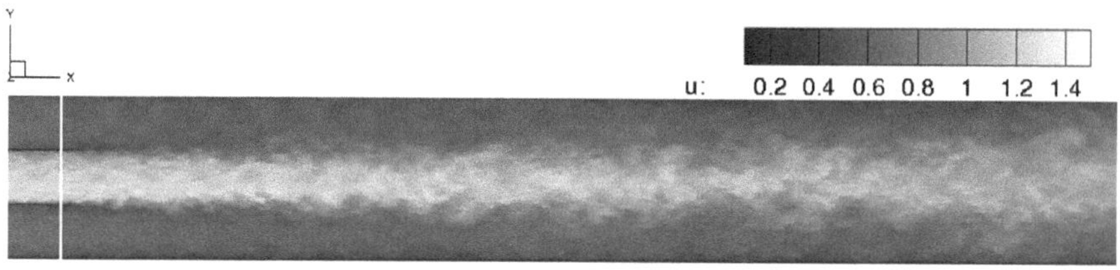

Fig. 3. Instantaneous velocity field

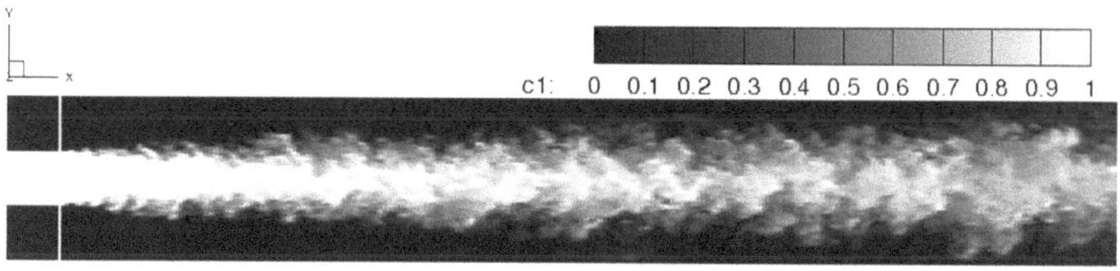

Fig. 4. Instantaneous scalar field

The two parts of the computational domain can be identified on the snapshots Figs. 3 and 4. On the left-hand side, one can see the inlet channels, one center channel and two lateral channels (where the bulk velocity is twice lower). The right-hand side consists of the long confined box representing the reactor itself, where the mixing layer develops and where the experimental measurements were carried out. In the region close to the nozzle, the jet expands. The confinement of the jet however stops this expansion at approximately $x/d = 12$.

Fig. 5 shows a first comparison between experimental data and numerical results for the streamwise velocity. The five observation stations S1 to S5 are displayed. Even though the numerical results show poor convergence, the agreement between experiment and simulation is satisfactory. The main problem that we encountered was the calibration of the inlet boundary condition for the simulation, since no reliable inlet condition could be used from the experimental results. For this reason, the value of the volume force imposed in the three inlet channels in the computation were tuned in accordance to the experiment, so that the velocity profile at the first station S1 match as close as possible. Unfortunately, the calibration of an instantaneous result with average quantities is not trivial, and since the required computational resource for the calculation of a single flow-through time is around 4600 CPU hours, a large amount of computational resource was used to solve this problem. One can observe, moreover that, close to the top and bottom walls, the velocity profile exhibits a large discrepancy with the experimental result. This can be due to the fact that the inlet velocities are slightly higher than in the experiment, however the presence of walls in the experiment can raise inaccuracy in the PIV measurements.

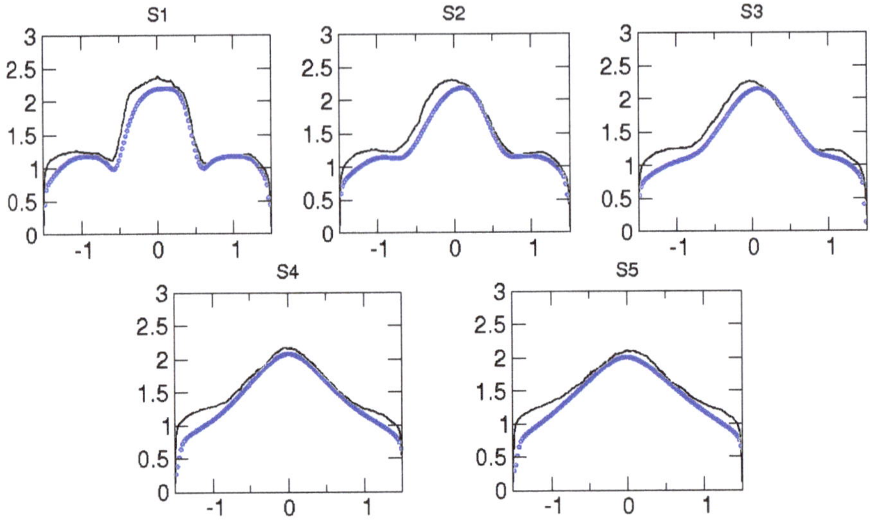

Fig. 5. Comparison of experimental (symbols) and numerical (solid) streamwise velocity at five stations

Fig. 6 shows a first comparison between experimental data and numerical results for the scalar concentration. The five observation stations S1 to S5 are displayed. Even though the numerical results show poor convergence, the agreement between experiment and simulation is satisfactory, raising our expectations on the comparison of numerical converged statistics with experimental data [1, 2]. The expansion of the jet is well represented and a next step will be to analyse the cross-correlations between scalar concentration and velocity.

8 Conclusions

We have presented a highly resolved Large-Eddy Simulation (LES) of the flow in a confined rectangular-jet reactor. The results presented are only preliminary results since the convergence of statistics is a very low process in this 3-dimensional configuration without homogeneity direction. Velocity profiles and Reynolds number are based on the experimental results from the group of Rodney Fox at Iowa State University [1]. The velocity and scalar concentration profiles show promising results. Statistics for this configuration are still accumulating. A further analysis with focus on the cross-correlation between scalar concentration and velocity will be available after convergence of the statistics.

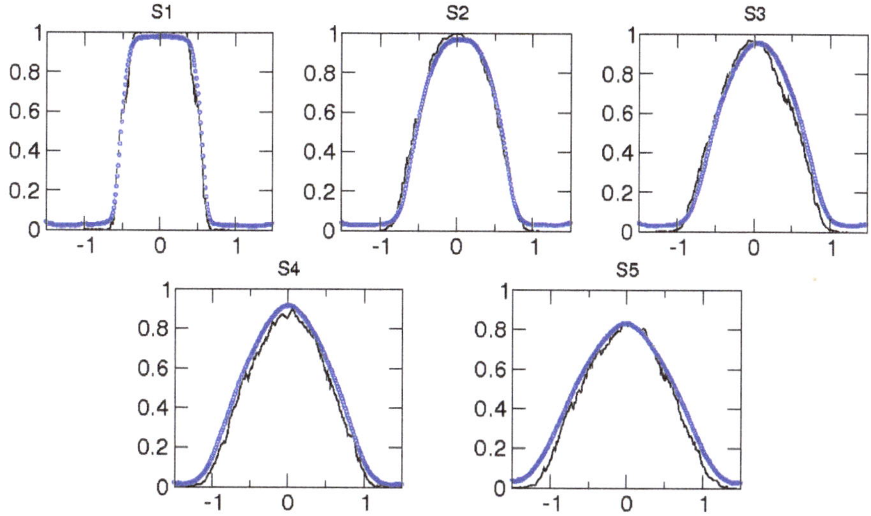

Fig. 6. Comparison of experimental (symbols) and numerical (solid) scalar concentration at five stations

References

1. H. Feng. *Experimental study of turbulent mixing in a rectangular reactor*. PhD thesis, Iowa State University, Ames, Iowa, 2006.
2. H. Feng, M.G. Olsen, Y. Liu, R.O. Fox, and J.C. Hill. Investigation of turbulent mixing in a confined planar-jet reactor. *AIChE J.*, 51:2649–2664, 2005.
3. A. Leonard. Energy cascade in large eddy simulations of turbulent fluid flows. *Adv. Geophys.*, 18A:237–248, 1974.
4. S. Ghosal. An analysis of numerical errors in large-eddy simulations of turbulence. *J. Comput. Phys.*, 125:187–206, 1996.
5. F.F. Grinstein and C. Fureby. From canonical to complex flows: Recent progress on monotonically integrated LES. *Comp. Sci. Eng.*, 6:36–49, 2004.
6. C.-W. Shu. Essentially non-oscillatory and weighted essentially non-oscillatory schemes for hyperbolic conservation laws. *Tech. Rep. 97-65, ICASE, NASA Langley Research Center, Hampton, Virginia*, 1997.
7. G. Batchelor. Small-scale variation of convected quantities like temperature in turbulent fluid. Part 1. general discussion and the case of small conductivity. *J. Fluid Mech.*, 5:113–133, 1959.
8. S. Corssin. On the spectrum of isotropic temperature fluctuations in an isotropic turbulence. *J. Appl. Phys.*, 22(4):469–473, 1951.
9. G. Batchelor, I. Howells, and A. Townsend. Small-scale variation of convected quantities like temperature in turbulent fluid. Part 2. the case of large conductivity. *J. Fluid Mech.*, 5:134–139, 1959.
10. R. Kraichnan. Small-scale structure of a scalar field convected by turbulence. *Phys. Fluids*, 11:945–953, 1968.
11. P. Yeung, S. Xu, and K. Sreenivasan. Schmidt number effects on turbulent transport with uniform mean scalar gradient. *Phys. Fluids*, 14(12):4178–4191, 2002.

12. S. Hickel, N.A. Adams, and N.N. Mansour. Implicit subgrid-scale modeling for large-eddy simulation of passive-scalar mixing. *Phys. Fluids*, 19:095102, 2007.
13. C.-W. Shu. Total-variation-diminishing time discretizations. *SIAM J. Sci. Stat. Comput.*, 9(6):1073–1084, 1988.
14. H.A. van der Vorst. Bi-CGSTAB: A fast and smoothly converging variant of Bi-CG for the solution of nonsymmetric linear systems. *SIAM J. Sci. Statist. Comput.*, 13:631–644, 1992.
15. S. Hickel and N.A. Adams. Efficient implementation of nonlinear deconvolution methods for implicit large-eddy simulation. In W.E. Nagel, W. Jäger, and M. Resch, editors, *High Performance Computing in Science and Engineering*, pages 293–306. Springer, 2006.
16. S. Hickel and N.A. Adams. On implicit subgrid-scale modeling in wall-bounded flows. *Phys. Fluids*, 19:105106, 2007.
17. S. Hickel and N.A. Adams. Large-eddy simulation of turbulent boundary-layer separation. In *5th International Symposium on Turbulence and Shear Flow Phenomena (TSFP5); Munich, Germany*, 2007.

Wing-Tip Vortex / Jet Interaction in the Extended Near Field

Frank T. Zurheide, Matthias Meinke, and Wolfgang Schröder

Institute of Aerodynamics, RWTH Aachen University,
Wüllnerstr. 5a, 52062 Aachen, Germany, f.zurheide@aia.rwth-aachen.de

Summary. The vortex wake behind a half wing is spatially simulated up to the extended near field. Instabilities of the wing tip vortex are analyzed. Results from wind tunnel measurements are used as inflow boundary conditions for an LES of the wake region. An aircraft engine is modeled in the experimental setup, where the jet is driven by pressurized air. The engine was mounted in two different positions under the wing model to investigate the influence of the location of the engine jet on the vortex wake. The numerical simulations of the wake were able to predict trajectories and instabilities of the vortex core. A position of the engine towards the root of the wing created a smaller deflections of the vortex.

1 Introduction

The problem of aircraft wake vortex hazard and its limitations on airport capacity is of growing importance due to the expected increase in passenger numbers and aircraft traffic. Heavy transport aircraft generate vortex wakes that endanger the following aircrafts, because they cannot be precisely detected. The decay of vortex wakes by diffusion and dissipation mechanisms is weak, such that the frequency of starting and landing aircrafts is limited by waiting times required for a sufficient decay of the vortex wake. Instabilities of wake vortex systems, however, can lead to a rapid decay of the vortex strength. Scientific research focuses on the evolution of instabilities inherent to or artificially introduced into the wake structure [1]. The aim is the excitation of mechanisms that lead to an amplification of different types of instabilities and therefore reduce the time required for the onset of rapid vortex decay.

In counter rotating vortex systems several three-dimensional instabilities are known. Crow [2] and Waleffe [3] performed stability analyses of a vortex pair with the vortex filament method resulting in basic findings for long and short-wave instabilities.

One possible strategy to influence the vortex wake is to use the jet from the engine. The influence of the jet on the tip vortex was analyzed with temporal simulations by Labbé et al. [4], Paoli et al. [5], Gago et al. [6] and Holzäpfel et al. [7]. In contrast to these temporal investigations, Fares and Schröder [8] simulated a spatially developing wake-jet interaction in the near field based on the Reynolds averaged Navier-Stokes equations.

In this paper we will present a Large-Eddy Smulation (LES) of a spatially developing wake-jet interaction in the near field and extended near field. Measurements from the wake are used as inflow data for the simulations. The wake for two different positions of the engine are examined and compared. This paper is organized as follows. The governing equations and the numerical procedure of the LES method are described in section 2. The inflow data for the simulations are presented in section 3. The results for the wake in the extended near field is given for two different positions of the engine and is discussed in detail in section 4. Finally, the findings of the present study are summarized in section 6.

2 Governing Equations

Since the continuum assumptions hold for the flows regarded here, the Navier-Stokes equations are an appropriate mathematical model for their simulation. Written in tensor notation and in terms of dimensionless conservative variables for a Cartesian coordinate system, they read:

$$\frac{\partial \boldsymbol{Q}}{\partial t} + (\boldsymbol{F}_\beta^C - \boldsymbol{F}_\beta^D)_{,\beta} = 0 \quad , \qquad \boldsymbol{Q} = [\varrho, \varrho u_\alpha, \varrho E]^T \quad , \tag{1}$$

where $\boldsymbol{Q}$ is the vector of the conservative variables, and $\boldsymbol{F}_\beta^C$ denotes the vector of the convective and $\boldsymbol{F}_\beta^D$ of the diffusive fluxes:

$$\boldsymbol{F}_\beta^C - \boldsymbol{F}_\beta^D = \begin{pmatrix} \varrho u_\beta \\ \varrho u_\alpha u_\beta + p\delta_{\alpha\beta} \\ u_\beta(\varrho E + p) \end{pmatrix} - \frac{1}{Re} \begin{pmatrix} 0 \\ \sigma_{\alpha\beta} \\ u_\alpha \sigma_{\alpha\beta} + q_\beta \end{pmatrix} \quad . \tag{2}$$

The stress tensor $\sigma_{\alpha\beta}$ is written as a function of the strain rate tensor $S_{\alpha\beta}$

$$\sigma_{\alpha\beta} = -2\,\mu(S_{\alpha\beta} - \tfrac{1}{3} S_{\gamma\gamma}\delta_{\alpha\beta}) \qquad \text{with} \qquad S_{\alpha\beta} = \tfrac{1}{2}(u_{\alpha,\beta} + u_{\beta,\alpha}) \quad . \tag{3}$$

The dynamic viscosity μ, is assumed here to be only a function of the temperature. Fourier's law of heat conduction is used to compute the heat flux q_β:

$$q_\beta = -\frac{k}{Pr(\gamma - 1)} T_{,\beta} \tag{4}$$

where Pr is the Prandtl number. The described system is closed with the equation of state for ideal gases

$$p = \frac{1}{\gamma}\varrho T \quad , \quad p = (\gamma - 1)\left(\varrho e - \frac{\varrho}{2} u_\beta u_\beta\right) \quad , \tag{5}$$

where γ is the ratio of the specific heat capacities and T the temperature.

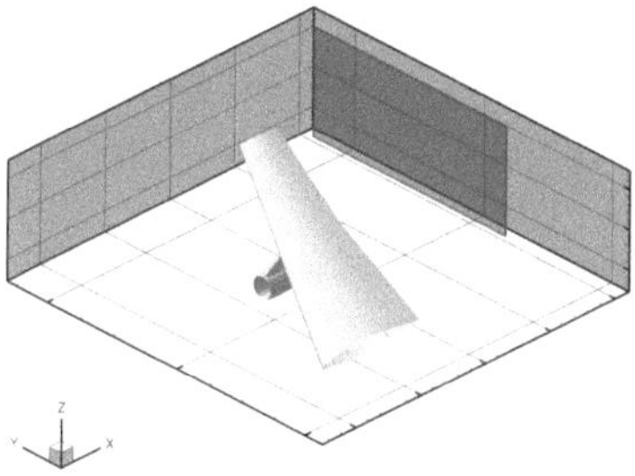

Fig. 1. Scheme of measurement set-up. BAC 3-11 wing with the engine mounted in position "A"/"C". The dark plane behind the wing indicates the location of the PIV measurements

For the LES, an implicit grid filter of width Δ is assumed to be applied to eqs. (1). A sub-grid scale model and the truncation error are coupled to the spatial steps of the computational grid in case of a finite volume scheme. Both terms lead to a damping of the smallest resolved scales and can be shown to be of the same order of magnitude for second-order accurate schemes, at least in the time averaged sense [9]. For the solution scheme presented in the next section, the influence of the sub-grid scale model has a negligible influence on accuracy of the turbulence statistics [10]. Therefore no sub-grid scale model was used for the simulations presented here.

3 Boundary Conditions

The wake is measured behind a half model of a wing below which an engine can be mounted at different positions. The results of the velocity measurements are used as inflow boundary conditions for the spatial computation of the wake in the near field and in the extended near field.

3.1 Experimental Set-Up

The experiments were carried out in the low-speed wind tunnel of the Institute of Aerodynamics, RWTH Aachen University [11, 12]. The flow speed is set to u_∞=27 m/s which results in a Reynolds number of Re_c=$2.8{\cdot}10^5$ based on the mean chord length c_m. The wing has a BAC 3-11/RES/30/21 profile and a three dimensional wing-tip geometry. The semi span is $s = 607\ mm$, and the mean chord length is $c_m = 150\ mm$. The angle of attack is set to $\alpha = 8°$. An engine model can be mounted at three different positions below the wing. The engine model works as an air jet apparatus and is operated with pressurized air. The jet speed is approximately $u_{engine}/u_\infty = 1.74$. The particle-image velocimetry (PIV) method is used for the measurement of the wake velocity field. The three velocity components of the wing and engine wake

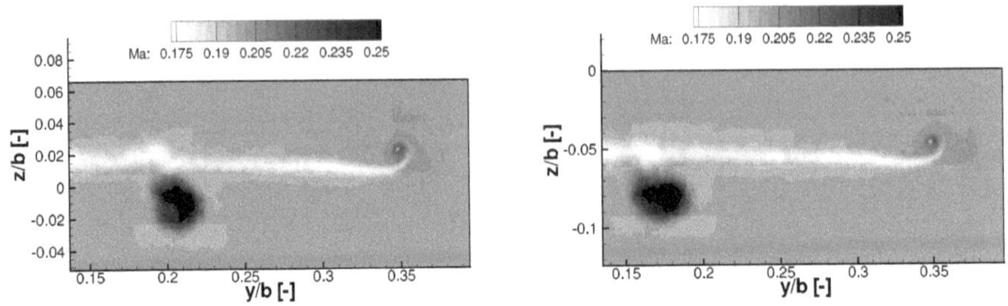

(a) Engine position "A" (b) Engine position "C"

Fig. 2. Mach number distribution for engine position "A" and "C"

are captured with stereo- or 3C-PIV one chord length c_m behind the wing, see Fig. 1. The measurements were carried out for two positions of the jet, position "A" towards the wing tip and position "C" towards the root of the wing. The results of the measurement are recorded on a cartesian mesh with 480×156 points. In Fig. 2 the Mach number distribution of the measurements is shown for the different engine positions. In Fig. 2(a) the jet can be recognized by the high velocity i.e. dark region centered at $y/b \approx 0.2, z/b \approx -0.01$. The shear layer of the wing can be identified by the lower velocity i.e. brighter thin region that starts at the left border at $z/b \approx 0.02$ and is entrained at the right by the tip vortex. The center of the vortex is located at $y/b = 0.35, z/b = 0.022$. For both engine positions the same features jet, shear layer, tip vortex can be seen, but the jet for engine position "C" is moved towards the root of the wing.

3.2 Inflow Boundary Conditions

The results of the PIV measurements are used as inflow boundary conditions of the numerical simulation. The extent of the measurement window in spanwise and wing normal direction is not large enough to permit a simulation of the interaction of the engine jet and the tip vortex, so the measurement data is interpolated to a larger area. For the interpolation, a parallel flow with Ma_∞=0.2 far away from the measurement window is used. The measured data comprise the three velocity components u, v, w and the fluctuating components u', v' and w'. From the velocities the initial pressure distribution p is computed by solving the Poisson equation for the pressure

$$\nabla^2 p = -\frac{\partial u_i}{\partial x_j}\frac{\partial u_j}{\partial x_i} \tag{6}$$

with the Laplacian operator ∇^2.

From the standard deviation σ of the fluctuating velocity components u', v', w', random fluctuations with the same amplitude are superimposed to the time averaged velocity components u, v, w at the inflow boundary during the simulation.

4 Results and Discussion

The wing span b (b=11.56 c_m) can be regarded as a typical length scale of the simulation, since the instabilities in the far field are dependent on the distance b between the vortex centers. The Reynolds number based on the wing span is Re_b=3.2368·10^6. The domain of computation has a size of $x/b \times y/b \times z/b = 12.48 \times 0.35 \times 0.34$ with the x-axis pointing downstream (see also Fig. 1). The length of the domain is 144 c_m=12.48 b. With that domain only one half of the wake of the airplane is simulated. The number of mesh points are 1249 × 225 × 225 equally split into 24 blocks leading to $64 \cdot 10^6$ points.

The radius of the wing tip vortex is r_c=1.93 ·10^{-3}b, which is 44 times smaller than the radius of the vortex used by Holzäpfel et al. [13, 7, 14] or Zurheide and Schröder [15], about one order of magnitude smaller than the radii of the vortices used by Fabre and Jacquin [16], Jacquin et al. [17] and nearly two orders of magnitude smaller than the vortices examined by Dizès and Laporte [18] and by Bristol et al. [19].

While Labbé et al. [4], Paoli et al. [5] and Gago et al. [6] use a temporal approach for the simulation of a jet/wake interaction, we use a spatial approach described by Fares and Schröder [8] and Stumpf [20]. In the temporal approach the velocity deficit in the vortex core in axial direction is neglected although it is known that it has a strong effect on the absolute and convective instability of a vortex [21]. Additionally, the deformation of the vortex is not convected downstream and parts of the wrapped engine jet around the tip vortex have an influence on the flow field upstream, since periodic boundary condition in axial direction are used for the temporal development. On the other hand the spatial approach allows only relatively short extents of the wake to be computed due to the large amount of mesh points for a sufficient resolution of the domain.

In Fig. 3(a) the wake of the simulation at engine position "A" is displayed for t=29.0 in dimensionless simulation time. The wake is shown from a position inside the wake towards the wing. The thick dark tube is a contour of the λ_2-criterion [22] that visualizes the vortex. Streamlines are used to illustrate the jet development. The vortex is bend towards the root of the wing, while the jet is wrapped around the vortex. The vortex reaches the maximal deflection behind the wing at $x/b \approx 4$. The deflection is y/b = 0.8 for engine position "A" and y/b=0.84 for position "C" (Fig. 10). This is close to the theoretical deflection behind a wing with elliptical circulation distribution of $y/b=\pi/4$=0.785 [23].

When the jet is wrapped a half rotation around the vortex small perturbations can be identified on the vortex tube (at x/b=3). These perturbations change their shape and grow to a maximum size at $x/b = 5\ldots7$ and are convected downstream while they are decaying. The perturbations have a 3-D shape and are turned clock wise with the rotation of the vortex. The change of the shape of the disturbances can be explained by looking at the incom-

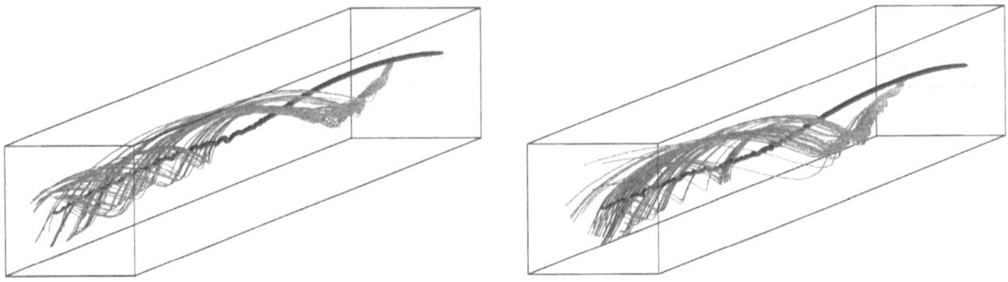

(a) Engine position "A" (b) Engine position "C"

Fig. 3. Wake of computed measurements with engine position "A" and "C". λ_2 vortex criterion is used to display the vortex (thick dark line) and streamlines to show the curvature and position of the jet. Dimensionless time t=29.0

pressible form of the azimuthal vorticity equation

$$\frac{D\omega_\theta}{Dt} = \omega_r \frac{\partial v_\theta}{\partial r} + \frac{\omega_\theta}{r}\frac{\partial v_\theta}{\partial \theta} + \omega_x \frac{\partial v_\theta}{\partial x}, \tag{7}$$

where $D(\cdot)/Dt$ is the total derivative. Shortly downstream of the wing the radial and tangential vorticity components are zero $\omega_r = \omega_\theta = 0$. Axial variations of the velocity field ($\partial v_\theta/\partial x \neq 0$) are induced when the jet is wrapped around the vortex. This leads to axial vorticity distribution that is transferred to an azimuthal vorticity $\partial\omega_\theta/\partial t \approx \omega_x \partial v_\theta/\partial x$.

In Fig. 3(b) the wake of the wing with the engine position "C" is shown. Since the initial distance between the vortex and the jet is larger than for position "A" the interaction between them occurs further downstream.

A comparison of the two engine positions shows, that the perturbations at position "A" have a slightly bigger amplitude and occur a little further upstream.

In Fig. 4 and 5 is the vorticity ω_x for several x values displayed. Fig. 4(a) contains the wing tip vortex and the wake. The wake and the jet are wrapped around the vortex for the slices downstream (Fig. 4(b)-4(f)). From x/b=2.0 one part of the jet can be seen at y/b=0.3, z/b=-0.07 for engine position "A" in Fig. 4(d). The jet is moved to y/b=0.3, z/b=0.016 at x/b=6.0 (Fig. 4(f)). For engine position "C" in Fig. 5 the vortex and the wake show a similar behavior as for the position "A". At x/b=2.0 in Fig. 5(d) the parts of the jet are at y/b=0.25, z/b=-0.1, which is outside the picture in contrast to Fig. 4(d). Then at $x/b = 6.0$ the jet is stretched to a line from $y/b = 0.37, z/b = -0.01$ to $y/b = 0.32, z/b = 0.0$.

To analyze the stability of the jet/vortex interaction the axial Fourier mode k is determined. To determine the growth of the modes of the instabilities the square root of kinetic energy $E^{1/2} = (1/2(u^2 + v^2 + w^2))^{1/2}$ is used to compute the discrete Fourier coefficients $E_k^{1/2}$ in the yz-plane for each grid point and average the values of each plane. The discrete Fourier coefficients are calculated using the Fast-Fourier Transformation (FFT).

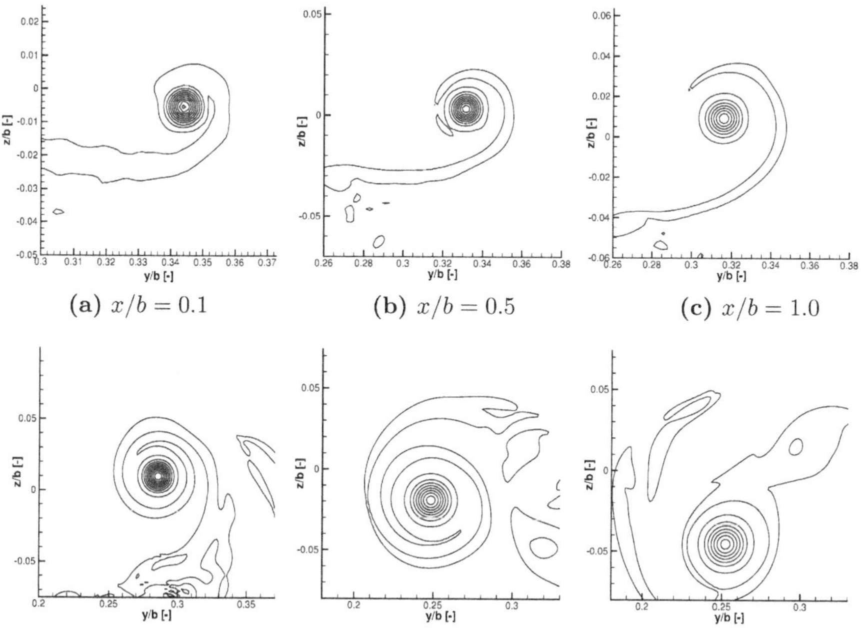

(a) $x/b = 0.1$ (b) $x/b = 0.5$ (c) $x/b = 1.0$

(d) $x/b = 2.0$ (e) $x/b = 4.0$ (f) $x/b = 6.0$

Fig. 4. Vorticity ω_x for engine position "A" at different x positions. 11 Levels for each plot. $x/b = 0.1 : 0.5 < \omega_x < 45$, $x/b = 0.5, 1.0 : 0.5 < \omega_x < 30$, $x/b > 2.0 : 0.2 < \omega_x < 16$

In Fig. 6 the time averaged Fourier coefficients for the modes $k = 2, 4$ are displayed for the two engine positions. The decay of the noise added at the inflow can be seen in the pictures 6(a) - 6(b) for $x/b < 2$. The decay of the kinetic energy of position "A" is stronger than that for position "C". In position "A" the second mode stops to decline and starts to grow at $x/b = 1.93$. This inflection point is at engine position "C" at $x/b = 2.36$ for the first small peak and at $x/b = 3.12$ for the start of the growth with the peak at $x/b = 4.08$. The growth rate of position "C" is larger than for position "A". Further downstream the Fourier coefficients of the energy for position "A" are larger than for position "C".

Compared to the temporal approaches of Paoli et al. [5], Laporte and Corjon [24], Jacquin et al. [17] or Zurheide [15] no dominant growth of a mode can be found in the simulated domain $x/b < 12$. Areas with lower Fourier coefficients change with areas of higher values of the Fourier coefficient. These values decline at higher modes but the main orientation is given by the local position x not by the Fourier mode k. Also a rise of the coefficients of several magnitudes cannot be found. There are several explanations for this behaviour. First, the growth of an elliptical instability requires two vortices

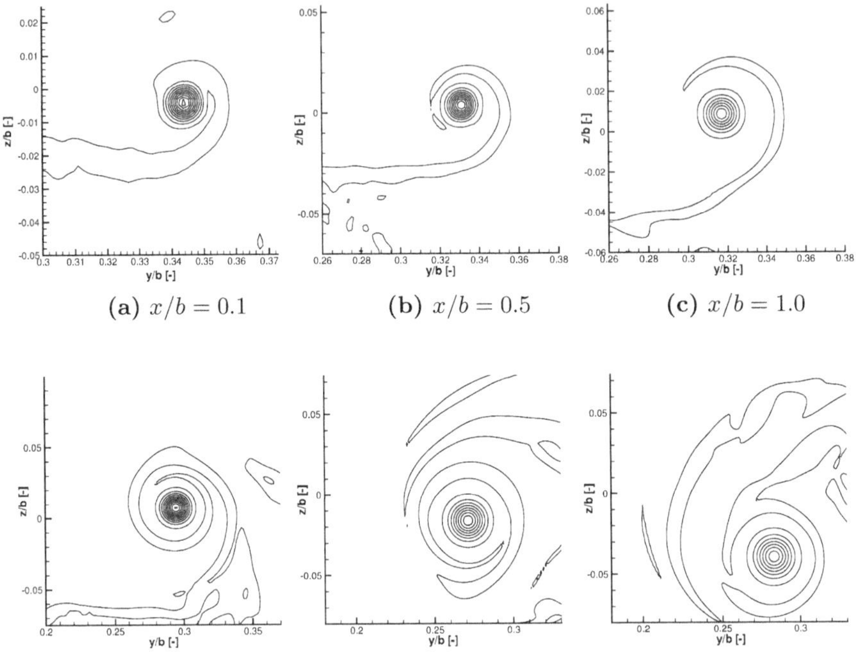

(a) $x/b = 0.1$ (b) $x/b = 0.5$ (c) $x/b = 1.0$

(d) $x/b = 2.0$ (e) $x/b = 4.0$ (f) $x/b = 6.0$

Fig. 5. Vorticity ω_x for engine position "C" at different x positions. 11 Levels for each plot. $x/b = 0.1 : 0.5 < \omega_x < 45$, $x/b = 0.5, 1.0 : 0.5 < \omega_x < 30$, $x/b > 2.0 : 0.2 < \omega_x < 16$

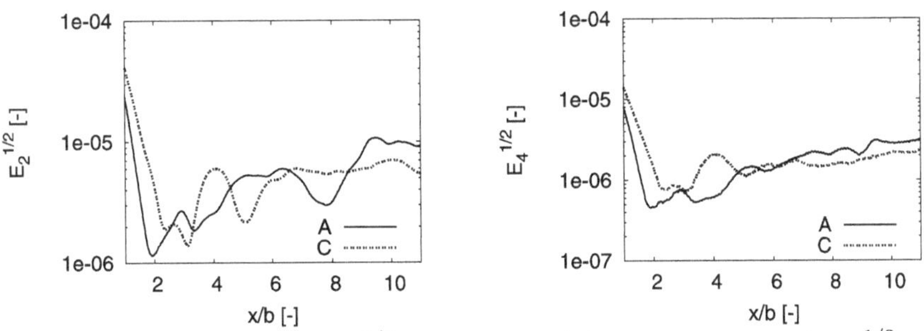

(a) Averaged values for $E_2^{1/2}$ (b) Averaged values for $E_4^{1/2}$

Fig. 6. Averaged Fourier coefficients $E_2^{1/2}$, $E_4^{1/2}$ along x-axis. Fourier coefficients are calculated with 64 points

that interact but we are computing only one half of the wake with one vortex. Note that a mirror condition at the symmetry plane y=0 would not result in a proper flow condition since the short wave instabilities have an offset of half the wave length [19, 25]. Secondly, the simulated extent in down stream

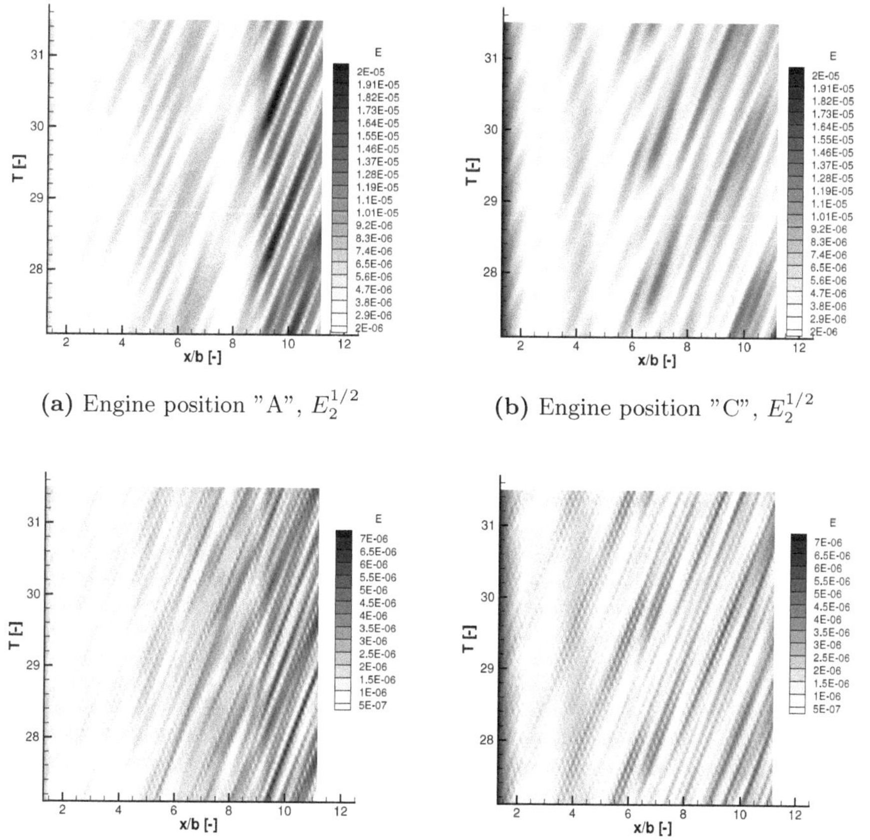

(a) Engine position "A", $E_2^{1/2}$

(b) Engine position "C", $E_2^{1/2}$

(c) Engine position "A", $E_4^{1/2}$

(d) Engine position "C", $E_4^{1/2}$

Fig. 7. Fourier coefficients $E_2^{1/2}$ and $E_4^{1/2}$ along x-axis for non dimensional time levels $t = 27.0 \ldots 31.4$. Fourier coefficients are calculated with 64 points

direction is too short. The long wave Crow instability appears several hundred wing span behind an airplane with large span [26, 27].

For the Fourier modes k=2 and 4 the temporal development is displayed in Fig. 7 for the two engine positions. In contrast to the averaged coefficients in Fig. 6 single perturbations of the flow field can be identified.

In Fig. 7(a) the Fourier coefficient has at $x/b = 8.0$, $t = 27.0$ a local maximum. The disturbance is convected down stream and is growing, e.g. at $x/b = 9.0$, $t = 28.0$ and at $x/b = 11.0$, $t = 30.0$. At a fixed x position the values show a heterogenous behaviour. The displayed values are striped – high values are followed by lower values. A comparison of the values of engine position “A” (Fig. 7(a)) and “C” (Fig. 7(b)) show, that position “A” is more uniform, while the changes between lower and higher values at position

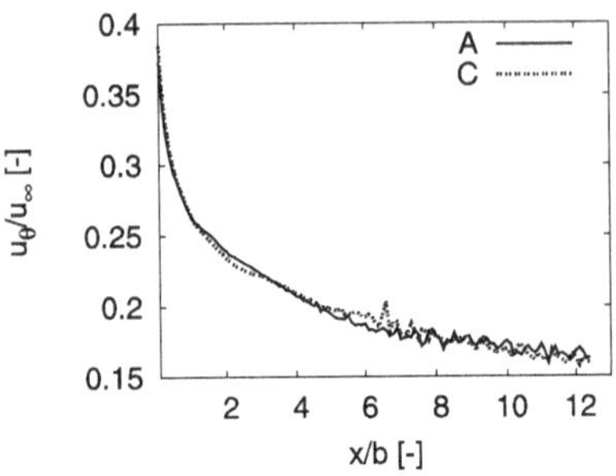

(a) Tangential velocity of the wing tip vortex

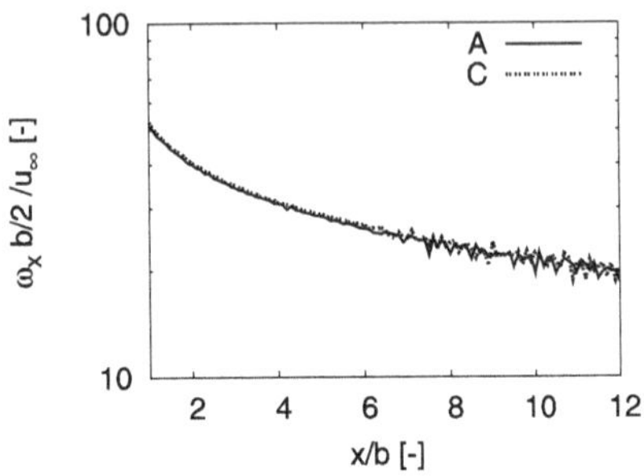

(b) Stream wise vorticity $(\omega_x b/2)/u_\infty$

Fig. 8. Comparison of values for engine position "A" and "C"

"C" are more random. Positions where the Fourier coefficients grow can be identified by transitions from lighter to darker regions, e.g. for case "A" $E_2^{1/2}$ at $x/b = 9.0$ in Fig. 7(a) or for "C" at $x/b = 4.0$ in Fig. 7(b). This peaks can also be found at Fig. 6(a). The Fourier coefficients for $k = 4$, $E_4^{1/2}$ in Fig. 7(c) and 7(d) depict a similar characteristic of the convection of flow features, but the growth is lower (see also Fig. 6(b)).

For further investigations of the wing tip vortex the flow field is analyzed at discrete x coordinates with a distance of $x/b = 0.1$. The values are taken at a constant dimensionless time step $t = 30.0$. The dimensionless tangential velocity u_θ/u_∞ of the wing tip vortex is compared for the two engine positions (Fig. 8(a)). The difference between the decay of the tangential velocity is negligible for the two cases. Özger et al. [28] measured the maximum tangential velocity u_θ/u_∞ in the wake of a BAC wing. The measured values decline from u_θ/u_∞=0.3 at x/b=0.0 to 0.2 at x/b=0.8, while the values of the simulation decline from 0.38 to 0.28 in the same distance.

The vortex has an axial velocity deficit u_D. The value at the inflow boundary is u_D/u_∞=0.8 for engine position "A" and u_D/u_∞=0.79 for position "C". The deficit decreases in downstream direction more rapidly than the axial deficit presented by Beninati and Marshall [29]. The initial velocity deficit of their simulation is also around u_D/u_∞=0.8. Fig. 8(b) depicts the non dimensional vorticity $(\omega_x b/2)/u_\infty$. The value is decreasing in downstream direction, but the difference for the two simulated cases is small. The swirl parameter q=$u_\theta/(u_\infty - u_D)$ is the ratio of the azimuthal to the axial velocity [30]. For a swirl parameter $q > 1.0$ a Batchelor vortex is stable, the rotation stabilizes all perturbations [30]. In our simulation the vortex is stable with q>1.0.

From $x/b = 5$ instabilities of the vortex have grown and become 3-D structures. These structures lead to strong perturbations in the flow field. The analysis summarized above in Fig. 8 was also used for the temporal development of vortical flows in [31, 4, 5]. In contrast to the results in those papers, no significant influence of the engine jet position on the wake development is

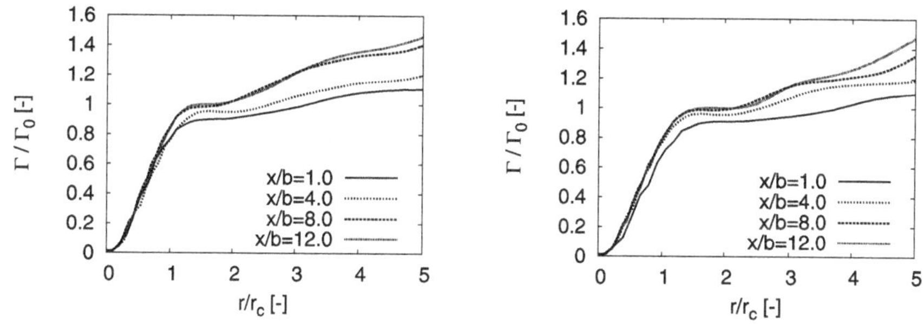

(a) Engine position "A" **(b)** Engine position "C"

Fig. 9. Circulation Γ/Γ_0 for selected positions x/b

obeserved here. One reason could be the unphysical coupling through the periodic boundary conditions in streamwise direction in the temporal developing simulations.

The circulation $\Gamma(r)$ is calculated for the radius r with the vorticity ω_x

$$\Gamma(r) = \int_0^{2\pi} \int_0^r \omega_x(\zeta)\zeta \mathrm{d}\zeta \mathrm{d}\theta \quad , \tag{8}$$

where the root circulation Γ_0 is defined as the circulation at $x/b = 12.0$. In Fig. 9(a) and 9(b) the non dimensional circulation Γ/Γ_0 is displayed for the non dimensional radius r/r_c. For a single vortex the circulation Γ grows to a constant and maximum value. At the initial position this is Γ_0. In Fig. 9(a) and 9(b) the circulation is growing beyond the maximum Γ_0, because the vorticity in the wake is also integrated at $r > r_c$, see Fig. 4 and 5 for the vorticity. The circulation reaches a local maximum at $1 < r/r_c < 2$. Then for higher values of r/r_c the vorticity of the wake is added to the circulation and so the value of Γ is increasing. For the x values down stream ($x/b \geq 8.0$) Γ is rising stronger than for smaller x values, because a larger part of the wake is already wrapped closer around the vortex core (compare Fig. 4(f) and 5(f)).

In Fig. 10 the positions of the center of the wing tip vortex are displayed. To get a more exact position of the vortex core, the position is interpolated with an ansatz function between the discrete positions defined by the mesh. The interpolation function is of second order in both directions:

$$c_0 + c_1 y + c_2 z + c_3 y^2 + c_4 z^2 = b_i \quad , \tag{9}$$

and is calculated for the position of the vortex core and the surrounding eight points at constant x values. This leads to a linear system of equations for each x position

$$\boldsymbol{A}\boldsymbol{c} = \boldsymbol{b} \quad . \tag{10}$$

For the nine points nine linear equations are obtained while there are only 5 unknowns ($c_0, \cdots, c_4$). This overdetermined system is solved with the least

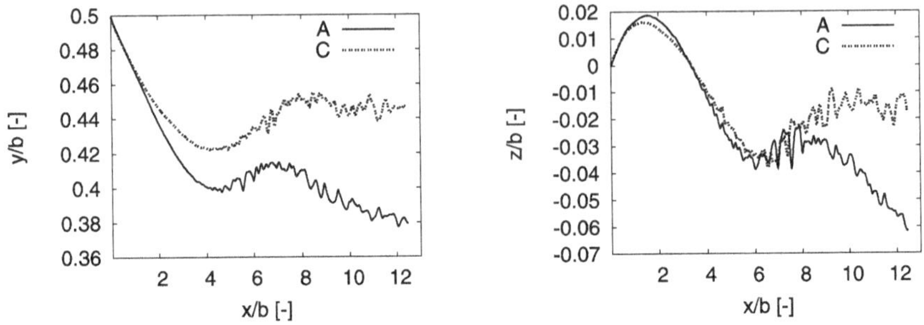

(a) y axis (b) z axis

Fig. 10. Refined position of wing tip vortex with least squares method. Comparison of bending for engine position "A" and "C"

squares method for the vector $\boldsymbol{c}$. Through derivation the interpolated position of the vortex core (y_c, z_c) is obtained

$$y_c = \frac{-c1}{2c_3} \quad , \quad z_c = \frac{-c2}{2c_4} \tag{11}$$

Figure 10(a) depicts the deflection on the y axis and Fig. 10(b) on the z axis. For engine position "A" - the one towards the wing tip - the deflection is larger than for position "C". At the local maximum at $x/b = 4.3$ the difference between the two trajectories is $\Delta y = 0.023b$. The differences in z direction are relative to the deflection even larger. For case "A" the vortex looses height from $x/b = 8$, while at position "C" the vortex position is still moving upwards.

One possible explanation for the different wing tip vortex positions is the interaction of the vortex with the shear layer and the jet of the wake. At position "A" the jet is rolled up at a shorter distance than at engine position "C". Therefore the moment needed for the rolling up and wrapping of the jet is smaller. Another explanation is the angle and position of the interaction of the vortex and jet. For "C" the angle is bigger than for "A" so the jet contains a larger impulse to move the jet towards a positive y direction.

5 Computational Resources

All simulations were carried out on the NEC SX-8 installed at the HLR Stuttgart. The results presented in chapter 4 were computed on a domain of integration that is divided into 24 blocks, while each block resides on a single CPU. Data between the blocks is exchanged via MPI (message passing interface). The cluster is arranged in nodes with 8 CPUs each, so three nodes are used for the simulation. The workload for each CPU, that is mesh points per block, is high, but since the scheduling system prefers jobs with fewer

Table 1. Sample performance on NEC SX-8

	Case "A"/"C"	Case "I"
Number of CPU	24	16
Number of Nodes	3	2
Mesh size	$63,2 \cdot 10^6$	$40.2 \cdot 10^6$
Meshpoints/CPU	$2.7 \cdot 10^6$	$2.6 \cdot 10^6$
Avg. User Time [s]	44500	43100
Avg. Vector Time [s]	41735	41100
Vector Operations Ratio [%]	99.5	99.6
Avg. Vector Length	240.0	242.3
Avg. MFLOPS/CPU	5035	5500
Max. MFLOPS/CPU	5090	5550
Memory/CPU [MB]	1750	1666
total GFLOPS	120.8	88.1
total Memory	41.2 GB	26.0 GB

nodes, the domain decompositioning was adapted to a rather small number of processors to achieve a small turnaround time and not to a massively parallelized simulation. In principle other distributions of the mesh points are possible.

Here, we present additionally computing statistics of another simulation which is not discussed elsewhere in this paper. This case "I" contains 40 million mesh points, that are distributed equally to 16 blocks on two nodes. Table 1 summarizes the performance of the two different mesh topologies.

For a realistic simulation of the wake, the whole wake with both wing tip vortices must be computed. A first estimation leeds to a six times larger number of mesh points for the same simulation length downstream ($12.68\,b$). For the simulation of the long wave Crow instability [2] the integration domain must be larger than 100 spans, since this slowly growing instability causes a breakdown of the vortex system [27] beyond this length. For this simulation more than 3 billion mesh points would be needed with approximately 2.5 TB RAM.

6 Conclusions

A successful LES of a wake-jet interaction in the wake of a wing was presented in this paper. The inflow boundary condition for the spatial simulation of the wake was determined from PIV measurements in the wake of a wing with an engine mounted in two different positions. The simulation showed that the vorticity distribution in the wake is influenced by the engine position, the Fourier coefficients of the Fourier modes k=2 and 4 depicted a different growth rate. The temporal development of the Fourier coefficients revealed, that the vortex instabilities are convected downstream. While the tangential velocity, the axial velocity deficit and the streamwise vorticity component in the tip vortex showed no significant difference for the two engine positions, the swirl parameter showed different values.

The main difference for the two cases could be shown to be the trajectory of the vortex core. Engine position "C" produces a larger deflection of the vortex core than the position "A".

Further investigation of the wake-jet interaction should focus on different cases like a fully turned off engine and also high lift devices like flaps, which produce additional vortices. Furthermore a simulation of the wake of the full wing with both wing tip vortices is required to predict the growth of instabilities in the near and extended near field.

Acknowledgement

The support of this research by the Deutsche Forschungsgemeinschaft (DFG) in the frame of SFB 401 is gratefully acknowledged.

References

1. Spalart, P.R.: Airplane trailing vortices. Annual Review of Fluid Mechanics **30** (1998) 107–138
2. Crow, S.C.: Stability theory for a pair of trailing vortices. AIAA J. **8** (1970) 2172–2179
3. Waleffe, F.: On the three-dimensional instability of strained vortices. Phys. Fluids **2** (1990) 76–80
4. Labbe, O., Maglaras, E., Garnier, F.: Large-eddy simulation of a turbulent jet and wake vortex interaction. Comput. & Fluids **36** (2007) 772–785
5. Paoli, R., Laporte, F., Cuenot, B., Poinsot, T.: Dynamics and mixing in jet/vortex interactions. Phys. Fluids **15** (2003) 1843–1860
6. Gago, C.F., Brunet, S., Garnier, F.: Numerical Investigation of Turbulent Mixing in a Jet/Wake Vortex Interaction. AIAA J. **40** (2002) 276–284
7. Holzäpfel, F., Hofbauer, T., Darracq, D., Moet, H., Garnier, F., Gago, C.F.: Analysis of wake vortex decay mechanisms in the atmosphere. Aerosp. Sci. Technol. **7** (2003) 263–275
8. Fares, E., Schröder, W.: Analysis of wakes and wake-jet interaction. In: Notes on Numerical Fluid Mechanics. Volume 84. (2003) 57–84
9. Ghosal, S., Moin, P.: The basic equations for the large eddy simulation of turbulent flows in complex geometry. J. Comput. Phys. **118** (1995) 24–37
10. Meinke, M., Schulz, C., Rister, T.: LES of spatially developing jets. In: Computation and Visualization of Three-Dimensional Vortical and Turbulent Flows. Notes on Numerical Fluid Mechanics. Vieweg Verlag (1997)
11. Huppertz, G., Klaas, M., Schröder, W.: Engine jet/vortex interaction in the near wake of an airfoil. In: AIAA 36th Fluid Dynamics Conference, San Francisco, CA, U.S. A (2006) AIAA-Paper 2006-3747.
12. Huppertz, G., Schröder, W.: Vortex/engine jet interaction in the near wake of a swept wing. In: 77th Annual Meeting of the GAMM, Berlin, Germany (2006)

13. Holzäpfel, F., Gerz, T., Baumann, R.: The turbulent decay of trailing vortex pairs in stably stratified environments. Aerosp. Sci. Technol. **5** (2001) 95–108
14. Holzäpfel, F., Hofbauer, T., Darracq, D., Moet, H., Garnier, F., Gago, C.F.: Wake vortex evolution and decay mechanisms in the atmosphere. In: Proceedings of 3rd ONERA–DLR Aerospace Symposium, Paris, France (2001) 10
15. Zurheide, F., Schröder, W.: Numerical analysis of wing vortices. In: New Results in Numerical and Experimental Fluid Mechanics VI. Volume 96 of Notes on Numerical Fluid Mechanics and Multidisciplinary Design., Springer Berlin, Heidelberg, New York (2008) 17–25
16. Fabre, D., Jacquin, L.: Stability of a four-vortex aircraft wake model. Phys. Fluids **12** (2000) 2438–2443
17. Jacquin, L., Fabre, D., Sipp, D., Theofilis, V., Vollmers, H.: Instability and unsteadiness of aircraft wake vortices. Aerosp. Sci. Technol. **7** (2003) 577–593
18. Le Dizès, S., Laporte, F.: Theoretical predictions for the elliptical instability in a two-vortex flow. J. Fluid Mech. **471** (2002) 169–201
19. Bristol, R.L., Ortega, J.M., Marcus, P.S., Savas, Ö.: On cooperative instabilities of parallel vortex pairs. J. Fluid Mech. **517** (2004) 331–358
20. Stumpf, E., Wild, J., Dafa'Alla, A.A., Meese, E.A.: Numerical simulations of the wake vortex near field of high-lift configurations. In: European Congress on Computational Methods in Applied Sciences an Engineering ECCOMAS 2004, Jyväskylä (2004)
21. Yin, X.Y., Sun, D.J., Wei, M.J., Wu, J.Z.: Absolute and convective instability character of slender viscous vortices. Phys. Fluids **12** (2000) 1062–1072
22. Jeong, J., Hussain, F.: On the identification of a vortex. J. Fluid Mech. **285** (1995) 69–94
23. Schlichting, H., Truckenbrodt, E.A.: Aerodynamik des Flugzeuges. Volume 2. Springer, Berlin, Heidelberg, New York, Barcelona, Hongkong, London, Mailand, Paris, Singapur, Tokio (2001)
24. Laporte, F., Corjon, A.: Direct numerical simulations of the elliptic instability of a vortex pair. Phys. Fluids **12** (2000) 1016–1031
25. Leweke, T., Williamson, C.H.K.: Cooperative elliptic instability of a vortex pair. J. Fluid Mech. **360** (1998) 85–119
26. Gerz, T., Holzäpfel, F.: Wing-tip vortices, turbulence, and the distribution of emissions. AIAA J. **37** (1999) 1270–1276
27. Saudreau, M., Moet, H.: Characterization of Extended Near-Field and Crow Instability in the Far-Field of a Realistic Aircraft Wake. In: European Congress on Computational Methods in Applied Sciences and Engineering ECCOMAS 2004. (2004)
28. Özger, E., Schell, I., Jacob, D.: On the Structure and Attenuation of an Aircraft Wake. J. Aircraft **38** (2001) 878–887
29. Beninati, M.L., Marshall, J.S.: An experimental study of the effect of free-stream turbulence on a trailing vortex. Experiments in Fluids **38** (2005) 244–257
30. Jacquin, L., Pantano, C.: On the persistence of trailing vortices. J. Fluid Mech. **471** (2002) 159–168
31. Gallaire, F., Chomaz, J.M.: Mode selection in swirling jet experiments: a linear stability analysis. J. Fluid Mech. **494** (2003) 223–253

Impact of Density Differences on Turbulent Round Jets

Ping Wang[1], Jochen Fröhlich[2], Vittorio Michelassi[3], and Wolfgang Rodi[1]

[1] Institut für Hydromechanik, Universität Karlsruhe, Kaiserstr. 12, D-76131 Karlsruhe wang@ifh.uka.de rodi@uka.de
[2] Institut für Strömungsmechanik, Technische Universität Dresden, George-Bähr-Str. 3c, D-01069 Dresden jochen.froehlich@tu-dresden.de
[3] Via Felipe Matteucci 2, 50127 Florence, Italy vittorio.michelassi@ge.com

Summary. Jets with different density than the fluid into which they are issued, constitute an important class of flows relevant for applications in technical and environmental situations. In this work three cases of variable-density turbulent round jets discharging from a straight circular pipe into a weakly confined low-speed co-flowing air stream are studied with the aid of large eddy simulation. The density ratios considered are 0.14 [Helium/air], 1.0 [air/air] and 1.52 [CO_2/air], with Reynolds numbers of 7000, 21000 and 32000, respectively. These computations closely correspond to experiments performed by F. Anselmet and co-workers. Detailed comparisons of the statistics show good agreement with the corresponding experiments. From a physical point of view it is observed that lower-density jet develop more rapidly than denser jets with the same exit momentum flux. Pseudo-similarity behavior in the three variable-density round jets is well reproduced in the simulation. The coherent structures of the three jets are investigated by visualization of the iso-surface of pressure fluctuations and vorticity. In the developing stage of the Kelvin-Helmholtz instability, large finger-shape regions of vorticity are observed for the helium jet close to the nozzle lip. This feature, however, is not found in the air and the CO_2 jet. The occurrence of strong streamwise vorticities across the shear layer in the helium jet is demonstrated by a characteristic quantity related to the orientation of the vorticity. The computations were all performed on the HP-XC cluster of the SSC in Karlsruhe.

1 Introduction

The present work has been undertaken within the framework of project A6 of SFB 606 "Unsteady Combustion: Transport phenomena, Chemical Reactions, Technical Systems", mainly located at the University of Karlsruhe. The final goal of this project is to study the oscillating flow in burner configurations with piloted premixed flames, with the aid of LES. During the first phase of the SFB 606 project A6, numerous simulations of the non-reactive, constant-density flows in combustor-related geometries have been performed.

The agreement achieved between the simulations and the reference experiment is outstanding [1]. During the second phase, the computational method is first extended to variable-density turbulent flows and will subsequently also account for chemical reactions. In this paper, the impact of density difference on the turbulent non-reactive round jets development is investigated numerically use large-eddy simulation. More details on the work summarized here can be found in a recent journal publication by the authors [2].

Turbulent jets with variable density are less well understood than the widely studied constant-density jets, although this type of flow widely occurs in technical applications as well as environmental situations. Relatively few experimental studies were reported for such cases. An experiment with helium/air mixture discharging into a confined swirling flow was carried out by Ahmed et al. [3]. Sreenivasan et al. [4] performed an experimental study on round jets of different densities issuing into the ambient air. Their different densities were obtained by premixing helium and air in various proportions. About the same time, Monkewitz et al. [5, 6] carried out an experimental investigation of entrainment and mixing in transitional axisymmetric jets, where density differences were achieved by heating the air. Panchapakesan and Lumley [7] conducted an experiment with helium injected into open quiescent air from a round nozzle. Later, Djeridane et al. [8] and Amielh et al. [9] performed experimental studies of variable-density turbulent jets, including helium, air and CO_2 jets exiting into a low-speed air co-flow. Numerical investigations of this type of flow are also relatively scarce. Jester-Zrker et al.[10] performed a numerical study of turbulent non-reactive combustor flow under constant- and variable-density conditions using a Reynolds-stress turbulence model. They obtained good agreement between simulation and experiment for the constant-density flow, whereas the results for the variable-density flow were less satisfactory. Some large-eddy simulation (LES) of variable-density round jets were also performed recently [11, 12]. To the authors' knowledge, however, detailed comparisons of LES results and experimental data for round jets, covering density ratios both lower and larger than unity, are not available in the literature.

The aim of the present work is to perform a detailed comparison of LES results and experimental data for three round jets with density ratios 0.14, 1.0 and 1.52 respectively, to gain a deeper understanding of the impact of density differences on the turbulent round jet development.

2 Numerical Method

The numerical method employed is based on the so-called low-Mach number version of the compressible Navier-Stokes equations. With this approach, the pressure P is decomposed into a spatially constant component $P^{(0)}$, interpreted as the thermodynamic pressure, and a variable component $P^{(1)}$, interpreted as the dynamic pressure. $P^{(0)}$ is connected to temperature and density, while $P^{(1)}$ is related to the velocity field only and does not influence the density. Due to this decomposition, sonic waves are eliminated from the flow, so that the time step is not restricted by the speed of sound.

Applying large eddy filtering to the low-Mach number equations, the corresponding filtered LES equations are obtained. The unclosed terms in these equations have to be determined by a subgrid scale (SGS) model. The variable-density dynamic Smagorinsky model by Moin et al. [13] is used to determine the SGS eddy viscosity, μ_T, in the momentum equations. The SGS scalar flux is modeled by the gradient diffusion model:

$$\overline{\rho\,\phi\,u_j} - \bar{\rho}\,\tilde{\phi}\,\tilde{u}_j = -\frac{\mu_T}{Sc_{T,SGS}}\frac{\partial\tilde{\phi}}{\partial x_j} \tag{1}$$

where $Sc_{T,SGS}$ is the subgrid-scale Schmidt number. $Sc_{T,SGS}$ may be determined with a dynamic procedure, as used for the SGS eddy viscosity [13]. In this work it is set to $Sc_{T,SGS} = 0.7$.

The simulations were performed with the in-house CFD code LESOCC2C, which is a compressible version of Finite Volume code LESOCC2 [14]. LESOCC2 is highly vectorized, and parallization is accomplished by domain decomposition and explicit message passing via MPI. It solves the incompressible Navier-Stokes equations on body-fitted curvilinear block-structured grids employing second-order central schemes for the spatial discretization and a 3-step Runge-Kutta method for the temporal discretization. The convection term of the species equation is discretized with the HLPA scheme [15]. Currently, LESOCC2 is employed for several LES projects at the Institute of Hydromechanics, University of Karlsruhe. In order to simulate reactive flows, LESOCC2C is designed to solve the above low-Mach-number version of the compressible Navier-Stokes equations, by means of a pressure-based method. Apart from the evaluation of additional non-linear terms, this approach amounts to solving a Poisson-type equation in each time step, similarly to the approach for constant density flows in LESOCC2, so that the algorithm of both versions is very much the same.

3 Computation Set Up and Computational Effort

Three jets issuing into a very slow co-flow of air are studied with density ratios equal to 0.14 [Helium/air], 1.0 [air/air], and 1.52 [CO_2/air], respectively (see Fig. 1). These cases correspond to situations studied experimentally by Djeridane et al. [8] and Amielh et al. [9]. The parameters employed are listed in Table 1, in which the Reynolds number is based on the centerline velocity at the jet exit and the jet nozzle diameter. The subscripts 'j' and 'e' relate to jet flow and external co-flow, respectively. It is worth noting that the momentum flux is the same for the three cases, $M_j = 0.1N$. The reason for comparing cases with identical momentum flux is that in the flow region investigated inertial forces dominate [8, 9]. The averaging time is given in units of D_j/U_j.

Table 1. Parameters of the three jets simulated

Jet	ρ_j/ρ_e	$U_j(m/s)$	$U_e(m/s)$	Re_j	t_{aver}
Helium	0.14	32	0.9	7000	1285
Air	1.0	12	0.9	21 000	803
CO_2	1.52	10	0.9	32 000	1078

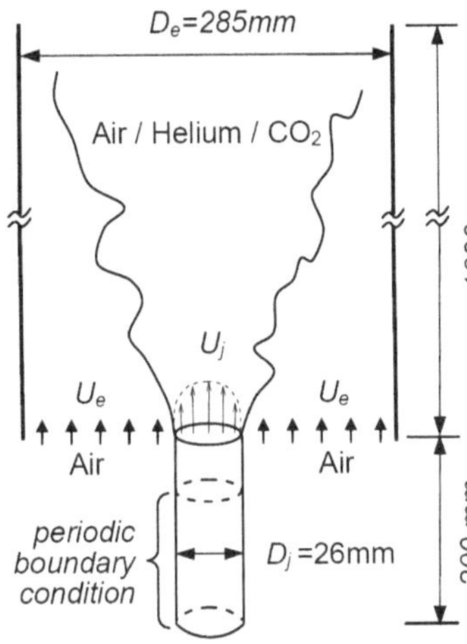

Fig. 1. Sketch of the computational domain

Additional averaging was performed in azimuthal direction. Note that this is not effective near the axis so that the quality of averaging is better remote from the axis. Fig. 1 shows a sketch of the flow configuration and the computational domain. The jet discharges from a long pipe into a coflow confined by an outer pipe. The ratio of the pipe diameters is $D_e/D_j = 11$, so that the confinement is weak.

A convective outflow boundary condition was imposed at the exit, and a uniform velocity profile without fluctuations at the inlet of the co-flow. Sensitivity studies were conducted imposing fluctuations in the inlet of the coflow and a small boundary layer configured according to the experimental setup, but this did not have any effect on the computed statistics. In order to obtain a fully developed turbulent flow in the pipe upstream of the jet exit, as indicated by the experimental data, a separate simulation of turbulent pipe flow with streamwise periodic conditions was performed simultaneously (see Fig. 1). The Werner-Wengle wall function [16] was used at the pipe walls.

4 Grid Resolution Study

The computational grid employed consists of 8 million cells, divided into 251 blocks. Fig. 2 provides related pictures. In the azimuthal direction 152 grid points are used and 385 grid points in axial direction from the outlet to the downstream end of the domain. The grid cells are stretched locally to reduce the cell size near the jet nozzle and across the shear layer. This resulted in a radial extent in wall units along the inner pipe wall of $\Delta r^+ = 3.0$, 8.0, and 10.5 for the helium, the air and the CO_2 jet, respectively. To perform a grid resolution study, a coarser grid consisting of 1 million cells was also employed.

Fig. 3 shows the radial profiles of streamwise rms-fluctuations and radial profiles of shear stress for the helium jet at three axial stations. In order to check the influence of grid resolution, two simulations are performed with the fine grid (8 mil. cells) and the coarse grid (1 mil. cells), respectively. It is seen that the difference is significant: the agreement between the experimental data and the computational results obtained with fine grid is excellent, whereas the prediction with the coarse grid is not satisfactory. Together with further results discussed below, this indicates that the fine grid employed is very well suitable for the helium jet simulation. For the air jet and the CO_2 jet, on the other hand, the relative resolution is not so high, due to the higher Reynolds number in these cases. This effect was quantified by means of the SGS eddy viscosity. Its time-averaged value $(\mu_T + \mu_l)/\mu_l$ is shown in Fig. 4. This ratio in the helium jet is always smaller than that in the CO_2 and the air jets, especially in the shear layer close to jet nozzle. For instance, at position $(x = 0.5D_j, r = 0.5D_j)$ a mean turbulent viscosity of $\mu_T \approx 2.4\mu_l$ is observed in the CO_2 jet, while in the helium jet $\mu_T \approx 0.5\mu_l$. On the other hand, an LES by definition is a simulation where the SGS terms overwhelm the molecular terms. To keep the required computational resources within a reasonable limit the same fine grid was used for all the three jets. The discussion below shows that this grid still yields satisfactory results for most of the quantities considered.

Generally, the computations with the fine grid divided into 251 blocks were carried out with 48-64 processors on the HP-XC4000 cluster. With this number of CPUs, good balance of work load could be achieved, and the com-

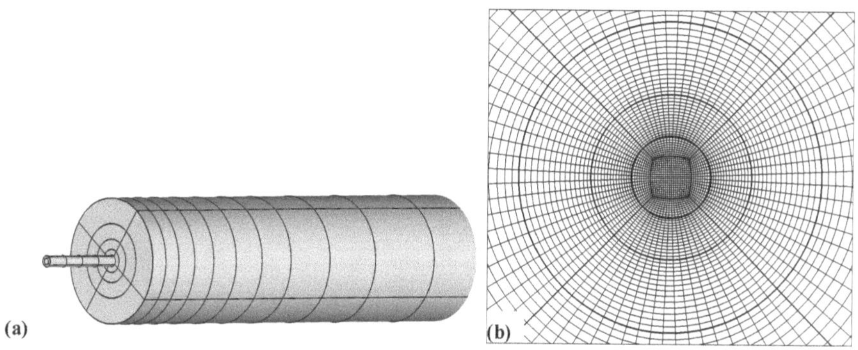

Fig. 2. (a) 3D view of the computational domain. Dark lines indicate block boundaries of the computational grid. (b) zoom of the grid around the symmetry axis in a plane x=const. (only every second point is shown)

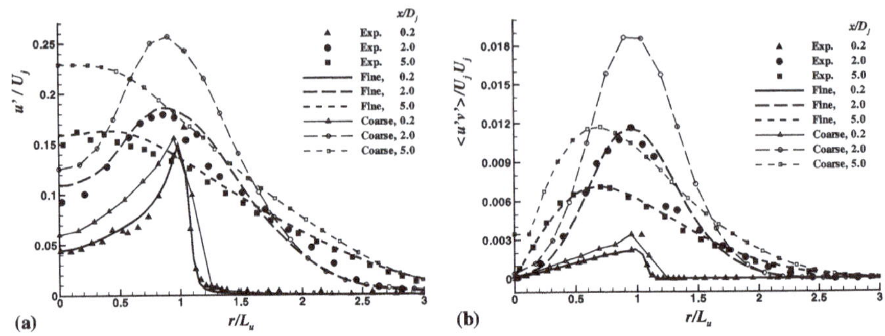

Fig. 3. Influence of grid resolution for the helium jet. (a) radial profiles of the streamwise rms-fluctuations. (b) radial profiles of the shear stress $\langle u'v' \rangle / U_j^2$

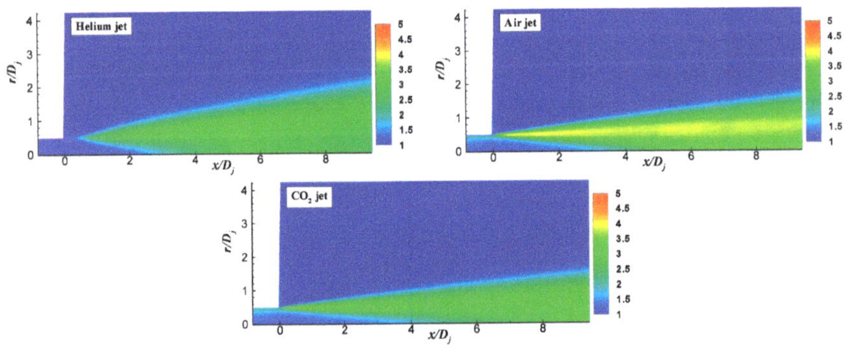

Fig. 4. Distribution of ratio $(\mu_T + \mu_l) / \mu_l$ with μ_T being the SGS eddy viscosity and μ_l the molecular viscosity. To provide an orientation, numerical values at position $(x = 0.5D_j, r = 0.5D_j)$ are reported here: the ratio is 1.5 for helium, 3.5 for air, and 3.4 for CO_2, respectively

munication costs were only about 5-10% of the total CPU time. On average, 43200 CPU hours were required to obtain converged second order statistics and velocity spectra for one jet with the fine grid. Consequently, the simulations of the three jets required a total of 129,600 CPU hours. In addition to this amount, numerous test were performed to investigate the grid resolution, to compare different inflow condition for the co-flow, and to check other issues of the simulation.

5 Statistical Results

Fig. 5 provides a comparison of the axial evolution of mean streamwise velocity and mean mass fraction along the jet axis, U_c and C_c, the subscript denoting 'c' values at the centerline. Analytical curves calculated from the similarity laws proposed for variable-density jets by Chen and Rodi [17] are also included.

The influence of the density difference is obvious: the centerline velocity and concentration of the helium jet decays much faster than in the air and the CO_2 jet. Light gas, helium, tends to mix more rapidly with the ambient air than the heavier gases do (recall that the momentum flux is the same for all flows). This faster mixing of helium is accompanied by a faster increase of turbulence intensity in the near-nozzle region (see Fig. 6). The potential core of the helium jet is only 3 diameters long, much shorter than that of the air and CO_2 jet. Very good agreement between the LES results and the experimental data is obtained for the helium jet. For the heavier jets, some differences in the decay of U_c are observed. They might be due to a small mismatch of inflow conditions.

The influence of the density ratio ρ_j/ρ_e on the decay of U_c and C_c is also described quite well by the similarity laws, but no perfect agreement between the LES and these laws should be expected because the laws are only approximate, and in fact 'x' in the laws should be replaced by the distance from the virtual origin of the jet as a point source.

The evolution of the rms-fluctuations of the streamwise velocity, u', along the jet axis is shown in Fig. 6. The influence of the density difference is obvious again: the helium jet decelerates much faster than the two heavier jets. At $x/D_j = 4.5$, the peak value of 17% for u'/U_j is already reached in the helium jet, whereas the peaks for the two heavier gases are lower, less pronounced, and attained further downstream. The fast increase of u'/U_j in the potential core region is due to the growth of the Kelvin-Helmholtz instability across the shear layer. Although the values of u'/U_j for the three jets at positions further downstream, for instance $x/D_j > 25$, are quite different, the values of $u'/(U_c - U_e)$, i.e. u' normalized with the local streamwise velocity difference, are close to each other and have a level of about 25-30%.

The radial profiles of the normalized mean streamwise velocity U/U_j are shown in Fig. 7 for the helium and the CO_2 jet. The results of the air jet are close to those of the CO_2 jet. Very good agreement between LES and

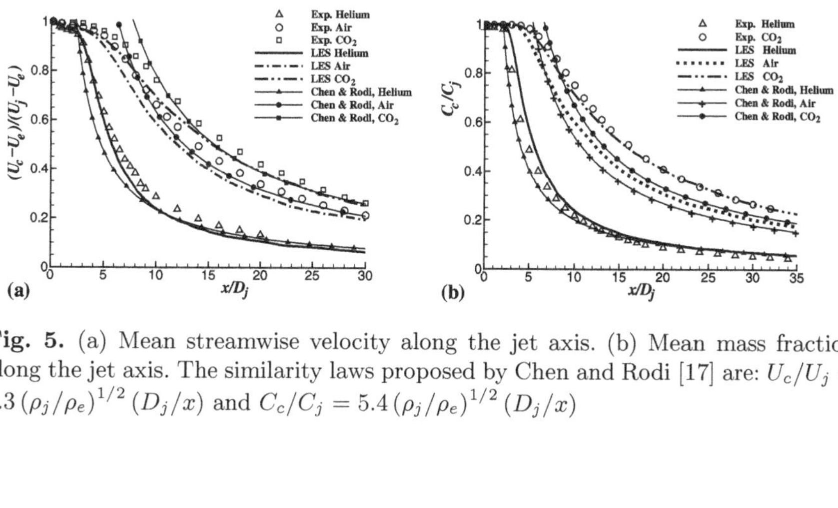

Fig. 5. (a) Mean streamwise velocity along the jet axis. (b) Mean mass fraction along the jet axis. The similarity laws proposed by Chen and Rodi [17] are: $U_c/U_j = 6.3\,(\rho_j/\rho_e)^{1/2}\,(D_j/x)$ and $C_c/C_j = 5.4\,(\rho_j/\rho_e)^{1/2}\,(D_j/x)$

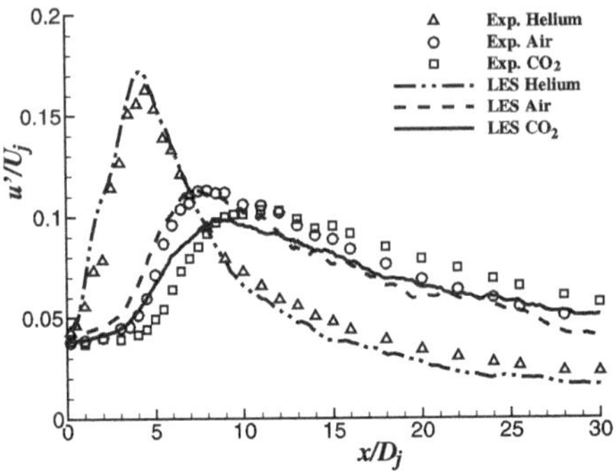

Fig. 6. Rms-fluctuations of the streamwise velocity component along the jet axis

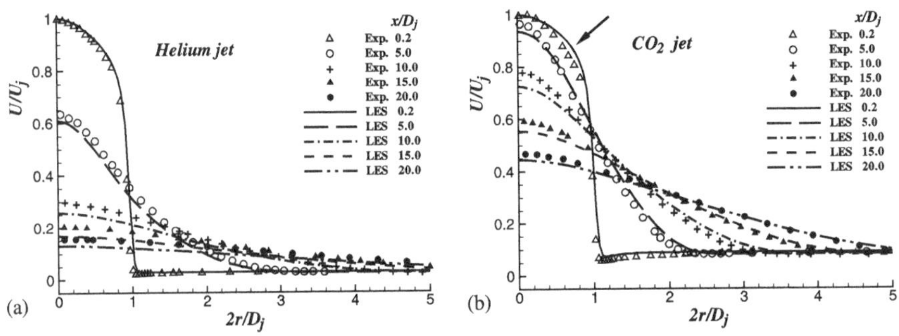

Fig. 7. Radial profiles of mean streamwise velocity at several axial positions

experiment is obtained for the helium jet. Starting from the typical turbulent pipe flow profile at $x/D_j = 0.2$, the centerline velocity of the jet decreases faster than for the CO_2 jet, as seen already from Fig. 3. At the first station $x/D_j = 0.2$ in Fig. 7(b), there is some disagreement for the CO_2 jet (see arrow). A reason may be that for this case the pipe flow was not fully developed at the exit in the experiment. Beyond $x/D_j = 0.2$, the agreement between LES and experiment is good also for the CO_2 jet.

Radial distributions of the normalized Reynolds shear stress $\langle u'v' \rangle / U_j^2$ are plotted in Fig. 8. With the two jets, different scales are used for the vertical axes for clarity. The agreement for the helium jet is excellent. In Fig. 8(a), the peak of $\langle u'v' \rangle / U_j^2$ at the first section $x/D_j = 0.2$ is only 0.003. Further downstream at $x/D_j = 2.0$, the maximum is almost four times as large and the shape of the curve is substantially smoother. At even larger x/D_j, the shear stress decays again due to the broadening of the jet and the resulting reduction of gradients. The results for the CO_2 jet in Fig. 8(b) show a substantially stronger peak in the shear stress right at the outlet which is extremely difficult to capture. At $x/D_j = 2.0$ and 5.0, the computed profiles have the correct shape but exceed the experimental data by about

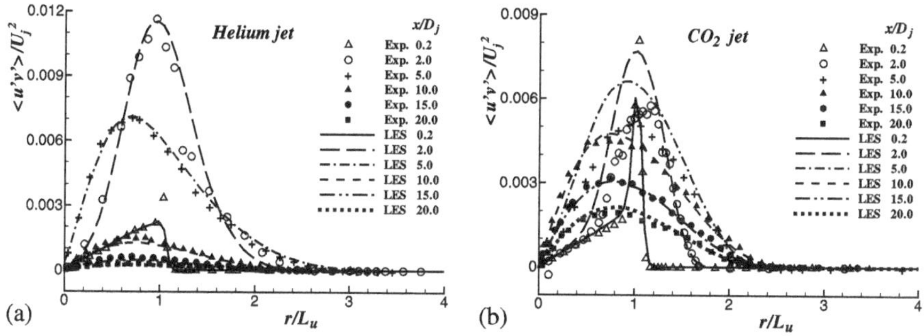

Fig. 8. Radial profiles of shear stress $\langle u'v' \rangle / U_j^2$

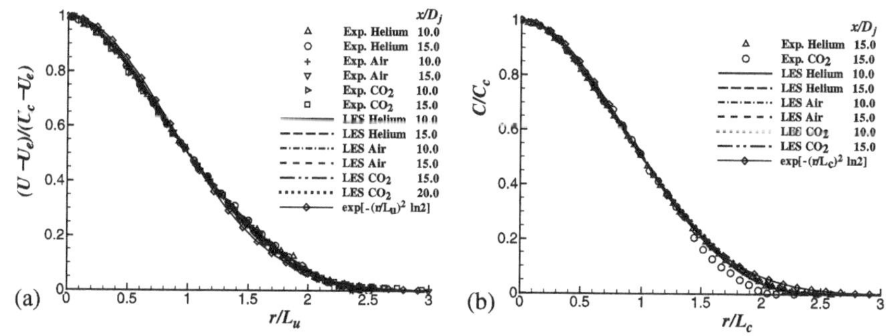

Fig. 9. Similarity profiles of mean streamwise velocity and mean mass fraction

25%. Further downstream, the agreement is very good. The reasons for the differences observed may reside in an under-resolution of the very first stages of the initial development (see Fig. 4 and its description).

Radial profiles of the mean streamwise velocity and the mean scalar concentration normalized by the values at the jet axis are shown in Fig. 9 for several axial positions. For each curve, the reference length is the local half-width of the respective quantity. The computed and the measured profiles collapse fairly well. These curves are well represented by a Gaussian function, except for the mean scalar fraction at the positions far downstream near the outer edge ($r/L_c > 1.4$). This is due to the confinement starting to have an effect [8]. The good collapse of the radial profiles and the approximately linear increase of the half-width (not shown in this paper) demonstrate the validity of pseudo-similarity in the present variable-density jets.

6 Coherent Structures

Many investigations have been carried out and aimed to explain the faster mixing and entrainment in the lighter jets [5, 6, 18]. By flow visualization in hot air jets, Monkewitz et al. [5] found a star-shaped jet cross section with two to six radial fingers in the potential core region. These fingers, which they called 'side jets' in this context, significantly increase the spreading angle, and it was concluded that the 'side jets' are generated by the so-called Widnall instability [19]. This mechanism is relevant for the azimuthal instability of vortex rings. In a later experimental investigation, however, Monkewitz and Pfizenmaier [6] revised the previous conclusion. By checking the phase-averaged axial and radial velocities, they concluded that strong streamwise vortex pairs on the braids between the vortex rings are as important for the generation of '*sidejets*' as the Widnall instability. Gharbi et al. [20] performed a statistical analysis of the jets considered here. They concluded that the structure of turbulence downstream and in the outer region is similar in all cases while in the near field the properties strongly depend on the density ratio.

The present LES data allow investigating the formation and evolution of coherent structures, which are responsible for the lager-scale exchange of mass and momentum. Iso-surfaces of the pressure fluctuation $p - \langle p \rangle$ are employed for this purpose, as successfully used already with swirling constant-density jets by Garca-Villalba et al. [1]. Like in this previous work, a 3D box filter of twice the step size of the grid is used for smoothing in a post-processing step in order to enhance the clarity of the picture. Fig. 10 shows such instantaneous iso-surfaces for the helium, the air and the CO_2 jet, respectively. For optimal presentation of the structures, the constant value for obtaining the iso-surface was slightly adjusted in each case, but all values are very close

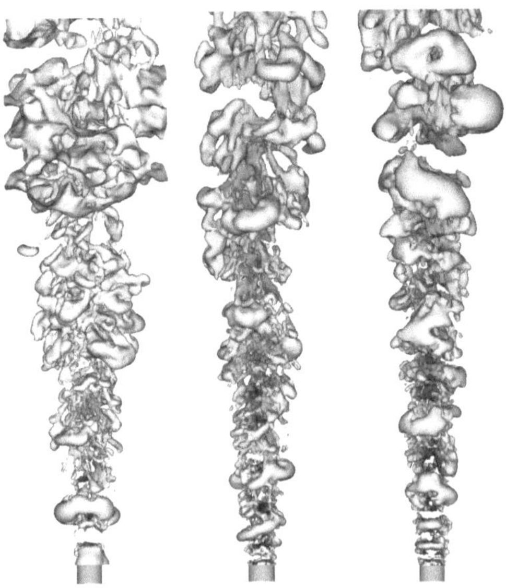

Fig. 10. Iso-surfaces of pressure fluctuation. From left to right: helium, air, CO2 jet, respectively

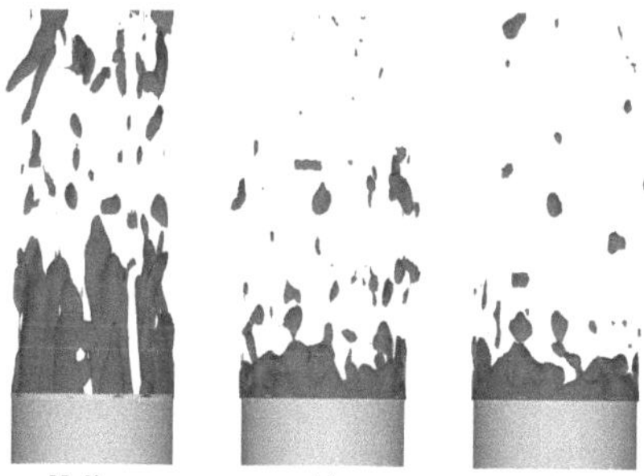

Fig. 11. Iso-surfaces of the vorticity modulus, $|\omega| = 9$

to zero. Shortly downstream of the nozzle, vortex rings are observed in all the three jets due to the Kelvin-Helmholtz instability. Further downstream in the air jet, larger structures form at the outer edge of the jet. In the CO_2 jet, the behavior is similar to the air jet, but the distance between the vortex rings appears to be shorter and the intermittency lower. For the helium jet in contrast, the vortex rings seem to have a larger distance and a larger size. These observations are made in the near-field region where an inner core and a surrounding flow can be distinguished. In Fig. 10 this would be up to $x/D_j \approx 10, 14, 18$ for the helium, air and CO_2 jet, respectively (lighter jets develop faster as discussed above). Beyond this region, the turbulent motion seems to be fairly homogeneous and vortex rings or similar structures can no longer be discerned.

Fig. 11 is concerned with the close vicinity of the jet outlet. It shows iso-surfaces of the vorticity modulus for the three jets in this region. In the following vorticity is normalized with the diameter of the pipe D_j and the center-line velocity of the jet at the outlet U_j. The maximum of the vorticity modulus in these units is approximately 15.0 for all cases. In the helium jet, large vortices are found which are regularly arranged along the lip. The length of these fingers is about one diameter. On the contrary, in the air and the CO_2 jet, only fairly continuous vortex sheets can be seen at the lip of the nozzle, the length of which is shorter. Since the finger-shape structures in the helium jet are approximately parallel to the jet axis, it is reasonable to believe that strong streamwise vorticity is present across the shear layer close to the lip. The streamwise vorticity in the helium jet is mainly generated by the Rayleigh-Taylor instability. This secondary instability developing in variable-density flows has been studied extensively [21, 22]. Schowalter et al. [21] showed that spanwise vortices generate streamwise vorticity if in a stratified shear layer a lighter fluid drives the heavier fluid.

Fig. 12 shows several snapshots of ω_x and $\left(\omega_y^2 + \omega_z^2\right)^{0.5}$ at three axial cross sections at the vicinity of the jet outlet. In each picture, a black circle is drawn to represent the location of the nozzle lip. It is found that the in-plane vorticity, denoted by $\left(\omega_y^2 + \omega_z^2\right)^{0.5}$, is generally slightly larger for the helium jet

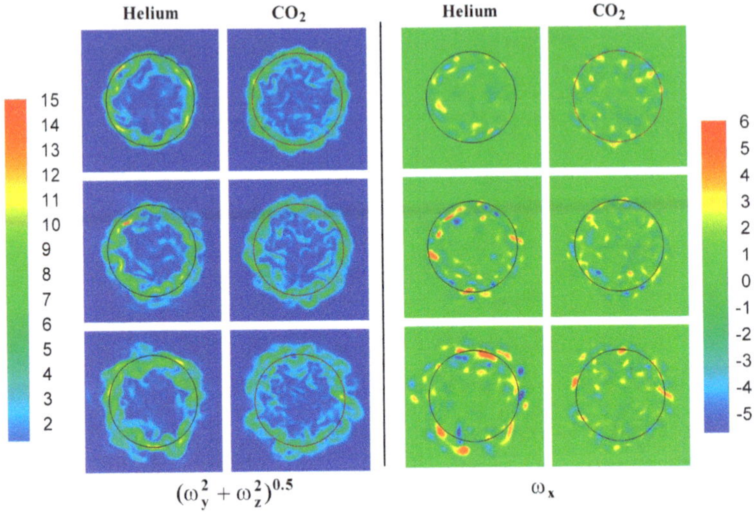

Fig. 12. Snapshots of the $\left(\omega_y^2+\omega_z^2\right)^{0.5}$ and ω_x at three axial cross sections, $x/D_j =$ 0.4, 0.8 and 1.2 from up to bottom, respectively

than for the CO_2 jet. The shear layer is fairly circular upstream, but develops more and more wrinkles further downstream. Together with the wrinkling of the shear layer streamwise vorticity is generated. It is also seen that these streamwise vorticities in the helium jet are stronger than in the CO_2 jet.

In order to investigate the orientation of vorticity with respect to the jet axis the following characteristic quantity is computed:

$$\Omega_{x/yz} = |\omega_x| \, / \, \left(\omega_y^2+\omega_z^2\right)^{0.5} = arc\,\mathrm{tg}(\phi) \tag{2}$$

where ϕ is the angle between the vorticity vector and the y-z-plane. Fig. 13(a) shows the average value of this quantity in the center plane (averaging is performed both in time and in azimuthal direction). For the helium jet, the change towards the free stream at the outer edge is somewhat broader which reflects the faster mixing of the helium jet. In the center and near the nozzle, a cone-shaped region with value around 0.8 exists in all three jets. It is closely related to the potential core region. An interesting difference among these figures is found in the initial developing region of the shear layer. Fig. 13 (d)-(f) show close ups of this region. It is seen that a region with high values exists just beside the shear layer in the helium jet (see Fig. 13(d)), but not in the air and the CO_2 jet. A local maximum of $\langle \Omega_{x/yz} \rangle$ of 1.15 is found at $x/D_j = 0.088$, $r/D_j = 0.54$ while for air the maximum is 0.73 and for CO_2 is 0.72, respectively, attained at the same radial position and slightly lower x. This reflects a higher contribution from streamwise vorticity to the whole vorticity modulus. It is worth pointing out that the vorticity and its components shown in Fig. 11-13 are all normalized with U_j/D_j. This excludes

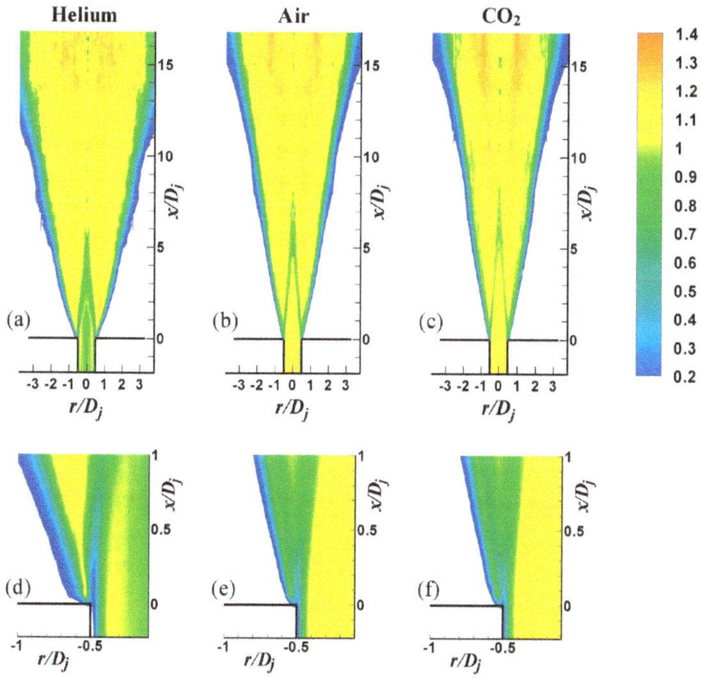

Fig. 13. Average value of $\Omega_{x/yz}$ in the center plane. (d), (e) and (f) is the zoom view of (a), (b) and (c), respectively

the possibility that the stronger contribution of streamwise vorticity for the Helium jet is due to the higher absolute inlet jet velocity.

7 Conclusions and Ongoing Work

Three weakly confined turbulent variable-density axisymmetric jets have been studied by LES. The density ratios range from 0.14 to 1.52. In the present paper, detailed comparison has been made for many statistical quantities, such as the axial evolution of mean streamwise velocity, turbulent intensities, as well as radial profiles of Reynolds stress. The agreement between simulation and experiment is generally very good. For the CO_2 jet the initial stages deviate somewhat, presumably due to imperfection in the experiment, but the downstream evolution is again well captured. Additionally, the effect of the density difference on the coherent structures in these jets is investigated. It is found that long finger-shaped streamwise vortices, regularly arranged along the jet nozzle lip, are produced in the helium jet, but not in the air and CO_2 jets. Furthermore, a characteristic quantity, which is used to detect the orientation of vorticity with respect to the jet axis is proposed and applied to the present cases. It reveals that there is a region with strong streamwise vorticity just beside the shear layer region in the helium jet, but not in the CO_2 jet.

In succession to study of non-reactive variable density jets presented here, the method is currently extended to compute turbulent premixed flames. The SGS model employed for the reaction term is the so-called Thickened-Flame model [23, 24]. It accounts for the Flame-turbulence interaction in a specific way by replacing the original flame with a thicker flame having the same turbulent flame speed as the fully resolved flame brush. Related simulations are under way, corresponding to experiments within SFB 606. Ultimately, a deeper understanding of the instabilities in turbulent reactive swirling flows is what is aimed at.

Acknowledgments

The authors gratefully acknowledge the support of the German Research Foundation (DFG) through the Collaborative Research Center SFB 606 'Unsteady Combustion'. The computations were performed on the HP-XC clusters of SSCK Karlsruhe, under project SFB606A6. F. Anselmet, Marseille, kindly provided the experimental data in electronic form.

References

1. M. Garcia-Villalba, J. Fröhlich, and W. Rodi. Identification and analysis of coherent structures in the near field of a turbulent unconfined annular swirling jet using large eddy simulation. *Phys. Fluids*, 18(5):055103, 2006.
2. P. Wang, J. Fröhlich, V. Michelassi, and W. Rodi. Large eddy simulation of variable-density turbulent axisymmertric jets. *Int. J. Heat and Fluid Flow*, 29:654–664, 2008.
3. S.A. Ahmed, R.M.C. So, and H.C. Mongia. Density effects on jet characteristics in confined swirling flow. *Exp. Fluids*, 3:231–238, 1985.
4. K.R. Sreenivasan, S. Raghu, and D. Kyle. Absolute instability in variable density round jets. *Exp. Fluids*, 7:309–317, 1989.
5. P.A. Monkewitz, B. Lehmann, B. Barsikow, and D.W. Bechert. The spreading of self-excited hot jets by side jets. *Phys. Fluids A*, 1:446–448, 1989.
6. P.A. Monkewitz and E. Pfizenmaier. Mixing by 'side jets' in strongly forced and self-excited round jets. *Phys. Fluids A*, 3:1356–1361, 1991.
7. N.R. Panchapakesan and J.L. Lumley. Turbulence measurements in axisymmetric jets of air and helium–part 2. helium jet. *J. Fluid Mech.*, 246:225–247, 1993.
8. T. Djeridane, M. Amielh, F. Anselmet, and L. Fulachier. Velocity turbulence properties in the near-field region of axisymmetric variable density jets. *Phys. Fluids*, 8:1614–1630, 1996.
9. M. Amielh, T. Djeridane, F. Anselmet, and L. Fulachier. Velocity near-filed of variable density turbulent jets. *Int. J. Heat Mass Transfer*, 39:2149–2164, 1996.
10. R. Jester-Zürker, S. Jakirlić, and C. Tropea. Computational modelling of turbulent mixing in confined swirling environment under constant and variable density conditions. *Flow, Turbulence and Combustion*, 75:217–244, 2005.

11. X. Zhou, K.H. Luo, and J.J.R. Williams. Study of density effects in turbulent buoyant jets using large-eddy simulation. *Theoret. Comput. Fluid Dynamics*, 15:95–120, 2001.
12. A. Tyliszczak and A. Boguslawski. LES of the jet in low mach variable density conditions. In *Direct and Large-Eddy Simulation*, volume 6, pages 575–582. Springer Netherlands, 2006.
13. P. Moin, K. Squires, W. Cabot, and S. Lee. A dynamic subgrid-scale model for compressible turbulence and salar transport. *Phys. Fluids A*, 3:2746–2757, 1991.
14. C. Hinterberger. *Dreidimensionale und tiefengemittelte Large–Eddy–Simulation von Flachwasserströmungen.* PhD thesis, Institute for Hydromechanics, University of Karlsruhe, 2004.
15. J. Zhu. A low diffusive and oscillation-free convection scheme. *Comm. Appl. Num. Meth.*, 7:225–232, 1991.
16. H. Werner and H. Wengle. Large eddy simulation of turbulent flow over and around a cube in a plate channel. In *8th Symp. on Turbulent Shear Flows.* Springer Verlag, 1993.
17. C.J. Chen and W. Rodi. Vertical turbulent buoyant jets–a review of experimental data. In *The Science and Application of Heat and Mass Transfer.* Pergamon Press, New York, 1980.
18. L. Fulachier, R. Borchi, F. Anselmet, and P. Paranthoen. Influence of density variations on the structures of low-speed turbulent flows: a report on euromech 237. *J. Fluid Mech.*, 203:577–593, 1989.
19. S.E. Widnall, D.B. Bliss, and C.Y. Tsai. The instability of short waves on a vortex ring. *J. Fluid Mech.*, 66:34–57, 1974.
20. A. Gharbi, M. Amielh, and F. Anselmet. Experimental investigation of turbulence properties in the interface region of variable density jets. *Phys. Fluids*, 7:2444–2454, 1995.
21. D.G. Schowalter, C.W. Van Atta, and J.C. Lasheras. Baroclinic generation of streamwise vorticity in a stratified shear layer. *Meccanica*, 29:361–371, 1994.
22. J. Reinaud, L. Joly, and P. Chassaing. Numerical simulation of a variable-density mixing-layer. In A. Giovannini et al., editor, *3rd Int. Workshop on Vortex Flows and Related Numerical Methods.* ESAIM, 1999.
23. O. Colin, F. Ducros, D. Veynante, and T. Poinsot. A thickened flame model for large eddy simulations of turbulent premixed combustion. *Phys. Fluids*, 12:1843–1863, 2000.
24. L. Selle, G. Lartigue, T. Poinsot, R. Koch, K.-U. Schildmacher, W. Krebs, B. Prade, P. Kaufmann, and Veynante D. Compressible large eddy simulation of turbulent combustion in complex geometry on unstructured meshes. *Combustion and Flame*, 137:489–505, 2004.

Thermal & Flow Field Analysis of Turbulent Swirling Jet Impingement Using Large Eddy Simulation

Naseem Uddin[1], Sven Olaf Neumann[1], Peter Lammers[2], and Bernhard Weigand[1]

[1] Institut für Thermodynamik der Luft- und Raumfahrt, Universität Stuttgart, Pfaffenwaldring 31, Stuttgart 70569, Germany
Naseem.Uddin@itlr.uni-stuttgart.de

[2] Höchstleistungsrechenzentrum (HLRS), Universität Stuttgart, Noblestrasse 19, Stuttgart 70596, Germany

Summary. Swirling jets are used in a variety of engineering applications like in chemical reactors, cyclone separators, mixing devices, drying and cooling applications. Good quality simulations of this highly complex flow field is a challenging task. In this work, the flow field and heat transfer of turbulent swirling and non-swirling impinging jets are computed using Large Eddy Simulation (LES). For the investigation of non-swirling jets, the ERCOFTAC recommended test case of an impinging jet at a Reynolds number of 23000 is simulated first. The agreement between experimental data and simulation gives encouragement for further investigation of complex flow of swirling jets impingement. Therefore, the swirling jets with Reynolds numbers of 21000 & 23000 and four different swirl numbers are investigated via LES. The results are compared with experimental data. The effect of inflow conditions and inlet temperature is investigated. The correlation between the heat transfer mechanism, flow kinematics and turbulence quantities is investigated. The numerical data computed within this investigation serve additionally for benchmarking results based on the solution of Reynolds Averaged Navier-Stokes Equations in complex flow configurations[3] at even higher Reynolds numbers. This comparison is not shown here, but could be found at [ITLR].

1 Introduction

Impinging jets are used in a variety of engineering applications like, drying, cooling, spraying and drilling. The accurate simulation of this complex flow field is a challenging task. The complexity of the turbulent impinging jet can be further enhanced by the addition of swirl. Because of high entrainment rates,

[3] WKS-Wärmeübertragung in komplexen Strukturen

swirling jets offer good mixing characteristics, which makes them desirable in chemical reactors, cyclone separators and mixing devices. The azimuthal velocity component plays an important role in swirling jets which introduces additional instabilities in the jet. The complex flow field of the swirling jet consists of flow features like spiral type flow instabilities, vortex breakdown and free shear layers. The differences between the impinging jet flow field of the swirling and non-swirling jet are depicted in figure 1.

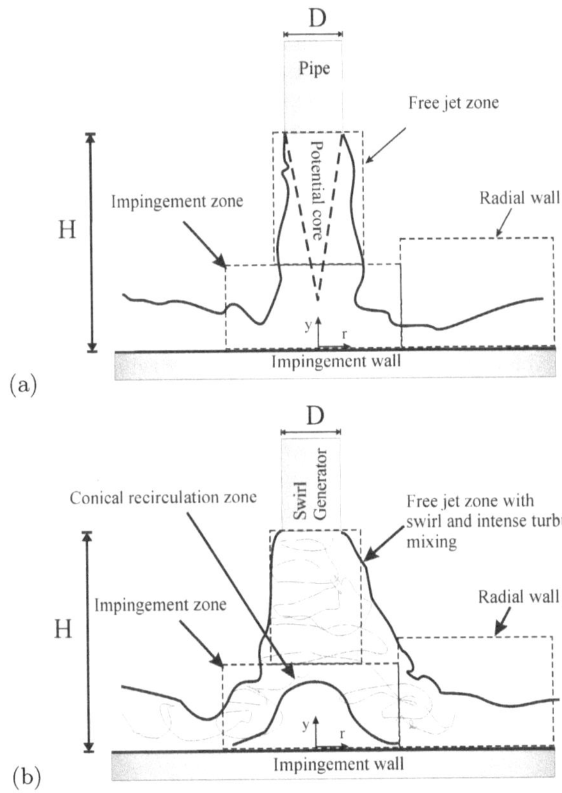

Fig. 1. The schematic representation of the complex flow field of a turbulent (a) non-swirling and (b) swirling jet impingement

The swirl in the incompressible jet is quantified through a swirl number defined by the following equation:

$$S = \frac{G_\theta}{RG_{ax}} = \frac{1}{R} \frac{\int_0^R r^2 u_{ax} u_\theta dr}{\int_0^R r u_{ax}^2 dr} \tag{1}$$

where, G_θ and G_{ax} are the tangential and axial momentum fluxes, u_θ and u_{ax} are the tangential and axial velocities of the swirling jet and R is the radius of the jet at the inlet.

The tangential velocity distribution in real swirling flows is intermediate between a forced vortex flow and a free vortex flow. Swirl with constant angular velocity is called forced vortex flow or solid body rotation. The tangential velocity in forced vortex flow is given by:

$$u_\theta = \Omega r, \tag{2}$$

where Ω is the angular velocity. The tangential velocity in free vortex flow is given by:

$$u_\theta = C/r. \tag{3}$$

A real swirling flow has normally a core region behaving like a solid body rotation and a region away from the center behaving like a free vortex. This combination is called Rankine vortex. The effect of swirl in impinging jet flow on heat transfer has been investigated experimentally by numerous researchers like [WAR82], [NOZ03], [SEN05]. There are two diverse opinions in general about the possibility of heat transfer enhancement. Ward & Mahmood [WAR82] have found that heat transfer is reduced because of the swirl in the jet. Nozaki [NOZ03] has proposed that there are two modes of swirling jets, one is called heat transfer enhancement mode and the other one is the heat transfer suppression mode. The main role is played by the dynamics of the recirculation zones produced due to swirling jet impingement. Senda et al. [SEN05] investigated the impingement heat transfer with swirl of 0.45 and 0.22 at Re=8100. It is found that for H/D less than and greater than one, the heat transfer at the target wall is reduced. However near H/D=1 the heat transfer can be improved. Abrantes and Azevedo [ABR06] have conducted an experimental investigation of heat transfer due to swirling jet impingement at a Reynolds number of 21000. The swirl number of the jet was 0.5. The purpose of their study was to investigate the effect of target-to-wall distance on the swirling jet heat transfer. They found that for the small H/D cases the swirl can increase heat transfer. Further, Lee et al. [LEE02b] and Yan et al. [YAN98a] have found in separate experimental investigations very different Nusselt number distributions for heat transfer due to swirling jet at a Reynolds number of 23000. Yan and Saniei [YAN98a] investigated experimentally the swirling jet impingement at Re=23000 and S=0.2, 0.35, 0.47 and H/D from 2 to 9.

There are very few numerical investigations of swirling jet impingement. Yan and Kalvakota [YAN98b] have used a Reynolds Averaged Navier-Stokes equations (RANS) based turbulence model for the investigation of the swirling jet impingement at Re=23000 with different swirl numbers (S= 0.2, 0.35, 0.47). They found that swirl affects the heat transfer only till r/D=2. In general, the RANS based models failed to capture the kinematics of the swirling jet. Direct Numerical Simulations (DNS) or Large Eddy Simulations (LES) can help in understanding the impinging jet flows. Hällqvist [HAL06] has conducted LES simulations of swirling jets at Reynolds number of 20000 and swirl numbers of 0.5 and 1. He found that a higher swirl number reduces the wall heat transfer. However, the LES results were presented without experimental validation.

In the present work, series of LES simulations of a swirling jet impingement are conducted. The Reynolds number of the jet is 23000 and the dimensionless jet outlet to target wall distance (H/D) is two. The heat transfer due to the swirling jet predicted by LES is validated by the experimental data of Yan and Saniei [YAN98a]. As a first step, the non-swirling jet is simulated at Re=23000 and H/D=2. This helps in understanding the set of boundary conditions required for a realistic simulation of turbulent jet impingement. Next, the swirl is added and boundary conditions are kept as close as described in original benchmark experiment [YAN98a, YAN98b]. The bulk axial Reynolds numbers are 21000 and 23000. The simulation conditions are tabulated below:

Table 1. LES Simulations

Cases	Swirl	Re	Jet's Inlet Temperature
Case-I	0	23000	293 K
Case-II	0.2	23000	293 K
Case-III	0.39	21000	293 K
Case-IV	0.47	23000	300 K
Case-V	0.5	21000	298 K

The swirling jet impingement features many characteristics which can be found in several heat transfer enhancement techniques like an amplification of free shear layer instabilities, a (re-)attachment point in the flow with a redeveloping boundary layer causing high heat transfer rates and a superimposed swirl to the main stream. Artificial protuberances in the boundary layer of a flow created by 3D vortex-generators mounted to the wall and their influence on the heat transfer have been recently studied by [HEN07] and [DIE07a]. The mentioned numerical analysis of the high Reynolds number flow based on RANS is especially concerned with the application of explicit non-linear scalar flux models, which can cover occurring anisotropies of the turbulent heat flux [DIE07b]. With its high resolution of turbulent structures the current Large Eddy Simulation provides validation data for the subsequent task of RANS modeling.

2 Computational Details

2.1 Numerical Grid

The geometry of the computational domain used for the non-swirling jet impingement is shown in Fig. 2(a). The domain consists of a circular impingement zone and a circular pipe of length 6D. The target to wall distance (H) is 2D. The domain is confined by an adiabatic, no-slip wall at the top. Also the boundary condition at the pipe wall is adiabatic. The domain is 16D in length. The computational domain is modeled via hexahedral structured O-grids, developed via ICEM-CFD of ANSYS, Inc. The most important domain region is the stagnation and wall jet region, in which the Δr^+ is ≈ 36.3 and $(r\Delta\theta)^+ \approx 20$, where $()^+ = (.)u_\tau/\nu$. The mean grid spacing inside the jet is 6.6 times length of the Kolmogorov scale. The grid used for the non-swirling jet has 5×10^6 control volumes.

Figure 2 (b) shows the schematic representation of the domain used for the investigation of the heat transfer due to impingement of the swirling jet at bulk Reynolds number of 21000 and 23000. The grid spacing in dimensionless units is $\Delta r^+ \approx 27$, $(r\Delta\theta)^+ \approx 20$ and $\Delta y^+/\eta^+$ is 6, where, η is the length of the Kolmogorov scale. In the investigations, a hexahedral structured grid has been used. The grid used for the swirling jet has 6.4×10^6 control volumes.

The heat flux of 1000 W/m^2 is applied at the impingement wall for both swirling and non-swirling jet cases.

2.2 Inflow Conditions

Non-swirling Jet Inflow Conditions

In the jet's axial direction, a fully developed turbulent pipe flow mean velocity profile, together with velocity fluctuations varying in time are prescribed at the inlet. The turbulence inflow conditions are generated, based on the procedure by Klein et al. [KLE03]. The velocity fluctuations generated are based on digital filtering of random data. In this procedure, the inflow turbulent velocity generated is based on the relation:

$$u_i = < u_i > + a_{ij} u_j', \tag{4}$$

where $< u_i >$ is the mean axial velocity, u_j' are velocity fluctuations and a_{ij} is a tensor related to the Reynolds stress tensor (see [KLE03]). According to this procedure, the prescribed auto-correlation function R_{uu} for homogeneous turbulence in later stages is used to describe the turbulence:

$$R_{uu}(\hat{r}) = exp\left(\frac{-\pi\hat{r}^2}{4L(t)^2}\right), \tag{5}$$

where $\hat{r}$ is the position vector and L(t), is the integral length scale at the inflow plane.

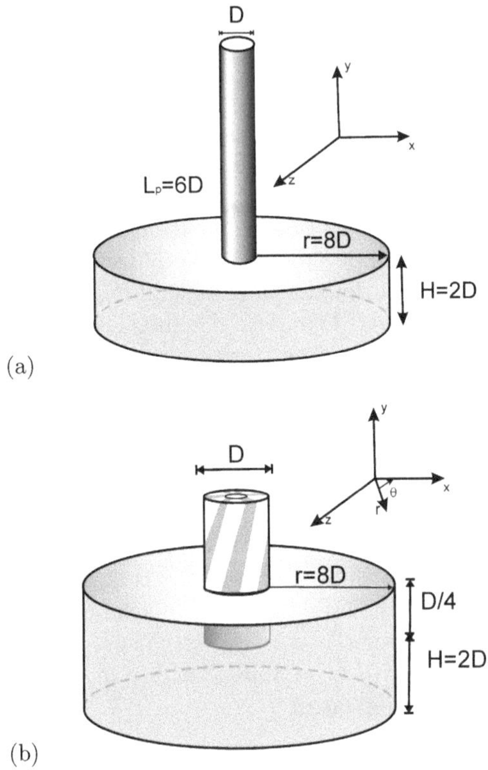

Fig. 2. Schematic diagram of the computational domain used for the simulation of (a) non-swirling and (b) swirling jet impingement cases. The origin is fixed at the geometric stagnation point. The jet enters at y=2D. A swirling device is shown for clarification but is not geometrically modeled in the simulation

Swirling Jet Inflow Conditions

Fahrokhi and Taghavi [FAH89] have found that the swirl number is an insufficient parameter to describe the characteristics of swirling flows. The correct prescription of the tangential velocity distribution is important for swirling jet investigations. In the experiment of Yan and Saniei, the swirl is imparted to the fully developed turbulent pipe flow by tangential jets [YAN98a]. The same approach is adopted here. The mean tangential velocity is superimposed on the fully developed pipe flow. The turbulent fully developed flow is prescribed in axial direction and described above for the non-swirling jet case. The empirical relation for the mean tangential velocity is reported by [KHA02]:

$$\frac{< u_\theta >}{u_\theta^*} = \left[\frac{2\vartheta}{1+\vartheta^2}\right]^{\kappa^*} \tag{6}$$

where, $\vartheta = r/r^*$, $r^* = 0.51$ R $\phi_*^{0.41}$ and $u_\theta^* = 2.04 < u_{ax} > \phi_*^{1.1}$. The κ^* is an index depending on the swirl conditions and R is the radius at the jet inlet. The parameter ϕ_* is comprised of swirl number and axial momentum flux S/G_{ax}. The turbulent flow fluctuations are prescribed in the same manner, as explained before.

2.3 Outflow Conditions

In transient simulations of the jet the correct choice of the outlet boundary conditions is a crucial problem. Convective boundary conditions are used for the outflow, as this allow a negligible influence on the evolution of the flow structures in the finite-size computational domain. It is experienced that the use of von Neumann type boundary conditions with a full three-dimensional domain simulation, for the impinging jet case, generates a non-physical feedback, which strongly affects the global stability of the simulation.

The convective boundary condition is spatially and temporally local, and defined as ([COL97])

$$\frac{\partial U}{\partial t} + \Lambda \frac{\partial U}{\partial r} = 0. \tag{7}$$

The quantity Λ is taken to be a constant convection velocity of the large-scale structures, which is set to a constant as per global mass conservation requirements.

3 Numerical Code & its Performance

3.1 FASTEST

The computations have been performed with the CFD code FASTEST (Flow Analysis Solving Transport Equations Simulating Turbulence). The code is based on a finite volume discretization with pressure velocity coupling done through a SIMPLE algorithm. For LES computations, a top hat filter has been used. The three dimensional filtered Navier-Stokes equations along with the subgrid model proposed by Germano [GER91] have been solved. The second order implicit Crank-Nicolson method is used for time discretization and for space discretization of the convective terms a second order central differencing scheme has been used. The resulting set of equations is solved using a SIP solver [STO88].

3.2 Code Performance & Solution Control

Computations are done on two different computing clusters available at Höchstleistungsrechenzentrum (HLRS), Stuttgart, Germany. For computations related to case I, II, III & V, the CRAY Opteron Cluster is used. The noteworthy improvements in the code are the vectorization of the subroutines related to LES subgrid scale model. The FASTEST has two different loop structures available in the code, which could be used for computations, depending on the architecture of the computing system. The routines related to LES had the three times nested loop version only. The use of a collapsed single loop is a natural choice for vector machines as it gives better vectorization. The single loop version is now also available for the LES-model. The code has attained 5.6 Gflops and high vectorization rates (>99.8%). For Case-IV, the vectorized version of the code is used and computations are done on the NEC SX-8.

The number of processors and computing platforms used are listed below:

Table 2. Computational Platforms

Cases	Re	Swirl	CPUs	Computing Platform
Case-I	23000	0	20	Cray Opteron
Case-II	23000	0.2	20	Cray Opteron
Case-III	21000	0.39	20	Cray Opteron
Case-IV	23000	0.47	4	NEC SX-8
Case-V	21000	0.5	20	Cray Opteron

The computational domain for non-swirling and swirling jet is divided into 42 and 90 blocks, respectively. This domain decompositions helps in computational-load balancing on the processors. The inter-processor communications was performed with standard Message Passing Interface (MPI). The simulation of the non-swirling jet requires 10000 CPU hours and on average the simulation of swirling jet requires 15000 CPU hours on Cray Opteron clusters. The computations on NEC SX-8 are much faster and require three to four times less computational time than the one on the Cray Opteron Platform.

On average total simulation time for each case is equivalent to 20 cycles, where one cycle corresponds to the natural frequency of a non-swirling jet. The flow becomes statistical stationary after about 9.3 cycles. The simulation is controlled in such a way that the CFL number remains less than one. The dimensionless time step used is $\Delta t D/U_{ax}$ equals to 7E-07.

4 Results

The Nusselt number distribution is defined as

$$Nu = \left(\frac{q_w^{''}}{T_w - T_j} \right) \frac{D}{\lambda}, \tag{8}$$

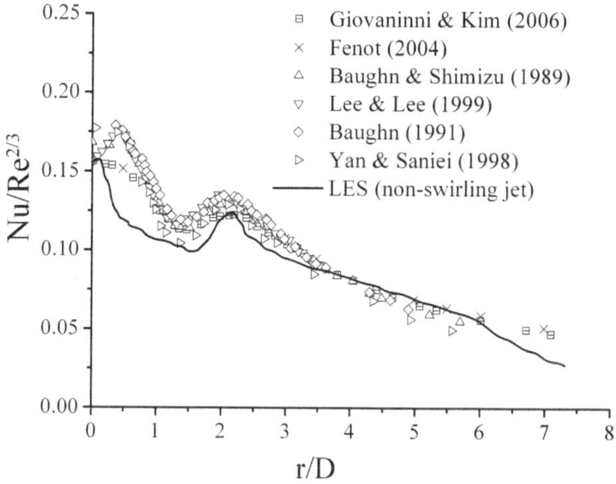

Fig. 3. The radial distribution of Nusselt number in case of non-swirling jet

where q''_w is the heat flux at the target wall. T_w is the temperature attained by target wall after jet impingement and T_j is the jet's inlet temperature. In Fig. 3 (a), the heat transfer for a non-swirling jet is compared with experimental data. As in some experiments, the Reynolds number is different from 23000, the Nusselt number distribution is therefore normalised by $Re^{2/3}$, which is recommended by Martin [MAR98]. Nusselt number from the LES simulation is compared with the experimental data of Giovannini & Kim ($Re = 23000$) [GIO06], Fenot ($Re = 23000$) [FEN04], Baughn ($Re = 23750$) [BAU89], Lee et al. ($Re = 20000$) [LEE99], Baughn & Shimizu ($Re = 23300$) [BAU91] and Yan & Saniei ($Re = 23000$) [YAN98a].

It is found that LES can correctly predict the stagnation point Nusselt number and also the secondary peak is well captured. However, for $r/D<1$ the LES under predicts the experimental Nusselt number distribution. One probable reason for this difference can be the artificial inflow turbulence conditions. Figure 3 shows the radial distribution of the Nusselt number. The case is used for a benchmark testing of the LES set-up.

The instantaneous velocity distribution in the swirling jet domain is shown in figure 4. The complexity of the flow field can be visualised through iso-velocity surfaces. Due to the swirl the jet breaks down before impingement. The breakdown is enhanced as the swirl increases.

The LES predictions of the radial distribution of the Nusselt number in case of swirling swirling jets at different swirl numbers, are plotted in figure 5 (a) and (b). The heat transfer distribution is affected not only by the Reynolds number but also the jet inlet temperature. The high inlet temperature of the jet increases the heat transfer. Yan and Saniei [YAN98a] have discovered that the heat transfer at the geometrical stagnation point is affected by the swirl

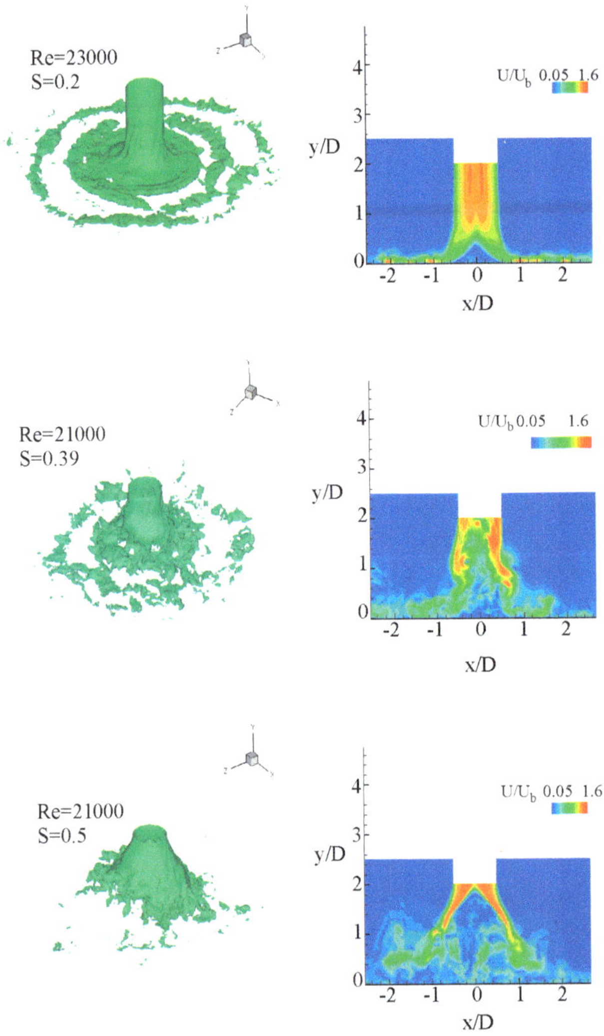

Fig. 4. The iso-velocity surfaces (U/U_b=0.77) and instantaneous velocity field contours in the swirling jet domain

rates. The series of LES simulations at different swirl rates, corroborates this finding.

Senda et al. [SEN05] have found that the Nusselt number is strongly correlated to maximum velocity approaching the wall. On the other hand, Abrantes and Azevedo [ABR06] have found that the peaks in the radial distribution of the Nusselt number are linked to peaks in the measured turbulent kinetic energy close to the target wall. The investigations at different swirl numbers,

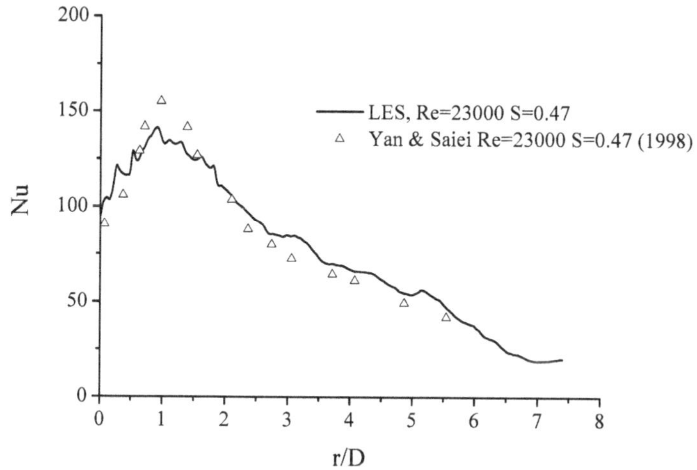

(a)

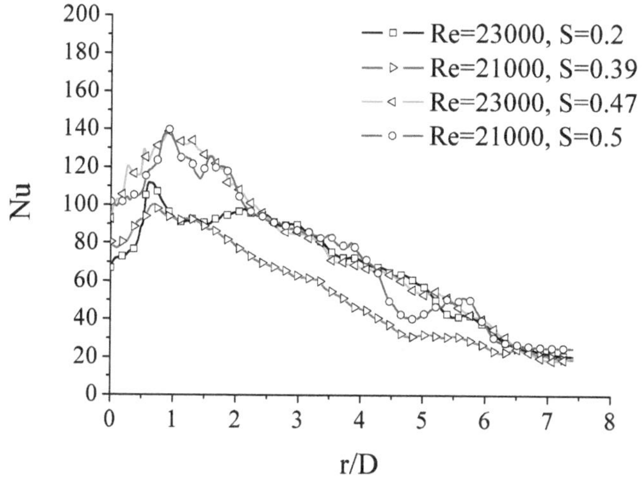

(b)

Fig. 5. Radial distribution of the Nusselt number in case of swirling swirling jets at (a) Re=23000, S=0.47, Experimental data from Yan and Saniei [YAN98a] (b) different swirl rates

Reynolds numbers and inlet temperature conditions help us further in understanding the correlation between heat transfer and swirling jet flow field.

Table 3 outlines the locations (r/D) at the wall where the said quantities appear. For example in case of jet's Reynolds number 21000 and swirl number 0.5, the maximum value of the production of turbulent kinetic energy occurs at r/D=0.66, the minimum value of the production of turbulent kinetic energy occurs at r/D=0.71, the maximum value of turbulent kinetic energy occurs at r/D=0.8. Similarly the maximum value of the Nusselt number oc-

Table 3. Location at the target wall (r/D)

Re	*S*	$P_{k_{max}}$	$P_{k_{min}}$	k_{max}	Nu_{max}	U_m
23000	0.2	1.38	1.92	1.93	0.644	0.52
21000	0.39	0.68	1.0	0.8	0.66	1
21000	0.5	0.66	0.71	0.8	0.90	1.2

Re=23000, S=0.2, $\omega_y = -294\ s^{-1}$ Re=21000, S=0.39, $\omega_y = -544\ s^{-1}$ Re=21000, S=0.5, $\omega_y = -1231\ s^{-1}$

Fig. 6. Coherent structures near the jet's outlet, visualised through iso-vorticity surfaces

curs at r/D=0.90 and the maximum value of the velocity approaching the target wall occurs at r/D= 1.2. Through the series of LES simulations, two different swirling jet behaviors are noted. It can be inferred from the information presented in table 3 that at small swirl numbers (here S=0.2) the jet velocity approaching the target wall is playing an important role in control of heat transfer. However, as the swirl increases (here S=0.39 & 0.5), the correlation between the turbulent kinetic energy and heat transfer is significant. LES thus confirms the result of Senda et al. [SEN05] but found to be valid at small swirl numbers. Correlation between turbulent kinetic energy and the Nusselt number is found to be strong when swirl number is high, as found be Abrantes et al. [ABR06].

Figure 6 shows the coherent structures visualised through iso-vorticity surfaces (ω_y) under different Reynolds number and swirl levels. The figure shows that at small swirl level the jet core is behaving like a rotating solid body. As the swirl increases the jet core exhibit helical shaped structures but their influence to the heat transfer at the wall is limited.

5 Conclusions

The series of LES investigations of turbulent swirling jet impingement on the HLRS computing platforms helps in understanding this complex flow field. It is found that:

- A detailed description of jet's inflow conditions are necessary for accurate Large Eddy Simulations.
- The peaks in radial distribution of the Nusselt number at the target wall are better correlated with the mean velocities approaching target wall at small swirl numbers. However at high swirl numbers, strong correlation exist between the heat transfer and the turbulent kinetic energy.
- The swirl rates alter the coherent structures in the swirling jet.

Acknowledgement

The authors would like to thank Prof. Dr. M. Schäfer and Dr.-Ing. D. Sternel, Fachgebiet für Numerische Berechnungsverfahren im Maschinenbau (FNB), Technische Universität, Darmstadt, Germany, for providing the FASTEST code and helpful discussions. The first author wants to thank the Higher Education Commission (HEC), Pakistan and DAAD, Germany for providing a PhD fellowship. The support provided by Höhstleistungsrechenzentrum (HLRS), Stuttgart is gratefully acknowledged.

References

[ABR06] Abrantes, J.K., Azevedo, L.F.A., Fluid flow and heat transfer charactersties of a swirl jet impinging on a flat plate, Annals of the Assembly for International Heat Transfer Conference 13 (2006)

[BAU89] Baughn, J.W., Shimizu, S., Heat transfer measurement from a surface with uniform heat flux and an impinging jet, Int. J. of Heat Transfer, **111**, 1096–1098 (1989)

[BAU91] Baughn, J.W., Hechanova, A.E., Yan, X., An experimental study of entrainment effects on the heat transfer from a flat surface to a heated circular impinging jet, Journal of Heat Transfer, **113**, 1023–1025, (1991)

[COO93] Cooper, D., Jackson, D.C., Launder, B.E., Liao, G.X., Impinging jet studies for turbulence model assessment- I. Flow field experiments, Int. J. Heat Mass Transfer, **36**, 2675–2684, (1993)

[COL97] Colonius, T., Numerically non-reflecting boundary and interface conditions for compressible flow and aeroacoustic computations, AIAA, 7, **35**, 1126–1133 (1997)

[DIE07a] Dietz, C., Henze, M., Neumann, S.O., von Wolfersdorf, J., Weigand, B., Flow and heat transfer investigations behind vortex inducing elements as benchmark for complex turbulence models. Sixth International Conference on Enhanced, Compact and Ultra-Compact Heat Exchangers: Science, Engineering and Technology, number CHE-0012, Potsdam (2007)

[DIE07b] Dietz, C., Neumann, S.O., von Wolfersdorf, J., Weigand, B., A comparative study of the performance of explicit algebraic models for the turbulent heat flux, Numerical Heat Transfer, Part A, **52**, 101–126, (2007)

[FEN04] Fenot, M., Etude du refroidissement par impact de jets. application aux aubes de turbines, Universite de Poitiers, France (2004)

[FAH89] Fahrokhi, S., Taghavi, R., Effect of initial swirl distribution on the evolution of a turbulent jet, AIAA Journal, 6, **27** (1989)

[GER91] Germano, M., Piomelli, U., Moin, P., Cabot, W.H., A dynamic subgrid-scale eddy viscosity model, Phy. Fluid, **3**, 1760–1765 (1991)

[GIO06] Giovannini, A., Kim, N.S., Impinging jet: Experimental analysis of flow field and heat transfer for assesment of turbulence models, Annals of the Assembly for International Heat Transfer Conference 13, TRB- 15 (2006)

[HAL06] Hällqvist, T., Large-eddy simulation of impinging jets with heat tansfer, Royal Institute of Technology, Department of Mechanics, Sweden, PhD thesis (2006)

[HEN07] Henze, M., Dietz, C., Neumann, S.O., von Wolfersdorf, J., Weigand, B., Heat transfer in complex internal flows - wedge-shaped vortex generators. Sixth International Conference on Enhanced, Compact and Ultra-Compact Heat Exchangers: Science, Engineering and Technology, number CHE-0011, Potsdam (2007)

[ITLR] http://www.uni-stuttgart.de/itlr/forschung/wks/ (Institut für Thermodynamik der Luft- und Raumfahrt)

[KLE03] Klein, M., Sadiki, A., Janicka, J., A digital filter based generation of inflow data for spatially direct numerical or large eddy simulations, Journal of Computational Physics, **18**, 652–665 (2003)

[KHA02] Khalatov, A.A., Avramenko, A.A., Shevchuk, I.V., Heat transfer and fluid flow in the fields of centrifugal forces, Swirl flows, vol-III, (russian edition), National Academy of Sciences of Ukraine, Institute of Engineering Thermophysics, Kiev (2002)

[LEE99] Lee, J., Lee, S., Stagnation region heat transfer of a turbulent axisymmetric jet impingement, Experimental Heat Transfer, **12**, 137–156 (1999)

[LEE02a] Lee, J., Lee, S.J., The effect of nozzle configuration on stagnation region heat transfer enhancement of axisymmetric jet impingement, Int. J. Heat Mass Transfer, **43**, 3497–3509 (2002)

[LEE02b] Lee, D.H., Won, S.Y., Kim, Y.T., Chung, Y.S., Turbulent heat transfer from a flat surface to a swirling round impinging jet, Int. J. Heat Mass Transfer, **45**, 223–227 (2002)

[LIU96] Liu, T., Sullivan, J.P., Heat transfer and flow structures in an excited circular impinging jet, Int. J. Heat and Mass Transfer, **17**, 3695–3706 (1996)

[MAR98] Martin, H., Wärmeübergang bei Prallströmung, VDI-Waermeatlas, VDI(1998)

[NOZ03] Nozaki, A., Igarashi, y., Hishida, K., Heat transfer mechanism of a swirling impinging jet in a stagnation region, Heat transfer Asian research, 8, **32** (2003)

[SEN05] Senda, M., Inaoka, K., Toyoda, D., Sato, S., Heat Transfer and Fluid Flow Characteristics in a Swirling Impinging Jet, Heat Transfer Asian Research, 5, **34** (2005)

[STO88] Stone, H.L., Iterative solution of implicit approximations of multidimensional partial differential equations, SIAM J. Numer. Anal., 3, **5**, 530–558 (1988)

[WAR82] Ward, J., Mahmood, M., Heat transfer from a turbulent swirling impinging jet, Proceedinds of 7th Int. Heat Transfer Conference, 3, 401–408 (1982)

[YAN93] Yan, X., A preheated wall transient method using liquid crystals for the measurement of heat transfer on external surfaces and in ducts University of California, Davis (1993)

[YAN98a] Yan, X., Saniei, N., Heat Transfer Measurements From a Flat Plate to a Swirling Impinging Jets, Proceedings of 11th International Heat Transfer Conference, Kyonju, Korea (1998)

[YAN98b] Yan, X., Kalvakota, R.S., Numerical analysis of local heat transfer from a flat plate to a swirling air impinging jet, Proceedings of IMECE2006 2006 ASME International Mechanical Engineering Congress and Exposition, November 5-10, 2006, Chicago, Illinois, USA, (1988)

Hybrid Techniques for Large–Eddy Simulations of Complex Turbulent Flows

Dominic A. von Terzi[1], Jochen Fröhlich[2], and Wolfgang Rodi[3]

[1] Institut für Thermische Strömungsmaschinen, Universität Karlsruhe, Kaiserstr. 12, D-76131 Karlsruhe vonterzi@its.uni-karlsruhe.de
[2] Institut für Strömungsmechanik, Technische Universität Dresden, George-Bähr-Str. 3c, D-01069 Dresden jochen.froehlich@tu-dresden.de
[3] Institut für Hydromechanik, Universität Karlsruhe, Kaiserstr. 12, D-76131 Karlsruhe rodi@uka.de

Summary. The paper presents developments for a segregated approach to the coupling of Reynolds-Averaged Navier–Stokes (RANS) calculations with a zone computed as Large-Eddy Simulation (LES). The mean velocity fields are matched at predefined interfaces and velocity fluctuations of the LES zone are treated according to the type of interface. If the RANS zone is downstream of the LES, fluctuations are allowed to leave the domain by employing a convective boundary condition. If the RANS zone is placed between the LES and a wall, the fluctuations at the interface are scaled to match the statistics predicted by the RANS computation. The proposed method was applied to turbulent channel flow and the flow over periodic hills. It was found that the method delivered an improvement over alternative techniques in the literature while removing the need for calibration constants. For incompressible flows, it is also necessary to prescribe conditions for the pressure or an equivalent variable. Several alternatives were tested. Decoupling of the pressure combined with explicitly enforced mass conservation at the interface yielded the best and, for the hill flow with aggressively placed interface, the only acceptable results.

1 Introduction

Large-Eddy Simulation (LES) has clear superiority over Reynolds-Averaged Navier–Stokes (RANS) methods for computation of complex flows, especially when large-scale structures dominate the turbulent transport, in transitional situations and when dynamic forces and noise generation must be calculated. However, due to the extremely high resolution and time-averaging requirements at high Reynolds numbers, LES still is too costly to be applied routinely to flows of practical interest. On the other hand, RANS can determine the mean flow with engineering accuracy at much lower cost in many cases, especially attached flows, which often prevail in subareas of complex flows. Hence, a hybrid method using RANS in regions with simpler flow behavior and LES only in the more critical areas appears to be the ideal approach for complex high Reynolds number flows. Thereby the key problem is how the RANS and LES zones are coupled. A detailed review of the literature and a classification of hybrid LES/RANS methods can be found in [1].

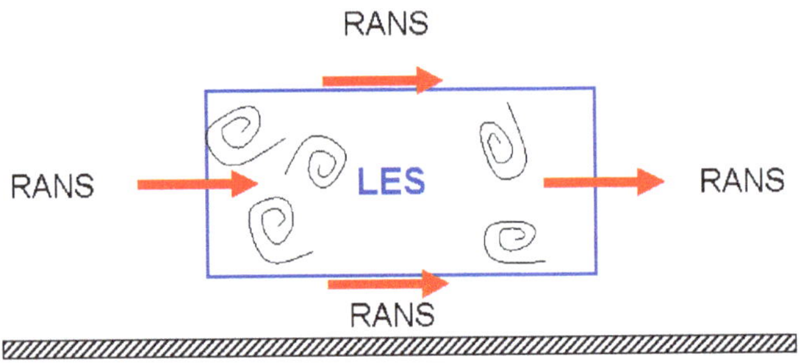

Fig. 1. Possible types of interfaces between an embedded LES and the surrounding RANS region; the red arrow indicates the direction of the largest mean velocity component

The approach followed here aims to develop a hybrid LES/RANS method based on the *segregated modeling* paradigm [1]: RANS and LES are applied in predefined subareas of the computational domain and linked by well-defined coupling conditions. A sketch of such a situation is shown in Fig. 1. The objective of segregated modeling is to compute all models in their regime of validity, i.e. steady RANS for flows with stationary statistics and unsteady LES with high resolution where it is needed. As a consequence, one can choose the best-suited method for each subdomain without considering their compatibility and without fear of inconsistencies in their use. Moreover, any so-called "grey zone" characterized by high uncertainties with respect to the generation and modeling of fluctuations can be avoided. Furthermore, for block-structured solvers the routines required for data exchange facilitate a straight-forward implementation and additional benefits in computational cost reduction can be reached by the possibility of sudden changes in mesh size at the subdomain boundaries. The price to pay is the need for comparatively complex coupling conditions, since inappropriate conditions lead to contamination of the results in the LES and/or the RANS subdomains. Three distinct types of interfaces can be discerned in Fig. 1: a RANS zone upstream of the LES ("inflow-type"), RANS downstream of LES ("outflow-type") and a tangential alignment of both regions. For all types, mean velocities are provided by the RANS calculation and can be coupled directly to the explicitly averaged LES velocities. However, in general, the LES domain needs to "see" an unsteady flow field at its boundaries. Depending on the type of the interface the requirements differ on how realistic the generation of fluctuations has to be and what kind of local flow physics need to be accounted for.

At inflow-type interfaces, if very strong instabilities exist inside the LES domain and if the upstream unsteadiness has only little impact on the downstream flow, it can suffice to omit the unsteadiness altogether. A successful example is the investigation of a supersonic baseflow in [2] where a thin, fully turbulent boundary layer separates at a sharp corner. In other cases, the LES requires the provision of fluctuations at the interface in order to avoid an artificial transition zone in the LES subdomain. To this end, methods already devised for pure LES can be used, such as those discussed in [3]. Due to the large body of existing literature inflow-type interfaces are excluded here.

For tangential and outflow-type interfaces, Quéméré and Sagaut [4] proposed to apply a strategy called *enrichment* for generating unsteadiness at the interface. This technique scales fluctuations from inside the LES domain and adds them to the mean values obtained from the RANS domain. The so-formed total flow quantity is then copied to ghost cells of the LES zone. An empirically determined calibration constant determines the amount of scaling for the fluctuations. Enrichment has been fairly successful for compressible flows where pressure coupling does not need to be considered. There is some sensitivity to the grid stretching at the boundary and the numerical method employed and the calibration constant must be close to but in most cases smaller than one [4]. Otherwise, the method causes reflections or the solution diverges. For the outflow-type interface, these shortcomings can be explained in the framework of the method discussed in Sect. 2.1 [5]. In the framework of segregated modeling, enrichment can be regarded as the state-of-the-art for tangential and outflow-type interfaces and it therefore serves as a benchmark here.

The present study is part of a project for which new coupling techniques have been devised that are more "physics-based" than enrichment, in particular with respect to obeying physical constraints and the elimination of *ad hoc* parameters. Since tangential and outflow-type interfaces are governed by different physical mechanisms, this necessitated distinct modeling approaches for each of these interfaces. In addition, the use of the above techniques (including the original enrichment) for incompressible flow requires the handling of pressure or an equivalent variable at the LES/RANS boundary [6, 7]. The resulting interface conditions are discussed in Sect. 2 and in more detail in [6, 7] and [8] for outflow-type and tangential interfaces, respectively. The numerical method and the performance of the particular flow solver employed are reported in Sect. 3. In Sect. 4, the developed techniques were scrutinized for turbulent channel flow and the flow over periodic hills. First, reference LES with the same flow solver were computed for the test cases in order to eliminate ambiguities due to different numerical methods, boundary conditions and resolution in the LES region. Then extensive parameter studies with the identical computational grid in the LES region, but less nodes in the RANS region, were performed for variations and combinations of the different coupling conditions. For brevity, only a few selected results can be shown in this report. For the channel flow, Direct Numerical Simulation (DNS) data from the literature [9, 10] are included for reference in some of the plots presented.

2 Coupling Techniques

2.1 Velocities at Outflow-Type Interfaces

The underlying idea for the outflow-type interface condition proposed in [6, 7] is that, for any RANS zone downstream of an LES zone, the primary task of a hybrid LES/RANS coupling is to propagate mean flow information upstream. At the same time, for flows with stationary statistics, the LES should provide only mean flow data to the RANS domain. Since the LES delivers unsteady data, the interface has to allow for the fluctuations to leave the LES domain without reflections. To this end, the proposed velocity interface condition couples the explicitly Reynolds-averaged velocity at the LES outflow directly to the RANS inflow boundary, whereas fluctuations are convected out of the LES domain using a one-dimensional, linear convection equation with a given convection speed U_c.

The inherent assumption for such a coupling is that the downstream transport of fluctuations across the interface is dominated by convection. For this to be true, U_c needs to be directed towards the RANS domain and its magnitude should be considerably larger than the amplitude of the fluctuations, hence restricting this method to outflow-type interfaces. In addition, laminar and modeled turbulent diffusion across the interface must be negligible which, however, is uncritical for turbulent flows and adequately resolved LES. This proposed "convective condition" for the velocity coupling is general and contains the original enrichment strategy as the limiting case of an infinite convection speed of the fluctuations [7]. Furthermore, no constant needs to be calibrated if the local mean velocity at the interface is used as the convection velocity for the fluctuations.

The convective coupling condition for a fluctuation ϕ is implemented in its discrete form using a first-order upwind difference in the n-direction (index j along a grid line normal to the interface) and a so-called θ–scheme with $0 \leq \theta \leq 1$ in time ($t = m\, \Delta t$):

$$\phi_j^{m+1} = C_1\, \phi_j^m + C_2\, \phi_{j-1}^m + C_3\, \phi_{j-1}^{m+1} \quad , \tag{1}$$

where C_1, C_2 and C_3 are coefficients depending on the convection speed U_c, the spatial mesh size, the time step and the choice of θ. For $\theta = 0.5$ and $\theta = 1$, as chosen for the simulations presented here, this results in the implicit second-order accurate trapezoid rule and the implicit first-order Euler method, respectively. In the following $\langle \ldots \rangle_{LES}$ designates an explicit averaging procedure in time and homogeneous directions applied in the LES domain. A double prime denotes fluctuations with respect to this average and the average inherent to the Reynolds–averaging procedure. The interface is located at the face between the cells with index j (RANS–side) and $j-1$ (LES–side). The resulting coupling conditions are then

$$U_{j-1}^{m+1} = \langle \overline{u}_{j-1}^{m+1} \rangle_{LES} \tag{2}$$

for the streamwise velocity U at the inflow boundary of the RANS calculation and

$$\overline{u}_j^{m+1} = U_j^{m+1} + \overline{u}_j^{\prime\prime m+1} \quad (3)$$

for the resolved streamwise velocity $\overline{u}$ at the LES outflow boundary, with the fluctuation $\overline{u}''$ obtained from (1). All other velocity components are computed accordingly. Setting $C_1 = C_2 = 0$ and $C_3 = C_E$ in (1), recovers the *ad hoc* formula for enrichment with its calibration constant C_E.

2.2 Velocities at Tangential Interfaces

For tangential interfaces, subdomain boundaries are more or less aligned with streamlines of the mean flow. If these interfaces are close to walls with the RANS region between the LES domain and the wall, the problem is analogous to near-wall modeling of LES using a two-layer approach. With segregated modeling, however, the velocities are discontinuous across the interface, since only mean values are directly coupled. Fluctuations need to be provided separately. Tangential coupling with segregated modeling has so far been proposed only in [4] using the enrichment strategy discussed in Sect. 1. In this reference, the method was applied to turbulent channel flow and the flow over a bluff body. Here, we follow their premise that copying fluctuations from inside the LES zones to ghost cells representing the other side of the interface provides physically realistic structures. But instead of scaling the structures with an *ad hoc* constant we use the information from the RANS model to appropriately scale the amplitude of the fluctuations. Hence, the solution in the ghost cells has the desired statistical properties determined by the RANS zone and is statistically consistent.

In this report a first step of the investigation is presented where the RANS solution is frozen. Suppose the interface is an $x-z$ plane located between the indices $j-1$ and j in the y-direction and that the RANS zone is below the LES region. In the present method, the values of the u-component of the instantaneous solution in the LES ghost cells are then determined as

$$u_{j-1}^{LES} = u_{j-1}^{RANS} + f_u\, u_j^{\prime\prime LES} \qquad \text{with} \qquad u_j^{\prime\prime LES} = u_j^{LES} - \langle u_j^{LES} \rangle_{LES} \quad . \quad (4)$$

The other velocity components are determined analogously. When using a turbulence model which is based on a K-equation, it is natural to scale the velocity fluctuations to the desired value of K. With $f_K = K_{j-1}^{RANS}/K_j^{LES}$ we use $f_u = f_v = f_w = \sqrt{f_K}$. For other models that do not provide a value K^{RANS}, an alternative scaling was designed to yield matching of the total turbulent shear stress which is the central component of most eddy-viscosity based statistical turbulence models [8].

Rescaling the amplitudes of the copied LES-fluctuations is meant to enforce chosen statistics provided by the RANS calculation, but it can not account for a phase shift in the fluctuation. A phase shift may occur due to the retardation of the faster moving fluid from above the interface when it is moved to the slower moving fluid below. In order to mimic this retardation effect, a relaxation was introduced to the fluctuation in (4)

$$u_j''^{LES} = \varepsilon \left(u_j^{LES}|^m - \langle u_j \rangle_{LES} \right) + (1-\varepsilon) \left(u_j^{LES}|^{m+1} - \langle u_j \rangle_{LES} \right) \quad , \qquad (5)$$

where m represents the time-level of the fluctuation and ε is the relaxation factor.

2.3 Pressure and Mass Conservation

For incompressible flows with well-posed boundary conditions, mass conservation inside the fluid domain is implicitly enforced through the pressure field or an equivalent constraint variable, e.g. the streamfunction. These variables are governed by a Poisson-type equation such that a convective condition or a scaled copying cannot be applied to their fluctuations and a different way of coupling the LES and RANS domains needs to be devised. In the following we restrict ourselves to formulations involving the pressure. Two distinct possibilities of handling this variable at the interface are scrutinized.

One possibility is to solve the instantaneous pressure globally in the union of the LES and the RANS domains. It is a strong coupling that enforces instant mass conservation in the complete fluid domain. If the algorithm for the Poisson solver employed uses a domain decomposition technique no adjustment to the algorithm is necessary making this a very attractive approach. However, complications may arise due to the so-called *modified pressure* caused by many turbulence models [7]. In addition, this method turned out to be unstable in the presence of large fluctuations [6]. An alternative approach is to decouple the pressure fields of the LES and RANS domains completely. In this case, both the velocity and pressure fields are discontinuous at the interface and mass conservation across this boundary is not guaranteed. Ignoring the mass conservation issue is the method identified as case P1 below. In cases where instantaneous mass conservation is violated, the boundary conditions for the Poisson solver become ill-posed and the solver converges poorly or not at all.

The problem of an integral mass flux imbalance is not limited to LES/RANS coupling, but occurs routinely with projection methods for incompressible flow. As a remedy, a global mass flux correction is applied computing the mass flux over all inflow boundaries $\dot{m}_{in}$ and calculating the actual mass flux over all exit boundaries $\dot{m}_{out}$ due to the uncorrected exit velocities u_i^*. All velocity components at all outflow cells can then be scaled with the same factor

$$u_i = f_m \, u_i^* \qquad \text{with} \qquad f_m = \frac{|\dot{m}_{in}|}{\dot{m}_{out}} \quad . \qquad (6)$$

In practical simulations, the mass flux ratio f_m in (6) is usually very close to one. This "global correction" can also be applied to the velocities constructed with (2) and (3) at a single outflow-type interface. For the LES boundary cells, one needs to replace $\dot{m}_{out}$ in (6) with the mass flux leaving the LES domain

$\dot{m}_{LES}$. Conversely, the velocities in the RANS boundary cell are then scaled using the magnitude of the mass flux entering the RANS domain $|\dot{m}_{RANS}|$. The simulation with decoupled pressure and a global mass flux correction is called case P2 below.

For coupling of LES and RANS in case of complex geometries, one might have multiple embedded LES domains or tangential interfaces making the global correction cumbersome, if not impossible. Hence, a local approximation to the global flux correction above is proposed with $|\dot{m}_{in}|$ in (6) being replaced, at each simply connected interface, by

$$\dot{m}_{interface} = 1/2\ (\dot{m}_{LES} + |\dot{m}_{RANS}|) \tag{7}$$

with $\dot{m}_{LES}$ and $\dot{m}_{RANS}$ determined by integration over the corresponding interface. A simulation using this "local correction" is denoted as case P3.

3 Computational Setup

3.1 Numerical Method

The simulations were performed with the Finite Volume Code LESOCC2 (Large Eddy Simulation On Curvilinear Coordinates) in FORTRAN 95 [11] that is an enhanced fully parallelized version of the code LESOCC [12], both developed at the Institute for Hydromechanics. It solves the incompressible, three-dimensional, time-dependent, filtered and/or Reynolds-Averaged Navier-Stokes equations on body-fitted, collocated, curvilinear, block-structured grids using second-order accurate central differences for the discretization of the convective and viscous fluxes. Time advancement is accomplished by either an explicit, low-storage Runge–Kutta or a second-order accurate implicit method. Conservation of mass is achieved by the SIMPLE algorithm with the pressure-correction equation being solved using the strongly implicit procedure (SIP) of Stone. The momentum interpolation method of Rhie and Chow is employed to prevent pressure–velocity decoupling and associated oscillations. Parallelization is achieved via a domain decomposition technique with the use of ghost cells and MPI for the data transfer.

3.2 Performance of Flow Solver

The flow solver LESOCC2 and its predecessors have been used in numerous studies and substantial experience has been gained with respect to its numerical properties and its performance on various hardware platforms. Versions of LESOCC2 have been used for several LES and DNS investigations at the Institutes for Hydromechanics and Technical Chemistry and Polymer Chemistry at the University of Karlsruhe. The program has been successfully employed for large-scale computations on various hardware such as the VPP-5000, IBM–SP, NEC–SX-8, SGI–Altix, HP–XC-6000 and HP–XC-4000 ("HP–XC-2" of the Steinbuch Centre for Computing). Scaling of LESOCC2 on the HP-XC-2 (the machine used for the present investigation) was tested for up to 128 processors and satisfactory results were obtained. For the test shown in Fig. 2, the problem size per processor was kept constant. As expected, results improved for larger numbers of grid points per processor, i.e. smaller relative communication overhead. For the study presented here, the problem sizes per processor were an order of magnitude larger than the largest case shown in Fig. 2 and typically 14 or 16 processors were used such that the efficiency of the parallelization was over 90%. A single run required between 2000 and 3000 CPU hours and almost 100 cases had to be run.

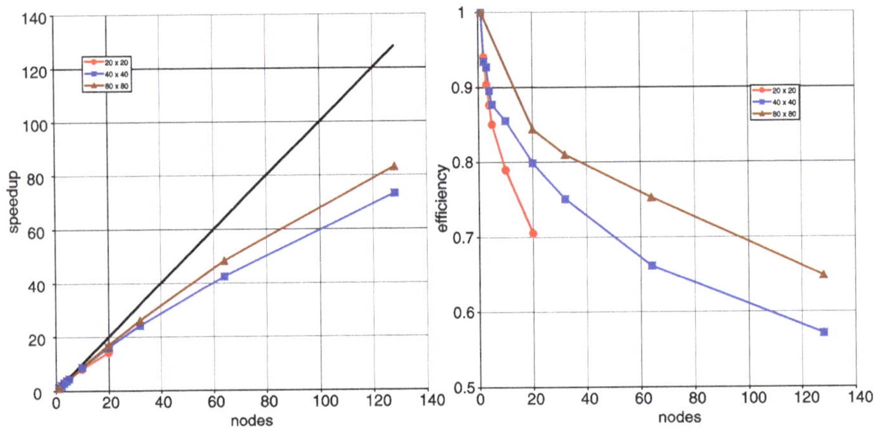

Fig. 2. Speedup (left) and efficiency (right) of the parallelization of the flow solver used on the HP–XC-2 for a model problem

4 Test Cases and Simulation Results

For all cases presented below, either the standard K–ω model of Wilcox (KO) or the Spalart–Allmaras (SA) model were used as RANS closures and the Smagorinsky model with van Driest wall damping was employed to determine the subgrid-scale stresses for LES.

4.1 RANS Downstream of LES

Turbulent Channel Flow

The performance of the convective velocity coupling was scrutinized for turbulent channel flow ($Re_\tau = 395$) by comparison with DNS data from [9]. Cases investigated are compiled in Tab. 1 and are reported in greater detail in [6, 7]. The domain is divided into three parts: the inflow generator, the principal three-dimensional LES zone and the two-dimensional RANS zone (Fig. 3).

Table 1. Investigated cases of velocity coupling strategies for LES with downstream RANS; computed in general with global pressure coupling except for cases P1–3 which use decoupled pressure and different methods for enforcing mass conservation at the interface as described in the text

case	method	C_E	case	method	U_c	θ	case	method	U_c	θ
E1	enrichment	0.0	C1	convective	1.0	0.5	P1	convective	$< u >$	0.5
E2	enrichment	0.1	C2	convective	72.	1.0	P2	convective	$< u >$	0.5
E3	enrichment	0.98	C3	convective	380.	1.0	P3	convective	$< u >$	0.5
E4	enrichment	1.0	C4	convective	$< u >$	0.5				

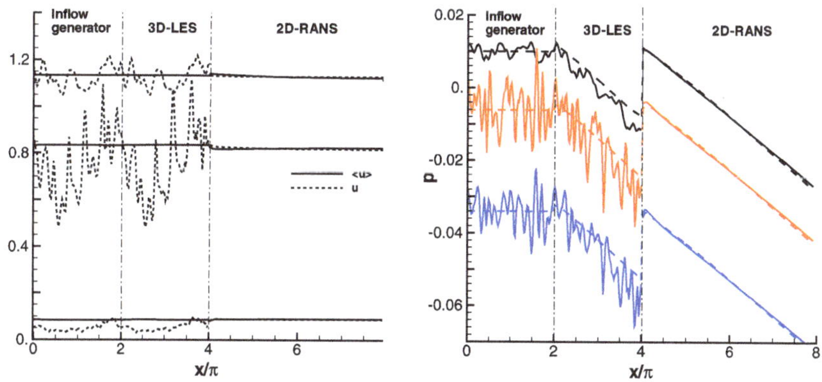

Fig. 3. RANS downstream of an LES zone with method P3 for turbulent channel flow at $Re_\tau = 395$; instantaneous and mean velocities (left) and pressure (right) for $y = 1$, 0.1 and 0.0037 (from top to bottom); pressure with arbitrary offset for clarity

All quantities are made dimensionless using the channel half-height δ and the bulk velocity U_b at the inlet of the principal LES zone. The inflow generator is a stand-alone LES with periodic boundary conditions in the streamwise direction and a mass flux enforced by volume forces and a controller. It provides planes of instantaneous velocities for the inflow of the principal LES zone. For each of the LES zones, the domain size is $2\,\pi \times 2 \times \pi$ in the streamwise (x), wall-normal (y), and spanwise (z) directions, respectively. Grid stretching is employed only in the wall-normal direction. Both domains are discretized using $80 \times 100 \times 80$ cells resulting in a near-wall scaling of $y_1^+ = 1.45$, $\Delta x^+ = 32$, and $\Delta z^+ = 16$. The RANS domain extends over a length of $4\,\pi$ on a stretched grid in the streamwise direction and was computed with the SA–model. The same wall-normal grid as for the LES zones is utilized with one cell in the spanwise direction. The time step was $\Delta t = 0.01$ and statistics were sampled over $t = 350\,\delta/U_b$ starting after a statistically steady state was obtained. All averages were taken in time and the lateral direction.

Fig. 3 illustrates that for the proposed coupling condition instantaneous velocity and pressure fluctuations can leave the domain without reflections.

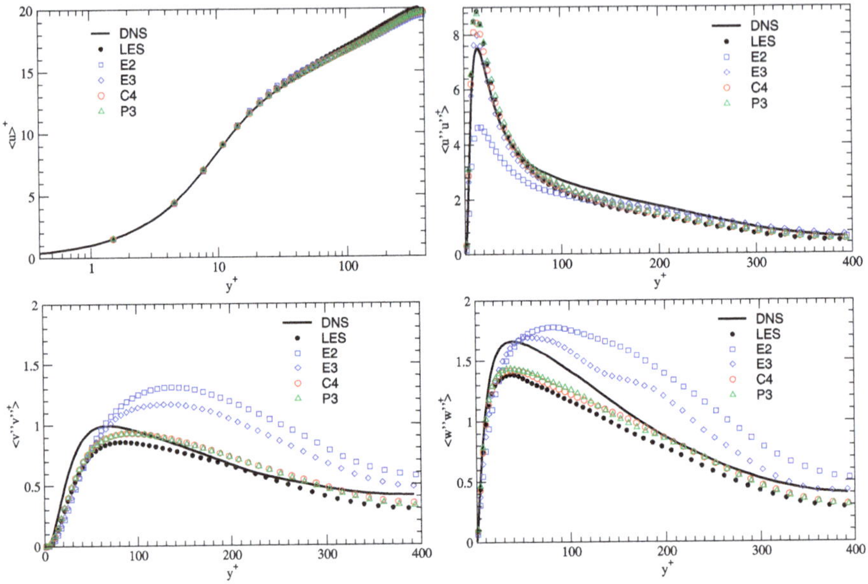

Fig. 4. Turbulent channel flow with outflow coupling: Comparison of mean streamwise velocity and resolved turbulent stresses in near-wall scaling for enrichment and the convective condition at the line adjacent to the interface; LES reference data from periodic simulation on same grid; DNS reference data from [9]

The flow structures are not altered even in closest proximity to the interface which is corroborated by contour plots and spectra at various locations [7]. Fig. 4 shows statistical data in the wall-normal direction for the reference DNS, for an LES without RANS coupling and at the interface plane for the proposed hybrid method (cases C4 and P3) and enrichment (cases E2 and E3). Cases C4 and P3 differ only in the pressure coupling or rather decoupling for P3 which is of no consequence for the channel flow, whereas the two enrichment cases refer to different values of the calibration constant C_E. The streamwise mean velocity is well represented by all methods, but the convective coupling conditions yield a clear improvement over enrichment with respect to the Reynolds stresses in the interface plane. While the longitudinal Reynolds stress can be captured by the well calibrated enrichment constant (E3), the other components still deviate considerably from the reference data. A more detailed comparison of enrichment and the proposed method (also for the hill flow below) can be found in [5, 7].

Flow Over Periodic Hills

The channel flow is a sensitive but uncritical test case since this flow is fully developed so that any modification in the streamwise direction results from changes in modeling, whereas no downstream information is really needed for the upstream LES. This is different in the flow over periodic hills. Again, all cases of Tab. 1 were tested, also with various locations of the LES-to-RANS interface [6, 7]. Only the most challenging setup is shown below for which a simulation without the RANS zone using a standard convective outflow condition diverges, illustrated by Fig. 5. The simulation is again divided into the three distinct zones used for the channel flow simulation. The first zone is computed with LES using wall functions and periodic boundary conditions in the downstream direction serving as inflow generator for the second zone. $200 \times 64 \times 92$ interior cells are used in the downstream, wall-normal and lateral direction, respectively. For the second zone, also LES is performed using the same resolution and wall-function as in Zone 1, however, before the crest of the next hill is reached, the simulation switches from LES to RANS (again using the SA–model). At the outflow of the RANS domain, Neumann boundary conditions are applied.

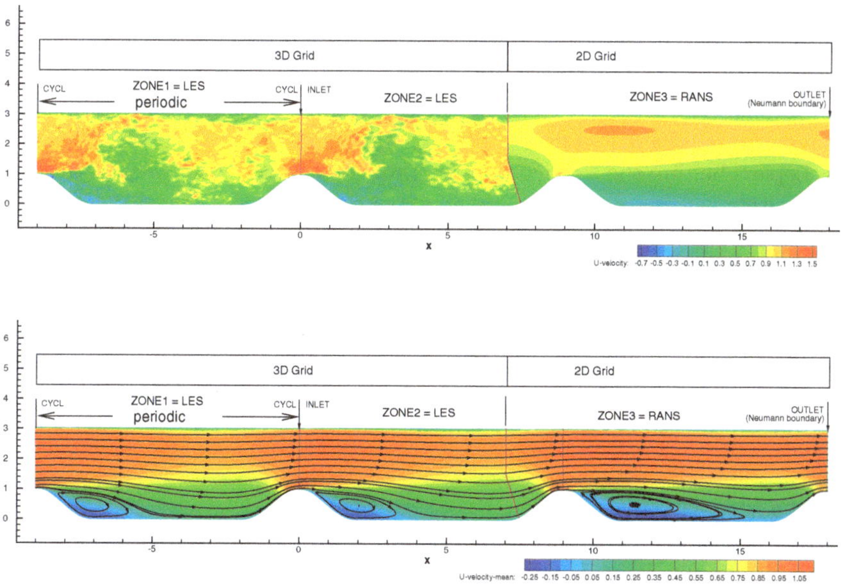

Fig. 5. Convective outflow coupling for the flow over periodic hills with the LES-to-RANS interface at $x \approx 7$; instantaneous streamwise velocity (top) and mean streamwise velocity (bottom) for method P3

Typical results obtained with convective coupling, decoupled pressure fields and explicit mass flux correction are displayed in Fig. 5. The instantaneous streamwise velocity contours show that the RANS flow field is completely steady. No reflections can be seen in the LES domain. The mean streamlines reveal that for the two-dimensional RANS solution reattachment occurs far too late, consistent with RANS results in the literature [13]. On the other hand, the LES in Zone 2 delivers results similar to the reference solution of Zone 1, albeit with a slightly longer recirculation region. Both reattachment lengths of $4.1\,h$ and $4.3\,h$ for Zone 1 and 2, respectively, are sufficiently close to the reference values of 4.6 to $4.7\,h$ in [14] obtained with a substantially finer grid. Note that for the setup shown here only cases P2 and P3 yielded acceptable results, whereas all other cases where fluctuations were generated (including enrichment) failed to obtain a result at all. This demonstrates the crucial importance of pressure decoupling in conjunction with mass conservation for complex situations.

4.2 RANS Between LES and Wall

Turbulent channel flow at two bulk Reynolds numbers ($Re = 7000$ and 20580) is used to assess the tangential coupling conditions for a RANS zone placed between LES and the wall. The investigated cases are compiled in Tab. 2. Here, only results for the highest Re are presented, for more details see [8]. Apart from the obvious naming for DNS, LES and RANS, the first letter in the case abbreviation identifies the method generating fluctuations, i.e. no fluctuations (N), classical enrichment with scaling constant 0.95 (E), scaling based on the turbulent kinetic energy (K) and the total turbulent shear stress (S). Letters behind the hyphen identify the RANS model used (KO or SA). For cases with relaxation, the letter R is appended after a second hyphen. Since the bulk velocity was enforced the Reynolds number based on friction velocity is a calculation result and reported in Tab. 2. For the hybrid simulations, the RANS zone next to the upper wall was frozen such that Re_τ of the RANS calculation is maintained whereas the corresponding value at the lower wall adjusts such that the total mass flux is maintained. The values obtained for Re_τ on both walls are included in Tab. 2. All simulations were computed with decoupled pressure fields and no explicit mass flux correction at the interface.

Table 2. Simulations reported and resulting Reynolds numbers Re_τ based on friction velocity for the top and the bottom wall, DNS reference data from [10]

method	$Re = 7000$		$Re = 20580$		method	$Re = 7000$		$Re = 20580$	
	top	bottom	top	bottom		top	bottom	top	bottom
DNS	395	395	934	934	E-KO	410	404	1084	952
LES	382	382	890	890	K-KO	410	396	1084	910
RANS-KO	410	410	1084	1084	S-SA	406	404	–	–
RANS-SA	406	406	1120	1120	K-KO-R (0.5)	–	–	1084	917
N-KO	410	383	1084	889	E-KO-R (0.5)	–	–	1084	950

The domain size is $2\,\pi \times 2 \times \pi$ in the streamwise (x), wall-normal (y), and spanwise (z) direction, respectively, with periodic boundary conditions in x and z. The bulk Reynolds number was imposed by a volume force in the

x−momentum equation and an appropriate controller. A grid of $64 \times 64 \times 64$ cells with stretching in the wall-normal direction was used throughout. For the low Reynolds number case this results in a near-wall scaling of $y_1^+ \approx 0.5$, $\Delta x^+ \approx 40$, and $\Delta z^+ \approx 20$ and for the high Re case in $y_1^+ \approx 1$, $\Delta x^+ \approx 90$, and $\Delta z^+ \approx 50$. The RANS domain extends over the last 16 points in the wall-normal direction ($1.885 \leq y \leq 2$). Here, the grid is two-dimensional with only one cell in the spanwise direction. The time step was $\Delta t = 0.01$ and statistics were sampled over $t_{aver} \geq 1500\,\delta/U_{bulk}$. All averages were taken in time and over wall-parallel planes. As initial condition for the hybrid simulations, the results of the RANS computation were used with the addition of random noise. The grid was chosen coarser than in Sect. 4.1 in order to more clearly see whether the RANS layer yields any improvements. For the results shown, the interface is placed within the logarithmic region of the U^+–profile ($y^+ \approx 120$).

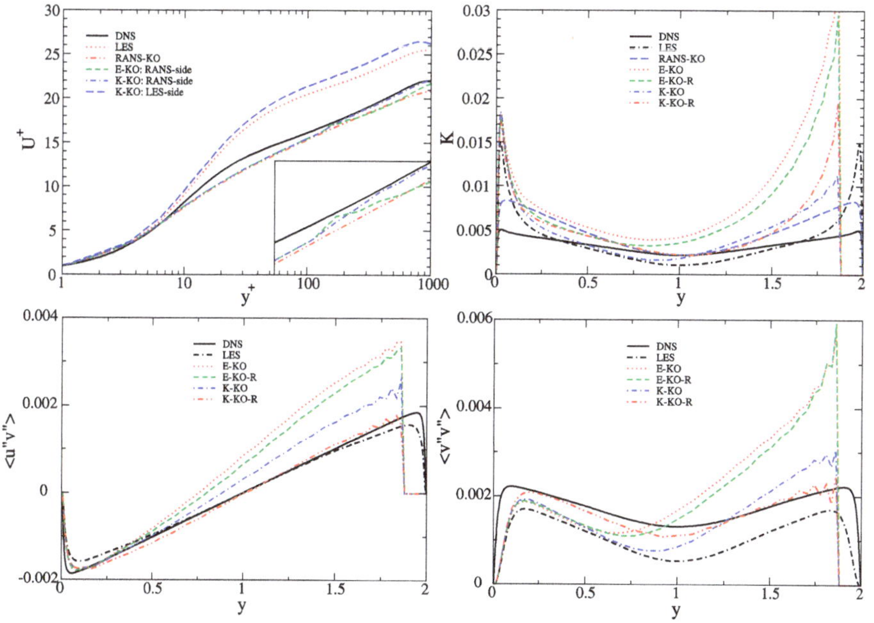

Fig. 6. Wall-normal profiles for turbulent channel flow simulation at $Re_\tau \approx 1000$ with tangential LES/RANS coupling: mean streamwise velocity (top left), resolved turbulent kinetic energy (modeled for RANS, top right) resolved Reynolds shear stress (bottom left) and resolved wall-normal Reynolds stress (bottom right); DNS reference data from [10]

Fig. 6 demonstrates that simulation K-KO delivered the best results with respect to the streamwise mean velocity improving over pure LES, pure RANS and enrichment in a smooth fashion. For K-KO, also the U^+–profile for the lower (LES) wall is shown. A slight shift in the constant of the log-profile can be discerned. This shift is due to the adjustment of the wall friction to the enforced mass flux, but otherwise the structure of the flow seems unperturbed. The resolved u−fluctuations however exhibit an overprediction close to the interface. Although the K–scaling represents a clear improvement over the original enrichment technique, this artificial increase of resolved fluctuations in the vicinity of the interface is still bothersome. It indicates that at least some of the flow structures are not represented accurately. A possible reason for this may be the strict vertical displacement of the fluctuations whereas, in particular for coarse grids, real flow structures might be transported to the location of the ghost cells at an angle leading to a time shift of the fluctuation. To assess whether such a mechanism might indeed be at work here, both the K–scaled and the classical enrichment technique were repeated with the relaxation of (5). Results are only shown for a relaxation factor of $\varepsilon = 0.5$ here, but other factors were tested as well and very similar results were observed. The relaxation has indeed a drastic effect on the flow structures as can be seen in the resolved Reynolds stresses plotted in Fig. 6. The shear stress profile is improved and, in particular for case K-KO-R, excellent agreement with the DNS data is obtained. This seems to be mainly due to a better prediction of the wall-normal velocity fluctuations, since the resolved wall-normal Reynolds stress also exhibits very good agreement with the DNS data near the interface. Also a considerable improvement can be seen in the lateral stresses (not shown here). The longitudinal Reynolds stress, however, is now overpredicted even more. Since it contributes most to the resolved turbulent kinetic energy this also attains higher values near the interface.

A more careful look at flow structures in the vicinity of the interface reveals that, for pure LES, predominately longitudinal structures of a velocity deficit are present. Enrichment leads indeed to unsteady flow structures but with shorter streamwise extension. The K–scaling improves on the resemblance and arguably realistic flow structures are generated, but this method still falls short, since the longitudinal structures appear to be shorter than in the LES. The relaxation destroys the high-frequency content of the fluctuations and only the large-scale structures survive [8].

5 Conclusions

Tangential and outflow-type interfaces for a segregated hybrid LES/RANS method were successfully tested. The methods were devised for incompressible flow for which the pressure coupling at the interface becomes critical in some situations. A decoupling of the pressure fields with an explicit mass flux correction performed best delivering a fairly robust hybrid method. Contrary to enrichment, for the method proposed here, the velocity coupling needs to distinguish outflow-type interfaces from tangential interfaces. For the first type, fluctuations are convected out of the LES domain, whereas for the second type the fluctuations are scaled to match the statistics predicted by the

RANS model. This more physics-based approach yielded better results than the original enrichment technique. However, the tangential interface condition can benefit from further improvements and a combined usage of outflow-type and tangential interfaces, in particular for application to more complex flows, is still awaiting further scrutiny.

Acknowledgements

Funding by the German Science Foundation under contract number Fr 1593 / 1-1,2 and provision of computer time by the Steinbuch Centre for Computing are gratefully acknowledged.

References

1. J. Fröhlich and D.A. von Terzi. Hybrid LES/RANS methods for the simulation of turbulent flows. *Prog. Aerospace Sci.*, 44:349–377, 2008.
2. J. Sivasubramanian, R.D. Sandberg, D.A. von Terzi, and H.F. Fasel. Numerical investigation of transitional supersonic base flows with flow control. *J. Spacecraft Rockets*, 44(5):1021–1028, 2007.
3. P. Sagaut, S. Deck, and M. Terracol. *Multiscale and Multiresolution Approaches in Turbulence*, pages 294–319. Imperial College Press, 2006.
4. P. Quéméré and P. Sagaut. Zonal multi–domain RANS/LES simulations of turbulent flows. *Int. J. Numer. Meth. Fluids*, 40:903–925, 2002.
5. D.A. von Terzi, W. Rodi, and J. Fröhlich. Scrutinizing velocity and pressure coupling conditions for LES with downstream RANS calculations. In S.-H. Peng and W. Haase, editors, *Advances in Hybrid RANS-LES Modelling*, volume 97 of *Notes on Numerical Fluid Mechanics and Multidisciplinary Design*. Springer, 2008. ISBN 978-3-540-77813-4.
6. D.A. von Terzi and J. Fröhlich. Coupling conditions for LES with downstream RANS for prediction of incompressible turbulent flows. In *Proc. of 5th Int. Symp. on Turbulence and Shear Flow Phenomena TSFP-5*, volume 2, pages 765–770. Elsevier, 2007.
7. D.A. von Terzi and J. Fröhlich. Zonal coupling of LES with downstream RANS calculations. 2008. in preparation.
8. D.A. von Terzi and J. Fröhlich. A statistically consistent approach to segregated LES–RANS coupling at tangential interfaces. In *Proc. of 7th Int. ERCOFTAC Symp. on Engineering Turbulence Modelling and Measurements ETMM-7*, 2008.
9. R.D. Moser, J. Kim, and N.N. Mansour. Direct numerical simulation of turbulent channel flow up to $Re_\tau = 590$. *Phys. Fluids*, 11:943–945, 1999.
10. J.C. Del Alamo and J. Jimenez. Scaling of the energy spectra of turbulent channels. *J. Fluid Mech.*, 500:135–144, 2004.
11. C. Hinterberger. *Dreidimensionale und tiefengemittelte Large–Eddy–Simulation von Flachwasserströmungen*. PhD thesis, Institute for Hydromechanics, University of Karlsruhe, 2004.
12. M. Breuer and W. Rodi. Large eddy simulation of complex turbulent flows of practical interest. In E.H. Hirschel, editor, *Flow simulation with high performance computers II*, volume 52 of *Notes on Numerical Fluid Mechanics*, pages 258–274. Vieweg, 1996.

13. S. Jakirlić, R. Jester-Zürker, and C. Tropea, editors. *9th ERCOFTAC / IAHR / COST Workshop on Refined Turbulence Modelling*. Darmstadt University of Technology, 2001.
14. J. Fröhlich, C.P. Mellen, W. Rodi, L. Temmerman, and M.A. Leschziner. Highly–resolved large eddy simulations of separated flow in a channel with streamwise periodic constrictions. *J. Fluid Mech.*, 526:19–66, 2005.

Vector Computers in a World of Commodity Clusters, Massively Parallel Systems and Many-Core Many-Threaded CPUs: Recent Experience Based on an Advanced Lattice Boltzmann Flow Solver

Thomas Zeiser, Georg Hager, and Gerhard Wellein

Regionales Rechenzentrum Erlangen (RRZE)
Universität Erlangen-Nürnberg
Martensstraße 1, 91058 Erlangen, Germany
thomas.zeiser@rrze.uni-erlangen.de

Summary. This report summarizes experience gained during the last year using the NEC SX-8 at HLRS and its wide range of competitors: commodity clusters with Infiniband interconnect, massively parallel systems (Cray XT4, IBM BlueGene L/P) and emerging many-core many-threaded CPUs (SUN Niagara2 processor). The observations are based on low-level benchmarks and the use of an advanced lattice Boltzmann flow solver developed in the framework of an international development consortium (ILBDC).

1 Preliminaries

The computational power of commodity cluster systems available e.g. at RRZE, the computing center of the University of Erlangen-Nuremberg, allowed to run many medium to large scale parameter studies on the local facilities using just 16–32 nodes of RRZE's Infiniband cluster with single- or dual-socket nodes equipped with Intel Core2-based dual-core CPUs and 2 GB of main memory per processor core. Using these cluster nodes seemed to be much more appropriate and economical than filling the NEC SX-8 at HLRS with just single-node jobs which would waste this expansive and scarce resources, i.e. misuse a capability computing resource for simple capacity computing.

Therefore, during the last year the NEC SX-8 was mainly used for (basic) performance studies only and to ensure that good vectorization of the used lattice Boltzmann flow solver is maintained despite major changes and algorithmic/physical extensions of the code. Continuous access to a multi-node

NEC SX installation therefore is mandatory to preserve the vector capabilities even though not much compute time is currently consumed on those systems.

2 Investigated Architectures

2.1 NEC SX-8

From a programmer's view, a NEC SX-8 CPU is a traditional vector processor with 4-track vector pipes running at 2.0 GHz. One multiply and one add instruction per cycle can be sustained by the arithmetic pipes, delivering a theoretical peak performance of 16 GFlop/s. The memory bandwidth of 64 GByte/s allows for one load or store of double precision floating point data ("word") per multiply-add instruction, providing a balance of 0.5 Word/Flop. The processor has 64 vector registers, each holding 256 64-bit words.

An SMP node comprises eight processors and provides a theoretical total memory bandwidth of 512 GByte/s, i.e. the aggregated single processor bandwidths should be saturated. The NEC SX-8 nodes are networked by a proprietary interconnect called *IXS*, providing a bidirectional bandwidth of *only* 16 GByte/s and a latency of about 5 microseconds.

The computational power of the 2006 introduced NEC SX-8R is (theoretically) more than twice that of the NEC SX-8 owing to the doubled number of ADD and MULT execution units and a slight increase of clock frequency from 2.0 to 2.2 GHz. The memory bandwidth also increased slightly, proportional to the increase of the clock frequency. The network interconnect was not changed at all.

The next generation, the NEC SX-9, will not only provide several times more performance (in terms of memory bandwidth, network interconnect and GFlop/s) but also introduce new cache-like memory concepts, called *fast buffering* and *assignable data buffers (ADB)*. Based on experience [29] with the Cray X1 system which was the first vector computer which included caches, we expected that much more effort will be required to optimally adapt existing (vectorized) codes to this new system design than was required when moving from the NEC SX-4 all the way to the NEC SX-8.

2.2 Cray XT4

The Cray XT4 is a successor in the tradition of the famous Cray T3D/T3E massively parallel processing (MPP) systems combining excellent network properties with state-of-the-art single-node processing power. It is based on single-socket compute nodes with commodity AMD Opteron processors, unbuffered DDR2 memory, a proprietary high-speed interconnect and a lightweight Linux-based operating system on the compute nodes or fully featured Linux on the service nodes. AMD's *HyperTransport* is used to directly connect the processor to Cray's *SeaStar2* router which provides six high speed network links (peak bidirectional bandwidth of 7.6 GB/s per link) to connect each node to six neighbors in the 3D torus topology.

2.3 IBM BlueGene L/P

The IBM BlueGene systems follow an alternative approach which was untypical for HPC before. Instead of trying to combine rather few but powerful compute nodes, the system is designed as massively parallel system where performance does not come from the individual processing element but their huge aggregation. This allows trading the speed of processors for lower power consumption. The building block of BlueGene/L are two 700 MHz PowerPC440 processors – well known from the embedded market and used e.g. as service processor in Cray XT4's *SeaStar2* router –, a rather low amount of memory (512 MB) per compute node and a lightweight OS supporting only one running process at a time but with minimal system overhead. The two processors are not cache coherent with one another and can be operated either in *virtual mode* where both CPUs are used for calculations or in *co-processor mode* where the second CPU is dedicated to handle the inter-node communication. Three different high-speed communication networks are available in the system: a 3D torus for peer-to-peer communication, one for collective communication, one for fast barriers. During operation, the BlueGene systems are partitioned into electronically isolated sets of nodes to allow multiple programs to run concurrently with the number of nodes in a partition always being a power of 2.

The IBM BlueGene/P system uses quad-core PowerPC450 processors running at 850 MHz, 2 or 4 GB of DDR2 memory per compute card and is designed to scale up to 884 736 processors (or 216 racks with 3 PFlop/s aggregated peak performance). As the PowerPC450 processors come with hardware cache coherence support, SMP features can now be used in the BlueGene/P, i.e. the possibility of OpenMP within the individual nodes complements the pure MPI approach previously known from BlueGene/L only.

2.4 Commodity Cluster with Dual-Core Intel *Core2* CPUs and Infiniband Interconnect

Commodity clusters are available in many different flavors, for price-performance reasons usually with two-socket nodes as standard building block. The reported benchmarks however were run on a bandwidth-optimized *low-cost MPP system*, i.e. an Infiniband cluster based on Intel's single-socket *Port Townsend* (S3000PT) server boards. Two half-sized boards, each with one dual/quad-core desktop or server Intel Core-based processor (2.66 GHz Xeon 3070 in the present case), (energy efficient) unregistered DDR2 memory and a single-port DDR Infiniband HCA, are assembled in standard 1U chassis. The relative costs for the network components (Infiniband cards, ports on the Infiniband switch and Infiniband cables) of this system are of course higher than in usual two-socket cluster nodes. However, owing to the simple system design, network balance and memory balance are much better than in usual commodity clusters. The sustained memory bandwidth (as e.g. measured by the STREAM TRIAD benchmark [19]) of these single-socket nodes with only one FSB-1066 connection to main memory is almost as high as the aggregated memory bandwidth of e.g. a two-socket HP DL140G3 node with two dual-core Intel Xeon 5160 CPUs (3.0 GHz, codenamed *Woodcrest*), two frond side buses with FSB-1333 to the main memory with FB-DIMMs and the Intel 5000X (*Greencreek*) chipset with snoop filter enabled. The *Port Townsend* successor, called *Melstone*, can fully parry the two-socket HP DL140G3 in terms of compute performance and memory bandwidth if quad-core processors are used.

The Intel *Core* micro-architecture is capable of performing a maximum of four double precision floating point operations (two multiply and two add) per cycle. The investigated dual-core CPUs have a small but private L1 cache and 4 MB of shared L2 cache.

2.5 The Sun UltraSPARC T2 Processor: *Niagara2* CPU

Trading high single core performance for a highly parallel single chip architecture is the basic idea of Sun's UltraSPARC T2 processor. Eight simple in-order SPARC cores (running at 1.16 or 1.4 GHz) are connected to a shared, banked L2 cache and four independently operating dual channel FB-DIMM memory controllers through a non-blocking switch, thereby providing UMA access characteristics with scalable bandwidth. Such features were previously only available in shared-memory vector computers like the NEC SX series. To overcome the restrictions of in-order architectures and long memory latencies, each core is able to support up to eight threads, i.e. there are register sets, instruction pointers, etc. to accommodate eight different machine states. There are two integer, two memory and one floating point pipeline per core. Although all eight threads can be interleaved across the floating point and memory pipes, each integer pipe is hardwired to a group of four threads. The CPU can switch between the threads in a group on a cycle-by-cycle basis, but only one thread per group is simultaneously active at any time. If a thread has to wait for resources like, e.g., memory references, it will be put in an inactive state until the resources become available which allows for effective latency hiding [21] but restricts each thread to a single outstanding cache miss. Running more than a single thread per core is therefore mandatory for most applications, and thread placement ("pinning") must be implemented. This can be done with the standard Solaris `processor_bind()` system call or, more conveniently but only available for OpenMP, using the `SUNW_MP_PROCBIND` environment variable.

Each memory controller is associated with two L2 banks. A very simple scheme is employed to map addresses to controllers and banks: Bits 8 and 7 of the physical memory address select the memory controller to use, while bit

6 determines the L2 bank [21, 22]. Consecutive 64-byte cache lines are thus served in turn by consecutive cache banks and memory controllers. Due to the fact that typical page sizes are at least 4 kB the distinction between physical and virtual addresses is of no importance here.

The aggregated nominal main memory bandwidth of 42 GB/s (read) and 21 GB/s (write) for a single socket is far ahead of most other general purpose CPUs and topped only by the NEC SX-8 vector series. Since there is only a single floating point unit (performing MULT or ADD operations) per core, the system balance of approximately 4 bytes/flop (assuming read) is the same as for the NEC SX-8 vector processor. In our experience only about one third of the theoretical bandwidth can actually be measured.

One should be aware that the T2 chip was not designed for the HPC market but geared towards commercial, database and typical server workloads. Therefore, on-chip PCIe-x8 and 10 Gb Ethernet connections are included as well as a cryptographic coprocessor.

3 Computational Method and Implementation Aspects

3.1 Basics of the Lattice Boltzmann Method

The lattice Boltzmann method (LBM) [4, 20] is a recent method from computational fluid dynamics (CFD) which has its roots in a highly simplified gas-kinetic description, i.e. a velocity-discrete Boltzmann equation with appropriate collision term. When properly applied, the results of LBM simulations satisfy the Navier-Stokes equations in the macroscopic limit with second order of accuracy [4, 20].

The simplest form is the lattice Boltzmann equation with BGK collision operator [2] which reads for the 3-D model with 19 discrete velocities (D3Q19 model) as follows if external forces are neglected:

$$f_i(\mathbf{x}+\mathbf{e}_i\Delta t, t+\Delta t) = f_i^{\mathrm{coll}}(\mathbf{x},t) = f_i(\mathbf{x},t) - \frac{1}{\tau}\left[f_i(\mathbf{x},t) - f_i^{eq}\left(\rho(\mathbf{x},t),\mathbf{u}(\mathbf{x},t)\right)\right] \qquad i = 0\ldots 18\,, \tag{1}$$

$$\text{with } f_i^{eq}\left(\rho(\mathbf{x},t),\mathbf{u}(\mathbf{x},t)\right) = \rho(\mathbf{x},t)w_i\left[1+\frac{3}{c^2}\mathbf{e}_i\cdot\mathbf{u}(\mathbf{x},t) + \frac{9}{2c^4}\left(\mathbf{e}_i\cdot\mathbf{u}(\mathbf{x},t)\right)^2 - \frac{3}{2c^2}\mathbf{u}(\mathbf{x},t)\cdot\mathbf{u}(\mathbf{x},t)\right]\,, \tag{2}$$

which describes the evolution of the single particle distribution function f_i. f_i^{coll} denotes the "intermediate" state after *collision* but before *propagation*. The macroscopic quantities, density ρ and velocity $\mathbf{u}$, are obtained as 0th or 1st order moments of f_i with regard to the discrete lattice velocities $\mathbf{e}_i$, i.e. $\rho(\mathbf{x},t) = \sum_0^{18} f_i(\mathbf{x},t)$ and $\rho(\mathbf{x},t)\mathbf{u}(\mathbf{x},t) = \sum_0^{18} \mathbf{e}_i f_i(\mathbf{x},t)$. The discrete equilibrium f_i^{eq} as given by Eq. 2 is a Taylor-expanded version of the Maxwell-Boltzmann equilibrium distribution function [16, 20]. w_i are direction-dependent constants [20], and $c = \frac{\Delta x}{\Delta t}$ with the lattice spacing Δx and lattice time step Δt. The pressure p is obtained locally via the equation of state of an ideal gas, $p(\mathbf{x}, t) = c_s^2 \rho(\mathbf{x}, t)$, using the speed of sound c_s. The kinematic viscosity of the fluid is determined by the dimensionless collision frequency $\frac{1}{\tau}$ according to $\nu = \frac{1}{6}(2\tau - 1)\Delta x c$ with $\tau > 0.5$ owing to stability reasons [4, 16, 20].

Alternative collision models, like the two-relaxation time (TRT) [7] or multi-relaxation time models (MRT) [5], may replace the single-relaxation time BGK operator, providing additional adjustable parameters and thus usually an improved stability while preserving the benefits of the explicit lattice Boltzmann equation.

In order to reduce the weak compressibility imposed by Eq. 2, the pressure p (and accordingly the density ρ) may be split into the constant contribution p_0 and a small deviation δp resulting in a slightly modified equilibrium distribution function [15],

$$f_i^{eq'}(\rho(\mathbf{x},t), \mathbf{u}(\mathbf{x},t)) = w_i \left[\rho + \rho_0 \left(\frac{3}{c^2} \mathbf{e}_i \cdot \mathbf{u}(\mathbf{x},t) + \right.\right. \tag{3}$$

$$\left.\left. + \frac{9}{2c^4} \left(\mathbf{e}_i \cdot \mathbf{u}(\mathbf{x},t) \right)^2 - \frac{3}{2c^2} \mathbf{u}(\mathbf{x},t) \cdot \mathbf{u}(\mathbf{x},t) \right) \right],$$

$$\text{with} \qquad \rho(\mathbf{x},t) = \rho_0 + \delta\rho(\mathbf{x},t) = \sum_0^{18} f_i(\mathbf{x},t)$$

$$\text{and} \qquad \rho_0 \mathbf{u}(\mathbf{x},t) = \sum_0^{18} \mathbf{e}_i f_i(\mathbf{x},t) .$$

Using $\rho_0 = 1$, no divisions are required any longer when calculating the local velocity for each cell update. If the modified equilibrium distribution function is now also shifted by $-w_i \rho_0$, we can easily transform numbers of order $\mathcal{O}(w_i \rho_0) \simeq \mathcal{O}(1)$ to small variations around zero (Eq. 4). The advantage of the latter is that certain operations are now carried out on numbers with the same order of magnitude as their result which should – in particular in the case of single precision – improve the numerical accuracy, i.e. no loss of digits when subtracting slightly varying numbers of the same magnitude.

$$f_i^{eq''}(\delta\rho(\mathbf{x},t), \mathbf{u}(\mathbf{x},t)) = w_i \left[\delta\rho(\mathbf{x},t) + \left(\frac{3}{c^2} \mathbf{e}_i \cdot \mathbf{u}(\mathbf{x},t) + \right.\right. \tag{4}$$

$$\left.\left. + \frac{9}{2c^4} \left(\mathbf{e}_i \cdot \mathbf{u}(\mathbf{x},t) \right)^2 - \frac{3}{2c^2} \mathbf{u}(\mathbf{x},t) \cdot \mathbf{u}(\mathbf{x},t) \right) \right],$$

$$\text{with} \qquad \delta\rho(\mathbf{x},t) = \sum_0^{18} f_i(\mathbf{x},t)$$

$$\text{and} \qquad \mathbf{u}(\mathbf{x},t) = \sum_{0}^{18} \mathbf{e}_i f_i(\mathbf{x},t)\,.$$

Lattice Boltzmann methods with an explicit lattice Boltzmann equation as outlined above are used on equidistant Cartesian meshes (if necessary with local mesh refinement [24] again using equidistant Cartesian cells). A *marker-and-cell* (MAC) approach is used to distinguish between fluid and solid regions. In the simplest case, solid wall boundary conditions are realized by the *bounce-back* rule [4,20], i.e. distributions hitting the wall which is assumed to be located half-way between the fluid and solid cell, return to their original cell but with inverted momentum, $f_{\bar{i}}(\mathbf{x},\, t+\Delta t) = f_i^{\mathrm{coll}}(\mathbf{x},t)$ with $\mathbf{e}_{\bar{i}} = -\mathbf{e}_i$ and $f_i^{\mathrm{coll}}(\mathbf{x},t)$ being the right hand side of Eq. 1.

Information about the geometric pore-scale structure can for example be directly taken from segmented X-ray micro-computed tomography images [6, 27] or magnetic resonance imaging (MRI) data sets [25].

If the staircase approximation of the geometry is not sufficient, 2nd order geometric boundary conditions [3,24] can be applied which resemble *cut-cell* techniques and inter- or extrapolate the required distribution functions using data from (over-next) neighbor cells. A general form of the corresponding update rule using common linear or quadratic schemes is [8]

$$\begin{aligned} f_{\bar{i}}(\mathbf{x},\, t+\Delta t) = {} & \kappa_1 f_i^{\mathrm{coll}}(\mathbf{x},t) + \kappa_0 f_i^{\mathrm{coll}}(\mathbf{x}-\mathbf{e}_i,t) + \kappa_{-1} f_i^{\mathrm{coll}}(\mathbf{x}-2\mathbf{e}_i,t) + \\ & + \bar{\kappa}_{-1} f_{\bar{i}}^{\mathrm{coll}},t) + \bar{\kappa}_{-2} f_{\bar{i}}^{\mathrm{coll}} - \mathbf{e}_i,t)\,, \end{aligned} \tag{5}$$

where the different κ values depend on the actual position of the wall and the interpolation scheme used [3,8,24].

3.2 General Implementation Aspects

The *collision* described by the lattice Boltzmann equation (right hand side of Eq. 1) is a purely local operation and involves arithmetic operations, whereas the *propagation* (left hand part of Eq. 1) only exchanges data with all direct neighbors of a cell. A usual way to work around the resulting data dependencies owing to the propagation step is the use of two arrays, i.e. one for the current and the other for the next time step, and toggling between them. To reduce the memory traffic, it is important that the collision and propagation are executed in a single loop and not independently of each other in separate loops or routines [23]. The D3Q19 lattice with either BGK or TRT collision operator requires about 180–200 floating point operations per cell and time step as well as reading 19 floating point values and writing to 19 different memory locations [23]. Assuming double precision floating point data and a cache based architecture with write-allocate strategy, this results in a balance of about 2.2–2.5 bytes/flop which cannot be sustained by most hardware. Therefore, the data layout (i.e. order of the different indexes) of the multi-dimensional arrays should be chosen in such a way that cache lines are efficiently used before they get replaced again. In most cases, a propagation optimized layout ("structure-of-arrays", in Fortran: xyzQ) is preferential with all entries of a single direction i being consecutive in memory [23].

3.3 ILBDC Solver: Sparse List-based Approach

Many lattice Boltzmann implementations use *full arrays* and a flag field to represent the data and the geometry. If only few cells are blocked out, this seems to be an optimal approach as the three nested spatial loops can be fused into just one, thus reducing loop overheads and ensuring large loop counts, and, moreover, the indexes of neighbor cells can be obtained directly via simple index shift arithmetics. However, if highly complex porous media with at least in some cases low porosities (e.g. packed bed reactors, vertebral bone or vascular system) are a main target, other implementation strategies become favorable, e.g. only *patches* of *full arrays* [9] or a true sparse representation which includes only the fluid cells. The patch approach resembles domain decomposition including only boxes which contain at least some fluid cells and requires clever "communication" between patches e.g. via halo-cells but it preserves the local order of cells. The sparse representation on the other hand sticks with the individual Cartesian fluid cells only and allows any order of them as the connectivity has to be stored anyway as adjacent cells no longer can be obtained by simple index arithmetics anyway. Within an *International Lattice Boltzmann Development Consortium* the latter approach has been chosen. The resulting data structure (cf. Fig. 1) is one 1-D list containing the density distribution values of the M fluid cells (for the current and next time step) as well as a second 1-D list with information about the adjacency of the cells used during propagation. For the present investigations on the PC cluster and the NEC SX-8, this 1-D list-based implementation with the propagation optimized *structure-of-arrays* data layout [23], i.e $f(1{:}M, 0{:}18, 0{:}1)$, and a combined collide–propagate algorithm is used unless otherwise noted. Stride-one access is always ensured for the adjacency list.

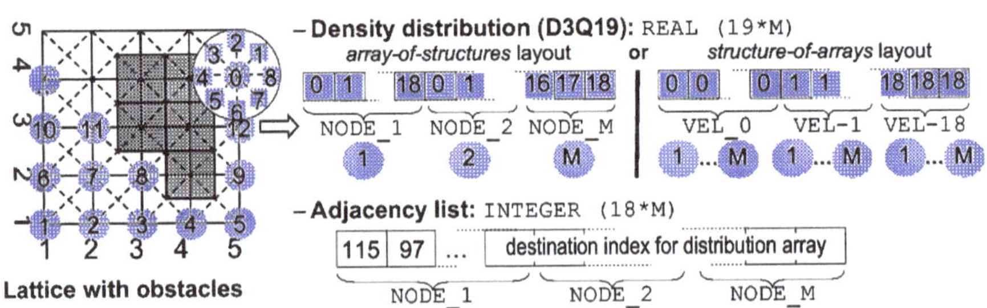

Fig. 1. Data structures of the sparse 1-D LBM solver

3.4 Partitioning of ILBDC's 1-D Node Lists

Lattice Boltzmann flow solvers are generally parallelized using domain decomposition *if* distributed memory systems are to be used – and even for ccNUMA systems it may be easier to use explicit domain decomposition instead of implementing proper local data placement with OpenMP throughout the code [13, 26].

For a *full arrays* approach with only few cells marked as solid (e.g. turbulent channel flow), a 3-D Cartesian domain decomposition using `MPI_Cart_Create` seems to be most appropriate leading to minimized communication interfaces and good load balance at the same time. The sparse list-based approach of ILBDC, however, resembles more an unstructured code although it originates from an equidistant Cartesian mesh. METIS [18] is a widely used library to partition arbitrary unstructured grids based on assigned weights for edges, nodes and links. As good scalability of lattice Boltzmann solvers not only depends on the minimization of the communication but moreover on load balance [1], METIS does not seem to be the optimal choice for ILBDC as the minimization of communication is overrated. Owing to the simple and regular Cartesian structure of the original grid which easily allows to enumerate all required fluid cells in arbitrary order, e.g. based on their lexicographic position, in spatially blocked fashion, or any type of space filling curve, the 1-D list can be build up in an *intelligent* way, reducing the partitioning to simply cutting the list into equally sized chunks. Load balance is automatically guaranteed in this way. The amount of communication between partitions is also low if the spatial locality of the selected ordering is high. A few examples for the index numbering and the resulting partitioning are schematically shown in Fig. 2 for a simple 2D case. The computational effort in terms of CPU-time and memory requirements for partitioning a realistic, complex 3D geometry of a packed bed is shown in Fig. 3. It can clearly be seen that the computational effort of directly cutting the 1-D is much lower than the effort for running METIS. However, good heuristics for finding an appropriate local numbering scheme still have to be found and are planed for a complementary project.

4 Scalability and Sustained Performance

Figure 4 shows the intra-node performance of the NEC SX-8 as function of the domain size. It is surprising that even for the single processor version rather large domains of 50–100 cells cord length (i.e. domains with 125k–1M fluid cells) are required to obtain full speed although all cells are processed in one single loop owing to the 1D list, thus, giving long vector length also for much smaller domains. In contrast to cache-based architectures, the array-of-structure data layout [23, 26] is more appropriate for the NEC SX-8 than the structure-of-array layout, in particular if many OpenMP threads are used. For

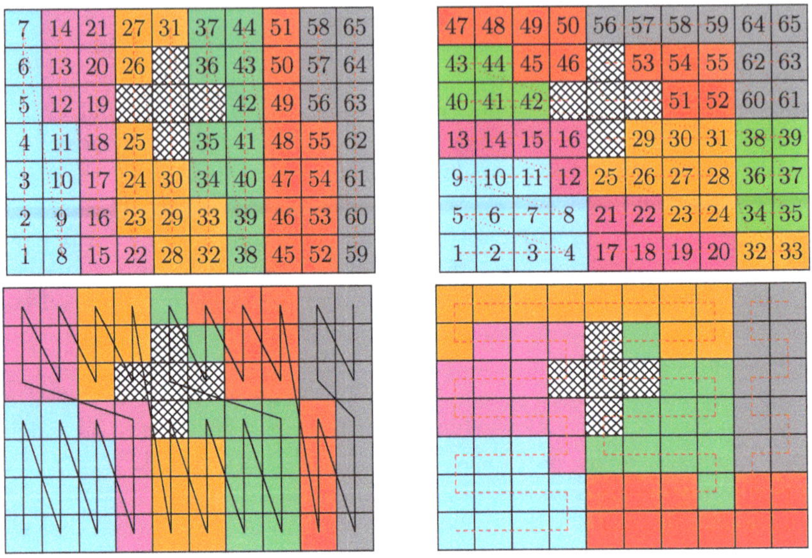

Fig. 2. Schematic representation of different approaches for the local numbering and the resulting partitioning in 6 domains

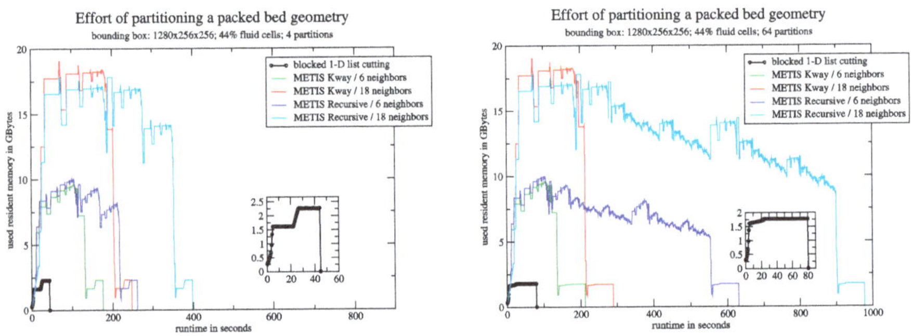

Fig. 3. Partitioning effort to generate 4 or 64 partitions: different METIS methods vs. simple 1-D list cutting. Packed bed of spheres with bounding box 1280x256x256

the structure-of-array layout bank conflicts are very pronounced if the leading dimension is a power of two, in particular in the case of a domain size of 128^3. This corresponds to cache trashing effects on other architectures.

Figures 5 and 6 show the scalability of different parallel computers for a fixed total domain size of only 100^3 as function of the number of MPI processes (or OpenMP threads in case of the NEC SX-8). On all clustered systems, the reported numbers are for the case where all cores of the nodes have been utilized – although using only some cores per node usually gave better numbers – as users usually have to pay on a per-node basis. It is remarkable how well all systems behave in this strong scaling experiment with only 1M cells in total

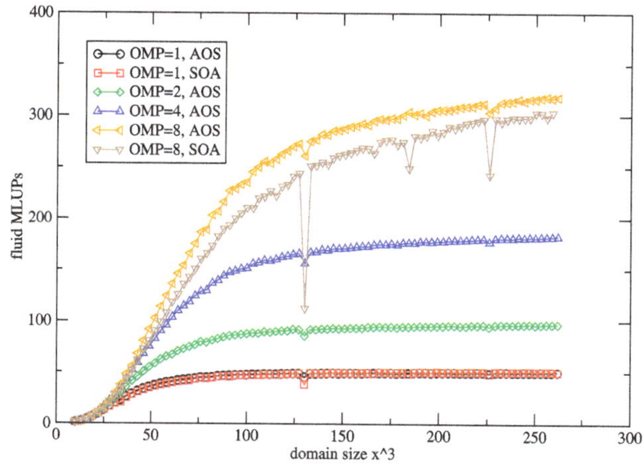

Fig. 4. Sustained performance of one single NEC SX-8 node as function of the domain size and the number of OpenMP threads

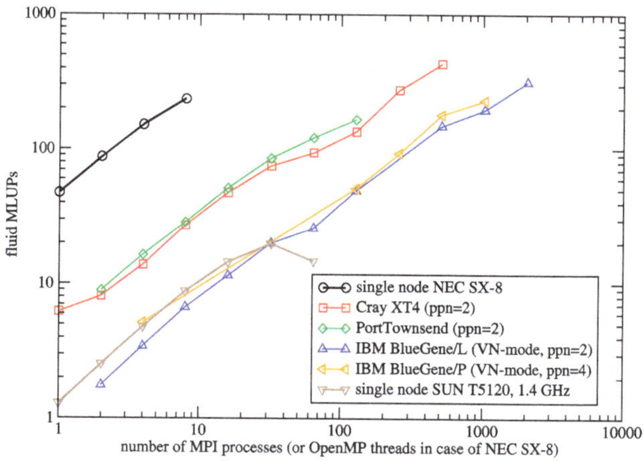

Fig. 5. Sustained aggregated performance on different systems as function of the number of MPI processes for a fixed global domain size of only 100^3

even for large number of processes or nodes, respectively. The straight lines in Fig. 5 might lead to the assumption that scalability is perfect. However, the per-core performance shown in Fig. 6 reveals that efficiency continuously drops as the number of cores is increased owing to the fixed total amount of work and additional efforts for exchanging partition boundaries.

Up to 32 MPI processes, the single one-socket Sun T5120 system with Sun's new multi-core multi-threaded Niagara2 processor can keep up with 16 BlueGene/L or 8 BlueGene/P nodes – which probably are much more expensive and consume in total much more power than this single high-end workstation server.

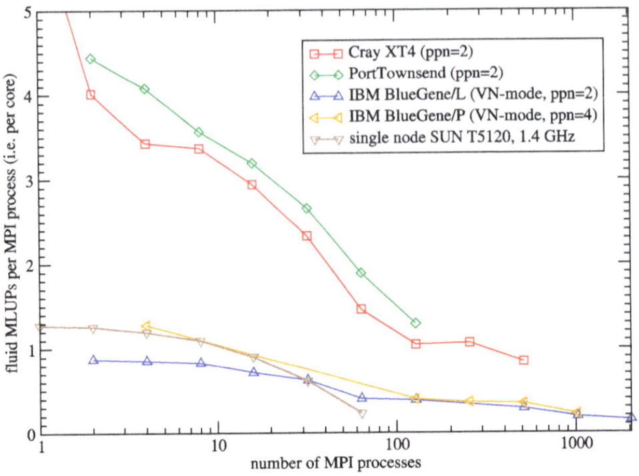

Fig. 6. Sustained performance per core on different systems as function of the number of MPI processes for a fixed global domain size of only 100^3

The performance of the *Port Townsend* cluster installed at RRZE is quite comparable with the Cray XT4. Unfortunately, RRZE's *Port Townsend* cluster consists only of 64 nodes, thus, performance studies could not be extended to larger partition numbers.

The NEC SX-8 is a class of its own. Multi-node experiments have not been carried out for this special case as the total domain size is far too small to still give reasonable performance numbers beyond one node. On the other hand, larger domains would not fit into the main memory of some of the clustered systems when using only few nodes.

5 Conclusions and Outlook

Scalability and sustained performance are not directly connected to each other. It is well known that *slow* code scales much better than a code with highly optimized single processor performance owing to the ratio of computation to communication time. Similar rules apply when comparing parallel computers with slow and powerful processors / single nodes. The results presented above clearly demonstrate for the ILBDC solver that the NEC SX-8 is a class of its own. However, clustered systems which state-of-the-art CPUs scale quite well if a good interconnect (DDR Infiniband or better) is used. For a fixed total amount of work, the per-core efficiency gradually drops as the number of cores is increased, although the sustained aggregated performance still increases. Other applications, e.g. the molecular dynamics code *Amber*, will not scale at all beyond just a few dozens of MPI processes. The single processor or single node performance of the IBM BlueGene systems is (as expected) very low. Using huge numbers of BlueGene nodes, the performance level of faster systems can be reached in the case of ILBDC, however, domain decomposition and load balancing becomes more and more complicated as the number of partitions increases. For other applications (e.g. *Amber*) a BlueGene system will never reach the performance level of even simple commodity clusters owing to scalability constraints. Except on the NEC SX-8, the experience with a hybrid approach (i.e. OpenMP within a node and MPI between nodes) was rather disappointing. Pure MPI with the explicit separation of memory accesses not only gave advantages in the case of ccNUMA nodes. It seems to be questionable if hybrid approaches are really the way to go for parallel computers consisting of SMP nodes with multi-core CPUs as e.g. suggested in [17].

Related studies, including many low level benchmarks can be found in complementary work of our group [10–12, 14, 26, 28].

Currently, the joint proposal *SKALB* of several partners is under review for the BMBF call *HPC-Software für skalierbare Parallelrechner* which hopefully allows to extend the present investigations on a large variety of systems, including the vector facilities at HLRS.

Acknowledgments

The scaling benchmarks presented in this report were measured on the NEC SX-8 system at the High Performance Computing Center Stuttgart (HLRS) within the project *lba-diff*, on the IBM BlueGene L/P systems at the Jülich Supercomputing Centre (JSC), the IBM BlueGene P system of the Rechenzentrum Garching (RZG) of the Max Planck Society, the Cray XT4 system of the National Energy Research Scientific Computing Center (NERSC) at the Lawrence Berkeley National Laboratory, the Sun Niagara2 cluster at RWTH Aachen and RRZE's own resources. We thank Rick Hetherington, Denis Sheahan and Ram Kunda from Sun Microsystems for valuable discussions and early access to pre-production Sun Niagara2-based systems.

This work is financially supported since 2000 through the framework of the *Competence Network for Technical, Scientific High Performance Computing* in Bavaria (KONWIHR). The 1-D list based code is the result of joined work of several partners within the *International Lattice Boltzmann Development Consortium* (ILBDC).

References

1. L. Axner, J. Bernsdorf, T. Zeiser, P. Lammers, J. Linxweiler, and A.G. Hoekstra. Performance evaluation of a parallel sparse lattice Boltzmann solver. *J. Comput. Phys.*, 227(10):4895–4911, 2008.
2. P. Bhatnagar, E.P. Gross, and M.K. Krook. A model for collision processes in gases. I. small amplitude processes in charged and neutral one-component systems. *Phys. Rev.*, 94(3):511–525, 1954.

3. M. Bouzidi, M. Firdaouss, and P. Lallemand. Momentum transfer of a Boltzmann-lattice fluid with boundaries. *Phys. Fluids*, 13(11):3452–3459, 2001.
4. S. Chen and G.D. Doolen. Lattice Boltzmann method for fluid flows. *Annu. Rev. Fluid Mech.*, 30:329–364, 1998.
5. D. d'Humières, I. Ginzburg, M. Krafczyk, P. Lallemand, and L.-S. Luo. Multiple-relaxation-time lattice Boltzmann models in three dimensions. *Phil. Trans. R. Soc. Lond. A*, 360(1792):437–452, 2002.
6. B. Ferreol and D.H. Rothman. Lattice-Boltzmann simulations of flow through Fontainebleau sandstone. *Transport in Porous Media*, 20:3–20, 1995.
7. I. Ginzburg. Equilibrium-type and link-type lattice Boltzmann models for generic advection and anisotropic-dispersion equation. *Advances in Water Resources*, 28(11):1171–1195, 2005.
8. I. Ginzburg and D. d'Humières. Multireflection boundary conditions for lattice Boltzmann models. *Phys. Rev. E*, 68(6):066614_30, 2003.
9. J. Götz. Numerical simulation of bloodflow in aneurysms using the lattice Boltzmann method. Master thesis, Lehrstuhl für Informatik 10 (Systemsimulation), Universität Erlangen-Nürnberg, 2006.
10. G. Hager, H. Stengel, T. Zeiser, and G. Wellein. RZBENCH: Performance evaluation of current HPC architectures using low-level and application benchmarks. In *HLRB/KONWIHR Results and Review Workshop, LRZ-Munich, Garching, Dec. 3-4, 2007, http://arxiv.org/abs/0712.3389*, Berlin, Heidelberg, in press, 2008. Springer-Verlag.
11. G. Hager and G. Wellein. Architecture and performance characteristics of modern high performance computers. In H. Fehske, R. Schneider, and A. Weiße, editors, *Computational Many-Particle Physics*, volume 739 of *Lecture notes in Physics*. pages 681–730, Berlin, Heidelberg, 2008. Springer-Verlag.
12. G. Hager and G. Wellein. Optimization techniques for modern high performance computers. In H. Fehske, R. Schneider, and A. Weiße, editors, *Computational Many-Particle Physics*, volume 739 of *Lecture notes in Physics*, pages 731–767, Berlin, Heidelberg, 2008. Springer-Verlag.
13. G. Hager, T. Zeiser, J. Treibig, and G. Wellein, Optimizing performance on modern HPC systems: Learning from simple kernel benchmarks. In E. Krause, Y. Shokin, M. Resch, and N. Shokina, editors, *Computational Science and High Performance Computing II: The 2nd Russian-German Advanced Research Workshop, Stuttgart, Germany, March 14 to 16, 2005*, volume 91 of *Notes on Numerical Fluid Mechanics and Multidisciplinary Design (NNFM)*, pages 273–287, Berlin, Heidelberg, 2006. Springer-Verlag.
14. G. Hager, T. Zeiser, and G. Wellein. Data access optimizations for highly threaded multi-core CPUs with multiple memory controllers. In *Workshop on Large-Scale Parallel Processing 2008 (IPDPS2008), Miami, FL, April 18, 2008, http://arxiv.org/abs/0712.2302*, 2008.
15. X. He and L.-S. Luo. Lattice Boltzmann model for the incompressible Navier-Stokes equation. *J. Stat. Phys.*, 88(3/4):927–944, 1997.
16. X. He and L.-S. Luo. Theory of the lattice Boltzmann method: From the Boltzmann equation to the lattice Boltzmann equation. *Phys. Rev. E*, 56(6):6811–6817, 1997.
17. V. Heuveline, M.J. Krause, and J. Latt. Towards a hybrid parallelization of lattice Boltzmann methods. *Proceedings of ICMMES2007*, 2008.
18. G. Karypis and V. Kumar. METIS: Serial graph partitioning and fill-reducing matrix ordering, 1998.

19. J.D. McCalpin. STREAM: Sustainable memory bandwidth in high performance computers, 1991-2007.
20. S. Succi. *The Lattice Boltzmann Equation – For Fluid Dynamics and Beyond.* Clarendon Press, 2001.
21. Sun Microsystems. OpenSPARC T2 core microarchitecture specification. Technical report, 2007.
22. Sun Microsystems. private communication, 2008.
23. G. Wellein, T. Zeiser, S. Donath, and G. Hager. On the single processor performance of simple lattice Boltzmann kernels. *Computers & Fluids*, 35(8-9):910–919, 2006.
24. D. Yu, R. Mei, L.-S. Luo, and W. Shyy. Viscous flow computations with the method of lattice Boltzmann equation. *Progr. Aero. Sci.*, 39:329–367, 2003.
25. T. Zeiser. Combination of detailed CFD simulations using the lattice Boltzmann method and experimental measurements using the NMR/MRI technique. In E. Krause, W. Jäger, and M. Resch, editors, *High Performance Computing in Science and Engineering '04, Transactions of the High Performance Computing Center, Stuttgart (HLRS) 2004*, pages 277–292, Berlin, Heidelberg, 2005. Springer-Verlag.
26. T. Zeiser. *Simulation von durchströmten Schüttungen auf Hochleistungsrechnern.* PhD thesis, Technische Fakultät, Universität Erlangen-Nürnberg, 2008.
27. T. Zeiser, M. Bashoor-Zadeh, A. Darabi, and G. Baroud. Pore-scale analysis of Newtonian flow in the explixit geometry of vertebral trabecular bone using lattice Boltzmann simulation. *J. Eng. Med., Proc. Inst. Mech. Eng. Part H*, 222(2):185–194, 2008.
28. T. Zeiser, J. Götz, and M. Stürmer. On performance and accuracy of lattice Boltzmann approaches for single phase flow in porous media. In *Computational Science and High Performance Computing – Russian-German Adwanced Research Workshop, Novosibirk, Russia, July 2007*, Berlin, Heidelberg, in press. Springer-Verlag.
29. T. Zeiser, G. Wellein, G. Hager, S. Donath, F. Deserno, P. Lammers, and M. Wierse. Optimized lattice Boltzmann kernels as testbeds for processor performance. Technical report, Regionales Rechenzentrum Erlangen, Universität Erlangen-Nürnberg, May 2004.

Numerical Modeling of Fluid Flow in Porous Media and in Driven Colloidal Suspensions

Jens Harting, Thomas Zauner, Rudolf Weeber, and Rudolf Hilfer

Institut für Computerphysik, Pfaffenwaldring 27, 70569 Stuttgart, Germany

Summary. This article summarizes some of our main efforts performed on the computing facilities provided by the high performance computing centers in Stuttgart and Karlsruhe. At first, large scale lattice Boltzmann simulations are utilized to support resolution dependent analysis of geometrical and transport properties of a porous sandstone model. The second part of this report focuses on Brownian dynamics simulations of optical tweezer experiments where a large colloidal particle is dragged through a polymer solution and a colloidal crystal. The aim of these simulations is to improve our understanding of structuring effects, jamming behavior and defect formation in such colloidal systems.

1 Resolution Dependent Analysis of Geometrical and Transport Properties of a Porous Sandstone Model

Geometrical characterization of porous media and the calculation of transport parameters present an ongoing challenge in many scientific areas such as petroleum physics, environmental physics (aquifers), biophysics (membranes) and material science. We perform large scale lattice-Boltzmann simulations to investigate the permeability of computer generated samples of quartzitic sandstone at different resolutions. To obtain these laboratory scale samples a continuum model is discretized at different resolutions. This allows to obtain high precision estimates of the permeability and other quantities by extrapolating the resolution dependent results.

1.1 Simulation Method and Implementation

The lattice-Boltzmann (hereafter LB) simulation technique is based on the well-established connection between the dynamics of a dilute gas and the Navier-Stokes equations [3]. We consider the time evolution of the one-particle velocity distribution function $n(\mathbf{r}, \mathbf{v}, t)$, which defines the density of particles with velocity $\mathbf{v}$ around the space-time point $(\mathbf{r}, t)$. By introducing the assumption of molecular chaos, i.e. that successive binary collisions in a dilute gas are uncorrelated, Boltzmann was able to derive the integro-differential equation for n named after him [3]

$$\partial_t n + \mathbf{v} \cdot \nabla n = \left(\frac{dn}{dt} \right)_{coll} , \tag{1}$$

where the right hand side describes the change in n due to collisions.

The LB technique arose from the realization that only a small set of discrete velocities is necessary to simulate the Navier-Stokes equations [4]. Much of the kinetic theory of dilute gases can be rewritten in a discretized version. The time evolution of the distribution functions n is described by a discrete analogue of the Boltzmann equation [10]:

$$n_i(\mathbf{r} + \mathbf{c}_i \Delta t, t + \Delta t) = n_i(\mathbf{r}, t) + \Delta_i(\mathbf{r}, t) , \tag{2}$$

where Δ_i is a multi-particle collision term. Here, $n_i(\mathbf{r}, t)$ gives the density of particles with velocity $\mathbf{c}_i$ at $(\mathbf{r}, t)$. In our simulations, we use 19 different discrete velocities $\mathbf{c}_i$. The hydrodynamic fields, mass density ϱ and momentum density $\mathbf{j} = \varrho \mathbf{u}$ are moments of this velocity distribution:

$$\varrho = \sum_i n_i , \qquad \mathbf{j} = \varrho \mathbf{u} = \sum_i n_i \mathbf{c}_i . \tag{3}$$

We use a linear collision operator,

$$\Delta_i = -\frac{1}{\tau}(n_i - n_i^{eq}) , \tag{4}$$

where we assume that the local particle distribution relaxes to an equilibrium state n_i^{eq} at a single rate τ [1]. By employing the Chapman-Enskog expansion [3, 4] it can be shown that the equilibrium distribution

$$n_i^{eq} = \varrho \omega^{c_i} \left[1 + 3\mathbf{c}_i \cdot \mathbf{u} + \frac{9}{2}(\mathbf{c}_i \cdot \mathbf{u})^2 - \frac{3}{2}u^2 \right] , \tag{5}$$

with the coefficients ω^{c_i} corresponding to the three different absolute values $c_i = |\mathbf{c}_i|$,

$$\omega^0 = \frac{1}{3} , \qquad \omega^1 = \frac{1}{18} , \qquad \omega^{\sqrt{2}} = \frac{1}{36} , \tag{6}$$

and the kinematic viscosity

$$\nu = \frac{\eta}{\varrho_f} = \frac{2\tau - 1}{6} , \tag{7}$$

properly recovers the Navier-Stokes equations

$$\frac{\partial u}{\partial t} + (u\nabla)u = -\frac{1}{\varrho}\nabla p + \frac{\eta}{\varrho}\Delta u \,, \qquad \nabla u = 0 \,. \tag{8}$$

We use LB3D [6], a highly scalable parallel LB code, to implement the model. LB3D is written in Fortran 90 and designed to run on distributed-memory parallel computers, using MPI for communication. It can handle up to three different fluid species and is able to model flow in complex geometries as it occurs for example in porous media. In each simulation, the fluid is discretized onto a cubic lattice, each lattice point containing information about the fluid in the corresponding region of space. Each lattice site requires about a kilobyte of memory per lattice site so that, for example, a simulation on a 128^3 lattice would require around 2.2GB of memory. The code runs at over $6 \cdot 10^5$ lattice site updates per second per CPU on a recent machine, and has been observed to have roughly linear scaling up to order $3 \cdot 10^3$ compute nodes (see below).

Simulations on larger scales have not been possible so far due to the lack of access to a machine with a higher processor count. The largest simulation we performed used a 1536^3 lattice and ran on the AMD Opteron based cluster in Karlsruhe. There, it is not possible to use a larger lattice since the amount of memory per CPU is limited to 4GB and only 1024 processes are allowed within a single job. On the NEC SX-8 in Stuttgart, typical system sizes are of the order of $256 \times 256 \times 512$ lattice sites. The output from a simulation usually takes the form of a single floating-point number for each lattice site, representing, for example, the density of a fluid at that site. Therefore, a density field snapshot from a 128^3 system would produce output files of around 8MB.

Writing data to disk is one of the bottlenecks in large scale simulations. If one simulates a 1024^3 system, each data file is 4GB in size. The situation gets even more critical when it comes to the files needed to restart a simulation. Then, the state of the full simulation lattice has to be written to disk requiring 0.5TB of disk space. LB3D is able to benefit from the parallel file systems available on many large machines today, by using the MPI-IO based parallel HDF5 data format [7]. Our code is very robust regarding different platforms or cluster interconnects: even with moderate inter-node bandwidths it achieves almost linear scaling for large processor counts with the only limitation being the available memory per node. The platforms on which our code has been successfully used include various supercomputers like the NEC SX-8, IBM pSeries, SGI Altix and Origin, Cray T3E, Compaq Alpha clusters, as well as low cost 32- and 64-bit Linux clusters.

During the last year, a substantial effort has been invested to improve the performance of LB3D and to optimize it for the simulation of flow in porous media. Already during the previous reporting period, we improved the performance on the SX8 in Stuttgart substantially by rearranging parts of the code and by trying to increase the length of the loops. These changes were proposed by the HLRS support staff. However, while the code scales very well with the number of processors used, the single CPU performance is still below what one could expect from a lattice Boltzmann implementation on a vector machine. The vector operation ratio is about 93%, but due to the inherent structure of our multiphase implementation, the average loop length is only between 20 and 30. Thus, the performance of our code stays below 1GFlop/s. For this reason, we are currently performing most of our simulations on the Opteron cluster *XC2* in Karlsruhe. Our code performs extremely well there and shows almost linear scaling to up to 1024 CPU's. Further, due to extensive improvements of the code during the last year, we were able to increase the per CPU performance by a factor of about two.

The lattice Boltzmann code can read voxel based 3D representations of porous media. Such data can either be obtained from XMT measurements of real samples or from numerical models. In order to compute the permeability of such a sample, the velocity field $v(x)$ and pressure field $p(x)$ as created by the LB simulations are required. For a liquid with dynamic viscosity η, the permeability κ is defined according to Darcy's law

$$\langle v(x)\rangle_{x\in S} = -\frac{\kappa}{\eta}\langle \nabla p\rangle_{x\in S} \tag{9}$$

and thus

$$\kappa = -\eta \frac{\langle v(x)\rangle_{x\in S}}{\langle \nabla p\rangle_{x\in S}}, \tag{10}$$

with $\langle v(x)\rangle_{x\in S}$ being the velocity component in flow direction averaged over the full pore space S. We approximate the average pressure gradient $\langle \nabla p\rangle_{x\in S}$ of the full sample like

$$\langle \nabla p\rangle_{x\in S} \approx \frac{\langle p(x)\rangle_{x\in OUT} - \langle p(x)\rangle_{x\in IN}}{aL}, \tag{11}$$

with IN/OUT representing the plane, perpendicular to the flow direction, in front of / behind the porous medium, the total length L (in voxel) and a the resolution. The accelerating force, in positive z-direction, is applied as a body force only in the first quarter of the "IN-flow" buffer before the sample, another buffed "OUT-flow" was added after the sample. Periodic boundary conditions in flow direction are being imposed. Fig. 1 shows the simulation setup for a sample with resolution $a = 10\mu m$ and voxel dimension 136x136x168 and the two "IN/OUT-flow" buffers.

1.2 Sample Creation and Simulation Setup

Because appropriately sized digital samples at sufficient resolutions are not available from experimental data, a continuum model of a quartzitic sandstone was discretized at different resolutions and sizes (Table 1) and then thresholded to generate digital voxel data. For details explaining the model see [2]. With increasing resolution a the microstructure becomes more and more resolved at the expense of increasing the amount of data and CPU time

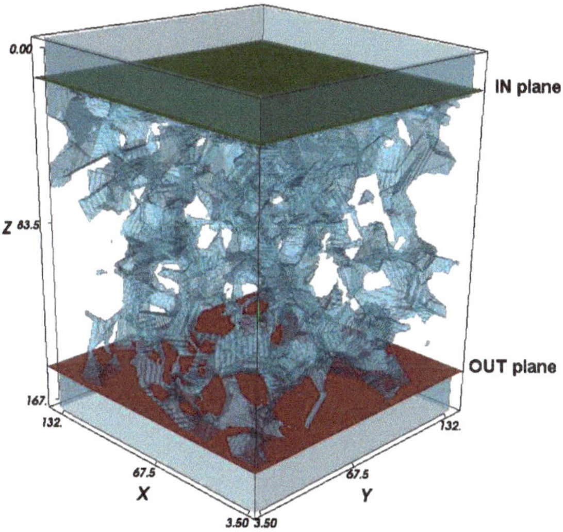

Fig. 1. Porous medium with pore space (blue) and the two IN/OUT-flow buffers shown. The flow is in positive z-direction, the sample size is 136x136x168 voxel, at a resolution of $a = 10\mu m$

in simulations. To find a good compromise between a high enough resolution and manageable systems for LB-simulations, geometrical characterizations for each of our samples were calculated (See Fig. 3,4).

Table 1. List of available digitized samples

Sidelength [voxel]	Resolution $a[\mu m]$	Number of samples
256	80	1
256	40	1
512/256	20	1/8
512/256	10	8/16
512/256	5	8/16
512/256	2.5	8/16

Typical lattice-Boltzmann simulation required 50,000–80,000 simulation steps to reach stationary flow within the pore space. Simulations were performed on 64 to 512 CPU's. For the total of 83 samples (including calibration test runs) more than 80,000 CPU hours were required. The LB3D code scales linearly and memory usage was moderate, making investigations of larger samples feasible and tempting. Fig. 2 shows the average time for one simulation step and per voxel in nano seconds (left axis). The average time was calculated from more than 100 runs on different systems with varying parameters and sizes. The speedup factor is defined as

$$speedup(n) = \frac{(total\ runtime\ on\ 4\ CPU's)}{(total\ runtime\ on\ n\ CPU's)}$$

The total runtime of a simulation is the time the program needs for it's full execution. The speedup factor was defined with reference to a run on 4 CPU's because on 4-core systems one physical node has 4 CPU's. The communication and I/O overhead for a simulation run on one physical node is negligible, compared to network communications.

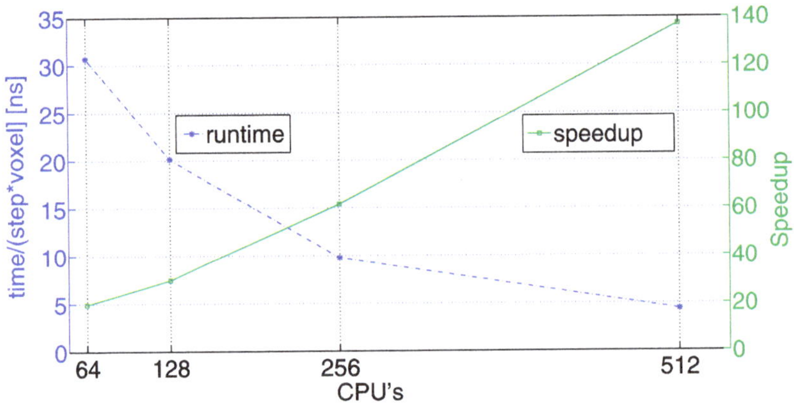

Fig. 2. LB3D code scaling behavior. The sequential lattice Boltzmann code is known to scale linearly with the sample size (voxel). The time per simulation step and per voxel is shown on the left axis. The speedup, shown on the right axis

1.3 Geometric Properties at Different Resolutions

For each sample geometrical properties such as the porosity, specific surface and mean/total curvature have been calculated to investigate their behavior with changing resolution and subsampling. From the resolution dependent porosity $\phi(a)$ (Fig. 3) conclusions can be drawn which resolution ranges approximate the true physical porosity well enough and thus, are suitable for simulations. Increasing the resolution (lower a) beyond a certain point (here approx. $a = 5 - 10\mu m$) will not justify the increase in data and CPU time but on the other hand low resolutions ($a = 50 - 80\mu m$) will most likely not yield relevant simulation data, because not even the sample porosity is close to the true porosity of the physical sandstone. As can be seen in (Fig. 3), subsamples much smaller than $256\mu m$ will not represent the full sample well. The data

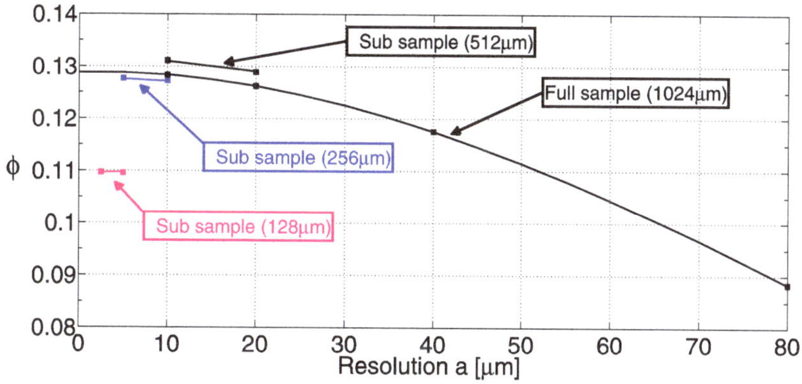

Fig. 3. Porosity ϕ for the full sample and subsamples at different resolutions a. The data for the full sample is extrapolated to the true physical porosity at a=0. With increasing resolution the porosity approaches the true physical porosity. Large enough subsamples, $512\mu m$ and $256\mu m$, approximate the full sample (size $1024\mu m$) porosity well. Subsamples at $128\mu m$ are not representative for the full sample

for the full sample (black line) can be extrapolated to approximate the true physical porosity as $a \to 0$.

Local porosity distributions [8, 9] are defined as

$$\mu(\phi, L, a) = \frac{1}{m} \sum_{x \in S} \delta(\phi - \phi(x, L, a))$$

where $\phi(x, L, a)$ is the local porosity of a measurement cell with sidelength L at the position x and a sample resolutions a. S is the full sample, m is the number of evaluated measurement cells and δ is the Dirac delta function.

To calculate this local porosity distribution a small cubic measurement cell, with sidelength L, is being moved through the whole sample and at all positions x, the local porosity $\phi(x, L, a)$ within the cell is calculated.

The local porosity distributions $\mu(\phi, L, a)$, as shown in Fig. 4, give the probability density that a randomly placed measurement cell with sidelength L has a porosity ϕ when the sample resolution is a. In our case they have a maximum close to the porosity of the full sample at that resolution. For resolutions $a = 40, 20, 10\mu m$ the distributions are well converged. Together with other geometric properties defined within Local porosity theory $\mu(\phi, L, a)$ can be used to estimate the permeability [8].

The lattice Boltzmann simulations calculated the velocity field and pressure field for all available samples with different resolutions. In Fig. 5 the z-Component v_z, component in flow direction, of the velocity field for a sample with resolution $a = 10\mu m$ and voxel size 136x136x168 is shown. The pore structure of the same sample is shown in Fig. 1. The "IN-flow" buffer, where the liquid is accelerated to a speed of approx 0.002 lattice units, is shown in

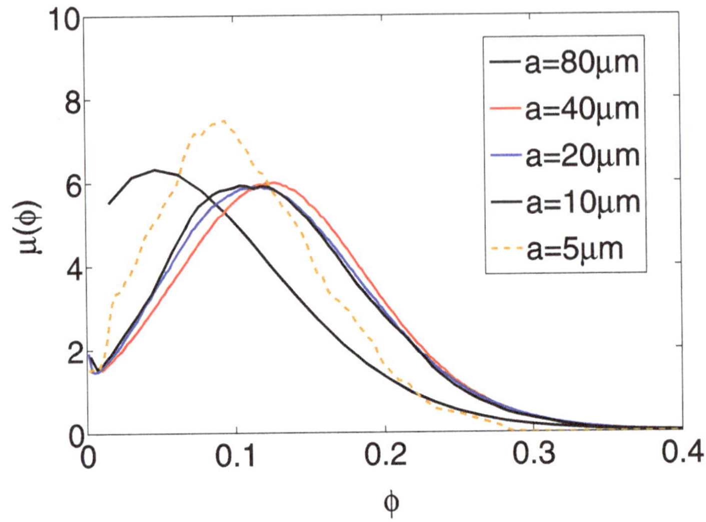

Fig. 4. Local porosity distributions for a measurement cell size $L = 320\mu m$ at different resolutions a. Very high resolutions ($a = 5\mu m$) yield bad statistics, at the size of the measurement cell used here. The distributions for resolutions $a = 40, 20, 10\mu m$ are very similar in shape, the mean porosity changes according to Fig. 3

green at the top. All voxel with $v_z > 0$ are shown in translucent blue. The brown isosurfaces depict areas where $v_z > 0.001$ in lattice units, being half of the maximum speed in the "IN-flow" buffer. They represent channels where the liquid flows fast. The flow through the porous medium is not homogeneous, even if the porous medium is quite homogeneous at this resolution.

In addition to the geometric characterizations and the velocity fields high precision permeability calculations (Tbl. 2) for all 83 samples have been performed and are now been critically analyzed for accuracy, to gain further insight into their resolution dependence and thereby into the nature of fluid transport within highly complex geometries.

Table 2. Selected permeability results of the full sample ($1024\mu m$) for different resolutions. An relative error of 0.05 has been estimated, resulting from the inaccuracy of the velocity field, calculated by the LB-simulation

Resolution $[\mu m]$	Permeability $[\mu m^2]$
40	1.7±0.043
20	2.2±0.055
10	1.9±0.047

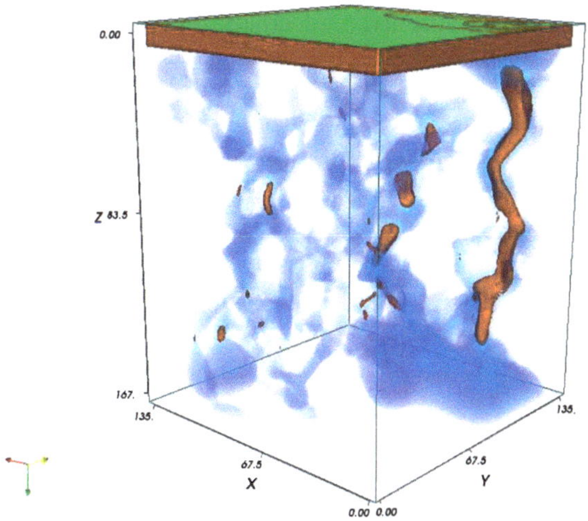

Fig. 5. Z-Component of the velocity field. Fluid is accelerated in the IN-flow buffer on top(green). Shown in translucent blue is the complete volume fraction with $v_z > 0$. Darker regions correspond to smaller velocities. All areas with a velocity with $v_z > 0.001$ are shown as isosurfaces. Channels with large connected isosurfaces thus carry the main flow through the porous medium

2 Simulation of Optical Tweezer Experiments

A colloidal suspension is a mixture of a fluid and particles or droplets with a length scale of some nanometers to micrometers suspended in it. Colloids are a common part of everyday life. Substances like paint, glue, milk, blood and fog are just some examples. Due to their technical applications, colloids are studied in several disciplines, among them physics, chemistry, and engineering.

Colloidal particles are too large to be affected directly by quantum mechanical effects. On the other hand, they are still small enough to be affected by thermal fluctuations. Therefore, colloidal suspensions are an interesting system to study thermodynamic phenomena like diffusion, phase transitions and more rare phenomena like stochastic resonance and critical Casimir forces. In solid state physics, colloidal crystals are used as model systems to study defect formation, crystal structures and melting. In contrast to many systems in these fields, which are on the nanometer scale, colloidal suspensions can be observed and manipulated directly using techniques like video microscopy, confocal microscopy, total internal reflection microscopy (TIRM) and optical tweezers. This offers numerous possibilities to control these systems on a per particle basis.

We study dynamical properties of colloidal suspensions using computer simulations. The advantage of simulations is that parameters can be controlled in ways that are not accessible in experiments. Also, in many cases, information is available that cannot be measured directly in a real system. Over the past decades, several simulation techniques have become available, which model different aspects of suspensions.

Methods like molecular dynamics and Brownian dynamics only model the dynamics of the suspended colloidal particles and handle the solvent implicitly by adding simplified forces to mimic the solvents behavior. Other approaches like lattice-Boltzmann models, dissipative particle dynamics and stochastic rotation dynamics simulate the complete fluid and couple it to suspended particles. While the first set of methods are computationally very efficient, more complicated hydrodynamic effects are usually not taken into account (polarization of the solvent). The methods that do simulate the full fluid field can reproduce hydrodynamic effects, but achieving quantitative accuracy far from equilibrium is still a challenge. Here, we focus on the Brownian dynamics technique in order to be able to simulate very large systems with an affordable amount of CPU time.

The dynamics of driven suspensions can be examined by dragging a colloidal particle through it. In this article we present our simulations of a particle dragged through a colloidal crystal and a suspension of coiled polymers using an optical tweezer. The focus of the optical tweezer is moved with time, thereby pulling the impurity along.

Optical tweezers trap a colloid (or even an atom) in the focus of a laser beam: this is because a dielectric is always driven along the field gradient. They are a very important tool in soft condensed matter physics: colloids can not only be trapped, they can also be moved around individually by moving the focus of the laser beam. Thereby the colloidal system can be controlled with an accuracy that would be impossible for an atomic system. The optical tweezer is modeled with a harmonic potential, i.e. as if the impurity were connected with the trap center by an ideal spring.

The simulations are motivated by experiments performed by R. Dullens in the group of C. Bechinger in Stuttgart and C. Gutsche in the group of F. Kremer in Leipzig, respectively. In both cases, the simulation parameters are chosen to reproduce the experimental conditions as closely as possible.

2.1 Simulation Setup

The experiments are simulated using a modified Brownian dynamics (BD) method which includes some of the hydrodynamics caused by the dragged colloid, as explained in Ref. [11]. The colloids/polymers and the probe particle are modeled as hard spheres with their respective radii.

We use a rectangular simulation volume with periodic boundary conditions in all three directions. Due to long range hydrodynamic interactions, large systems are required in order to reduce finite size effects. Thus, we typically handle several hundred thousand particles in a single simulation.

The probe particle is trapped in a moving parabolic potential $V(r) = \frac{1}{2}\,a\,r^2$, mimicking the optical tweezer. In the case of the polymer suspension, the potential has a spring constant of $a = 7.5 \times 10^{-5}$ pN/nm, which gives a better signal to noise ratio than the experimental value of 8.5×10^{-2} pN/nm. Figure 6 shows a cut out of a snapshot of our simulation setup used to describe the experiments performed by the group in Leipzig.

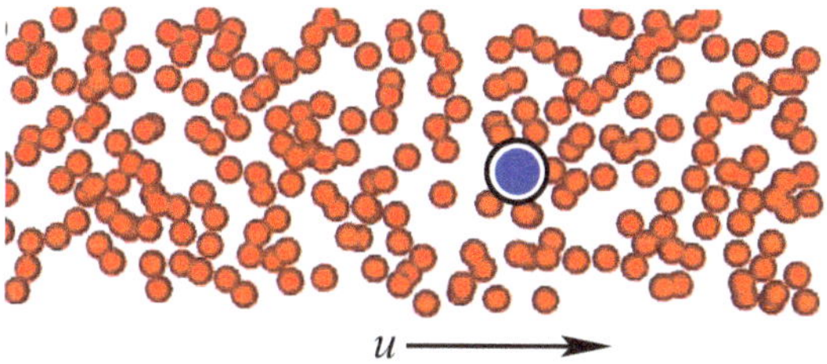

Fig. 6. A cut through the simulated system, where a probe particle is dragged through a suspension (from [5]). The arrow indicates the direction of motion of the probe particle

In conventional BD, the two most important aspects of hydrodynamics felt by the suspended particles are taken into account, namely the Stokes friction and the Brownian motion. Correspondingly, this is done by adding to a molecular dynamics simulation two additional forces. The Langevin equation describes the motion a Brownian particle with radius R at position $\mathbf{r}(t)$ as

$$m\,\ddot{\mathbf{r}}(t) = 6\,\pi\,\eta\,R\,\dot{\mathbf{r}}(t) + \mathbf{F}_{\mathrm{rand}}(t) + \mathbf{F}_{\mathrm{ext}}(\mathbf{r},t), \tag{12}$$

where the first term models the Stokes friction in a solvent of viscosity η, $\mathbf{F}_{\mathrm{ext}}(\mathbf{r},t)$ is the sum of all external forces like gravity, forces exerted by other suspended particles, and, for the colloid, the optical trap. $\mathbf{F}_{\mathrm{rand}}(t)$ describes the thermal noise which gives rise to the Brownian motion. The random force on different particles is assumed to be uncorrelated, as well as the force on the same particle at different times. It is further assumed to be Gaussian with zero mean. The mean square deviation of the Gaussian (i.e., the amplitude of the correlator) is given by the fluctuation-dissipation theorem as

$$\langle |\mathbf{F}_{\mathrm{rand}}|^2 \rangle = 12\,\pi\,\eta\,R\,k_B T. \tag{13}$$

This conventional BD scheme is widely used to simulate suspensions because it is well understood, not difficult to implement, and needs much less computational resources than a full simulation of the fluid. However, this simulation method does not resolve hydrodynamic interactions between particles. In particular, the long-ranged hydrodynamic interactions between the dragged colloid and the surrounding particles are not modeled. However, in the system we consider, these interactions are important, as the dragged colloid moves quickly and has a strong influence on the flow field around it. Therefore, the BD scheme is modified such that the effect caused by the flow field around the dragged colloid is included. This is achieved by calculating the friction force on the small particles with radius R_c not with respect to a resting fluid ($\mathbf{F} = 6\pi\eta R_c\mathbf{u}$), but with respect to the flow field caused by the moving colloid. The friction force then is

$$\mathbf{F} = 6\pi\eta R_c(\mathbf{u} - \mathbf{v}(\mathbf{r})), \tag{14}$$

where $\mathbf{v}(\mathbf{r})$ is the flow field around the moving colloid at a position $\mathbf{r}$ with respect to the colloid's center. This correction leads to the inclusion of two hydrodynamics-mediated effects. Due to the large component of the flow field along the direction of motion, both, in front and behind the probe, small colloids are dragged along. Also the flow of particles is advected around the moving probe particle, i.e., obstacles are moved out of the way to its sides. Both these effects lead to a reduction of drag force on the driven colloid.

2.2 Investigating the Stiffness and Occurrence of Defects in Colloidal Crystals

To analyze the effects of the disturbance due to the dragged probe particle, we can either measure the distance that the probe stays behind the focus of the optical tweezer - and thereby the force required to drag the impurity through the crystal - or examine the reaction of the crystal itself. With our simulations we show that in a colloidal crystal, the velocity-force relation for the dragged colloidal particle is close to linear despite the complicated surrounding. It is also shown that the inter-colloid potential does have an influence on the drag force, though not as strong as velocity. This is one example of a result that could not be obtained easily in experiments. Using maps of average particle density and defect distributions, we illustrate the effect of the dragged colloidal particle on the crystal structure.

An example of a rather small system is given in Fig. 7. For large tweezer velocities, much larger crystals are needed and for studying the relaxation of the crystal, we also have to simulate for at least a 100 to 300 seconds of real time causing these simulations to cost up to a few thousand CPU hours each.

2.3 Dragging a Colloidal Probe Through a Polymer Suspension

The second system we consider is a suspension of coiled polymers which are modeled as hard spheres. A colloidal particle is dragged through the suspension at high velocities. In the experiment and in the simulation, a drag force is measured that is higher than that calculated from the suspension's viscosity as obtained from a shear rheometer. This increase in drag force can be explained

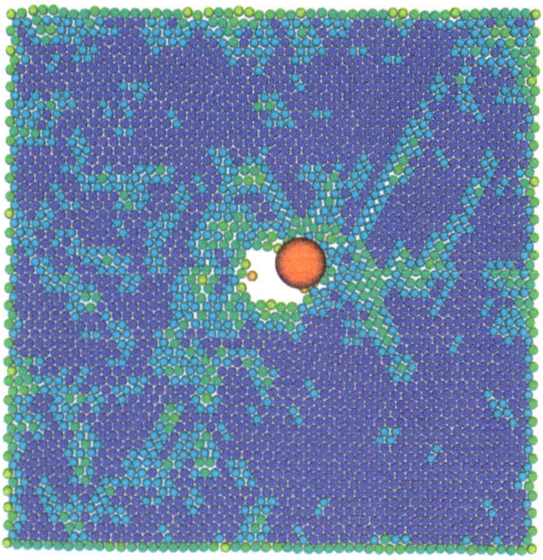

Fig. 7. Simulation of a large colloidal particle dragged through a crystal consisting of smaller colloids by means of an optical tweezer. The coloring denotes defects occurring due to the distortion of the crystal

by a jamming of polymers in front of the moving colloidal particle. In contrast to experiments, this jamming can be observed directly in computer simulations as the positions of the polymers are available. The simulation results are compared to dynamic density functional theory calculations by Rauscher et. al and experimental results by Gutsche et al.. A very good quantitative agreement between experiment, theory and simulation is observed [5].

From the simulation data, it is possible to measure the effective polymer concentration around the dragged colloid. To accomplish this, we take about 2000 two dimensional slices of the simulation and move each snapshot such that the position of the colloid coincides in each snapshot. We calculate the probability for each of the 200×200 bins to be occupied by a polymer by averaging over all snapshots. Polymers accumulate in front of the colloidal particle and the concentration in the back is reduced due to the finite Peclet number of the polymers. For high polymer concentrations, the probability to find a polymer in front of the colloid is close to one. The region right behind the colloid is almost clear of polymers, because the polymers get advected away from the colloid before they can diffuse into this region.

Our model reproduces very well the experimentally found linear relation between drag force and the drag velocity for different polymer concentrations. As higher drag velocities require a larger system (even with periodic boundary conditions) and short numerical time steps, we are limited to about $80\,\mu\mathrm{m/s}$ by the available computational resources and time. However, the linearity of the drag force with respect to the velocity at high enough velocities allows to extrapolate to the higher velocities used in the experiments.

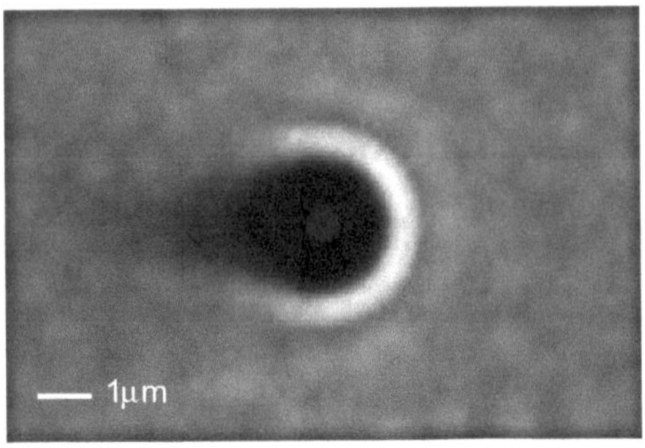

Fig. 8. Polymer density around the colloidal particle averaged over 2000 snapshots of the system. Lighter colors correspond to higher polymer densities. Also visible are density oscillations in front of the colloid, which are characteristic for hard sphere systems

3 Conclusion

In this report we have presented results from lattice Boltzmann simulations of fluid flow in porous media and the simulation of optical tweezer experiments. In particular the AMD Opteron cluster in Karlsruhe has been found to perform particularly well with our simulation codes. In the case of porous media simulations we have demonstrated that we are able to systematically determine the permeability of digitized quartzitic sandstone samples – even if the resolution of the samples is very high resulting in the need of substantial computational resources.

In the second part of this article we reported on our Brownian dynamics simulations of optical tweezer experiments, where a large probe particle is trapped by a laser beam and dragged either through a colloidal crystal or through a polymer suspension. In both cases, quantitative agreement with experimental data was observed.

Acknowledgments

We are grateful to the High Performance Computing Center in Stuttgart and the Scientific Supercomputing Center in Karlsruhe for providing access to their NEC SX-8 and HP 4000 machines. We would like to thank Bibhu Biswal, Peter Diez, and Frank Raischel for fruitful discussions. This work was supported by the collaborative research center 716 and the DFG program "nano- and microfluidics".

References

1. P.L. Bhatnagar, E.P. Gross, and M. Krook. Model for collision processes in gases. I. small amplitude processes in charged and neutral one-component systems. *Phys. Rev.*, 94(3):511, 1954.

2. B. Biswal, P.E. Oren, R. Held, S. Bakke, and R. Hilfer. Stochastic multiscale model for carbonate rocks. *Phys.Rev. E*, 75:061303, 2007.
3. S. Chapman and T.G. Cowling. *The mathematical theory of non-uniform gases.* Cambridge University Press, second edition, 1952.
4. U. Frisch, D. d'Humiéres, B. Hasslacher, P. Lallemand, Y. Pomeau, and J.P. Rivet. Lattice gas hydrodynamics in two and three dimensions. *Complex Systems*, 1(4):649, 1987.
5. C. Gutsche, F. Kremer, M. Krüger, M. Rauscher, J. Harting, and R. Weeber. Colloids dragged through a polymer solution: experiment, theory and simulation. *Submitted for publication*, arXiv:0709.4142, 2007.
6. J. Harting, M. Harvey, J. Chin, M. Venturoli, and P.V. Coveney. Large-scale lattice Boltzmann simulations of complex fluids: advances through the advent of computational grids. *Phil. Trans. R. Soc. Lond. A*, 363:1895–1915, 2005.
7. 2003. HDF5 – a general purpose library and file format for storing scientific data, http://hdf.ncsa.uiuc.edu/HDF5.
8. R. Hilfer. Transport and relaxation phenomena in porous media. *Adv. Chem. Phys.*, XCII:299, 1996.
9. R. Hilfer. Local porosity theory and stochastic reconstruction for porous media. In K. Mecke and D. Stoyan, editors, *Statistical Physics and Spatial Statistics*, volume 554 of *Lecture Notes in Physics*, page 203, Berlin, 2000. Springer.
10. A.J.C. Ladd and R. Verberg. Lattice-boltzmann simulations of particle-fluid suspensions. *J. Stat. Phys.*, 104(5):1191, 2001.
11. M. Rauscher, M. Krüger, A. Dominguez, and F. Penna. A dynamic density functional theory for particles in a flowing solvent. *J. Chem. Phys.*, 127:244906, 2007.

Numerical Characterization of the Reacting Flow in a Swirled Gasturbine Model Combustor

A. Widenhorn, B. Noll, and M. Aigner

Institut für Verbrennungstechnik der Luft- und Raumfahrt, Universität Stuttgart, Pfaffenwaldring 38-40, 70569 Stuttgart, Germany, axel.widenhorn@dlr.de

Summary. In this work the three-dimensional reacting turbulent flow field of a swirl-stabilized gas turbine model combustor was analysed with compressible CFD. For flow analysis the Scale Adaptive Simulation (SAS) turbulence model in combination with the Eddy Dissipation/Finite Rate Chemistry combustion model (EDM/FRC) was applied. The simulations were performed using the commercial CFD software package ANSYS CFX-11.0. Both the numerically achieved time-averaged values of the velocity components and their appropriate turbulent fluctuations (RMS) show a good agreement with the experimental values obtained by Laser Doppler Anemometry (LDA). The same excellent results were found for other flow quantities such as the temperature which was compared to Raman measurements that are obtained for the time-averaged temperature distributions as well as for the appropriate temperature fluctuations. Furthermore, experiments and simulations reveal a precessing vortex core (PVC) with a frequency of 1594Hz at the entry of the combustion chamber. The simulations have been performed on the HP XC4000 system of the High Performance Computing Centre Karlsruhe.

1 Introduction

In order to achieve low levels of pollutants modern combustion systems of industrial gas turbines are operating in swirl stabilized lean premixed mode. The combustor design is usually based on the common injection of air and gaseous fuel in the form of a swirling jet. The central recirculation zone, which arises due to the swirl of the incoming flow, serves to anchor the flame within the combustion zone. However, especially under lean premixed conditions often severe self-excited combustion oscillations arise. These unwanted oscillations are in conjunction with high amplitude pressure oscillations which for example can decrease the lifetime and availability of the gas turbine. Depending on the swirl number swirling flows can exhibit different topologies [1, 2, 3, 4]. A typical flow instability at a high swirl number is the precessing vortex core (PVC). This flow phenomenon can be detected at the outlet of the injector system and exhibits a rotation around the swirl flow axis at a certain frequency. Further typical flow instabilities of swirling flow can arise due to an unsteady vortex breakdown which can have a significant impact on combustion dynamics.

In future the design process of modern gas turbines combustion systems shall rely more and more on numerical simulation. In order to allow a reliable design the CFD methods have to predict accurately the aerodynamics and the combustion driven dynamics. However, there is still a large need for improvements in the field of turbulence and combustion modelling as well as in the definition of appropriate boundary conditions [5, 6, 7, 8]. To achieve this aim unsteady 3D simulations are mandatory. Nowadays different approaches are available to capture unsteady flow fields. These are the Unsteady Reynolds Averaged Simulation (URANS), Large Eddy Simulation (LES) and hybrid RANS/LES methods. The URANS approach, which is commonly used in practical applications, uses complete statistical averaging [9, 10]. This allows the prediction of the time-mean or ensemble-averaged quantities for velocity, temperature and species distribution of non-reacting and reacting flow fields. However, experience shows that the URANS approach can lead to an excess of turbulent dissipation and thus there is the risk that important flow structures are dissipated. LES methods on the other side allow the resolution of part of the turbulent structures. Thus only the smaller structures have to be modelled. However, these models for practical purposes in general require prohibitive high computation times. In order to get high quality results at an acceptable computational effort in the present work hybrid RANS/LES turbulence models like the Detached Eddy Simulation (DES) and the Scale Adaptive Simulation (SAS) were used. Here the computational domain is divided into RANS and LES domains. The potential of the hybrid RANS/LES turbulence models in the simulation of the non-reacting flow fields in a model gas turbine combustion chamber was demonstrated by Widenhorn et al. [11, 12]. Compared to LES which has a similar capability the computational effort is reduced drastically, because the strengths of LES and RANS are combined.

For the simulation of reacting flows the complex combustion processes have to be modelled accurately. This is necessary since for example in a partially premixed flame the changes in the flow and the fuel to air ratio have a direct influence on the rate of combustion. The change in the heat release causes pressure waves which under certain conditions may lead to combustion oscillations. Furthermore, due to chemical-kinetic effects the heat release rate and the location of the flame can be influenced. This has an effect on the pressure field and time lag between the flame front and the air and fuel inlets thus influencing the stability map of the flame. Therefore the accurate prediction of heat release rate and flame location by the combustion model is important. Up to now it is not clear which combustion model can be used in conjunction with SAS or DES turbulence models. In the present work the reacting dynamic turbulent flow field of a model gas turbine combustor was investigated using the SAS turbulence model in combination with different combustion models. The first goal was to analyse the capability and the limits of the combination of the SAS turbulence model with the EDM/FRC combustion model by comparing the simulation data against the experimental data set. The obtained results are reported in this work. Furthermore, the required computational resources were assessed. For the numerical simulations the commercial CFD package ANSYS CFX 11.0 was used.

2 Physical Model

2.1 Conservation Equations

The initial set for the numerical simulation of reacting flows includes the continuity, momentum, energy, turbulence and species equations. In this paper the compressible formulation is used. The equations are given by:

$$\frac{\partial Q}{\partial t} + \frac{\partial (F - F_v)}{\partial x} + \frac{\partial (G - G_v)}{\partial y} + \frac{\partial (H - H_v)}{\partial z} - S \qquad (1)$$

The conservative variable vector Q consists of the density, the velocity components, the total specific energy, the turbulent kinetic energy, the specific dissipation rate and the species mass fractions and is defined as:

$$Q = \left[\bar{\rho}, \bar{\rho}\tilde{u}, \bar{\rho}\tilde{v}, \bar{\rho}\tilde{w}, \bar{\rho}\tilde{E}, \bar{\rho}k, \bar{\rho}\omega, \bar{\rho}\tilde{Y}_i\right]^T \qquad i = 1, 2, \ldots, N_{k-1} \qquad (2)$$

Here, Favre-averaged quantities are used.

F, G and H are the inviscid and F_v, G_v and H_v are the viscous fluxes in x-, y- and z-direction, correspondingly. The vector S in eq. (1) contains the source terms and is defined as:

$$S = [0, S_u, S_v, S_w, S_E, S_k, S_\omega, S_{Y_i}]^T \qquad i = 1, 2, \ldots, N_{k-1} \qquad (3)$$

2.2 Turbulence Modelling

For the closure of the above system of partial differential equations for turbulent flows the Boussinesq hypothesis is used. The required values for the eddy viscosity can be obtained by appropriate turbulence models.

Scale Adaptive Simulation Model (SAS)

In the applied SAS-model the SST-RANS model [13] is used to cover the boundary layer. Depending on the grid used the SAS model switches to a LES-like mode which is desired for example in detached regions. Thus the model allows resolving partially the turbulent spectrum. In contradiction to the standard turbulence models which provide a length scale proportional to the thickness of the shear layer, SAS adjusts dynamically to the length scale of the resolved structures. The length scale of the resolved eddies is taken into account by the introduction of the von Karmann length scale into the turbulence scale equation. This information allows the SAS model to operate in LES-like mode. Thus, usually, in attached boundary layers the RANS model is in operation. The SAS model is based on the k-kL formulation given in Menter and Egorov [14, 15]. Menter [16] transformed the term containing the von Karmann length scale according to the SST model. This transformation results in a modified transport equation for the specific dissipation rate ω of the SST model. The new source term contains two independent scales. In addition to the standard velocity gradient tensor the von Karmann length scale, which is computed from the second derivative of the velocity field is introduced.

$$\begin{aligned}\frac{\partial(\rho\omega)}{\partial t}+\frac{\partial(\rho\bar{U}_j\omega)}{\partial x_j}&=\tilde{\alpha}\rho S^2-\tilde{\beta}\rho\omega^2+\frac{\partial}{\partial x_j}\left[\frac{\mu_t}{\tilde{\sigma}_\omega}\frac{\partial\omega}{\partial x_j}\right]\\&+\frac{2\rho}{\sigma_\Phi}\frac{1}{\omega}\frac{\partial k}{\partial x_j}\frac{\partial\omega}{\partial x_j}+F_{SAS-SST}\end{aligned} \tag{4}$$

The additional term is given by eq. (5):

$$F_{SAS-SST}=-\frac{2\rho}{\sigma_\Phi}\frac{k}{\omega^2}\frac{\partial\omega}{\partial x_j}\frac{\partial\omega}{\partial x_j}+\bar{\zeta}_2\kappa\rho S^2\frac{L}{L_{\nu k}} \tag{5}$$

In order to preserve the SST model in the RANS regions a modified formulation of eq. (5) is used.

$$\begin{aligned}F_{SAS-SST}=\rho F_{SAS}\max&\left[\bar{\zeta}_2\kappa S^2\frac{L}{L_{\nu k}}\right.\\&\left.-\frac{2}{\sigma_\Phi}k\max\left(\frac{1}{\omega^2}\frac{\partial\omega}{\partial x_j}\frac{\partial\omega}{\partial x_j},\frac{1}{k^2}\frac{\partial k}{\partial x_j}\frac{\partial k}{\partial x_j}\right);0\right]\end{aligned} \tag{6}$$

Since the grid spacing is not an explicit term in (6) the SAS model can operate in scale resolving mode without explicit grid information. The issue of grid induced separation of the flow in the boundary layer as it can appear in DES model is eliminated. The model constants are given in [16].

2.3 Combustion Modelling

To take combustion processes into account the respective chemical production rate of all species has to be modelled. The required values for this quantity can be obtained by appropriate combustion models. In the present work the combined Eddy Dissipation/Finite Rate Chemistry combustion model is used applying a one step global reaction mechanism for methane combustion.

Combined Eddy Dissipation/Finite Rate Chemistry Model

The Eddy Dissipation combustion model concept which was introduced by Magnusson [17] is based on the hypothesis that the chemical reaction is fast in relation to the transport processes of the flow. The reaction rate is assumed to be proportional to a mixing time defined by the turbulent kinetic energy k and the specific dissipation rate ϵ. The reaction rate is not kinetically controlled which may result in poor predicts for processes where the chemical kinetics limit the reaction rate. Thus the source term of the species conservation equation of a species i is calculated by eq. (7).

$$S_i = M_i \sum_{r=1}^{Nr} (\nu''_{ir} - \nu'_{ir}) \, R_i \tag{7}$$

where R_r defines the reaction rate and ν' and ν'' are the stoichiometric coefficients. A more detailed explanation of the model and the model constants can be found in [17].

The Finite Rate Chemistry combustion model is based on the assumption that the mixing is much faster than the kinetically controlled processes. Here, the chemical production rate of the species i is defined by

$$S_i = M_i \sum_{r=1}^{Nr} \left[(\nu''_{ir} - \nu'_{ir}) \cdot \left(k_{fr} \prod c_\beta^{\eta'_{\beta r}} - k_{br} \prod c_\beta^{\eta''_{\beta r}} \right) \right] \tag{8}$$

In eq. (8) k_{fr} and k_{br} are the forward and backward reaction rates of the reaction r. The reaction can be described by the Arrhenius function:

$$k_r = A_r T^{n_r} exp \left(\frac{-E_r}{R_m T} \right) \tag{9}$$

A detailed explanation is given in [18].

In the combined model that is applied in this work both reaction rates are computed first independently from each other. Then, for the calculation of the effective chemical production rate the minimum value of both models is used. Thus the chemical production rate is either limited by the chemical kinetics or by the turbulent mixing. Therefore this model potentially is applicable for a wide range of turbulent reacting flows from low to high Damköhler numbers.

3 Numerical Method

The simulations were performed applying the commercial software package ANSYS CFX 11.0. The fully implicit solver is based on a finite volume formulation for structured and unstructured grids. Mulitgrid strategy is used to solve the linear set of coupled equations. For the spatial discretization a second order scheme is used except for the species and energy equations. For these equations a high order resolution scheme which is essentially second order accurate and bounded is used. For SAS simulations a non-dissipative second order central difference scheme is applied in the detached regions. This is necessary to avoid excessive numerical diffusion which would interfere the resolution of the turbulent structures. In RANS regions the computational method switches back to the second order accurate upwind based scheme. For the time discretization an implicit second order time differencing scheme is used. The parallelisation in CFX is based on the Single Program Multiple Data (SPMD) concept. The numerical domain is decomposed into separate tasks which can be executed separately. The communication between the processes is realized using the Message Passing Interface (MPI) utility. The partitioning process is fully automated and the memory usage is equally distributed among all processors.

4 Results and Discussion

The aim of the present work was to elaborate the strengths and weaknesses of different turbulence and combustion models for the numerical prediction of gas turbine combustor flows. To this end numerical simulation runs and comparisons of calculated and measured values were done for an aero engine model combustor. Within the present project numerical calculations were set up and non-reacting and reacting flow simulations were accomplished. Exemplarily the results of a reacting flow simulation will be reported in the subsequent chapters.

4.1 Test Case

The simulated gas-turbine model combustor is schematically illustrated in Fig. 1 [19, 20, 21]. The burner is a modified version of a burner of an aero engine combustor. Air at room temperature and atmospheric pressure is supplied from a common plenum and admitted through an annular nozzle to the flame. Non-swirling gaseous fuel is injected between the two co-rotating air flows. The annular fuel injection slot is divided in 72 sections with an area of each segment 0.5x0.5mm^2. The exit plane of the outer air nozzle is taken as the reference height x=0mm. The combustion chamber, which permits a good optical access, consists of 4 quartz windows held by four posts in the corners. The square cross section of the chamber is 85x85mm^2 and its height is 114mm. The combustion chamber is connected via a conical top plate to a central exhaust pipe (diameter 40mm, length 50mm). Microphones are installed in the plenum and the combustion chamber to detect the pressure fluctuations. Depending on the load conditions the combustion within this combustor can be stable or strongly oscillating. In the present work the reacting case with a thermal load of 35kW is investigated.

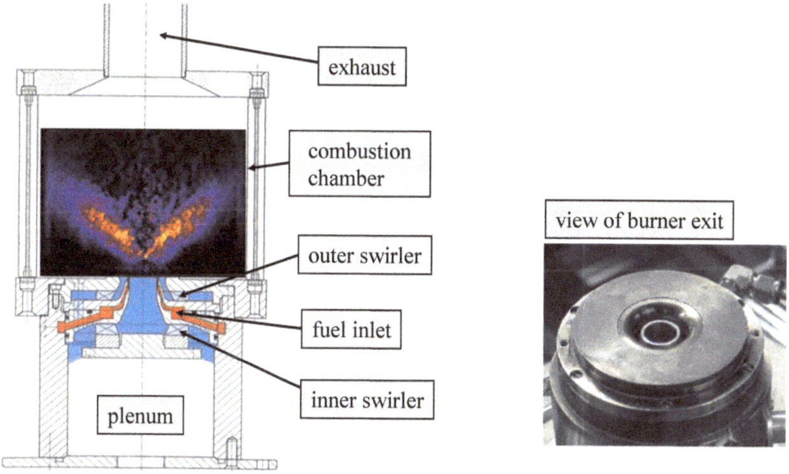

Fig. 1. Schematic of the gas turbine model combustor

4.2 Numerical Setup

The computational grid consists of 1.91 million grid points. For the nozzle and the combustion chamber an unstructured hexaeder grid with 1.6 million grid points was created. In the regions of potential turbulence generation and large velocity gradients a fine mesh was used in order to fulfil the LES requirements. Furthermore, the growth of the adjacent cells was limited to 10% in these zones. For the plenum an unstructured tetrahedral mesh was used. It consists of 1.79 million tetrahedral elements and 0.31 million grid points. At the air inflow boundary condition a mass flow of 0.01762 kg/s is set. The air temperature is set to 330K. The turbulent quantities are defined by using the medium intensity option of 5% in ANSYS CFX. The numerical boundary condition at the fuel inlet specifies a pure methane mass flow of 0.000696 kg/s at 330K. Here, the turbulent intensity is denoted by 15% and the eddy length scale by 0.0005m. The wall of the plenum and the combustor is assumed to be adiabatic. The wall of the combustion chamber bottom is set to 600K and of the combustion chamber to 1050K. For the outlet a static pressure boundary condition is used. The relative static pressure is set to 0Pa. The reference pressure of the computational domain is defined by 101325Pa. Furthermore Fig. 2 shows the location of positions where a comparison of measured and calculated averaged velocity profiles will be presented.

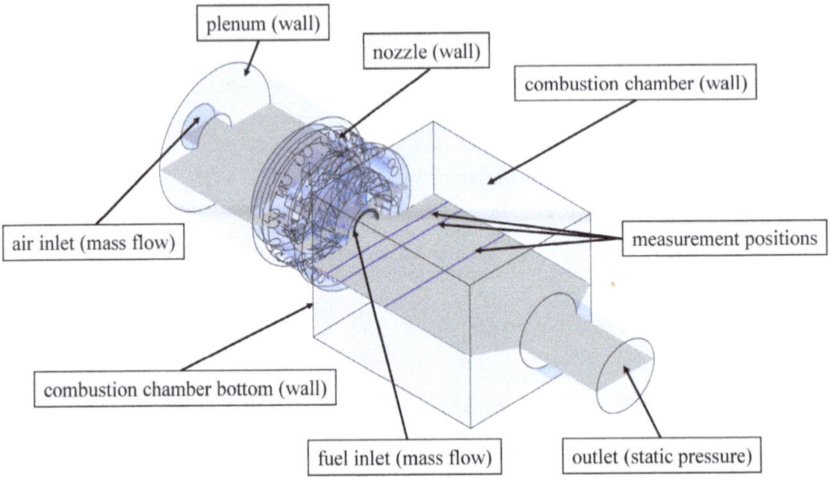

Fig. 2. Computational domain of the gas turbine model combustor with numerical boundary conditions

4.3 Time-Averaged Results

Velocity

Figure 3 shows the whole geometry including contour plots of the calculated flow pattern of the time-averaged and the instantaneous axial velocity. The black lines represent the locations of zero velocity. In Fig. 4 two-dimensional streak line plots are used to visualise the time-averaged and instantaneous velocity field within the combustion chamber. The time-averaged plots show a flow field which is typical for enclosed swirl burners with a concentrically shaped inflow. The negative velocities in the centre indicate the inner recirculation zone (IRZ) which occurs due to vortex break down. The length of the IRZ is about 81mm and the maximum width of the IRZ is about 40mm. The outer recirculation zone (ORZ) is developing in the corners of the combustion chamber. The IRZ and ORZ zones are unsteady and their positions and sizes oscillate with time. One shear layer is located between the IRZ and the inflowing stream and a second one towards the ORZ. Both recirculation zones are characterised by high velocity gradients and strong fluctuations. Note that in the complex flow inside the plenum further recirculation zones are formed. The instantaneous streak line plot shows clearly the existence of small non-stationary vortices close to the inner and outer shear layer. These phenomena are not visible in the time-averaged images. Along with the high turbulence intensity in the shear layers these small scale vertical structures are significantly contributing to the intensive mixing between the cold fresh gas and the hot burning gas coming from the IRZ and the ORZ. These processes

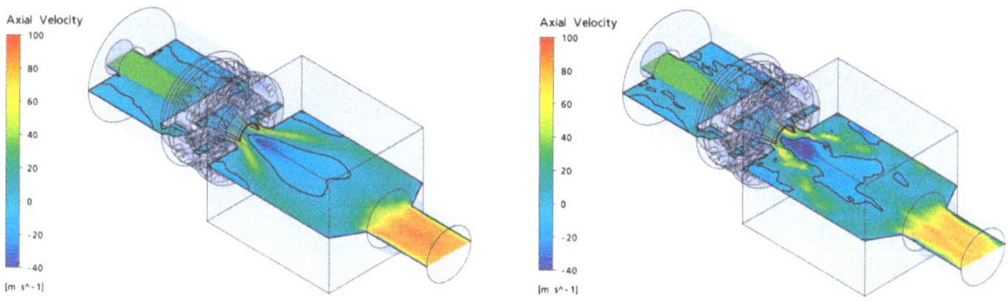

Fig. 3. Time-averaged (left) and instantaneous (right) axial velocity plot on a cutting plane

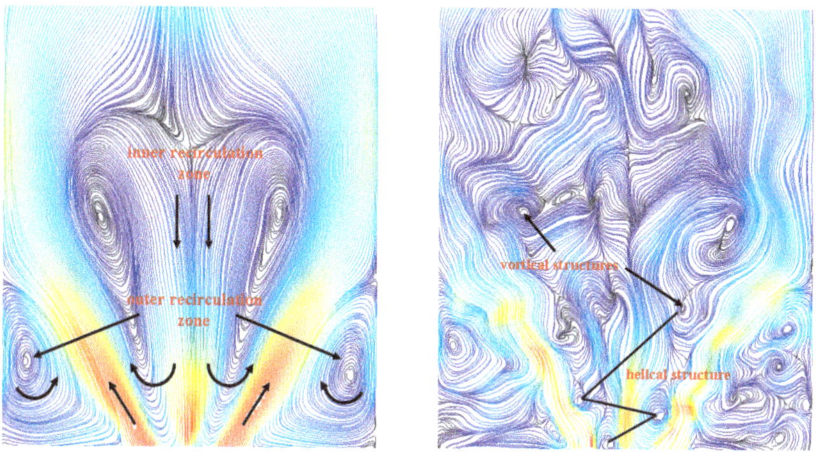

Fig. 4. Time-averaged (left) and instantaneous (right) streamline plots of the axial velocity on a cutting plane

play an important role in the mixing and ignition of the fresh gas and therefore for the stabilisation of the flame. The transient simulation was started from a steady state solution. After a start-up phase of 2 combustor residence times the statistical averaging of the velocities was started. This averaging was performed over four residence times. Figure 5 shows the comparison of the numerically obtained time-averaged axial, radial and tangential velocity profiles with appropriate LDA measurements at three axial positions. At all positions the simulated time-averaged radial and tangential velocity profiles agree very well with the experiment. A deviation occurs at x=10mm which is close to the burner mouth for the time-averaged axial velocity profile. Here, the positive peak values which belong to the incoming stream of fresh gas are predicted well whereas the strength of the recirculation zone is too high. Nevertheless at the other positions the time-averaged axial velocity is predicted well. Figure 6 visualizes the turbulent fluctuations of the three velocity

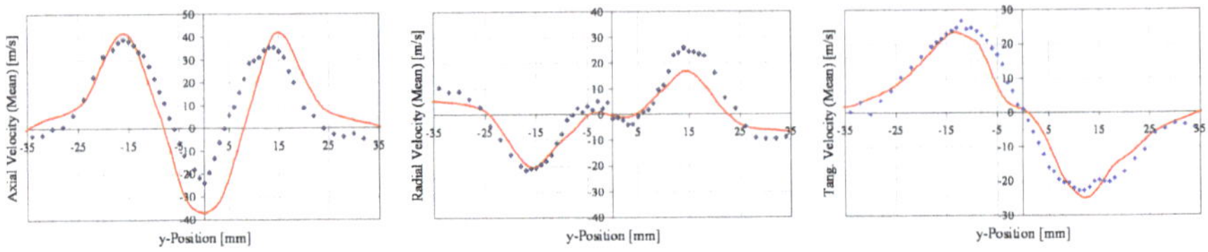

(a) x=10mm

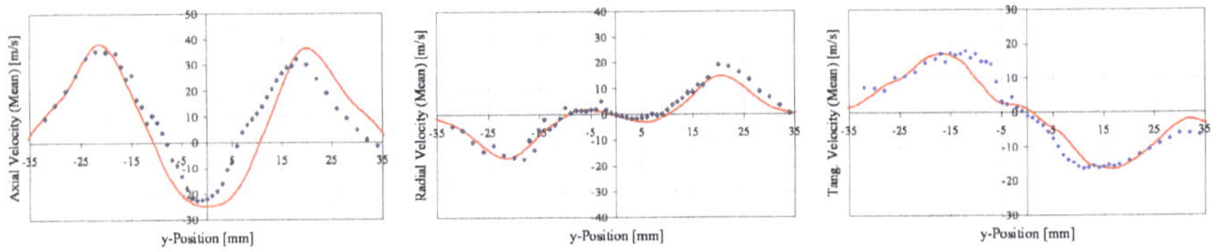

(b) x=20mm

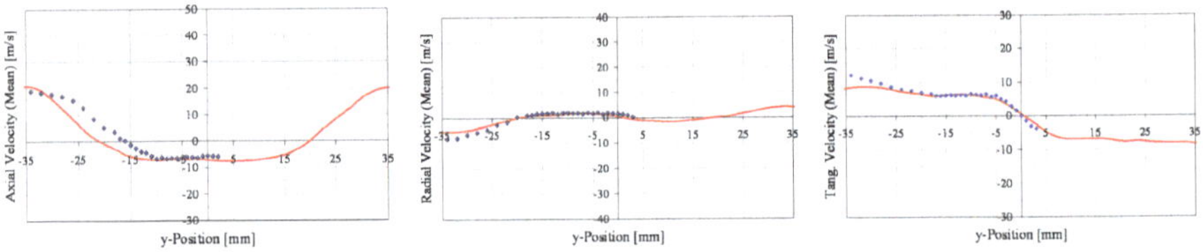

(c) x=50mm

Fig. 5. Time-averaged axial (left), radial (middle) and tangential (right) velocity profiles at x=10mm, x=20mm, x=50mm

components. This quantity has an important impact on the mixing processes of the cold fresh gas and the hot burning gas coming from the IRZ and the ORZ and therefore on ignition and flame stability. To predict these values accurately CPU intensive LES or hybrid RANS/LES turbulence models have to be used instead of RANS models. The turbulent fluctuations induced in the shear layers are clearly visible while in the central region the level of turbulence is much smaller for the axial and radial velocity components. Generally the comparison of the simulated profiles obtained by using the SAS turbulence model shows an excellent agreement with the experiment at the most positions. The calculation slightly under predicts the turbulent intensities of the axial velocity component at the centreline at x=10mm. At x=50mm the simulations predicts the tangential velocity components at the centreline very well but deviates towards the combustion chamber wall.

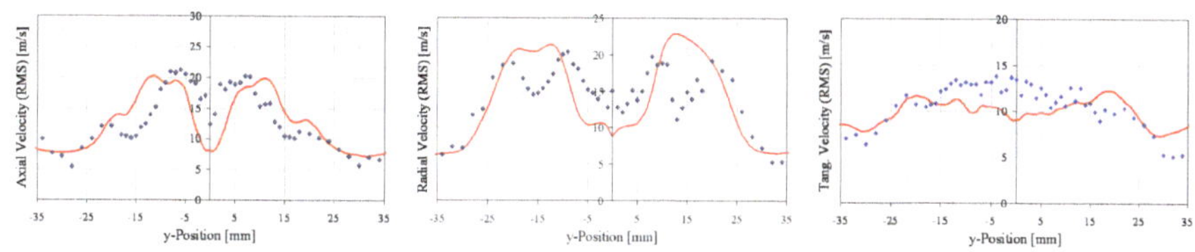

(a) x=10mm

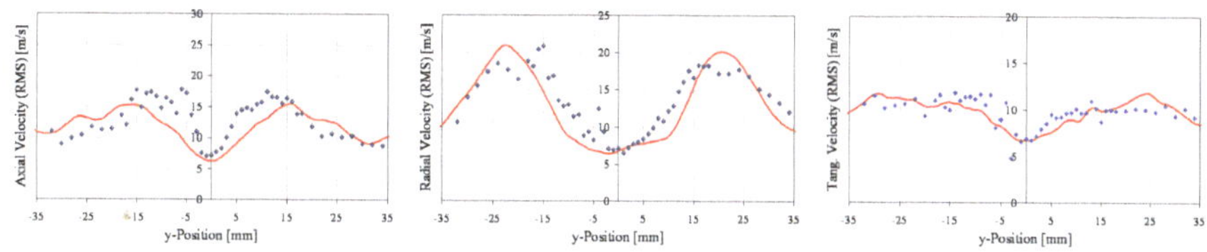

(b) x=20mm

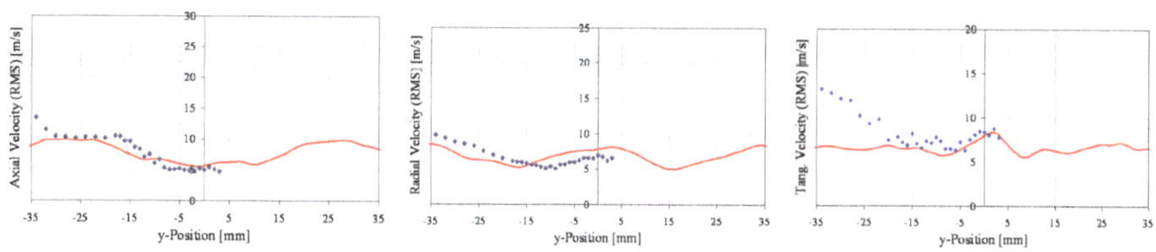

(c) x=50mm

Fig. 6. RMS values of the axial (left), radial (middle) and tangential (right) velocity components at x=10mm, x=20mm, x=50mm

Temperature

Figure 7 shows the simulated time-averaged and an instantaneous temperature pattern. The distribution of the mean temperature shows the conical shape of the heat release zone. Furthermore, the flame root lies approximately 2-5mm above the nozzle exit, which implies that the flame is partially premixed and a high level of mixing has occurred before combustion takes place. Figure 8 show the comparison of the numerically obtained time-averaged temperature profiles with appropriate Raman measurements. At all positions the simulated time-averaged temperatures profiles agree excellent with the experiment. At a height of x=10mm the flame exhibits a temperature of T=1800K in the centre. Between a height of x=10mm and x=20mm the highest temperatures are reached and decrease slowly afterwards. The low temperature regions reflect the incoming stream of the fresh gas. The temperature level in the ORZ is lower compared to the IRZ. This is due to the leaner mixture in these regions and the heat losses to the wall. Figure 9 visualizes the turbulent temperature fluctuations. The comparison of the simulated profiles shows a very good agreement with the experiment at most positions. The temperature fluctuations at x=10mm reach a level of 550-580K and decreases at larger heights.

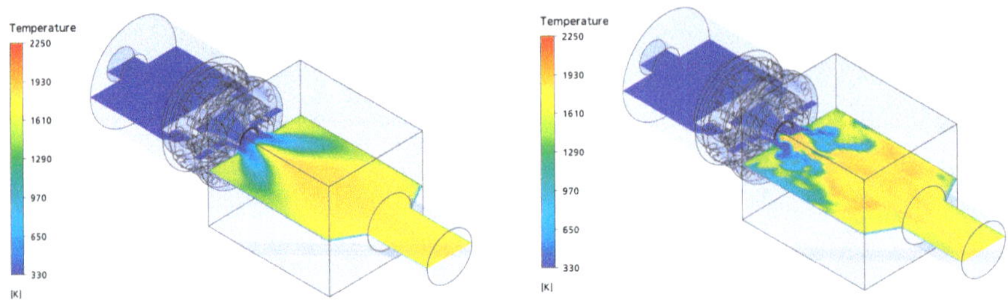

Fig. 7. Time-averaged (left) and instantaneous (right) temperature plots on a cutting plane

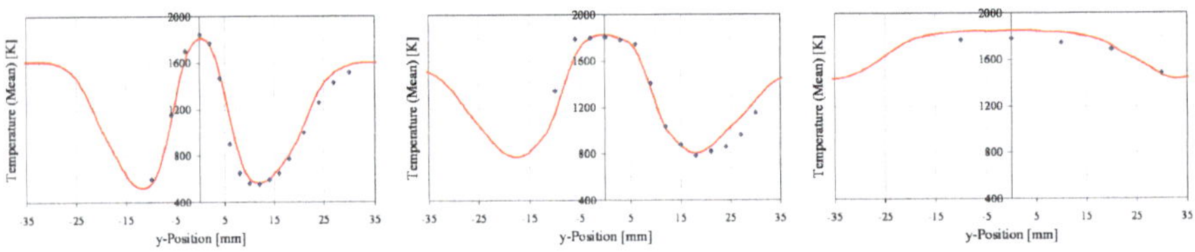

Fig. 8. Time-averaged temperature profiles at x=10mm (left), x=20mm (middle) and x=50mm (right)

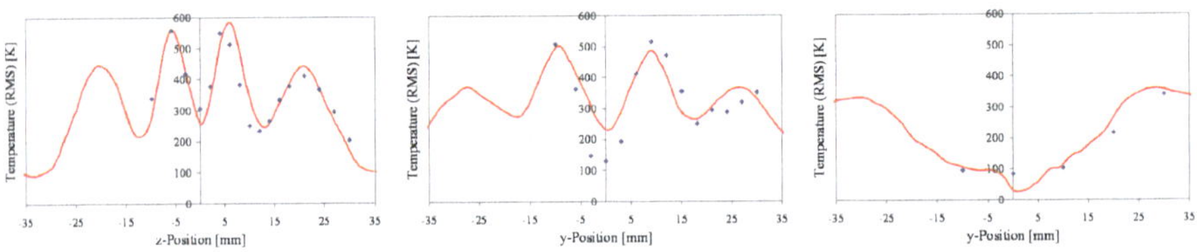

Fig. 9. RMS temperature profiles at x=10mm (left), x=20mm (middle) and x=50mm (right)

At x=20mm the calculation slightly over predicts the turbulent intensities of the temperature at the centreline. Nevertheless the peak values and the shape agree very well with the measured data. The high temperature fluctuations within the IRZ evidences additionally to the velocity fluctuations that the IRZ is a highly unsteady flow region.

4.4 Instantaneous Flow

Precessing Vortex Core (PVC)

Time-averaged data are informative to assess the general feature of the flow, but do not describe the full flow characteristics which is highly unsteady. In

Fig. 4 it is clearly visible that the vortex structures in the inner shear layer which propagate in time are ordered in a zig-zag arrangement. These vortices are formed inside the fuel/air nozzle in the inner air flow section. This phenomenon is a strong indication of the formation of a Precessing Vortex Core (PVC) which is a typical instability in non-reacting and reacting swirling flows at high swirl numbers. The occurrence and role of PVCs under combustion conditions is a complex issue and strongly depends on quantities like the equivalence ratio and combustor geometry. This vortex system can be extracted from the results of a hybrid LES/RANS or LES calculation. Figure 10 shows the instantaneous views of a PVC using a low pressure isosurface which is coloured with the axial velocity component. To show the evolution of the PVC in time the same isosurfaces are displayed at two different times. The PVC is rotating around the burner axis in the same direction as the imposed swirl. The PVC which is a hydrodynamic phenomenon oscillates with time at a frequency of 1594Hz. Due to these vortices the turbulent mixing inside the combustor is enhanced significantly.

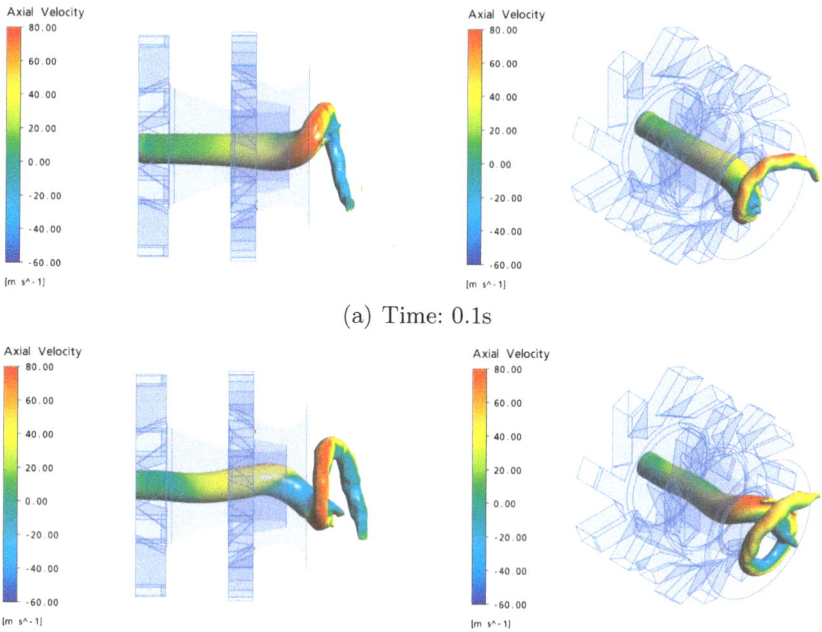

(a) Time: 0.1s

(b) Time: 0.138s

Fig. 10. Precessing vortex core visualised by a low pressure iso-surfaces and coloured by the axial velocity

5 Computational Resources

The simulations have been performed on the HP XC4000 system. According to our experience to obtain a statistically converged solution a total integration time of at least 4 combustor residence times is required. This corresponds for the applied test case to a simulated real time of 0.12s. Additionally for the start-up phase 2 combustor residence times are needed. Therefore the overall simulated real time is 0.18s. Furthermore, since the SAS turbulence model is applied in combination with a combustion model a relatively fine mesh has to be used to achieve low CFL numbers and hence the numerical effort is very large. A typical grid which is applied for the simulation has 1.91million grid points. In order to perform the calculations within adequate turnaround times the CPU numbers varies between 20 and 24. The typical total CPU time required for such runs is about 60 days and the total RAM requirements are 1040MB. In all simulations one time step consists of four inner iterations loops. The time step is set to 1e-5s. This leads to 18000 time steps and 72000 inner iteration loops per run.

6 Conclusions

The potential of the SAS turbulence models in combination with the EDM/FRC combustion model for the simulation of the reacting flow in gas turbine combustion chambers has been worked out. In general the results show a remarkable predictive capacity of these methods. Compared to the classical LES approach the SAS models can save one order of magnitude of computing power, since the large portions of the computational domain can be treated by a RANS method thus partly enabling larger grid sizes. Another advantage is the stationary RANS boundary conditions formulation which can be applied more easily than unsteady LES boundary conditions. Nevertheless, high performance computing is necessary to perform the calculations within adequate turnaround times. The numerically obtained time-averaged velocity components as well as the time-averaged temperature profiles match very well the LDA and Raman measurements. Likewise the turbulent velocity and temperature fluctuations agree very well with the experimental data. The dynamic behaviour of the reacting flow is visualised. A precessing vortex core (PVC) with a frequency of 1594Hz could be found in the combustion chamber.

Acknowledgement

The authors would like thanks the High Performance Computing Centre Karlsruhe for the always helpful support and the computation time on the high performance computers.

References

1. Liang, H., Maxworthy, T.: An Experimental Investigation of Swirling Jets. J. Fluid Mech., **525**, pp. 115–159 (2005)
2. Cala, C.E., Fernandez, E.C., Heitor, M.V., Shtork, S.I.: Coherent Structures in Unsteady Swirling Jet Flows. Exp. Fluids, **40**, pp. 267–276 (2006)
3. Fernandez, E.C., Heitor, M.V., Shtork, S.I.: An Analysis of Unsteady Highly Turbulent Swirling Flow in a Model Vortex Combustor. Exp. Fluids, **40**, pp. 177–187 (2006)
4. Midgley, K., Spencer, A., McGuirk, J.J.: Unsteady Flow Structures in Radial Swirler Fed Fuel Injectors. ASME Turbo Expo 2004, GT2004-53608
5. Thompson, K.W.: Time Dependent Boundary Conditions for Hyperbolic Systems. J. of Comput. Phys., **68**, pp. 1–24 (1987)
6. Poinsot, T., Lele, S.: Boundary Conditions for Direct Simulations of Compressible Viscous Flows. J. of Comput. Phys., **101**, pp. 104–129 (1992)
7. Widenhorn, A., Noll, B., Aigner, M.: Accurate Boundary Conditions for the Numerical Simulation of Thermoacoustic Phenomena in Gas-Turbine Combustion Chambers. ASME Turbo Expo 2006, GT2006-90441
8. Widenhorn, A., Noll, B., Aigner, M.: Impedance Boundary Conditions for the Numerical Simulation of Gas-Turbine Combustion Systems. ASME Turbo Expo 2008, GT2008-50445
9. Noll, B.: Numerische Strömungsmechanik. Springer Verlag (1993)
10. Noll, B., Schütz, H., Aigner, M.: Numerical Simulation of High Frequency Flow Instabilities Near an Airblast Atomizer. ASME Turbo Expo 2001, GT2001-0041
11. Widenhorn, A., Noll, B., Stöhr, M., Aigner, M.: Numerical Investigation of a Laboratory Combustor Appling Hybrid RANS-LES Methods. DESider 2007, Second Symposium on Hybrid RANS-LES Methods, Corfu, Greece GT2008-50445
12. Widenhorn, A., Noll, B., Stöhr, M., Aigner, M.: Numerical Characterization of the Non-Reacting Flow in a Swirled Gasturbine Model Combustor. The 10th Results and Review Workshop of the HLRS
13. Menter, F.R.: Two Equation Eddy Viscosity Turbulence Models for Engineering Applications. AIAA Journal **32(8)**, pp. 269–289 (1995)
14. Menter, F.R., Egorov, Y.: Re-visting the Turbulent Scale Equation. IUTAM Symposium (2004)
15. Menter, F.R., Egorov, Y.: A Scale-Adaptive Simulation Model Using Two-Equation Models. AIAA-Paper, 2005-1095
16. Menter, F.R., Kuntz, M., Bender, R.: A Scale Adaptive Simulation Model for Turbulent Flow Prediction. AIAA Paper, 2003-0767
17. Hjertager, B.H., Magnusson, B.F.: On Mathematical Modelling of Turbulent Combustion with Special Emphasis on Soot Formation and combustion. Sixteenth Symposium on Combustion, 1976
18. Gerlinger, P.: Numerische Verbrennungssimulation, Springer Verlag, 2005
19. Giezendanner, R., Keck, O., Weigand, P., Meier, W., Meier, U., Stricker, W., Aigner, M.: Periodic Combustion Instabilities in a Swirl Burner Studied by Phase-Locked Planar Laser-Induced Fluorescence. Combust. Sci. Technol., **175**, pp. 721–741 (2003)

20. Weigand, P., Meier, W., Duan, X.R., Stricker, W., Aigner, M.: Investigation of swirl flames in a gas turbine model combustor; I. Flow field, structures, temperatures and species distributions, Combustion and Flame, **144**, pp. 205–224 (2006)
21. Weigand, P., Meier, W., Duan, X.R., Stricker, W., Aigner, M.: Investigation of swirl flames in a gas turbine model combustor; II. Turbulence-chemsitry interactions, Combustion and Flame, **144**, pp. 225–236 (2006)

Numerical Simulation of a Transonic Wind Tunnel Experiment

Benedikt König, Thorsten Lutz, and Ewald Krämer

Institut für Aerodynamik und Gasdynamik, Universität Stuttgart
Pfaffenwaldring 21, 70569 Stuttgart, Germany

Summary. Numerical studies were undertaken to assess the consistency of a wind tunnel experiment and of numerical simulations performed to investigate the potential of shock control devices in transonic flow. Behaviour of the shock control bumps under consideration heavily depends on the local flow condition, which on the other hand is considerably influenced by wall interferences experienced in the wind tunnel experiment. The study presented herein uses a CFD-RANS method to numerically model the wind tunnel and simulate the complete channel flow with all relevant interference effects. Comparisons of experimental and numerical results for a clean airfoil mounted in the wind tunnel are presented.

1 Introduction

The Institut für Aerodynamik und Gasdynamik (IAG) has a tradition of several years of research on shock control in transonic flow. Special emphasize is thereby placed on the so-called shock control bump (SCB) which was introduced by Ashill [1] in 1992. Early studies conducted at the IAG by Sommerer [9, 10] started numerical investigations and optimizations of the potential benefits of adaptive shock control bumps on two-dimensional airfoils. Kutzbach extended this work to two-dimensional SCBs in quasi-two-dimensional and three-dimensional flows on swept wings [4]. Three-dimensional effects of finite shock control bumps (3D-SCBs), with a width-to-length ratio of around one, were numerically studied by Pätzold [7, 8] with an emphasis on the application on swept wings. Ogawa and Pätzold showed in [6] a comparison of experimental measurements of a three-dimensional bump mounted on a wind tunnel wall with numerical results of the same bump on a transonic airfoil with comparable flow conditions upstream of the shock.

However, until recently no experimental verification of the numerically predicted drag reductions by 3D-SCBs were performed. Such experiments were finally conducted at the Transonic Wind Tunnel Göttingen (TWG) for a transonic airfoil equipped with an array of especially designed three-dimensional shock control bumps. The experimental results confirmed the predicted overall drag reductions but, at the same time, showed considerable differences compared to the numerical simulations conducted beforehand [2]. These differences could be attributed to interference effects of the wind tunnel walls on the local flow field at the model position in the experiments.

To assess the wall interferences and increase the trustworthiness of the numerical results a study of the flow effects present in the wind tunnel was conducted. This paper will present results obtained for the clean airfoil by numerical RANS simulation of the wind tunnel environment present during the experimental campaign.

2 Experimental Setup

The work presented in this paper is related to an experimental measurement campaign conducted in the Transonic Wind Tunnel Göttingen. The TWG is a continuous, pressurisable wind tunnel with three exchangeable test sections for subsonic, transonic, and supersonic speeds. The experiments were carried out in the 1 m $\times$ 1 m adaptive walls test section at a Mach number of $Ma_\infty = 0.79$ and a Reynolds number of $Re_c = 5.0 \times 10^6$. Adaptation of the upper and lower test section walls to the momentary flow condition was applied to minimize wall interference effects on the wind tunnel measurements as much as possible. Besides measurements on the clean 40 cm chord metal airfoil model, an overall of eleven bumps were manufactured from glass fiber composite and glued onto the airfoil model for measurements.

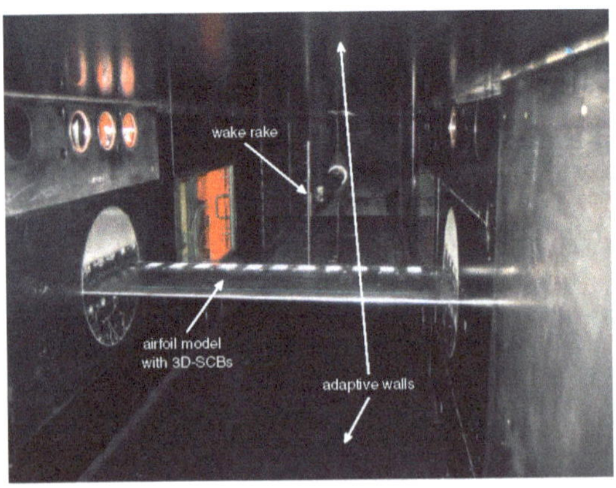

Fig. 1. Model with 3D-SCBs mounted in adaptive wall test section of TWG

The clean airfoil model was equipped with pressure orifices which, however, were partly covered by the additionally fitted composite bumps for the SCB measurements. Therefore, these pressure taps could not be used for lift measurement in the SCB-on configuration and no lift measurements were provided in the clean cases either. Drag was measured by means of a wake rake with 128 pitot tubes at a distance of 1.5 chord lengths behind the model. Additional pressure taps were located along the adaptive walls as well as the side walls of the test section. Actuator positions of the 22 jacks of the upper flexible wall and the 20 jacks of the lower flexible wall were recorded to recover the full state of the test section for the numerical simulations now presented herein.

3 Numerical Method

The CFD simulations were performed using the well established FLOWer code [3] developed by DLR and German universities and industry. FLOWer is a finite-volume RANS solver for block-structured grids and a well established code.

3.1 Governing Equations

FLOWer solves the three-dimensional Reynolds Averages Navier-Stokes equations (RANS) in integral form

$$\frac{\partial}{\partial t}\int_V \mathbf{W} dv + \int_{\partial V} \mathsf{F} \cdot \mathbf{n} ds = 0 \tag{1}$$

with the vector of the conservative variables

$$\mathbf{W} = [\rho, \rho u, \rho v, \rho w, \rho E]^T \tag{2}$$

on block-structured finite volume meshes. The conservative variables are given in a Cartesian coordinate system with ρ, u, v, w, E denoting the density, the Cartesian velocity components of the velocity vector $\mathbf{v}$ and the specific total energy, respectively. V represents the control volume and ∂V its closed outer surface. The flux tensor F is divided in a convective inviscid part F^c and a viscous part F^v such that

$$\mathsf{F} = \mathsf{F}^c - \mathsf{F}^v \tag{3}$$

with

$$\mathsf{F}^c = \begin{bmatrix} \rho u & \rho v & \rho w \\ \rho u^2 + p & \rho uv & \rho uw \\ \rho uv & \rho v^2 + p & \rho vw \\ \rho uw & \rho vw & \rho w^2 + p \\ \rho uE + up & \rho vE + vp & \rho wE + wp \end{bmatrix} \quad \text{and} \quad \mathsf{F}^v = \begin{bmatrix} 0 & 0 & 0 \\ \sigma_{xx} & \sigma_{xy} & \sigma_{xz} \\ \sigma_{yx} & \sigma_{yy} & \sigma_{yz} \\ \sigma_{zx} & \sigma_{zy} & \sigma_{zz} \\ \psi_x & \psi_y & \psi_z \end{bmatrix} \tag{4}$$

The ψ_i are abbreviations of the type

$$\psi_i = \left(u\sigma_{ix} + v\sigma_{iy} + w\sigma_{iz} + K\frac{\partial T}{\partial x_i} \right) \quad \text{for } i = x, y, z \tag{5}$$

The pressure p is calculated by the equation of state of the perfect gas

$$p = (\gamma - 1)\rho \left(E - \frac{u^2 + v^2 + w^2}{2} \right) \tag{6}$$

with the specific heats ratio γ. The temperature T is defined by

$$T = \frac{p}{\rho R}. \tag{7}$$

The Reynolds stress tensor σ_{ij} in (4) and (5), that represents correlations between fluctuating velocities and which depends on the fluid viscosity, is given through

$$\sigma_{ij} = \mu(\mathbf{v}_{i,j} + \mathbf{v}_{j,i} - \frac{2}{3}\delta_{ij}\mathbf{v}_{k,k}) \tag{8}$$

where $(.)_{i,j}$ denotes derivation of the i-th component with respect to x_j. Likewise the heat conductivity K in (5) depends on the viscosity μ, as is shown by the relation

$$K = \frac{\gamma}{\gamma - 1}\frac{\mu}{\mathrm{Pr}} \tag{9}$$

with Pr being the Prandtl number. For laminar flow μ in (8) and (9) is set to $\mu = \mu_L$, which is according to the Sutherland law

$$\mu_L = \mu_0 \left(\frac{T}{T_\infty} \right)^{3/2} \frac{T_\infty + S}{T + S}. \tag{10}$$

where $\mu_0 = 1.716 \times 10^{-5}\ \mathrm{kg/(m\,s)}$ and $S = 110.4$ K.

3.2 Turbulence Models

For turbulent flows FLOWer provides several algebraic and one- or two-equation turbulence models as well as seven-equation Reynolds Stress models. All these models, with exception of the Reynolds Stress models, make use of the *Boussinesq assumption*, which relates the Reynolds stresses to the velocity gradients and a so-called turbulent or eddy viscosity μ_T. Subsequently, μ in (8) and (9) is replaced by $\mu_L + \mu_T$ for turbulent flows, where the eddy viscosity μ_T is provided by either an algebraic or a transport equation turbulence model. The Boussinesq assumption is not used for the Reynolds Stress models but instead a set of additional transport equations derived from the Navier-Stokes equations is solved to model the components σ_{ij} directly.

3.3 Spatial and Time Discretization

The fluxes F^c and F^v are approximated using either a central or an upwind spatial discretisation scheme. The standard central scheme is based on the formulation according to Jameson with a blend of artificial dissipative terms added for high frequency damping and sufficiently sharp shock resolution. Additionally, several flux vector upwind schemes like AUSM are available. Time integration to steady state is carried out by an explicit five-stage Runge-Kutta scheme featuring convergence acceleration techniques like multigrid, local time stepping and implicit residual smoothing.

4 Computational Setup

The simulation of a wind tunnel necessitates several new aspect to be considered besides the general requirements for reliable CFD computations in external aircraft aerodynamics. The following section will give an overview of the modelling procedure applied to simulate the wind tunnel.

4.1 Numerical Simulation of the Wind Tunnel

The suspected influence of the wind tunnel walls on the flow motivated the numerical simulation of the experiment in the wind tunnel environment. At the beginning of this work it was not clear to what extent the wind tunnel should be modelled to capture its main interference effects. Preliminary studies comparing simulations using far-field boundary conditions with a two-dimensional model of the center plane of wind tunnel test section, with the upper and lower wall modelled, showed already a tendency of improving the agreement between numerical and experimental results. It was obvious that considerable interactions between the shock wave on the airfoils upper surface and the upper wind tunnel walls boundary layer took place and influenced the shock position itself. However, it also became clear that the presence of the horizontal walls could not explain all of the differences between measurements and numerical free stream simulations. It was therefore reasoned that additional modelling of the side walls would be necessary to reproduce all interference effects.

Furthermore, the preliminary studies showed a noticeable influence of the thickness and condition of the wall boundary layer at the airfoil position. To simulate a realistic boundary layer thickness at the model position in the wind tunnel, the straightforward approach would have been to extend the test section walls upstream and thereby to produce flat plate boundary layers along the walls. This method would have required information about the thickness of the boundary layer that should be simulated. These information were not available for the experiments under consideration. Furthermore, it could not be ruled out that not only the thickness but also the velocity profile of the boundary layer would influence the shock boundary layer interactions experienced in the experiment. In general, the boundary layer in a wind tunnel will have different characteristics than a flat plate boundary layer.

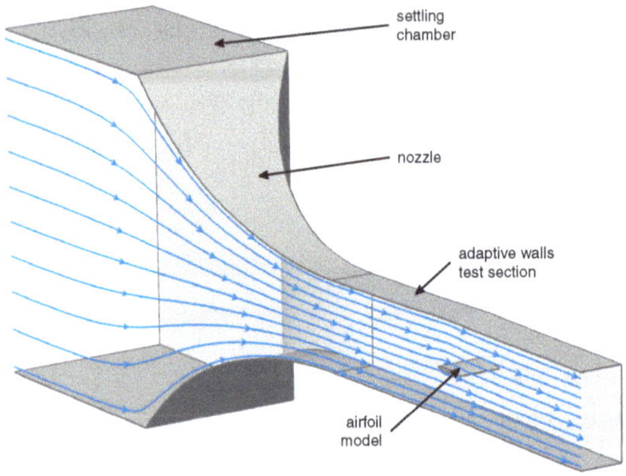

Fig. 2. CFD half-model of the adaptive walls test section of the TWG with nozzle and settling chamber

The problem of providing a realistic boundary layer history was solved by adding the nozzle and the settling chamber to the simulated test section. Figure 2 shows how the flow in a wind tunnel passes through a flow straightener before it enters the settling chamber and then passes through the nozzle into the test section. The flow straightener effectively removes the boundary layer and, thus, a new boundary layer develops from the entry into the settling chamber. Starting the numerical simulation with the flow in the settling chamber offered the possibility to simulate the boundary layer development through the nozzle and into the test section as realistic as possible. A similar approach was also described by Melber et al. [5].

The contour of the adaptive walls test section was modelled following the actual test section settings during the experiment. The actuator positions of the upper and lower adaptive walls, which were recorded at the TWG for each flow condition, were applied to the baseline geometry of the test section. An additional correction of the wall positions in the rear part of the test section was introduced to account for the displacement effect of the wake rake. Modelling the wake rakes true shape was dismissed as it would have led to excessive grid generation efforts and increased computation times. Instead, a volume distribution of the rake and its associated boundary layer was estimated and the tunnel cross section reduced by the respective amount. The influence of this correction will be shown in Sect. 5.1.

4.2 Boundary Conditions

All of the walls of the wind tunnel and the airfoil model surface were modelled with no-slip wall boundary conditions. A symmetry plane was specified along the wind tunnel center to reduce the computational costs. Inflow to the settling chamber and outflow of the test section were modelled using special inflow and outflow boundary conditions. The boundary conditions made use of two versions of the characteristic theory specialised to subsonic inflow and outflow. At the inflow plane the total pressure p_0 and the total temperature T_0 were specified according to the recorded values from the measurements for each flow condition separately. The total values p_0 and T_0 in the simulation corresponded to the fluid state in the reservoir, which equals the ambient pressure and temperature in the case of this campaign. Flow velocity in the numerical wind tunnel was adjusted by specifying the static pressure p_{exit} at the outflow plane. The FLOWer code offers an automatic routine to iteratively adjust p_{exit} until a target flow speed is achieved at a prescribed position in the flow domain. Though, to avoid the time consuming iteration process of p_{exit} the outflow static pressure was set to the mean value of the most downstream upper and lower pressure orifices in the test section.

4.3 Computational Grid

The block-structured finite volume grids used for the simulations were generated automatically and individually for each flow condition. The repeated grid generation was necessary, as not only the angle of attack of the airfoil in the tunnel was varied for each measurement but also the shape of the upper and lower adaptive walls was adjusted. Wall adaptation parameters were recorded for all on-flow conditions during the measurements. A script-based geometry generator was implemented to automatically model the wind tunnel walls and the airfoil based on the recorded parameters.

A study on different sized grids was conducted to find the grid resolution necessary to capture all relevant effects. Three grids with a refinement ratio of approximately 1.5 in each index direction were produced leading to overall numbers of grid cells of 5.6×10^6, 18.8×10^6, and 68.3×10^6 for the coarse, fine and super fine grid, respectively.

Based on the results of the grid refinement study a final regular grid with the following characteristics was set up. To model both the wall interferences of the wind tunnel and the shock boundary layer interactions at the shock control bumps on the airfoil, an overall high grid resolution was necessary in the computational domain. All wind tunnel walls were modelled as no-slip walls with full boundary layer resolution, featuring y^+-values of $y^+ \leq 1$ in the inflow and test section and $y^+ \leq 2$ in the outflow region, to reproduce the boundary layer and its influence on the channel flow as accurate as possible. Very fine grid resolution was applied to the airfoil model in order to capture the complex three-dimensional flow patterns due to the control applied. A total of 384×256 cells were placed on the half span of the airfoil model in streamwise and spanwise direction, respectively. In case of the control bumps applied the innermost SCBs close to the symmetry plane were reproduced using 24×32 cells on their surface. The overall grid size was therefore 22×10^6 cells for the half model using a symmetric boundary condition in the wind tunnel center plane. The grid was divided into eight blocks to make use of the parallel computation capability of FLOWer on the NEC SX-8 vector computer of the High Performance Computing Center Stuttgart (HLRS).

5 Results

In order to assess the potential of the wind tunnel simulations a wealth of simultions was performed first for the clean wing configuration and later for the airfoil with control bumps applied. Parameter variations were conducted for dedicated on-flow conditions and polars over a range of angles of attack were computed. This section presents results for the clean airfoil simulations with flow conditions at the design point of the experiment.

5.1 Simulation of the Wind Tunnel

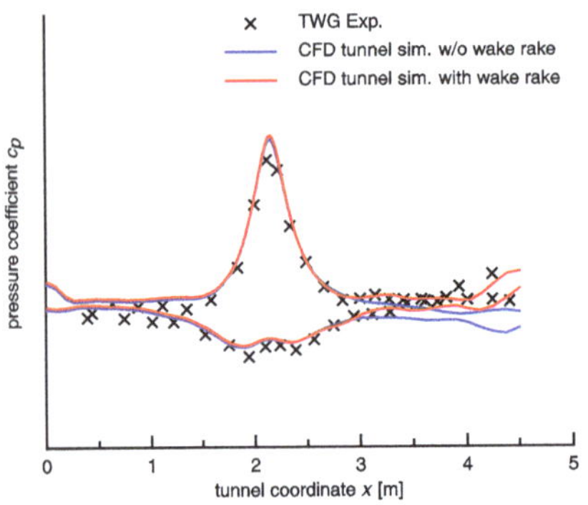

Fig. 3. Pressure distribution along the center line of the test section

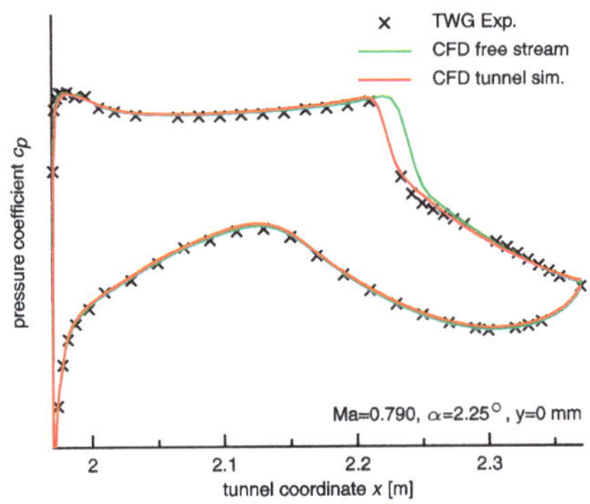

Fig. 4. c_p-distributions at the center of the airfoil surface with and w/o simulation of the tunnel walls at Ma=0.79

Figure 3 shows the c_p-pressure distribution along the wind tunnel walls in the center plane. The simulations without accounting for the displacement effects of the wake rake were performed using an iteration of the exit pressure p_{exit} with a target Mach number at the test section entry of Ma $= 0.79$. Agreement of the pressure distribution along the tunnel walls up to the model position from $x = 1.97$ m to $x = 2.37$ m in particular was good. However, at the rear part of the test section the influence of the missing wake rake became obvious. Modelling of the wake rake displacement effect, as it was described in Sect. 4.1, gave comparable results at the test section front part and at the model position. Again the agreement with the measurements was good. But in this case also the c_p-levels at the rear part of the test section matched the TWG data. The approximately constant mean pressure level, which was achieved through the adaptation of the test section walls, was clearly reproduced by this method.

Besides the better agreement in the pressure distribution the rough modelling of the wake rake has the additional benefit that p_{exit}-values obtained from the measurements could be prescribed for the simulations. No iteration of the exit pressure was therefore necessary which led to a considerable reduction of computation time.

Like a real wind tunnel the numerical wind tunnel needed to be started up and the flow speed increased slowly up to the final velocity in the test section. Initialisation of the whole flow domain with the final reference flow conditions, as it is usually done for CFD simulations, caused an over-acceleration through the nozzle with peak velocities reaching a multiple of the desired value. This did not only produce numerical stability problems but also led to unphysical flow solutions. The non-realistic occurrence of shock waves, separations, and resulting blockages, all associated with very high velocities, established stable flow solutions for some cases that would not have been reached through a more realistic start-up of the tunnel. Therefore, a start-up procedure was applied that initialised the flow domain with a very low Ma-number and then gradually increased the velocity in the test section by decreasing the exit pressure p_{exit} until it reached its desired value.

5.2 Comparison to Free Stream Simulation

The original goal of the test campaign in the TWG was to verify the applicability of the numerical models to the flow physics of the shock-boundary layer interactions associated with shock control bumps. It was mentioned before that the experiments confirmed the overall drag reduction by the 3D-SCBs as it was expected by CFD computations. However, beyond that, there are considerable differences between the drag polars in general and the pressure distributions (Fig. 4) in particular. Interferences of the flow with the wind tunnel walls were identified as a source of these differences.

Simulation of the wind tunnel environment gave a considerable improvement compared to the CFD results with free stream boundary conditions. Figure 4 shows the influence of the wall interferences at the wind tunnel center plane. The pressure distribution obtained by simulation of the wind tunnel is in very good agreement with the experimental results. Of special interest for the investigations on shock control was the correct prediction of the shock location. The CFD simulations without consideration of the wind tunnel featured a streamwise offset of approximately 5% of the airfoil's chord length for the flow condition shown in this example. The preliminary two-dimensional tunnel simulations indicated that this effect is mainly due to the horizontal walls influence but a certain impact of the side walls may not be ruled out, as will be described in the following section 5.3.

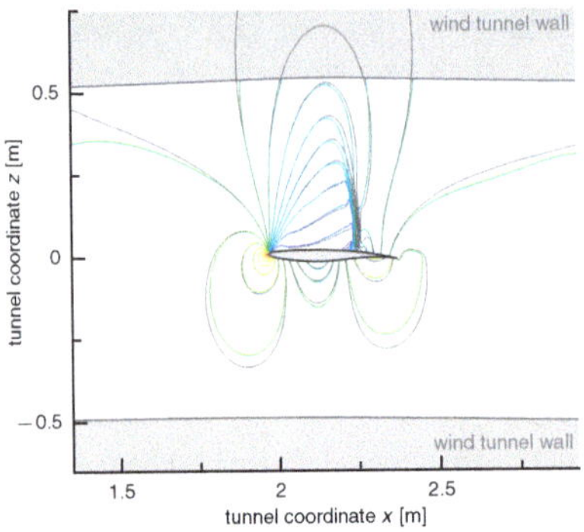

(a) overview of the test section

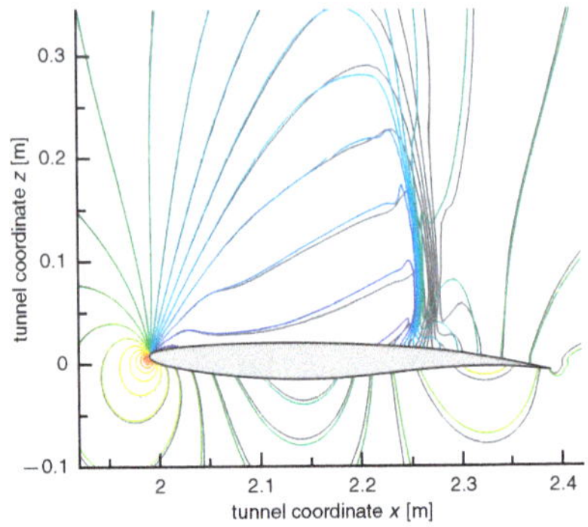

(b) close-up of the airfoil

Fig. 5. Comparison of c_p-distributions in the flow field with (*coloured lines coded with local pressure*) and w/o (*black lines*) simulation of the tunnel walls

Figure 5 shows pressure iso-lines with and without simulation of the wind tunnel walls. There were two main areas of influence of the tunnel walls discernible, namely the aforementioned shock region on the suction side above and the pressure side below the airfoil. While the flow field below the airfoil showed noticeable differences between the two simulations, these pressure variations did hardly propagate to the surface pressure distribution shown in Fig. 4. On the other hand the pressure iso-lines on the upper surface were, for the most part, only slightly bent by the wall interference. The wall adaption was obviously working as expected. However, the streamwise extent of the supersonic flow region was slightly reduced due to the wind tunnel wall. Through the non-linear characteristic of the transonic flow this small deterioration of the supersonic region led to the significant upstream relocation of the shock front of about 5% in the tunnel compared to the free stream flow.

5.3 Three-Dimensional Wall Interference Effects

The interferences described in the previous section were mostly due to interactions of the flow with the upper and lower wind tunnel walls. The upper and lower walls influence can be assumed to have a spanwise constant effect on the flow and, thus, the flow and the interferences can be considered as two-dimensional. Flow interactions with the tunnel side walls on the other hand lead to spanwise varying and therefore three-dimensional perturbations of the flow field.

One considerable effect that occurred during the simulations of the wind tunnel were flow separations along the model–side wall junction. These separations produced massive blockages of the tunnel cross section in some cases and thereby heavily influenced the complete flow characteristics in the tunnel. However, measurement data from the wind tunnel did not show signs of such significant flow separation and it was concluded that this was merely a numerical effect. It is well known that state-of-the-art turbulence models have difficulties to correctly predict the occurrence or non-occurrence of separations for certain flow conditions. It shall not be the goal of this work to improve turbulence model behaviour and therefore an engineering approach was chosen to prevent the obviously unphysical effects. The boundary layers susceptibility to separation was reduced by decreasing the grid resolution in the region of the model–side wall junction.

Another three-dimensional effect, which was indeed clearly visible in the measured data, was the development of oblique show waves. These shock waves stemmed from the joint of the airfoil leading edge and the wind tunnel side walls and ran across approximately one third of the models span towards the center of the tunnel. The influence on the pressure distribution is clearly visible in Fig. 6 at a streamwise and spanwise position of $x \approx 2.14$ m and $y = 0.28$ m. This effect was certainly not captured in the CFD free stream simulations without the wind tunnel walls influence. Figure 7 gives an illustration of the three-dimensional shock configuration and the resulting spanwise pressure distribution. It can be seen that even though the oblique shock waves led to inherently three-dimensional flow characteristics they reached only about one third from each tunnel wall into the flow domain. In the center of the tunnel the flow could indeed be considered as two-dimensional but still the normal shock wave may have been influenced in position and strength by the oblique shocks. Hence, there might be a coupling of the three-dimensional flow structures to the two-dimensional shock-dislocation described before in Sect. 5.2 that needs further research.

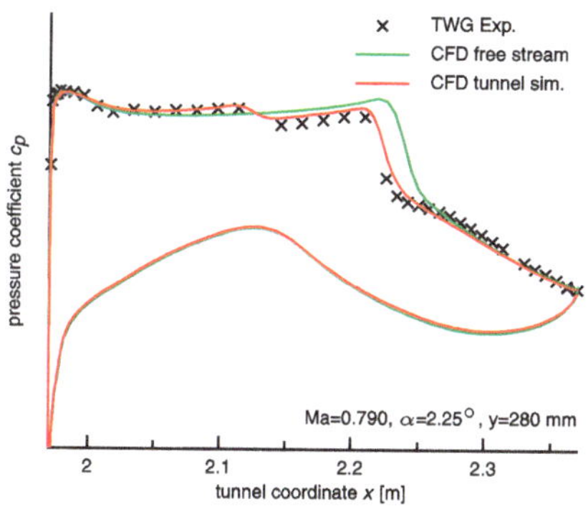

Fig. 6. c_p-distributions of the airfoil surface at $y = 280$ mm with and w/o simulation of the tunnel walls at Ma=0.79

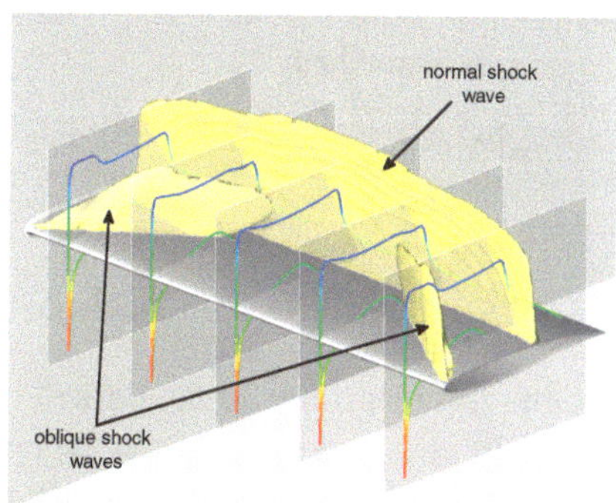

Fig. 7. Illustration of the 3D shock in the wind tunnel (*shock front coloured yellow*). Slices show surface pressure levels (*colour-coded with local c_p*)

6 Computational Resources

All computations presented in this work were conducted using the DLR FLOWer code on NEC SX-8. FLOWer is very well suited for vector operations with its structured grid approach. Parallelisation is done using either OpenMP or MPI, or an hybrid approach of both. OpenMP parallelisation however seems to somewhat reduced vector lengths and is therefore considered to be not optimal. The computations presented herein where thus conducted using MPI, even on the shared memory architecture of a single node. The highly optimised implementation of FLOWer allows for average vector operation ratios of more than 99% and a performance of around 5 GFLOPS per CPU. Table 1 gives an overview of the performance for simulations using the regular grid and the super fine grid from the grid refinement study mentioned in the text. The regular grid was used to conduct computations for single onflow conditions as well as complete polars. For a polar a series of around 15 single computations for increasing angles of attack were performed. Restarting from previous solutions and the small angular increments allowed for a reduced number of iterations per step. The differences in the time/iteration ratio between single point and polar computations stem from the extensive file output that was done in the example shown here.

Table 1. Computational performance of the TWG wind tunnel simulations

	regular grid		super fine grid
	single point	polar (15 points)	single point
Number of cells	22020096	22020096	68296704
Number of iterations	3000	16000	3000
GFLOPS	40	40	38
Average Vector Length	220	220	230
Vector operation ratio [%]	99.3	99.3	99.4
Global Memory size [GB]	30.3	30.3	95.0
Wall clock time [h]	4.5	32	14.5

7 Conclusions and Outlook

Results from the ongoing numerical research on shock control bumps were validated by experiments. Discrepancies between CFD and experimental results could be attributed to interference effects in the wind tunnel. In this work an approach was described to investigate and resolve these discrepancies for the case of a clean airfoil without shock control. Good agreement between experimental results and numerical simulations of the wind tunnel environment was reached. Currently, this method is applied to the test cases of the airfoil with shock control applied, and comparisons with detailed surface pressure measurements are performed. This work will identify the potential and the limitations of current CFD methods to simulate the complex flow phenomena involved in the design and investigation of shock control bumps.

Acknowledgements

This work has partly been funded by the Deutsche Forschungsgemeinschaft (DFG). The authors would like to express their gratitude to Dr. Rosemann and Dr. Rein from the German Aerospace Research Center (DLR) in Göttingen for affording the opportunity of the wind tunnel tests. Furthermore, we would like to thank H. Uhlemann and the team of TWG for supporting the tests. Discussions with S. Melber-Wilkending from DLR provided helpful insights into the numerical simulation of wind tunnel environments. Finally, we thank the system administrators of HLRS for competent support whenever support was needed.

References

1. P.R. Ashill, J.L. Fulker and A. Shires: *A Novel Technique for Controlling Shock Strength of Laminar-Flow Aerofoil Sections.* In: Proceedings 1st European Forum on Laminar Flow Technology, March 16–18, 1992, Hamburg, Germany, DGLR, AAAF, RAeS, 1992, S. 175–183.
2. B. König, M. Pätzold, Th. Lutz, E. Krämer, H. Rosemann, K. Richter and H. Uhlemann: *Numerical and Experimental Validation of Three-Dimensional Shock Control Bumps.* Paper 2008-4001, AIAA, 2008.
3. N. Kroll, C.C. Rossow, D. Schwamborn, K. Becker and G. Heller: *MEGAFLOW – a Numerical Flow Simulation Tool for Transport Aircraft Design.* In: Proceedings of the 23rd International Congress of Aeronautical Sciences, ICAS, 2002. Toronto.
4. M. Kutzbach, Th. Lutz and S. Wagner: *Investigations on Shock Control Bumps for Infinite Swept Wings.* Paper 2004-2702, AIAA, 2004.
5. S. Melber-Wilkending, A. Heidebrecht and G. Wichmann: *CFD Validation by Wind Tunnel Measurements: Uncertainty Assessment by Numerical Simulation of Complete Wind Tunnel Flows.* NATO-RTO, Applied Vehicle Technology, AVT 147, 2007, Greece
6. H. Ogawa, H. Babinsky, M. Pätzold and Th. Lutz: *Numerical Optimization of Finite Shock Control Bumps.* Paper 2007-324, AIAA, 2007.
7. M. Pätzold, Th. Lutz, E. Krämer and S. Wagner: *Numerical Optimization of Finite Shock Control Bumps.* Paper 2006-1054, AIAA, 2006.
8. M. Pätzold: *Auslegungsstudien von 3-D Shock-Control-Bumps mittels numerischer Optimierung.* PhD Thesis, Institut für Aerodynamik und Gasdynamik, Universität Stuttgart, 2008, to be published
9. A. Sommerer, Th. Lutz and S. Wagner: *Design of Adaptive Transonic Airfoils by Means of Numerical Optimization.* In: Proceedings of ECCOMAS 2000, European Congress on Computational Methods in Applied Sciences and Engineering, 11–14 September, 2000, Barcelona, Spain, 2000.
10. A. Sommerer: *Numerische Optimierung adaptiver transsonischer Profile.* PhD Thesis, Institut für Aerodynamik und Gasdynamik, Universität Stuttgart, 2005, Shaker Verlag, ISBN 3-8322-3691-O

Numerical Simulation of Helicopter Aeromechanics in Slow Descent Flight

M. Embacher, M. Keßler, F. Bensing, and E. Krämer

Institut für Aerodynamik und Gasdynamik (IAG)
Universität Stuttgart, Pfaffenwaldring 21, 70550 Stuttgart, Germany

Summary. In this paper we present numerical simulation results for a generic helicopter configuration in slow descent flight. The well known HART-II test case has been chosen as the experimental reference, especially the baseline case. This test case is characterized by the occurrence of Blade-Vortex Interactions (BVI) and can thus be considered as very ambitious with respect to the aerodynamic simulation. The HART-II test case has been subject to previous investigations at the Institute. A local mesh refinement technique using tube-shaped vortex-adapted Chimera grids was developed in order to improve the vortex conservation in the numerical simulation and thus to allow for the quantitative reproduction of BVI induced airloads. In contrast to earlier results, the vortex grids are extended to fill the entire rotor disk and the trim including the fuselage is run to convergence. A comparison of the numerical simulations with wind tunnel experiments shows the improvement in the prediction of the aeromechanics of the helicopter. The numerical simulations incorporating the enhancements mentioned above were run on the vector computer NEC SX-8 located at the high performance computing center Stuttgart.

1 Introduction

For more than two decades helicopter aerodynamics has been one of the main research interests at the Institut fr Aerodynamik und Gasdynamik (IAG). Starting from inviscid flow around main rotors in hover, we have made progress in the direction of forward flight, viscous simulations, aeroelastic coupling and recently full helicopter configurations including fuselage and tail rotor.

Fluid-structure coupling between Computational Fluid Dynamics (CFD) and Computational Structural Dynamics (CSD) for the main rotor has been proven to be indispensable to obtain viable results, at least in forward flight conditions. The trim of the rotor is absolutely mandatory as well, in order to allow for a meaningful comparison of the numerical results to experimental or flight test data. This way, at least the global structure of the rotor wake is reproduced by the simulation, regardless of subtle differences between the numerical and experimental setup, e.g. wind tunnel intricacies.

Slow descent flight belongs to the most complex flight conditions of the helicopter. This is due to the fact that blade tip vortices shed from the rotor blades remain within the rotor disc area and thus do interact closely with other rotor blades. This undesirable phenomenon is denoted as Blade-Vortex Interaction (BVI, [8,10]) and has a severe impact on the acoustic behaviour of the rotor, as well as on vibrations. This so-called BVI-induced impulsive noise is the main noise source on a helicopter in slow descent flight. The accurate prediction of the corresponding airloads in the numerical simulation is a prerequisite in order to compute the acoustics and thus to investigate possible noise reduction strategies.

Using a field method the reproduction of BVI-induced airloads is a difficult task due to the fact that the tip vortices are subject to the dissipation of the numerical scheme. In order to reduce the numerical dissipation of the CFD method, a Chimera-based local mesh refinement method utilizing vortex-adapted grids has been developed at the IAG. Previous results have been published in [2,4,13].

Fig. 1. HART-II hingeless rotor model in the DNW wind tunnel

Again, we use a weakly coupled aeroelastic simulation method. The aerodynamic modelling is carried out by the CFD method FLOWer of DLR [7] and the dynamic modelling including rotor trim capability is performed by the flight mechanics method HOST of Eurocopter [1]. In the following sections we will present the numerical methods, the weak coupling scheme between FLOWer and HOST and the results obtained.

2 Mathematical Formulation and Numerical Scheme

2.1 Computational Setup

The numerical setup used for all our helicopter simlations has been described already in detail (e.g. [2]). In short, the CFD part consists of a standard structured multiblock Finite Volume scheme of second order accuracy on smooth meshes, solving the Reynolds-averaged Navier-Stokes equations including some turbulence model for closure. Runge-Kutta time integration of second order for our instationary results is done using implicit dual-time stepping. The Chimera technology of overlapping grids together with an Arbitrary Lagrangian-Eulerian (ALE) formulation allows for grid deformations and motions, as needed for the complex blade dynamics. Additional information on the code can be found in References [7, 11].

The CSD problem of the blade is formulated as a quasi one–dimensional Euler-Bernoulli beam, that allows for deflections in flap and lag direction as well as elastic torsion along the blade axis. In addition to the assumption of a linear material law, tension elongation and shear deformation are neglected. However, possible offsets between the local cross-sectional center of gravity, tension center and shear center are accounted for, thus coupling bending and torsional degrees of freedom. Rigid segments are connected through virtual joints, allowing for geometrical nonlinearities. The large number of local rotations as degrees of freedom is reduced by a modal Rayleigh-Ritz approach, incorporating only a limited number of lower frequency modes and corresponding eigenforms. The computation is carried out by the Eurocopter flight mechanics tool HOST [1].

Coupling between CFD and CSD is done in a weak fashion, exchanging periodic coefficients once per revolution. These consist of forces and moments from CFD as inputs to CSD, and resulting deformations in the other direction. At the same time, control angles (collective and cyclic) are updated in order to reach the trim objectives, namely rotor thrust and the two mast moments. Trim is carried out by HOST, which uses its internal blade element formulation together with airfoil tables to calculate the required trim jacobian.

Grid deformation is done analytically according to the surface deformations defined by CSD. In order to keep mesh quality, surface rotations and motion are damped out towards the far field using Hermite polynomials. The multiblock capable grid deformation process is not completely general, but sufficient for all blade grids encountered so far. Different blocks of the blade grid may thus be distributed onto different processors, allowing for an effective parallelization of the flow computation.

2.2 Vortex-Adapted Grid Technique

One prerequisite for successful modeling of BVI is to preserve the strength of the blade tip vortices, such that decay and dispersion occur at realistic rates, and sufficient interaction with a following blade is generated also after a wake age of one or several rotor revolutions. In order to minimize the dissipation of vorticity within the otherwise highly dissipative coarse background grid, a Chimera-based local mesh refinement strategy utilizing vortex-adapted grids was developed at the IAG and incorporated into the FLOWer code.

The method adds tube-shaped Chimera child grids to the existing grid setup, which are generated from a nearly circular, distorted Cartesian mesh that is evolved along a predefined trajectory. The length of the grids is determined by the wake age ψ_W to be covered. The convection of the tip vortices is anticipated by analytically describing the trajectory according to Egolf and Landgrebe [5]. The vortex adapted grids emanate from within the blade grids to ensure a loss free transfer of tip shed vorticity. The accurate flow exchange with the remaining grids of the configuration is ensured by Chimera hole cutting. As the section of the helical tip vortex trajectory covered by the vortex grid changes with the physical time step the vortex grids have to be treated as deformable. Further details of the method are given in [3].

3 Results

3.1 Description of the Test Case

The HART-II test ([9, 12, 14]) uses a 40% Mach scaled, four-bladed hingeless Bo105 model rotor with a diameter of 4.00 m. It is dynamically scaled in order to match the natural frequencies of the full-scale blade in the first three flap, first two lead-lag and first torsion modes. The rotor blades are rectangular with a $-8°$ linear twist and a precone angle of $2.5°$. The blade features a modified NACA23012 airfoil with a chord aligned trailing edge tab. The hover tip Mach number is 0.64, and the advance ratio is 0.15 for the present case. The rotor was installed with a rotor shaft angle of $5.3°$ (nose up). For the computation we prescribed a shaft angle of $4.5°$, taking $0.8°$ of wind tunnel interference effect into account. For the present paper we restricted our investigations to the HART-II baseline case, which features a pure monocyclic pitch input without any higher harmonic control (HHC).

3.2 Computational Setup

In the present report we will show results obtained from three different grid arrangements:

a) Near-field grids for rotor and fuselage.
b) Near-field grids for rotor, fuselage and a truncated blade tip vortex system
c) Near-field grids for rotor, fuselage and the complete blade tip vortex system

A cartesian background grid embeds all near-field grids. The blade grids make use of the multi-block capability of the grid deformation tool and consist of 10 blocks each. A C-topology is used in the chordwise direction and an H-topology in the spanwise direction. The fuselage mesh consists of 6 blocks and uses a CO-topology. The geometry of the generic fuselage is part of the HART-II database. It has been slightly simplified by neglecting the cylindrical rotor shaft fairing on its upper surface. This simplification is justified, as in the present paper we just want to capture the dominating blockage effect of the fuselage and its impact on the rotor solution, and not the accurate prediction of some fuselage pressure distribution. The cartesian background grid features a successive grid refinement utilizing hanging grid nodes towards the fuselage mesh overlap region. This allows for a grid spacing in the overlap region comparable to the fuselage grid, while saving grid cells in the far field. The situation is depicted in Figure 2 showing slices through the rotor/fuselage grid system.

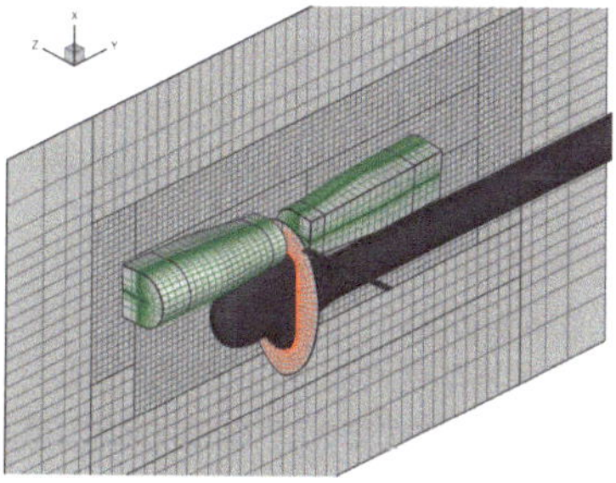

(a) Slice in longitudinal direction

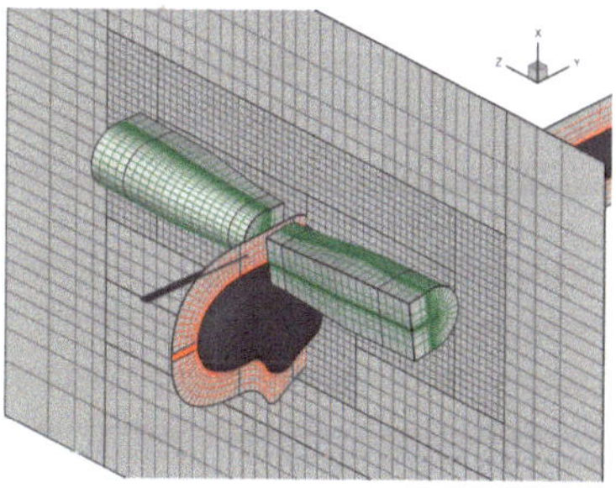

(b) Slice in lateral direction

Fig. 2. Illustration of the grid system (vortex adapted grids not included, every 4th grid line shown)

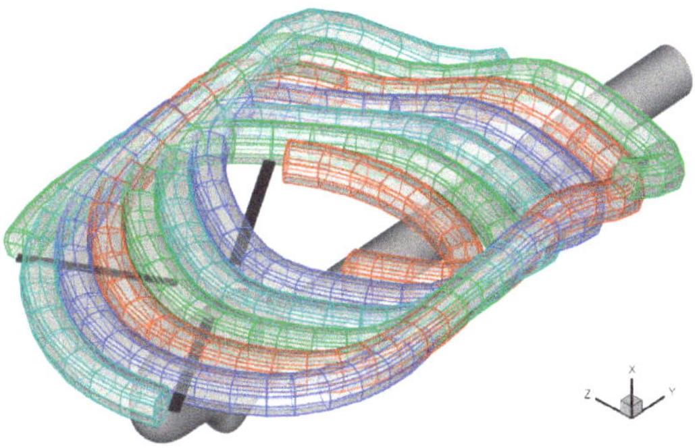

Fig. 3. Vortex adapted grids extending over a wake age of $\psi_W = 544°$

In grid arrangement b), the mesh refinement by vortex adapted grids is applied over a fixed wake age of $\psi_W = 544°$. Thus, the tip vortex reaches the rear border of the tube shaped near-field grid when it becomes older than approximately 1.5 rotor revolutions and is hereupon computed in the background grid. As it becomes visible in Figure 3, showing the vortex adpated grids in arrangement b) for $\Psi = 30°$, this setup fills the rotor disk region with near-field grids except for a small section at its rear border. To effectively cover the entire rotor disk area by vortex adapted grids, their wake age can be estimated by an approximate rule to be $1/\mu\pi$ rotor revolutions. Hence, for the present flight case at the relatively low advance ratio of $\mu = 0.15$ a prolongation of the vortex adapted grid age to at least 2.12 rotor revolutions is appropriate, and accordingly in grid arrangement c) the age is extended to $\psi_W = 768°$. The locally refined vortex system now is complete in the sense that coverage of the entire rotor disk is ensured for all blade azimuths. However, since the total number of grid cells is increased by 21% compared to arrangement b), the computationally intensive setup c) was employed only for one of the simulations presented here. Simultaneously with extending the vortex-adapted grids to $\psi_W = 768$, also their trajectory was slightly adjusted. Based on the observation of the vortex paths in preceeding simulations with grid arrangement b), the shape and envelope functions S and E, which describe the offset of the grid centerline from a helical path, were determined by a least mean squares fit to the observed vortex core locations at several wake ages. Figure 4 compares the trajectory of the two vortex adapted grids in terms of the axial and radial offset, showing that mainly the radial component was modified at wake ages older than $\psi_W = 270°$.

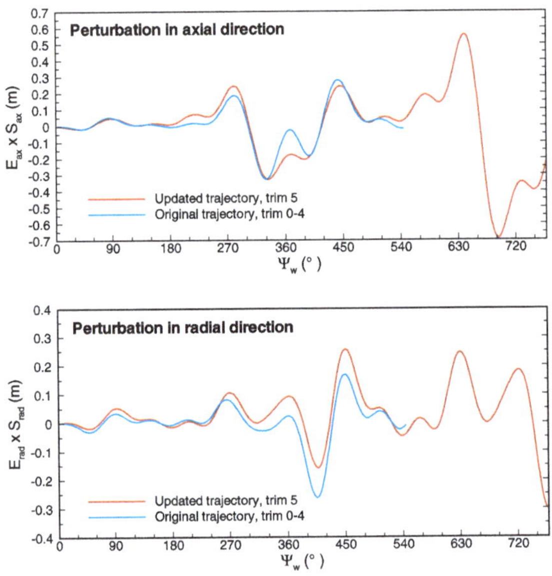

Fig. 4. Offset of vortex adapted grid trajectory from helical path

The grid resolutions are summarized in Table 1. As it can be seen from the table two different blade grids were generated. The standard blade grid features 1.2 million grid cells and was used for the rotor/fuselage computations. If vortex-adapted grids are added to the setup in order to capture the BVI, a comparable grid resolution should be provided in the overlapping regions between the blade grids and the vortex grids. In this case the fine blade grid featuring 5.5 million grid cells is used. As shown in Table 1 the rotor/fuselage setup amounts to approximately 20 million cells in total, while the rotor/fuselage/vortex grid setup uses approximately 73 million cells in case of grid arrangement b). When the vortex grids are prolonged to a wake age of $\psi_W = 768°$, the number of grid cells rises to 87.2 million.

Table 1. Grid resolution

Grid	Number of cells	a)	b)	c)
background	10,780,672	X	X	X
fuselage	4,030,464	X	X	X
coarse blade	4 x 1,202,688	X	—	—
fine blade	4 x 5,534,208	—	X	X
standard vortex	4 x 8,912,896	—	X	—
long vortex	4 x 12,582,912	—	—	X

All CFD computations were carried out using RANS modelling. The $k-\omega$ turbulence model according to Wilcox was used for the closure of the RANS equations. A time step of 1° azimuth was used for all simulations, and to achieve a converged state of the flow three rotor revolutions were computed.

Using the weak coupling method by adaption of the collective and cyclic pitch input, the rotor is trimmed towards the experimental trim objectives. Previous trim simulations with the isolated rotor, i.e. not including the generic fuselage, provided a trimmed dynamic state of the isolated rotor during iterations no. 0 until no. 2. Along this process, the trailing edge tab extending 0.045 chordlengths from the rear edge of the blade was deflected upwards by 1°. This corrective measure aims at a reduction of nose-down pitching moment, since earlier investigations indicated an overprediction of the blade's elastic twist. The modification results in a spanwise redistribution of lift and slightly improves the reproduction of the experimental reference values for blade tip flapping and elastic torsion.

Serving as a starting point for the present investigations, the blade dynamic state and the control angles obtained with the isolated rotor simulations were prescribed for two computations with grid arrangements a) and b). This initial step in the trim process of the complete configuration is referred to as trim 3 in the following. While the trim 3 computation with grid arrangement a) provides the updated blade dynamics for further trimming, another trim 3 computation with identical setup besides the omission of vortex adapted grids was used in comparison in order to assess the impact of the mesh refinement technique. The trim process is continued by the trim 4 computation with grid arrangement b), and completed by the trim 5 computation using the grid arrangement c) with elongated vortex grids. Trim 6 just proves convergence using the CFD forces and moments from the last step.

3.3 Trim Convergence and Blade Dynamics

Figures 5 and 6 show the convergence properties of the weakly coupled method using three different representations. The first figure plots the development of the collective and cyclic control angles θ_0, θ_C and θ_S during the trim process. As mentioned in the previous section, an isolated rotor trim which converges during three retrim cycles is used as an initial solution for trim 3. The changes from trim 3 to trim 4 mark the transition towards a grid setup including the generic fuselage. Due to the displacement effect of the fuselage the rotor thrust and moments depart from the trim objectives, and as a consequence the angles are readjusted in the subsequent trim 4. Since the fuselage introduces mainly a fore/aft disturbance to the flow through the rotor disk, θ_S that controls the $\sin(\Psi)$-variation in blade pitch is less affected. A converged trim state for the full configuration, i.e. including the fuselage, is obtained after two further retrim cycles in trim 5. This is verified by performing one further trim calculation with the flight mechanics tool, indicating that the changes in all three control angles do not exceed 0.025° from trim 5 to trim 6. The corresponding development of the blade dynamic state is shown in Figure 6a) in terms of blade tip flapping. Note that the curve pertaining to trim 3 represents both the isolated rotor simulation and the computation including the fuselage. As expected, the changes in the flap motion reduce from one retrim cycle to the next and approach a converged state. The same statement holds for the blade tip torsion distribution, as it becomes evident from Figure 6b). In comparison to the experiment, both flapping and torsion distributions reproduce well the measured azimuthal variations. However, a clear difference persists in the mean tip flap deflection despite the tab angle modifications. With respect to the torsional motion, reasonable quantitative results are obtained. The difference between the dynamic equilibrium state for the isolated rotor and for the complete configuration indicates that including the generic fuselage into the simulation setup significantly improves the prediction of the measured values, which can be attributed to the changes in control angles as well as to rotor-fuselage interference effects.

3.4 Aerodynamic Loads and BVI

Figure 7 shows the three-dimensional flow field with the λ_2 criterion used to visualize the tip vortex pattern at the instant $\Psi = 70°$. Only the solution calculated within the vortex adapted grids is shown. The isosurfaces at $\lambda_2 = -0.0008$ are colored by vorticity magnitude, with red color denoting high vorticity and thus a low age of the vortex. As can be seen from the small but nearly constant diameter of the isosurfaces that vaguely represent the vortex cores, dispersion is minimized and the vorticity remains concentrated. Equally, conservation of the tip vortices is ensured beyond a wake age of more than one rotor revolution, recognizable from the persistence of tip vortices

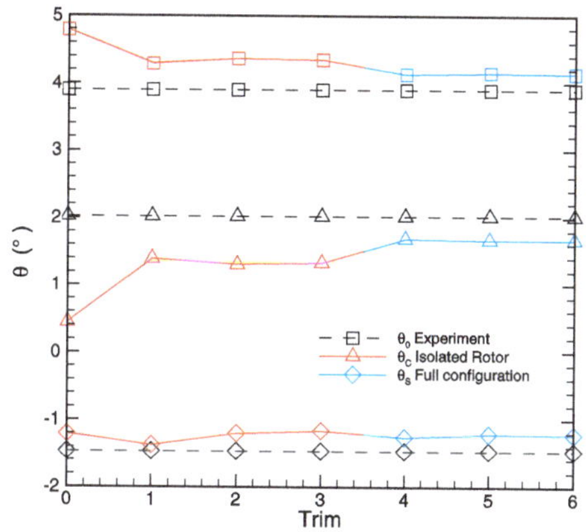

Fig. 5. Convergence of the control angles

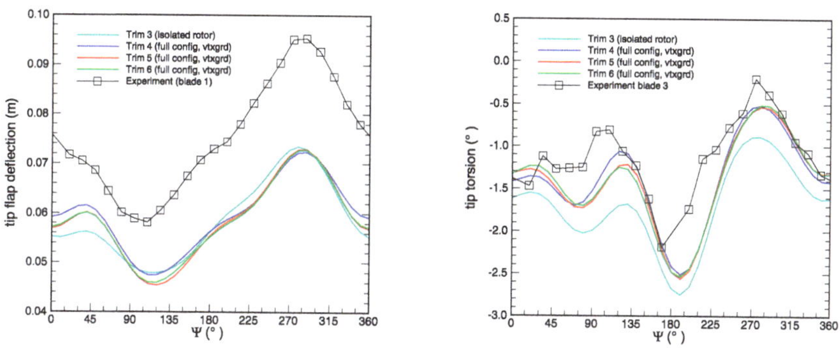

(a) Flap deflection

(b) Elastic torsion

Fig. 6. Tip motion versus azimuth

downstream of the rotor in the top right corner of Figure 7. Numerous small-scale structures and packets of several aligned vortex cores occur particularly in the rear half of the rotor area. These structures represent sheets of vorticity emanating from the blade trailing edges, or are the outcome of blade vortex interactions.

The vertical lift distribution on the rotor disk is shown in Figure 8(a) for the trim 3 computation using grid arrangement a) without vortex adapted grids, while Figure 8(b) plots the same value in case of trim 5 and grid arrangement c). The azimuthal load distribution in Figure 8(a) is comparably smooth, whereas a first indication of BVI can be recognized from Figure 8(b), as high-frequency azimuthal variations in lift become visible around a blade azimuth of $\Psi = 300°$. By forming the difference of the loadings given in Fig-

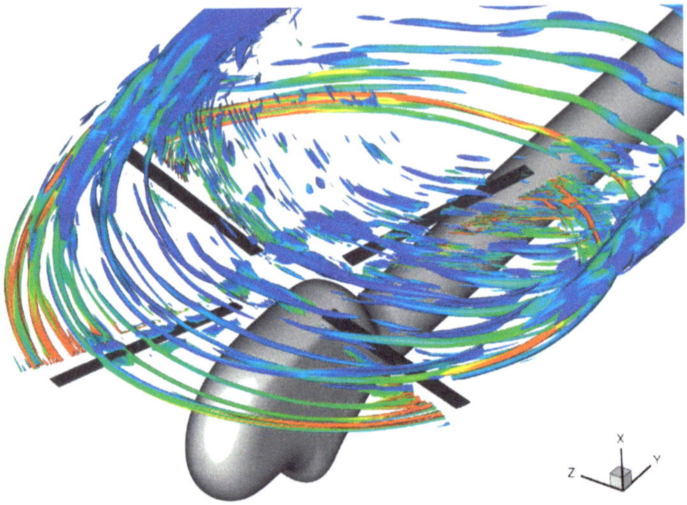

Fig. 7. λ_2 isosurface of the flow, colored by vorticity magnitude

ure 8(a) and (b), these load fluctuations become clearly evident in Figure 8(c). Accordingly, BVI occurs in the numerical simulation primarily for a blade azimuth between $\Psi = 280°$ and $\Psi = 330°$, where three distinct pairs of local maxima and minima in the loading of the outer portion of the rotor blade are registered. A fluctuation amplitude of approximately 130 N/m was determined at $0.87\,r/R$. At these blade positions tip vortices are encountered that are oriented nearly parallel to the blade, as can be verified from Figure 7 and Figure 3. Further peaks in the loading are detected close to the blade tip at $\Psi = 350°$ and $\Psi = 80°$, and smaller load fluctuations with an amplitude of approximately 50 N/m are experienced by the blade when it passes through the first quadrant of the rotor disk. Besides demonstrating the impact of using a dedicated vorticity conservation technique, Figure 8(c) also shows the redistribution of lift effected by the control angle changes from trim 3 to trim 5 (cf. Figure 5). Due to the fuselage displacement or blockage effect on the flow through the rotor disk, the angle of attack is temporarily increased when the blade passes by the fuselage nose, and decreased during the passage above the mounting fairing. Readjustment of θ_C compensates this effect by redistributing lift from the front to the rear part of the rotor disk, such that the rotor integral pitching moment objective is restored.

3.5 Comparison to the HART-II Experiment

In Figure 9 the sectional load $C_n Ma^2$ measured at $r/R = 0.87$ in the experiment is compared to the simulation results. Comparing the green and blue

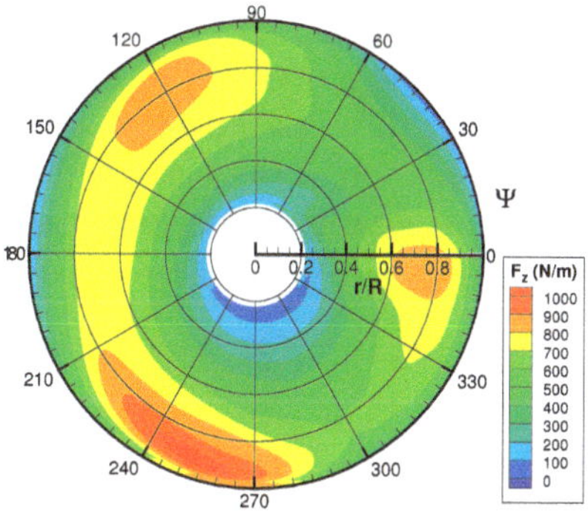

(a) Trim 3 without vortex adapted grids

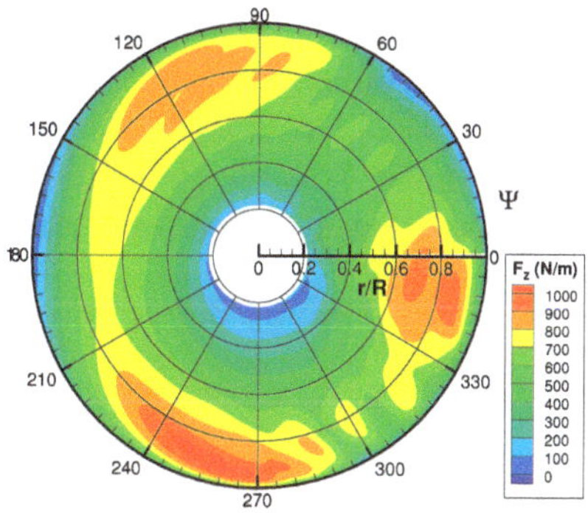

(b) Trim 5 with vortex adapted grids

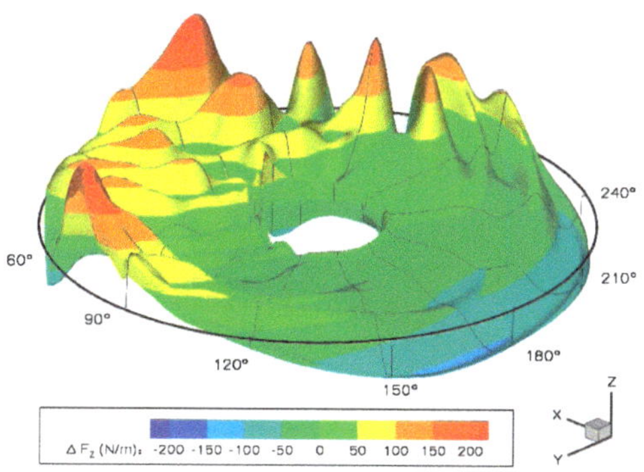

(c) Difference between Trim 5 and Trim 3

Fig. 8. Distribution of normal sectional load over the rotor disk

curves for the trim 3 simulations with grid arrangements a) and b), again the role of the vortex adapted grids for the prediction of BVI oscillatory loads becomes apparent. At the retreating blade side between $\Psi = 225°$ and $\Psi = 330°$ the result from trim 3 using vortex grids compares fairly well with the observations of the HART-II experiment in terms of both fluctuation amplitude and phase. This statement however does not apply to blade azimuth angles around $\Psi = 45°$, where both the fluctuation magnitude as well as the $C_n Ma^2$ stripped of the BVI event clearly underpredict the experimental values. A cause of the mismatch in the BVI oscillations might be found in a not sufficiently accurate analytical description of the vortex trajectory in this region. Concerning the unperturbed component, the situation is improved in trim 5 by the fore/aft redistribution of load mentioned in the previous section. Also at the retreating side, the rotor trim with the fuselage present shifts the $C_n Ma^2$ curve to higher values in closer proximity of the experiment. The slight reduction of BVI oscillatory load around $\Psi = 300°$ from trim 3 to trim 5 is caused by weaker tip vortices encountered by the blade in this region. This reduction in vorticity strength can be directly attributed to the fact that during generation of the respective tip vortices about half a rotor revolution before the blade creates less lift in trim 5. An influence from the vertical separation of the blade and vortex positions can be ruled out since neither blade nor vortex locations are changed significantly during the trim process.

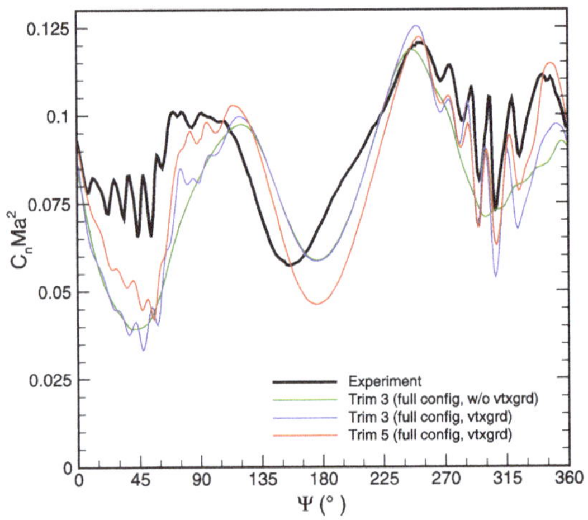

Fig. 9. Lift coefficient C_nMa^2 at the radial position $r/R = 0.87$

In order to examine whether the simulated tip vortex pattern is consistent with the experiment, vortex core positions on two planes cutting vertically through the rotor disk were evaluated. The planes are oriented longitudinally and are determined by a constant lateral distance of $z = 1.4$ m and $z = -1.4$ m. Relevant experimental flow data is based on PIV measurements. From the velocity field data obtained with PIV, vortex core locations were extracted and published by the HART-II consortium as part of the test documentation [6]. Due to the restriction of the PIV technique to record data only in one sector between two blades, vortex core positions are given in this reference in two different qualities. For a blade positioned at $\Psi = 110°$, vortex core locations extracted directly from the PIV data at $z = 1.4m$ are readily available for all vortices located behind the blade, i.e. in the first quadrant of the rotor disk. The position of vortices above and in front of the blade, however, have to be reconstructed from a complementary measurement at $\Psi = 70°$ and are referred to as interpolated positions in the following. Figures 10 and 11 show contours of vorticity magnitude of the numerical simulation at $z = 1.4$ m and

$z = -1.4\,\mathrm{m}$, with the boundary of the blade and vortex adapted grids marked as a black line. In both plots view is from right to left, such that in Figure 10 the blade at $\Psi = 110°$ travels to the right and advances in flight direction. In Figure 11 the plane $z = -1.4\,\mathrm{m}$ at the retreating blade side is shown with the blade at $\Psi = 290°$ travelling to the left in wind direction. Besides the tip vortices, which appear as distinct spots of concentrated vorticity mainly above the rotor plane, also the vortex sheet between the blade trailing edge and its corresponding tip vortex is visible. The location of the vortex cores in the experiment is marked by red circles, or red squares in case of interpolated positions. The overall agreement between simulation and experiment can be considered good, with all locations separated by less than one chordlength. Particularly the vertical position is predicted with high accuracy, which is beneficial in light of the sensitivity of oscillatory loads to the interaction distance between the blade and a passing vortex.

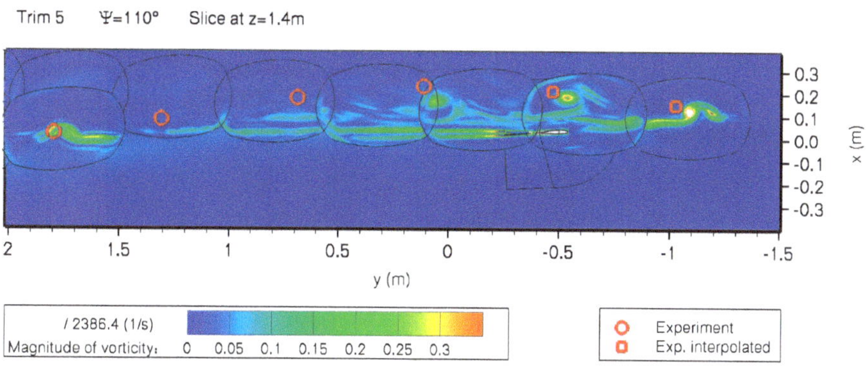

Fig. 10. Vorticity magnitude in the plane $z = 1.4\,\mathrm{m}$, flight direction to the right

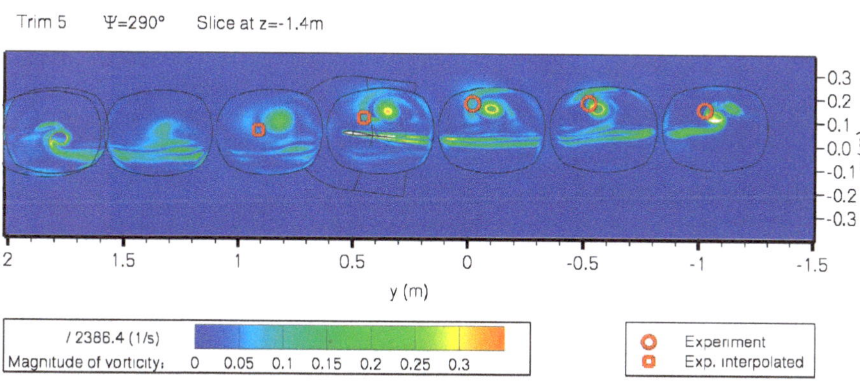

Fig. 11. Vorticity magnitude in the plane $z = -1.4\,\mathrm{m}$, flight direction to the right

3.6 Computational Performance

All CFD computations presented in this paper were performed on the NEC SX-8 as multi-node computations using 64 processors. The choice of eight nodes is due partly to memory requirements and partly to the wish to converge the flow state for one trim iteration within reasonable time. The multiblock grid deformation tool allowed for an effective parallelization of the computation. The computational performance data for trim 4 and trim 5 are given in Table 2. The vector operation ratio is highly influenced by the extent of the Chimera overlap regions, since at the beginning of every physical time step the Chimera connectivities have to be set, involving a costly search and interpolation process which is only partially vectorized. As can be seen from Figures 2 and 3, the Chimera setup is rather complex for this configuration involving near-field grids for fuselage, blade as well as vortex grids. Recalling from table 1 that the grid arrangement for trim 5 exceeds the trim 4 setup by 20% in size, it is interesting to note that the difference in computational effort is only 11% of trim 4 wall clock time. This is rooted in a simplified Chimera exchange policy for vortex adapted grids in certain parts of the computational domain, a measure that is also reflected in a higher number of GFLOPS during trim 5. The average vector length is comparable in both cases.

Table 2. Computational performance

	Trim 4	**Trim 5**
Grid arrangement	b)	c)
Number of blocks	107	107
Number of cells	72,599,522	87,279,616
Number of CPUs	64	64
GFLOPS	74.6	84.7
Vector operation ratio	97.7%	97.8%
Average vector length	185.7	182.2
Wall clock time, 3 rotor revolutions	77.1h	85.7h
Required memory	701 GB	701 GB

4 Conclusions and Outlook

We presented results of numerical aeroelastic simulations of the HART-II experimental test case. In contrast to a previous publication [3] on this setup, the vortex adapted grids have been extended and the trim procedure has been continued. These modifications resulted in an improved agreement of the numerical prediction and the experimental reference.

The inclusion of the fuselage necessitated a significant change in control angles (collective as well as cyclic) to counter the blockage effect. This change partly reduced discrepancies with experimental values, especially an improvement in tip torsion can be observed. However, while seeing some BVI effects in the first quadrant, there still is a significant underprediction unexplained up to now. At the retreating side, vortex positions obtained by PIV measurements coincide well with the simulated results, and the corresponding BVI fluctuations can be considered adequate for quantitative acoustic postprocessing.

This acoustic evaluation of the obtained aerodynamic results is the next step on our roadmap to a complete analysis of helicopter configurations. Further cases of the HART-II experimental database as well as the upcoming GOAHEAD measurements will enable us to further validate the established tool chain and provide qualitative insight into the physics as well as quantitative figures for forthcoming developments. Clearly, at the complexity of the problem and the stated resources required, helicopter simulation will be a challenging task for high performance computers in the foreseeable future. The NEC SX-8 is an ideal platform for the parallelized, vectorized approach taken here.

Acknowledgement

This work has been funded in part by the European Union. The authors would like to thank all GOAHEAD partners for their support and cooperation. Special thanks go to the HART consortium for providing substantial experimental data. Furthermore we would like to thank the system administrators of HLRS for their technical support.

References

1. Benoit, B., Dequin, A-M., Kampa, K., Grünhagen, W. v., Basset, P-M., Gimonet, B.: **HOST: A General Helicopter Simulation Tool for Germany and France.** American Helicopter Society, 56th Annual Forum, Virginia Beach, Virginia, May 2000.
2. Dietz, M., Kessler, M., Krämer, E.: **Trimmed Simulation of a Complete Helicopter Configuration Using Fluid-Structure Coupling.** High Performance Computing in Science and Engineering, pp. 487–501, Springer Verlag, 2007.
3. Dietz, M., Kessler, M., Krämer, E., Wagner, S.: **Tip Vortex Conservation on a Helicopter Main Rotor Using Vortex-Adapted Chimera Grids.** AIAA Journal, Vol. 45, No. 8, August 2007, pp. 2062–2074.
4. Dietz, M., Krämer, E., Wagner, S.: **Tip Vortex Conservation on a Main Rotor in Slow Descent Flight Using Vortex-Adapted Chimera Grids.** Proceedings of the AIAA 24th Applied Aerodynamics Conference, San Francisco, June 2006.
5. Egolf, T.A., Landgrebe, A.J.: **Helicopter Rotor Wake Geometry and Its Influence in Forward Flight.** NASA Contractor Report 3726, 1983.

6. HOFFMANN, F., VAN DER WALL, B.: **The vortex trajectory method applied to HART II PIV data.** Institute Report IB 111-2005/34, Institute of Flight Systems, DLR, Braunschweig, 2005.
7. KROLL, N., EISFELD, B., BLEECKE, H.M.: **The Navier-Stokes Code FLOWer.** Notes on Numerical Fluid Mechanics, Vol. 71, Vieweg, 1999, pp. 58–71.
8. LIM, J.W., NYGAARD, T.A., STRAWN, R., POTSDAM, M.: **BVI Airloads Prediction Using CFD/CSD Loose Coupling.** AHS Fourth Vertical Lift Aircraft Design Conference, San Francisco, January 2006.
9. LIM, J.W., TUNG, C., YU, Y.H., BURLEY, C., BROOKS, T., BOYD, D., VAN DER WALL, B., SCHNEIDER, O., RICHARD, H., RAFFEL, M., BEAUMIER, P., BAILLY, J., DELRIEUX, Y., PENGEL, K., MERCKER, E.: **HART-II: Prediction of Blade-Vortex Interaction Loading.** 29th European Rotorcraft Forum, Friedrichshafen, Germany, September 2003.
10. RÖTTGERMANN, A., BEHR, R., SCHÖTTL, C., WAGNER, S.: **Calculation of Blade-Vortex Interaction of Rotary Wings in Incompressible Flow by an Unsteady Vortex-Lattice Method Including Free Wake Analysis.** Notes on Numerical Fluid Mechanics, Vol. 33: Numerical Techniques for Boundary Element Methods, W. Hackebusch (Ed.), Vieweg, 1991, pp. 153–166.
11. SCHWARZ, T.: **The Overlapping Grid Technique for the Time-Accurate Simulation of Rotorcraft Flows.** 31st European Rotorcraft Forum, Florence, Italy, September 2005.
12. VAN DER WALL, B., JUNKER, B., BURLEY, C., BROOKS, T., YU, Y.H., TUNG, C., RAFFEL, M., RICHARD, H., WAGNER, W., MERCKER, E., PENGEL, K., HOLTHUSEN, H., BEAUMIER, P., PRIEUR, J.: **The HART-II Test in the LLF of the DNW – a Major Step towards Rotor Wake Understanding.** 28th European Rotorcraft Forum, Bristol, England, September 2002.
13. WAGNER, S. DIETZ, M., EMBACHER, M., SCHNEIDER, C., KRÄMER, E.: **Influence of Grid Arrangements and Fuselage on the Numerical Simulation of the Helicopter Aeromechanics in Slow Descent Flight.** 15th International Conference on Computational & Experimental Engineering and Sciences, Honolulu, Hawaii, USA, March 2008.
14. YU, Y.H., TUNG, C., VAN DER WALL, B., PAUSDER, H., BURLEY, C., BROOKS, T., BEAUMIER, P., DELRIEUX, Y., MERCKER, E., PENGEL, K.: **The HART-II Test: Rotor Wakes and Aeroacoustics with Higher-Harmonic Pitch Control (HHC) Inputs – The Joint German/French/Dutch/US Project.** American Helicopter Society, 58th Annual Forum, Montreal, Canada, June 2002.

Partitioned Fluid-Structure Coupling and Vortex Simulation on HPC-Systems

Felix Lippold, Eugen Ohlberg, and Albert Ruprecht

Institute of Fluid Mechanics and Hydraulic Machinery, Universität Stuttgart, Pfaffenwaldring 10, 70550 Stuttgart ruprecht@ihs.uni-stuttgart.de

Summary. In this article two examples for applications in high-performance computing are presented. One is the fluid-structure interaction (FSI) in a tidal turbine. The other one is the simulation of unsteady vortices in hydraulic turbine draft tubes. The architecture for fluid-structure interaction with the in-house CFD code FENFLOSS and a commercial structural code is presented. Furthermore, issues of installing the environment on a NEC SX-8 vector computer is discussed and performance data is presented. Additionally, the issues of turbulence and the impact of turbulence models on the accuracy of the vortex capturing are investigated. All simulations are performed on a cluster (structural mechanics) and a NEC SX-8 vector computer (CFD).

1 Introduction

In many fields of engineering complex vortex dominated flows play an important role. Often the flow behaviour gets unstable and a complex unsteady vortex movement occurs. These vortices lead to pulsating pressure field and consequently also to unsteady loads on the structure.

In case of slim structures the unsteady loading can lead to a deformation of the structure. A relatively large deformation of the structure on the other hand can again influence the flow behaviour. For an accurate prediction of these behaviour at first the unsteady vortex motion have to be simulated very accurately. For example by using a wrong turbulence model one can completely suppress the unsteady vortex motion. Here different sophisticated models are analysed for the prediction of an unsteady vortex movement. The used application is the part load vortex rope in the draft tube of a water turbine. This is a very critical flow situation for hydro turbines, because the rotating vortex rope can lead to severe vibrations which can even result in a operation restriction of the machine.

In a second step a coupled solution of the fluid and of the resulting structural behaviour is necessary. For this fluid structure interaction the coupling of a flow simulation code and a structure simulation code is carried out. As an example the flow through a tidal current turbine is taken into account. This type of turbine is characterized by rather slim and long turbine blades, where the deformation cannot be neglected.

In this paper an overview over the used turbulence models and its performance is given. Also the fluid structure coupling is described in detail.

2 Basic Equations

Incompressible, turbulent flows are described by the continuity equation for constant fluid density and the Reynolds-averaged Navier-Stokes equations. If the fluid domain is changing, e.g. in fluid-structure interaction, the mesh movement has to be accounted for. This yields the Arbitrary-Lagrange-Euler (ALE) formulation. Furthermore, turbulent flow plays an important role in engineering applications. Nevertheless, a direct simulation of turbulent fluctuations is not possible for complex problems, today. Hence, the influence of the turbulent fluctuations on the mean flow is modelled by an additional viscosity. Two-equation models solve two additional transport equations. They are widely used today even though they have certain drawbacks. But, due to sophisticated enhancements to the standard models a wide range of flow phenomena can be represented with acceptable accuracy.

An additional topic in fluid-structure interaction is the representation of the structure. In hydraulic machinery the added fluid mass has a great influence on the vibrational behaviour. The transformation of the fluid pressure to the strucutral surface models this additional mass. Usually, a linear material behaviour with small deformations may be assumed.

Flow Equations

In order to simulate the flow of an incompressible fluid the momentum equations and mass conservation equation are derived in an infinitesimal control volume. Including the turbulent fluctuations yields the incompressible Reynolds-averaged Navier-Stokes equations [FP02]. Considering the mesh deformation introduces a velocity to the grid nodes U_{G} and results in the Arbitrary-Langrange-Euler (ALE) formulation [DO03, HUG81]

$$\frac{\partial U_i}{\partial x_i} = 0 \tag{1}$$

$$\left.\frac{\partial U_i}{\partial t}\right|_{\mathcal{X}} + (U_j - U_{\mathrm{G}})\frac{\partial U_i}{\partial x_j} + \frac{1}{\rho}\frac{\partial P}{\partial x_i} + \frac{\partial}{\partial x_j}\left[\nu\left(\frac{\partial U_i}{\partial x_j} + \frac{\partial U_j}{\partial x_i}\right) - \overline{u_i' u_j'}\right] = f_i \tag{2}$$

The components of the flow velocity are U_i and the pressure is P. On the right hand side there are loads due to rotational and body forces f_i. In this case we use an ALE-formulation of the Navier-Stokes equations, hence, the time derivatives are computed in the ALE frame of reference $\mathcal{X}$. In case of a non-moving mesh, $U_{\mathrm{G}} = 0$, equation (2) complies with the regular Euler formulation of the Navier-Stokes equations.

Turbulence Modelling

Usually, Reynolds Stresses are modelled following Boussinesq's eddy viscosity principle. To model the resulting turbulent viscosity, for many engineering problems k-ε and k-ω-models are combined with logarithmic wall functions. Alternatively, Low-Reynolds formulations are applied. For both models and their enhanced variations two additional transport equations have to be solved. The turbulent viscosity ν_{t} is computed from the turbulent kinetic energy k and its dissipation ratio ε or the turbulent vorticity ω, respectively. Basically, the k-equation stays the same for both models and reads for the standard versions

$$\frac{\partial k}{\partial t} + \hat{U}_i \frac{\partial k}{\partial x_i} - \frac{\partial}{\partial x_i}\left[\left(\nu + \frac{\nu_t}{\sigma_k}\right)\frac{\partial k}{\partial x_i}\right] = G - \varepsilon \quad . \tag{3}$$

The equations for the dissipation ratio and the turbulent vorticity read

$$\frac{\partial \varepsilon}{\partial t} + \hat{U}_i \frac{\partial \varepsilon}{\partial x_i} - \frac{\partial}{\partial x_i}\left[\left(\nu + \frac{\nu_t}{\sigma_\varepsilon}\right)\frac{\partial \varepsilon}{\partial x_i}\right] = c_{1\varepsilon}\frac{\varepsilon}{k}G - c_{2\varepsilon}\frac{\varepsilon^2}{k} \tag{4}$$

$$\frac{\partial \omega}{\partial t} + \hat{U}_i \frac{\partial \omega}{\partial x_i} - \frac{\partial}{\partial x_j}\left[\left(\nu + \frac{\varepsilon}{\sigma_\omega}\right)\frac{\partial \omega}{\partial x_j}\right] = \alpha\frac{\omega}{k}R_{ij}\frac{\partial u_i}{\partial x_j} - \beta\omega^2 \quad . \tag{5}$$

The model constants may vary depending on the model. A good reference is given in [CE04].

Vortex simulations usually require more sophisticated turbulence models than those shown above. For this reason there are certain enhancements available. Most of them only add further expressions to the transport equations which minimizes the additional computational effort. In this paper we use the Standard k-ε model [LAU74] for most computations. In case of the draft tube vortex the results are compared to those obtained with the SST-Model [ME94] and an enhanced k-ε (Kim-Chen) [CH87, HEL07]. Both models are based on versions of the k-ε and k-ω model. Hence, the compuational effort is comparable to the Standard models.

Structural Equations

The discretised linear structural equations with mass, damping and stiffness matrices $\boldsymbol{M}$, $\boldsymbol{D}$, and $\boldsymbol{K}$, load vector $\boldsymbol{f}$, and displacments $\boldsymbol{u}$ can be written as

$$\boldsymbol{M}\ddot{\boldsymbol{u}} + \boldsymbol{D}\dot{\boldsymbol{u}} + \boldsymbol{K}\boldsymbol{u} = \boldsymbol{f} \quad , \tag{6}$$

see Zienkiewicz [ZI89]. For fluid-structure coupled systems the load vector $\boldsymbol{f}$ represents the fluid loads due to surface pressure and inertia.

3 Fluid-Structure Coupling

Seen from the physical point of view fluid-structure interaction is a two field problem. But numerically, the problem has three fields, where the third field is the fluid grid that has to be updated after every deformation step of the structure to propagate the movement of the fluid boundaries, the wetted structure, into the flow domain.

The solution of the numerical problem can be arranged in a monolithic scheme, which solves the structural and flow equations in one step. This method is applied for strongly coupled problems, e.g. for modal analyses. Another scheme, suitable for loosely and intermediately coupled simulations, is the separated solution of flow and structural equations with two independent codes and models. In the separated scheme well proven codes and models can be used. However, data has to be exchanged between the codes, including the interpolation between the two computational meshes. This is the reason why for engineering problems usually the second approach is employed. In order to account for the coupling and to avoid unstable simulations, some coupling schemes have been developped, e.g. see Farhat et. al. [FAR98, MOK01].

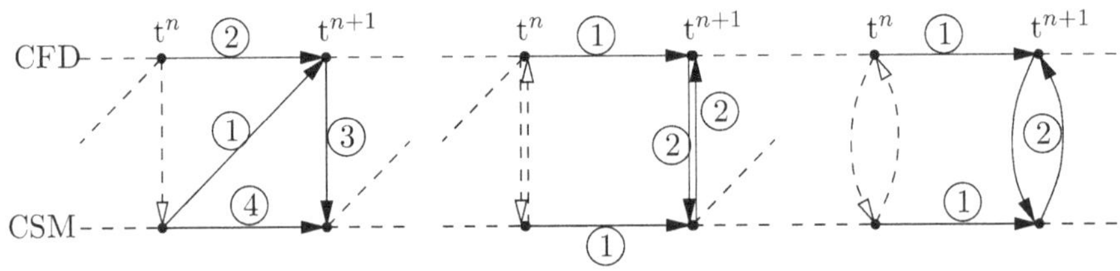

Fig. 1. Partitioned schemes for loose and strong coupling

All schemes shown in figure 1 exchange data at each time-step. In the first scheme, a sequentially staggered scheme, one code waits until the other one is done with it's time-step. Hence, one code always uses data of the remote code's previous step for the time-step integration. This gives the scheme explicit character. The second explicit scheme is a parallel staggered one. Here, both codes do the time-step integration simultanously. If the solution requires approximately the same real time in both codes, this scheme will reduce waiting times.

Explicit coupling schemes will exhibit instabilities if the moving fluid mass is high compared to the structural mass [FOE07]. The high density of water leads to a high added fluid mass in hydraulic machinery applications. Due to the iteration between the codes these schemes increase the computational effort.

In this paper the explicit schemes are applied for quasi-static analyses of coupled problems. For time dependent problems an implicit scheme is used. Furthermore, the difference between the two explicit schemes will be examined.

4 Dynamic Mesh Approach

The mesh update method applied in the computations in this paper uses an interpolation between the nodal distance between moving and fixed boundaries to compute the new nodal position after a displacement step of the moving boundary. The simplest approach is to use a linear interpolation value $0 \leq \kappa \leq 1$. Here we use a modification of the parametre $\kappa = \frac{|\boldsymbol{s}|}{|\boldsymbol{r}|+|\boldsymbol{s}|}$ proposed by Kjellgren and Hyvärinen [KH98].

$$\tilde{\kappa} = \begin{cases} 0, & \kappa < \delta \\ \frac{1}{2} \cdot \cos\left(\left(1 - \frac{\kappa-\delta}{1-2\delta}\right) \cdot \pi\right) + 1, & \delta \leq \kappa \leq 1 - \delta \\ 1, & 1 - \delta < \kappa \leq 1 \end{cases} \quad (7)$$

This parameter is found from the nearest distance $\boldsymbol{r}$ to the moving boundary and the distance to the fixed boundary $\boldsymbol{s}$ in the opposite direction. To use this approach for parallel computations the boundaries have to be available on all processors. Since a graph based domain decomposition is used here this is not implicitly given. Hence, the boundary exchange has to be implemented, additionally. Usually the number of boundary nodes is considerably small, i.e. the additional communication time and overhead is negligible. The algorithm is discussed and compared with other methods regarding computational performance, parallelisation and robustness in [LI06].

5 Software and Implementation

All simulations shown in this paper require a considerable amount of computing power. In case of the vortex simulations a fine mesh is necessary to resolve the high gradients in the flow. Furthermore, several periods of the periodic vortex rope have to be computed. For the fluid-structure simulations the CFD part is computationally expensive compared to the structural simulation. Especially for dynamic FSI simulations a great number of steps has to be solved. Hence, the CFD runs will all be performed on an HPC system, NEC SX-8. This means to verify the performance of the CFD-code FENFLOSS as well as to install the coupling on the vector system.

5.1 FENFLOSS - CFD on Vector Systems

The flow simulation (CFD) code FENFLOSS[1], developped at the IHS, is an unsteady Navier-Stokes solver based on the finite element method. A stabi-

[1] **F**inite **E**lement based **N**umerical **FL**ow **S**imulation **S**ystem

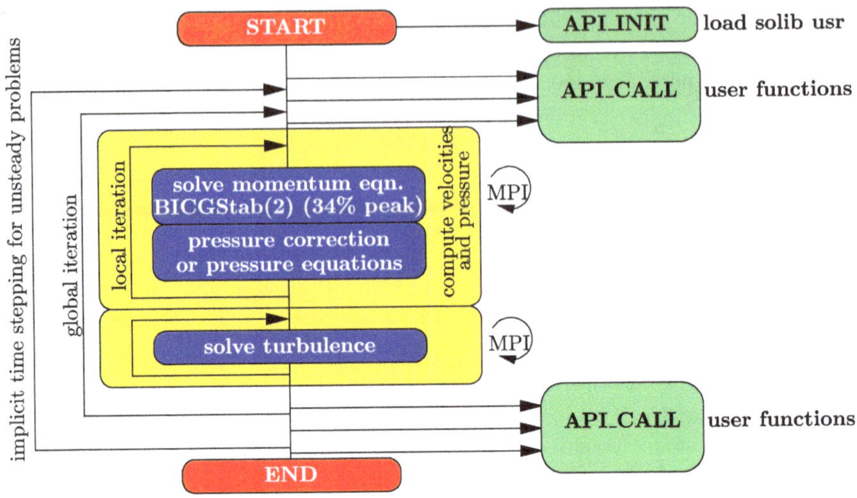

Fig. 2. FENFLOSS solver scheme - including user-function calls

lized Petrov-Galerkin Finite Element approach [ZI89, GR99] is applied for the discretization of equation (1,2). The nonlinear system is solved by a fixed-point iteration scheme. In each nonlinear iteration loop the equations are linearised and solved by an iterative BICGStab(2) solver [VDV92, VDV94], figure 2 The three velocity components can be solved coupled or decoupled followed by a modified UZAWA pressure correction, see Ruprecht [RU89]. Working on parallel architectures, MPI is applied in the precoditioner and the matrix-vector and scalar products, see Maihoefer [MAI02]. FENFLOSS supports all unstructured grids with hexahedral (3D) or quadrilateral (2D) elements.

Furthermore, FENFLOSS provides a programming interface (API), which allows to implement user-defined subroutines. These will be called at certain points in the solver loops. The user-subroutines are implemented by a shared object library that is loaded on demand during the initialisation of the CFD solver. This guarantees a very flexible method to enhance the FENFLOSS solver kernel or even exchange data to external programs.

FENFLOSS has been successfully optimised for vector architectures [BOR06]. Despite the indirect addressing the matrix vector product in the linear solver reaches a ratio of almost 50% of the peak performance (16 GFLOPS) on a SX-8 vector cpu for vector lengths near to the optimum of 256. The whole solver still has a relative performance of nearly 35% and the total application, including start-up and I/O gets close to 30%.

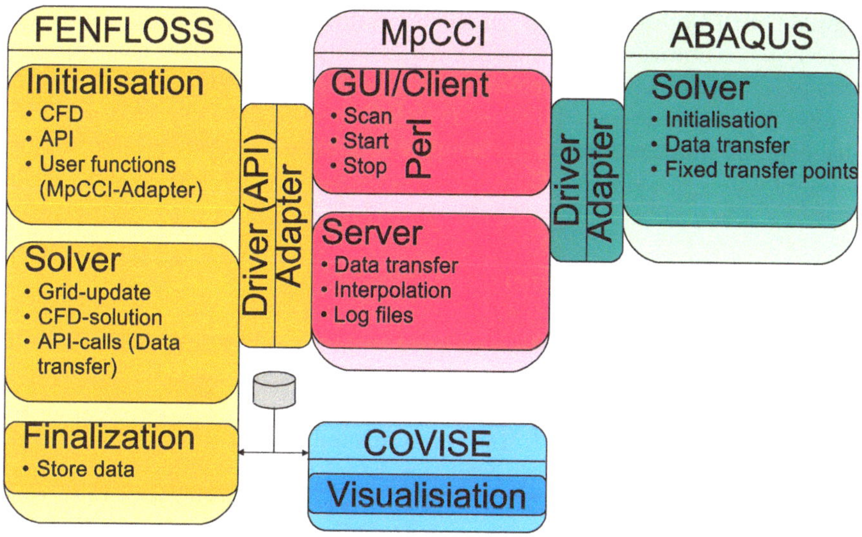

Fig. 3. FENFLOSS coupled with MpCCI and ABAQUS

5.2 FSI Coupling with FENFLOSS on SX-8

For fluid-structure simulations the code coupling software MpCCI[2], developped at the Fraunhofer Institute SCAI, is applied. It provides a coupling manager that controls the simulation codes and interpolates the data to the different meshes. To communicate data with the codes MpCCI needs a special code adapter.

The MpCCI-adapter for FENFLOSS is available as shared object library. It has access to the pressure solution in FENFLOSS and the geometry data. For the explicit coupling scheme, see section 3, data is transferred once at the beginning of the time-step. A data transfer sends the current pressure values and receives the deflections on the coupling region. If the coordinates of the coupling surface have changed the mesh is updated automatically. For implicit schemes an additional transfer is made in the global iteration loop.

In this paper we use the commercial structural simulation program ABAQUS$^{(C)}$. The adapter is included in MpCCI and can not be changed. The visualisation software COVISE is used for post-processing of the CFD results. Figure 3 shows the complete set-up for the simulation. During our simulations ABAQUS will run on a single cluster CPU that is allowed to connect to the SX-8 frontend to exchange data with the CFD solver via MpCCI.

For the installation of the MpCCI-coupled version of FENFLOSS on the NEC SX-8 vector system we have to make some modifications. The compiler

[2] Mesh based parallel Code Coupling Interface$^{(C)}$

installed with Super-Unix 15.1 on the SX-8 does not support the concept of shared object libraries. Hence, we have to give up the flexibility this approach gives us on regular Linux systems and link the user-defined functions statically to the FENFLOSS kernel. This means to have a special version that only runs with MpCCI. If something in the adapter is changed, the whole program has to be rebuilt.

6 Tidal Turbine Runner

Tidal current turbines are similar to wind turbines. They only use the kinetic energy of the flow. As a consequence they have rather long, slim blades. In Figure 4 a prototype of a tidal current turbine is shown. The diameter of the rotor is 16 m. The flow field to the runner usually is not uniform. Because of a boundary layer on the bottom and because of the wakes behind the support structure the flow to a turbine blade is unsteady. As a consequence this non-uniform flow field leads to a severe periodic loading on the blades and finally to a noticeable deformation of blade. To calculated the resulting vibrations of the runner blades is the final task of this project.

After computing a first solution for steady inflow boundary conditions the pressure loads are used to evaluate the static behaviour of one turbine blade. Since the change of the flow due to the deflection is not examined this analysis

Fig. 4. Computer rendered picture of tidal current turbines

is called a one-way coupled analysis. This analysis is made to verify the further results. The second step is a two-way coupled analysis, which involves a direct interaction between the flow and the structural solution. Both codes exchange data until an equilibrium is reached.

6.1 Computational Model and Performance

Here we examine the structural behaviour of a three-blade tidal turbine runner on the fluid loads. Figure 5 shows the mesh around one blade. Since we assume a constant flow distribution at the inlet the problem is modelled as a periodic section of the whole runner. The open boundaries at the outlet and in radial direction have to placed sufficiently far away from the runner to avoid influences on the flow solution around the runner.

Pressure loads to and deflections from the structure are exchanged at the blade surface. First, a CFD analysis without structural interaction is performed on the coarse (0.8 mio. nodes) and a refined mesh (2.57 mio. nodes). Table 1 shows a comparison of the performance numbers on a NEC SX-8 vector node (8 CPUs). MPI is applied for parallelisation. Due to the better, but not yet optimal, exploitation of the vector pipes the refined mesh shows a slightly better performance than the coarse one. All timing values are relative in % to the user-time. Since the code uses unstructured meshes the Bank conflict ratio of about 7% is acceptable. Also the overall code performance of

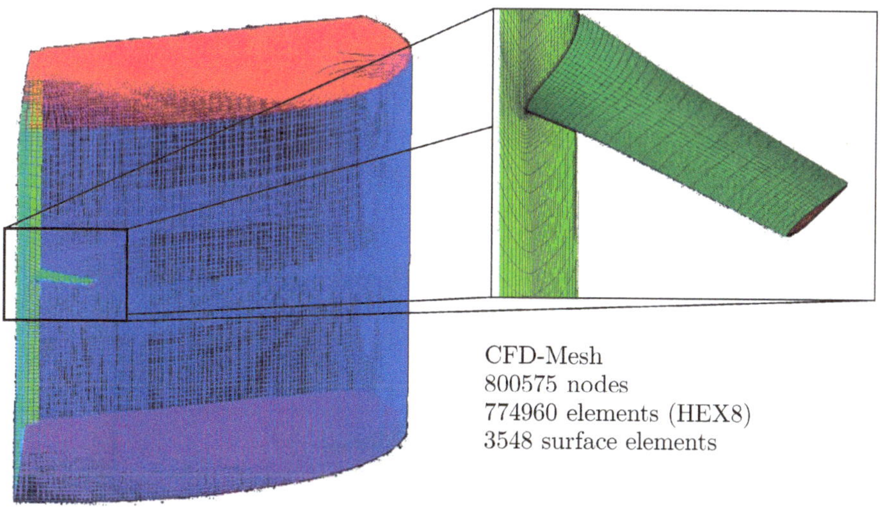

Fig. 5. Coarse periodic CFD mesh (800575 nodes) and detail view of blade surface mesh

Table 1. Overall average performance data for tidal turbine runner. All times in % relative to user-time

			Vectorisation		Cache Miss	
Mesh size	% peak	Bank conflict	Vec.len.	Vec.ratio	Instruction	Operation
800575	28.06	6.55	248.94	99.44	1.42	1.29
2574740	29.77	7.08	251.55	99.46	1.71	0.87

almost 30% peak performance (16 GFLOPS * 8 CPUs) including all I/O and the initialisation part is satisfying.

6.2 Fluid-Structure Interaction Results

First, a one-way coupled analysis is conducted, i.e. solving the flow equations and then solving the structure separately with the pressure loads obtained from the fluid solution. The results are used to verify the two-way coupling. Here a quasi-static equilibrium between the flow solution with constant inflow boundary conditions and the structural deflection is sought. For this analysis the sequential explicit staggered scheme, see figure 1, left, is applied. The blade material is steel with a Young's modulus of 2.1E+11 N/m^2, Poisson ratio of 0.3 and a density of 7850 kg/m^3.

Figure 6 shows the results of both analyses. In the coupled computation case the pressure distribution (on deflected surface) in the equilibrium state is almost identical to the CFD-only (one-way) solution. In reality the deflection of the blade is rather small; here it is scaled by a factor of 10. This is the

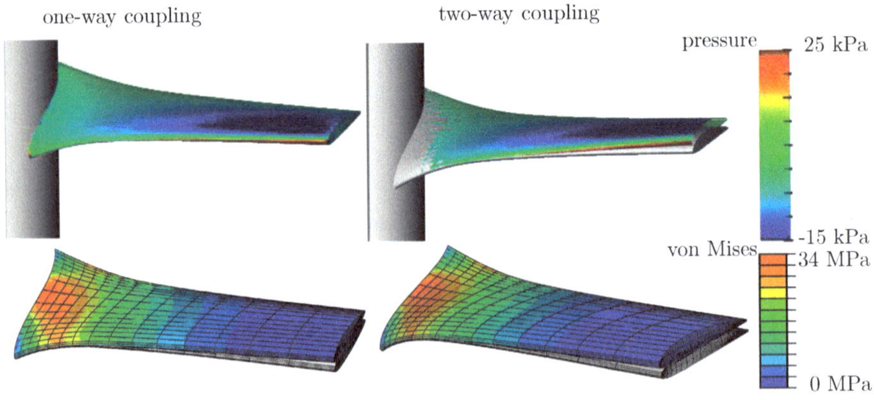

Fig. 6. Pressure field and v. Mises stresses for a one-way coupled solution and a quasi-static two-way coupled solution

reason why the pressure field does not change in the two-way coupled analysis case. In further simulations unsteady boundary conditions will be applied to study the dynamic behaviour of the blade.

Regarding computational time the coupled FSI analysis takes twice the time of the CFD simulation. Looking at the time needed by the mesh update algorithm this might be surprising. The profiling shows an additional effort of 5.5% of the total simulation time for the new routines. About 3.5% is needed for the initial node search, 2% for the mesh-update after each data transfer. But, there are about 50% MPI_WAIT on the four processors that do not participate in the coupling. On the other four processors that do exchange data, there is almost no waiting time issued. This shows the drawback of the sequentially coupled scheme. The time needed for one step is the same for both codes, but FENFLOSS has to wait until ABAQUS is done with its time-step. To reduce the wasted MPI_WAIT time either the structural code has to run even faster or the coupling scheme has to be changed. Applying the parallel coupling scheme the MPI_WAIT-time is reduced to 50% in comparison to the sequential scheme. The total ratio is about 33% of the whole computation. This is due to the fact that there is still some time needed for the data exchange and interpolation, and furthermore the codes still sometimes have to wait. But the improvement is clear. Obviously, for this kind of simulation the parallel scheme is the best regarding the computation time. It is furthermore desirable to run both codes on platforms that guarantee almost identical step times between the data transfer.

7 Vortex Rope in a Straight Diffusor

The flow in turbo machinery is mostly characterized by complex geometries and by the presence of a rotating runner or impeller. Additionally it shows a very high Reynolds number, which means that a highly turbulent flow exists.

Whereas under design conditions the flow in turbo machinery behaves quite smooth, in off-design, however, the flow behavior is characterized by the existence of strong vortices. These vortices can be unstable and move around in the flow domain. A typical vortex structure in hydro turbines is the draft tube vortex rope. In the design point the velocity distribution downstream of the runner is nearly free of swirl. In part load, however, this distribution shows a high swirling component. This leads to a strong vortex in the draft tube. This vortex can get unstable and form a rotating helical structure, which typically rotates with approximately 30-50% of the runner speed.

To understand these phenomena experimental studies have been made on a special test rig. Now, the numerical models have to be tested and improved. Figure 7 shows the vortices in a straight diffusor obtained with three different turbulence models. The Standard k-ε model fails to reproduce any fluctuations or vortex breakdown. Visually, the SST model gives the best results. Measurements yield a frequency of the pressure fluctuations of 15.7 Hz. Obviously, the

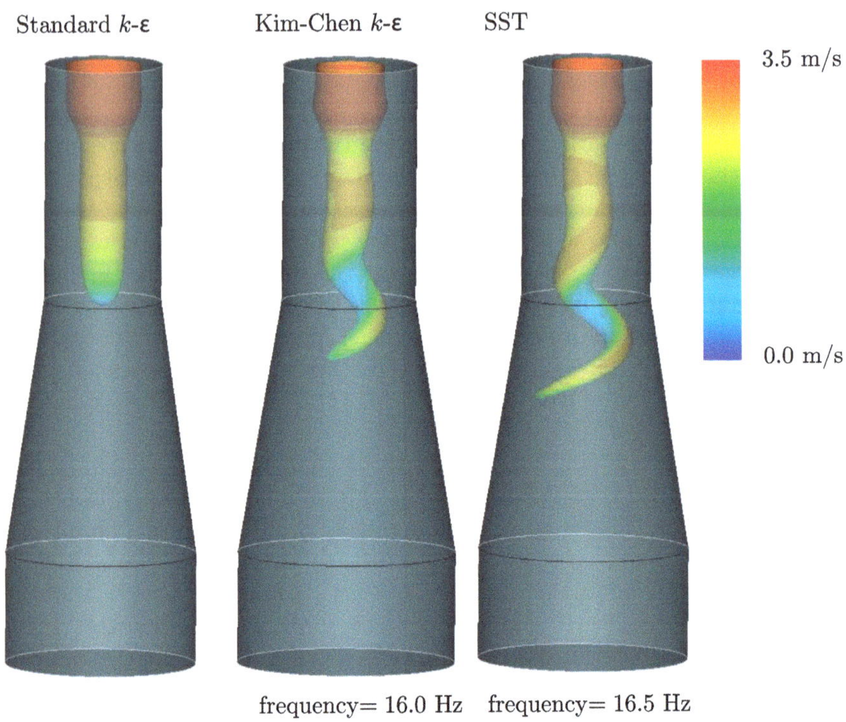

Fig. 7. Vortex ropes obtained for different turbulence models. Pressure iso-surface at -3500Pa relative pressure

results of the Kim-Chen k-ε model complies better with the measurements than the SST-model.

8 Summary

In this paper two different applications in high-performance computing are discussed. One is the simulation of unsteady vortices in a straight diffusor. Experimental studies for this example are available and the simulation results can be verified. The second application is the fluid-structure interaction of the flexible blades of a tidal current turbine with the surrounding fluid. Especially the coupled application is examined regarding computational performance.

Due to the increased number of necessary iterations unsteady simulations require more computational power, which is available on vector machines. In case of unsteady fluid-structure interactions in hydraulic machinery implicit partitioned compling schemes require also higher number of solution steps.

Here, we examine the performance of the coupling schemes in a quasi-static FSI analysis. The in-house CFD-code FENFLOSS is running on an SX-8 vector node that provides the required computational power, the commercial structural mechanics solver ABAQUS on a cluster. Both codes are coupled with the code coupling software MpCCI that is installed on the SX-8 frontend.

After the verification of the sustained vector performance of FENFLOSS a single CFD analysis is conducted. The pressure loads are used to obtain static deflections of a tidal current turbine blade as reference for a quasi-static coupled analysis. For a sequential fluid-structure interaction scheme the simulations show a high ratio of waiting time for the CFD-processes that do not participate in the coupling. A parallel scheme reduces the idle time by 50%.

The vortex rope simulations show certain differences between different turbulence models. The Standard k-ε model is not able to reproduce vortex instability. A modified k-ε model and a SST model both show good agreements with the measurements.

Future work will be, first, about the improvement of the performance of coupled simulations. Second, dynamic analyses with unsteady boundary conditions will be made to examine the vibrational behaviour of the tidal current blades. Further simulations of the vortex in the straight diffusor will be made with refined meshes and more sophisticated turbulence models.

References

[BOR06] Borowski, S., Tiyyagura, S.R., Küster, U.: Matrix Assembly without Coloring on Vector Machines. In: Proc. of the Int. Conference of Numerical Analysis and Applied Mathematics (ICNAAM 2006), Crete, Greece, 2006.

[CE04] Cebeci, T.: Turbulence Models and Their Applications. Horizons Publishing, Long Beach, CA, 2004.

[CH87] Chen, Y.S., Kim, S.W.: Computation of turbulent flows using an extended $k - \epsilon$ closure model. NASA CR-179204.

[DO03] Donea, J., Huerta, A.: Finite element methods for flow problems. Wiley, Chichester, 2003.

[FAR98] Farhat, C., Lesoinne, M., le Tallec, P.: Load and motion transfer algorithms for fluid-structure interaction problems with non-matching interfaces. Computer Methods in Applied Mechanics and Engineering, **157**, 95–114 (1998)

[FP02] Ferziger, J.H., Perić, M.: Computational Methods for Fluid Dynamics (third Ed.). Springer (2002).

[FOE07] Förster, C.: Robust methods for fluid-structure interaction with stabilised finite elements. PhD-Thesis, Universität Stuttgart, Institut für Baustatik, 2007.

[GR99] Gresho, P.M., Sani, R.L.: Incompressible Flow and the Finite Element Method (Vol. I). John Wiley & Sons (1999)

[HEL07] Helmrich, T.: Simulation instationärer Wirbelstrukturen in hydraulischen Maschinen. PhD-Thesis, IHS, Universität Stuttgart, 2007.

[HUG81] Hughes, T.J.R., Liu, W.K., Zimmermann, T.K.: Lagrangian-Eulerian Finite Element Formulation for Viscous Flows. Comp Methods in Applied Mech. and Eng. **29**, 329–349 (1981)

[KH98] Kjellgren, P., Hyvärinen, J.: An Arbitrary Langrangian-Eulerian Finite Element Method. Computational Mechanics, **21**, 81–90 (1998)

[LAU74] Launder, B.E. Spalding, D.B.: The Numerical Computation of turbulent Flows. Comp. Methods in Applied Mech. Eng., 3, 1974.

[LI06] Lippold, F., Fluid-structure interaction in an axial fan. HPC-Europa report (2006)

[MAI02] Maihoefer, M.: Effiziente Verfahren zur Berechnung dreidimensionaler Stroemungen mit nichtpassenden Gittern (PhD-Thesis). University of Stuttgart, (2002)

[ME94] Menter, F.R.: Two-Equation Eddy-Viscosity Turbulence Models for Engineering Applications. AIAA Journal, 32:1598–1605, 1994.

[MOK01] Mok, D.P.: Partitionierte Lösungsansätze in der Strukturdynamik und der Fluid-Struktur-Interaktion. PhD-Thesis, Universität Stuttgart, Institut für Baustatik, 2001.

[RU89] Ruprecht, A.: Finite Elemente zur Berechnung dreidimensionaler turbulenter Stroemungen in komplexen Geometrien (PhD-Thesis). University of Stuttgart (1989)

[VDV92] van der Vorst, H.A.: BI-CGSTAB: A fast and smoothly converging variant of BI-CG for the solution of nonsymmetric linear systems. SIAM Journal of Scientific Stat. Computing, **13**, 631-644, (1992)

[VDV94] van der Vorst, H.A. (1994): Recent Developments in Hybrid CG Methods, Proc. High Performance Computing & Networking, München.

[ZI89] Zienkiewicz, O.C., Taylor, R.L.: The Finite Element Method (Vol. I). McGraw-Hill (1989)

FEASTSolid and FEASTFlow: FEM Applications Exploiting FEAST's HPC Technologies

Sven H.M. Buijssen, Hilmar Wobker, Dominik Göddeke, and Stefan Turek

Institute for Applied Mathematics and Numerics,
Dortmund University of Technology, 44227 Dortmund, Germany
feast@math.tu-dortmund.de

1 Introduction and Motivation

Finite Element (FE) codes typically operate on sparse matrices and feature low arithmetic intensity, resulting in their performance being limited by the available memory bandwidth rather than the peak compute performance. FEAST (Finite Element Analysis and Solution Tools) is our toolkit providing FE discretisations and corresponding optimised parallel multigrid solvers for PDE problems, addressing this *memory wall problem* with what we call "hardware-oriented numerics" [11]. These techniques allow FEAST to exploit a significant share of modern processors' peak performance for FE applications while maintaining numerical efficiency, robustness and flexibility. Last year we reported on our efforts solving Poisson problems with FEAST on NEC SX-6 and SX-8, JUMP Jülich and commodity based clusters [4]. In this paper, we address our progress in solving problems from solid mechanics and fluid dynamics with FEAST. We only briefly summarise the main ideas here (cf. Fig. 1), and refer to previous publications for related work and more details.

The two main principles underlying our approach are:

Logical tensorproduct structure: In FEAST, the discretisation is closely coupled with the domain decomposition for the parallel solution. The computational domain $\bar{\Omega}$ is covered with a collection of quadrilateral subdomains $\bar{\Omega}_i$. The subdomains form an unstructured coarse mesh and are hierarchically refined such as to preserve a logical tensorproduct structure of the mesh cells within each subdomain. Consequently, FEAST maintains a clear separation of globally unstructured and locally structured data. The resulting mesh is used for the discretisation with Finite Elements, and linewise numbering of the unknowns leads to band structured matrices.

SBBLAS: Since the underlying data structures store matrix bands as sequential vectors, there is no need for general storage formats such as CSR.

Consequently, matrix-vector multiplication can be implemented bandwise, entirely without indirect addressing (*Sparse Banded BLAS*). On cache-based architectures, only slices of the complete diagonals (cf. Fig. 1) are operated on simultaneously which allows for a greater part of the result vector being held in cache. On non von Neumann architectures such as the NEC SX-8, matrix-vector multiplication can be efficiently vectorised due to this blocking strategy.

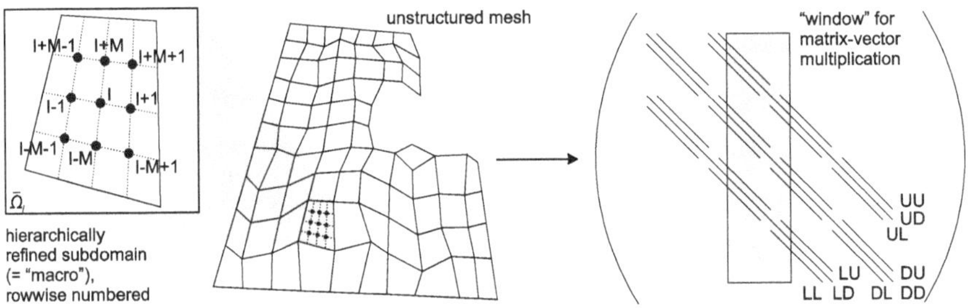

Fig. 1. Grid with logical tensorproduct structure, exemplary one coarse grid cell hierarchically refined. Matrix vector multiplication in SBBLAS operates only on slices of the corresponding FE band matrix

2 Computational Solid Mechanics

In Computational Solid Mechanics (CSM), the deformation of solid bodies under external loads is simulated. In this report, we prototypically consider a two-dimensional body covering a domain $\bar{\Omega} = \Omega \cup \partial\Omega$, where Ω is a bounded, open set with boundary $\Gamma = \partial\Omega$. The boundary is split into two parts: the Dirichlet part Γ_{D} where displacements are prescribed and the Neumann part Γ_{N} where surface forces can be applied ($\Gamma_{\mathrm{D}} \cap \Gamma_{\mathrm{N}} = \emptyset$). Furthermore the body can be exposed to volumetric forces, e. g. gravity. We treat the simple, but nevertheless fundamental, model problem of elastic, compressible material under static loading, assuming small deformations. We use a formulation where the displacements $\boldsymbol{u}(\boldsymbol{x}) = \big(u_1(\boldsymbol{x}), u_2(\boldsymbol{x})\big)^{\mathsf{T}}$ of a material point $\boldsymbol{x} \in \bar{\Omega}$ are the only unknowns in the equation. The strains can be defined by the linearised strain tensor $\varepsilon_{ij} = \frac{1}{2}\Big(\frac{\partial u_i}{\partial x_j} + \frac{\partial u_j}{\partial x_i}\Big), \quad i,j = 1,2$, describing the linearised kinematic relation between displacements and strains. The material properties are reflected by the constitutive law, which determines a relation between the strains and the stresses. We use Hooke's law for isotropic elastic materials, $\boldsymbol{\sigma} = 2\mu\boldsymbol{\varepsilon} + \lambda\,\mathrm{tr}(\boldsymbol{\varepsilon})\boldsymbol{I}$, where $\boldsymbol{\sigma}$ denotes the symmetric stress tensor and μ and λ are the so-called Lamé constants.

The basic physical equations for problems of solid mechanics are determined by equilibrium conditions. For a body in equilibrium, the inner forces (stresses) and the outer forces (external loads $\boldsymbol{f}$) are balanced:

$$-\mathbf{div}\boldsymbol{\sigma} = \boldsymbol{f}, \qquad \boldsymbol{x} \in \Omega.$$

Using Hooke's law to replace the stress tensor, the problem of linearised elasticity can be expressed in terms of the following elliptic boundary value problem, called the Lamé equation:

$$-2\mu\,\mathbf{div}\boldsymbol{\varepsilon}(\boldsymbol{u}) - \lambda\,\mathbf{grad}\,\mathrm{div}\,\boldsymbol{u} = \boldsymbol{f}, \qquad \boldsymbol{x} \in \Omega \tag{1a}$$

$$\boldsymbol{u} = \boldsymbol{g}, \qquad \boldsymbol{x} \in \Gamma_{\mathrm{D}} \tag{1b}$$

$$\boldsymbol{\sigma}(\boldsymbol{u}) \cdot \boldsymbol{n} = \boldsymbol{t}, \qquad \boldsymbol{x} \in \Gamma_{\mathrm{N}} \tag{1c}$$

Here, $\boldsymbol{g}$ are prescribed displacements on Γ_{D}, and $\boldsymbol{t}$ are given surface forces on Γ_{N} with outer normal $\boldsymbol{n}$. For details on the elasticity problem, see for example Braess [5].

3 Computational Fluid Dynamics

We model problems from Computational Fluid Dynamics (CFD) with the Navier–Stokes equations, which describe the flow of incompressible Newtonian fluids (e. g. water and many other liquids) in a domain Ω.

Confining the domain and imposing boundary conditions, i. e. in- and outflow conditions on the "artificial" boundaries and no-slip conditions at rigid walls, yields the following system of equations under the assumption of constant temperature ϑ and constant kinematic viscosity $\nu > 0$:

$$-\nu\Delta\boldsymbol{u} + (\boldsymbol{u} \cdot \mathrm{grad})\boldsymbol{u} + \mathrm{grad}\,p = \boldsymbol{f}, \qquad \boldsymbol{x} \in \Omega \tag{2a}$$

$$\mathrm{div}\,\boldsymbol{u} = \boldsymbol{0}, \qquad \boldsymbol{x} \in \Omega \tag{2b}$$

$$\boldsymbol{u} = \boldsymbol{g}, \qquad \boldsymbol{x} \in \Gamma_{\mathrm{D}} \tag{2c}$$

$$\nu\partial_n \boldsymbol{u} + p \cdot n = 0, \qquad \boldsymbol{x} \in \Gamma_{\mathrm{N}} \tag{2d}$$

where p denotes pressure, n the outer normal vector and Γ_{D} and Γ_{N} the boundary parts with, respectively, Dirichlet and Neumann boundary conditions (i. e. inflow, outflow and adhesion conditions). For more details on the theoretical background of this, see for example Ferziger and Perić [7].

4 Solution Strategy

4.1 Parallel Multigrid Solvers in FEAST

For the problems we are concerned with in the (wider) context of this report, multigrid methods are obligatory from a numerical point of view. When parallelising multigrid methods, numerical robustness, numerical efficiency and (weak) scalability are often contradictory properties: A strong recursive coupling between the subdomains, for instance by the direct parallelisation of ILU-like smoothers, is advantageous for the numerical efficiency of the multigrid solver. However, such a coupling increases the communication and synchronisation requirements significantly and is therefore bound to scale badly. To alleviate this high communication overhead, the recursion is usually relaxed to the application of *local smoothers* that act on each subdomain $\bar{\Omega}_i$ independently. The contributions of the separate subdomains are combined in an additive manner only after the smoother has been applied to all subdomains, without any data exchange during the smoothing. The disadvantage of such a (in terms of domain decomposition) *block-Jacobi* coupling is that typical local smoothers such as Gauss-Seidel are usually not powerful enough to treat, for example, local anisotropies. Consequently, the numerical efficiency of the multigrid solver is dramatically reduced [9, 11].

To address these contradictory needs, FEAST employs a generalised multigrid domain decomposition concept called ScaRC (Scalable Recursive Clustering). The basic idea is to apply a *global*, data-parallel multigrid algorithm which is smoothed in an additive manner by *local* multigrids acting on each subdomain independently. In the nomenclature of the previous paragraph, this means that the application of a local smoother translates to performing few iterations – sometimes even only one iteration – of a local multigrid solver, and we can use the terms *local smoother* and *local multigrid* synonymously. This cascaded multigrid scheme is very robust as local irregularities are 'hidden' from the outer solver, the global multigrid provides strong global coupling (as it acts on all levels of refinement), and preserves the scalability of data-parallel multigrid methods by design. Obviously, this cascaded multigrid scheme is prototypical in the sense that it can only show its full strength for reasonably large local problem sizes and locally ill-conditioned systems [3].

Instead of keeping all data in one general, homogeneous data structure, FEAST stores only the local FE matrices and vectors, corresponding to the local subdomains (which is common in domain decomposition methods). Global matrix-vector operations are performed by a series of local operations on matrices representing the restriction of the 'virtual' global matrix on each subdomain. These operations are *directly* followed by exchanging information via MPI over the boundaries of neighbouring subdomains. There is only an implicit subdomain overlap; the domain decomposition is implemented via special boundary conditions in the local matrices [3]. Several subdomains are typically grouped into one MPI process, exchanging data via shared memory. All global and local coarse grid problems are solved exactly by a tuned direct LU decomposition solver taken from UMFPACK [6].

Figure 2 illustrates a typical solver in FEAST. The notation 'local multigrid (C 4+4, S, UMFPACK)' denotes a multigrid solver on a single subdomain, configured to perform the cycle $C \in \{V,F,W\}$ with 4 pre- and postsmoothing steps with the smoothing operator $S \in \{\text{Jacobi, Gauss-Seidel, ILU}, \ldots\}$. To improve solver robustness, the global multigrid solver is used as a precondi-

global BiCGStab
preconditioned by
  global multigrid (V 1+1)
  additively smoothed by
    for all $\bar{\Omega}_i$: **local multigrid** (C 4+4, S, UMFPACK)
  coarse grid solver: UMFPACK

Fig. 2. Illustration of the family of cascaded multigrid solver schemes in FEAST

tioner to a Krylov subspace solver such as BiCGStab which executes on the global fine grid. As a preconditioner, the global multigrid performs exactly one iteration without convergence control.

We finally emphasise that the entire concept – comprising domain decomposition, solver strategies and data structures – is independent of the spatial dimension of the underlying problem. Implementation of 3D support is tedious and time-consuming, but conceptually straightforward.

4.2 Scalar and Vector-Valued Problems

The guiding idea to treating vector-valued problems with FEAST is to rely on the modular, reliable and highly optimised scalar operations, in order to formulate robust schemes for a wide range of applications rather than using the best suited numerical scheme for each application and go through the optimisation and debugging process over and over again. Vector-valued PDEs as they arise, for instance, in the application domains in the focus on this report, can be rearranged and discretised in such a way that the resulting discrete systems of equations consist of blocks that correspond to scalar problems (for the CSM case see beginning of Sect. 4.3, for CFD see Sect. 4.4). Due to this special block-structure, all operations required to solve the systems can be implemented as a series of operations for scalar systems, taking advantage of the highly tuned linear algebra components in the SBBLAS library. To apply a scalar local multigrid solver, the set of unknowns corresponding to a global scalar equation is restricted to the subset of unknowns that correspond to the specific subdomain.

To illustrate the approach, consider a matrix-vector multiplication $\mathbf{y} = \mathbf{A}\mathbf{x}$ with the exemplary block structure:

$$\begin{pmatrix} \mathbf{y}_1 \\ \mathbf{y}_2 \end{pmatrix} = \begin{pmatrix} \mathbf{A}_{11} & \mathbf{A}_{12} \\ \mathbf{A}_{21} & \mathbf{A}_{22} \end{pmatrix} \begin{pmatrix} \mathbf{x}_1 \\ \mathbf{x}_2 \end{pmatrix}$$

As explained above, the multiplication is performed as a series of operations on the local FE matrices per subdomain $\bar{\Omega}_i$, denoted by superscript $(\cdot)^{(i)}$. The global scalar operators, corresponding to the blocks in the matrix, are treated individually:

For $j = 1, 2$, do

1. For all $\bar{\Omega}_i$, compute $\mathbf{y}_j^{(i)} = \mathbf{A}_{j1}^{(i)} \mathbf{x}_1^{(i)}$.
2. For all $\bar{\Omega}_i$, compute $\mathbf{y}_j^{(i)} = \mathbf{y}_j^{(i)} + \mathbf{A}_{j2}^{(i)} \mathbf{x}_2^{(i)}$.
3. Communicate entries in $\mathbf{y}_j$ corresponding to the boundaries of neighbouring subdomains.

4.3 Solving the Elasticity Problem

In order to solve vector-valued linearised elasticity problems with the application FEASTSOLID using the FEAST intrinsics outlined in the previous paragraphs, it is essential to order the resulting degrees of freedom corresponding to the spatial directions, a technique called separate displacement ordering [2]. In the 2D case where the unknowns $\boldsymbol{u} = (u_1, u_2)^\mathsf{T}$ correspond to displacements in x and y-direction, rearranging the left hand side of equation (1a) yields:

$$-\begin{pmatrix} (2\mu + \lambda)\partial_{xx} + \mu\partial_{yy} & (\mu + \lambda)\partial_{xy} \\ (\mu + \lambda)\partial_{yx} & \mu\partial_{xx} + (2\mu + \lambda)\partial_{yy} \end{pmatrix} \begin{pmatrix} u_1 \\ u_2 \end{pmatrix} = \begin{pmatrix} f_1 \\ f_2 \end{pmatrix} \tag{3}$$

We approximate the domain $\bar{\Omega}$ by a collection of several subdomains $\bar{\Omega}_i$, each of which is refined to a logical tensorproduct structure as described in Sect. 4.1. We consider the weak formulation of equation (3) and apply a Finite Element discretisation with conforming bilinear elements of the Q_1 space. The vectors and matrices resulting from the discretisation process are denoted with upright bold letters, such that the resulting linear equation system can be written as $\mathbf{Ku} = \mathbf{f}$. Corresponding to representation (3) of the continuous equation, the discrete system has the following block structure,

$$\begin{pmatrix} \mathbf{K}_{11} & \mathbf{K}_{12} \\ \mathbf{K}_{21} & \mathbf{K}_{22} \end{pmatrix} \begin{pmatrix} \mathbf{u}_1 \\ \mathbf{u}_2 \end{pmatrix} = \begin{pmatrix} \mathbf{f}_1 \\ \mathbf{f}_2 \end{pmatrix}, \tag{4}$$

where $\mathbf{f} = (\mathbf{f}_1, \mathbf{f}_2)^\mathsf{T}$ is the vector of external loads and $\mathbf{u} = (\mathbf{u}_1, \mathbf{u}_2)^\mathsf{T}$ the (unknown) coefficient vector of the FE solution. The matrices $\mathbf{K}_{11}$ and $\mathbf{K}_{22}$ of this block-structured system correspond to scalar elliptic operators (cf. Equation (3)), i.e. FEAST's tuned solvers can be applied to the corresponding subsystems.

We illustrate the details of the solution process with a basic iteration scheme, a preconditioned defect correction method:

$$\mathbf{u}^{k+1} = \mathbf{u}^k + \omega \tilde{\mathbf{K}}_{\mathrm{B}}^{-1}(\mathbf{f} - \mathbf{K}\mathbf{u}^k), \qquad k = 1, \ldots \tag{5}$$

This iteration scheme acts on the global system (4) and thus couples the two sets of unknowns $\mathbf{u}_1$ and $\mathbf{u}_2$. The *block-preconditioner* $\tilde{\mathbf{K}}_{\mathrm{B}}$ explicitly exploits the block structure of the matrix $\mathbf{K}$. In this report, we use a *block-Gauss-Seidel* preconditioner $\tilde{\mathbf{K}}_{\mathrm{BGS}}$. One iteration of the global defect correction scheme consists of the following three steps:

1. Compute the global defect (cf. Sect. 4.2):
$$\begin{pmatrix} \mathbf{d}_1 \\ \mathbf{d}_2 \end{pmatrix} = \begin{pmatrix} \mathbf{f}_1 \\ \mathbf{f}_2 \end{pmatrix} - \begin{pmatrix} \mathbf{K}_{11} & \mathbf{K}_{12} \\ \mathbf{K}_{21} & \mathbf{K}_{22} \end{pmatrix} \begin{pmatrix} \mathbf{u}_1^k \\ \mathbf{u}_2^k \end{pmatrix}$$
2. Apply the block-preconditioner
$$\tilde{\mathbf{K}}_{\text{BGS}} := \begin{pmatrix} \mathbf{K}_{11} & \mathbf{0} \\ \mathbf{K}_{21} & \mathbf{K}_{22} \end{pmatrix} \tag{6}$$
by approximately solving the system $\tilde{\mathbf{K}}_{\text{BGS}}\mathbf{c} = \mathbf{d}$. This is performed by two global scalar solves and one global (scalar) matrix-vector multiplication:
   a) Solve $\mathbf{K}_{11}\mathbf{c}_1 = \mathbf{d}_1$.
   b) Update RHS: $\mathbf{d}_2 = \mathbf{d}_2 - \mathbf{K}_{21}\mathbf{c}_1$.
   c) Solve $\mathbf{K}_{22}\mathbf{c}_2 = \mathbf{d}_2$.
3. Update the global solution with the (eventually damped) correction vector: $\mathbf{u}^{k+1} = \mathbf{u}^k + \omega\mathbf{c}$

Instead of the illustrative defect correction scheme outlined above, our full solver is a preconditioned BiCGStab solver. Figure 3 summarises the entire scheme. Note the similarity to the general template solver in Fig. 2, and that this specialised solution scheme is entirely constructed from FEAST intrinsics.

global BiCGStab
with block-Gauss-Seidel preconditioner (Equation (6)):
1) solve $\mathbf{K}_{11}\mathbf{c}_1 = \mathbf{d}_1$ by
  global multigrid (V 1+1), additively smoothed by
    for all $\bar{\Omega}_i$: **local multigrid** (V 4+4, Jacobi, UMFPACK)
  coarse grid solver: UMFPACK
2) update RHS: $\mathbf{d}_2 = \mathbf{d}_2 - \mathbf{K}_{21}\mathbf{c}_1$
3) solve $\mathbf{K}_{22}\mathbf{c}_2 = \mathbf{d}_2$ by
  global multigrid (V 1+1), additively smoothed by
    for all $\bar{\Omega}_i$: **local multigrid** (V 4+4, Jacobi, UMFPACK)
  coarse grid solver: UMFPACK

Fig. 3. Our solution scheme for the elasticity equations, scalar solvers are highlighted

4.4 Solving the Navier-Stokes Equation

The application FEASTFLOW solves the Navier-Stokes equations, a nonlinear and vector-valued problem, in a similar way as FEASTSOLID solves elasticity problems. Again, we approximate the domain $\bar{\Omega}$ by a collection of quadrilateral subdomains $\bar{\Omega}_i$, each of which is refined to a logical tensorproduct structure. We consider the weak formulation of equation system (2) and apply a Finite Element discretisation with conforming bilinear elements of the Q_1 space. It is well-known that this pure Galerkin approach exhibits numeric instabilities which stem from dominating convection and from the violation of the discrete inf-sup or LBB-condition. For stability on arbitrary meshes, pressure-stabilisation (PSPG) and streamline-upwind stabilisation (SUPG) is applied, choosing the mesh-dependent parameters in accordance with Apel et al. [1].

The resulting discrete nonlinear equation system reads

$$\begin{pmatrix} \mathbf{A}_{11} & \mathbf{A}_{12} & \mathbf{B}_1 \\ \mathbf{A}_{21} & \mathbf{A}_{22} & \mathbf{B}_2 \\ \mathbf{B}_1^\top & \mathbf{B}_2^\top & \mathbf{C} \end{pmatrix} \begin{pmatrix} \mathbf{u}_1 \\ \mathbf{u}_2 \\ \mathbf{p} \end{pmatrix} = \begin{pmatrix} \mathbf{f}_1 \\ \mathbf{f}_2 \\ \mathbf{g} \end{pmatrix}, \tag{7}$$

with

$$\mathbf{A}_{11} := \nu \mathbf{L}_{11} + \mathbf{N}_{11}(\mathbf{u}) + \tilde{\mathbf{C}}_{11} \qquad \mathbf{A}_{12} := \tilde{\mathbf{C}}_{12}$$
$$\mathbf{A}_{12} := \tilde{\mathbf{C}}_{21} \qquad \mathbf{A}_{22} := \nu \mathbf{L}_{22} + \mathbf{N}_{22}(\mathbf{u}) + \tilde{\mathbf{C}}_{22}$$

where the matrix $\mathbf{L}_{ii}$ corresponds to the Laplacian operator and $\mathbf{N}_{ii}(\mathbf{u})$ to the convection operator. $\mathbf{B}$ and $\mathbf{B}^\top$ are discrete analogues of the gradient and divergence operator while $\mathbf{C}$ and $\tilde{\mathbf{C}}_{ij}$ stem from the discretisation of the PSPG and SUPG stabilisation terms, respectively. For the case of an isotropic mesh, we notice the following: $\mathbf{C}$ is identical to a discretisation of the pressure Poisson operator, scaled with the mesh size h^2.

The nonlinear problem is reduced to a sequence of linear problems by applying a fixed point defect correction method, which can be written in a manner similar to equation (5), where $\tilde{\mathbf{K}}_{\mathrm{B}}$ can be identified with the solution of linearised subproblems.

The linearised, but still vector-valued subproblem is subsequently tackled by a pressure Schur complement approach: We illustrate it with the following basic iteration, but – as in the elasticity case – prefer a Krylov subspace solver such as BiCGStab for increased numerical efficiency in the tests in Sect. 5.2:

$$\begin{pmatrix} \mathbf{u}_{n+1} \\ \mathbf{p}_{n+1} \end{pmatrix} = \begin{pmatrix} \mathbf{u}_n \\ \mathbf{p}_n \end{pmatrix} + \mathbf{K}_{\mathrm{S}}^{-1} \left[\begin{pmatrix} \mathbf{f} \\ \mathbf{g} \end{pmatrix} - \begin{pmatrix} \mathbf{A} & \mathbf{B} \\ \mathbf{B}^\top & \mathbf{C} \end{pmatrix} \begin{pmatrix} \mathbf{u}_n \\ \mathbf{p}_n \end{pmatrix} \right] \tag{9}$$

Here, $\mathbf{A}$ is a block-structured matrix consisting of the linearised matrices $\mathbf{A}_{ij}$, $\mathbf{B}^\top$ is defined as $(\mathbf{B}_1^\top, \mathbf{B}_2^\top)$ and the vectors $\mathbf{u}_n$ and $\mathbf{f}$ as the iterates of the solution $(\mathbf{u}_1, \mathbf{u}_2)^\top$ and right hand side $(\mathbf{f}_1, \mathbf{f}_2)^\top$, respectively. The preconditioner $\mathbf{K}_{\mathrm{S}}$ is defined as the block-structured matrix:

$$\mathbf{K}_{\mathrm{S}} := \begin{pmatrix} \tilde{\mathbf{A}} & 0 \\ \mathbf{B}^\top & -\tilde{\mathbf{S}} \end{pmatrix} \tag{10}$$

where $\tilde{\mathbf{A}}$ denotes a preconditioner for the matrix $\mathbf{A}$ and $\tilde{\mathbf{S}}$ a preconditioner for the pressure Schur complement matrix

$$\mathbf{S} := \mathbf{B}^\top \mathbf{A}^{-1} \mathbf{B} - \mathbf{C}.$$

We note that solving equation (9) requires the solution of linear systems with the matrix $\mathbf{A}$ as well as some matrix-vector multiplications.

fixed point iteration
solving linearised subproblems with
global BiCGStab (reduce initial residual by 1 digit)
preconditioned by block preconditioner (Equation (10))
1) realise preconditioner $\tilde{\mathbf{A}}$ as **global BiCGstab** (1 digit)
with block-Gauss-Seidel preconditioner (Equation (6)):
a) solve $\mathbf{A}_{11}\mathbf{c}_1 = \mathbf{d}_1$ by
global multigrid (V 1+1), additively smoothed by
for all $\bar{\Omega}_i$: **local multigrid** (V 4+4, Jacobi, UMFPACK)
coarse grid solver: UMFPACK
b) update RHS: $\mathbf{d}_2 = \mathbf{d}_2 - \mathbf{A}_{21}\mathbf{c}_1$
c) solve $\mathbf{A}_{22}\mathbf{c}_2 = \mathbf{d}_2$ by
global multigrid (V 1+1), additively smoothed by
for all $\bar{\Omega}_i$: **local multigrid** (V 4+4, Jacobi, UMFPACK)
coarse grid solver: UMFPACK
2) realise preconditioner $\tilde{\mathbf{S}}$ as
$\mathbf{c}_3 = \mathbf{M}_p^{-1}(\mathbf{d}_3 + \mathbf{B}_1^\mathsf{T}\mathbf{c}_1 + \mathbf{B}_2^\mathsf{T}\mathbf{c}_2)$

Fig. 4. Excerpt of our solution scheme for the Navier–Stokes equations, scalar solvers are highlighted

Murphy et al. [8] have pointed out that the square of the iteration matrix of the preconditioned system

$$\mathbf{K} := \mathbf{I} - \mathbf{K}_\mathrm{S}^{-1}\begin{pmatrix}\mathbf{A} & \mathbf{B}\\ \mathbf{B}^\mathsf{T} & C\end{pmatrix} \tag{11}$$

vanishes. The associated Krylov space, $\mathrm{span}\{\mathbf{r}, \mathbf{Kr}, \mathbf{K}^2\mathbf{r}, \mathbf{K}^3\mathbf{r}, ...\}$, has hence dimension 2 which implies that – with exact numerics – any Krylov subspace iterative method terminates in at most two iterations with the solution to the linear system arising in system (9) if the preconditioner $\mathbf{K}_\mathrm{S}$ is used. A few iterations with a "good" approximation of $\mathbf{K}_\mathrm{S}$ therefore suffice to solve (9).

The off-diagonal parts of $\mathbf{A}$ stem from the SUPG stabilisation terms only, they are of order $O(h^2)$ (and vanish for isotropic grids in the Stokes case). Taking this into account, the approximation of the upper left block of $\mathbf{K}_\mathrm{S}$ is straightforward: The diagonal block matrices $\mathbf{A}_{11}$ and $\mathbf{A}_{22}$ correspond to scalar elliptic operators, FEAST's tuned solvers can be applied as in the elasticity case. Turek has shown that – neglecting the convective terms – the lumped pressure mass matrix $\mathbf{M}_p$ is a good preconditioner for the diffusive part of the Schur complement matrix $\mathbf{S}$ [10], i.e. $\tilde{\mathbf{S}} = \nu\mathbf{M}_p$. Solving in this case is reduced to scaling a right hand side with a diagonal matrix. Figure 4 summarises the solver scheme we use throughout this report.

5 Experimental Results

The questions we address in this report are: To what extend do FEASTSOLID and FEASTFLOW benefit from the highly tuned scalar solvers from the FEAST library? What MFLOP/s rates do we achieve on NEC SX-8? How well do these FE applications scale? The following paragraphs are dedicated to these questions.

5.1 Scalar Performance of Feast

Table 1 lists our results of solving a Poisson problem with FEAST using 2–16 CPUs[1] on the BLOCK configuration grid (Fig. 5 (a)), applying Dirichlet conditions on the whole boundary, though. These results are a significant improvement compared to those we presented in last year's report. Every CPU is assigned four subdomains which are subsequently hierarchically refined. We employ the same scalar solver as used for solving scalar sub-problems in the CSM and CFD test cases later on, i. e. the global multigrid (used as a preconditioner to a global BiCGStab iteration) performs one pre- and postsmoothing step in a V cycle, and the local multigrids use a V cycle with four smoothing steps. The goal in all tests is to reduce the initial residuals by six orders of magnitude.

For FE codes, the available system bandwidth is the dominant factor for performance. On the NEC SX-8, each CPU can access main memory at 64 GB/s, while the IXS crossbar switch provides 16 GB/s per node (each node has 8 CPUs). The results obtained on 2 and 4 CPUs – on a single node – thus illustrate a monotonic increase in MFLOP/s rates for increasing level of refinement. For smaller problem sizes, the performance is inhibited by the sequential parts of the code, while for larger problem sizes the performance is mainly determined by the fully vectorised, throughput-oriented operations, and the tuned matrix-vector multiplication pays off. This is further underlined by the fact that with increasing level of refinement (four times the amount of unknowns), the time per iteration (T_{solve}/iter) increases by less than a factor of four. We reach a stable performance of more than 5 GFLOP/s per CPU, which is roughly 30% of the peak performance. Detailed analysis reveals that the single-node performance is close to the peak memory bandwidth. The results obtained on 8 and 16 CPUs include communication via the interconnect, and thus exhibit a significant decrease in performance.

5.2 Performance of FeastSolid and FeastFlow

With FEASTSOLID, we evaluate four configurations that are prototypical for practical applications. Figure 5 shows the coarse grids, the prescribed boundary conditions and the partitioning for the parallel execution of each configuration. The BLOCK configuration (Fig. 5 (a)) is a standard test case in CSM, in which a block of material is vertically compressed by a surface load. The PIPE configuration (Fig. 5 (b)) represents a circular cross-section of a pipe clamped in a bench vise. It is realised by loading two opposite parts of the

[1] There is always one master process which we do not list explicitly.

Table 1. Efficiency tests. Solving a scalar Poisson problem on NEC SX-8

# CPU	level	DOF	T_{solve}/iter (s)	MFLOP/s	MFLOP/s/CPU
2	7	131,841	0.27	1,111	556
	8	525,825	0.40	2,806	1,403
	9	2,100,225	0.80	5,930	2,965
	10	8,394,753	1.94	9,098	4,549
	11	33,566,721	5.89	10,959	5,480
4	7	263,425	0.27	2,116	529
	8	1,051,137	0.43	5,336	1,334
	9	4,199,425	0.74	11,049	2,762
	10	16,787,457	2.08	17,091	4,273
	11	67,129,345	6.36	20,269	5,067
8	7	525,825	0.30	3,685	461
	8	2,100,225	0.50	9,059	1,132
	9	8,394,753	1.00	18,390	2,299
	10	33,566,721	2.57	27,645	3,456
	11	134,242,305	8.07	31,959	3,995
16	7	1,051,137	0.33	6,989	437
	8	4,199,425	0.53	17,701	1,106
	9	16,787,457	1.07	34,408	2,151
	10	67,129,345	3.03	46,888	2,931
	11	268,476,417	8.96	57,600	3,600

outer boundary by surface forces. With the `CRACK` configuration (Fig. 5 (c)) we simulate an industrial test environment for assessing material properties. A workpiece with a slit is torn apart by a device attached to the two holes. In this configuration the deformation is induced by prescribed horizontal displacements at the inner boundary of the holes, while the holes are fixed in the vertical direction. For the latter two configurations we exploit symmetries and consider only sections of the real geometries. Finally, the `STEELFRAME` configuration (Fig. 5 (d)) models a section of a steel frame, which is fixed at both ends and asymmetrically loaded from above.

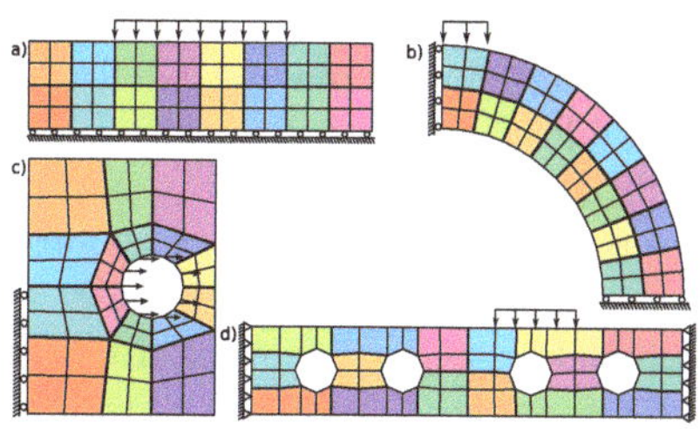

Fig. 5. Coarse grids, boundary conditions and static partition into subdomains for the configurations (a) `BLOCK`, (b) `PIPE`, (c) `CRACK` and (d) `STEELFRAME`

In all the CSM tests we configure the solver scheme (cf. Fig. 3) to reduce the initial residuals by 6 digits, the global multigrid performs one pre- and postsmoothing step in a V cycle, and the inner multigrid uses a V cycle with four smoothing steps.

For CFD, we use a standard benchmark case: a lid driven cavity at Reynolds number $Re = 100$. The solver scheme (cf. Fig. 4) is configured to reduce the residuals for velocity and pressure to below 10^{-8}.

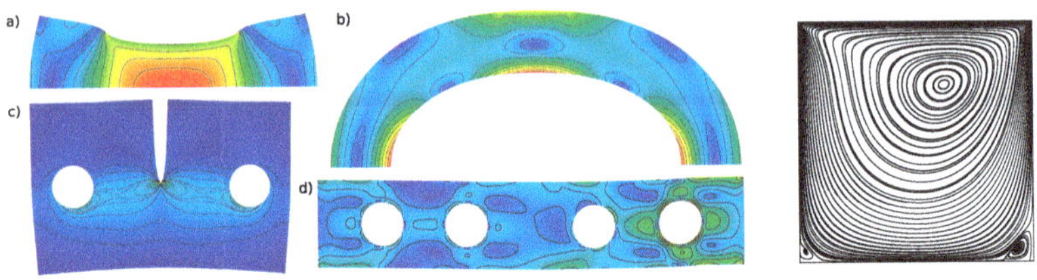

Fig. 6. Computed displacements and von Mises stresses for the CSM configurations, velocity streamline plot for Driven Cavity

Figure 6 shows the computed deformations of the four CSM geometries and the von Mises stresses, which are an important measure for predicting material failure in an object under load, as well as a streamline plot of the driven cavity velocity solution.

Absolute Performance of NEC SX-8 and a Commodity Based Cluster

To compare performance across different architectures and to be able to assess the performance of our code on NEC SX-8, we execute the elasticity solver (cf. Fig. 3) on 16 nodes of a commodity based Opteron cluster (LiDO, Dortmund) and the NEC SX-8. The (slightly outdated) cluster consists of two Opteron DP 250 CPUs with 8 GB DDR-400 memory, and is fully connected via Infiniband. For these tests, we use the four prototypical test cases illustrated in Fig. 5, and refine each subdomain 10 times for a problem size of 134,258,690 degrees of freedom.

Table 2 contains our timing measurements. We first note that the obtained results are consistent for all four configurations. The longer computation times of the PIPE and the STEELFRAME configuration result from the fact that they exhibit a relatively long and thin geometry, while only a relatively small portion of the boundary is fixed (cf. Axelsson [2]). As expected from the experiments with the scalar Poisson problem (cf. Table 1), the Opteron cluster outperforms the NEC SX-8 on small levels of refinement. For larger problem sizes, the NEC SX-8 executes FEASTSOLID roughly 5 times faster than the Opteron cluster. The comparison between the total solving time T_{solve} and the accumulated times of the scalar solves T_{scalar} verifies that our approach of reducing vector-valued problems to sequences of scalar solves works as expected. Independent of the level of refinement, more than 95% of the total time to solution is spent inside scalar solvers. Consequently, the MFLOP/s rates for the solver scheme (cf. Fig. 3) are in the same range as in the scalar case depicted in Table 1. We expect the gap between the Opteron cluster

Table 2. Performance results for four prototypical test cases, computed with FEASTSOLID on 16+1 CPUs on LiDO and NEC SX-8

Configuration	level	NEC SX-8 T_{solve} (s)	NEC SX-8 T_{scalar} (s)	NEC SX-8 MFLOP/s	NEC SX-8 MFLOP/s/CPU	LiDO T_{solve} (s)	LiDO T_{scalar} (s)	LiDO MFLOP/s	LiDO MFLOP/s/CPU
BLOCK	7	6.4	6.3	3,533	221	2.6	2.4	8,709	544
	8	9.5	9.3	9,434	590	10.9	10.1	8,156	509
	9	14.4	14.0	22,008	1,375	39.3	36.3	8,062	503
	10	31.3	29.9	40,256	2,516	157.2	145.3	8,025	502
CRACK	7	6.7	6.6	3,532	221	2.8	2.6	8,582	536
	8	9.2	9.1	10,122	633	10.8	10.0	8,634	540
	9	15.9	15.5	23,250	1,453	46.5	43.5	7,970	498
	10	33.5	32.1	39,693	2,481	166.7	155.5	7,966	498
PIPE	7	11.7	11.5	3,570	223	4.5	4.1	9,281	580
	8	16.2	15.9	10,155	635	18.6	17.2	8,827	552
	9	25.2	24.4	24,454	1,528	77.1	71.7	7,989	499
	10	66.4	63.4	41,498	2,594	345.7	321.7	7,969	498
STEEL-FRAME	7	13.8	13.6	3,544	221	5.5	5.0	8,914	557
	8	20.3	20.0	10,442	653	23.7	21.9	8,958	560
	9	35.7	34.6	24,647	1,540	103.6	96.4	8,487	530
	10	92.1	92.1	41,287	2,580	470.2	438.5	8,090	506

and NEC SX-8 to widen further when refining the coarse grids more than 10 times (cf. Sect. 5.1), which we could not do on LiDO due to lack of local memory.

Weak Scalability of FEASTSOLID

We evaluate weak scalability on NEC SX-8 and LiDO with FEASTSOLID. For these tests, we employ modifications of the BLOCK configuration (Fig. 5 (a)) such that each CPU is assigned four subdomains. The number of CPUs (and hence, DOF) is increased from 2 to 64 (16 Mi to 537 Mi DOF, refinement level $L = 10$). Due to the different geometries, the number of solver iterations until convergence varies, so we normalise the timings with the iteration numbers to emphasise the scalability of our approach rather than presenting obscured results due to the elasticity solver's dependence on the geometry (see previous paragraph).

Figure 7 illustrates good scalability of FEASTSOLID on both architectures. On the NEC SX-8, the bump in performance as soon as more than a single node is involved is clearly visible (cf. Sect. 5.1), and for larger CPU numbers, performance reaches a stable 7.9 seconds per iteration. We thus attribute the linear increase in runtime from 8 to 32 CPUs to granularity effects of the IXS crossbar switch.

When comparing performance between the two architectures, we see that the speed-up achieved by the NEC SX-8 over the Opteron cluster LiDO slowly decreases from a factor of 7.7 to a stable factor of 5 when increasing the number of CPUs, consistent with the results in the previous paragraph. We also attribute this to the interconnects.

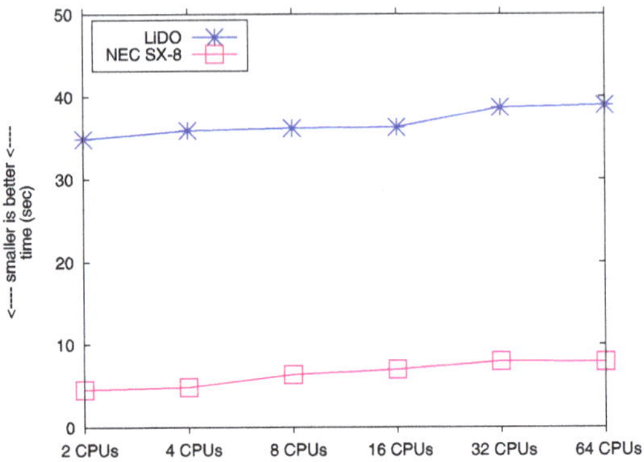

Fig. 7. Weak scalability tests for FEASTSOLID on LiDO and NEC SX-8

Performance Results with FEASTFLOW

Table 3 presents results obtained by FEAST's CFD application FEASTFLOW, using the solver shown in Fig. 4 applied to the lid driven cavity benchmark problem. We only list the time spent in solving the linearised problems, as the assembly process has not been fully tuned yet and obfuscates the solver timings we are interested in. The values labeled "T_{scalar} %" demonstrate that, similar to the elasticity solver above, more than 90% of the total time is spent inside FEAST's optimised scalar solvers. This confirms the feasibility of our solution approach; and accordingly, the MFLOP/s rates obtained by the simple Poisson solver are preserved (see last column of Table 3, copied from Table 1).

Table 3. Performance results of FEASTFLOW

# CPU	level	DOF	T_{solve} (s)	T_{scalar} (s)	T_{scalar} %	MFLOP/s	MFLOP/s/CPU	MFLOP/s/CPU poisson
2	7	395,523	35.1	33.4	95	1,155	583	556
	8	1,577,475	59.0	55.5	94	2,894	1,456	1,403
	9	6,300,675	117.7	108.2	92	5,931	2,977	2,965
	10	25,184,259	345.2	309.1	90	9,137	4,576	4,549
4	7	789,507	40.7	38.4	94	2,145	541	529
	8	3,151,875	66.9	62.8	94	5,359	1,348	1,334
	9	12,595,203	128.3	118.2	92	11,078	2,779	2,762
	10	50,356,227	404.5	363.9	90	17,197	4,306	4,273

We did not perform tests on more CPUs and higher problem sizes on the NEC SX-8 yet. However, these preliminary results convince us that FEASTFLOW will scale just as well as FEASTSOLID or the simple Poisson solver.

6 Conclusions and Future Work

We have demonstrated the feasibility of our approach to reduce the solution of vector-valued problems from CSM and CFD to sequences of scalar problems to be treated with optimised and architecture-aware scalar multigrid solvers. For several prototypical applications from linearised elasticity and fluid dynamics, the approach maintains weak scalability and node performance of the prototypical Poisson problem. In particular, FEAST applications on NEC SX-8 execute significantly faster than on commodity based clusters. In future work, we will focus not only on tuning solvers, but also on the assembly process which turned out to be a bottleneck during our experiments.

Acknowledgements

We would like to thank Christian Becker for initial work with FEAST on the NEC SX-8, and his help and support. This work has been supported by DFG, under grants TU 102/22-1, TU 102/27-1, TU 102/11-3.

References

[1] Th. Apel, T. Knopp, and G. Lube. Stabilized finite element methods with anisotropic mesh refinement for the Oseen problem. In G. Lube and G. Rapin, editors, *Proceedings of the International Conference on Boundary and Interior Layers (BAIL 2006), Göttingen*, pages 1–8, 2006.

[2] Owe Axelsson. On iterative solvers in structural mechanics; separate displacement orderings and mixed variable methods. *Mathematics and Computers in Simulations*, 50:11–30, 1999. doi: 10.1016/S0378-4754(99)00058-0.

[3] Christian Becker. *Strategien und Methoden zur Ausnutzung der High-Performance-Computing-Ressourcen moderner Rechnerarchitekturen für Finite Element Simulationen und ihre Realisierung in FEAST (Finite Element Analysis & Solution Tools)*. PhD thesis, Universität Dortmund, Fachbereich Mathematik, May 2007. http://www.logos-verlag.de/cgi-bin/buch?isbn=1637.

[4] Christian Becker, Sven H.M. Buijssen, and Stefan Turek. FEAST: Development of HPC technologies for FEM applications. In *High Performance Computing in Science and Engineering '07, Transactions of the High Performance Computing Center, Stuttgart (HLRS) 2007*, pages 503–516. Springer, Berlin, 2007. doi: 10.1007/978-3-540-74739-0_34.

[5] Dietrich Braess. *Finite Elements – Theory, fast solvers and applications in solid mechanics*. Cambridge University Press, 2nd edition, 2001.

[6] Timothy A. Davis. A column pre-ordering strategy for the unsymmetric-pattern multifrontal method. *ACM Transactions on Mathematical Software*, 30(2):165–195, 2004. doi: 10.1145/992200.992205.

[7] Joel H. Ferziger and Milovan Perić. *Computational Methods for Fluid Dynamics*. Springer, Berlin, 1996.

[8] Malcolm F. Murphy, Gene H. Golub, and Andrew J. Wathen. A note on preconditioning for indefinite linear systems. *SIAM J. Sci. Comput.*, 21(6):1969–1972, 1999. doi: 10.1137/S1064827599355153.

[9] Barry F. Smith, Petter E. Bjørstad, and William D. Gropp. *Domain Decomposition: Parallel Multilevel Methods for Elliptic Partial Differential Equations.* Cambridge University Press, 1996.

[10] Stefan Turek. *Efficient Solvers for Incompressible Flow Problems: An Algorithmic and Computational Approach.* Springer, Berlin, 1999.

[11] Stefan Turek, Christian Becker, and Susanne Kilian. Hardware–oriented numerics and concepts for PDE software. *Future Generation Computer Systems,* 22(1):217–238, 2003. doi: 10.1016/j.future.2003.09.007.

Overview on the HLRS- and SSC-Projects in the Field of Transport and Climate

Prof. Dr. Ch. Kottmeier

Institut für Meteorologie und Klimaforschung - Forschungsbereich Troposphäre, Universität Karlsruhe, Wolfgang-Gaede-Straße 1, 76131 Karlsruhe

Currently six high-performance computing projects make use of the HLRS in Stuttgart and of the SSC in Karlsruhe in the field of "Transport and Climate". All of these deal with models and data related to processes in the climate system. The topics cover a broad range of scales, ranging from cloud microphysics to large ocean circulation systems. The CPU time requirements of such models are continuously increasing, since in oceanic and atmospheric models there are energy-containing processes in the mesoscale (10–1000 km), the convective scale (100 m – 10 km), and turbulent scales (down to mm). Processes on such scales interact considerably and progress in grid resolution means that the fields can be directly resolved better. Approximations such as parametrizations can be avoided and net effects of small scales are calculated at grid resolution.

The HPC requirements for simulations of natural systems in general are still increasing. This is reflected by, e.g., the large storage amount and CPU times of regional climate simulations in the project "Modelling Regional Climate in Southwest Germany" (IMK-TRO at KIT), not included in the Annual Report. A model restart possibility needed to be implemented to allow for decadal runs. The study aims at an assessment of the capabilities of regional climate models in simulating the observed climate of the last decades in a highly structured mountaineous region. The project AMMA (African Monsoon Multidisciplinary Analysis; IMK-TRO at KIT) focusses on the sensitivity of the rare precipitation events in Westafrica on soil moisture. Since the analysis is still in an early stage, it is not included in the report. On the other hand, the Global Long-Term MIPAS Processing (IMK-ASF at KIT) is rather advanced, but not included since it was presented extensively in 2007.

The three projects chosen for oral presentation and for the HLRS report reflect very well the high importance of the HLRS and SSC computing facilities for highly visible research programmes in actual research in meteorology and oceanography. The project GICS (Effects of Intentional and Inadvertent Hygroscopic Cloud Seeding; IMK-TRO at KIT) considers an innovative approach of cloud seeding in Israel, namely to shift the rain from the ocean closer the land by reducing the speed of cloud droplet growth. This requires a very sophisticated treatment of microphysical processes in clouds as well as a full threedimensional atmospheric model (COSMO). The project AGULHAS (The Agulhas system as a key region of the global oceanic circulation, IFM-Geomar Kiel) applies two-way coupling of a high-resolution regional model with a coarse-resolution ocean model. It is shown that in nested mode the Agulhas mesoscale processes cause significant variability even in the tropical and North Atlantic, which is missing without the high resolution Agulhas nest. Another innovative approach in ocean modelling is realized in the project DYNE ("Simulating El Nino in an eddy-resolving coupled ocean-ecosystem model", IFM-GEOMAR Kiel). In their numerical experiments, a biological ecosystem model is coupled to a regional ocean circulation model in the tropical Pacific ocean. The variable phytoplankton affects the absorption of solar radiation. The results show, that the sea surface temperature as a critical parameter of ENSO changes considerably, but there is also an increased variability of the ocean current systems.

In general, more and coupling between different complex models or model submodules for, e.g., the atmosphere, the ocean, and ecosystems is realized. There is also a trend towards nested models, where either 1-way coupling and, still rarely, 2-way coupling is realized between a coarsely resolving model being applied for a large domain such as a global model and a limited area model that is run at high resolution. Developing such coupling tools requires substantial efforts in adaption and testing with many model test runs. Related projects therefore need relatively small HPC in the beginning of projects, but have much large CPU and storage requirements in a later stage when "production runs" for scientific results are performed.

Effects of Intentional and Inadvertent Hygroscopic Cloud Seeding

Heike Noppel and Klaus D. Beheng

Institut für Meteorologie und Klimaforschung, Universität Karlsruhe / Forschungszentrum Karlsruhe, Kaiserstr. 12, 76128 Karlsruhe, Germany, heike.noppel@kit.edu

Summary. In order to study possible effects of intentional cloud seeding and air pollution on clouds and precipitation, investigations were performed with the numerical weather prediction model COSMO (formerly called LM) using a sophisticated cloud microphysics scheme developed at the Institut für Meteorologie und Klimaforschung. Simulations with this model system on the high performance computer (HP XC400) operated by the computing center at the University of Karlsruhe show that in a land-sea breeze situation, typical for wintertime in the Eastern Mediterranean, aerosols have a considerable impact on the amount and spatial distribution of precipitation at the ground. The results suggest that it might be possible to gain a significant amount of freshwater shifting precipitation from sea to land by seeding the clouds with hygroscopic particles. It could also be shown that the model system is able to reproduce the spatial and temporal evolution of a hailstorm that was observed in South-West Germany, producing realistic radar reflectivities and amounts of precipitation and hail. Sensitivity studies reveal that CCN concentration has a significant impact on the severity of the hailstorm.

1 Introduction

In this report cloud seeding means the emission of substances into the air that serve as cloud condensation nuclei (CCN) or ice nuclei (IN) and may change the amount or type of precipitation or the dynamics of a cloud or a storm. Usually the term is used if this is done intentionally in order to enhance precipitation or to suppress hail. However, emissions from traffic, industry, biomass-burning and other human activities also change CCN and IN conditions and may lead to similar modification of clouds and precipitation. This is, what we call *inadvertent* cloud seeding.

In the following, we will study the impact of CCN conditions only, i.e. the effect of hygroscopic seeding (in contrast to glaciogenic seeding which deals with IN). It is well known, that high number concentrations of aerosol particles (strictly speaking, CCN) lead to high numbers of cloud droplets. As they compete for the available water vapor, the mean size of the nucleated drops will be smaller than for lower CCN concentrations.

Cloud droplets grow into raindrops by coalescence, which will take longer when starting from small droplets. Additionally, if all cloud droplets have about the same size they move with about the same velocity and collisions between them are less likely than if their size distribution is broad. As a consequence, small aerosol particles in high number concentrations will produce small cloud droplets with a narrow size distribution and finally lead to a delay in precipitation formation and often to a decrease in rain amount. On the other hand, large hygroscopic particles may lead to large cloud droplets and a wider spectrum, and therefore may accelerate and enhance the formation of rain.

In an environment where precipitation is formed via the ice phase more complicated effects and feedbacks occur. Initially smaller droplets may shift precipitation formation to higher and thus colder altitudes leading to enhanced freezing, stronger updrafts and finally to increased precipitation rates. Consequently, air masses with low aerosol loading (clean, maritime) and air masses with high aerosol concentration (polluted, continental air) show different precipitation characteristics [7].

Since shortage of fresh water is an issue for many countries, the prospect of being able to enhance precipitation has attracted many scientists to seek for and apply specific materials for cloud seeding. After years of experiments with uneven results [3] recent findings hint to the possibility that rain production can be initiated and/or enhanced by some sort of hygroscopic particles released into clouds during their early stages of formation [9].

An even more recent field of investigation is aerosol impact on convective storms. Studies, performed mostly on a conceptual level, show that pollution aerosols can invigorate convective storms. The rather fast rain formation in pristine air invokes early downdrafts and hinders the lifting of water to the supercooled levels, so that the cloud dies early with a moderate amount of rainfall. Due to the slower rain formation in polluted air, the amount of supercooled water in the mature stage of the cloud is increased. This leads to enhanced growth of ice particles by collision with drops, in the extreme case to the production of large hail, and finally to high precipitation rates and strong downdrafts.

Unfortunately, experimental evidence of the efficacy of intentional (and unintentional) cloud seeding is extremely difficult to obtain [3] and for the time being, the most efficient tool for estimating aerosol effects on precipitation is cloud-resolving numerical modeling using advanced cloud microphysical schemes.

In the following, we will give a short introduction into the model system used. Section 3 deals with rain enhancement by hygroscopic cloud seeding in Israel and section 4 with the effect of CCN on a hailstorm. After a summary of the scientific part an overview of the computational resources used for the simulations is given in section 6.

2 Model Description

All simulations were performed with the non-hydrostatic numerical weather prediction model COSMO developed by the German Weather Service (and others) for their operational weather forecasts. It is based on the fully compressible Navier-Stokes equations using finite differences with a terrain following grid and rotated spherical coordinates. It allows for a wide range of spatial resolutions, provided that adequate parameterizations for the subgrid-scale processes are considered.

The standard COSMO model uses a one moment scheme for cloud physics, i.e. only the bulk masses of the different hydrometeors (cloud droplets, raindrops, cloud ice, snow, graupel) are predicted at each grid point. For the number concentration of the particles or their sizes, respectively, certain assumptions have to be made. Therefore, such a scheme is not appropriate to study the impact of aerosols on clouds, as it works via particle numbers and sizes.

More advanced microphysical schemes take into account balance equations not only for mass concentrations but also for number concentrations (so called two-moment schemes). In this way, two parameters are predicted in space and time which allow to reconstruct hydrometeor spectra comprising the bulk quantities. Therefore, a sophisticated two-moment scheme developed by Seifert and Beheng [6; 8] has been implemented into the COSMO model.

Originally, the particle types were the same as in the standard COSMO scheme. In the course of the project the two-moment scheme has been continously improved and extended. Two of the most important extensions are an additional hail class that has been implemented [1] and a new scheme to parameterize cloud droplet nucleation based on look-up tables by Segal and Khain [4]. A multitude of processes and interactions between the particles are considered (cf. [5]) and further parameters as, e.g., radar reflectivity, can be calculated within the model.

For different purposes different configurations are used. In the following, horizontal resolution is mainly 1 km or 2 km with a model domain of up to 291×291 grid points and 40 to 64 vertical levels.

For time integration COSMO uses a time-splitting technique. The time step for the slow processes is set at the beginning of the run which, for the 1 km resolution, amounts to 6 s. For certain fast processes (sound and gravity wave propagation, some microphysical processes) a smaller, and for very slow processes a larger time step is applied. Different time integration schemes can be chosen within COSMO. We used the two time-level, third order Runge-Kutta method.

3 Precipitation Enhancement by Cloud Seeding

Many regions around the world are subject to severe water shortage, where agriculture alone is responsible for over 70 percent of freshwater use. One possible measure to improve water supply, especially for agriculture, is to try to enhance precipitation by cloud seeding. Although their efficacy has not been proved, more than 24 countries carry out operational cloud seeding activities [3], among them the US and Israel. Most often, the aim is to accelerate and enhance precipitation production.

In our joint German-Israeli project on cloud seeding we follow a new and unique approach, namely a gain of water by hygroscopic cloud seeding with the aim to *delay* the formation of precipitation.

In Israel rain falls mainly during the cold season and often due to cyclones approaching Israel from the Mediterranean Sea or the formation of a convergence zone off the coast where the easterly land-sea breeze meets the large scale westerly wind. In these situations, a significant fraction of precipitation does not reach the land but falls over the sea.

Fig. 1 shows the land-sea-breeze situation typical for winter time. The large scale wind blows from the west. Due to the temperature difference between the air above the warm sea and the cold land surface a land-sea breeze forms at the coast. This leads to convergence of mass in some distance from the coast and the air is forced to rise. The ascending air cools down, condensation takes place and clouds form. Whether a large number of small droplets form or a lower number of larger droplets depends, as mentioned above, on the CCN content. The nucleated droplets grow mainly by collision with other droplets and when they are large enough precipitate. Meanwhile, the large scale wind is driving the clouds towards the coast. Usually, maritime air is rather clean and contains a lot of sea salt. Consequently, large cloud droplets with a wide spectrum form, precipitation formation is fast and most of the precipitation falls over the sea (Fig. 1, left panel) The idea is, that by seeding the clouds with many small CCNs, it might be possible to slow down precipitation formation, shift precipitation to the East and hopefully increase the amount of precipitation accumulated over land (Fig. 1, right panel).

This opportunity was investigated by 3D-simulations with COSMO and the 2-moment microphysics scheme.

3.1 Model Setup for Cloud Seeding Experiments

3D-simulations (with idealized initial and boundary conditions but real topography) have been performed in order to study the impact of aerosols (CCN) in the described land-sea-breeze situation. COSMO with the 2-moment scheme for cloud microphysics is used with a horizontal resolution of about 2 km (0.018°) and a time step of 10 s for the slow processes. Total simulation time is 5 h.

The size of the model domain is 201 × 201 × 60 gridpoints and comprises Israel and adjacent regions. The model was initialized with a modified radio sounding from Bet Dagan (Israel) assuming that the sea is 5 K warmer than the land, and that this temperature deviation decreases exponentially with

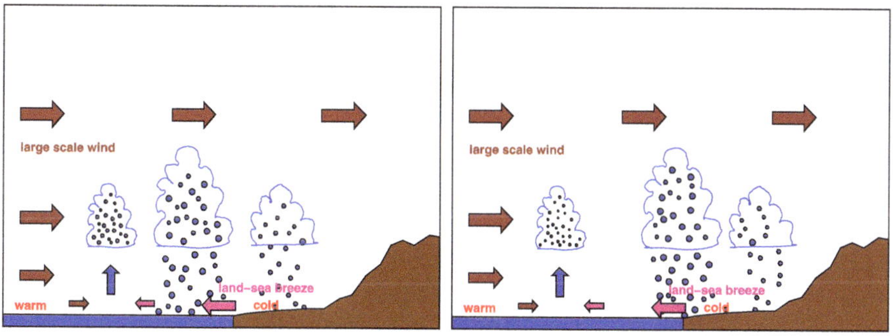

Fig. 1. A sketch of the typical land-sea breeze situation. *Left:* for clean, maritime air with low CCN concentration. *Right:* higher CCN concentration

height. Relative humidity over land is set 25% lower than over sea. Soil and sea temperatures are held constant.

As mentioned above, cloud droplet nucleation is parameterized by using the look-up tables given in [4], where the number of nucleated droplets is determined by vertical velocity at cloud base and aerosol properties represented by CCN number concentration N_{ccn}, mean radius of the larger aerosol mode R_2 (a bi-modal aerosol distribution is assumed), logarithm of the standard deviation of the size distribution $\log\sigma$, and a factor ε representing the effect of the soluble fraction. To study the impact of aerosols and thus possible effects of cloud seeding, simulations for different aerosol scenarios were performed.

What cloud droplet concentrations result from these values depends on the vertical velocities that develop near cloud base and therefore on simulation time.

As mentioned above, the time it takes to form raindrops out of cloud droplets not only depends on the size of the nucleated droplets but also on the width of the cloud droplet size distribution.

In the 2-moment scheme a general gamma distribution with four parameters is assumed for all size distributions (with x = hydrometeor mass)

$$f(x) = A x^{\nu} \exp\left(-\lambda x^{\mu}\right) . \tag{1}$$

A and λ are determined by the predicted number and mass densities (the 0th and 1st moment of the distribution), μ and ν have to be fixed.

In the model experiments cloud seeding was simulated by

1. using three scenarios of CCN concentration leading to three different typical maximum cloud droplet concentrations N_{drop} at cloud base: (a) low CCN concentration, $N_{\mathrm{drop}} = 100\,\mathrm{cm}^{-3}$, (b) intermediate CCN concentration, $N_{\mathrm{drop}} = 300\,\mathrm{cm}^{-3}$, (c) high CCN concentration, $N_{\mathrm{drop}} = 1000\,\mathrm{cm}^{-3}$.
2. decreasing the width of the gamma distribution from broad ($\nu = 0$, $\mu = 1/3$) to narrow ($\nu = 6$, $\mu = 1$).

For the control run, clean maritime air is assumed, i.e. low CCN concentration and a broad cloud droplet spectrum.

3.2 Model Results

Fig. 2 shows the simulated fields of temperature at 2 m agl and wind at 10 m agl after a simulation time of 3 h. In some distance from the coast the large scale westerly winds prevail. Temperature is higher over sea and lower over land where it decreases with the height of topography. Near the coast a land-sea breeze has developed leading to a clearly defined convergence zone about 40 km off the coast, where the air is forced to rise and clouds form. They move towards the mainland where the mountains form an additional trigger for vertical movement.

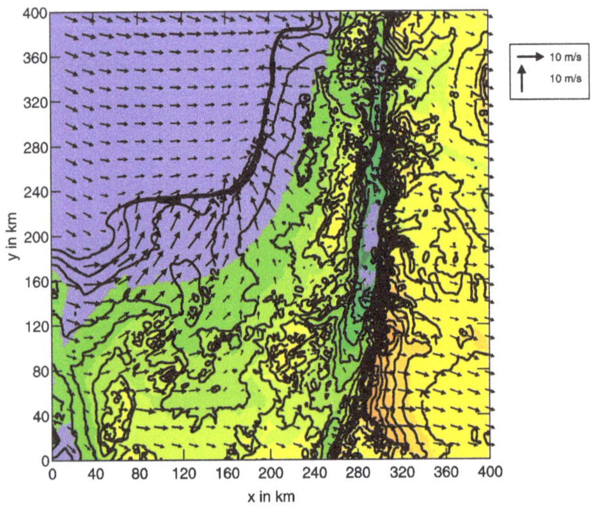

Fig. 2. Horizontal wind field at 10 m agl and air temperature at 2 m agl (isolines, spacing: 1°C) after 3 hours of simulation time for the control run

After 3 h of simulation time, maximum accumulated precipitation over sea is about 5 mm (Fig. 3, top panel). Compared to that over sea, precipitation over land is more structured and reaches higher maximum values. Precipitation occurs up to the Jordan valley, i.e. about 80 km inland. In total about 31 mio m^3 are accumulated over sea and about 35 mio m^3 over land during that time.

This can also be seen in Fig. 4 and 5. Fig. 4 shows the temporal evolution of total accumulated precipitation for the whole model domain (P_{tot}) on the left panel and over land only (P_{land}) on the right panel. The ratio P_{land}/P_{tot} is shown in Fig. 5. Precipitation starts after about 1 h of simulation time. In the beginning less then 20% of the precipitation falls over land. After 2 h the ratio increases to more than 40% and after 3 h about 50% are reached.

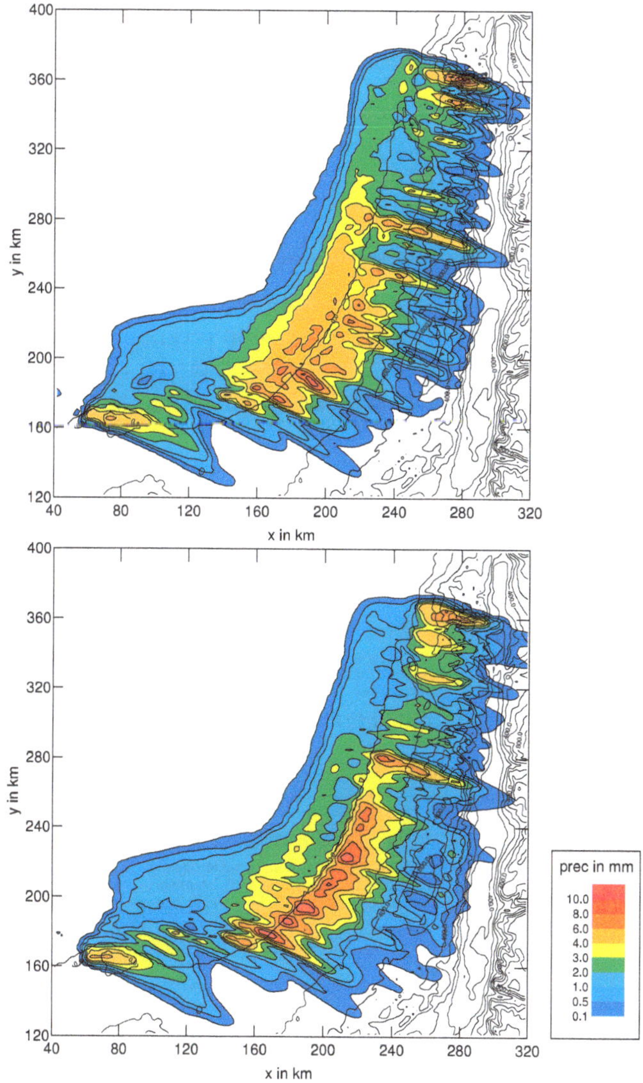

Fig. 3. Accumulated precipitation after 3 hours of simulation time. *Top:* for the control run (low CCN concentration, broad cloud droplet spectrum); *bottom:* for the run with intermediate CCN concentration and a narrow cloud droplet spectrum

As can be seen in Fig. 4 (solid lines) an increase in CCN, and thus droplet number concentration, hardly affects total accumulated precipitation with a slight decrease for high CCN concentration. However, precipitation formation is slowed down a little, the rain is shifted towards the land and $P_{\text{land}}/P_{\text{tot}}$ increases (Fig. 5). Together with the decrease in P_{tot} this leads to a slight enhancement in P_{land} for the first 2 h and a decrease afterwards (Fig. 4).

What would happen if the width of the cloud droplet size distribution was narrowed considerably, e.g. by seeding with many particles of same size? For high CCN concentration this leads to a significant reduction in precipitation, while for low CCN concentration a small increase in P_{tot} results (Fig. 4). In the intermediate case, the narrower spectrum leads to a slight decrease, especially in the first 2 h.

However, precipitation formation is much slower for the narrow spectrum and therefore $P_{\text{land}}/P_{\text{tot}}$ increases for all CCN scenarios (Fig. 5). For intermediate and high CCN concentration more than 60% of the total precipitation fall over land almost from the beginning. In the high CCN case this positive effect is compensated by the general decrease in precipitation, so that the effect on P_{land} is only small (Fig. 4, right panel). For the low and intermediate CCN case though, we get the desired effect, namely a significant increase in precipitation over land. According to these results the most effective seeding scenario is to increase CCN concentration from low to intermediate and at the same time narrow the width of the cloud droplet spectrum.

Fig. 3 (bottom panel) shows the field of accumulated precipitation for this optimum case after 3 h. One can see the clear spatial shift of precipitation compared to the control run. Now about 24 mio m^3 of rainwater are accumulated over sea and 41 mio m^3 over land, which means an increase of 6 mio m^3 (17%). Assuming that 10% of this amount could be stored or used directly for irrigation or other purposes, and estimating desalination costs of 1 \$/m^3, this means a monetary gain of 600 000 \$ in only one event.

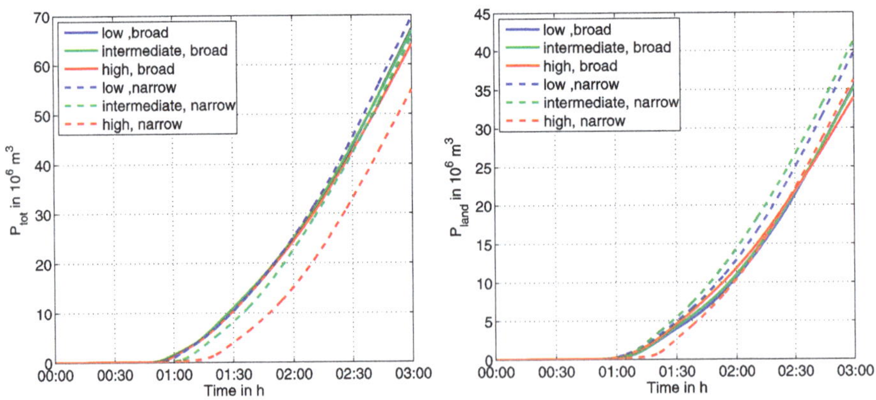

Fig. 4. Time series of accumulated precipitation for different model runs. (Blue: low CCN conc.; green: intermediate CCN; red: high CCN; solid lines: broad cloud droplet size distribution; dashed lines: narrow size distribution). *Left:* over the whole model domain; *right:* over the mainland only

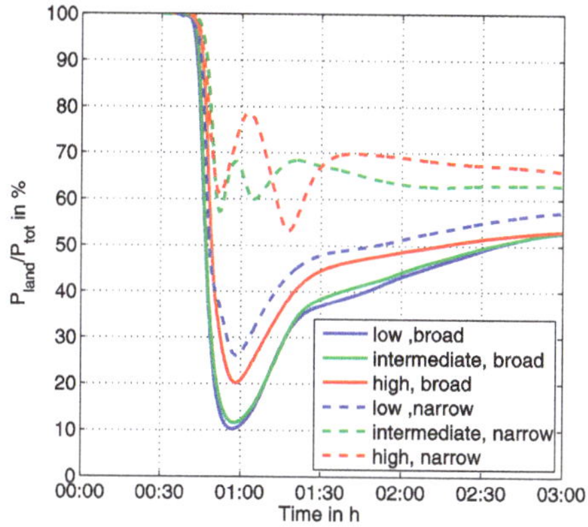

Fig. 5. Time series of the ratio $P_{\mathrm{land}}/P_{\mathrm{tot}}$ for different CCN concentrations and cloud droplet size distributions (line colors and styles as in 4)

4 Impact of Aerosols on the Severity of a Hailstorm

In the situation described above the question was, if by changes in CCN conditions, a spatial shift and thus an enhancement in precipitation over land might be achieved. A different issue, though physically closely related, is, if and to which extent CCN conditions, and hence inadvertent cloud seeding by anthropogenic emissions, may alter the severity of convective storms in Europe. In the following, some results for a case study – a hailstorm in South-West Germany – will be presented.

The hailstorm with up to tennisball sized hailstones occured on 28/06/2006 and caused significant damage in the vicinity of Villingen-Schwenningen (VS), a small town in the Black Forest, South-West Germany.

Radar observation showed the splitting of a convective cell at the western edge of the Black Forest. The right moving cell moved directly east, intensified, hit VS at about 17:30 UTC and moved further east (Fig. 6, right column).

4.1 Model Setup for the Hailstorm Experiment

In the following, one has to bear in mind, that convection that is not triggered by large scale processes like a convergence zone or a front is difficult to predict because (a) it is quite sensitive to environmental conditions, i.e. stratification, humidity and wind shear, and (b) it can be triggered by rather stochastic processes. Nevertheless, we tried to simulate the hailstorm of 28/06/2006 in a realistic setup.

Main differences to the setup for the seeding experiment are a higher spatial resolution (1 km horizontal grid spacing, 64 vertical levels) and the use of operational COSMO-DE forecasts with 2.8 km horizontal resolution as initial and boundary values. The COSMO-DE forecasts, provided by the German Weather Service had been performed with the standard one-moment microphysical scheme and the output was available for every hour. Simulations started at 12 UTC and integration time was 13 to 18 hours with a "slow" time step of 6 s.

Four different CCN conditions are assumed, leading to different maximum cloud droplet concentrations: (a) low CCN concentration, $N_{\text{drop}} = 100\,\text{cm}^{-3}$, (b) intermediate CCN concentration, $N_{\text{drop}} = 300\,\text{cm}^{-3}$, (c) high CCN concentration, $N_{\text{drop}} = 1000\,\text{cm}^{-3}$, (d) very high CCN concentration, $N_{\text{drop}} = 2000\,\text{cm}^{-3}$. A cloud droplet spectrum with moderate width is assumed in all model runs ($\nu = 1/3$,$\nu = 2/3$).

4.2 Results

Fig. 6 shows that in the model simulations (left column) a right moving cell develops at about the same place as the observed one (right column) and passes Villingen-Schwenningen about 20 km to the North. Though there are some differences, e.g. there is a temporal shift of about 5 h and the left mover survives in the simulation, the general development of the simulated storm is in good agreement with the observed one. In CCN scenario (b), the simulated storm produces precipitation rates of more than 200 mm/h in total and up to 100 mm/h for hail (not shown). Accumulated precipitation by the right mover reaches about 50 mm, 15 mm of which is due to precipitating hail (Fig. 7).

The comparison of the results for the different CCN concentrations makes clear that it has a significant impact and that dynamics-microphysics interactions play an important role. For CCN scenario (b) a second, earlier hailstorm passes east of VS from the South-West bringing 4 mm of hail. For other CCN concentrations this convective storm still exists but is much weaker and produces only little precipitation and no hail at the ground at all. Generally, higher CCN concentration tends to lower the amount of rain and hail at the ground. However, there is no monotonous decrease of precipitation with increasing CCN loading, and there are exceptions like the cell east from VS. One can conclude that CCN concentration has a significant impact on the severity of a hail storm and/or the number of hailstorms, respectively, but a general statement whether an increase of CCN loading, e.g. by anthropogenic emissions, will lead to an invigoration of storms or not is not possible.

5 Summary and Conclusions

The simulations with the COSMO model showed that CCN conditions may have a significant impact on the development of clouds and and precipitation. In case of the eastern Mediterranean this means, that with an appropriate cloud seeding strategy it could be possible to shift precipitation from sea to

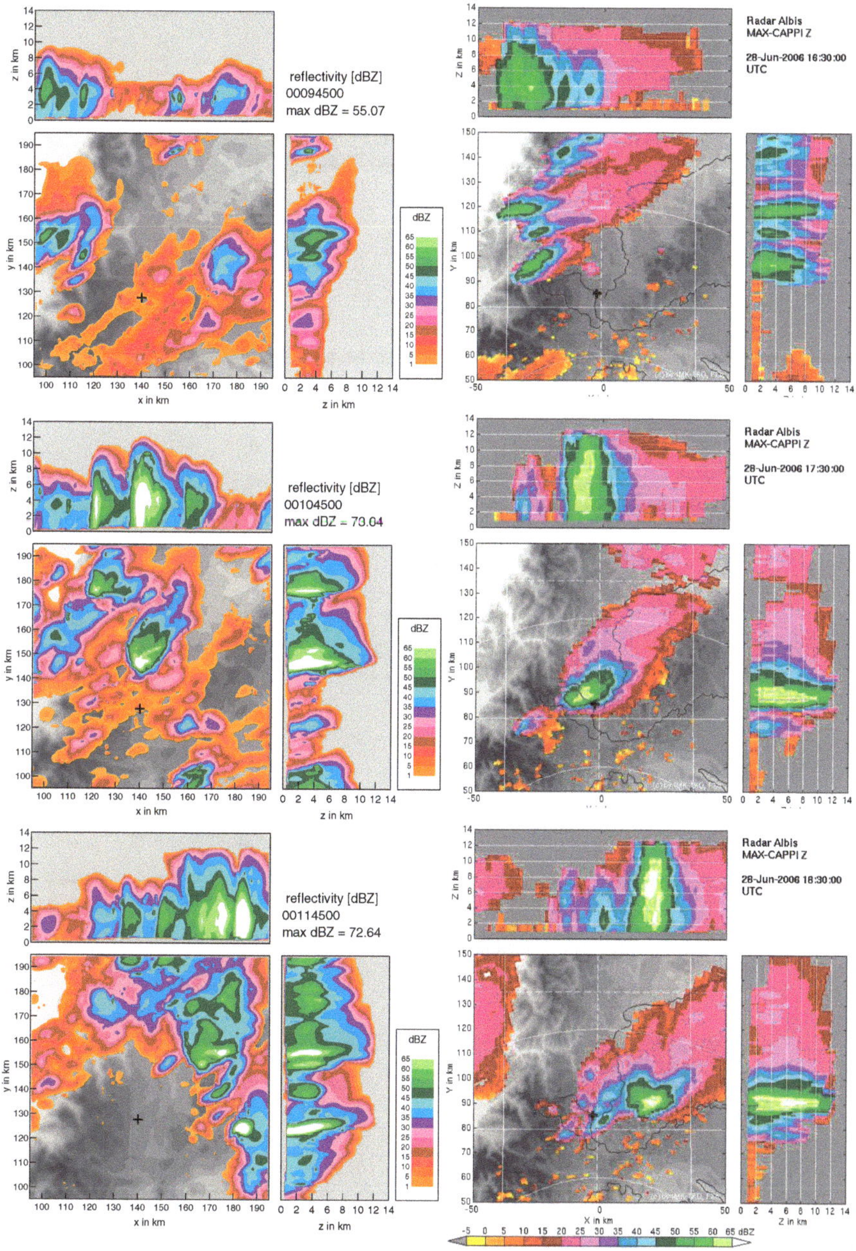

Fig. 6. Maxcappis of radar reflectivity in dBZ (cf. color bar). *Right:* Observations by the radar at Albis, Switzerland. The black cross denote the location of Villingen-Schwenningen. From top to bottom: 1630, 1730, 1830 UTC. *Left:* Calculated from the model simulations with intermediate CCN concentration. From top to bottom: 2145, 2245, 2345 UTC. Topography in grey shades

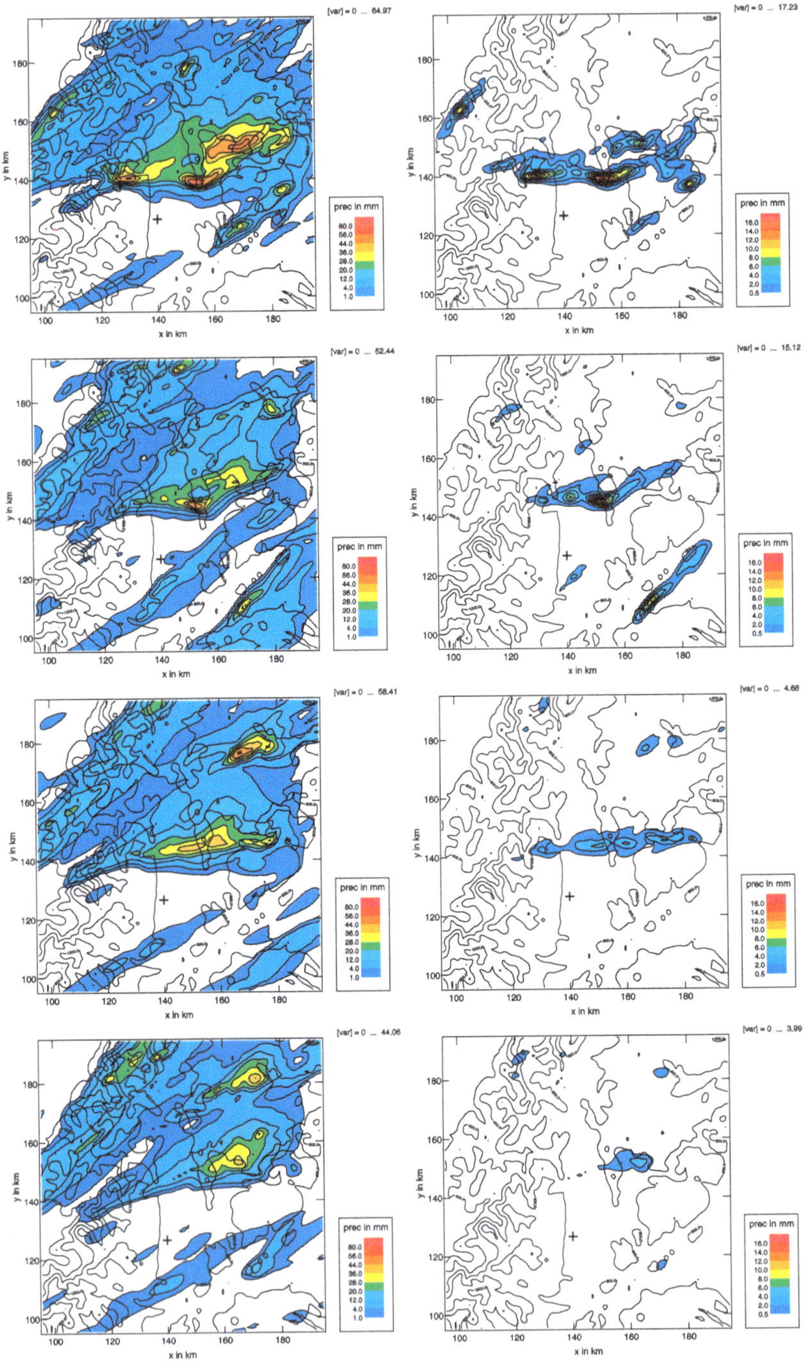

Fig. 7. Accumulated precipitation after 12 h of simulation time. *Left*: total precipitation; *right:* precipitation by hail and graupel. From top to bottom: low, intermediate, high, very high CCN concentration

the land which would be especially beneficial for agriculture but also for other kinds of water use. The simulations also made evident, that in doing this, one has to be careful because cloud seeding may also lead to a decrease in precipitation over land, for example if too many particles or particles of the wrong size are added.

So far, quite simple assumptions have been made in these studies, e.g. idealised initial and boundary conditions were used and "cloud seeding" took place over the whole model domain. In further studies, changes in CCN concentration will be restricted to a certain area off the coast and it is planned to perform simulations with realistic boundary and initial conditions. To provide these, additional simulations with a lower resolution but a larger model domain will be necessary.

For severe storms in Europe the model simulations also indicate a significant effect of CCN conditions. However, it is difficult to draw clear conclusions in this case, except that the model system is able to reproduce a storm similar to the observed one and that further investigations will be necessary. In the hailstorm simulations presented above, only CCN concentration was varied, but as in the cloud seeding experiment further simulations show that the effect of CCN depends on the shape of the assumed cloud size distribution. Another issue is, that in reality, a change in CCN conditions will usually be accompanied by a change in ice nuclei concentration which may enhance or counteract the impact of CCNs. Unfortunately, so far only little is known about how heterogenous freezing of cloud and raindrops actually takes place.

6 Computational Resources

All numerical experiments described in this paper were performed on the high performance computer (HP XC4000) operated by the computing center of the University of Karlsruhe (Steinbuch Centre for Computing).

The weather forecast model COSMO uses dynamic memory allocation for all diagnostic and prognostic arrays, most of them being allocated at the beginning and deallocated at the end of the run. In addition local work space is allocated and deallocated automatically by every routine.

According to the COSMO-LM-User-Guide [2], taking into account the arrays needed, the data space required for version COSMO 2.1 for DOUBLE PRECISION values can be estimated from:

$$8 \times [75\, i \times j \times k + 120\, i \times j + 20.000 \times k]$$

where i and j are the number of gridpoints in east-west and north-south direction and k is the number of vertical layers..

CPU requirements depend on the number of grid points as well as on the lengths of the large and small time steps. The User Guide gives an estimation of

$$i \times j \times k \times \left[1600\,\text{flop}\frac{T_{\text{sim}}}{\Delta t} + 100\,\text{flop}\, i_{\text{ts}}\frac{T_{\text{sim}}}{\Delta t} + 35000\,\text{flop}\left(\frac{T_{\text{sim}}}{1h} + 1\right)\right]$$

with $\Delta t=$ large time step. i_{ts} is the number of small time steps that the acoustic solver needs within one large time step. This value depends on the large time step and on spatial resolution. For the model setups we used $i_{\text{ts}} \approx 7$.

We use a more recent version of COSMO with more cost intensive algorithms which increases the necessary memory and the number of floating point operations. We started with version 2.13 and use 2.19 now. In addition, we apply a more sophisticated scheme for cloud microphysics with additional prognostic and diagnostic variables requiring much more floating point operations than the standard cloud scheme of the model. As the microphysics scheme is quite costly and clouds are often distributed inhomogeneously within the model domain, load leveling is performed in some of the runs.

Numbers of processors used for the simulations described in these paper varied between 16 and 64. For the cloud seeding studies presented in section 3, radiation calculations and the soil model were switched off, which in turn saves some computational costs. In general, the "real" 3-dimensional simulations of section 4 are much more demanding concerning computational resources. These simulations were performed using 291 × 291 × 64 grid points, a large time step of 6 s, and a simulation time of up to 18 h, giving a total of 5 419 584 grid points and 518 400 time steps.

One of these recent "real" model runs with an integration time of 15 h on HP XC4000 demanding 36 CPUs resulted in 36 tasks running on 9 nodes using 328.0 h (user) + 17.4 h (system) total CPU-time and a maximum physical memory of 1463 MB (1949 MB virtual memory) by any process.

According to the timing tools implemented in COSMO most CPU time was used for the dynamical part (i.e. resolved processes) of the model (200 h) and more than half of this for the fast waves alone. Less than 10% of the time was used for the physical part of the model (i.e. sub-scale processes like turbulence), where cloud microphysics is not included. A significant amount of CPU time was spent on output (11 h). For dynamics as well as for physics, communication takes less than 3% of CPU time. In the physical part, however, the barrier waiting time is somewhat larger, namely almost 8% (1.68 h). In the dynamical part it is quite small (about 1%).

Since the further development of COSMO, as well as of the 2-moment cloud microphysics scheme has been progressing during the course of the project, several updates of the model system have been necessary and may be necessary in the future. Every update also means that some new tuning of the cloud module has to be done, which means that many model runs may be required to test the sensitivity of the simulations to different parameters or settings of the model and to get reliable results.

Acknowledgments

We gratefully acknowledge the support of the German Federal Ministry of Education and Research (BMBF) funding this project under grant 02WT0536 in the framework of the joint German-Israeli Cooperation in Water Technology as well as the funding of the ANTISTORM project by the EC-STREP project as part of FP6-2003-NEST-B1. We are also thankful to Profs A. Khain and D. Rosenfeld from the Hebrew University, Jerusalem, Israel, who provided some of the ideas underlying these investigations as well as computer programs for comparison studies.

References

[1] Blahak, U.: Towards a better representation of high density ice particles in a state-of-the-art two-moment bulk microphysical scheme. International Conference on Clouds and Precipitation, July 7–11 2008, Cancun, Mexico. Extended abstract available online at: http://convention-center.net/iccp2008/abstracts/Program_on_line/Poster_07

[2] Doms, G. and U. Schättler: A Description of the Nonhydrostatic Regional Model LM, Part I-VII. Deutscher Wetterdienst, Offenbach, Germany (2005)

[3] National Research Council: Critical Issues in Weather Modification Research. The National Academies Press, Washington, D.C., 2003

[4] Segal, Y., Khain, A.: Dependence of droplet concentration on aerosol conditions in different cloud types: Application to droplet concentration parameterization of aerosol conditions. J. Geophys. Res., **111**, D15240, doi:10.1029/2005JD006561 (2006)

[5] Seifert, A.: Parametrisierung wolkenmikrophysikalischer Prozesse und Simulation konvektiver Mischwolken. Dissertation (in German). Institut für Meteorologie und Klimaforschung, Universität Karlsruhe (TH) / Forschungszentrum Karlsruhe

[6] Seifert, A., Beheng, K.D.: A two-moment cloud microphysics parameterization for mixed phase clouds. Part 1: Model description. Meteorol. Atmosph. Phys., **92**, 45–66 (2006)

[7] Seifert, A., Beheng, K.D.: A two-moment cloud microphysics parameterization for mixed phase clouds. Part 2: Maritime vs. continental deep convective storms. Meteorol. Atmosph. Phys., **92**, 67–82 (2006)

[8] Seifert, A., Khain, A., Pokrovsky, A., Beheng, K.D.: A comparison of spectral bin and two-moment bulk mixed-phase cloud microphysics. Atmos. Res. **80**, 46–66 (2006)

[9] Woodley, W.L., Rosenfeld, D., Axisa, D., Lahav, R., Bomar, G.: On the Documentation of Microphysical Structures Following the Base-Seeding of Texas Convective Clouds Using Salt Micro-Powder. 16th Conference on Planned and Inadvertent Weather Modification, San Diego, Jan. 2005

The Agulhas System as a Key Region of the Global Oceanic Circulation

A. Biastoch[1], C.W. Böning[1], M. Scheinert[1], and J.R.E. Lutjeharms[2]

[1] Leibniz-Institut für Meereswissenschaften, Düsternbrooker Weg 20, 24106 Kiel
abiastoch@ifm-geomar.de
[2] University of Cape Town, 7700 Rondebosch, South Africa

Summary. The Agulhas system at the interface between the Indian and Atlantic Ocean is an important region in the global oceanic circulation with a recognized key role in global climate and climate change. The simulation of the Agulhas system was performed by a high-resolution regional model nested in a global coarse-resolution ocean model. It is shown that this model simulates all characteristics of the Agulhas regime in a highly realistic manner. Due to the two-way coupling of both models the importance of the Agulhas leakage on the large-scale thermohaline circulation was demonstrated.

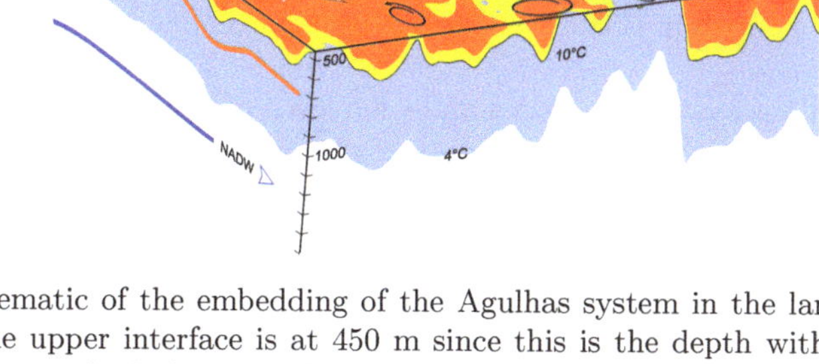

Fig. 1. Schematic of the embedding of the Agulhas system in the large-scale circulation. The upper interface is at 450 m since this is the depth with the largest contrast between the Indian and Atlantic Ocean. Circulation features as the Agulhas Current and Return Current, Agulhas rings are marked as black arrows. The grey arrow indicates the general drift of Agulhas rings. The upper (Antarctic Intermediate Water and Thermocline Water) and lower (North Atlantic Deep Water) limbs of the meridional overturning circulation are shown as orange and blue arrows. Modified from [1]

1 Introduction

The flow of warm and salty waters from the Indian Ocean to the Atlantic Ocean around the southern tip of Africa is an important element of the global ocean circulation [2]. Under present climate conditions this interoceanic flux provides the bulk [3] of the upper limb of the meridional overturning circulation (MOC) in the Atlantic Ocean, highly affected by the nonlinear constituents of the Agulhas Current system [4].

Paleo observations and model studies have linked variations in the MOC to changes in the Agulhas leakage [5, 6]. Similar arguments have been advanced for future climate trends, so that the Agulhas region is acknowledged to play a key role in global climate and climate change [7]. Factors causing changes in the Agulhas leakage are still under investigation. Variations in the wind fields, such as latitudinal shifts of the southern hemisphere mid-westerlies [8], would directly impact on the retroflection of the Agulhas Current and in consequence the interoceanic transport of heat and salt. Especially the supply of salt has repercussions for the large-scale global ocean circulation by its influence on the deep water formation in the subpolar North Atlantic (and therefore the lower limb of the MOC) via advective processes [9]; in consequence this feeds also back to climate [10].

What factors determine the intensity of the Agulhas leakage? How will it react on changes in the atmospheric conditions, in particular a southward shift of the westerlies as projected by some climate scenario calculations for the recent IPCC report [11]? Addressing such questions does require an improved quantitative understanding of the dynamics of the Agulhas region and its interplay with the global circulation. The goal of this project is to significantly advance this understanding by a sequence of model studies based on a newly developed "nested" model configuration which combines the global ocean circulation with a high-resolution representation of the Agulhas leakage. The studies focus on some key aspects of the Agulhas regime: the role mesoscale dynamics around South Africa, their interplay with the mean current system and their effect on the interocean exchange and the large-scale circulation.

The Agulhas system is not a simple conveyor of heat and salt from the Indian to the Atlantic Ocean. It consists of a strong western boundary current, the Agulhas Current [4], which flows southward along the African Coast. After shooting over the southern tip of the continent it abruptly retroflects back into the Indian Ocean, forming the Agulhas Return Current, which then gradually closes the return flow of the subtropical gyre in the South Indian Ocean. The retroflecting Agulhas Current intermittently sheds the largest mesoscale eddies in the World Ocean [12]. These Agulhas rings are the dominating vehicles transporting and gradually releasing the Indian Ocean waters into the Atlantic. However, they are only one constituent in a vast range of mesoscale features in the "Cape Cauldron" [13], complicating the exact quantification of the interocean exchange [14].

The retroflection process appears to be affected by mesoscale perturbations upstream of the Agulhas Current. For the first time described by results from a numerical model [15], later verified by analyses from satellite observations [16], these eddies are being formed in the Mozambique Channel and drift pole-ward. Further south the Mozambique eddies do strongly interact with the Agulhas Current, possibly in combination with offshore displacements of the Agulhas Current, called Natal Pulses [17]. Observationally this interplay [18] has only been verified for a single instance [19]. Eddies that appear east of Madagascar [20] are even more under debate since there are contradicting theories on the question whether the South East Madagascar Current is retroflecting [21] or not [22], which then has also consequences for its feeding of the Agulhas Current.

This project aims to realistically simulate this complex current system and its effect on the interoceanic transport with the highest spatial resolution to date. Using a hierarchy of global ocean models with realistic atmospheric forcing, the effect of inter-ocean transport on the large-scale circulation in the Atlantic is established on time scales up to several decades. This includes the variability of the meridional overturning.

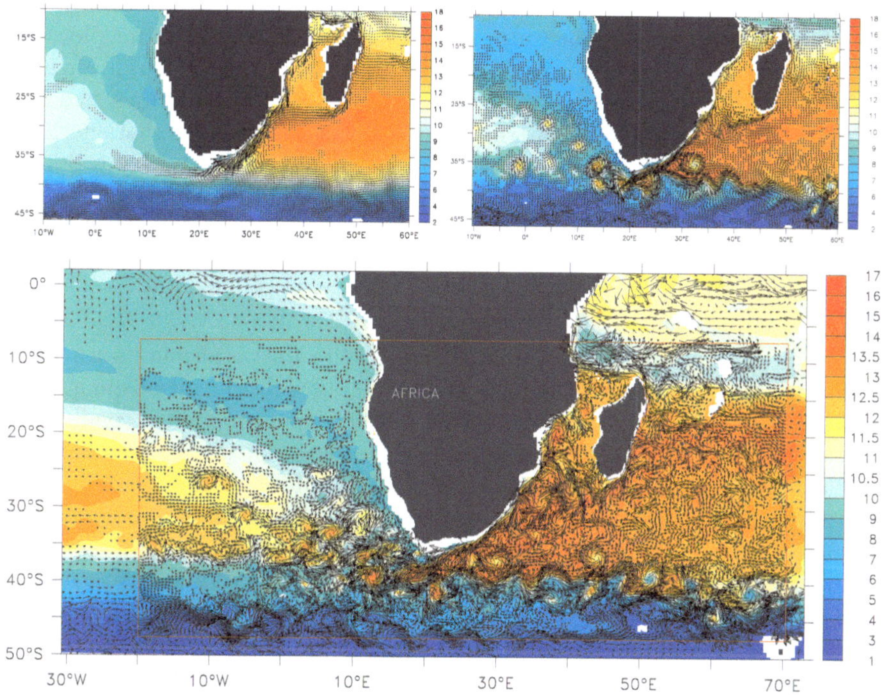

Fig. 2. Circulation around South Africa: Shown are 5-day snapshots of temperature and velocity at 450 m depth for (a) ORCA05 and (b) ORCA025 and (c) AG01 (note that for ORCA025 every second vector is shown, for AG01 every 4th)

2 The Model Hierarchy

The model used in this study is based on the "Nucleus for European Modelling of the Ocean" (NEMO, v.2.3) [23], consisting of the C-grid primitive equation ocean model OPA [24] and the LIM2 sea-ice model [25]. The global ORCA version used here is part of a model hierarchy developed by the European model collaboration DRAKKAR [26].

The ocean component is based on a derivation of the Navier-Stokes Equations (the "primitive equations"), stepping velocity, temperature and salinity forward in time. A free surface formulation (e.g. by a conjugate gradient solver), a high-order polynominal fit of the density equation and lots of parametrizations of different ocean physics let the complete package appear as a wide range of different numerical methods, though flexible in its use due to the modular formulation. It is written in FORTRAN90, has a finite differences layout with a horizontal (geographical) domain decomposition for MPI parallelization. Traditionally the grid space layout leads to a good performance on vector systems.

The global configurations used in this project have nominal horizontal resolutions of 1/2° (*ORCA05*) and 1/4° (*ORCA025*) respectively. The first one is "coarse-resolving" (Fig. 2a), not resolving the important mesoscale at O(100 km). Without Agulhas rings it therefore misses an important physical component of the Agulhas system. The second resolution (Fig. 2b) is already eddy-permitting [27], simulating Agulhas rings and eddy structures to some degree. However, these Agulhas rings appear too regular in its characteristic structures and shedding frequencies. In addition, this model configuration is not able to properly simulate mesoscale perturbations occurring in the source regions of the Agulhas Current.

The full representation of the Agulhas system can only be represented at high spatial resolution (grid scales less than 10 km). This can be reached by two ways: Either further refining the global grid (e.g. [28]), leading to enormous computational costs and an inflexibility for performing sensitivity studies. Or by reducing the model domain; an approach which leads to a large dependence of the circulation on the prescribed open boundary conditions; effectively removing a flexibility of the system and not always leading to a realistic representation of the Agulhas system in all its facets [15, 29].

Although the NEC SX-8 at HLRS would be powerful enough to deal with a global high-resolution configuration, we have used a third alternative: embedding a high-resolution Agulhas nest in a coarser resolved global base model (ORCA05). Apart from having a smaller configuration (and therefore allowing more sensitivity studies) this approach also is the only means to isolate the effects from the Agulhas mesoscale from other dynamics. Comparing the base model outside of the Agulhas region (e.g. in the North Atlantic) with a solution without a high-resolution nest unequivocally allows one to trace back its differences to mesoscale Agulhas dynamics.

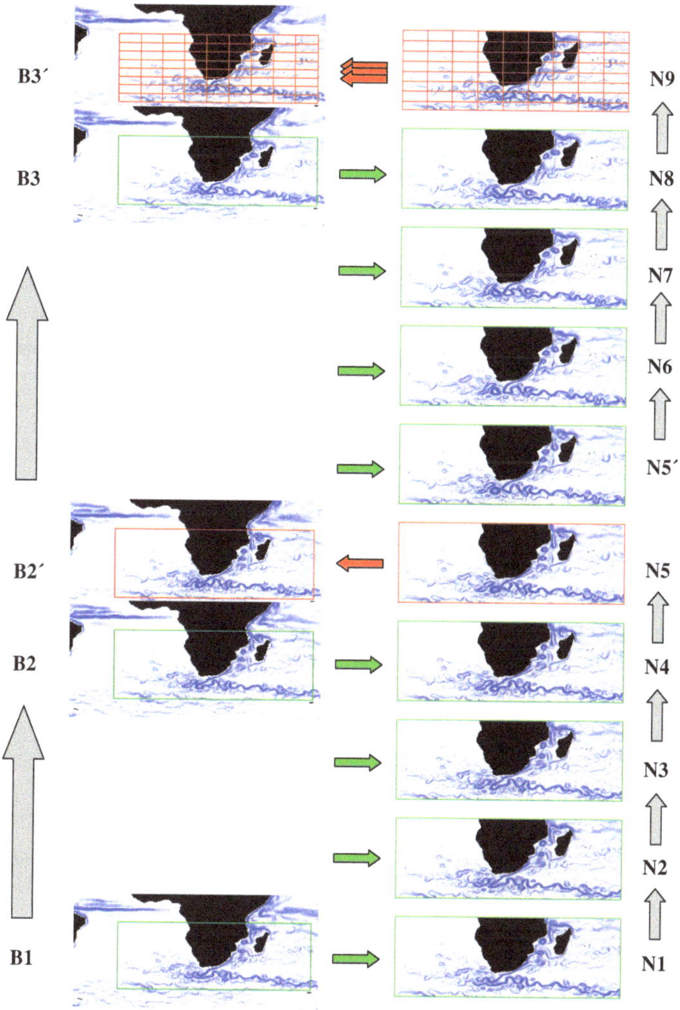

Fig. 3. Time-stepping of the base (left) and nested (right) grids. The green boxes and arrows indicate an interpolation from the base grid onto the outer boundaries of the nest, the red ones an averaging of the outer and surface boundaries of the nest onto the base grid; the mesh indicates an averaging of the whole nest onto its base grid points in the Agulhas region. Gray arrows and numbers indicate the timesteps of base (Bn) and nest (Nn) and their respective updates (Bn', Nn') (Note that this is a schematic representation, the actual circulation will not evolve so fast within one timestep.)

This nesting capability is provided by AGRIF (Adaptive Grid Refinement In Fortran, [30]), directly embedded in the NEMO code. It is realized via a preprocessing step, inserting additional subroutines into the model code, allowing both models to communicate in a two-way coupling behavior (Fig. 3), where

- the host updates the boundaries of the nest
- the nest updates the boundaries and the sea surface height of the host

Coupling both model grids takes place at every baroclinic time step of the base model (2160 s in this case). At any given timestep the base grid provides its prognostic data along the boundary of the nest (green box in Fig. 3), interpolated in time between two base model time steps, e.g. B1 and B2. Then the nest is integrated some time steps (4 times in this configuration, each 540 s, N1 to N5), afterwards all data are averaged onto the base grid along the boundary of the nest (red box in Fig. 3); in addition all coarse-resolution grid points in the Agulhas region are updated with the sea surface height from the nest. Both averaging processes feed the baroclinic and barotropic states back to the base model. Using this updated time step B2' the base model is integrated for another timestep and the cycle starts again. Every few timesteps (3 in this configuration) the full three-dimensional, baroclinic state vector of the nest is averaged onto the base model grid points and fed back to the base model (red mesh in Fig. 3). The interpolation and averaging between both grids is conservative, so that mean model fields are almost maintained during the full length of integration (the drift in global mean temperatures and salinities due to the nesting approach are less than 0.02 °C and 0.02 psu). This is a significant improvement over earlier attempts [31] where hydrodynamic conditions in base and nest diverged after a few model weeks. Due to its success the AGRIF technique embedded in OPA has been used in applications of the Labrador Sea [32] and in the Caribbean [33].

As a first configuration on the NEC SX-8 a 1/10° nest within ORCA05 has been set up (Figs. 4 and 2c) which simulates the full variability of the Agulhas regime, including a realistic variety of scales and paths of Agulhas rings and the important upstream perturbations. It was integrated over a hindcast period using prescribed atmospheric fields for the period 1958-2004. In addition a sensitivity experiment was performed to examine the importance of upstream perturbations in the Mozambique Channel and east of Madagascar on the Agulhas dynamics. For this approach the high-resolution nest was reduced, leaving those key regions at the lower resolution of the base model. This, again, demonstrates the usefulness of the nesting approach for the physical interpretation.

2.1 Computational Requirements

Base model and nest have similar physical dimensions (Tab. 1), but since the time step of the base model is four times larger than that of the nest most

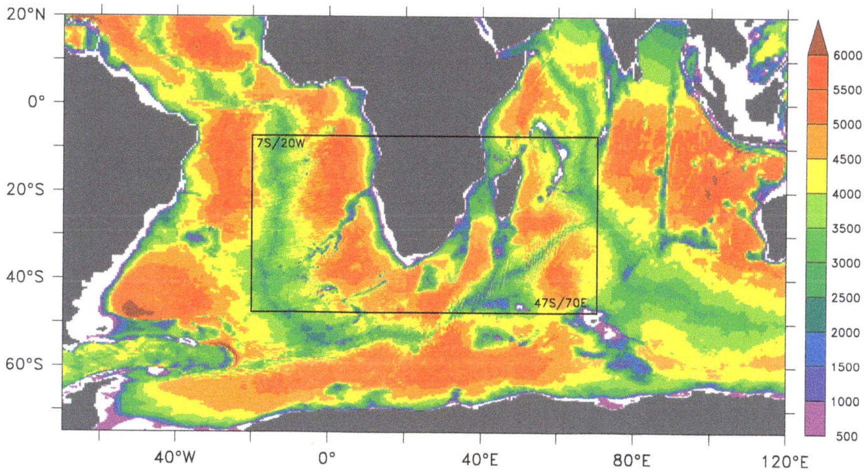

Fig. 4. Part of ORCA05 bathymetry hosting the 1/10° Agulhas nest (The color key is denoting water depth in meters, white areas are shallower than 500 m.)

of the computational cost is spent in the nested model. When the integration was started AGRIF itself was very ineffective on vector computers, leading to an overall performance of just 14.5 GFLOP/s per node (in comparison to 27 GFLOP/s per node = more than 20% of the peak performance for a stand alone configuration). In the meantime significant manpower by the NEMO and AGRIF teams in collaboration with the Teraflop Project went into the optimization, which led to an increase to about 24 GFLOP/s, which is a reasonable number for a GFD code.

Table 1. Characteristic numbers of the base model and nest

	dimensions (lon × lat × depth)	grid points	time step
base	$722 \times 511 \times 46$	$\approx 17 \times 10^6$	36 min
nest	$909 \times 474 \times 46$	$\approx 20 \times 10^6$	9 min

Beside the "pure number crunching" figure, the massive output is a typical bottleneck for this type of high-resolution ocean models: The 5-daily output of both base model and nest together produced more than 100 GB per model year, summing up to a total of up to 5 TB for a 47-year experiment. A significant amount of work load had to go in the transfer of these data to servers outside the HLRS. A data reduction is hereby not possible since those hindcast runs have a realistic structure that can be directly compared with observations. Therefore, they serve as sort of "community runs", where different groups (e.g. those at the University of Cape Town) will analyse individual components of the Agulhas system.

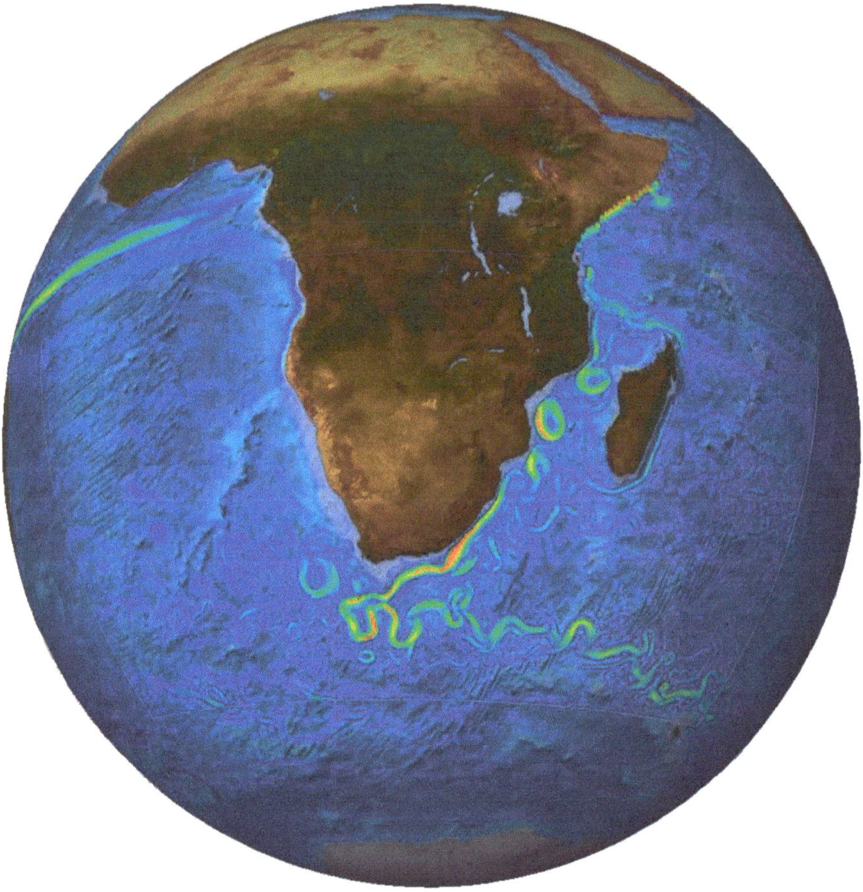

Fig. 5. Snapshot of the high-resolution Agulhas model nested in the global coarse resolution model. Shown are speeds (5-day average around 12 Oct 2001) at 100 m depth (in m^{-1}). Speeds below 0.05 m^{-1} are transparent, so that the bottom topography shines through

3 Results

Analysis showed that the nested Agulhas model provides a realistic picture of the system [34]: An Agulhas Current with a realistic transport; a highly variable Agulhas Undercurrent; Agulhas rings with a correct diameter rapidly spinning down in the Cape Basin and then moving off into the South Atlantic and an Agulhas Return Current with a faithful representation of its average location, variability and shedding of eddies to either side. About 5±1 anti-cyclonic Mozambique eddies are formed per year; Natal Pulses are evident on the landward border of the Agulhas Current at least 2±1 times per year and quickly move downstream. Upstream retroflections are seen in the expected location south of Port Elizabeth. In consequence the Agulhas model simulates an interoceanic transport that is comparable to observational estimates, in contrast to the over-estimation in lower resolved models.

A reduction of the nested grid, excluding the Mozambique Channel effectively prevented Mozambique eddies being formed. In consequence no Natal Pulses were generated and no upstream retroflection of the Agulhas Current occurred. However, in contrast to former speculations, this did not lead to a reduction of the interoceanic transport. Instead, Agulhas rings occurred more regular, similar to those in the eddy-permitting model (ORCA025) that also was not properly simulating the Mozambique eddies and Natal Pulses due to its limited resolution.

By comparing the circulation in the base model with a solution not containing a high-resolution Agulhas nest it was possible to isolate the effect of the high-resolved Agulhas area on the Atlantic meridional overturning. The mesoscale dynamics in the Agulhas regime hereby appear as an important source of decadal variability in the meridional overturning [1]. Propagating by boundary waves along the South American shelf transport signals do quickly communicate northward into the North Atlantic. In the tropical and sub-tropical Atlantic the Agulhas induced variability has similar amplitudes as the variability introduced by sub-polar deepwater formations events [35]. This underlines the importance for studying the Agulhas regime and it associated interoceanic transport as another prominent key region of the Atlantic thermohaline circulation.

Further analyses are currently in progress, both shedding light on the behavior of individual components of the greater Agulhas system (as, for instance, the Agulhas Undercurrent or the retroflection of the East Madagascar Current) or on the embedding this system in the global oceanic circulation under current and changing climate conditions. Therefore supporting sensitivity experiments will be performed with the optimized code version.

References

1. A. Biastoch, C.W. Böning, and J.R.E. Lutjeharms. Agulhas leakage dynamics affects decadal variability in Atlantic overturning circulation. *Nature*, doi:10.1038/nature07426, in press, 2008.
2. A.L. Gordon. Oceanography: The brawniest retroflection. *Nature*, 421:904–905, 2003.
3. S. Speich, B. Blanke, and G. Madec. Warm and cold water routes of an OGCM thermohaline conveyor belt. *Geophys. Res. Lett.*, 28(2):311–314, 2001.
4. J.R.E. Lutjeharms. *The Agulhas Current.* Springer, Berlin, 2006.
5. G. Martínez-Méndez, R. Zahn, I.R. Hall, L.D. Pena, and I. Cacho. 345,000-year-long multi-proxy records off South Africa document variable contributions of Northern versus Southern Component Water to the Deep South Atlantic. *Earth and Planetary Science Letters*, 2007.

6. G. Knorr and G. Lohmann. Southern Ocean origin for the resumption of Atlantic thermohaline circulation during deglaciation. *Nature*, 424(6948):532–6, 2003.
7. T.F. Stocker, G.K.C. Clarke, H. Le Treut, R.S. Lindzen, V.P. Meleshko, R.K. Mugura, T.N. Palmer, R.T. Pierrehumbert, P.J. Sellers, K.E. Trenberth, and J. Willebrand. Physical climate processes and feedbacks. In J.T. Houghton, Y. Ding, D.J. Griggs, M. Noguer, P.J. van der Linden, X. Dai, K. Maskell, and C.A. Johnson, editors, *Climate Change 2001: The Scientific Basis. Contribution of Working Group I to the Third Assessment Report of the Intergovernmental Panel on Climate Change*. Cambridge University Press, 2001.
8. W. Cai. Antarctic ozone depletion causes an intensification of the Southern Ocean super-gyre circulation. *Geophys. Res. Lett.*, 33(3), 2006.
9. W. Weijer, W.P.M. de Ruijter, H.A. Dijkstra, and P.J. Van Leeuwen. Impact of interbasin exchange on the Atlantic overturning circulation. *J. Phys. Oceanogr.*, 29:2266–2284, 1999.
10. R. Marsh, W. Hazeleger, A. Yool, and E.J. Rohling. Stability of the Thermohaline Circulation under millennial CO2 forcing and two alternative controls on Atlantic salinity. *Geophys. Res. Lett.*, 34(3), 2007.
11. G.A. Meehl, T.F. Stocker, W.D. Collins, P. Friedlingstein, A.T. Gaye, J.M. Gregory, A. Kitoh, R. Knutti, J.M. Murphy, A. Noda, S.C.B. Raper, I.G. Watterson, A.J. Weaver, and Z.-C. Zhao. Global Climate Projections. pages 747–846. Cambridge University Press, Cambridge, United Kingdom and New York, NY, USA, 2007.
12. D.B. Olson and R.H. Evans. Rings of the Agulhas Current. *Deep-Sea Res.*, 33:27–42, 1986.
13. O. Boebel, J. Lutjeharms, C. Schmid, W. Zenk, T. Rossby, and C. Barron. The Cape Cauldron: a regime of turbulent inter-ocean exchange. *Deep-Sea Res. II*, 50:57–86, 2003.
14. W.P.M. de Ruijter, A. Biastoch, S.S. Drijfhout, J.R.E. Lutjeharms, R. Matano, T. Pichevin, P.J. van Leeuwen, and W. Weijer. Indian-Atlantic inter-ocean exchange: Dynamics, estimation and impact. *J. Geophys. Res.*, 104:20,885–20,910, 1999.
15. A. Biastoch and W. Krauss. The role of mesoscale eddies in the source regions of the Agulhas Current. *J. Phys. Oceanogr.*, 29:2303–2317, 1999.
16. M.W. Schouten, W.P.M. de Ruijter, P.J. van Leeuwen, and H. Ridderinkhof. Eddies and variability in the Mozambique channel. *Deep-Sea Res. II*, 50:1987–2004, 2003.
17. J.R.E. Lutjeharms and H.R. Roberts. The Natal Pulse: an extreme transient on the Agulhas Current. *J. Geophys. Res.*, 93:631–645, 1988.
18. W.P.M. de Ruijter, P.J. van Leeuwen, and J.R.E. Lutjeharms. Generation and evolution of Natal Pulses, solitary meanders in the Agulhas Current. *J. Phys. Oceanogr.*, 29:3043–3055, 1999.
19. M.W. Schouten, W.P.M. de Ruijter, and P.J. van Leeuwen. Upstream control of Agulhas Ring shedding. *J. Geophys. Res.*, 107(10.1029), 2002.
20. W.P.M. de Ruijter, H.M. Aken, E.J. Beier, J.R.E. Lutjeharms, R.P. Matano, and M.W. Schouten. Eddies and dipoles around South Madagascar: formation, pathways and large-scale impact. *Deep-Sea Res. I*, 51(3):383–400, 2004.
21. J.R.E. Lutjeharms. Remote sensing corroboration of retroflection of the East Madagascar Current. *Deep-Sea Res.*, 35:2045–2050, 1988.

22. G.D. Quartly, J.J.H. Buck, M.A. Srokosz, and A.C. Coward. Eddies around Madagascar: The retroflection re-considered. *J. Mar. Res.*, 63(3-4):115–129, 2006.
23. G. Madec. Nemo ocean engine. Technical Report 27, Note du Pole de modelisation, Institut Pierre Simon Laplace (IPSL), France, 2006.
24. G. Madec, P. Delecluse, M. Imbard, and C. Levy. Opa 8.1 ocean general circulation model reference manual. Technical report, Institut Pierre Simon Laplace des Sciences de l'Environment Global, 1999.
25. T. Fichefet and M.A. Morales Maqueda. Modelling the influence of snow accumulation and snow-ice formation on the seasonal cycle of the Antarctic sea-ice cover. *Climate Dynamics*, 15:251–268, 1999.
26. The DRAKKAR Group. Eddy-Permitting Ocean Circulation Hindcasts of Past Decades. *Clivar Exchanges*, 12:8–10, 2007.
27. B. Barnier, G. Madec, T. Penduff, J.-M. Molines, A.-M. Treguier, A. Beckmann, A. Biastoch, C. Böning, J. Dengg, J. Gulev, S. Le Sommer, E. Remy, C. Talandier, S. Theetten, and M. Maltrud. Impact of partial steps and momentum advection schemes in a global ocean circulation model at eddy permitting resolution. *Ocean Dynamics*, 56:doi: 10.1007/s10236 006 0082 1, 2006.
28. K. Matsumoto, H. Sasaki, T. Kagimoto, N. Komoro, A. Ishida, Y. Sasai, T. Miyama, T. Motoi, H. Mitsudera, K. Takahashi, H. Sakuma, and T. Yamagata. A fifty-year eddy-resolving simulation of the world ocean: preliminary outcomes of OFES (OGCM for the Earth Simulator). *J. Earth Simulator*, 1:35–56, 2004.
29. S. Speich, J.R.E. Lutjeharms, P. Penven, and B. Blanke. Role of bathymetry in Agulhas Current configuration and behaviour. *Geophys. Res. Lett.*, 33, 2006.
30. L. Debreu, C. Vouland, and E. Blayo. AGRIF: Adaptive grid refinement in Fortran. *Computers and Geosciences*, 34(1):8–13, 2008.
31. A.D. Fox and S.J. Maskell. A nested primitive equation model of the Iceland-Faeroe front. *J. Geophys. Res.*, 101:18259–18278, 1996.
32. J. Chanut, B. Barnier, L. Debreu, W. Large, L. Debreu, T. Penduff, J.-M. Molines, and P. Mathiot. Mesoscale eddies in the Labrador Sea and their contribution to convection and re-stratification. *J. Phys. Oceanogr.*, page doi: 10.1175/2008JPO3485.1, 2008.
33. J. Jouanno, J. Sheinbaum, B. Barnier, J.M. Molines, L. Debreu, and F. Lemarié. The mesoscale variability in the Caribbean Sea. Part I: Simulations and characteristics with an embedded model. *Ocean Modelling*, page doi:10.1016/j.ocemod.2008.04.002, 2008.
34. A. Biastoch, C.W. Böning, J.R.E. Lutjeharms, and M. Scheinert. Mesoscale perturbations control inter-ocean exchange south of Africa. *Geophys. Res. Lett.*, doi:10.1029/2008GL035132, in press, 2008.
35. A. Biastoch, C.W. Böning, J. Getzlaff, J.-M. Molines, and G. Madec. Causes of interannual-decadal variability in the meridional overturning circulation of the mid-latitude North Atlantic Ocean. *J. Climate*, doi:10.1175/2008JCLI2404.1, in press, 2008.

HLRS Project Report 2007/2008: "Simulating El Nino in an Eddy-Resolving Coupled Ocean-Ecosystem Model"

Ulrike Löptien and Carsten Eden

IFM-GEOMAR, Dsternbooker Weg 20, 24105 Kiel

Key Findings

- By including a radiatively active biological component we find considerable improvements of the circulation in a realistic eddy-resolving circulation model of the Pacific Ocean.
- The long-term changes in the dynamically adjusted circulation and watermass distribution differ clearly from the first order response and are mediated by a damped wave response.
- A similar, but somewhat weaker model improvement is obtained using a fixed non-flow-interactive chlorophyll concentration.

Methods and Results

Ocean components of current climate models still suffer from difficulties in simulating the tropical Pacific concerning both, mean state and variability [2]. Recently, considerable reseach effort has been devoted to reduce these model biases. One idea is to account for variable phytoplankton concentrations in the solar absorption scheme of the ocean models, which can lead to a rather large change in the vertical distribution of solar heating in the upper hundred meters of the ocean. This effect was shown to be in particular important for the tropical ocean (e.g. [10]. During the HLRS report period 2007, we explored the influence of changes in the solar absorption scheme due to inclusion of a biological component and its associated impact on solar heating on the simulated circulation and the El Nino/Southern Oscillation phenomenon in a regional, eddy-resolving model of the Pacific Ocean (FLAME, see also http://www.ifm-geomar.de/~spflame).

A simple nitrogen-based ecosystem model consisting of four biological components (NPZD) is coupled to a regional ocean general circulation model as described by [3]. The solar absorption scheme accounts for the simulated variable phytoplankton concentration as described in [5] and replaces the fixed solar absorption scheme in the circulation model. For the formulation of non-solar (turbulent) surface heat fluxes we use an Haney-type surface thermal boundary conditions following [1]. All simulations described here use interannually varying NCEP/NCAR-wind stress forcing during the period 1948-2003 [4]. The influence of different surface boundary conditions was tested successfully during the report period 2006/2007 and in 2007/2008 we concentrated on the effect of the variable solar absorption scheme on the circulation.

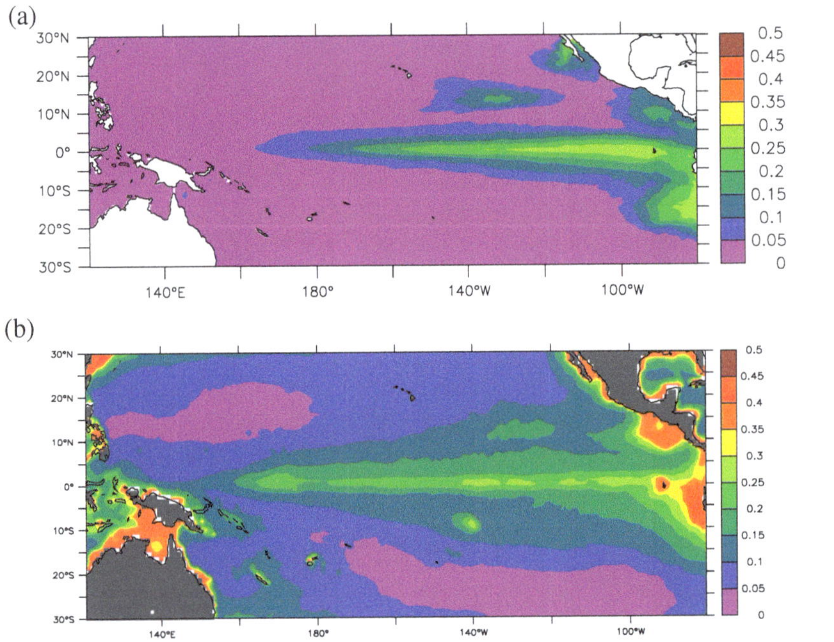

Fig. 1. Chlorophyll concentration ($mgchl/m^3$) averaged over the period 1997-2001 (a) as modeled in the upper 20m and (b) as observed from satellite observatory (SeaWIFS, [6], [7]

Three major sensitivity ocean hindcast experiments are performed and discussed in this report: In the first one (experiment BIO hereafter) we include the full biogeochemical model and the simulated phytoplankton affects the attenuation of light in the water column, i.e. in case of higher phytoplankton concentration more incoming shortwave solar radiatation is absorped locally in the water column. Since running a biological model in parallel to ocean circulation models is computationally time and cost intensive we performed a second experiment (PREBIO hereafter) where the chlorophyll concentration is kept constant in time using a fixed pattern. This fixed pattern was derived from the first experiment as a long-term average such that the approach resembles a cheap alternative to include the biological radiative forcing without running the complete biological model. The third (control) experiment (NO-BIO) uses a spatially and temporally constant value for the attenuation depth, i.e. no phytoplankton is included in the model.

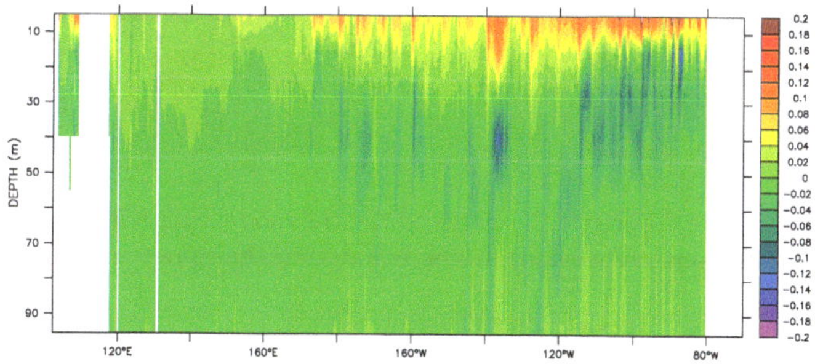

Fig. 2. Temperature difference [K] between the the model runs with biology included and ithout biological component at the equator in the upper 150m averaged over the first three days

The mean simulated chlorophyll concentration in experiment BIO in comparison to satellite data is shown in Figure 1 (SeaWIFS, [6], [7]) agreeing resonable well. Note that the regions with vanishing chlorophyll concentrations, in particular in the interior of the subtropical gyre are a well known problem of NPZD models. Note also that the large ocean color value near the continental margins in the satellite observations should be viewed with caution since they might be related to observational biases. The influence of the biological component on the physical model due to its changes in differential heating by absorption of solar irradiance is striking. The expected first order direct thermal response of the biologically-induced changes in the solar absorption scheme is clearly visible within the first days of our model integration (Figure 2).

Due to the presence of chlorophyll more light is absorbed in the top layers of the ocean, which leads to an enhanced warming of the upper ocean of up to 0.2 K, while the deeper ocean remains cooler than without radiatively interactive biology. Note that the simulated 'patchiness' in the figure is due to mesoscale eddies that are connected with anomalous chlorophyll concentrations. Note also that the non-uniform distribution of chlorophyll leads to an enhanced east-west gradient radiant heating gradient which would in a coupled model lead to a reduction of the trade winds and in consequence to a reduction of the upwelling in the east Pacific. Hence, for a coupled model we would expect an amplification of the initial temperature signal. In our hindcast experiments, however, this effect is not included such that we can focus here on the effect of the temperature redistribution by oceanic dynamics only.

The temperature anomaly is amplified during the following days and the warming signal is transported to the west (not shown). Within the first month

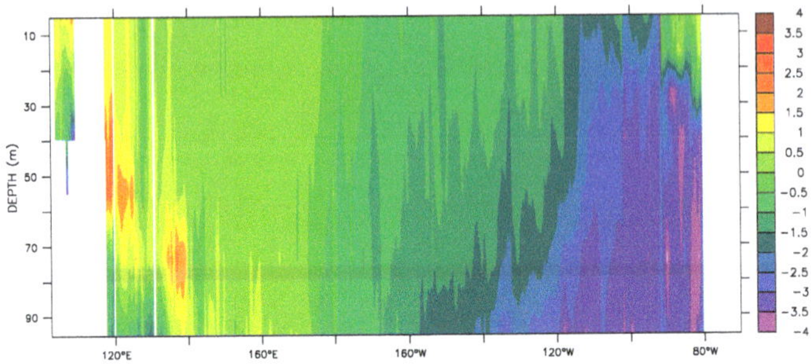

Fig. 3. Temperature difference [K] between the the model runs with biology included and without biological component at the equator in the upper 150m averaged over the first month

of the hindcast simulation the direct thermal response weakens considerably and during the following months the steady-state difference between the BIO and NOBIO experiments finally establishes (1948-2003). For this equilibrium response the upwelling in the east Pacific is enhanced especially at the equator which leads to a strong signal in the subsurface temperature (Figure 3). In this new equilibrium the slope of the thermocline is more pronounced. Also, the current undergo considerable changes. Figure 4 shows a strongly enhanced meridional component of the near equatorial surface currents compared to the experiment with no biological component and a strengthening of the tropical and subtropical meridional cells. Also, the zonal currents change (not shown): The South Equatorial Current (SEC) is considerable enhanced in the eastern Pacific by up to $10 - 15\%$. While the SEC strengthens the Equatorial Undercurrent (EUC) is getting shallower coming along with the shallowing of the thermocline. However, the EUC is also considerable stronger. Except for a proper representation of the Tsuchiya jets, the simulations including a biological component reproduce many of features of the observed zonal currents [9] quite accurately and more realistically than the experiment with no biological component included (not shown). Experiment PREBIO with the prescribed constant phytoplankton concentration shows similar but somewhat weaker changes compared to the full biogeochemical model (BIO).

Coming along with the new equilibrium of the Pacific ocean, the variability of the circulation also undergoes striking changes. Figure 5 shows departures of the sea surface temperature (SST) from the mean annual cycle averaged in the Nino3.4-region ($5°S - 5°N, 170°W - 120°W$). The black line shows the observed values while the colored lines depict the model results obtained form the three described model experiments (BIO, PREBIO and NOBIO). Since the surface wind stress is prescribed and identical in all experiments it is not surprising that the general agreement between simulated (NOBIO, standard model without biology) and observed ENSO events is very high and the correlation between the modeled and observed Nino3-time series is 0.96. However, in NOBIO, i.e. without radiatively active biological component, the inter-

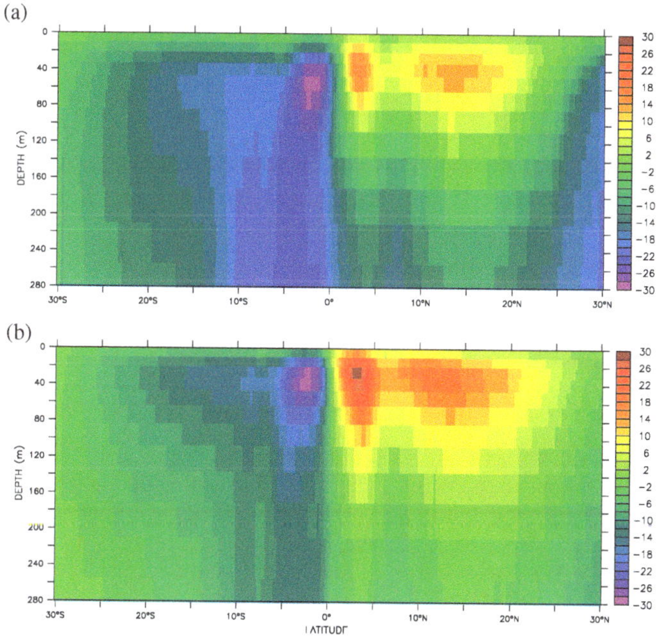

Fig. 4. Meridional overturning streamfunction in $10^6\ m^3/s$ of the upper 300m for the experiments (a) without biological component and (b) with the full biogeochemical model included

annual SST-variability of our model is too weak compared to observations. Including a biological component (BIO, red line) enhances the variability significantly. However, a similar effect is obtained when considering the time constant phytoplankton concentration (PREBIO, green line). The standard deviation of the Nino3-index increases from 0.49 (NOBIO) to 0.8 (BIO) when including biology. Using a time constant biology leads to a standard deviation of 0.76 in experiment PREBIO. Again, the circulation changes point into the same direction when including a prescribed, non-flow-interactive chlorophyll concentration instead of a full biogeochemical model.

Future Work

In addition to the discussed hindcast runs, experiments using a hybrid coupled model with and without biological component are presently carried out. This hybrid coupled model setup is described in the HLRS project report 2006/2007 and all initial tests of the hybrid coupled model are completed by now. The results of the coupled system will be compared with the hindcast runs described above. In particular a detailed investigation of the influence of a biological component in a coupled model will be investigated to gain

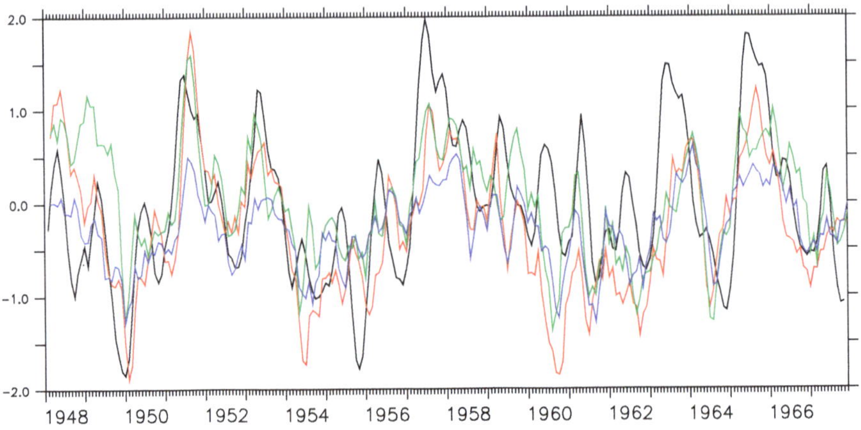

Fig. 5. Nino3.4-timeseries (°C) for different surface forcing with (BIO) and without biological component (NOBIO). The black lines corresponds to observed values (Reynolds-SST) while the blue line corresponds to the hindcast experiment with the ocean model only (NOBIO). In the other two experiments a full biogeochemical model is included (BIO, red line) and additionally a time constant biological component is tested (PREBIO, green line)

practical benefits for the development of future (less biased) climate models. Futhermore, we aim to explore the influence of westerly windburst on ENSO in the coupled system in detail during the remaining project-period. Our investigations are part of the EU-project DYNAMITE.

Computing Resources

During the year 2007, we used only a fraction of the requested computer resources since our project period at HLRS was extended until 2009 due to initial problems concerning the hybrid coupled model. However, all initial setup problems are resolved by now. Our model is running as a chain-job. The need for computer resources is variable and depends on the number of time steps calculated until restarting the model. In 2007/2008, we usually used two to six processors with a cpu-time of 10-30 hours per job. The performance (MFLOPS) varied between 3200 and 4300, with a mean value typically around 3600-4100.

References

1. Barnier B., L. Siefridt, P. Marchesiello, 1995: Thermal forcing for a global ocean circulation model using a three-year climatology of ECMWF analyses, J. Mar. Sys. 6, 363-380, doi:10.1016/0924-7963(94)000.

2. Bony S., and Coauthors, 2007: Climate Models and their Evaluation. In: Climate Change 2007: The Physical Science Basis. Contribution of Working Group I to the Fourth Assessment Report of the Intergovernmental Panel on Climate Change.
3. Eden, C., A. Oschlies, 2006: Adiabatic reduction of circulation-related CO_2 air-sea flux biases in North Atlantic carbon-cycle models. Glob. Biogeochem. Cycles, 20 (GB2008) doi:10.1029/2005GB002521.
4. Kalnay E., and Coauthors, 1996: The NCEP/NCAR 40-Year Reanalysis Project, Bull. Am. Met. Soc. 77, 437-471.
5. Marzeion, B., A. Timmermann, R. Murtugudde, F.F. Jin, 2005: Biophysical feedbacks in the tropical Pacific, J. Clim., 18, 58-70.
6. McClain, C.R., G.C. Feldman, S.B. Hooker, 2004: An overview of the SeaWiFS project and strategies for producing a climate research quality global ocean bio-optical time series, Deep-Sea Research II 51: 5-42 2004
7. McClain, C.R., J.R. Christian, S.R. Signorini, M.R. Lewis, I. Asanuma, D. Turk, C. Dupouy-Douchement, 2002: Satellite ocean-color observations of the tropical Pacific Ocean, Deep-Sea Research II 49, pp 2533-2560.
8. Reynolds, R.W., 1988: A real-time global sea surface temperature analysis. J. Climate, 1, 75-86.
9. Schott, F.A., J.P. McCreary, G.C. Johnson, 2004: Shallow overturning circulations of the tropical-subtropical oceans, Geophys Monogr. Geophys Union 137, 261-304.
10. Timmermann, A., and F.-F. Jin, 2002: Phytoplankton influences on tropical climate, Geophys. Res. Lett., 39, doi: 10.10129/2002GL15434.

Structural Mechanics

Prof. Dr.-Ing. Peter Wriggers

Institut für Kontinuumsmechanik, Leibniz Universität Hannover. Appelstr. 9A, 30167 Hannover, Germany

With the development of computer simulations, a powerful tool was created in the last 50 years which enhances engineering science and design in many ways. With the availability of fast and distributed computing facilities large scale problems can be analysed and such numerical simulations can then be sued within the industrial design process. Virtual testing is nowadays used in research institutions and in industry at large scale. Most designs of modern cars, buildings and aircraft are obtained using simulation based approaches. However there is still a long way to go in order to develop robust and reliable simulation tools for virtual testing of real materials and structures. This is especially true for nonlinear processes and when validated results of numerical simulations are needed. Due to the mathematical and physical complexity of real structures, robust simulation methods do not exist for general nonlinear applications. Among these are inadequate physical models where e.g. the influence of the microstructure of materials or of material surfaces was neglected which however might influence the structural behaviour eminently. But also modelling of connections of different structural parts can lead to simulation models which do not reflect the real physical behaviour. Furthermore damage and failure of new structural engineering design based on composite materials is often not predicted accurately with classical techniques based on continuum models.

Due to the fact that industrial problems are large, consider e.g. the wing of an aircraft made of laminated material, but the failure within the material occurs on a micro- to millimetre length scale, it is not possible to solve such structural problems using standard numerical methods. Here the challenge is develop new methodologies for virtual testing like multi-scale approaches. These have to be able to reproduce correct physical behaviour and to treat complex material and structural behaviour with the necessary efficiency. For this new methods and algorithms are currently worldwide under development and still have to be developed further and to be refined.

The contributions which are contained in this section tackle also new problems in which high performance computing is necessary to obtain meaningful results which can be validated by experiments. The first two contributions are concerned with classical engineering applications. The first contribution tackles a problem from mechanical engineering in which the influence of thickness on development of residual stresses due to shot peening is investigated. The numerical simulation is performed by using standard commercial software, however due to the size of the problem computing on a fast high performance computer was necessary. While the first paper is concerned with a problem stemming from surface treatment of metals the second one is related to a detailed problem in structural analysis. Here the transient behaviour within a welding process in multiple layers is investigated when applied to large structures. Again a standard finite element program is applied in which the author apply special formulations to model the process. Finally the last application is concerned with high performance computation using the massive parallel structure of the computional facilities for an application of discrete dislocation simulations.

All three contributions underline the need of high performance computing tools when real three-dimensional solid mechanics problems have to be solved numerically. Here the number of unknowns grow very fast and the investigated complex nonlinear problems demand a high level of computational speed in order to complete the iterative solution algorithms in a reasonable time frame.

Numerical Studies on the Influence of Thickness on the Residual Stress Development During Shot Peening

Marc Zimmermann, Manuel Klemenz, Volker Schulze, and Detlef Löhe

Institut für Werkstoffkunde I
Universität Karlsruhe (TH)
Kaiserstraße 12
76131 Karlsruhe
volker.schulze@iwk1.uni-karlsruhe.de

Summary. Shot peening is an important and in industrial production widely-used mechanical surface treatment with the purpose to improve fatigue life of components subjected to cyclic loading by inducing compressive residual stresses in the surface near region of the treated component. Finite Element simulation models are being developed since more than three decades in order to investigate, understand, explain, and predict the correlation between the influencing factors of shot peening and the post process residual stress state. All of the FE models proposed in literature have in common that the component thickness can not be taken into account realistically as an influencing parameter. To overcome this shortcoming a more sophisticated type of boundary condition was developed and investigated. With this boundary condition effects of thickness on the residual stress state were studied and the differences to the boundary conditions known from literature were analyzed. Obtained simulation results were compared with experimental x-ray stress measurements. The strain rate dependent deformation behavior of the investigated and aged material IN718 was taken into account using an elasto-viscoplastic material model with combined isotropic and kinematic hardening. An important finding was that a small thickness has no influence on the compressive residual stresses in the surface region but great influence on the tensile residual stresses present in deeper regions.

Key words: Shot Peening, simulation, material modeling, IN718, curved surfaces

1 Introduction

Shot peening is a mechanical surface treatment with the purpose to induce compressive residual stresses and work hardening in the surface near region of the treated component in order to suppress and/or decelerate crack initiation and propagation in components subjected to cyclic loading. The changes in the material state of the treated component are produced due to the interaction of accelerated peening media with the treated component surfaces. Experimental investigations showed that the compressive residual stresses induced by shot peening in the surface near regions are balanced by an almost constant tensile residual stress field in greater depths [1]. In the case of bulky components the magnitude of the tensile stresses is small due to the available thickness on which the compressive stresses are balanced. However since shot peening is also applied on rather thin components such as turbine blades, it can be assumed that larger tensile residual stresses can be found in the inner component regions representing a potential risk to crack initiation during service. Hence thickness has to be taken into account as a geometric parameter influencing the residual stress state of a component. Since costly experimental trials should be minimized in the design process of new components FE simulation of shot peening is an interesting approach of analyzing the influence of component thickness.

As the residual stress development during shot peening can not be captured adequately by a single shot impact most of the recently proposed shot peening models take into account multiple shot impacts [2, 3, 4, 5]. In order to minimize the model size and computing time it is a common approach to constrain the lateral faces of the body by mirror symmetry boundary conditions expanding the simulated body laterally to infinity (cp. Fig. 1). The

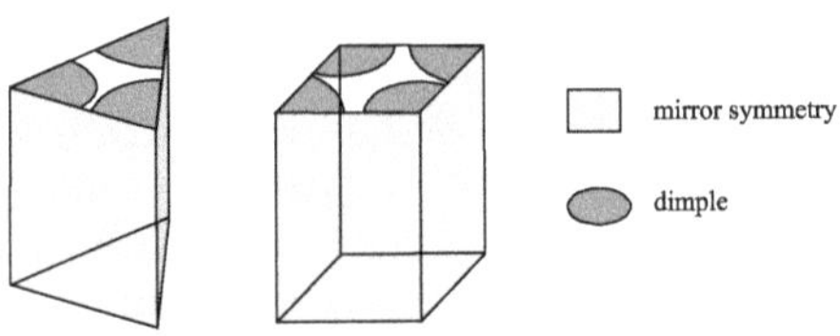

Fig. 1. Schematic presentation of boundary conditions used in literature for 3D shot peening simulation models [2, 3, 4, 5]

major shortcoming of this type of boundary condition is that thickness effects like deflection due to the successive induction of compressive stresses in the surface layer can not be taken into account. In order to resolve this shortcoming this work addresses the development and application of a boundary condition with the capability to incorporate thickness effects in FE modeling of shot peening.

2 Methods

2.1 Experimental Procedure

In order to analyze the geometry parameter thickness on the residual stress state experimentally and to provide experimental data for the comparison with the simulation model a test specimen featuring three sections of different thickness (cp. Figure 2) was treated with an air blast machine (type Baiker) with a shot medium of type CCW31 with a measured mean diameter of 0.89 mm and a hardness of 550 HV, an impact angle of 80°, and a coverage of 98 %. The mean shot velocity was determined by means of an optical measuring system provided from KSA [6] to 23 m/s. The impact angle is defined according to Figure 2. The residual stresses after shot peening

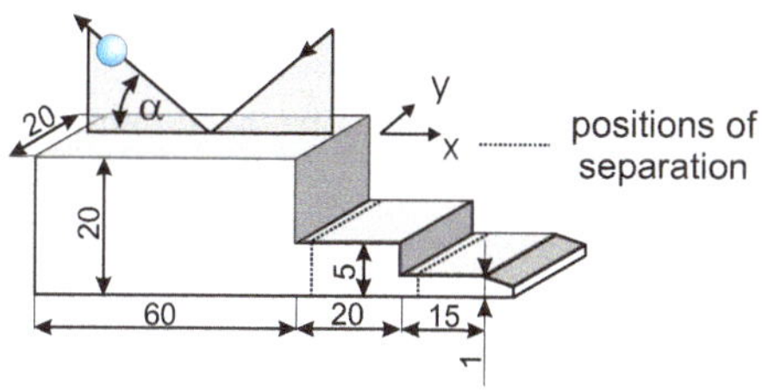

Fig. 2. Test specimen geometry, coordinate system and positions of separation

were measured by x-ray diffraction on the {311}-interference line of the age hardened material IN718 using Mn-Kα-radiation. The measured interference peaks were evaluated according to the $sin^2\psi$ method, using a Young's modulus of $E_{311} = 200000MPa$ and a Poisson's ratio of $\nu_{311} = 0.32$. Measurements in depth were performed by successive electro polishing of a circular area with a diameter of 5 mm. Stress relaxation due to the material removal by electro polishing was not taken into account, since the removed area was small. The test specimens were separated after the shot peening treatments at the positions shown in Figure 2 in order to enable the x-ray diffraction measurement in the x-direction at the test specimen sections with a thickness of 5 and 1 mm. After the treatments the test specimen section with a thickness of 1 mm was slightly bent around the y-axis and after the separation also around the x-axis.

2.2 Shot Peening Model

The shot peening model consists of a 3-dimensional rectangular body of arbitrary thickness and quadratic base and is based on the work of [7]. ABAQUS/Explicit is used for the dynamic analysis of the shot impacts on the surface taking into account inertia effects. The mesh is set up by 400000 8-node linear brick elements with reduced integration and hourglass control. To assure a good mesh quality the element size of the surface elements were adjusted to 1/15th of the dimple diameter produced by a single shot impact for the given shot diameter and velocity. Stress waves induced by the shot impacts and being reflected inside the model are damped using so-called infinite elements surrounding the faces of the rectangular body. These special types of elements lead to a minimization of the reflection of dilatational and shear waves into the body during the analysis and do not affect the stress state of the model when static equilibrium is reached [8]. The steel shots are modeled by half spherical rigid surfaces with parameterized diameter, velocity, and flight direction featuring the physical properties of a full sphere. Isotropic Coulomb Friction between the shots and the surface of the plate is assumed with a constant friction coefficient $\mu = 0.4$. The stochastic impact order of the shots on the treated surface is modeled by a dimple pattern of full coverage of the entire model surface with the impact order and shot arrangement shown in Fig. 3 a) providing an axis-symmetric residual stress state in the case of an impact angle $\alpha = 90°$. The residual stresses from the simulation are determined

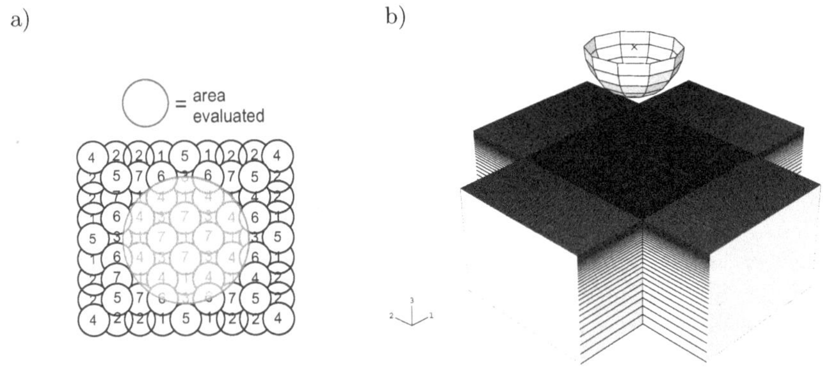

Fig. 3. a) Model geometry and b) shot arrangement with evaluated area to calculate the residual stress profile; numbers indicate the impact order of the shots

according to [7] with an averaging technique where the residual stresses of the elements lying within the gray marked circular area are averaged for every depth layer. Fig. 3 b) shows the model geometry for a thickness of 1 mm after the impact of 81 shots according to the dimple pattern shown in Fig. 3 a).

2.3 Boundary Conditions

Three different types of boundary conditions were investigated in this study. The first type is the classic approach of applying mirror symmetry to the lateral faces of the model. The second type corresponds to the boundary condition used by [7] where the lateral faces of the model are not constrained and the model's base is fixed in z-direction (cp. Figure 4 a). The third and new type of boundary condition aims to model a small section of an initially flat plate with a defined thickness where deflection can occur depending on the thickness and the peening intensity. The lateral faces of the model are forced to be perpendicular to the surface as well to the base by a kinematic coupling constraint of the lateral faces in normal direction with 4 reference points surrounding the model according to Figure 4 b). Control of deflection is established by constraining the rotational degree of freedom of the reference points. This is important when deflection is omitted due to geometrical constraints like for instance the deflection of the 1mm thick section of the test specimens around the x axis during the shot peening treatment.

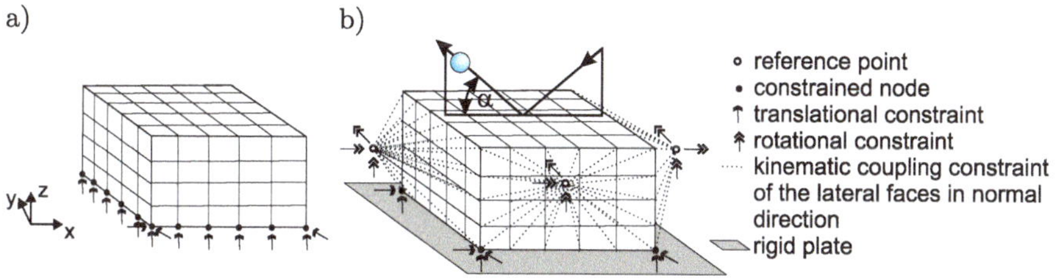

Fig. 4. Boundary conditions with a) free and b) constraint lateral faces

2.4 Material Model

Theoretical, experimental, and numerical studies showed that the Bauschinger effect occurring during cyclic deformation [9, 10, 11, 12] and the materials strain rate sensitivity [9, 13, 3] exhibit a serious impact on the residual stress development during shot peening. Hence these material effects can not be neglected in a realistic shot peening simulation and are taken into account by the usage of a combined isotropic/kinematic elasto-viscoplastic material model based on the work of [14]. This material model is implemented into ABAQUS/Explicit as a user defined subroutine (a so called VUMAT) written in FORTRAN code, which calculates the material response in terms of stress changes $\dot{\sigma}_{ij}$ on basis of total strain changes $\dot{\epsilon}_{ij}$ calculated by the FEM solver for every element and time increment. The stress changes $\dot{\sigma}_{ij}$ are calculated by means of equation 1 combining elastic and inelastic strain to total strain and Hook's law (equation 2) assuming isotropic elasticity.

$$\dot{\epsilon}_{ij} = \dot{\epsilon}_{ij}^{E} + \dot{\epsilon}_{ij}^{I} \tag{1}$$

$$\dot{\sigma}_{ij} = \lambda \dot{\epsilon}_{kk}^{E} \delta_{ij} + 2\mu \dot{\epsilon}_{ij}^{E} \tag{2}$$

Here λ and μ are the Lamé constants, which are calculated by the elastic modulus E and the Poisson's ratio ν.

$$\lambda = \frac{\nu E}{(1+\nu)(1-2\nu)} \quad , \quad \mu = \frac{E}{2(1+\nu)} \tag{3}$$

The inelastic strain rate is calculated using the flow equation proposed by [14]

$$\dot{\epsilon}_{ij}^I = D_0 \exp\left[-\frac{1}{2}\left(\frac{Z^2}{3K_2}\right)^n\right]\frac{S_{ij}-\Omega_{ij}}{\sqrt{K_2}} \tag{4}$$

with the material model parameters D_0 correlating with the maximum inelastic strain rate and n controlling strain rate sensitivity. K_2 is the second invariant of the overstress tensor.

$$K_2 = \frac{1}{2}\left(S_{ij}-\Omega_{ij}\right)\left(S_{ij}-\Omega_{ij}\right) \tag{5}$$

S_{ij} is the deviatoric stress and Ω_{ij} and Z are two implicit internal state variables tracking microstructural development and associated hardening effects. Ω_{ij} is the so called back stress, being a second order tensor that models orientation dependent strain hardening and Bauschinger effects associated with dislocation pileups at barriers like grain boundaries, other dislocations, and precipitates. Isotropic hardening effects are taken into account by the scalar drag stress Z. For a better agreement with experimental data the back stress Ω_{ij} is decomposed into two back stress terms according to equation 6.

$$\dot{\Omega}_{ij} = \dot{\Omega}_{1,ij} + \dot{\Omega}_{2,ij} \tag{6}$$

Each of the back stress terms develops according to its own evolution equation.

$$\dot{\Omega}_{1,ij} = \frac{2}{3}a_1\dot{\epsilon}_{ij}^I - a_1\frac{\Omega_{1,ij}}{\Omega_{1,m}}\dot{\epsilon}_e^I \tag{7}$$

$$\dot{\Omega}_{2,ij} = \frac{2}{3}a_2\dot{\epsilon}_{ij}^I - a_2\frac{\Omega_{2,ij}}{\Omega_{2,m}}\dot{\epsilon}_e^I \tag{8}$$

Both evolution equations of the two back stress terms $\Omega_{1,ij}$ and $\Omega_{2,ij}$ characterize the effects of strain hardening (first term) and dynamic recovery (second term). Thereby a_1, a_2, $\Omega_{1,m}$ and $\Omega_{2,m}$ are material model parameters, $\dot{\epsilon}_{ij}^I$ is the inelastic strain rate tensor and $\dot{\epsilon}_e^I$ is the effective inelastic strain rate defined as:

$$\dot{\epsilon}_e^I = \sqrt{\frac{2}{3}\dot{\epsilon}_{ij}^I\dot{\epsilon}_{ij}^I} \tag{9}$$

The drag stress Z develops according to equation 10, where m and Z_1 are material model parameters.

$$\dot{Z} = m\left(Z_1 - Z\right)\dot{\epsilon}_e^I \tag{10}$$

The numeric time integration scheme used for the solution of the partial differential equation system of the material model is a forward Euler estimate for the first iteration loop and the trapezoidal rule for subsequent iterations.

The material model parameters were determined in a two step procedure. In the first step the parameters n, governing the strain rate sensitivity, and the initial flow stress of the material represented by the parameter $Z_0 = Z(\epsilon_e^I = 0)$, were determined by means of the method of least squares on the basis of compression tests carried out at strain rate regimes ranging from 10^{-3} to 10^{+4} $1/s$. Therby D_0 was set to 10^6 s^{-1}. The remaining material model parameters a_1, a_2, $\Omega_{1,m}$, $\Omega_{2,m}$, m, and Z_1 describe the deformation and hardening behavior and are fitted in a second step of the material model parameter determination procedure by means of the commercial software FitIt [15] on basis of strain controlled push pull tests, which provide information about the deformation behavior during cyclic loading conditions. In order to realize the material model parameter determination procedure on basis of the uniaxial test data the material model was reduced to its uniaxial case. The determined material model parameters as well as the elastic properties are summarized in Table 1.

Table 1. Elastic material properties and determined material model parameters

parameter	unit	value
E	MPa	207000
ν	–	0.284
D_0	s^{-1}	10^{+6}
n	–	1.34
a_1	MPa	12390
$\Omega_{1,m}$	MPa	393

parameter	unit	value
a_2	MPa	123070
$\Omega_{2,m}$	MPa	560
Z_0	MPa	3095
Z_1	MPa	1900
m	–	200

3 Results and Discussion

In Fig. 5 a) the capability of the material model to describe strain rate sensitivity is compared to the experimental data. It can be seen that the moderate yield stress increase for strain rate regimes from 10^{-3} s^{-1} to 10^2 s^{-1} is being well captured by the material model. Small deviations between experiment and model are present in the strain rate regime from 10^3 s^{-1} to 10^4 s^{-1}. Since no experimental data is available for strain rates larger than 10^4 $1/s$ no conclusion can be drawn about the reliability of the material model in these strain rate regimes. However the predicted yield strength increase at very high strain rates is qualitatively in accordance to experimental data found for copper [16]. A qualitative comparison with copper is reasonable since both copper and the nickel based alloy IN718 are face centered cubic (fcc) metals and the mechanisms governing strain rate sensitivity are the same. In Fig. 5 b) the predicted material response of a cyclic test with changing total strain maximum in the first three cycles is compared with the experimental results. This comparison validates the capability of the material model of predicting the material response for complex loading paths as they can be assumed to occur in the surface near regions of a shot peened component during shot peening. It can be seen that the material response is well predicted by the

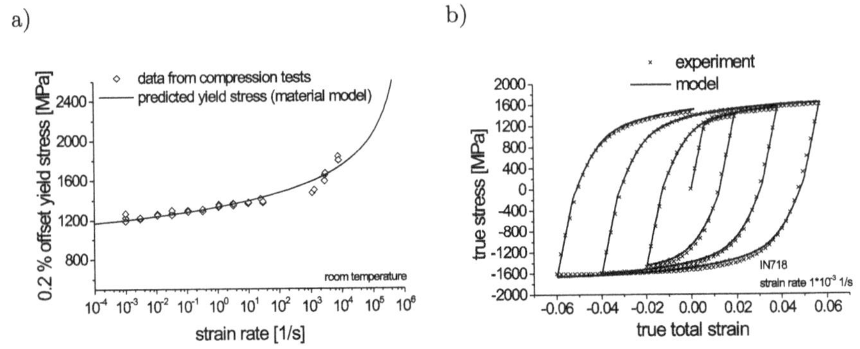

Fig. 5. Comparison of the predicted and experimentally determined strain rate dependent yield stress and the stress strain curve for a strain controlled cyclic test with increasing total strain amplitude after each cycle for age hardened IN718 at room temperature; experimental compression test data obtained from the European research project MMFSC

material model. Consequently a material model is provided to the shot peening simulation that should meet the requirements for a realistic description of the component material behavior during shot peening.

The capability of the three types of boundary conditions to realistically constrain the simulated body was investigated for a model thickness of 1 and 5 mm. For a model thickness of 1 mm the time between the successive shot impacts was chosen to be sufficiently large, in order to capture deflection effects occurring after each shot impact when the kinematic coupling constraint was used. Fig. 6 shows the comparison of the calculated residual stress depth profiles with the experimental measurements. Analyzing the experimental data

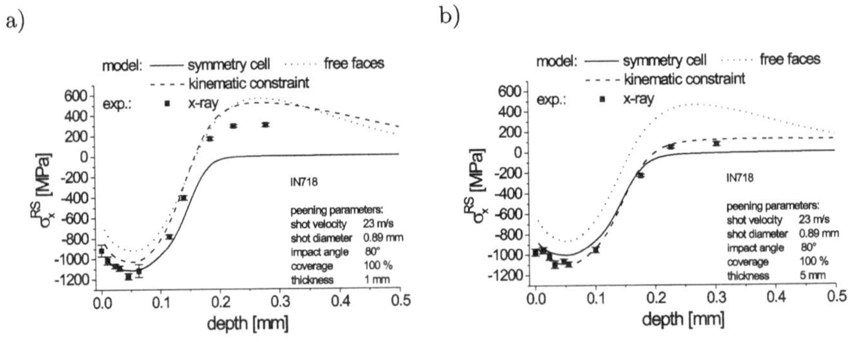

Fig. 6. Influence of boundary conditions on the calculated residual stress depth profiles for a thickness of a) 1 and b) 5 mm

it can be seen that for a thickness of 1 mm large and for 5 mm small tensile residual stresses are present in depths of about 0.2 mm with almost identical compressive residual stress values in the surface near region. The difference in the magnitude of the tensile stresses can be attributed to the different thicknesses available to balance the compressive stresses. The symmetric boundary conditions yield to a good prediction of the compressive residual stresses close to the surface. However the calculated residual stress depth profile is unbalanced since no tensile residual stresses are present. This shortcoming is due to external nodal forces on the lateral faces representing the infinite lateral extension. Hence the calculated stress state with symmetric boundary conditions on the lateral faces is not internally equilibrated and therefore unrealistic if small thicknesses are investigated. The simulation with unconstraint and therefore free lateral faces yields to a qualitatively good accordance with the experimental results for a thickness of 1 mm. However for a thickness of 5 mm the compressive residual stresses in the surface region are strongly under- and the tensile residual stresses strongly overestimated. This effect is due to the smaller stiffness in the edge near regions of the models surface leading to an "out squeezing", which is shown in Figure 7. Overcoming these disadvantages

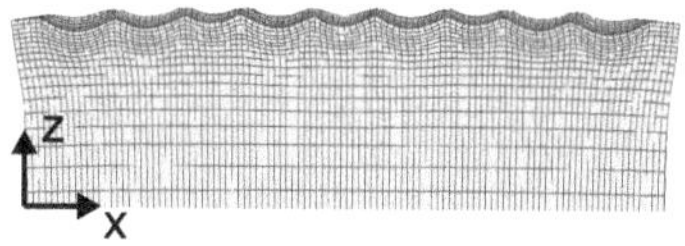

Fig. 7. Scaled mesh deformations with free faces (cut view)

the kinematic coupling constraint suppresses the "squeeze out" effect and enforces the induced compressive residual stresses to be balanced with tensile residual stresses.

In Figure 8 the capability of the new type of boundary condition describing the influence of thickness on the residual stress state after shot peening is shown. In order to represent the geometrical stiffness of the test specimen sections with a thickness of 5 and 1 mm the simulations of these thicknesses were carried out suppressing the deflection of the rectangular body around the y axis during the shot impacts. The separation of the 5 and 1 mm thick test specimen sections and the corresponding stress relaxation was taken into account by releasing the rotational constraint of the reference nodes linked to the lateral faces normal to the y-direction after the shot impacts. The results show that in x-direction no influence of the thickness on the surface and maximum compressive residual stresses can be observed. However, in y-direction the surface and maximum compressive residual stresses for a thickness of 1 mm are smaller. This effect can be attributed to the stress relaxation due to the separation of the test specimen. In both directions increasing tensile residual stresses can be observed with decreasing thickness. The predicted tensile residual stresses for a thickness of 1 mm are overestimated in comparison to the x-ray measurements. This deviation can likely be caused due to strong stress rearrangement effects resulting from electro polishing of such a thin test specimen decreasing the measured tensile residual stresses.

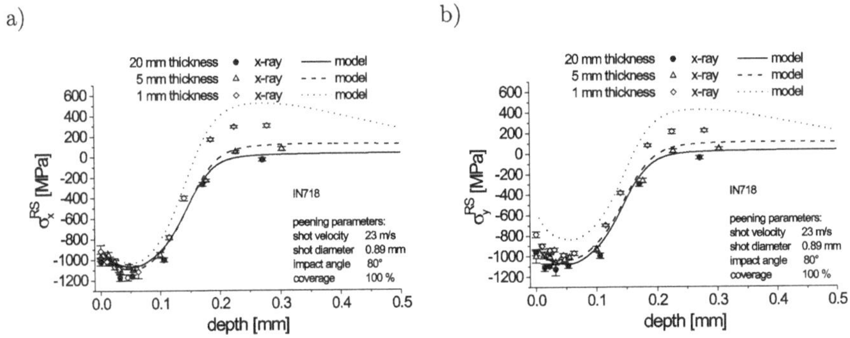

Fig. 8. Measured and simulated residual stress depth profiles for different thicknesses

4 Computational Requirements and Computing Time

The shot peening FEM calculations were performed on the HP-XC4000 high performance computer located at the computational center of the University of Karlsruhe. MPI-based parallelization was used in combination with the domain-level decomposition method of ABAQUS splitting the model into a number of topological domains being computed individually by each CPU involved in the analysis. The physical memory required for each simulation job was about 4000 MB. Typical wall clock computing times for a shot peening simulation job parallelized over 8 cpus ranged from 2 to 4 hours.

Due to the high numbers of simulation runs needed for the shot peening model development and for the parametric study carried out the accumulated cpu time for this study laid in the order of several months. Only due to the enormous computational resources of the HP-XC4000 this study could be realized within a reasonably short time period.

5 Conclusions

Experimental and numerical investigations were carried out in order to analyze the influence of component thickness on the residual stress state after shot peening. The numerical simulations were carried out with ABAQUS/Explicit and a user defined material model capturing the Bauschinger effect and strain rate sensitivity of the investigated material IN718. Three different approaches for constraining the model boundaries were tested. The results showed that with the classic approach of symmetric boundaries at the lateral model faces the influence of thickness is completely neglected and the calculated residual stress depth profile is not balanced. Leaving the lateral faces free leads to squeezing effects at the edges of the model surface and to unrealistic stress states. Constraining the lateral faces to be perpendicular to the surface as well to the base of the model leads to the incorporation of deflection effects and to realistic simulation results being in very good accordance with the experiment. The numerical and experimental results showed that a decrease of the component thickness does not affect the compressive residual stresses in the surface region but leads to a shift of the zero crossing to the surface and to higher tensile residual stresses in greater depths. Since the simulated plastic deformations were always identical independent of the model thickness it can be concluded that the residual stress distribution not only depends on the plastic deformation field but also on the geometrical constraints.

Acknowledgments

The authors gratefully acknowledge the financial support from the European 6th Framework Program through the research project VERDI (Virtual Engineering for Robust manufacturing with Design Integration) and Hans Uwe Baron from MTU Aero Engines for the manufacturing and peening of the test specimens.

References

1. R. Menig, L. Pintschovius, V. Schulze, and O. Vöhringer. Depth profiles of macro residual stresses in thin shot peened steel plates determined by x-ray and neutron diffraction. *Scripta Materialia*, 45:977–983, 2001.
2. K. Schiffner and C. Droste gen. Helling. Simulation of residual stresses by shot peening. *Computers and Structures*, 72(1):329–340, 1999.
3. S.A. Meguid, G. Shagal, and J.C. Stranart. 3d fe analysis of peening of strain-rate sensitive materials using multiple impingement model. *Int J Impact Eng*, 27(2):119–134, 2002.
4. G.H. Majzoobi, R. Azizi, and A. Alavi Nia. A three-dimensional simulation of shot peening process using multiple shot impacts. *J Mater Process Tech*, 164-165:1226–1234, 2005.
5. S.A. Meguid, G. Shagal, and J.C. Stranart. Development and validation of novel fe models for 3d analysis of peening of strain-rate sensitive materials. *J. Eng. Mater. Tech.*, 129(2):271–283, 2007.
6. F. Wüstefeld, W. Linnemann, S. Kittel, and A. Friese. On-line process control for shot peening applications. In V. Schulze and A. Niku-Lari, editors, *Proc. of the 9th Int. Conf. on Shot Peening*, pages 366–372, Noisy-le-Grand, 2005. iitt.

7. J. Schwarzer, V. Schulze, and O. Vöhringer. Evaluation of the influence of shot peening parameters on residual stress profiles using finite element simulation. In T. Chandra, J.M. Torralba, and T. Sakai, editors, *Proc. Thermec 03*, volume 462-432, pages 3951–3956, Uetikon, 2003. Materials Science Forum, Trans Tech Publications.
8. Dassault Systèmes. *ABAQUS Manuals*, 2007.
9. S.T.S. Al-Hassani. Mechanical aspects of residual stress development in shot peening. In A. Niku-Lari, editor, *Proc. of the 1st Int. Conf. on Shot Peening*, pages 583–602, Oxford, 1981. Pergamon Press.
10. E. Rouhaud, A. Ouakka, C. Ould, J.L. Chaboche, and M. François. Finite elements model of shot peening, effects of constitutive laws of the material. In V. Schulze and A. Niku-Lari, editors, *Proc. of the 9th Int. Conf. on Shot Peening*, pages 107–112, Noisy-le-Grand, 2005. iitt.
11. M. Klemenz, V. Schulze, O. Vöhringer, and D. Löhe. Finite element simulation of the residual stress states after shot peening. In W. Reimers and S. Quander, editors, *7th European Conference on Residual Stresses*, volume 524-525, pages 349–354, Uetikon-Zürich, 2006. Materials Science Forum, Trans Tech Publications Ltd.
12. M. Klemenz, M. Zimmermann, V. Schulze, and D. Löhe. Numerical prediction of the residual stress state after shot peening. *High Performance Computing in Science and Engineering: Transactions of the High Performance Computing Center, Stuttgart (Hlrs)*, pages 437–448, 2006.
13. M. Kobayashi, T. Matsui, and Y. Murakami. Mechanism of creation of compressive residual stress by shot peening. *Int J Fatigue*, 20(5):351–357, 1998.
14. V.G. Ramaswamy, D.C. Stouffer, and J.H. Laflen. A unified constitutive model for the inelastic uniaxial response of rené 80 at temperature between 538 °c and 982 °c. *Journal of Engineering Materials and Technology*, 112:280–286, 1990.
15. Fitit. https://www.fitit.fraunhofer.de/fitit/help.htm.
16. P.S. Follansbee, G. Regazzoni, and U.F. Kocks. The transition to drag controlled deformation in copper at high strain rates. In J. Harding, editor, *3rd Conf. Mech. Prop. High Rates of Strain*, pages 71–80, 1984.

A Transient Investigation of Multi-Layered Welds at Large Structures

Tobias Loose

Ingenieurbüro Tobias Loose GbR, Haid-und-Neu-Straße 7, D-76131 Karlsruhe
loose@tl-ing.de

1 Introduction

Since welding processes are developed and used welding distortions have been known as well. The distortions influence the dimensional accuracy of the welded construction and the remaining residual stresses can effect the load carring capacity and the stiffness of a member negatively. Therefore, residual stresses, welding distortions and their mitigation are the focus of many projects.

In the last few years growing calculation capacity enables a welding simulation of small welds on conventional PCs. Powerful parallel computing stations e.g. the HP-XC4000 of the Rechenzentrum of the Universität Karlsruhe which was launched in 2007 permit a transient welding simulation of members for in civil engineering common dimensions.

One interesting question in welding engineering is the different behavior of single- and multi-layered welds. With the powerful utility of supercomputing residual stresses and distortions of the mentioned members and their influence can be investigated.

The investigation of this question was performed using the geometry of a cylinder segment which is welded circumferentially. The chosen dimensions are a radius R of 6000 mm, a sheet thickness t = 6 mm and a cylinder segment of 11,25; the chosen cylinder segment has an arc length of 1178 mm.

The numerical model is using a combination of shell and volume elements. The welded area with its highly non linear behavior with high gradients across the thickness is represented by a fine mesh of volume elements and the boundary areas with shell elements.

Although the model uses a lot tolerable simplifications a mesh with a high value of nodes and elements is needed.

The finite-element calculations are performed on the HP-XC4000 using the program SYSWELD from the ESI Company. The simulations include phase transformations of steels and the resulting important effects.

2 Heat Sources for Two-Layered Welds

The two-layered welds are investigated for two different kind of joints, a V-butt joint and a X-butt joint. The first layer is welded from inside of the cylinder.

To piont out the difference between single- and multi-layered welds, the V-butt and X-butt joint is modelled as a single joint with backing run as well. The criterion for the chosen parameters for the single joint with the backing run is the creation of a weld bead similar to the weld beads of the conventional V-butt and X-butt joint. The temperature distribution in the cross-section of the investigated two-layered welds can be seen in figure 1.

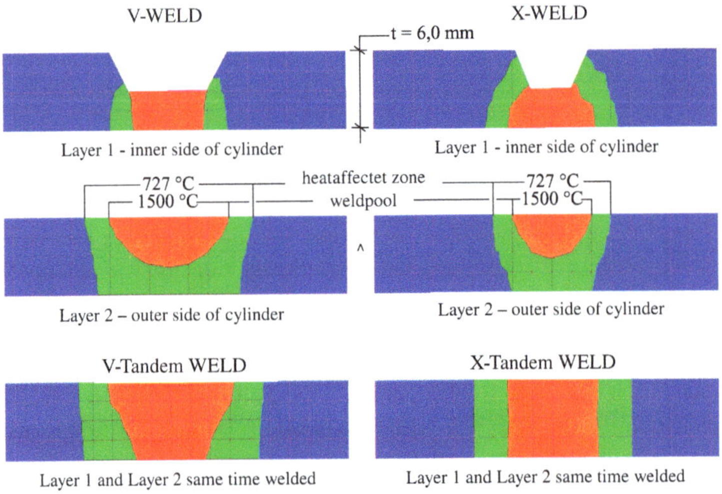

Fig. 1. Temperature distributions of the two-layered welds

3 Overview of the Investigated Cases

Using shell-volume-element models, cylinders made of the steel grade S355 and with the dimensions listed in table 1 are used. The temperature fields due to the different welding orders are varied. The cylinder slenderness $\frac{R}{t}$ is 1000. The expected critical stresses are elastic stresses.

Table 1. Dimensions and welding orders of the invested cases

Material	R	t	Segment	L	Tacks	Welding order
	mm	mm		mm		1th layer - 2nd layer
S355	6000	6	11,25	2400	5	4 - 2

The tack welds are made with five equidistant tackings with 1 cm length each. The distance between tackings is 295 mm. Only the first inner layer is tacked. The tack welds start a 0 s.

Two geometries of a weld line are investigated: X-butt joint and V-butt joint. The welding gap is filled from alternating sides in case of the choosen X-butt joint. This causes a nearly symmetrical heat input. Therefore, this case is similar to the single-layer joint of the previous chapter, the only difference is a reheating of the start and the end of previously welded layers. The filling of the welding gap in case of a V-butt joint is onesided and unsymmetrical. The angular distortion known from V-butt joint of plates leads in case of a perimeter weld line on a cylinder and for the desired opening of the welding gap outwards to an anlargement of radial deformation inwards. The choosen forms of joints describe favorable and inappropriate orders of weld line geometries referring to the distortions.

The first layer - in case of a multi-layered joint - is welded following welding order 4, the second layer with welding order 2. Welding line 2 of the first layer starts at 1000 s, welding line 1 of the first layer starts at 2000 s and the welding line of the second layer at 3000 s

In models with single-layered weld line the welding order 4 is used with in previous paragraph mentioned times for layer 1. The heat source in single-layer models is modelled in the way, that the melting pool of a single layer joint equates to molten area of a multy-layer joint.

Proper parameters of the investigated models are shown in table 2.

Table 2. Weldform realisations of the investigated cases

Cylinder Name	Weld
V2	two-layered V-weld
V1	single-layered V-weld
X2	two-layered X-weld
X1	single-layered X-weld

4 Axial Stresses

For the model with a double-layer welded V-butt joint (V2) the axial stresses at the equator in the middle of the shell after tacking, after the welding of the first layer and after the welding of the second layer are shown in picture 2.

An axial stress bulb known from single-layer welded cylinders occures after the layer 1 between the weld joint 1 and 2. This is visible in the middle of the segment where there is a tensile-compression-stress-change of the blue curve in picture 2.

The compression stresses prevailing in this curve are a result of the excentricity of the middle area of the shell in case that, after layer 1, the welding gap is filled only to a half.

Axial residual stresses disappear after the overwelding of butt joint change from layer 1 to layer 2. this is shown with the green curve in picture 2.

The situation respected to axial stresses for X-butt joint is the same as for V-butt joints.

In case of a X-butt joint the situation referable to axial stresses equals to the situation in case of a V-butt joint.

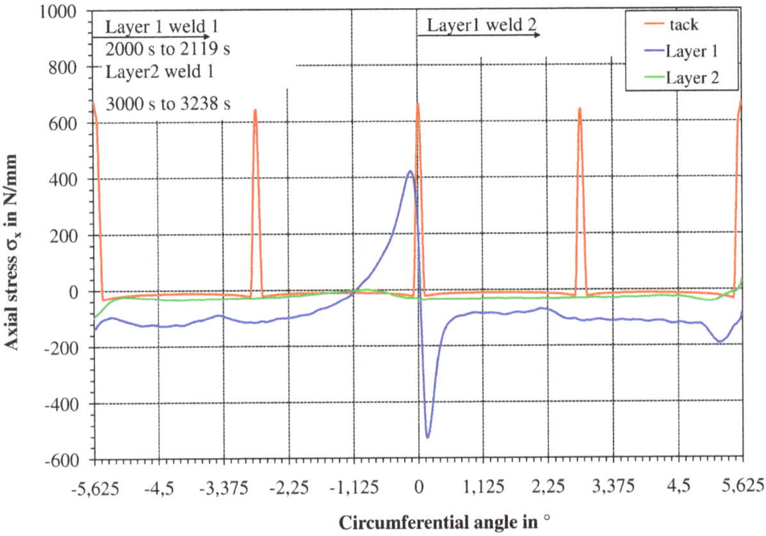

Fig. 2. Axial stress σ_x in $\frac{N}{mm^2}$ at the equator of the shell midth, two-layered v-butt joint (V2)

5 Welding Distortions

The highest radial deformations w outwards and inwards for the investigated models are arranged in table 3. Deformations normalized to the section thickness are presented in table 4. The highest radial deformations inwards are located - as well as in cases of single-layer weld lines - in the weld line adjacent area.

Picture 3 shows the radial deformation w after separate steps for cases with X-butt weld (X1, X2) at the meridian VL = -2,8125. The weld line 2 is on the right side of the segment and not on the examined Meridian. This is the reason why the deformation in the middle of the second weld line is oriented outwards.

The inwards oriented radial deformation of the two-layer welded cylinder segment is significant lower than in case of a single-layer welded segments. This is visible in picture 3 for a X-butt weld and is also valid vor a V-butt weld (table 3).

Table 3. Maximum values of the radial deformation w in mm

Weld	V 2-layered	V 1-layered	X 2-layered	X 1-layered
after Tacking:				
outwards	0,314	0,314	0,314	0,314
inwards	0,0672	0,0672	0,0672	0,0672
after Layer 1:				
outwards	0,479	-	0,379	-
inwards	0,682	-	0,841	-
after Layer 2:				
outwards	0,170	0,273	0,136	0,257
inwards	1,611	2,00	1,24	1,95

In the picture 4 we can see the radial deformation at the equator for a double-layer V-butt weld and in picture 5 for a double-layer X-butt-weld after single process steps.

The second layer of the V-butt weld has a clearly larger volume of the molten pool than the second layer of the X-butt weld. This is the reason for a much larger weldseam shrinkage compared to a model using X-butt weld.

Welding deformations are demonstrated in pictures 4 to 4 for investigated models.

Table 4. Maximum values of the normalized radial deformation $\frac{w}{t}$

Weld	V 2-layered	V 1-layered	X 2-layered	X 1-layered
after Tacking:				
outwards	0,0523	0,0523	0,0523	0,0523
inwards	0,0112	0,0112	0,0112	0,0112
after Layer 1:				
outwards	0,0798	-	0,0632	-
inwards	0,114	-	0,140	-
after Layer 2:				
outwards	0,0283	0,0455	0,0227	0,0428
inwards	0,269	0,333	0,207	0,325

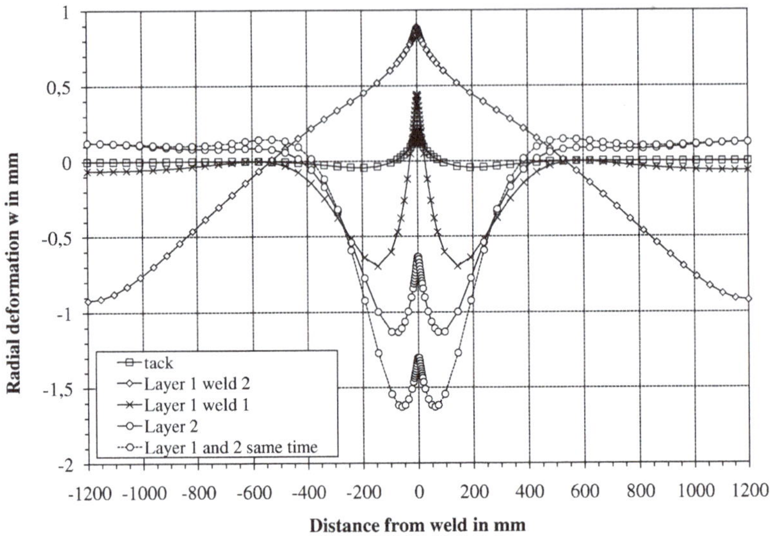

Fig. 3. Radial deformation w in mm at the meridian VL = -2,8125, two-layered (X2) and single-layered (X1) X-butt weld

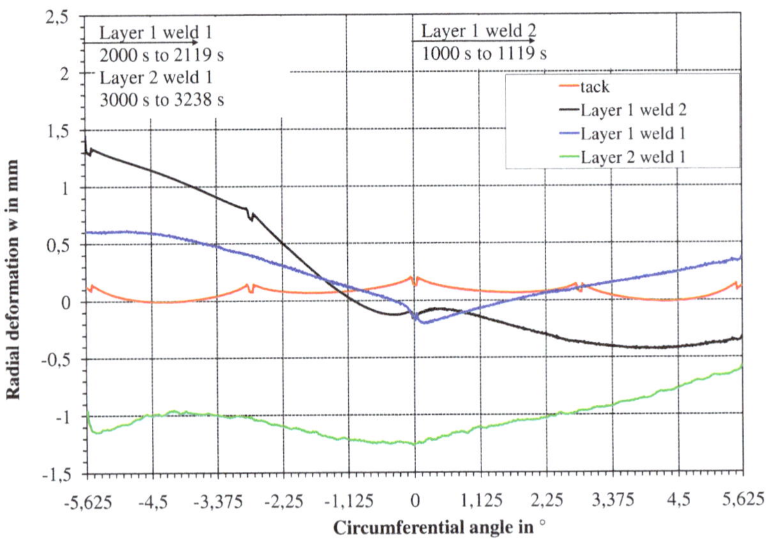

Fig. 4. Radial deformation w in mm at the equator, two-layered V-butt weld (V2)

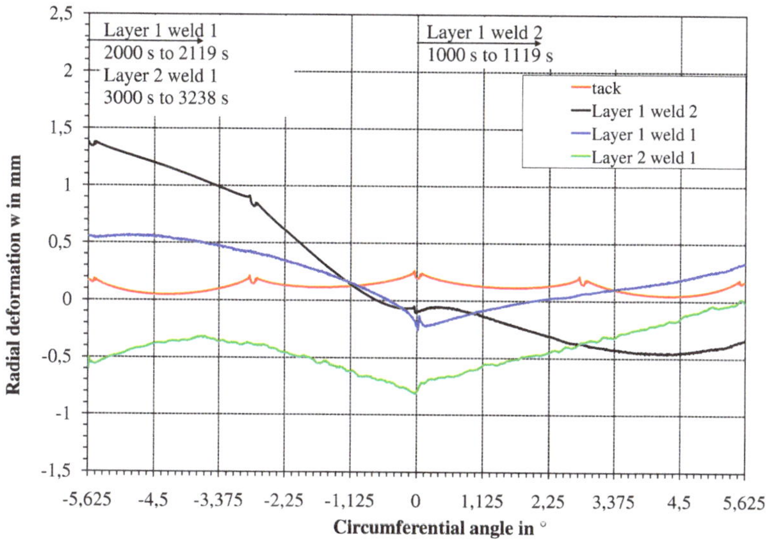

Fig. 5. Radial deformation w in mm at the equator, two-layered X-butt weld (X2)

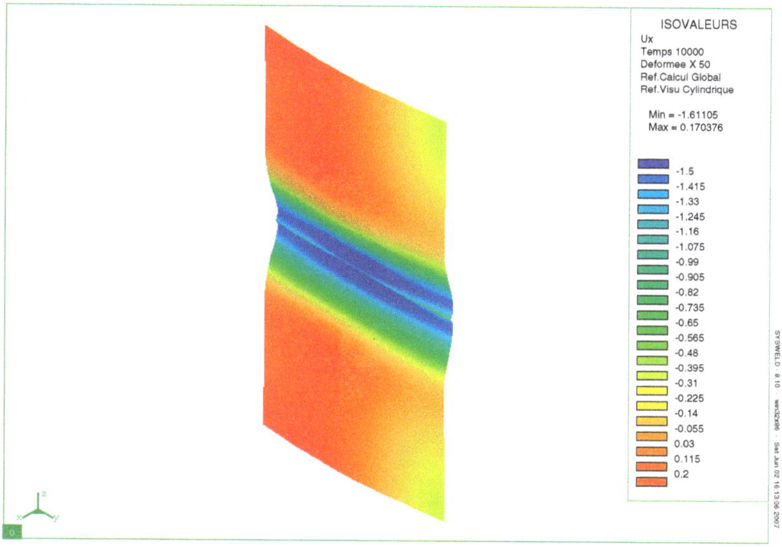

Fig. 6. Radial deformation w in mm after welding, 50-times deformed, two-layered V-butt weld (V2)

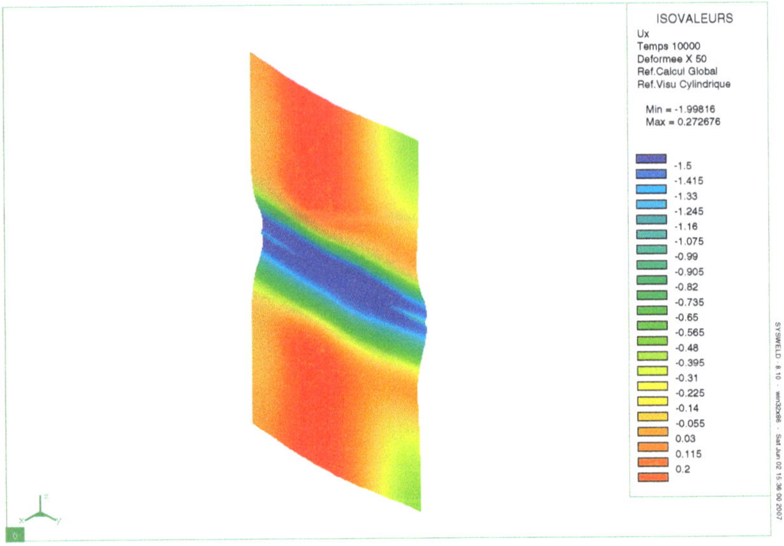

Fig. 7. Radial deformation w in mm after welding, 50-times deformed, single-layered V-butt weld (V1)

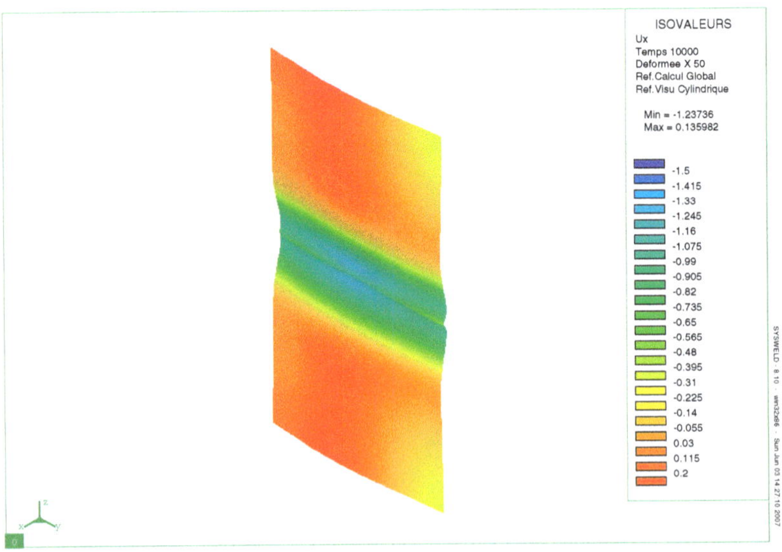

Fig. 8. Radial deformation w in mm after welding, 50-times deformed, two-layered X-butt weld (X2)

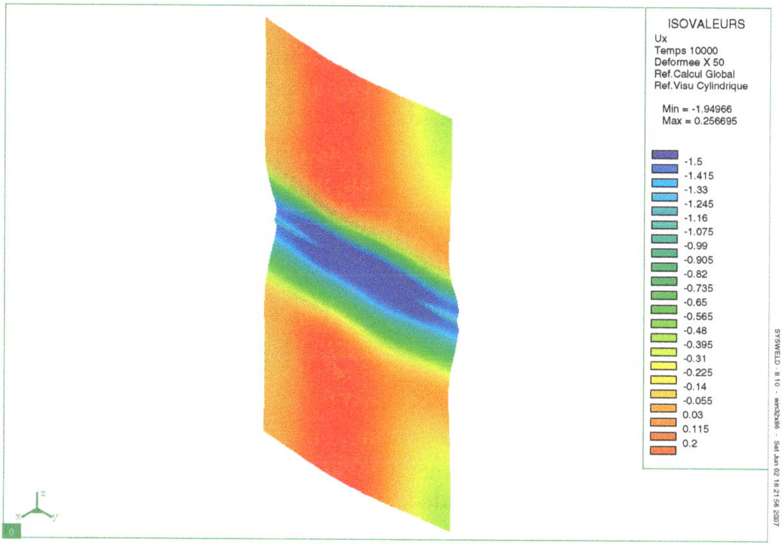

Fig. 9. Radial deformation w in mm after welding, 50-times deformed, single-layered X-butt weld (X1)

6 Critical Stresses Under Axial Load

Critical stresses under axial load of investigated models are described in table 5. Deformations belonging to the critical stresses are shown in pictures 10 to 13.

Calculated critical stresses are larger compared to models with single-layer weld line and comparable slenderness of the cylinder, which is a result of the choosen size of a segment, which is 11,25 and very small. This causes a high perimeter spindle count and a high eigenvalue. For this reason, the critical stresses are only to be considered as comparative values of different models.

Critical stresses in models with single-layer butt welds and double-layer V-butt weld are in the same order of magnitude. The critical stress in case of a double-layer X-Butt weld is significant larger. The reason for this effect is a much smaller radial deformation in this model.

Table 5. Critical stress σ_{gr} in $\frac{N}{mm^2}$

Cylinder Name	Weld	critical stess	normed critical stress $\frac{\sigma_{gr}}{\sigma_{kl}}$
V2	V 2-layered	84,5	0,67
V1	V 1-layered	86,6	0,68
X2	X 2-layered	99,7	0,79
X1	X 1-layered	86,5	0,68

The clearly reduced distortion in case of the two-layers X-butt weld compared to a single-layer welding causes a significant augmentation of critical stresses. It can be stated for this welding geometry, that a simplified calculation of single-layer welding is a conservative estimation.

The difference of distortion between a double-I V-butt weld and a single-layer V-butt weld is less distinctive than in case of an X-butt weld. The buckling is however clearly different. The buckling at the beginning and the end of the weld line in the middle of the segment is more developed. In case of a two-layers weld line lies the distinctive buckling at the left border of the segment. This is the reason for the following conclusion:

There are significant geometrical and structural imperfections in the transition area begin of the welding line - end of the welding line that are leading to a distinctive buckling in ultimate state. This is shown in models with single-layer perimeter weld line as well as the model with a single-layer V-butt weld (Picture 11). If, in case of a multi-layer weld line, the beginning and the and areas of the weld line are overwelded, the imperfections of the first layer are diminished, the buckling takes effect in some other area, in investigated model (Picture 10) on the left border of the segment.

The calculated critical stresses are equal in both models with a V-Butt weld. It can be observerd that the simplified calculation with a single-layer weld line leads to equal critical stresses as a multi-layered calculation.

We can conclude that multi-layered welding does not lead to larger imperfections compared to a single-layerd welding. The results of a parametric study with single-layered welding can be transferred to cylinders with greater sheet thickness that are welded multi-layered for reasons of production technology.

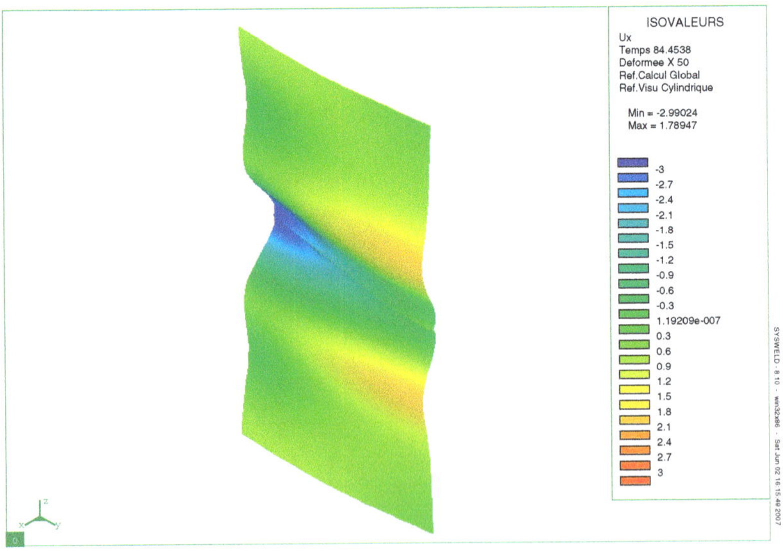

Fig. 10. Radial deformation w in mm under critical load, 50-times deformed, two-layered V-butt weld (V2)

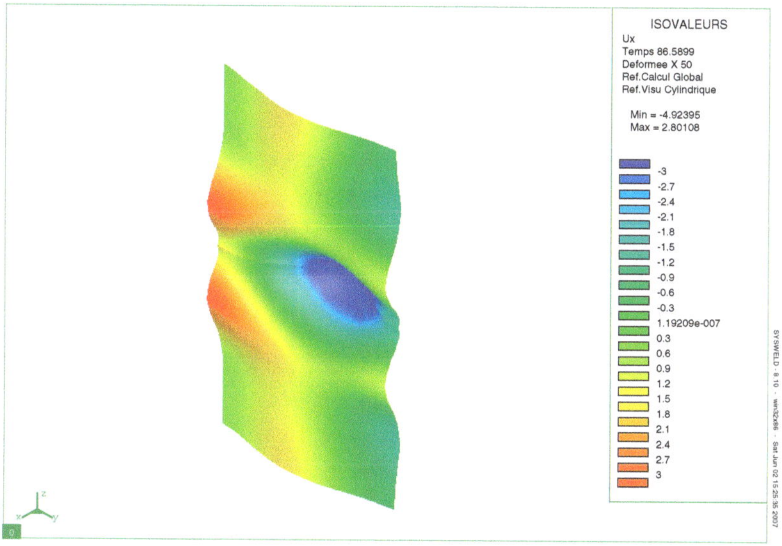

Fig. 11. Radial deformation w in mm under critical load, 50-times deformed, single-layered V-butt weld (V1)

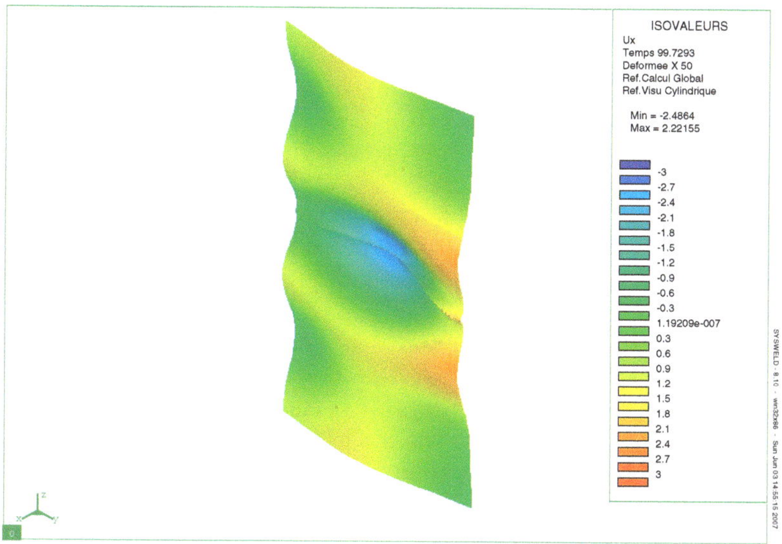

Fig. 12. Radial deformation w in mm under critical load, 50-times deformed, two-layered X-butt weld (X2)

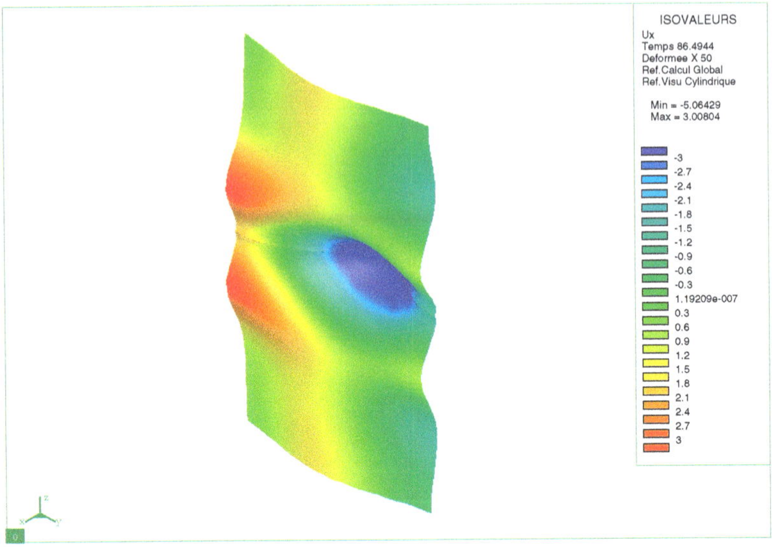

Fig. 13. Radial deformation w in mm under critical load, 50-times deformed, single-layered V-butt weld (V1)

7 Calculation Times

A particular difficulty of the welding simulation lies in the calculation of large gradients that appear during the calculation of the temperature field as well as in the mechanical calculation. It is necessary to use a fine mesh in the weld line area and small time steps during the calculation of welding. This causes a high calculation complexity for finite elements models.

The calculations of multi-layered weld lines were made at scientific supercomputer HP XC4000 at the electronic data processing center of the Universität Karlsruhe.

The calculations using Sysweld were done - depending of the certain model - on a node with 4 CPUs. A calculation duration of one day results in a needed calculation complexity of 4 CPU-days. It is allowed to start up to 10 calculations per user simultaneous, so the models for the parametric study were calculated at the same time. Because of a long calculation duration for one model a huge time saving compared to a total calculation time for all models was reached.

Geometry, mesh size, welding time and the calculation times divided in the calculation of the temperature field, mechanical calculations and the calculation of axial critical stresses for a Shell-Volume-Model with a double-layer weld line are shown in the table 6.

Table 6. Calculation time for a shell-volume model of a cylinder segment with multi-layered circumferrential welds

Cylindername	V2
R in mm	6000
t in mm	6
L in mm	2400
Segment in degrees	11,25
Segment in mm	1178
Area in m^2	2,83
Number of nodes	77 823
Number of shellelements	12 288
Number of volumeelements	55 296
Welding time in s	474
Thermal analysis:	
Number of timesteps	710
Computingtime in days	1,45
Computingtime in CPU-days	5,82
Mechanical analysis:	
Number of timesteps	859
Computingtime in days	25,2
Computingtime in CPU-days	101
calculation of the axial critical stress:	
Number of timesteps	51
Computingtime in days	0,35
Computingtime in CPU-days	1,41

High Performance Computing and Discrete Dislocation Dynamics: Plasticity of Micrometer Sized Specimens

D. Weygand[1], J. Senger[1], C. Motz[1], W. Augustin[2], V. Heuveline[2], and P. Gumbsch[1]

[1]KIT, Institut für Zuverlässigkeit von Bauteilen und Systemen, 76128 Karlsruhe
[2]KIT, Steinbuch Computing Centre, 76128 Karlsruhe

Summary. A parallel discrete dislocation dynamics tool is employed to study the size dependent plasticity of small metallic structures. The tool has been parallelised using OpenMP. An excellent overall scaling is observed for different loading scenarios.

The size dependency of the plastic flow is confirmed by the performed simulations for uniaxial loading and micro-bending tests. The microstructural origin of the size effect is analysed. A strong influence of the initial microstructure on the statistics of the deformation behaviour is observed, for both the uniaxial and bending scenario.

1 Introduction

Miniaturisation of metallic systems becomes more and more relevant in technical applications, e.g. micro electro mechanical system (MEMS) for sensors, filters or actuator. These devices have structural elements in the sub- and micrometer range and show pronounced size effect in the mechanical properties, as increasing flow stress with decreasing sample size for torsion experiments [1]. The size effect in these first experiments gave rise to the development of so-called strain gradient theories, which were considered to be the main origin of the observed behaviour [1]. Classical continuum theories are unable to describe this behaviour as the underlying assumptions that the dislocation microstructure can be averaged out does not hold anymore at this length scale. The details of dislocation microstructure play a paramount role at these scales. Therefore modelling of individual dislocations is required for understanding the plasticity at these small scales.

Dedicated experiments to study the plasticity of nominally uniformly loaded samples, such as compression tests at the micrometer range, showed a size dependency of the plastic flow for all tested face-centered cubic (fcc) materials like nickel [2] or gold [3]. Furthermore a strong statistical variation of the deformation behaviour of samples with nominally identical internal and external characteristics is observed. As general trend, the smaller the sample the more pronounced the scatter in the mechanical response, which clearly points to the strong influence of the initial dislocation structure and possibly experimental uncertainties. Very recently, tension tests on single crystal copper samples complemented the compression results [4]. This recent work has pinned down differences between compression and tension tests, as here also an influence of the aspect ratio is observed. In compression, this ratio plays no significant role [5]. In contrast, in tension tests the flow stresses and the hardening behaviour is reduced with increasing aspect ratio [4].

The size effect in micro pillars cannot be explained by the presence of grain boundaries that constrain the motion and nucleation from Frank-Read sources inside the grains, a successful model to explain the size effect in thin films [6, 7], as the pillars are single crystals and the side surfaces are traction free. A number of possible explanations of the size effect are discussed at the moment: dislocation starvation in small samples is assumed, where dislocation leave the volume prior multiplication can take place [8, 9]; a further model to explain is the source truncation model where longer dislocation sources are cut by the surface during glide or fabrication of the pillars with a focused ion beam [10, 11] effectively decreasing the dislocation source length L in the sample. These shorter sources require a higher activation stress as this stress scales with $1/L$.

A further important aspect for the characterisation of stress strain curves consists in a statistical analysis of the so-called strain burst [12] or stress drops, which could also be found with our discrete dislocation dynamics model [13].

Discrete dislocation dynamics (DDD) codes are nowadays used by different groups to investigate the size dependent plastic properties of micro-samples [14, 15, 16, 17]. These codes allow an analysis of the evolving dislocation structure in terms of dislocation reactions and densities evolution during the total deformation process. The unique feature of the Karlsruhe code is the approach to solve mechanically consistent the elastic boundary value problem, proposed in [18]. A different very powerful code is ParaDis, which has been developed for simulating bulk properties of body-centered cubic (bcc) on massiv parallel machines at LLNL [16]. The Karlsruhe code is parallelised using OpenMP, which allows an efficient exploration of plastic flow of the deformation at small scales treating both microstructural analysis and statistical aspects, for which many different realisations have to be performed.

The paper starts with an overview of the simulation results on the mechanical properties of uniaxially deformed pillars. Both the dependency on the diameter and the length of the sample is analysed [17, 19] and the underlying dislocation mechanism are sketched. In order to investigate the influence of stress and strain gradients on the mechanical response of metallic materials, simulations on micro-bending beams are performed and compared to experimental results [20]. The influence of the initial dislocation microstructure is discussed in a separate chapter. The last section concerns brief technical overview on the parallelisation of the ddd model and an efficiency analysis.

2 Size Effects in Single Crystal Micro Pillars

Uniaxial deformation experiments of single crystalline aluminum (Al) pillars with diameters between 0.5 and 2.0 μm are simulated. The loading direction is <123>, where single slip is expected. The aspect ratio (AR), defined as *diameter:high*, is varies between AR 1:1.5 and 1:5. These ratios cover the experimental range in compression tests [3]. The ratio 1:2 is used for a detailed size effect study in section 2.1

The displacements along the long pillar axis is prescribed at the top and bottom surface. The remaining degrees of freedom at the surfaces are set to be traction free. A strain rate of 5000 s^{-1} is imposed and a constant initial dislocation density of about $2.1 \cdot 10^{13}$ m^{-2} is chosen. Randomly distributed Frank-Read sources (FRS) with a source length of 220 nm are placed within the volume. To reduce statistical scatter, the FRS are uniformely distributed on the 12 *fcc* glide systems.

The results are analysed based on the following quantities: (i) yield stress at 0.2% plastic deformation; (ii) internal structural length defined as the maximum distance between pinning points of dislocations and (iii) the analysis of the elastic stress inhomogeneities.

2.1 Results of the Micro Pillars with an Aspect Ratio 1:2

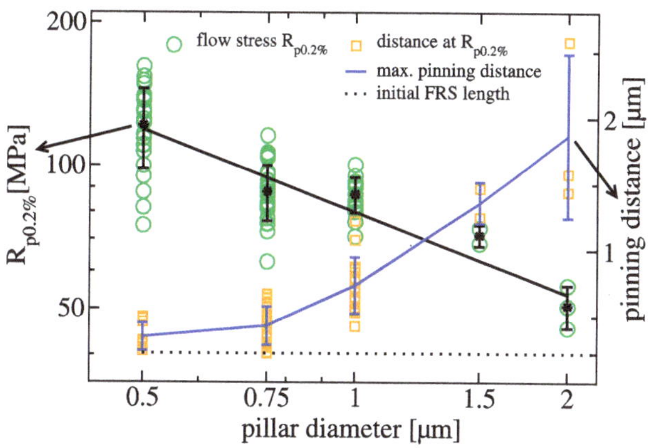

Fig. 1. Size effect: increasing flow stresses at 0.2 % plastic deformation with decreasing diameter (left axis). Pictured are the results of the individual flow stresses (green circles) and the mean value per sample diameter (black stars with standard deviation). The mean values are fitted by a power law. On the right axis the maximum distance between two pinning points are presented (orange squares = individual lengths; blue = mean values and standard deviation). The dotted line represents the initially chosen Frank-Read source length of 220 nm

In fig. 1 flow stresses at 0.2 % plastic strain are plotted versus the diameter. Similar to the experiments, the average flow stress increases with decreasing diameters and can be fitted to a power law (black line in fig. 1). A pronounced size effect is found for these samples, with an exponent of -0.55 ± 0.1, which is in the range of the experimental observations [3].

For a given size, a nominally identical initial dislocation microstructure is used. Statistical effect due to the low number of dislocation sources present in the small volumes cause the large scattering in the observed flow stresses. Furthermore, during deformation dislocation reactions or surfaces may deactivate sources and force less favourably oriented dislocation sources or even sources on glide systems with a lower Schmid factor to carry on the plastic flow (details in [17]). The influence of individual sources is reduced with increased pillar diameters (volume) as more sources on each glide system are available.

The analysis of the weakest dislocation source, defined as the maximum pinning distance found in the dislocation network is plotted on the right axis in fig. 1. With increasing sample diameter, longer dislocation segments can be created by multiplication and reactions like cross slip or glissile junctions. For comparison, the length of the prescribed FRS (220 nm) is shown with a dotted line in fig. 1. These longer segments require lower stresses resulting in the overall observed size dependence of the flow stress.

2.2 Influence of the Aspect Ratio

In a next step, the aspect ratio of the specimens is varied as tension and compression tests show differences in the size dependency of the flow stress. Tensile experiments show a pronounced size effect only for very low aspect ratios. Possible reasons for this behaviour are pure volume effects, due to the even lower number of sources and the influence of the boundary conditions on the sample giving rise to back-stresses.

In fig. 2 several stress strain curves of pillars with the smallest simulated diameter $d = 0.5\ \mu m$ and aspect ratio of 1:1.5 are presented. The yield stresses and overall stress strain curves scatter much more in the sample with AR 1:1.5 (70-140 MPa) than the ones for the higher AR. Less initial FRS lowers the probability of a well oriented source which lead to a higher probability for increased flow stresses. With increasing pillar diameter the difference between the aspect ratios concerning yield stress and hardening are found to disappear. In pillars with $d = 1.0\ \mu m$, the flow stress is at a similar level as the yield stress. The largest simulated pillars with $d = 2.0\ \mu m$ show the same stress strain response and the yield stress is found to be independent of the aspect ratio.

The aspect ratio influences mainly the stress strain response of the smallest pillars with a diameter $d = 0.5\ \mu m$. Here, the yield stress and the hardening behaviour are extremely dependent on the AR. One reason is the volume effect as only 1-2 sources (AR 1:1.5) on the glide system with the highest

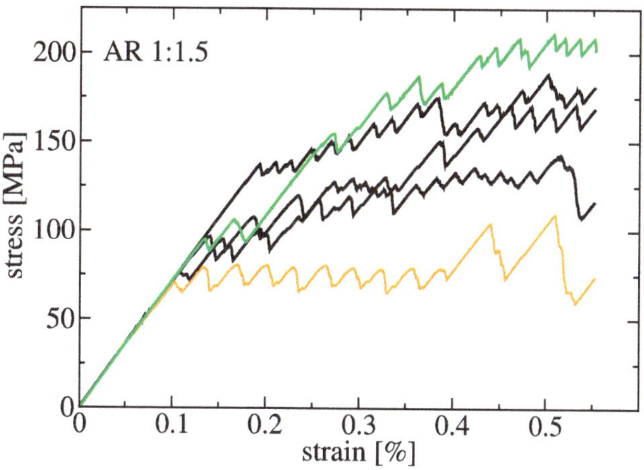

Fig. 2. Different stress strain curves for pillars with diameter $d = 0.5\ \mu m$ and aspect ratio 1:1.5

Schmid factor are available. This leads to higher yield and flow stresses at the chosen reference deformation of 0.2 % plastic strain. The consequence is a more pronounced size effect for pillars with aspect ratios 1:1.5 than for AR 1:3 and above. As lower aspect ratios show a larger size effect, other factors like as the influence of the boundary conditions and the geometry of the sample must be important. In low aspect ratio samples, the plastic slip is likely to concentrate on slip planes, which intersect the top and bottom surfaces of the sample. Back stresses are generated by the surface steps which act on the dislocation sources and which require increasing activation stress or force other - less favoured oriented - sources to activation. This is observed in the sample corresponding to the green stress strain curve in fig. 2, where the pillar shows a high hardening rate. The dislocation structure as well as the stress distribution in tensile axis at 0.55 % total strain is presented in fig. 3. The tensile stress varies between 100 and 300 MPa inside the sample. The high stress close to the bottom surface near the origin is caused by a surface step at the surface which is forced to be flat. A homogeneous and almost constant tensile stress profile is expected for dislocations leaving purely on the traction free side faces of the pillars, e.g. for the sample belonging to the orange curve in fig. 2. Increasing the AR favours the probability of the second scenario (dislocation slip trace at the traction side faces only).

3 Micro-Bending Tests

3.1 Specimen Set-Up

For the simulation of micro-bending tests the current 3D DDD code was extended for two new boundary conditions: (i) a free standing beam with one

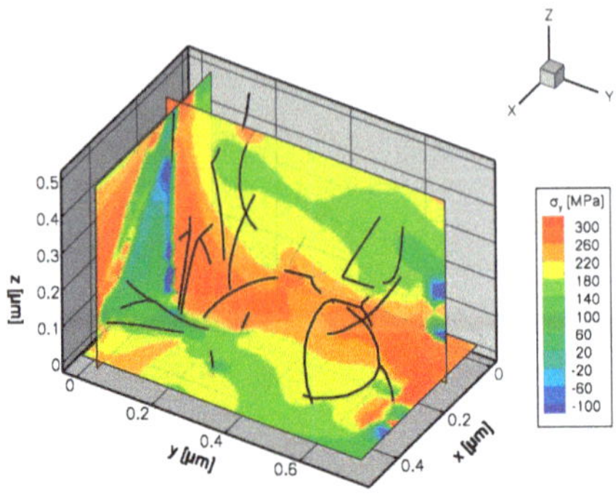

Fig. 3. Stress distribution in tensile direction of the sample corresponding to the green curve in Fig. 2 at 0.55% total strain. Thick black lines represent the current dislocation structure

end fixed, which mimics the experimental set-up [20], where one end is fixed in the bulk material and the other end is loaded with an indenter, which gives a linearly increasing bending moment, and (ii) a free beam with opposite bending moments applied on both ends, which gives a constant bending moment (details can be found in [21]).

The thicknesses t of the bending beams is varied between 0.5 and 1.5 μm with an aspect ratio of thickness to width to length of $t : B : l = 1 : 1 : 3$. For the finite element framework $8 \times 8 \times 24$ 20-node brick elements are used. The initial dislocation structure consists of FRS with a fixed length L of 220 nm. These sources are randomly distributed on the glide systems giving an initial dislocation density of $\rho_0 \approx 2 \cdot 10^{13}$ m^{-2}. The simulations are carried out on a single slip [010](001) and a multi-slip [123]$(7\bar{2}\bar{1})$ crystal orientation. For further details on the set-up and the used parameters the reader is referred to Motz et al. [21].

3.2 Results and Discussion

Compared to uniaxial compression or tension tests (chapter 2 or in detail [17]) a strong increase in dislocation density with deformation is observed. This increase in density is caused by the buildup of dislocation pile-ups around the neutral axis of the beam.

The observed dislocation pile-ups have a significant influence on the mechanical properties. Fig. 4 shows the normalised bending moment versus normalised displacement for the different beam thicknesses ([010](001) orientation) using boundary condition (ii). A pronounced size effect is visible, where the thinner bending beams exhibit higher bending moments and vice versa, which is in accordance with experimental findings. Furthermore, the thinner beams show a significant work hardening - a strong increase in bending moment with ongoing plastic deformation - compared to the thicker beams where only a slight increase in bending moment is observed. These key features do not change with different initial dislocation configurations. The dislocation

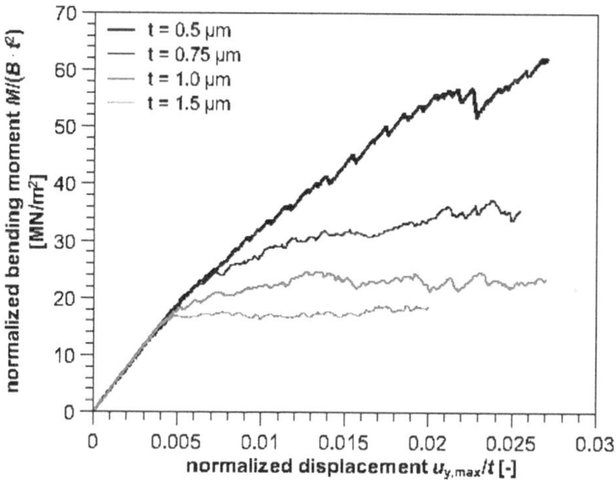

Fig. 4. Normalised bending moment vs. normalised displacement response for different beam thicknesses t using boundary condition (ii) and [010](001) orientation

density evolution for the simulations depicted in Fig. 4 is shown in Fig. 5. As already mentioned a strong increase in density is observed with ongoing deformation. The density increases from $2 \cdot 10^{13}$ m^{-2} (initial density) up to almost 10^{14} m^{-2}, which is a huge increase compared to uniaxial tension or compression tests. If the dislocation density is plotted versus the plastic deformation (dashed lines in Fig. 5) a linear increase in density with plastic deformation can be observed. The slope of the density increase scales inversely with the beam thickness here. This is in good agreement with strain gradient plasticity approaches (see for example [22]), where the density of the so-called geometrically necessary dislocations (GND) should scale inversely with the beam thickness. This implies that the major part of the dislocation density in the bending simulations consists of GNDs. For the thinnest beams distinct jumps in the dislocation density can be identified which are associated with load drops in the corresponding moment vs. displacement curve (see Fig. 4). Under load control these jumps would result in strain bursts presumably caused by intensive dislocation source operation [13].

In Fig. 6 the size dependence of the normalised bending moment is shown for the [010](001) orientation and the two boundary conditions at a relative plastic deformation of 0.01. For both boundary conditions a strong size effect is evident where the normalised bending moment scales about inversely with with beam thickness: $M^* \propto t^{-1}$. The same behaviour is found for the single slip $[123](7\bar{2}\bar{1})$ orientation. The bending moment level is higher for the

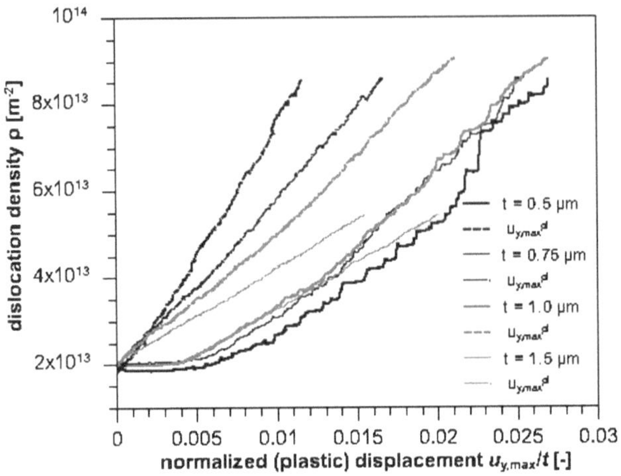

Fig. 5. Dislocation density vs. total displacement (solid lines) and plastic displacement (dashed lines) for the same simulations as depicted in Fig. 4

boundary condition (i) – the cantilever like bending – which is caused by the additional constraint in beam axis direction (additional strain gradient in beam axis direction). The overall scaling behaviour is similar as found in the experiments [20], leading to the conclusion that the principle dislocation mechanisms are the same. Details in the influence of the initial dislocation structure on the mechanical properties can be found in [21].

In general, plastic deformation starts with the activation of sources close to the surface, triggered by the high stress state there. Dislocation segments emitted from these sources glide into the volume and pile-up around the neutral plane of the bending beam. Multiple activation of the same source increases the pile-up until the back stress becomes too high and the dislocation source ceases operating. Further loading is necessary to re-activate the sources against the pile-up or to activate new sources located in less favourable positions. The activation of new sources is usually accompanied by immediate multiple operation and results in the formation of a new pile-up. In thin beams, these pile-ups are a predominant feature because the number of favourably located near surface dislocation sources is limited. For thicker beams the pile-up is less pronounced and the spacing of the dislocations in the pile-ups is larger. The difference between thin and thick beams is mainly due to volume effect leading to a higher number of initial dislocation sources for the thicker beams. More sources near the surface can be activated, which leads to a more homogenous overall slip distribution.

Usually, size effects in the plastic properties of samples subjected to strain gradients during loading (e.g. bending, torsion or indentation) are explained by the use of strain gradient plasticity approaches. This approach is not ap-

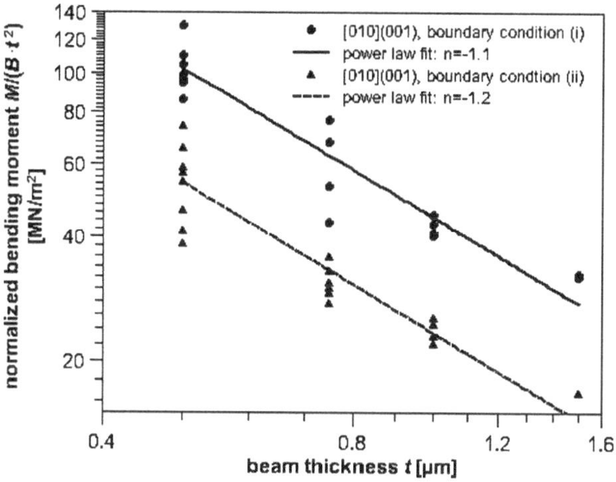

Fig. 6. Size dependence of the normalised bending moment at a plastic deformation of $u_{y,max}^{pl}/t = 0.01$ for the [010](001) orientation and the two boundary conditions

plicable for bending beams in the investigated size regime, because the dislocation density increase cannot explain the increase in bending moment by the Taylor relation $\sigma \propto \sqrt{\rho}$ (compare Fig. 4 and Fig. 5). It is obvious that the observed dislocation pile-up around the neutral plane of the beams may be responsible for the strong size effect with the scaling exponent of about -1. These pile-ups are a persistent feature in the structure independent on the initial dislocation structure.

4 Advanced Initial Dislocation Structures

Usually FRS are used in 3d DDD simulations as initial dislocation structures. Lengths, orientations and the distribution on the glide systems may be varied for the FRS to obtain different initial structures. These structures are highly artificial because "real" dislocation arrangement never consists of isolated FRS. Furthermore, persistent pinning points - the end-nodes of the FRS - are introduced into the dislocation structure, which does not allow to assess mechanisms like dislocation starvation. This was the cause for the investigation of alternative starting dislocation structures.

In a first study sub-volumes were cut out from a larger volume and loaded subsequently. For this purpose large samples of several micrometer in size were loaded to some extend and fully unloaded subsequently to develop a deformation dislocation structure. Then sub-volumes were cut out. After cutting, relaxation calculations were preformed on that sub-volume to bring the structure into equilibrium. These samples can be used in a subsequent simulation (e.g. tension, bending, etc.). The obtained starting dislocation structure consists of old FRS, truncated FRS (spiral sources), dislocation reactions and free dislocation segments. Such structures are more realistic compared to pure FRS structures. Simulations, both in tension and bending loading, have shown that there is now significant difference in the mechanical response for these starting configurations compared to FRS. Further details can be found in [21] and [19].

To get rid of the persistent pinning points in the initial dislocation structure a different approach was used. The initial structure closed circular dislocation loops with a certain diameter range were randomly distributed in the sample volume. The initial dislocation density was chosen in that way that after the relaxation process the standard density of $\rho_0 = 2 \cdot 10^{13}$ m^{-2} was reached. Dislocation loops which did not lie entirely in the sample volume were cut by the sample surfaces. Fig. 7 shows an example of the initial structure with the closed loops and the resulting structure after the relaxation process.

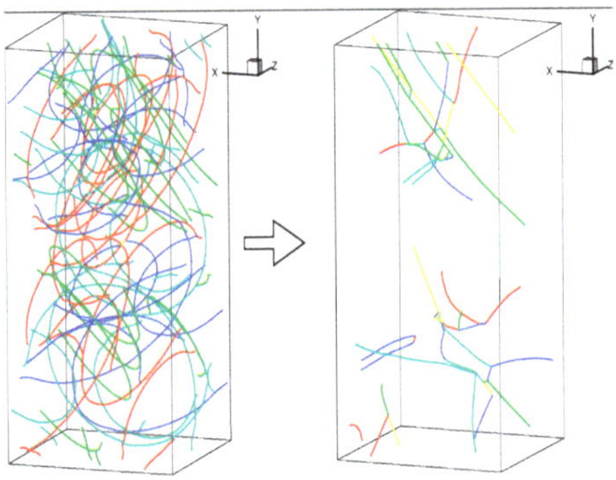

Fig. 7. Initial dislocation structure consisting of closed circular loops used for the relaxation process (left side) and the resulting dislocation structure (right side) for an $0.5 \times 0.5 \times 1.0$ μm sample

Fig. 8 shows first stress vs. strain curves of samples with such an initial dislocation structure for different sample sizes and different initial dislocation densities loaded in tension. It is obvious that the initial dislocation density has a much stronger influence on the mechanical response compared to samples with FRS as starting structure used in section 2. In these structures no artificial pinning points are present. A full study of the consequences of this choice is currently ongoing work.

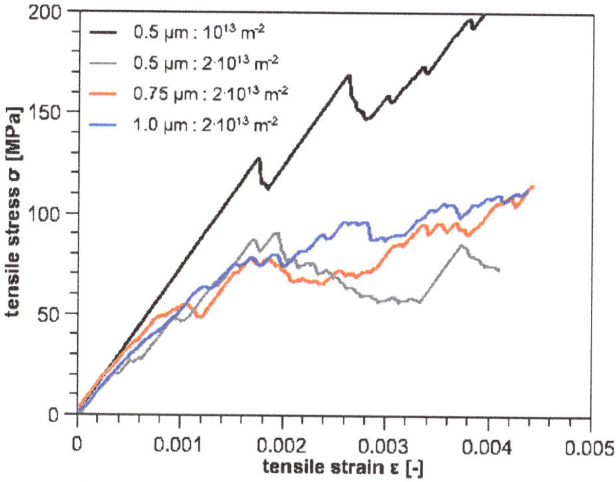

Fig. 8. Stress strain curves for specimens where the relaxed structure was used as starting structure in tension tests. The AR of the samples was 1:2

5 Parallelisation and Scaling

An overview on the implementation of a discrete dislocation dynamics (DDD) model, parallelised using OpenMP compiler directives, is described in this section. In the first part the overall program structure is presented. In the second part the scaling behaviour of the ddd tool is presented for two representative test cases with realistic parameters. The two "benchmark" loading cases, "uniaxial tensile test" and "bending test" lead to quite different dislocation microstructures, e.g. dislocation network and formation of pile-ups.

5.1 Dislocation Dynamics Model in a Nutshell

The objective of the DDD tool is the simulation of the time evolution of a dislocation microstructure during plastic flow of small and finite crystals. The tool is splitted in two parts, a discrete dislocation dynamics part, describing the plastic flow of the sample, and a boundary value problem for the elastic properties of the sample [14, 18]. The dislocation dynamics description consists of a geometrical part, where the bookkeeping of the evolution of dislocation network and connectivity is handled and a physical part, where the constitutive rules for the motion and the physics of intersecting dislocations is described. The following points are implemented:

I Discrete Dislocation part:

- Dislocations are line defects within a crystalline structures. The geometrical description includes the discretisation of the dislocations, based on straight segments connected at nodes, local refinement upon contact or approach, contact detection between dislocation, junction formation and a book-keeping algorithm to discriminate between short and long range interactions. The data structure needed to describe the dislocation network is adapted to an OpenMP environment: locking mechanisms for handling dynamic lists for dislocation objects (nodes, segments and loops) and controlled access to information stored on the basic objects are implemented.

- physical description: the elastic interactions between dislocation is long range ($\propto 1/r$, where r is the distance to the dislocation). This dependency does not allow to use a cut-off to reduce the interaction calculation. For a faster calculation a near field interaction scheme was implemented: In sub time steps only the interactions of nearby dislocations were recomputed, whereas the far field contribution is kept. A detection scheme has been implemented which detects the hot spots in the dislocation microstructure. "Hot spots" are regions, where dislocations are closeby and strong variations of the structure occur. This necessitates a parallel implementation of a linked list scheme, similar to the one commonly used in MD simulations, but extended to the handling of finite segments. This list is regularly updated, triggered through topological and rediscretisation events, where segments are added or removed from the system. Only in global time steps a complete update of all interactions (far and near field) is done. The boundary conditions are updated during the global time step.
- equation of motion for dislocations leads to a coupled system of equations for the discretisation nodes of the dislocation microstructure. This system is solved using a parallel conjugate gradient.

II Elastic boundary value problem:

- the simulation volume is discretised on a regular grid.
- the isotrapic elastic problem is solved with the finite elements methods using a quadratic approach with 20-node cubic elements [14, 18].
- in collaboration with the HPC group of Prof. Heuveline at the SCC in Karlsruhe, the conjugate gradient solver with LU-preconditioning was replaced by a multigrid algorithm which takes full advantage of the regular grid, has a better convergence order and shows a much better parallel performance. To increase the computationally feasible number of elements the sparse-matrix implementation of the linear transformation was replaced by a very compressed transformation which needs only a fraction of the former memory; for the $24 \times 72 \times 24$ elements grid mentioned in the next section one single sparse-matrix with about 535 thousand degrees of freedom would require approx. 750 MB while the multigrid solver for the same problem has a total memory footprint of about 143 MB.

To bring these two part together, an interface is defined, which allows to transfer the information between the two parts: (i) access to the displacements and stresses caused by the dislocation microstructure is needed in the elastic boundary value problem; (ii) the information on the elastic stresses from the boundary conditions are needed along the dislocation lines.

5.2 Scaling Behaviour

The computational time for evaluating the overall scaling of the implemented model is measured using the intrinsics omp_get_wtime() routine. The output presented here is given units of [s]. The scaling is measured on different AMD64 systems: (i) XC2 node 2*2 Core Opteron 2.6 GHz; (ii) 2*2 Core Xeon 2.6 GHz (iii) 2*4 Core Xeon Penryn 2.8 GHz and (iv) 4*4 Core Tigerton Xeon X7350 2.97 GHz.

Uniaxial tensile test:

A simulation configuration of a (100) oriented pillar with the size $2 \times 6 \times 2\ \mu m^3$ and approximately 20000 segments is used to benchmark the different parts of the program. A FEM mesh with $24 \times 72 \times 24$ elements is used. A total speedup of ≈ 3.9 on four CPUs (XC2 node: two socket system with dual core Opteron 2.6 GHz, absolute time: 1 CPU: 8345; 4 CPU: 2490) and ≈ 9.6 on 16 CPUs (4 socket Xeon 2.93 GHz, absolute time: 1 CPU: 7700; 16 CPU: 804), respectively, is obtained.

Micro - bending beam:

A bending configuration (size: $2 \times 6 \times 2\ \mu m^3$) with 24000 segments and a FEM mesh of $8 \times 24 \times 8$ elements are used to test also the second loading condition. Here, the overall speedups are 8.6 on 16 CPUs (4*4 Core Xeon – Tigerton) and 4.8 (8 CPUs, 2*4 Xeon 2.8 GHz Penryn). Both loading conditions are compared in fig. 9(b).

The speedup of the different parts of the DDD tool:

The calculation of global and local interaction shows an excellent scaling (green line in fig. 9 (force)). The interaction calculation of N segments require $N \times N$ calculations in the global time step, and is microstructural dependent in the subtime steps, where local "hot spots" that require local recalculation are collected. The computational time is despite of the excellent scaling important: (absolute time: XC2 Opteron: 1 CPU: 2749; 4 CPU: 703; speedup: 3.9; 4*4 Xeon: 1 CPU: 2780; 16 CPU 162; speedup: 17). The scaling of the local force calculation scheme is included in the above measurement. In these benchmarks the dislocation microstructure evolves and the number of nodes, degrees of freedom changes during the simulation.

The equation of motion for the dislocation is solved to calculate the new nodal positions. A parallel CG-solver is implemented in combination with the Störmer-Verlet algorithm. For isolated dislocations a tridiagonal solver is used, which reduces the degrees of freedom to be solved by the CG solver. This reduces the calculation time during the elastic loading. The initialisation of the CG solver, needed for every time step, performs a mapping of the dynamic dislocation data structure, which uses heavily Fortran 90 pointers, on indexed arrays. This mapping introduces a serial overhead in the CG part. Therefore an ideal scaling is not possible and an early saturation is observed. The blue line in fig. 9 (velocity) shows the scaling behaviour for solving the equations of motion (absolute time: XC2 Opteron: 1 CPU: 321; 4 CPU: 104; speedup: 3.1; 4*4 Core Xeon: 1 CPU: 255; 16 CPU: 50; speedup: 5.1). The array mapping has reduced the absolute time for the solver by one order of magnitude and takes now less than 10% of the total time.

Boundary conditions in the FEM-part require the tractions calculation on the surfaces. The parallelisation is done at the level of the surface elements, which can be calculated independently. An excellent scaling is observed (absolute times: XC2 Opteron: 1 CPU: 2298; 4 CPU: 595; speedup: 3.9; 4*4 Xeon: 1 CPU: 2006; 16 CPU: 123).

Refined FEM meshes are needed for the elastic boundary value problem to allow for a good resolution of the surface tractions and spacial resolution of the stress inhomogeneities as shown in Fig. 3. The multigrid solver described above fullfills these requirements. An excellent scaling behaviour is observed: (absolute time: XC2 Opteron: 1 CPU: 3706; 4 CPU: 960; speedup: 3.9; Xeon: 1 CPU: 3198; 16 CPU: 220; speedup: 14.5).

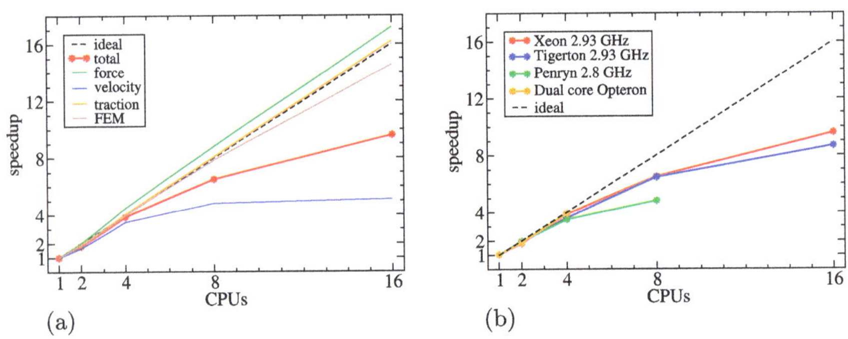

Fig. 9. Speedup analysis: (a) for uniaxial loading: overall and subdivided to the main calculation tasks. Benchmarks are performed on a Xeon 2.93 GHz (SUN X4450, 2.93 GHz); (b) Total speedup for uniaxial loading (marked as Xeon, Opteron on XC2) and bending configuration (Tigerton, Penryn) on different platforms

Computational time for the simulated configurations

Reasonable analyses of the size effect in the flow stress or the stress distribution in single crystals require a plastic deformation of the samples of 0.5 % and a large number of simulations. Simulations for micro-pillars of $0.5 \times 1.0 \times 0.5\ \mu m^3$ take $\approx$ 3 to 4 days of calculation time on a XC2 node with 4 CPUs. In order to capture the statistical properties 20 to 40 individual simulations have to be performed. The simulation time increases with the number of segments, a total force calculation scales with V^2 and only the use of the sub-incremental scheme which near field recalculations allows to study reasonably large systems. A simulation of a sample with a size of $2.0 \times 4.0 \times 2.0\ \mu m^3$ needs up to 1-2 months. This is reached using a restart facility of the program.

The dislocation microstructure evolution may lead to quite different responses, which are characterised by quite different dislocation densities and flow stresses as shown in Fig. 2 and the calculation time varies accordingly. The behaviour of dislocations have been shown to be chaotic [23] and therefore the dislocation microstructure is likely to be different even for simulations with different numbers of threads, e.g. some dislocation reactions may occur in a different order. The scaling may be affected by this considerably.

6 Conclusion

The presented parallel discrete dislocation tool allows for insight into the deformation process of small scale samples. Statistical properties and microstructural analysis require the use of HPC facilities as the computation time for these simulations is important.

Comparison of the simulation results to experimental findings show an excellent agreement in the investigation of flow stresses in micro pillars and bending moments for micro-bending tests. The influence of the boundary conditions, leading to non uniform elastic stresses, for the micro tensile experiments with low AR is demonstrated. The formation of pile-ups in micro-bending is proposed to be the origin of the thickness dependency of the normalised bending moment.

This work is currently extended to include grain boundaries, which will increase the computational costs even more.

Acknowledgement

The financial support of the European Commissions NANOMESO project under contract number NMP3-CT-2006-016710, as well as the Landesstiftung Baden-Württemberg HPC-project No. 667 is gratefully acknowledged. Part of the calculations were performed at the Baden-Württemberg HPC facilities in Karlsruhe within the HPC_DDD project.

References

1. N.A. Fleck; G.M. Muller; M.F. Ashby; J.W. Hutchinson. Strain gradient plasticity: Theory and experiment. *Acta Metall. et Mater.*, 42:475–487, 1994.
2. M.D. Uchic; D.M. Dimiduk; J.N. Florando; W.D. Nix. Samples dimensions influence strength and crystal plasticity. *Science*, 305:986–989, 2004.
3. C.A. Volkert; E.T. Lilleodden. Size effects in the deformation of sub-micron au columns. *Phil. Mag.*, 86(33-35):5567–5579, 2006.
4. D. Kiener; W. Grosinger; G. Dehm; R. Pippan. A further step towards an understanding of size-dependent crystal plasticity: In situ tension experiments of miniaturized single-crystal copper samples. *Acta Mater.*, 56:580–592, 2008.
5. W.D. Nix; J.R. Greer; G. Feng; E.T. Lilleodden. Deformation at the nanometer and micrometer length scales: Effects of strain gradients and dislocation starvation. *Thin Solid Films*, 515:3152–3157, 2007.
6. B. von Blanckenhagen; P. Gumbsch; E. Arzt. Dislocation sources and the flow stress of polycrystalline thin metal films. *Phil. Mag. Lett.*, 83(1):1–8, 2003.
7. B. von Blanckenhagen; E. Arzt; P. Gumbsch. Discrete dislocation simulation of plastic deformation in metal thin films. *Acta Mater.*, 52:773–784, 2004.
8. J.R. Greer; C.R. Weinberger; W. Cai. Comparing the strength of f.c.c. and b.c.c. sub-micrometer pillars: Compression experiments and dislocation dynamics simulations. *Mat. Sci. Eng. A*, doi:10.1016/j.msea.2007.08.093, 2008.
9. A.S. Budiman; S.M. Han; J.R. Greer; N. Tamura; J.R. Patel; W.D. Nix. A search for evidence of strain gradient hardening in au submicron pillars under uniaxial compression using synchrotron x-ray microdiffraction. *Acta Mater.*, 56:602–608, 2008.
10. D.M. Dimiduk; M.D. Uchic; S.I. Rao; C. Woodward; T.A. Parthasarathy. Overview of experiments on microcrystal plasticity in fcc-derivative materials: selected challenges for modelling and simulation of plasticity. *Mod. Sim. Mat. Sci. Eng.*, 15:135–146, 2007.
11. T.A. Parthasarathy; S.I. Rao; D.M. Dimiduk; M.D. Uchic; D.R. Trinkle. Contribution to size effect of yield strength from the stochastics of dislocation source lengths in finite samples. *Scripta Mater.*, 56:313–316, 2007.
12. D.M. Dimiduk; M.D. Uchic; T.A. Parthasarathy. Size-affected single-slip behavior of pure nickel microcrystals. *Acta Mater.*, 53:4065–4077, 2005.
13. F.F. Csikor; C. Motz; D. Weygand; M. Zaiser; S. Zapperi. Dislocation avalanches, strain bursts, and the problem of plastic forming at the micrometer scale. *Science*, 318:251–254, 2007.
14. D. Weygand; L.H. Friedman; E. van der Giessen; A. Needleman. Aspects of boundary-value problem solutions with three-dimensional dislocation dynamics. *Mod. Sim. Mat. Sci. Eng.*, 10:437–468, 2002.
15. H.D. Espinosa; M. Panico; S. Berbenni; K.W. Schwarz. Discrete dislocation dynamics simulations to interpret plasticity size and surface effects in freestanding fcc thin films. *Int. J. Plast.*, 22:2091–2117, 2006.
16. A. Arsenlis; W. Cai; M. Tang; M. Rhee; T. Oppelstrup; G. Hommes; T.G. Pierce; V.V. Bulatov. Enabling strain hardening simulations with dislocation dynamics. *Mod. Sim. Mat. Sci. Eng.*, 15:553–595, 2007.
17. J. Senger; D. Weygand; P. Gumbsch; O. Kraft. Discrete dislocation simulations of the plasticity of micro-pillars under uniaxial loading. *Scripta Mater.*, 58:587–590, 2008.

18. E. van der Giessen; A. Needleman. Discrete dislocation plasticity: a simple planar model. *Mod. Sim. Mat. Sci. Eng.*, 3:689–735, 1995.
19. J. Senger; D. Weygand; C. Motz; P. Gumbsch; O. Kraft. Uniform loading of micro-pillars with variable aspect ratio simulated with discrete dislocation dynamics. *in preparation*, 2008.
20. C. Motz; T. Schöberl; R. Pippan. Mechanical properties of micro-sized copper bending beams machined by the focused ion beam technique. *Acta Mater.*, 53:4269–4279, 2005.
21. C. Motz; D. Weygand; J. Senger; P. Gumbsch. Micro-bending tests: A comparison between three-dimensional discrete dislocation dynamics simulations and experiments. *Acta Mater.*, 56:1942–1955, 2008.
22. H. Gao; Y. Huang; W.D. Nix; J.W. Hutchinson. *Jour. Mech. Phys. Sol.*, 47:1239, 1999.
23. V.S. Deshpande; A. Needleman; E. van der Giessen. Dislocation dynamics is chaotic. *Scripta Mater.*, 45:1047–1053, 2001.

Miscellaneous Topics

Univ.-Prof. Dr.-Ing. Wolfgang Schröder

Aerodynamisches Institut, RWTH Aachen, Wüllnerstr. 5a, 52062 Aachen, Germany, office@aia.rwth-aachen.de

In completing the research topics which have been addressed before such as fluid mechanics, structural mechanics, aerodynamics, thermodynamics, chemistry, combustion, and so forth the interdisciplinary breadth of numerical simulations is emphasized in the following contributions. The articles clearly show the link between applied mathematics, fundamental physics, computer science and the ability to develop certain models such that a closed mathematical description can be achieved which can be solved by highly sophisticated algorithms on an up-to-date high performance computer. In other words, it is the collaboration of several scientific fields which on the one hand, defines the area of numerical simulations on the other hand, drives the progress in fundamental and applied research. The subsequent papers, which represent an excerpt of various projects linked with HLRS, will confirm that numerical simulations are not only used to compute some quantitative results but to corroborate basic physical models and to even develop new theories. However, it goes without saying that numerical simulations will always be completed by experimental investigations and analytical solutions. To rely on just one of these approaches appears to be no successful scientific route.

The first manuscript from the Institut für Technische Thermodynamik und Thermische Verfahrenstechnik of the Universität Stuttgart addresses the problem of molecular modeling of hydrogen bonding fluids. It is fair to state that molecular modeling and simulation becomes more and more important to predict thermophysical properties of pure fluids and mixtures. This is true in research and industry for several reasons. Superior to classical methods the predictive power of molecular models yields results with technically relevant accuracy over a wide range of state points. A given molecular model provides access to the full variety of thermophysical properties, such as thermal, caloric, transport, or phase equilibrium data. Through the event of cheaply available powerful computing infrastructure, reasonable execution times for molecular simulations can be achieved which are of particular importance for industrial applications. Molecular modeling and simulation are based on statistical thermodynamics which directly links the intermolecular interactions to the macroscopic thermophysical properties. This sound physical background also supports the increasing acceptance compared to classical phenomenological modeling. The objective of this study of the University of Stuttgart is to demonstrate the ability of molecular models to accurately predict transport properties. The research is based on rigid, united-atom type Lennard-Jones based models with superimposed point charges for methanol and ethanol. These models had been optimized using experimental data of vapor pressure and saturated liquid density.

The next contribution on the investigation of process specific size effects is from the Institute of Material Science and Engineering and the Institute of Reliability of Components and Systems of the University of Karlsruhe. A process-specific size effect is defined as a nonlinear scaling behavior of output quantities for a linear and self-similar scaling of a process. It is particularly important for the miniaturization and design of industrial cutting processes. To improve the knowledge of size effects, an orthogonal micro turning process is investigated, in which the nonlinear increase of the specific cutting force with a linearly decreasing cutting depth is a significant process-specific trait. In the Karlsruhe study a systematic analysis of the impact of the geometric dimensions of the cutting tool and the process parameters cutting depth and cutting velocity on the specific reaction forces, i.e., the specific cutting and passive force components, is performed. As to similarity mechanics, the geometrically similar scaling of the process is given, which allows to identify material inherent influences on size effects such as strain rate hardening. 3D-finite-element-simulations were performed for the material AISI 1045 in a normalized state. The parameters were adapted to micro cutting experiments such that plastic deformation fields and chip shapes could be compared.

Andean orogeny and plate generation is the title of the final contribution. It is from the Institut für Geowissenschaften of the University of Jena. The basic conception of a novel fluid dynamic and geodynamic project on the Andean orogeny is presented. In the beginning a kinematic analysis of the entire orogeny is performed and then different numerical options are tested to explain these systematized observations by a physical model. Therefore, partly kinematics and partly dynamic regional models as well as purely dynamic models are considered. An existing concept is used to embed a regional model into a global spherical-shell model to determine the boundary conditions of the regional model as a function of time such that the artificially simplified boundary conditions of some published models of the Andean mechanism are avoided. An iteration concept is applied allowing the regional model to retroact upon the global surrounding model. Furthermore, two kinds of spherical-shell convection models, i.e., circulation models and forward models, exist. A spherical-shell model of mantle convection with thermal evolution and generation of continents and the depleted mantle reservoir is presented. The numerical result shows plate tectonics to occur only if at least the lithosphere deviates from purely viscous rheology and if there is a low-viscosity layer beneath it. The numerical regional Andean model has to be embedded into a global circulation model which is why an improvement of the basic code Terra is discussed.

Molecular Modeling of Hydrogen Bonding Fluids: New Cyclohexanol Model and Transport Properties of Short Monohydric Alcohols

Thorsten Merker, Gabriela Guevara-Carrión, Jadran Vrabec, and Hans Hasse

Institut für Technische Thermodynamik und Thermische Verfahrenstechnik, Universität Stuttgart, D-70550 Stuttgart, Germany
vrabec@itt.uni-stuttgart.de

1 Introduction

Currently, molecular modeling and simulation gains importance for the prediction of thermophysical properties of pure fluids and mixtures, both in research and industry. This is due to several reasons: Firstly, the predictive power of molecular models allows for results with technically relevant accuracy over wide range of state points that is superior to classical methods. Secondly, a given molecular model provides access to the full variety of thermophysical properties, such as thermal, caloric, transport or phase equilibrium data. Finally, through the advent of cheaply available powerful computing infrastructure, reasonable execution times for molecular simulations can be achieved which are of particular importance for industrial applications. Molecular modeling and simulation are based on statistical thermodynamics which directly links the intermolecular interactions to the macroscopic thermophysical properties. That sound physical background also supports the increasing acceptance compared to classical phenomenological modeling.

Modeling thermophysical properties of hydrogen bonding systems remains a challenge. Phenomenological models often fail to describe the interplay between the energetics of hydrogen bonding and its structural effects. Molecular force field models, however, are much better suited for solving that task as they explicitly consider this interplay. Most of the presently available molecular models use crude assumptions for the description of hydrogen bonding which can, for instance, be simply modeled by point charges eccentrically superimposed to Lennard-Jones (LJ) sites. One benefit of this simple modeling approach for hydrogen bonding is the comparably small number of adjustable model parameters. Furthermore, the approach is compatible with numerous

LJ based models from the literature and it can successfully be applied to mixtures. This simple modeling approach emerged to be fruitful in many ways, although many of the molecular models proposed in the literature lack in the quantitatively sound description of thermophysical properties. The aim of this project is to tackle that problem and to show that a thorough modeling and parameterization does indeed yield quantitatively correct results.

Molecular models which accurately describe vapor-liquid equilibria over the full temperature range usually exhibit a good predictive power throughout the whole fluid region. The molecular model for ethanol developed in the first period of the MMHBF project [1] has these characteristics and excellently performed in the prediction of vapor-liquid equilibrium properties of mixtures, i.e. Henry's law constants [2, 3]. To study the chosen modeling approach also regarding other strongly associating fluids, a new molecular model for formic acid [4] was developed earlier in the present MMHBF project [5]. Formic acid is the simplest carboxylic molecule and has exceptional thermophysical properties due to its ability to act both as hydrogen bond donor and acceptor. Since both hydrogen atoms of formic acid can act as proton donors and both oxygen atoms provide proton acceptance, four unlike hydrogen bond types yield the basis for a complex self-association which is the reason for its exceptional thermophysical behavior. The developed molecular model for formic acid also excellently describes the vapor-liquid equilibrium properties.

The Collaborative Research Centre 706 (SFB706) offers attractive applications for the present work. There, novel octahedral molecular sieves for the heterogeneously catalyzed selective oxidation of cyclohexane are investigated. Supercritical fluids and carbon dioxide-expanded liquids are used as innovative reaction media. For a rational planning of catalytic experiments and process design, especially at higher pressures, reliable thermodynamic data are needed. Most groups used in the past nitrogen instead of oxygen to predict the phase behavior of the reacting system. For predictive applications, e.g. the Peng-Robinson equation of state has been used, as true experimental vapor-liquid equilibira of binary mixtures containing oxygen for this reaction system are rare, especially at elevated temperatures and pressures. Molecular modeling and simulation is an excellent approach to bypass the lack of experimental data. Cyclohexanol is the central component in this reaction system. Therefore, a new molecular model for cyclohexanol was developed in the present period of the MMHBF project with the aim to accurately describe the vapor-liquid equilibrium. Cyclohexanol is the largest hydrogen bonding molecule which was modeled within the present project to date.

Another interesting application of molecular modeling and simulation are the transport properties of liquids. Due to the complexity of the involved physical mechanism, only molecular methods offers promising predictive approaches to this problem, especially for hydrogen bonding fluids. In this project, the self-diffusion coefficient of pure methanol and ethanol as well as in the mixture of both components was regarded.

Two fundamentally distinct methods for calculating transport properties by molecular dynamics simulation are available. The equilibrium methods (EMD), using either the Green-Kubo formalism or the Einstein relations, determine the time dependent response of a fluid system to spontaneous fluctuations. With non-equilibrium molecular dynamics (NEMD), on the other hand, the system response to an externally applied perturbation is analyzed. The later method was developed in order to increase the signal to noise ratio and to improve statistics and convergence. Both methods exhibit different advantages and disadvantages, but are comparable in efficiency, as shown, e.g. by Dysthe et al. [6].

The present work is based on rigid, united-atom type Lenard-Jones based models with superimposed point charges for methanol and ethanol developed earlier [4, 2]. It should be pointed out that these models were optimized using experimental data of vapor pressure and saturated liquid density. The goal of this study is to demonstrate the ability of molecular models, adjusted to these vapor-liquid equilibrium data only, to accurately predict transport properties.

Results of this work are consistently published in peer-reviewed international journals. The following publications contribute to the present project:

- T. Merker, J. Vrabec and, H. Hasse: Comment on "An optimized potential for carbon dioxide"[J. Chem. Phys. 122, 214507 (2005)]. J. Chem. Phys., submitted (2008).
- T. Merker, J. Vrabec and, H. Hasse: Molecular models for carbon dioxide and cyclohexanol. In preparation
- G. Guevara-Carrión, C. Nieto-Draghi, J. Vrabec and, H. Hasse: Prediction of the Transport Properties for Short Monohydric Alcohols: Methanol, Ethanol and their Binary Mixture. In preparation

This report is organized as follows: Firstly, the new molecular model for cyclohexanol is introduced. Subsequently, the self-diffusion coefficient of methanol and ethanol is presented. Finally, remarks on computational details are given.

2 Molecular Model for Cyclohexanol

A new cyclohexanol model was developed based on quantum mechanical calculations and optimized using experimental vapor pressure, bubble density and heat of vaporization. As the complexity of a molecular model determines the required computing time in molecular simulation, it was attempted to find an efficient solution balancing accuracy and simplicity. A rigid model with seven LJ sites plus one point quadrupole and three point charges was chosen. The assumption of rigidity was chosen as the cyclohexane ring predominantly forms the energetically favorable chair-conformation.

The geometric parameters of the molecular model were taken directly from quantum mechanical calculations. For this purpose, initially a geometry optimization was performed with the GAMESS (US) package [7], employing the

Table 1. Coordinates and parameters of the LJ sites and the point charges in the principal axes system of the new molecular model for cyclohexanol. Bold characters indicate represented atoms

Interaction Site	x Å	y Å	z Å	σ Å	ε/k_B K	q e
$\mathbf{CH_2}(1)$	-2.16883	-0.55100	0	3.412	102.2	—
$\mathbf{CH_2}(2)$	-1.30893	0.44982	-1.57594	3.412	102.2	—
$\mathbf{CH_2}(3)$	0.56919	-0.40155	-1.56168	3.412	102.2	—
$\mathbf{CH_2}(4)$	0.56919	-0.40155	1.56168	3.412	102.2	—
$\mathbf{CH_2}(5)$	-1.30893	0.44982	1.57594	3.412	102.2	—
CH	1.06798	0.33582	0	3.234	60.0	0.256184
OH	2.45979	0.00085	0	3.150	85.1	-0.638767
H-O	2.50948	-0.97164	0	—	—	0.382583

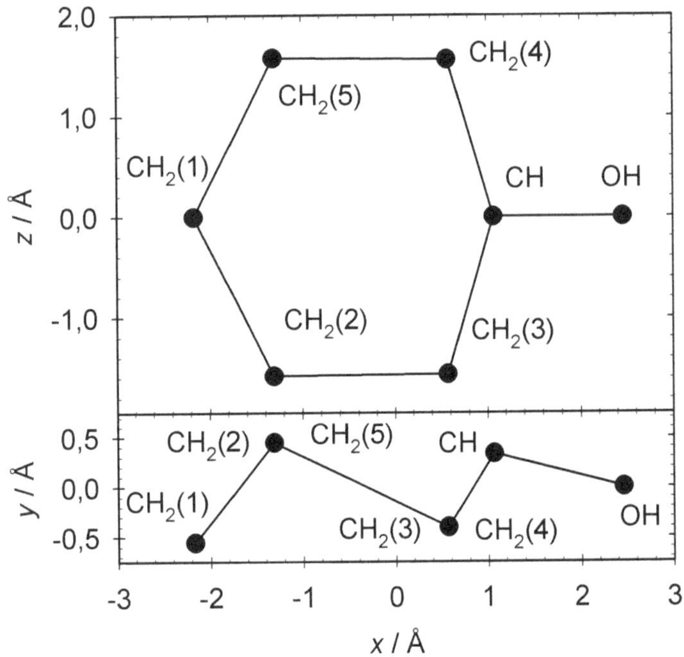

Fig. 1. Coordinates of the LJ sites for the present cyclohexanol model

Hartree-Fock method and the basis set 6-31G. For the quantum chemical calculations, the symmetry of the molecule was exploited and only half of it was regarded. A LJ site was located exactly at all resulting nuclei positions, except for the hydrogen atom. The ethyl and methyl group was modeled by a single LJ site, i.e. the united-atom approach was used. The coordinates of the seven LJ sites are given in Table 1 and in Figure 1.

Table 2. Orientation and moment of the point quadrupole placed in the center of mass of the new molecular model for cyclohexanol. Orientations are defined in standard Euler angles, where φ is the azimuthal angle with respect to the $x-z$ plane and θ is the inclination angle with respect to the z axis

Site	x Å	y Å	z Å	φ deg	θ deg	Q
Quadrupole	0	0	0	90	90	0.795561 B

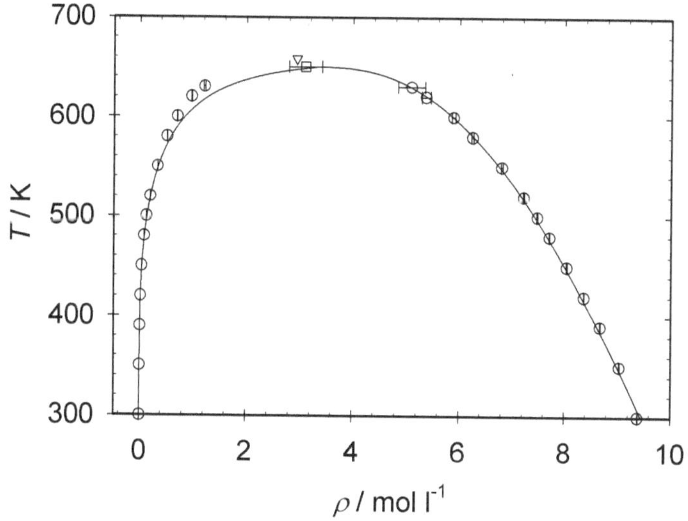

Fig. 2. Saturated densities of cyclohexanol: ○, present simulation data; –, experimental data [8]; ▽, critical point derived from simulated data; □, experimental critical point [8]

A subset of the parameters for the LJ sites, the point quadrupole and the point charges were optimized to fit correlations of experimental saturated liquid density and vapor pressure of pure cyclohexanol [8] in the range from 325 to 635 K. As a starting set for optimization, LJ parameters for the CH_2 site and electrostatic parameters for the point quadrupole were taken from a recently developed cyclohexane model. The remaining LJ parameters for the CH and OH group and the point charges were taken from Schnabel et al. [3]. During the present optimization, the LJ parameters of the CH_2 sites, the point charges and the point quadrupole were adjusted. The optimization followed the procedure presented by Stoll [9].

Vapor-liquid equilibria of the new cyclohexanol model are presented together with experimental data [8] in Figures 2 to 4 and in Table 3. The agreement between the molecular model and the experimental data is good. The mean unsigned errors in vapor pressure, bubble density and heat of va-

Table 3. Vapor-liquid equilibria of cyclohexanol: simulation results (sim) are compared to experimental data (DIPPR) [8] for vapor pressure, saturated densities and enthalpy of vaporization. The number in parentheses indicates the statistical uncertainty in the last digit

T K	p_{sim} MPa	p_{DIPPR} MPa	ρ'_{sim} mol/l	ρ'_{DIPPR} mol/l	ρ''_{sim} mol/l	$\Delta h^{\text{v}}_{\text{sim}}$ kJ/mol	$\Delta h^{\text{v}}_{\text{DIPPR}}$ kJ/mol
300	—	—	9.378(3)	9.440	—	—	—
350	0.002(4)	0.003	9.032(8)	9.000	0.00076(0)	54.06(9)	56.05
390	0.023(2)	0.022	8.674(6)	8.625	0.00704(0)	49.96(3)	51.16
420	0.062(3)	0.065	8.371(4)	8.328	0.01807(2)	46.71(3)	47.31
450	0.145(6)	0.159	8.043(6)	8.014	0.04062(6)	43.39(3)	43.27
480	0.31 (1)	0.331	7.71 (1)	7.677	0.0842 (2)	40.03(4)	39.01
500	0.47 (2)	0.504	7.48 (1)	7.439	0.1267 (3)	37.80(3)	36.02
550	1.23 (3)	1.202	6.80 (2)	6.762	0.335 (1)	31.42(4)	27.81
600	2.43 (2)	2.394	5.89 (3)	5.883	0.704 (4)	23.92(6)	17.88
620	3.11 (5)	3.046	5.37 (9)	5.396	0.972 (8)	19.78(7)	12.92

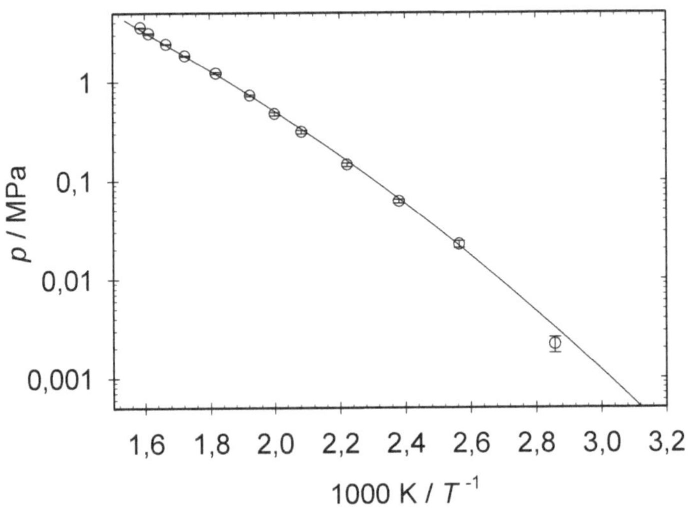

Fig. 3. Vapor pressure of cyclohexanol: ○, present simulation data; –, experimental data [8]

porization are 4.1, 0.4, and 13.4 %, respectively, in the temperature range from 325 to 635 K, which is about 50 to 97 % of the critical temperature. The seemingly high unsigned error in the heat of vaporization is due to the lack of experimental data for cyclohexanol. In fact, the only available experimental data for the heat of vaporization are at low temperatures, where the present molecular model shows a very good agreement, cf. Figure 4.

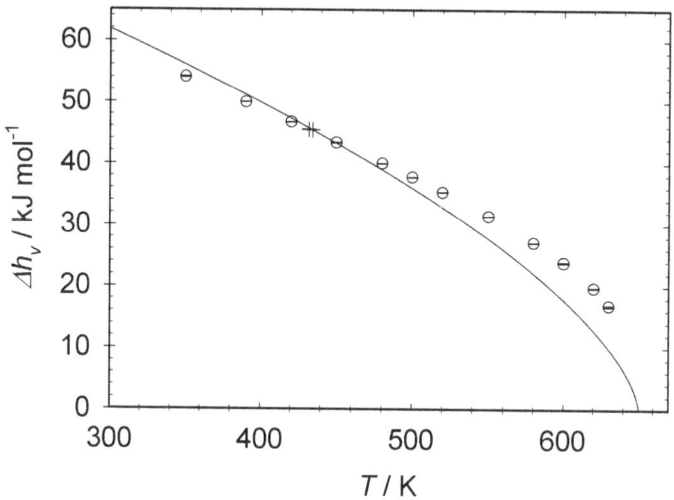

Fig. 4. Heat of vaporization of cyclohexanol: ○, present simulation data; –, predicted experimental data [8]; +, experimental data [8]

3 Transport Properties

Dynamic properties can be obtained from EMD simulations by means of the Green-Kubo equations [10, 11]. These equations are a direct relationship between a transport coefficient and the time integral of an autocorrelation function of a particular microscopic flux in a system at equilibrium. This method was used in this work to calculate the self-diffusion coefficient.

3.1 Diffusion Coefficient

The self-diffusion coefficient D_i is related to the mass flux of single molecules within a fluid. Therefore, the relevant Green-Kubo formula is based on the individual molecule velocity autocorrelation function as follows

$$D_i = \frac{1}{3N} \int_0^\infty \mathrm{d}t \, \left\langle \mathbf{v}_k(t) \cdot \mathbf{v}_k(0) \right\rangle, \tag{1}$$

where $\mathbf{v}_k(t)$ is the center of mass velocity vector of molecule k at some time t, and <...> denotes the ensemble average. Eq. (1) is an average over all N molecules in a simulation, since all contribute to the self-diffusion coefficient.

3.2 Simulation Results

The self-diffusion coefficient of pure methanol and ethanol was predicted at atmospheric pressure in the temperature range between 200 and 340 K. Present numerical data are given in Table 4. Figure 5 shows the self-diffusion coefficient for both alcohols as a function of temperature. The statistical error of the simulation data was estimated to be in the order of 1%.

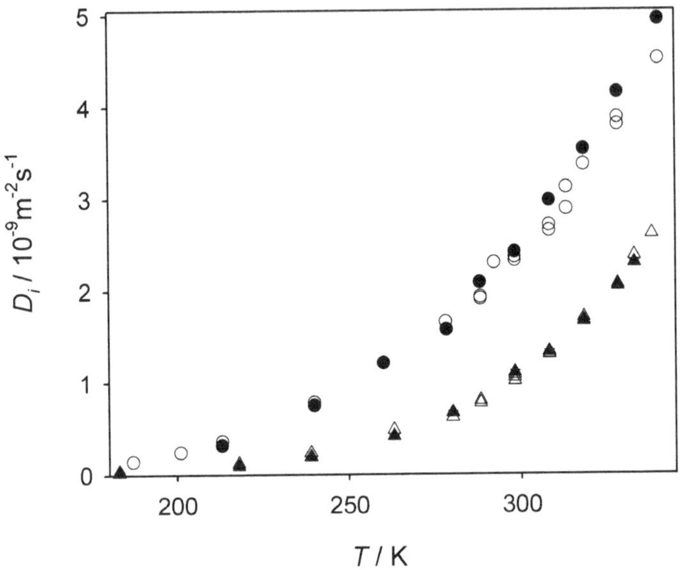

Fig. 5. Temperature dependence of the self-diffusion coefficient at 0.1 MPa. Present simulation results for methanol (•) and ethanol (▲) are compared to experimental data (○) [12, 13, 14, 15, 16] and (△) [12, 13, 14, 15, 17]. Simulation error bars are within symbol size

Table 4. Density and self-diffusion coefficient of pure liquid methanol and ethanol at 0.1 MPa from present molecular dynamics simulations. The number in parenthesis indicates the statistical uncertainty in the last digit

T K	ρ mol/l	D_i 10^{-9}m^2/s	T K	ρ mol/l	D_i 10^{-9}m^2/s
Methanol			Ethanol		
213	27.02	0.174(4)	239	18.11	0.192(6)
240	26.22	0.549(7)	263	17.74	0.415(9)
260	25.63	1.01 (1)	280	17.44	0.67 (1)
278.15	25.14	1.57 (1)	298.15	17.08	1.11 (2)
288	24.81	2.09 (2)	308.15	16.88	1.33 (2)
298.15	24.51	2.41 (2)	318.15	16.65	1.71 (1)
318.15	23.91	3.47 (2)	328.15	16.45	2.04 (2)
328.15	23.58	4.16 (2)	333	16.31	2.30 (2)
340.15	23.21	4.95 (3)	—	—	—

The self-diffusion coefficient of methanol shows a good agreement with the experimental values. Below ambient temperature, the mobility of the model molecules decreases more rapidly than in real methanol. Hence, the self-diffusion coefficient from simulation is underestimated by about 5%. At temperatures above 298 K the self-diffusion coefficient is overestimated by up to 12% at 340 K. However, the significant scatter of the experimental data should be noticed.

The self-diffusion coefficient of ethanol shows an excellent agreement with the experimental data. The predicted data also reproduce the temperature dependence correctly over the whole regarded temperature range from 239 to 333 K. Deviations are on average 5%.

Selected normalized velocity autocorrelation function (VACF) of methanol and ethanol are shown in Figure 6. In agreement with findings in the literature, VACF may have two minima at short times ($t < 1$ ps) [18]. At low temperatures and high densities, the VACF decrease rapidly and assume negative values. After this first minimum, the VACF increase slightly to drop again into a deeper minimum, for methanol, or to a higher minimum, in the case of ethanol. Beyond the second minimum, the VACF converge to zero following a characteristic path. The observed negative values of the VACF are related to backscattering collisions, also known as cage-effect, which governs short time dynamics. Michels and Trappeniers [19] attribute this phenomenon to the formation of "bound states" since this effect can only be observed for molecular models that include attractive forces.

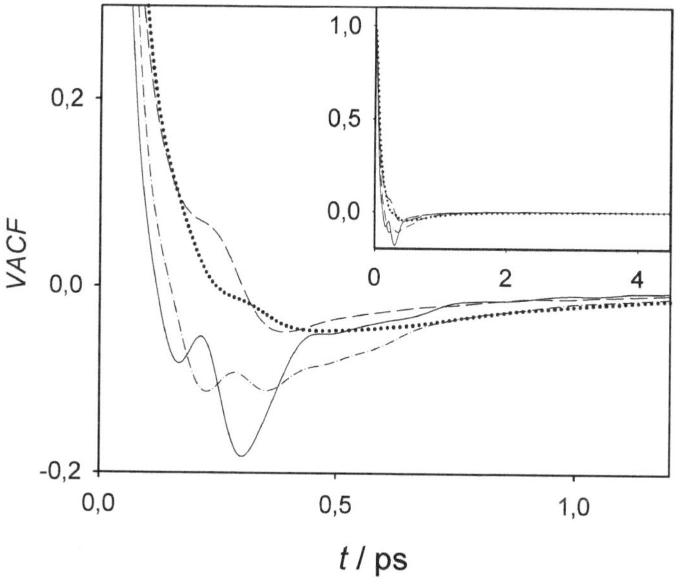

Fig. 6. Velocity autocorrelation functions at 0.1 MPa of methanol at 180 K (–) and 340 K (– –) as well as of ethanol at 173 K (– · –) and 333.15 K (· · ·)

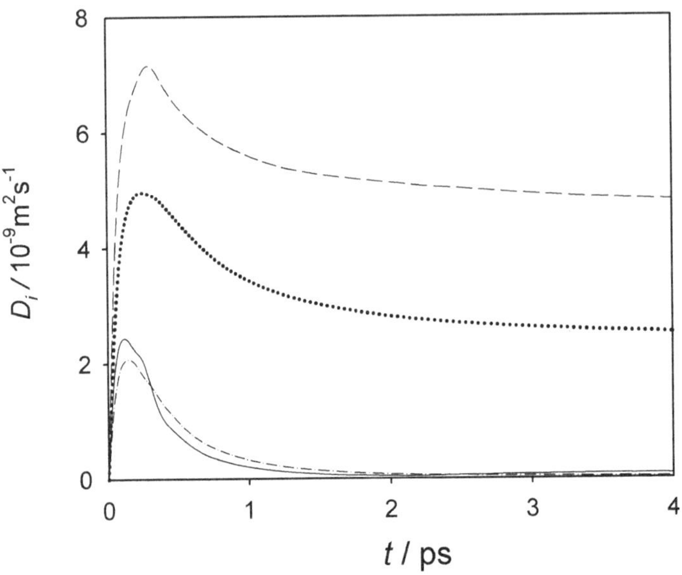

Fig. 7. Integral of the velocity autocorrelation functions at 0.1 MPa of methanol at 180 K (–) and 340 K (– –) as well as of ethanol at 173 K(– · –) and 333.15 K (· · ·)

For increasing temperatures at atmospheric pressure, where the density is lower, molecular collisions are less frequent and the depth of the minima is gradually reduced until they almost disappear, cf. Figure 6. As can be assumed, the VACF decay faster at lower temperatures, which is clearly visible in Figure 6. Moreover, the initial drop of the VACF of methanol occurs earlier than the corresponding VACF for ethanol. Thus, methanol VACF exhibit a more pronounced backscattering than those of ethanol, particularly at low temperatures approaching its melting temperature ($T_m = 175.37$ K) [20]. Note that the absolute minimum of the ethanol velocity autocorrelation function occurs in between the two minima of the methanol VACF.

Figure 7 shows the behavior of the integrals of the VACF given by Eq. (1). It can be observed that all VACF converge to their final value after less than 4 ps, i.e. no long time tail contributes significantly to the value of the self-diffusion coefficient at the regarded state points.

The changes of the self-diffusion coefficient of methanol and ethanol in that binary mixture were predicted at 298.15 K and 0.1 MPa for the entire composition range. Numerical results are presented in Table 5. Fig. 8 shows the self-diffusion coefficient of methanol and ethanol as a function of the methanol mole fraction in the mixture. As can be observed, there is a very good agreement between the present self-diffusion coefficients and the experimental data

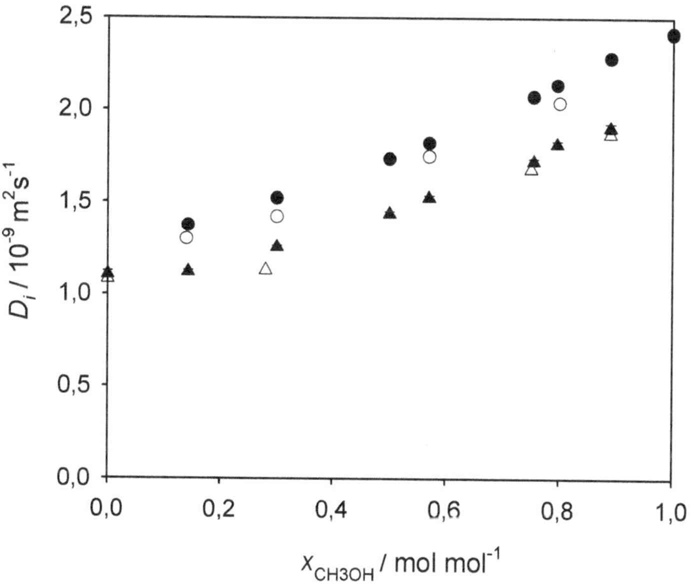

Fig. 8. Composition dependence of the self-diffusion coefficient for the mixture methanol + ethanol at 298.15 K and 0.1 MPa. Present simulation results for methanol (•) and ethanol (▲) are compared to experimental data (○) and (△) [21]

Table 5. Density and self-diffusion coefficient for the mixture methanol(1) + ethanol(2) at 298.15 K and 0.1 MPa from present molecular dynamics simulations. The number in parenthesis indicates the statistical uncertainty in the last digit

x_1 mol/mol	ρ 1/mol	D_1 $10^{-9}m^2/s$	D_2 $10^{-9}m^2/s$
0	27.02	—	1.11(2)
0.14	26.22	1.38(2)	1.12(1)
0.30	25.63	1.52(1)	1.25(1)
0.50	25.14	1.73(1)	1.44(1)
0.57	24.81	1.82(1)	1.53(1)
0.75	24.51	2.07(1)	1.72(2)
0.80	23.91	2.14(2)	1.82(2)
0.89	23.58	2.28(1)	1.90(2)
1	23.21	4.95(3)	—

for both alcohols. The present simulation results overestimate slightly the self-diffusion of both alcohols by 5% on average. Furthermore, the predicted data reproduces the composition dependence of both self-diffusion coefficients.

4 Computing Performance

All MD simulations were carried out on the NEC SX-8 with the MPI based molecular simulation program *ms*2 developed in our group. The parallelization of the MD part of *ms*2 is based on Plimpton's particle-based decomposition algorithm [22].

A typical MD simulation on the NEC SX-8 using one node with 8 processors is compared here to simulations on two other platforms available on our institute. Firstly, LEO, a workstation with 4 dual core Opteron 8216 processors, was considered and secondly, S11, a workstation with 2 Athlon MP 2800+ processors. In all cases, the same simulations were carried out with 864 methanol molecules in the liquid state over 100.000 time steps. It should be noted that for transport properties, as presented in section 3, typically 2.000.000 time steps have to be performed. The comparison is presented in Table 6. As can be assumed, both NEC SX-8 and LEO are much faster than the Athlon based workstation. Nevertheless, it can be confirmed that NEC SX-8 is clearly the most suited platform for MD.

Table 6. Run time on NEC SX-8, LEO and S11 simulating 864 methanol molecules over 100.000 time steps

System	run time minutes
NEC SX-8	30
LEO	55
S11	447

References

1. W.E. Nagel, W. Jäger, M. Resch: High Performance Computing in Science and Engineering '05. Springer, Berlin (2005).
2. T. Schnabel, J. Vrabec, H. Hasse: Henry's law constants of methane, nitrogen, oxygen and carbon dioxide in ethanol from 273 to 498 K. Fluid Phase Equilib., **233**, 134 (2005).
3. T. Schnabel, J. Vrabec, H. Hasse: Erratum to "Henry's law constants of methane, nitrogen, oxygen and carbon dioxide in ethanol from 273 to 498 K". Fluid Phase Equilib., **239**, 125 (2006).
4. T. Schnabel, M. Cortada, J. Vrabec, S. Lago, H. Hasse: Molecular Model for Formic Acid adjusted to Vapor-Liquid Equilibria. Chem. Phys. Lett., **435**, 268 (2007).
5. T. Schnabel, B. Eckl, Y.-L. Huang, J. Vrabec, H. Hasse: Molecular Modeling of Hydrogen Bonding Fluids: Formic Acid and Ethanol + R227ea. in W.E. Nagel,

W. Jäger, M. Resch, High Performance Computing in Science and Engineering '07. Springer, Berlin (2007).
6. D.K. Dysthe, A.H. Fuchs, B. Rousseau: Fluid Transport properties by equilibrium molecular dynamics. I. Methodology at extreme fluid states. J. Chem. Phys., **110**, 4047 (1999).
7. M.W. Schmidt, M.W. Baldridge, J.A. Boatz, et al.: General atomic and molecular electronic structure system. J. Comput. Chem., **14**, 1347 (1993).
8. DIPPR Project 801 - Full Version. Design Institute for Physical Property Data/AIChE, 2005.
9. J. Stoll: Molecular Models for the Prediction of Thermophysical Properties of Pure Fluids and Mixtures, Fortschritt-Berichte VDI, Reihe 3, 836, VDI Verlag, Düsseldorf, (2005).
10. M.S. Green: Markoff Random Processes and the Statistical Mechanics of Time-Dependant Phenomena 2. Irreversible Processes in Fluids. J. Chem. Phys., **22**, 398 (1954).
11. R. Kubo: Statistical-Mechanical Theory of Irreversible Processes I. General Theory and Simple Applications to Magnetic and Conduction Problems. J. Phys. Soc. Jpn., **12**, 570 (1957).
12. P.A. Johnson, A.L. Babb: Liquid Diffusion in Non-electrolytes. Chem. Rev., **56**, 587 (1956).
13. F.A.L. Dullien: Predictive Equations for Self-Diffusion in Liquids: a Different Approach. AIChE Journal, **18**, 62 (1972).
14. R.L. Hurle, A.J. Eastel, L.A. Woolf: Self-diffusion in monohydric Alcohols under Pressure. J. Chem. Soc. Far. Trans, **81**, 769 (1985).
15. N. Karger, T. Vardag, H.D. Lüdemann: Temperature dependende of self-diffusion in compressed monohydric alcohols. J. Chem. Phys., **93**, 3437 (1990).
16. N. Asahi, Y. Nakamura: Nuclear magnetic resonance and molecular dynamics study of methanol up to supercritical region. J. Chem. Phys., **109**, 9879 (1998).
17. S. Meckl, M.D. Zeidler: Self-diffusion measurements of ethanol and propanol, Mol. Phys., **63**, 85 (1988).
18. J. Alonso, F.J. Bermejo, M. Garcia-Hernandez, J.L. Martinez, W.S. Howells: H-Bond in methanol: a molecular dynamics study. J. Mol. Struc., **250**, 147 (1991).
19. P.J. Michels, N.J. Trappeniers: Molecular Dynamics calculations of the self-diffusion coefficient below the critical density, Chem. Phys. Lett., **33**, 195 (1975).
20. P. Sindzingre, M.L. Klein: A molecular dynamic study of methanol near the liquid-glass transition. J. Chem. Phys., **95**, 4681 (1992).
21. P.A. Johnson, A.L. Babb: Self-diffusion in liquids. I. Concentration dependence in ideal and non-ideal binary solutions, J. Chem. Phys., **60**, 14 (1956).
22. S. Plimpton: Fast Parallel Algorithms for Short-Range Molecular Dynamics. J. Comp. Phys., **117**, 1 (1994).

Investigation of Process-Specific Size Effects by 3D-FE-Simulations

H. Autenrieth[1], M. Weber[2], V. Schulze[1], and P. Gumbsch[2]

[1] KIT, Institute of Material Science and Engineering I (iwk I)
[2] KIT, Institute for Reliability of Components and Systems (izbs)

Summary. The miniaturization of cutting processes into the micrometer regime shows process-specific size effects like the nonlinear increase of the specific cutting force for decreasing cutting depth. In order to investigate these size effects, the mechanics of the material as well as the operation have to be investigated. A turning process was chosen to study the influence of process parameters like cutting depth h, cutting width b, cutting edge radii r, and cutting velocity v_c on the specific reaction force by 3D-Finite-Element-Simulations for normalized AISI 1045. For an adequate numeric reproduction of the material behavior, a physically based rate-dependent plasticity law was used in combination with a failure criterion describing the material damage and chip separation. The characteristics of the influences of the different parameters were analyzed mathematically precisely by similarity mechanics. The characteristics of the chip shapes determined by numerical simulations were compared with experimental results and a good correlation was found. The finite element simulations were executed on the high end mainframe XC4000 for a significant improvement of the run time of the simulations.

1 Introduction

A nonlinear scaling behavior of output quantities for a linear and self-similar scaling of a process is referred to as process-specific size effect. It is particularly important for the miniaturization and design of industrial cutting processes. To expand on the knowledge about size effects, an orthogonal micro turning process is investigated exemplarily, in which the nonlinear increase of the specific cutting force with a linearly decreasing cutting depth is a significant process-specific trait. This size effect is classically ascribed to the relative increase of friction energy or to the reduced probability for the existence of stress relaxing micro-structural defects in the shear zone [18]. Furthermore, hardening effects due to strain gradients have been investigated in [8], [15]. Previous studies [7] showed that strain rate hardening, as observed for many metals including steels, also yields to an increase of the specific cutting force during miniaturization. At higher cutting velocities, a further contribution to the occurring size effect is caused by the changing heat distribution in the workpiece, with increasing relevance for very small dimensions [13]. It was further shown by finite element simulations in [2] that a higher friction coefficient between the cutting tool and workpiece leads to an intensified size effect. While the size effect of the specific cutting force is widely discussed, the influence of geometrically similar scaling on the passive and feed force components was neglected in many studies. These quantities characterize the cutting tool deflection, which is important for the residual geometry of the workpiece and therefore of great importance for the optimization of the machining process. In this study, a systematic analysis of the influence of the geometrical dimensions of the cutting tool and the process parameters cutting depth and cutting velocity on the specific reaction forces, namely the specific cutting and passive force components, is performed. With respect to similarity mechanics, the geometrically similar scaling of the process is given, which allows for the identification of material inherent influences on size effects, e.g. strain rate hardening. For these investigations, 3D-Finite-Element-Simulations were performed using ABAQUS/Explicit. The investigated material was AISI 1045 in a normalized state. The investigated parameter ranges in the simulation were adapted to micro cutting experiments, described in [14], allowing for the final comparison of the plastic deformation fields and chip shapes.

2 Applied Methods and Simulation Model

2.1 Basic Equations and Simulation Model

The Finite Element Method (FEM) is a mathematical method for solving partial differential equations derived by the constitutive equations of continuum mechanics. Details can be found in [4], [22]. The equation system, obtained by the derivation, can be written as follows:

$$\frac{\partial \rho}{\partial t} = (\rho\, v_k)_{,k} = 0 \tag{1}$$

$$\sigma_{lk,l} + \rho\, b_k = 0 \tag{2}$$

$$\frac{\mathrm{d}\Pi^{\mathrm{i}}}{\mathrm{d}t} = \frac{1}{\rho}\,\sigma_{kl}\, D_{kl} - \frac{1}{\rho}\, c_{k,k} + z \tag{3}$$

In these equations (1-3), ρ is the material density, v is the velocity of the material point, σ denotes the stress tensor and b the sum of the volume forces per norm volume in the material point. Furthermore, Π^{i} is the specific inner energy, σD is the inner work performed by the inner stress, c denotes the heat flux vector, and z is the term which takes local heat sources into account. The differential equation system of the thermo-mechanically coupled FE simulation is solved by ABAQUS/Explicit by discretization of space and

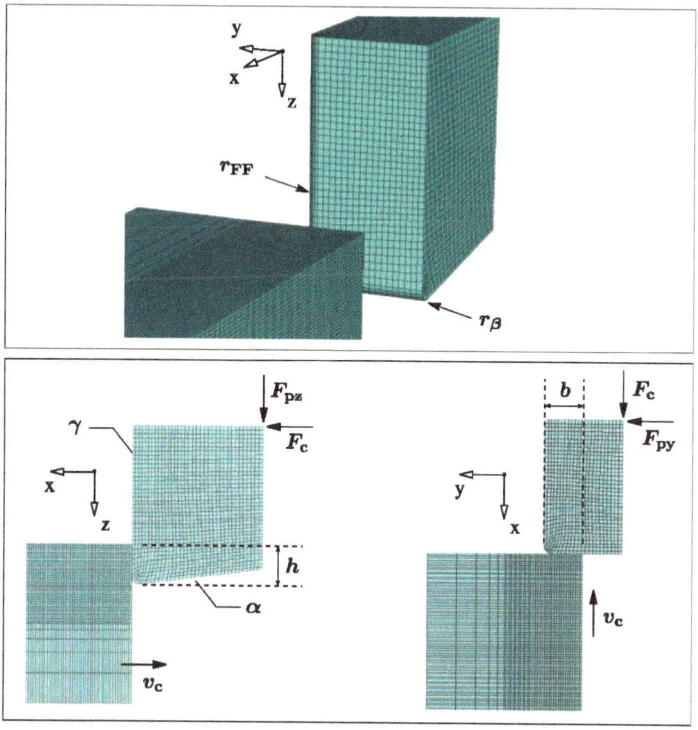

Fig. 1. 3D-Finite-Element-Simulation model with relevant process parameters and reaction forces

time. The yield stress has to be defined as the function of strain, strain rate and temperature which was done in the applied user subroutine VUMat.

The micro cutting process is three-dimensionally simulated. The applied simulation model is depicted in Figure 1 with the process-specific parameters. The workpiece is moved with the cutting velocity v_c, indicated in Figure 1, while the cutting tool is spatially fixed. The cutting depth h is defined as the constant feed of the cutting tool relative to the workpiece in z-direction while the constant feed in y-direction is denoted as cutting width b. The cutting tool is characterized by the rake angle γ, clearance angle α and the radii at the major and minor cutting edge, cutting edge radius r_β and flank face radius r_{FF}, respectively. A symmetric simulation model is used for the case of $h = b$ and $r_\beta = r_{FF}$ to reduce the number of degrees of freedom and therefore computation time. The discretization of the workpiece is homogeneous in front of the cutting tool with an element edge size of 3 μm for a cutting depth h of 50 μm, a cutting width b of 50 μm, and a cutting edge radius r_β of 10 μm. For variation of the cutting width b, the element edge length is scaled along b holding the ratio of the element edge length to the cutting width constant. In the workpiece, heat generation by plastic deformation and heat conduction is taken into consideration. The workpiece was meshed using a thermo-mechanically coupled element of a bi-linear type, whereas the cutting tool was meshed with discrete rigid elements.

2.2 Material Model

Due to the complex loading history the workpiece material has undergone during the machining process, and the mutual interacting physical parameters influencing the process, the modeling of the material behavior became complex. Basically, the flow stress of the material model consists of a thermal and an athermal contribution, σ^* and σ_{G}, respectively, shown schematically in Figure 2:

$$\sigma_{\mathrm{y}} = \underbrace{\left(\sigma_{\mathrm{G}_0} + \kappa\, \bar{\varepsilon}_{\mathrm{p}}^{r}\right) \cdot \frac{G(T)}{G(0\mathrm{K})}}_{\sigma_{\mathrm{G}}} \cdot g(T, T_{\mathrm{t}}) + \underbrace{\sigma_0^* \cdot \left(1 - \left(\frac{T}{T_0}\right)^n\right)^m}_{\sigma^*} \tag{4}$$

$$\text{with:} \quad T_0 = \frac{\Delta G_0}{k_{\mathrm{B}}} \ln\left(\frac{\dot{\varepsilon}_0}{\dot{\bar{\varepsilon}}_{\mathrm{p}}}\right) \tag{5}$$

The flow stress is described in dependence on temperature T, plastic strain rate $\dot{\bar{\varepsilon}}_{\mathrm{p}}$ and accumulated plastic strain $\bar{\varepsilon}_{\mathrm{p}}$. The effect of long-range stress fields, induced by dislocations, grain boundaries, precipitations and solute atoms on the mobility of dislocations, is taken into consideration by the athermal part of the flow stress σ_{G}. This part is proportional to the shear modulus G, which can be calculated from the Young's modulus E and the Poisson ratio ν. These temperature dependent elastic material parameters are given by Richter [17]:

$$E(T) = E(273\,\mathrm{K}) + e_1 \cdot (T - 273\,\mathrm{K}) + e_2 \cdot (T - 273\,\mathrm{K})^2 \tag{6}$$

$$\nu(T) = \nu(273\,\mathrm{K}) + \Delta\nu \cdot (T - 273\,\mathrm{K}) \tag{7}$$

The athermal flow stress for $\bar{\varepsilon}_{\mathrm{p}} = 0$ is denoted as σ_{G_0}. Strain hardening is modeled by the power law based on Ludwik [16] with the hardening coefficient κ and the hardening exponent r. The thermal part σ^* is based on the model of thermally activated dislocation motion. The model can be found in [20] and is similar to the theory proposed by Follansbee and Kocks [9] for copper. For temperatures $T > T_0(\dot{\bar{\varepsilon}}_{\mathrm{p}})$, the thermal contribution of flow stress is zero because small range obstacles are overcome by dislocations without additional mechanical stresses. The thermal part of the yield stress at zero Kelvin is denoted as σ_0^*, the free activation enthalpy necessary to overcome the decisive obstacles is ΔG_0. The critical strain rate is written as $\dot{\varepsilon}_0$ and the shape of the obstacles is described by m and n. The Boltzmann constant is denoted as k_{B}. Strong softening of the material at high temperature, induced by diffusion processes is observed and described in [5], and is realized by the function g in equation (8), which is applied to the athermal part of the flow stress σ_{G}.

$$g(T, T_t) = \begin{cases} 1 & \text{for } T \leq T_t \\ \left(1 - \left(\dfrac{T - T_t(\dot{\varepsilon}_p)}{T_{melt} - T_t(\dot{\bar{\varepsilon}}_p)}\right)^{\xi}\right)^{\zeta} & \text{for } T > T_t \end{cases} \tag{8}$$

$$\text{with:} \quad T_t(\dot{\bar{\varepsilon}}_{pl}) = \vartheta_0 + \Delta\vartheta \ln\left(1 + \frac{\dot{\bar{\varepsilon}}_p}{\dot{\varepsilon}_n}\right) \tag{9}$$

In equation (9) T_t denotes the transition temperature for the initiation of strong thermal softening. The transition temperature T_t can be adapted to the material by the parameters ϑ_0 and $\Delta\vartheta$. The slope of flow stress decrease is adjusted to experimental data by the parameters ξ and ζ. The melting temperature is denoted as T_{melt}.

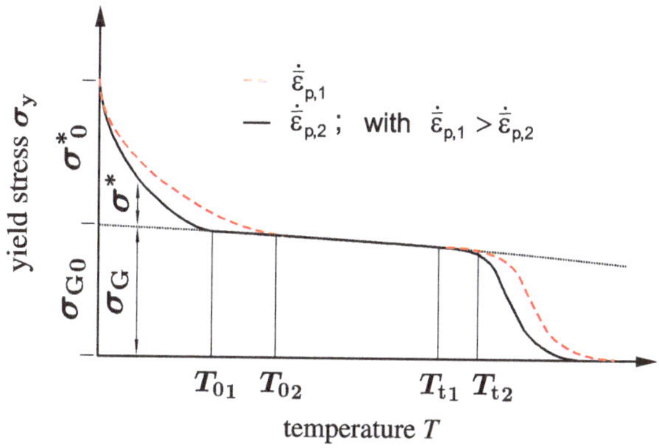

Fig. 2. Schematic dependence of the physically based material law for the yield stress σ_y on the temperature T and the plastic strain rate $\dot{\bar{\varepsilon}}_p$

The failure behavior of the material, induced by the evolution of the local material damage, is described by the phenomenological model proposed by Johnson and Cook [12], which describes the failure of ductile materials. The advantage of this method is the material inherent description of the damage evolution as a function of the local load history. Any finite element is deleted, when its damage parameter D reaches unity. The evolution of the damage parameter D is given by:

$$D(\tau) = \int_0^{\tau} \frac{\dot{\bar{\varepsilon}}_p(t)}{\varepsilon_f(t)} \, dt \tag{10}$$

with

$$\varepsilon_f = \left(d_1 + d_2 \exp\left(d_3 \frac{p}{\bar{\sigma}}\right)\right) \cdot \left(1 + d_4 \ln \frac{\dot{\bar{\varepsilon}}_p}{\dot{\varepsilon}_{jc}}\right) \cdot \left(1 + d_5 \frac{T - T_{room}}{T_{melt} - T_{room}}\right) \tag{11}$$

The hydrostatic pressure is defined as $p = tr(\sigma)/3$, while the von Mises equivalent stress is denoted as $\bar{\sigma}$. Further T_{melt} and T_{room} are the melting and room temperature, the coefficients $d_1, \ldots, d_5$ are dimensionless material parameters and $\dot{\varepsilon}_{\mathrm{jc}}$ is a characteristic strain rate introduced for dimensional reasons. The failure parameters for AISI 1045 were determined by tension-torsion tests for positive stress triaxialities $p/\bar{\sigma}$. The material parameters, used in the simulation for AISI 1045, are summarized in [21].

2.3 Similarity Mechanics

A process-specific size effect is defined as the nonlinearity of quantities of the process result with moderate change of the input parameters while scaling the process geometrically self-similarly and linearly. Therefore, similarity mechanics is the appropriate method for the identification of such nonlinear influence functions. As derived mathematically precisely by Buckingham and postulated in the π-theorem [6], any (influence) function preliminary named as f_i, which describes the influence of the physical input parameters x_i on output quantities, has to map the physical problem from the input space into the identical output space. Therefore, the basis dimensions of the parameter space have to be conserved during the mapping process, which leads to further implications as presented in [7], and finally to the derivation of the so-called similarity numbers ξ_j. These similarity numbers ξ_j are free of dimensions and allow for the identification of the mutual interaction of model parameters, as exemplarily shown in Table 1 for the process-specific parameters. Furthermore, they have to be kept constant in case of a self-similar geometrical process scaling while one parameter is varied for the identification of its influence function f_j. The identification of influence functions is done separately for each process parameter by finite element simulations. Finally, these functions are combined in the so-called product ansatz to obtain one closed formula for all of the separate influencing parameters.

Table 1. Similarity numbers for the investigated process parameters

Process Parameter	similarity number	Definition
Cutting depth h	Size Number Si	$Si = \frac{\dot{\varepsilon}_0 \cdot h}{\sqrt{E/\rho}}$
Cutting width b	Width Number Wi	$Wi = \frac{b}{h}$
Cutting edge radius r_β	Relative Sharpness Rs	$Rs = \frac{r_\beta}{h}$
Flank face radius r_{FF}	Relative Sharpness Rs_{FF}	$Rs_{\mathrm{FF}} = \frac{r_{\mathrm{FF}}}{b}$
Cutting velocity v_c	Cauchy Number Ca	$Ca = \frac{v_c}{\sqrt{E/\rho}}$

3 Simulation Results and Discussion

3.1 Specific Reaction Forces

For the investigation of the process the specific reaction forces k_c, k_{py} and k_{pz} acting in x-, y- or z-direction, respectively, are used. These output quantities are determined by dividing the reaction forces F_c, F_{py} and F_{pz}, depicted in Figure 1, by the uncut chip area. The forces are averaged under steady state conditions. The uncut chip area is defined by the product of the cutting depth h and the cutting width b. For the systematic identification of influence functions of the result quantities on the process parameters, standard conditions, indicated by "0", are defined in Table 2.

Table 2. Standard conditions for the simulation models

Cutting depth	$h_0 = 50\ \mu m$
Cutting width	$b_0 = 50\ \mu m$
Cutting edge radius	$r_{\beta 0} = 10\ \mu m$
Flank face radius	$r_{FF0} = 10\ \mu m$
Rake angle	$\gamma_0 = 0°$
Clearance angle	$\alpha_0 = 7°$
Cutting velocity	$v_{c0} = 200$ m/min
Initial workpiece temperature	$T_{i0} = 293$ K

The specific reaction forces for standard conditions k_{c0}, k_{py0} and k_{pz0} were determined to be 2200 N/mm^2, 620 N/mm^2 and 620 N/mm^2, respectively. The determined result quantities for a parameter variation are normalized by the results for standard conditions and are denoted as k_c/k_{c0}, k_{py}/k_{py0} and k_{pz}/k_{pz0} in the following. The dependences of the normalized specific reaction forces on the Si number, or the cutting depth h for geometrical similar scaling, are shown in Figure 3. For a decreasing cutting depth h , the normalized specific cutting force k_c/k_{c0} increases nonlinearly. The normalized specific passive forces k_{py}/k_{py0} and k_{pz}/k_{pz0} show a less pronounced, but also nonlinear dependence for smaller cutting depth h. The simulation results are approximated by an exponential equation. The nonlinear scaling behavior of k_c/k_{c0} can be explained by a nonlinear increase in the strain rate in the shear plane – which follows $\dot{\bar{\varepsilon}}_p \propto 1/h$ – at the major and minor cutting edge resulting in a higher thermal part of the flow stress σ^* [7]. This leads to a higher average shear flow stress $\bar{\tau}$ in the primary shear zones evolving at the major and minor cutting edge. The higher strain rate is explained by the smaller lateral dimension of the shear plane resulting from the miniaturization of the process. On the other hand, the temperature in front of the rake face decreases because the heat flux $\mathbf{q} = \lambda \cdot \partial T/\partial \mathbf{x}$ depends nonlinearly on the geometrical dimensions. This reduces the effect of temperature dependent softening and leads to an increase of flow stress in the secondary deformation zones at the rake face of the cutting tool at the major and minor cutting edge. Furthermore, the more homogeneous temperature distribution results in a decrease of the shear angle representing the orientation of the plane, where the maximum plastic strain rates are present, to the cutting velocity vector. The lower shear angle leads to a higher shear plane length causing an extension of the shear zone and therefore an increase of the specific cutting force. The occurring size effect is reduced by the lower failure strain at the point of separation. The lower failure strain is caused by higher strain rate and lower temperature in front of the cutting tool for lower cutting depth h. For a similar stress state in front of the cutting tool, the maximum plastic work up to failure is reduced. This causes a decrease of the specific cutting force for decreasing cutting depth h. The increase of the normalized deflection forces k_{py}/k_{py0} and k_{pz}/k_{pz0} for small cutting depths may be explained by the higher ploughing forces induced by the higher material strength in front of the cutting edge radii.

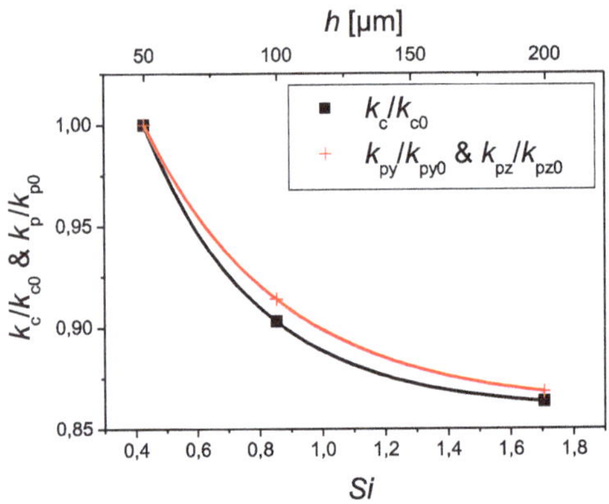

Fig. 3. Dependences of the normalized reaction forces on the Si number or the cutting depth h respectively

The dependences of the normalized specific reaction forces on the Wi number are shown in Figure 4. The reaction forces decrease with increasing cutting width b. The normalized specific cutting force can be described by a linear influence function while the specific passive forces are approximated by power laws for the investigated parameter range. The normalized specific cutting force k_c/k_{c0}, representing the energy per machined volume, can be separated into the energy needed for the machining process along the major and the minor cutting edge. The specific energy per machined volume stays constant along the major cutting edge while increasing the cutting width b, as discussed for the cutting depth variation, which results in a decrease of the plastic strain rate and higher temperature in the shear zones at the minor cutting edge. This causes a lower strength of the material at the minor cutting edge, reducing the ploughing force in y-direction, characterized by k_{py}/k_{py0}; further leading to a decrease of k_c/k_{c0}. As the cutting conditions do not change along the cutting width b for the Wi variation, k_{pz}/k_{pz0} should stay constant. This cannot be verified by the simulation results. The decrease found is referred to the changing chip geometry for higher cutting width b and the low actual value of k_{pz} but has to be investigated in more detail.

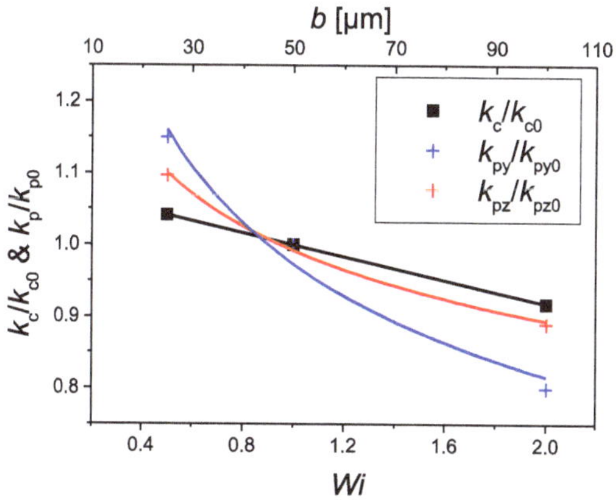

Fig. 4. Dependences of the normalized reaction forces on the Wi number or the cutting width b respectively

There is a significant influence by the edge radii of the tool on the specific reaction forces for micro cutting as the radii r_β and r_{FF} become of the same order as the cutting depth h and the cutting width b. The dependences of the reaction forces on the relative sharpness $Rs = r_\beta/h$ are depicted in Figure 5 for $0.2 \leq Rs \leq 1$. The dependences of the normalized specific reaction on Rs are described by linear influence functions. The slope for k_c/k_{c0} is negative while k_{pz}/k_{pz0} increases for increasing Rs and k_{py}/k_{py0} stays constant. The decrease of the specific cutting force k_c/k_{c0} for higher Rs conflicts with simulation results under plain strain conditions [2] where k_c/k_{c0} increases linearly for increasing Rs. This was ascribed to the more pronounced ploughing process [1]. The decrease of k_c/k_{c0} found here is explained by the lateral squeezing of the material in y-direction for higher Rs, where stress free conditions are present. The strong increase of k_{pz}/k_{pz0} for higher Rs can be explained by the higher amount of normal forces, acting in z-direction. As the edge radius r_{FF}, at the flank face, is held constant, there is no significant change of the ploughing process at the minor cutting edge and therefore k_{py}/k_{py0} stays nearly constant for the Rs variation. The determined parameters for the influence function of the cutting edge radius at the minor cutting edge r_{FF} on k_{py}/k_{py0} are the same as the parameters for the r_β variation on k_{pz}/k_{pz0} because of symmetrical reasons.

The dependences of the normalized specific reaction forces on the Ca number, or the cutting velocity v_c, are depicted in Figure 6. The simulation results are approximated by a power law. The increase of the normalized specific cut-

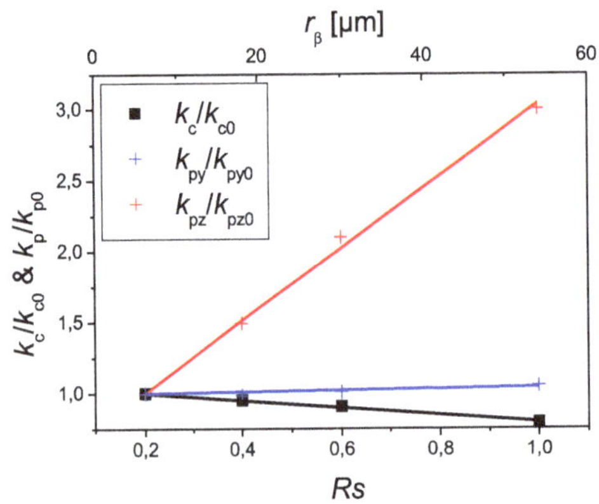

Fig. 5. Dependence of the normalized specific reaction forces on the relative sharpness Rs

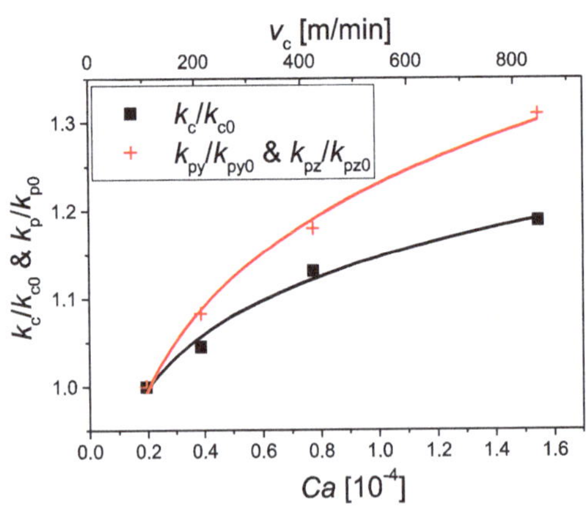

Fig. 6. Dependences of the normalized reaction forces on the Ca number or the cutting velocity v_c respectively

ting force k_c/k_{c0} with increasing cutting velocity can be explained by a higher thermal flow stress σ^* because of the growing plastic strain rate in the shear plane [7]. The effect of strain rate hardening of the material dominates the material softening caused by the more localized temperature distribution in the shear zones for higher cutting velocities v_c. In earlier simulations [21], an important result was the minimum in the cutting forces, which was found for a material-specific characteristic cutting velocity $v_{c,\,spec} \leq 100$ m/min. Due to the experimentally chosen process parameters, in this simulation model the range of cutting velocity is above this characteristic velocity $v_{c,\,spec}$ and therefore it cannot show the characteristic minimum in the specific forces. The higher strength of the material in front of the cutting edges, caused by strain rate hardening, leads also to an increase of the specific deflection forces k_{py}/k_{py0} and k_{pz}/k_{pz0} for higher cutting velocities v_c. The stronger nonlinear increase of k_p/k_{p0} compared with the increase of k_c/k_{c0} is caused by the lower reference value of k_{p0}.

Finally, a product ansatz is used to describe the dependences of the normalized specific reaction forces on the similarity numbers Si, Wi, Rs and Ca, in a closed formulation. The equation for the specific cutting forces follows:

$$k_c/k_{c0} = f_c(Si) \cdot f_c(Wi) \cdot f_c(Rs) \cdot f_c(Ca) \tag{12}$$

The validity of this approach was shown in [21] for the cutting depth h and the relative sharpness Rs. The approximation functions, describing the effect of different similarity numbers, or process parameters, on the result quantity, are given by:

$$f_c(Si) = a_c + b_c \exp(c_c \cdot Si) \tag{13}$$

$$f_c(Wi) = 1 + d_c \cdot (Wi - 1) \tag{14}$$

$$f_c(Rs) = f_c(Rs_{FF}) = 1 + e_c \cdot (Rs - 0.2) \tag{15}$$

$$f_c(Ca) = g_c \cdot Ca^{h_c} \tag{16}$$

The specific passive forces can be described analogously. A validation of the simulation model by comparing the specific reaction forces from simulation and experiment will be provided in the near future as the number of performed experiments increases.

3.2 Chip Shape

The chip shapes' simulation results and experimental measurements will be compared to verify the simulation and material model. There exist two main characteristics of the chips. On one hand, there is tearing that occurs periodically and can be seen at the border of the chip in experiments and simulations, as shown in Figure 7. This tearing is formed in the simulation by material failure caused by high accumulated plastic strains at both cutting edges. The deletion of an element leads to a higher deformation degree close to this position resulting in the progress of element deletion in the surrounding region. This mechanism may be of special importance for the burr formation at the newly generated edge. The correlation between chip shape and burr formation will be investigated in more detail in the future.

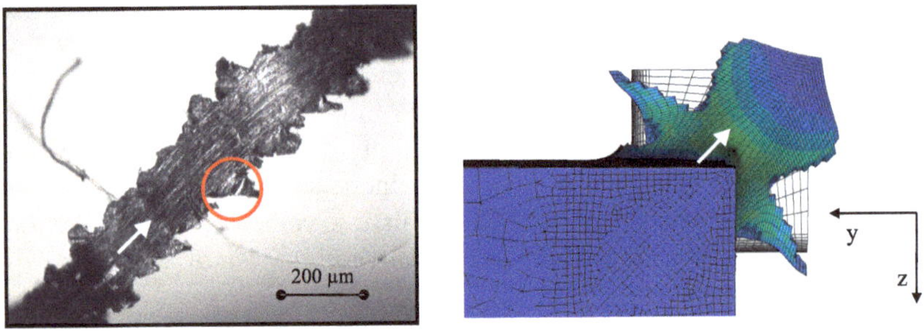

Fig. 7. Lower side of the chip for experiment (left) and simulation (right) for similar process conditions

On the other hand, another type of chip damage can be found at the upper side of the chip in the simulation and the experiment as depicted in Figure 8. In literature, several reasons are given for the formation of segmented chips. One, there is thermal dependent softening of the material, caused by quasi adiabatic conditions in the shear plane [3], and two, decreasing flow stress caused by the high plastic strains, induced by ductile damage [19]. In [10], chip segmentation was modeled in finite element simulation by using the Johnson-Cook failure criterion. The deletion of elements therein leads to failure paths along the shear plane for the hard machining process similar to the results presented herein.

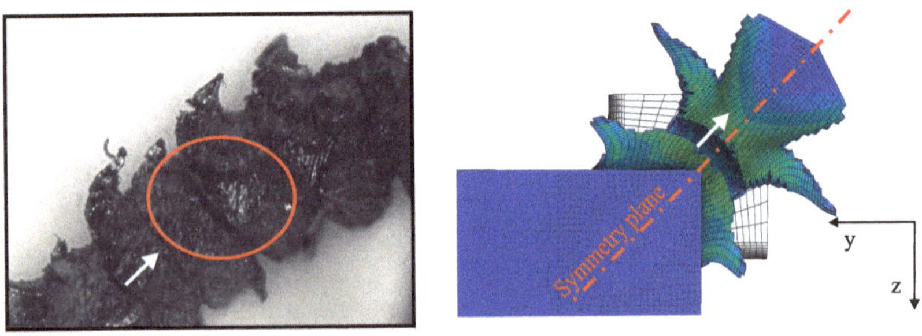

Fig. 8. Upper side of the chip for experiment (left) and simulation (right) for similar process conditions

In the simulations presented here, chip segmentation is induced by material failure beginning at the upper side of the chip at an equivalent plastic strain of about 1.6. The initiation of material failure is depicted in Figure 9 (top left) in the symmetry plane introduced in Figure 8 (right). The deletion of the element at the surface acts as the starting point of the failure path depicted in Figure 8 at different time steps (I-IV). This mechanism leads to a complete separation of the chip and the workpiece.

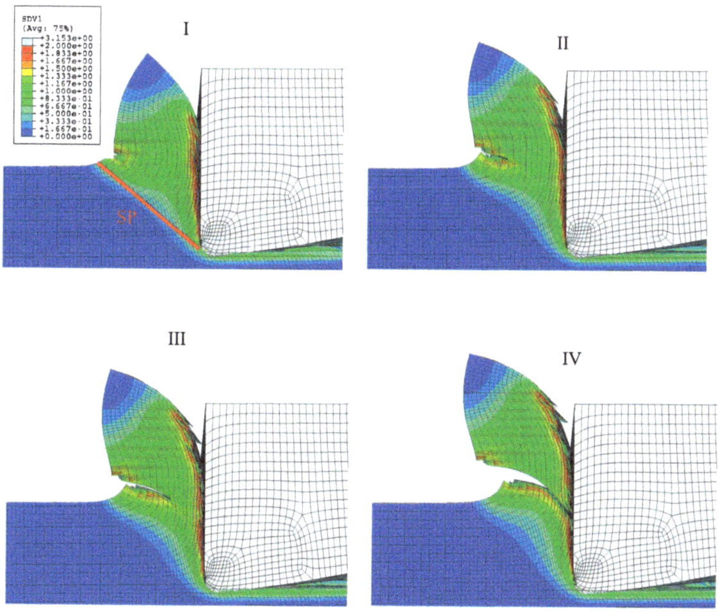

Fig. 9. Temporal development of the failure path for the chip formation process in the symmetry plane of the micro cutting process; SDV1 corresponds to the accumulated plastic strain

The development of a failure path is caused by the higher plastic deformation and a changed stress state close to the position of a deleted element. The starting point of the failure path lies above the shear plane (SP), marked in Figure 9 (top left). The initiation of failure is caused by the deformation history and therefore by the accumulated local material damage. It can further be seen in Figure 8 for simulation and experiment that the regions for tearing and segmentation at the upper side of the chip are close and therefore the tearing is assumed to be coupled with the segmentation mechanism. A complete separation of chip and workpiece cannot be found in the experiments in contrast to simulation, so the material's failure behavior has to be evaluated for negative stress triaxialities, appearing at the upper side of the chip. If the failure strain is underestimated for the appearing stress state, the chip segmentation will be significantly influenced. We are also concerned with the discretization of the workpiece, as chip segmentation is significantly influenced by the mesh density and mesh orientation [11]. These aspects will be part of further investigations of the chip formation process.

4 Computational Requirements and Computing Time

The computational cost for the 3D simulation of the micro cutting process by the Finite Element Method is quite high. This is caused by the fine discretization of the workpiece leading to a low critical time increment, which depends on the element edge length, for the explicit algorithm and a high number of degrees of freedom. Furthermore, the heat equilibrium equation has to be solved, in parallel to the stress equilibrium equation, what therefore takes into account the mutual influences of the temperature distribution with the local flow stress. Finally, the strongly nonlinear material behavior also leads to high computational requirements. For comparison, two methods of calculations are offered by ABAQUS, namely the "Domain Decomposition" and "Loop" algorithms. For the parallelization of the simulation, the "Loop" algorithm is used as for the "Domain Decomposition" algorithm one big partition is created in the workpiece, consisting of all elements which may get into contact with the cutting tool, resulting in no performance benefit by parallelization. The maximum number of CPUs for the parallelization is restricted to four for the "Loop" algorithm in ABAQUS. The dependence of computational time up to a definite simulation time on the number of parallel CPUs in use is summarized in Table 3. The computational time can be significantly decreased by the parallelization of the micro cutting process and shows the need for an account on the xc4000. The computational time for the model for standard conditions is about nine days for four CPUs to get steady state conditions which can be evaluated. The computation time can be reduced to six days by applying the symmetrical model. The maximum computation time for one simulation is restricted to 4,320 minutes so the simulations have to be re-started.

Table 3. Dependences of the computational time on the number of parallel CPUs in use up to a definite simulation time

Number of CPUs	1	2	4	8
Computational time [days]	19	12.3	9	–

5 Summary and Conclusion

For a detailed investigation of size effects occurring in micro cutting, 3D-Finite-Element-Simulations were performed with ABAQUS/Explicit. For a realistic material response, a physically based material model was implemented in a user subroutine VUMat. Similarity mechanics were used for the description of the micro cutting process. The dependences of the normalized specific reaction forces on the process parameters cutting depth, cutting width, cutting edge radius, and cutting velocity, were discussed. The nonlinear increase of the specific cutting and passive forces for decreasing cutting depths, named as size effect, can be assigned to the strain rate hardening in the shear zones and to changing heat distribution in the workpiece. Finally, a product ansatz is presented for the description of the dependences of the specific reaction forces on the process parameters. For the verification of the simulation model, the characteristic shape of the chip is compared with experiments and shows a good correlation. This can be attributed to the fact that the failure behavior of the material is taken into consideration. It was shown that computation time can be significantly decreased by parallelization of the simulation justifying the account on the XC4000.

Acknowledgement

The authors gratefully acknowledge the support of the Deutsche Forschungsgemeinschaft in the DFG Priority Program Process Scaling (SPP 1138). The authors greatly appreciate the support and supply of computational time on the high performance computer XC4000.

References

1. P. Albrecht. New development in the Theory of the Metal-Cutting Process Part I. The ploughing Process in Metal Cutting. *Journal of Engineering for Industry*, pages 348–358, 1960.
2. H. Autenrieth, M. Weber, J. Kotschenreuther, V. Schulze, D. Löhe, P. Gumbsch, and J. Fleischer. Influence of friction and process parameters on the specific cutting force and surface characterisitcs in micro cutting. *Proceedings of the 10th CIRP International Workshop on Modeling of Machining Operations*, pages 539–546, 2007.
3. M. Bäker. An investigation of the Chip Segmentation Process Using Finite Elements. *Technische Mechanik*, 23:1–9, 2003.
4. K.J. Bathe. *Finite-Elemente-Methoden.* Springer, 1990.
5. Frank Biesinger. *Experimentelle und numerische Untersuchung zur Randschichtausbildung und Spanbildung beim Hochgeschwindigkeitsfräsen von CK45.* PhD thesis, Universität Karlsruhe (TH), 2005.
6. E. Buckingham. On physically similar systems; illustrations of the use of dimensional equations. *Phys. Rev.*, 4:345–376, 1914.

7. L. Delonnoy, T. Hochrainer, V. Schulze, D. Löhe, and P. Gumbsch. Similarity considerations on the simulation of turning processes of steels. *Zeitschrift für Metallkunde*, 96:761–769, 2005.
8. D. Dinesh, S. Swaminathan, S. Chandrasekar, and T.N. Farris. An intrinsic size-effect in machining due to the strain gradient. *ASME/MED-IMECE*, 12:197–204, 2001.
9. P.S. Follansbee and U.F. Kocks. A constitutive description of the deformation of copper based on the use of mechanical threshold stress as an internal state variable. *Acta Metallurgica*, 36:81–93, 1988.
10. Y.B. Guo and D.W. Yen. A FEM study on mechanisms of discontinuous chip formation in hard machining. *Journal of Materials Processing Technology*, 155-156:1350–1356, 2004.
11. C. Hortig and B. Svendsen. Simulation of chip formation during high-speed cutting. *Journal of Materials Processing Technology*, 186:66–76, 2007.
12. G.R. Johnson and W.H. Cook. Fracture characteristics of three metals subjected to various strains, strain rates, temperatures and pressures. *Engineering Fracture Mechanics*, 21 (1):31–48, 1985.
13. E.M. Kopalinsky and P.L.B. Oxley. Size effects in metal removal processes. *Institute of Physics Conference Series*, pages 389–396, 1984.
14. J. Kotschenreuther, L. Delonnoy, T. Hochrainer, M. Weber, J. Schmidt, J. Fleischer, V. Schulze, D. Löhe, and P. Gumbsch. Modellierung und experimentelle Untersuchungen zu Gröseneffekten beim Stirndrehen von 90MnCrV8 im vergüteten Zustand. *Strahltechnik*, 27:219–240, 2005.
15. K. Liu. *Process Modeling of Micro-Cutting including strain gradient effects*. PhD thesis, Georgie Institute of Technology, 2005.
16. P. Ludwik. *Elemente der technologischen Mechanik*. Springer Verlag, Berlin, 1909.
17. F. Richter. *Physikalische Eigenschaften von Stählen und ihre Temperaturabhängigkeit*. Verlag Stahleisen M.B.H., Düsseldorf, 1983.
18. M. Shaw. The size effect in metal cutting. *Sadhana*, 28:875–896, 2003.
19. R. Sievert, H.-D. Noack, A. Hamann, P. Löwe, K.N. Singh, G. Künecke, R. Clos, U. Schreppel, P. Veit, E. Uhlmann, and R. Zettier. Simulation der Spansegmentierung beim Hochgeschwindigkeitszerspanen unter Berücksichtigung duktiler Schädigung. *Technische Mechanik*, 23:216–233, 2003.
20. O. Vöhringer. Temperatur- und Geschwindigkeitsabhängigkeit der Streckgrenze von Kupferlegierungen. *Zeitschrift für Metallkunde*, 65:32–36, 1974.
21. M. Weber, T. Hochrainer, H. Autenrieth, L. Delonnoy, J. Kotschenreuther, P. Gumbsch, V. Schulze, D. Löhe, and J. Fleischer. Investigation of Size-Effects in Machining with Geometrically Defined Cutting Edges. *Journal of Machining Science and Technology*, 11:447–473, 2007.
22. O.C. Zienkiewicz. *Methode der finiten Elemente*. Carl Hanser Verlag München Wien, 1984.

Andean Orogeny and Plate Generation

Uwe Walzer, Roland Hendel, Christoph Köstler, and Jonas Kley

Institut für Geowissenschaften, Friedrich-Schiller-Universität, Burgweg 11, 07749 Jena, Germany, u.walzer@uni-jena.de

Summary. We present the basic conception of a new fluid-dynamic and geodynamic project on the Andean orogeny. We start with a kinematic analysis of the entire orogeny and test different numerical options to explain these systematized observations by a physical model. Therefore we consider partly kinematic, partly dynamic regional models as well as purely dynamic models. Because of stochastic effects which are unavoidable in purely fluid-mechanical mechanisms of this kind and which influence the specific form of the Andes and because of the, to a large extend, unknown initial conditions, the partly kinematic, partly dynamic models have their right to exist. A purely dynamic model would be, of course, much more satisfactory. Therefore we want to approach nearer to the purely dynamic models prescribing a less number of parameters and dropping some artificial constraints. We have a concept to embed a regional model into a global spherical-shell model to determine the boundary conditions of the regional model as a function of time. So we avoid the artificially simplified boundary conditions of some published models of the Andean mechanism. On the other hand, the regional model has to retroact upon the global surrounding model. So, we have an iteration concept. For the two mentioned reasons there are, analogously to the two kinds of regional models, also two kinds of spherical-shell convection models, namely circulation models and forward models. As a first step, we present a spherical-shell model of mantle convection with thermal evolution and generation of continents and, as a complement, the depleted mantle reservoir. Our presented numerical result is that plate tectonics occurs only if at least the lithosphere deviates from purely viscous rheology and if there is a low-viscosity layer beneath of it. We suppose especially a viscoplastic yield stress for the lithosphere and a mainly temperature-independent asthenosphere which is determined, e. g., by the intersection points of water abundance and water solubility curves. The number of plates, at a certain fixed time of evolution, depends on Rayleigh number and, to a minor degree, on yield stress. We discuss our new efforts to improve the basic code Terra. The numerical regional Andean model has to be embedded into a global circulation model. Therefore we need an improved Terra for the latter one.

1 Introduction

A physical process is thought to be clearly understood if the numerical model succeeded to reproduce the essential features of this process where the model is based on the solution of the balance equations, if this solution is stable in a certain range of parameters and if the parts of a more complex model are understood separately.
a) The definition of the boundaries of a natural system is often very difficult. The same applies for the specification of the *boundary conditions*. The Earth as a whole has natural boundaries. The matter transfer from and to space is rather small for the geological time except for the accretion period at the beginning. The energy exchange is essentially known and can be described by the boundary conditions. If, however, a system, as e.g. the Andes, has no natural boundaries we have to take into account that the artificially supposed boundaries have boundary conditions that are neither temporally constant nor known at all.
b) For regional models neither the beginning age nor the *initial conditions* are known.
c) The Earth's crust and mantle are polycrystalline solids but the internal heating generates solid-state convection, the mathematical description of which is fluid dynamics. So, *stochastic processes* are unavoidable so that a specific final configuration cannot be forecasted in a fully deterministic manner.
d) If we set us the task of designing a numerical model of the evolution of a specific mountain range, e.g. the Andes, then we also have to *systematize* a multitude of *observations*. Of course, we should expect that only some essential observational features can appear in the model. We want to give some examples for such kind of questions. What is the essential condition for the appearance of orogenesis with its stock-work tectonics? How to explain *episodic* orogenesis by *continuous* subduction? What induces the eastward migration of crustal shortening?
e) We search for a numerical model of the *generation of plate tectonics* taking into account the effects of the endogenic water cycle and want to present first results. The connection with the Andean model is produced by the idea to determine the boundary and initial conditions of an *embedded regional model* by a global spherical-shell model.

2 Observations and Conceptions of Modeling of the Andean Orogeny and Surrounding Circulation Models of the Earth's Mantle

This Section is divided into five parts:
a) geological description of dynamic problems of the Andean orogeny
b) models which are partially kinematic and partially dynamic. In this kind

of models, essential features are prescribed in order to gain a large adaptation to geological and geophysical observations (tectonic movements, magmatism, seismic fault-plane solutions etc).
c) geochemical models of growth and differentiation of continents which do not contain any dynamic modeling
d) self-consistent dynamic models of the subduction process to understand the physical mechanism behind subduction
e) fully dynamic circulation models.
We intent to use such a global dynamic model to define the time-dependent boundary conditions of an embedded regional dynamic model of Andean subduction and orogeny.

Up to now, the model types b) to e) are only loosely connected. A principal aim of this paper is to search for a better integration of these model types and to better understand b) and possibly d).

a) *Conception.* The problems of this subsection are specified in items.

- It would be important to understand why a plateau-type orogen formed between a purely oceanic lithospheric plate (Nazca plate) and a continent (South America) [53]. As a rule, such elevated plateaus are formed by underthrusting a continental plate beneath another one. E. g., the plateaus of Tibet and Iran have been created in such a way. During the Cenozoic, however, the Altiplano and the Puna Plateau developed as 4 km high plateaus during uninterrupted subduction of the Nazca plate.
- It would be necessary to explain why, simplified expressed, except in the western Altiplano, the deformation starts in the West and migrates to and finishes in the East [53]. Simultaneously also the volcanism moves eastward [71]. There is no time lapse between onset of magmatism and onset of shortening in the central Andes [42]. Fig. 2.3 of [71] shows a pronounced eastward migration of arc volcanism but only for the segment between 20° S and 28° S. In the segment between 14° S and 20° S, the volcanism starts only at $\tau = 25..30$ Ma and is broadly distributed, so there is no distinct migration. The age is denoted by τ. Fig. 2.7 of [71] demonstrates the strong increase of arc volcanism and of ignimbrites as a function of time starting at $\tau = 30$ Ma for the segment between 14° S and 28° S. This phenomenon could be connected with the 30 Ma Africa-Eurasia collision [62]. In the early period of the Andean orogeny, phases of high convergence rates as the Incaic and Quechua phases coincide with phases of tectonic shortening in the overriding plate. For the time span $\tau = 25$ Ma .. 0 Ma, however, plate convergence and Andean strain rates are decoupled [42]. This seems to be a further hint toward a connection with the 30 Ma Africa-Eurasia collision. In Section 3, we propose an investigation on fluid-dynamic mechanisms which could possibly explain phenomena of that kind.
- In the Andes, there are two two-sided orogens, the Western and Eastern Cordilleras, the wings of which are differently strong developed. The Altiplano with a low degree of deformation is situated between them. It is

remarkable that the Altiplano has a high heat flow. This is in contradiction to the hypothesis that the continental lithosphere is extraordinarily thick in this region since the thermal lattice conductivity would allow only a less efficient heat transfer. Other attempts of explanation are to be found in paragraph b).

- In spite of more than 200 Ma continuous subduction of the Nazca plate [51], the high topography did not begin to grow earlier than an age of about 35 Ma [42]. During the early stages of orogeny, the strain-rate variations in the orogen reflect changes in plate convergence rate. But for ages smaller than 20 Ma, the two rates are obviously decoupled [42]. In this connection we should think about which process has thinned the South American lithosphere. Geological data show that the central Andean plateau was at most at half of its present-day height when the eastern marginal thrust belt began to grow [42]. It would be desirable if we could explain these observations by a physical model.
- Also in the Andes, mountain building has an episodic nature and is spatially limited in spite of essentially continuous subduction taking place simultaneously everywhere at the 7500 km long front of the Andes [42]. This is not really explained up to now. We would like to evaluate the different attempts of explanation possibly in search of an appropriate secondary mechanism.
- The decreasing convergence rate is *not* mimicked by the Andean strain rates that even increased, especially in the Eastern Cordillera and the Subandean belt [39, 42, 53]. Although the Nazca-South American convergence reached its climax at an age of 25-20 Ma and decreased after this time, the shortening rates rise considerably, especially in the Subandean belt, so that they have their highest values between 10 and 0 Ma of age. It is necessary to explain this delay.
- Also the following problem has not been understood from a physical point of view. The Nazca plate has rather small dip angles for latitudes between 2° S and 15° S as well as between 27° S and 33° S. But the active volcanism is mainly restricted to the segment in between with a more steeply dipping subduction zone [1, 37, 42, 61]. This and the arc shape is probably connected with the observation that the shortening is stronger in the center than on the wings [31, 36, 40]: The shortening rate of the central Andes is 1 - 1.75 cm/a, that of the southern Andes is 0 - 0.5 cm/a [47].
- A further open question is whether the lower crust of the central Andes has a felsic composition [42, 79]. This has probably a connection with the question of an initial weakening of the central base and with the anomalously high mantle heat flow of the plateau region (Altiplano) [42]. In the case of an affirmative answer we should check the models of Babeyko et al. [2, 3] and Sobolev et al. [63] in order to avoid their artificial boundary conditions. This idea would probably lead to a two-stage evolution of the central Andean plateau.

The hitherto enumerated questions seem to be relevant but only a few of them could be resolved in the project since more basic questions of the 3D subduction mechanics are unresolved yet (see d)).

- Some authors consider the global asymmetry of subduction zones as a problem: Only the western Pacific margins and the northern and southern Antilles have present-day extensional basins. Attempted explanations [12, 60] seem to be unconvincing from a physical point of view. Furthermore, a switch from backarc extension to backarc contraction in mid-Cretaceous was observed [42].

b) *Partially dynamic models.* Medvedev et al. [47, 48] propose a thin-sheet numerical model for the deformation of the central and southern Andes. They assume that the continental and oceanic crust is a much stronger viscous fluid than the rocks in an assumed slanting subduction channel. They invented this channel as a thin sliding layer above the subducting plate. The model ignores deformation in the continent and in the oceanic crust and assumes that the stresses associated with deformation have the deforming effect in the subduction channel and in the crust between ocean and Brazilian shield only. Although the model is three-dimensional, it contains rather strongly simplified assumptions. Poiseuille flow is assumed with respect to the channel-normal velocity. A lateral movement of the two plates causes a Couette flow in the subduction channel. The movements of the plates are prescribed: The present movement of the Nazca plate with 5-6 cm/a in eastward direction and the westward velocity of the Brazilian shield with 3 cm/a are prescribed by the boundary conditions. The sideward boundaries act as indenting plates. The inclination of the subducting slab, dipping 15-30° to the east, is prescribed, too. Realistic shortening rates were found when the orogenetic lithosphere was assumed to be 20-100 times weaker than the foreland lithosphere. The incorporation of a weak and easily flowable middle crust of the upper plate generates the necessary reduction of the topographic relief and provides a mechanism to explain the flatness of the Andean plateau. Vietor [72] presented a 2D rectangular-box model with a noncohesive Navier-Coulomb rheology driven by kinematic boundary conditions applied as fixed velocities. He showed that the lateral expansion of a weak zone at the base of the plateau can switch the tectonics of the plateau from vertical thickening to lateral expansion. This kind of switch has been observed also in the Basin and Range province of North America.

Babeyko et al. [2] calculate a 2D thermo-mechanical model with prescribed velocities at the right and left sideward boundaries of the box model. Therefore the box is growing higher and narrower as a function of time. So, the model contains neither the subducting slab nor the Brazilian shield but the main topics of the model are the high heat flow at the Altiplano-Puna plateau and the peak of ignimbrite activity in the late Miocene and Pliocene. Radiogenic heat production of the crust with growing thickness, shear heating and heat brought by intrusions prove to be too small to explain the high heat flux of the plateau. Therefore Babeyko et al. [2] calculate the effects of a hot lower crust with internal convection. In doing so, they use a quartz-dominated rheology and apply a high heat flow from below. In this model, the hypothesis of a delamination of the continental lithospheric mantle and of the lower crust due to eclogitization [3] plays a role in the background conception. We remark that the latter conception fits well to the fact that it is impossible to explain the *episodicity* of the Andean orogenesis or orogenesis in general by a *continuous* subduction, only. It remains to be seen whether it is really necessary to assume a laterally connected layer of thermal convection in the remaining lower crust since the heat flow of the Altiplano-Puna plateau is not only high but also laterally strongly variable. So, also the option of a multitude of intrusions should be checked again. Babeyko et al. [3] propose a similar 2D model containing the plateau *plus* the Brazilian shield but not the subducting slab. Ongoing eclogitization of the lower mafic crust beneath the plateau is proposed to be responsible for the orogenetic episodes.

However, Sobolev et al. [63, 64] investigate the total mechanism of Andean orogeny. They use a viscoelastic rheology supplemented by Mohr-Coulomb plasticity for the layered lithospheres. The drift of the overriding plate and the pulling of the slab is prescribed by the velocities at the boundaries of the 2D model area and it is not calculated by solution of the balance equations though. What drives Andean orogeny? Sobolev et al. [63, 64] answer this question by numerical experiments using their 2D model and varying only one influence parameter each. They conclude that the major factor is the westward drift of the South American plate. Furthermore they use alternatively a stronger thin crust (35-40 km) or a thick crust (40-45 km) in the backarc as well as different friction coefficients. Both additional parameters produce considerable effects. The model, however, did not confirm that climate-controlled changes of the sedimentary trench-fill has a significant influence on the shortening rate. We think highly of these results [3, 63, 64] and want to refer to them in Section 3.

Burov and Toussaint [17] apply the same mixed FE/FD code Parovoz (Poliakov et al. [54]), which is based on the FLAC technique (Cundall [20]), and use it for the India-Eurasia collision and especially for the Himalayan mountain belt. They conclude that the total amount of subduction may *largely* vary as a function of the denudation rate. Sedimentation helps down-thrusting of the lower plate. However, very strong or very slow sedimentation augment the probability of plate coupling. They conclude that there is an optimum sedimentation to support subduction. This is in contrast with the conclusion of Sobolev et al. [63], regarding the Andes, that sedimentation plays only a *minor* role for subduction. These opposing conclusions are the more remarkable since the two groups of authors used virtually the same code. It is not probable that the different geographical regions are the cause for this disagreement. The cause of this contradiction still has to be clarified.

The last mentioned models are two-dimensional. However, the structure of the Andes varies as a function of latitude. The thrust belts at the east flank of the Andes show, e. g., large differences [41]. Furthermore, Gerbault et al. [23] refer to the relatively low elevation and the thick crust in the Altiplano, in comparison to the higher elevation, but thinner crust in the Puna plateau. They speculate whether orogen-parallel lower crustal flow could play a role. Problems of this kind can, of course, be investigated only in a three-dimensional model.

A 3D model of the Andean dynamics of the last 10 Ma has been presented by Heidbach et al. [29]. Iaffaldano et al. [35] couple the global circulation model of Bunge et al. [16] with the global lithospheric model of Bird [11] which, however, uses a thin-sheet approximation. Heidbach et al. [29] use the results of the circulation model to calculate the creeping velocities of the asthenosphere at the bottom of the plates. The boundary conditions are rather realistic since the circulation models prescribe the velocities of the plates as a function of time. A dislocation creep rheology of olivine is used inside the lithosphere. Heidbach et al. [29] determine the geographical distribution of the maximum horizontal compressional stress, S_H, for the ages of 10 Ma, 3.2 Ma, and 0 Ma where the topographies are a priori prescribed in the model. They obtain a good coincidence of the computed S_H with the observed one for $\tau = 0$ Ma and conclude that the growth of the central Andes controls the overall slow down of the Nazca/South American plate convergence. So, we close the paragraph on the partly dynamical models which directly refer to the Andes.

c) Now, there are at least five *geochemical/geological models* for the origin of the continental crust [21]. One of them postulates an additional fractionation of the arc crust. A delamination of cumulate layers beneath the seismological Moho back into the mantle could play a role.

d) *Dynamic subduction models which do not refer explicitely to the orogeny of the Andes.* Many efforts have been made to find self-consistent solutions to the generation problem of oceanic lithospheric plates. The oceanic lithosphere was generated by a strong temperature dependence of the shear viscosity in most of such numerical experiments (Christensen [19]), Hansen and Yuen [28], etc.). It proved to be impossible to produce plate-like solutions *and* subduction by a purely viscous rheology. Therefore the viscous creep has been supplemented by different constitutive laws and mechanisms [8, 9, 52, 58, 66, 67, 70]. Gerya et al. [26] treat the slab breakoff triggered by thermal diffusion using a 2D finite-difference and marker-in-cell technique. The temperature- and pressure-dependent thermal conductivity proves to have a significant effect on thermal weakening of the slab. Gerya et al. [24] present a 2D high-resolution petrological-thermomechanical model of the slab using a fine-scale oceanic crust with 1 km sediment, 2 km hydrothermally altered basalts and lower 5 km of gabbro whereas the mantle is supposed to be either anhydrous or contains about 2 wt. % water. In this 2D model, the water is entrained by the slab and generates not only a hydration front above the subducting slab but also small unmixed and mixed plumes which rise from the upper surface of the slab. This model contains certain features of our started 3D Andean backarc model (cf. Section 3). Gerya et al. [25] search for the reason why subduction is one-sided. They achieve one-sided convection assuming that the subducting plate is considerably thicker than the overriding plate. They start from the assumption that not only the subducting plate has oceanic crust and oceanic subcrustal lithosphere but also the overriding plate which is considerably thinner, yet. But we observe a perpendicular downgoing slab in the case of an ocean-ocean collision. Subduction angles smaller than 90° are observed, however, for ocean-continent collisions where the slabs subduct obliquely under the continent, the lithosphere of which is considerably thicker than and chemically different from the lithosphere of the subducting plate. Therefore we believe that for modeling the oblique subduction, the existence of a continent on the overriding plate and its very thick subcontinental lithospheric mantle should be crucial for the model. Further numerical models were published by [6, 18, 22, 30, 69]. They could be relevant for the conception of our regional model of Andean orogenesis. Schellart et al. [57] show that slab width controls the curvature of subduction zone and the tendency to retreat backwards as a function of time. So it is understandable that the shortening is very large in the case of the central Andes. Billen [10] reports on slab dynamics, especially on models showing that spatial and temporal variations in slab strength determine whether slabs subduct into the lower mantle or remain in the transition zone.

e) *Circulation models.* Not only because we do not know the initial conditions but also because of some stochastic features of mantle convection [73,74], it is necessary to introduce mantle circulation models in order to reconstruct the Mesozoic and Cenozoic history. Based on seismic tomographic models and reconstructions of plate motions successful circulation models have been derived for the last 100 Ma [13, 15] which could be used as a spherical-shell model which should define the time-dependent boundary conditions of our special Andean model. Other circulation models have been discussed [7, 81] and should be compared with [13, 15].

3 Our New Model of Andean Orogenesis

a) General features of the model. We do not want to treat the problem of Andean orogenesis by purely regional models only, as the papers described in Section 2, b), except Heidbach et al. [29] do since, in this case, the temporally varying boundary conditions are unknown. Therefore it is often assumed, for reasons of simplicity, that there are no or only very simple effects from outside of the regional computational domain. As mentioned in Section 2, however, some changes in the arc volcanism and in the tectonic-shortening behavior are evidently in connection with the 30 Ma-Africa-Eurasia collision. Therefore we intend to embed a regional 3D model into a 3D spherical-shell model. So, we want to solve the balance equations of momentum, energy and mass in the spherical-shell model using somewhat larger time steps on a coarser whole-mantle grid, coarser than in the regional model. The values of creeping velocity, temperature and pressure, determined in that way and lying at the boundaries of the *regional* computational domain, serve then as temporarily fixed boundary conditions for some smaller time steps for which the balance equations are solved in the regional computational domain. Then, the final values of this regional computation will be transferred to the coarser grid of the whole mantle where the balance equations will be solved for the next time step. So, we want to jump iteratively between global and regional domain.

b) The spherical-shell model. We continue with considerations on the spherical-shell model, only afterwards we report on the regional model. As a base of the spherical-shell model, we intend to use the code Terra which was developed by Baumgardner [4], has been parallelized by Bunge et al. [14] and has been rearranged by Yang et al. [77] to allow for steep viscosity gradients. Walzer et al. [75] developed a convection model of the 4500 Ma of thermal evolution of the Earth's mantle with stable but temporally variable oceanic lithospheric plates, using Terra. Walzer et al. [73] present a spherical-shell model for 4500 Ma of mantle evolution with convection, plate tectonics *and* chemical differentiation. Origin and growth of the continents and of the depleted mantle are modeled by the interplay of differentiation and convection. They could numerically show that today, also in the model, different mantle reservoirs exist in spite of 4500 Ma of stirring convection. The latter two papers are based on forward computations starting with certain initial conditions and fulfilling certain very simple boundary conditions. Number, size, form, distribution and velocities of plates and continents are not constrained or even prescribed. Such a procedure is appropriate for more fundamental investigations since, in this case, it is senseful to compare the computed distributions of the mentioned quantities as well as Rayleigh number, *Ra*, Urey number, *Ur*, laterally averaged heat flow, *qob*, etc, as a function of time with observations.

It is a different matter in case of the 3D circulation models where number, size, form, distribution and velocities of the plates are prescribed (for the younger geological past). In this case we have to take care that the balance equations are really fulfilled. But this kind of spherical-shell model is necessary if we want to study *definite* plates, continents and orogens like Nazca plate, South American plate, South American continent and the Andes. We have to decide on the kind of a global 3D spherical-shell *circulation* model. Furthermore we have to adopt a global model of lithospheric motions, e. g., that of Bird [11] or another one. In no way, we intend to repeat or modify the procedure of Heidbach et al. [29] since the thin-sheet approximation used there is not able to solve our intended questions. But we think highly of this valuable and stimulating paper.

c) Numerical improvements. It is necessary to invent further improvements in Terra to solve the questions of the geodynamic modeling. Partly we already began to work in such a direction. Müller [50] and Köstler [43] work on improvements of Terra which, however, could not be used yet in our latest simulations. Müller analyzed the discretization of the Stokes problem and found a local grid refinement with hanging nodes which is inf-sup stable. Beate Sändig augmented the block size of the Jacobi smoother in the solver and Köstler adapted the grid transfer because of the small irregularities of the icosahedric grid. However, the influence of these alterations on the convergence behavior proved to be small. Further studies on the Krylov subspace method and multigrid method have been carried out and are continued to apply the latest results of numerical research to the Stokes solver in Terra. Especially a multigrid solver of the coupled Stokes problem, proposed by Larin and Reusken [44], promises to be successful for an augmentation of the convergence rates and an improved pressure correction. We intend to implement such a solver and a MINRES procedure and we agreed on cooperation with John Baumgardner (San Diego, USA) regarding these problems. As a latest result, Baumgardner got the new free-slip boundary treatment debugged and working correctly. It is to be expected that he can move forward now on a set of benchmarks and get a Terra benchmark paper.

d) The regional model. Before we sketch the *regional* model, we want to outline the idea behind of it. This idea has to be considered like a tentative diagnosis. It is well possible that we will be compelled to modify it. Apart from the details described in Section 2, a), the model should contain or allow to derive the following items for the geological present. The present-day average heat flow of the backarc, consequently the area between Western and Eastern Cordillera, should be 85 ± 16 mW/m^2, the heat flow of the Brazilian shield should be 42 ± 7 mW/m^2. The deformation ages across the southern central Andes migrated from the west to the east [53]. At a latitude of 21° S, the following ages, τ, have been observed: western flank/Precordillera $\tau = 47..38$ Ma, central Altiplano $\tau = 32..12$ Ma, Eastern Cordillera/Interandean $\tau = 42..7$ Ma. We suppose that the thickness of the Brazilian shield was and is 200 km. The present backarc lithosphere is 50-60 km thick. The present high backarc asthenosphere below of it shows a very low shear viscosity. We presume that the strength structure of the craton can be described by the right-hand side of Fig. 1.

Using these assumptions, we can sketch the following mechanism. The subduction of the slab is rendered possible by the presence of large quantities of water. The absorption of large amounts of water by the oceanic crust essentially lowers the melting point [27, 38, 55]. So, the deflection of the oceanic lithospheric plate will be made possible in the first place. The upper and lower boundary surfaces of the asthenosphere as a low-viscosity layer are determined by the intersecting points of the curves of water solubility and water abundance [49]. Therefore the thickness of the asthenosphere is a function of location and time. If the slab is dipped into the mantle, it carries *continuously* water into the upper mantle. Many tectonic problems cannot be explained by plate collision or friction. E. g., the subduction of the Nazca plate essentially takes place *continuously*, but the orogenetic events of the Andes are *episodically* distributed on the time axis. We propose to solve this problem as follows.

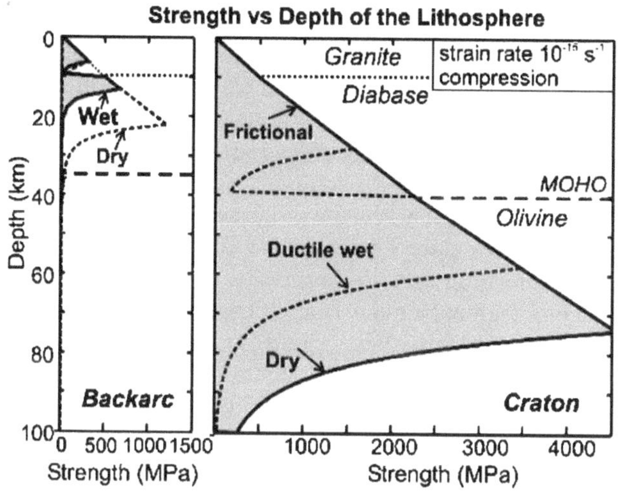

Fig. 1. Strength versus depth for hot backarc belts and cold cratons according to Hyndman et al. [34]

According to Model V of Davidson and Arculus [21], the cumulate complement of the evolved continental crust beneath the seismological Moho and the continental lithospheric mantle below of it will be so flowable by absorption of water that they *episodically* detach from the middle and upper crust. Then they sink down into the mantle because of a Rayleigh-Taylor instability. In this process, the eclogitization of the lower crust plays a part. From the Nb/Ta-Nb plot for primitive mantle, depleted mantle and continental crust [56] it follows that there is such an eclogitic reservoir in the mantle. To express it in another way, although it would be possible to reestablish the primitive mantle by a total stirring of the depleted-mantle and continental-crust reservoirs regarding many chemical elements, for the Nb/Ta-Nb plot this is not possible without the assumption of an additional eclogite reservoir. The proposed mechanism would also explain the felsic composition of the present Altiplano and Puna crust [80].

So we think that the high backarc asthenosphere with particularly low viscosity evolves gradually by the continuous water transfer from the slab. Small-scale thermal convection takes place continuously in the backarc asthenosphere explaining the high heat flow at the surface of the only 50-60 km thick lithosphere of the Altiplano-Puna region. This heat flow is considerably higher than that of the Brazilian shield as well as that of the neighboring parts of the Nazca plate. By the enduring growth of the quantity of water in the backarc produced by the slab movement, the backarc asthenosphere spreads piecewise eastward by episodic detachment of pieces of the mantle lithosphere and of the eclogitisized lower crust. This explains the *episodicity* of the orogenetic events and their *eastward migration*. The mobile belt of the backarc asthenosphere is exclusively generated and maintained by the water transfer of the slab. The word exclusively refers, of course, only to our model. If the slab mechanism vanishes then this continental margin will loose its fertility. Therefore *continuous* slab subduction is necessary to produce the *episodes* of orogenesis. The orogenesis was particularly active at the two margins of the backarc asthenospheric zone. These boundaries are characterized by the largest lateral viscosity contrasts between backarc and craton or between backarc and Nazca slab. Therefore the Eastern and Western Cordilleras are situated at the leading and trailing edges of the fertile belt. This means for the code of our regional model that it has to be designed to tackle also high *lateral* viscosity gradients. Walzer and Hendel [73] emphasized that there have to exist *considerable* viscosity gradients in radial direction at the well-known mineral phase boundaries since activation energies and volumes jump there. The activation enthalpy determines the exponent of the e-function of the viscosity. The same applies for the intersecting points of the curves of water solubility and water abundance which define top and bottom boundaries of the asthenosphere.

From the previous considerations, it follows that *our* regional computational domain has to be considerably smaller than that of Heidbach et al. [29]. So, the quantities of the larger parts of the South American and the Nazca plates have to be taken from the results of the global circulation model, in contrast [29]. Furthermore in contrast to the coarser grid of the surrounding circulation model, the grid-point distances of the regional domain have to be much smaller. A further grid refinement is necessary at the surface of the subducting slab, at the boundary between craton and backarc asthenosphere and at the Earth's surface. The time window, however, may not be restricted to the last 10 Ma as in the [29]-model but at, e. g., 119 Ma [45]. The depth of the regional computational domain is about 900 km in order to include all influences of the phase boundaries on the slab plus a small additional domain. Regarding the phase boundaries of the upper mantle, only the effects of the phase boundary distortion by advection of thermal anomalies should be taken into account. In comparison to this, the effects of the phase boundary distortion due to release or absorption of latent heat and the effects of expansion or contraction due to release or absorption of latent heat are small. Although the transition from basalt/gabbro to eclogite includes only minor volumes it has a density increase of 15 %. Therefore this transition is relevant for modeling the delamination or a less viscous detachment of the lower crust. That is why the basalt-eclogite transition should be taken into account.

We intend to describe the transport of water by a tracer approach. In [73], we had modified the tracer module of Dave Stegman and used it to describe the transport of heat-producing elements. In the *first* numerical cases of the model, we should simply use a bulk-silicate-earth (BSE) distribution of the heat-producing elements, U, Th and K, temporally declining but spatially simply homogeneous heating from within [32,46]. As usual [73,75], we intend to add a core-cooling model [33,65]. The CMB temperature, T_c, is *laterally* constant in this procedure. But T_c of the surrounding spherical-shell model is adapted to the temporally variable heat flow through the CMB after each time step. In *later* numerical experiments, tracers are to be used to take into account the different abundances of heat-producing elements in the different reservoirs. This is probably already necessary to show that the high heat flow of the Altiplano-Puna region is not *mainly* caused by the higher abundances of U, Th, K in the felsic crust which has a *present* thickness of 50-60 km but by small-scale convection in the high backarc asthenosphere.

We do not want to criticize the partly dynamic, partly kinematic models described in Section 2, b). But we intend to produce an Andean model with *less* restrictive assumptions. In spite of the higher number of degrees of freedom, the new model should allow to derive the essential features of the Andean orogenesis. In doing so, the catalog of unresolved or only partly resolved problems of Section 2 a) ought to be our guide. The models of Sobolev et al. [63,64] and Babeyko et al. [2,3] are good and very stimulating. However, we intend to alter not only the computational method but also the treated schedule of problems. We suppose that the westward migration of the Brazilian shield is *not* the main reason for the entirety of the Andean orogenetic processes but the water subducted by the Nazca plate. We believe that this water generates the orogenetically active backarc belt. Nevertheless, the very active thrust faults of the Subandean Ranges show that the westward movement of the Brazilian shield has a large influence especially in the last 10 Ma. We continue with some numerical considerations on the construction of the computational code to realize the outlined model.

e) Numerical methods: Coupling a regional with a global mantle convection model. The computational model Terra uses a discretization of the sphere which consists of 10 diamonds evolving from the projection of the regular icosahedron onto the sphere [5]. Because domain decomposition is done in every diamond separately by explicit message passing, it is worthwhile to use one diamond as a regional domain and to couple this regional to the global model of Terra. The lateral extension of one diamond edge on the Earth's surface is 7000 km and the depth of this domain can be chosen arbitrarily. The size of such a segment is very appropriate to model the Andean subduction zone as Schellart et al. [57] found that subduction zone width strongly influences trench migration and crustal extension, respectively shortening. Because parallelization in Terra is done in lateral directions, it can be applied very efficiently to a regional domain which is a wide aspect ratio spherical shell segment. For instance on an SGI Altix 4700 machine with Intel Itanium2 Montecito Dual Core processors we could achieve an efficient cache usage with subdomains of 33x33x129 grid points which would give an overall regional grid spacing of 7 km at the surface and 6 km at the bottom boundary if we use 1024 processors on one diamond with a depth of 900 km. This is 4 times less the spacing we could achieve in the whole spherical shell. Communication overhead would be much less than 10 % for this configuration. A similar approach has been done with the convection code CitcomS in [68], where a 2 times finer regional grid is used to model the ascent of a plume through the surrounding mantle.

Nevertheless, at the boundaries of the subducting slab an even smaller grid spacing is desirable and very strong gradients in material parameters, in particular viscosity, have to be taken into account. Therefore we have further advanced the Terra code with respect to stability, local grid refinement, and solver techniques. Precisely we have successfully investigated an inf-sup-stable grid modification that enables us to use grids with hanging nodes. Thus we are able to adapt the grid resolution rapidly towards a heavily refined region with only a few successively refined layers. We have changed the smoother of the multigrid solver, the preconditioner, the algorithm for the solution of an included Stokes problem and the method for the time integration. We have started refactoring the solver to achieve greater flexibility with regard to algorithm changes. We created a test framework that automates the build process, starts automatically a series of test cases on different machines in different resolutions and checks the results. This way the response time to changes and errors in the code will be noticeably reduced. That means, we will have facilities assisting even radical changes of the code.

Moving towards the coupling of the two mentioned models we first want to implement a regional model containing all small scale parameters of interest, but without dynamic coupling to the whole mantle convection. The first step to achieve this is to modify the existing convection code by simply constraining

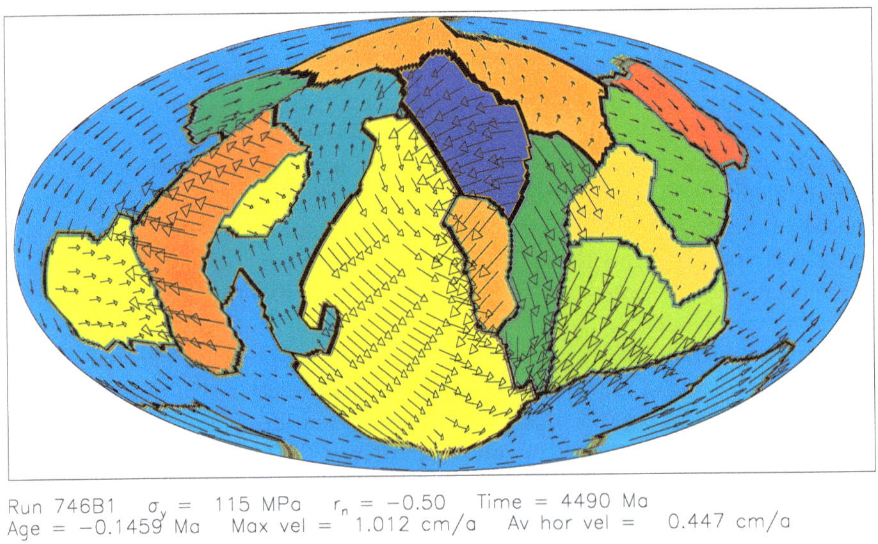

Fig. 2. An equal-area projection of the lithospheric plates (colors) and creep velocities (arrows) of our 3D spherical-shell convection-evolution model *with* chemical differentiation and continents (not shown here) for the present time. The viscoplastic yield stress is $\sigma_y = 115$ MPa

the computational domain to one diamond. This results in additional vertical boundaries and the necessity to be able to prescribe appropriate conditions on them, since up to now the code is concerned with the whole shell which has only two spherical boundaries. We can easily provide a fully featured parallel multigrid solver for the convection in the domain of interest, including parallel implemented markers for chemical differentiation by a slight modification of the existing code.

In addition to the simple adaption described we plan to implement refined solving strategies to meet the requirements of strongly varying viscosity due to differences in temperature, pressure and water content between subducted material and its surroundings. We will use the techniques we investigated for whole mantle convection like the local grid refinement, a special preconditioning technique as well as our existing test framework. In particular, the above-mentioned preconditioned MINRES algorithm will be implemented. The benefits achieved this way would also pay off for the global model.

Having obtained a small scale method that uses essentially the same solver we have very good preconditions to integrate the small high resolution area seamlessly in the global convection. The spatial grid will be refined with the described technique of hanging nodes towards the boundaries of the small scale model with only a few successive refinement steps. One has however also to couple the time stepping scheme of both methods in a very flexible manner. We will have to address both, stability and optimality with respect to the load balancing. The plan to achieve these goals is very simple, because we do not need full adaptivity in space and time, but prescribe the spatiotemporal resolution according to optimal load balancing.

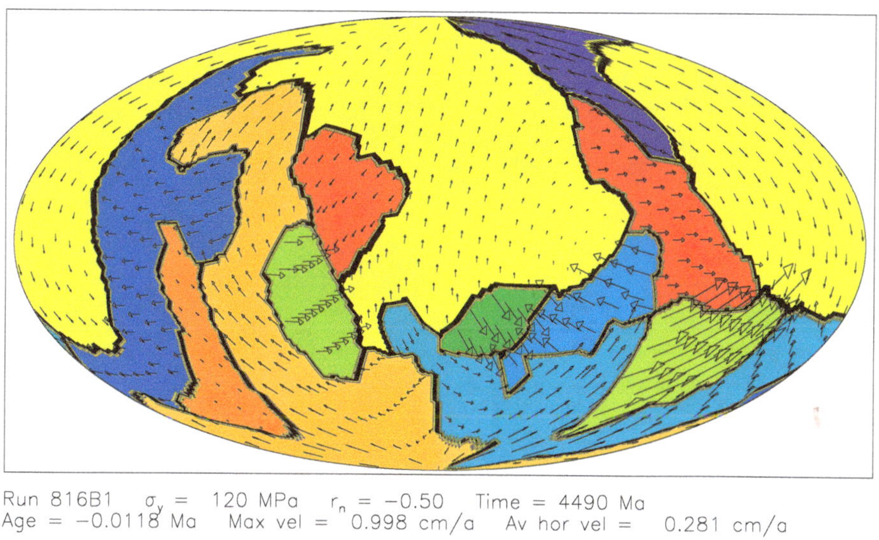

Fig. 3. Text see Fig. 2, but $\sigma_y = 120$ MPa

f) Numerical methods: Using a 2D model for rheology influence on subduction. Furthermore, we mention that there exists a 2D version of Terra which can already model viscosity contrasts of up to 10^{10} between adjacent points [78]. This code is currently used by the group of Slava Solomatov at Washington University in St. Louis to study mantle convection with realistic rheologies and related issues such as kinetics of phase transformations and grain growth. As we also use this code to study different solving strategies in two dimensions first, there is a benefit to be expected also in modeling the influences of composition and grain size on subduction. Moreover, the matrix-dependent transfer technique in the 2D code still promises an improvement of the convergence of the multigrid in the 3D code. Therewith it will be possible also to model lateral viscosity variations of several orders of magnitude between adjacent points in the global and in the regional 3D model.

4 First Results: Plate Generation

These results do not yet refer to a circulation model but to forward computations where we solve the balance equations to a spherical shell starting from assumed initial conditions at an age of 4500 Ma. The included tracer code is constructed in such a way as to conserve the four sums of the numbers of atoms of the pairs ^{238}U-^{206}Pb, ^{235}U-^{207}Pb, ^{232}Th-^{208}Pb, and ^{40}K-^{40}Ar. The decay of these four radionuclides and the primordial heat are the principal energy sources driving the Earth's solid-state mantle convection and lateral plate movements by heating from within. There is only a small contribution

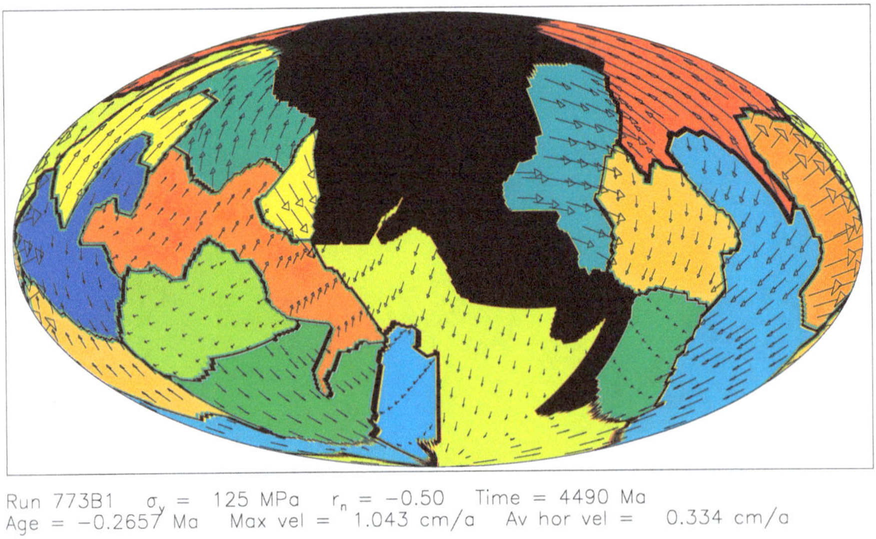

Fig. 4. Text see Fig. 2, but $\sigma_y = 125$ MPa

by heating from the core-mantle boundary (CMB) but it is not negligible. In this model, there is a certain partial random influence [74] on the formation of oceanic lithospheric plates and of continents. The two essential conditions of *our* model for the generation of plates are the assumption of a viscoplastic yield stress, σ_y, for the lithosphere and the existence of a low-viscosity asthenosphere below it [73, 75]. It is evident that it is impossible to produce plate tectonics without a deviation from the purely viscous rheology and without asthenosphere. The viscosity, η, of mantle and crust is calculated by

$$\eta(r,\theta,\phi,t) = 10^{r_n} \cdot \frac{\exp(c\,\overline{T_m/T_{av}})}{\exp(c\,\overline{T_m/T_{st}})} \cdot \eta_3(r) \cdot \exp\left[c_t \cdot T_m \left(\frac{1}{T} - \frac{1}{T_{av}}\right)\right] \quad (1)$$

where r is radius, θ colatitude, ϕ longitude, t time, r_n viscosity-level parameter, T_m melting temperature, T_{av} laterally averaged temperature, T_{st} initial temperature profile, T temperature as a function of r, θ, ϕ, t; c and c_t are constants. The quantity $\eta_3(r)$ denotes the viscosity profile at initial temperature and for $r_n = 0$. The parameter r_n serves for a stepwise shift of the viscosity profile to vary the time-averaged Rayleigh number from run to run. For an idealized plate tectonics with sharp boundaries, the divergence divh $\mathbf{v}$ would form mountain crests of a pseudo-topography drawn on the spherical surface where

$$\mathrm{divh}\,\mathbf{v} = \nabla_h \cdot \mathbf{v} = \frac{1}{r_0}\left[\cot\theta \cdot v_\theta + \frac{\partial v_\theta}{\partial \theta} + \frac{1}{\sin\theta} \cdot \frac{\partial v_\phi}{\partial \phi}\right] \quad (2)$$

and r_0 is the Earth's radius. The velocity components v_r, v_θ and v_ϕ are assigned to r, θ and ϕ. Only the divergent and convergent plate boundaries

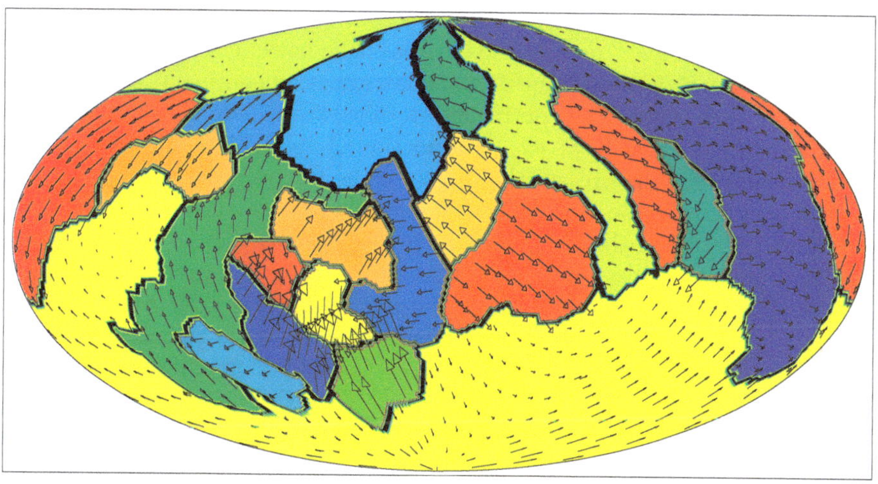

Run 796B1 σ_y = 130 MPa r_n = −0.50 Time = 4490 Ma
Age = −0.0246 Ma Max vel = 0.873 cm/a Av hor vel = 0.274 cm/a

Fig. 5. Text see Fig. 2, but $\sigma_y = 130$ MPa

appear as hogbacks of divh $\mathbf{v}$. For transform faults, the pseudo-topography of the curl

$$\mathrm{roth}\,\mathbf{v} = (\nabla \times \mathbf{v})_h = \frac{1}{r_0}\left[\cot\theta \cdot v_\phi + \frac{\partial v_\phi}{\partial \theta} + \frac{1}{\sin\theta} \cdot \frac{\partial v_\theta}{\partial \phi}\right] \tag{3}$$

shows mountain crests. The surface expression of the square root of the second invariant of the strain-rate tensor, invh $\mathbf{v}$, will produce mountain crests for *all* plate boundaries,

$$\mathrm{invh}\,\mathbf{v} = \dot{\epsilon}_{surf} = \frac{1}{r_0}\left[\left(\frac{\partial v_\theta}{\partial \theta}\right)^2 + \left(\frac{1}{\sin\theta} \cdot \frac{\partial v_\phi}{\partial \phi} + v_\theta \cdot \cot\theta\right)^2 \right.$$
$$\left. + \frac{1}{2}\left(\frac{1}{\sin\theta} \cdot \frac{\partial v_\theta}{\partial \phi} + \frac{\partial v_\phi}{\partial \theta} - v_\phi \cdot \cot\theta\right)^2\right]^{1/2} \tag{4}$$

Therefore we used the crest lines of invh $\mathbf{v}$ to determine the exact plate boundaries. A successive grid refinement toward the surface could improve the procedure, yet. The Figures 2 to 5 present the plate distribution and the plate velocities for the geological present time where r_n and the viscosity-profile factor, $\eta_3(r)$, were kept constant and only the yield stress, σ_y, was varied in equal steps from case to case. Deriving the viscosity profile, we assumed that the oceanic lithosphere is produced not only by the temperature dependence of viscosity but also by devolatilization and secondary chemical segregation generating the layering of oceanic lithosphere. For higher σ_y, the number of plates is larger than for low σ_y. We observed a similar, but stronger effect for the variation of the temporal average, Ra, of the Rayleigh number. Keeping the other parameters constant, we obtain more plates for low Ra and less plates for high Ra. The arrows in Figures 2 to 5 denote the lateral creep velocities at the surface. These plate velocities have not only different directions but also different absolute values. Some plates contain continents, other plates are free of continents. The continents are not shown here. Fig. 6 shows the laterally averaged heat flow, *qob*, as a function of age. This quantity arrives at realistic values for the present time close to the laterally averaged, measured surface heat flow of the Earth. It is remarkable that the decrease of *qob* as a function of time is rather moderate in comparison to that of parameterized models [59]. In [76], we emphasized that this result is in agreement with the results of komatiite research. This slow decrease of *qob* is a further indication that not only the temperature dependence of viscosity is the reason for the generation of oceanic lithosphere but also devolatilization and other chemical effects.

5 Conclusions

We began to research into models of Andean orogenesis and present some basic considerations. We obtained explicit results in the further development of

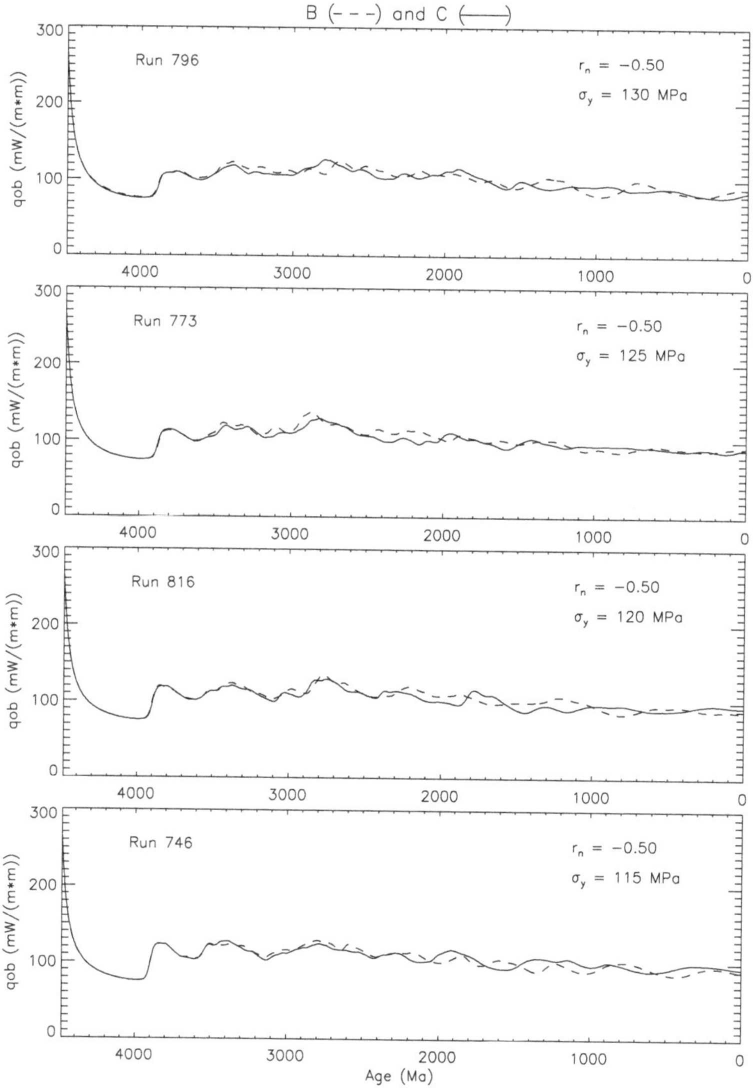

Fig. 6. The evolution of the laterally averaged surface heat flow, *qob*, for neighbor runs. The viscosity-level parameter, r_n, is kept constant at -0.50. Dashed lines belong to basic runs with about 10.5 million tracers, solid lines to comparative runs with about 84 million tracers

the code Terra and regarding the generation of plate tectonics on a spherical-shell mantle. Here, only plate generation is presented in some detail. Referring to the items a) to d) of the Introduction, we summarize the following considerations, regarding item e) also specific results. Because of the limitations of available computing capacity and the necessary fine grid of the Andean dynamic model, we have to cut out only a small part of the spherical shell as a regional model with unknown time-dependent boundary conditions. For the same reasons and because of the restriction of knowledge of observed global plate motions to the younger Phanerozoic, also the evolution time of the regional model has to be restricted. Therefore, the initial conditions of the regional model are unknown, too. We intend to embed the regional model into a global circulation model to determine the initial conditions and the time-dependent boundary conditions of the Andean model. We have to prefer a circulation model in order to eliminate the majority of stochastic processes. If we would not do so, the model could not arrive at the *specific* form of plates and continents. We developed a hierarchy of observational facts which ought to be explained by the embedded Andean evolution model. In Section 4, we present a specific model of generation of plate tectonics showing only a few cases with a systematic variation of the yield stress. The number of plates depends mainly on the temporally averaged Rayleigh number, Ra, but to a minor degree also on the yield stress, σ_y. A variation of the parameters shows the existence of a central area in the $Ra - \sigma_y$ plot where stable plate-like solutions on the sphere have been found.

Acknowledgements

We gratefully acknowledge the stimulating discussions with John Baumgardner and Markus Müller. We thank the High Performance Computing Center Stuttgart (HLRS) for support and supply of computational time under grant *sphshell* /12714. This work was partly supported by the Deutsche Forschungsgemeinschaft under grant WA 1035/5-3.

References

1. R. Allmendinger, T. Jordan, S. Kay, and B. Isacks. The evolution of the Altiplano-Puna plateau of the Central Andes. *Annu. Rev. Earth Planet. Sci.*, 25:139–174, 1997.
2. A. Babeyko, S. Sobolev, R. Trumbull, O. Oncken, and L. Lavier. Numerical models of crustal scale convection and partial melting beneath the Altiplano-Puna Plateau. *Earth Planet. Sci. Lett.*, 199:373–388, 2002.
3. A. Babeyko, S. Sobolev, T. Vietor, O. Oncken, and R. Trumbull. Numerical Study of Weakening Processes in the Central Andean Back-Arc. In O. Oncken et al., editors, *The Andes. Active subduction orogeny*, Frontiers in Earth Sciences, pages 495–512. Springer, 2006.
4. J.R. Baumgardner. *A Three-Dimensional Finite Element Model for Mantle Convection.* PhD thesis, Univ. of California, Los Angeles, 1983.
5. J.R. Baumgardner and P.O. Frederickson. Icosahedral discretization of the two-sphere. *SIAM J. Numer. Anal.*, 22:1107–1115, 1985.
6. T.W. Becker and C.Faccenna. A review of the role of subduction dynamics for regional and global plate motions. In S. Lallemand and F. Funiciello, editors, *Subduction Zone Geodynamics*, Frontiers in Earth Sciences. Springer, 2008.

7. T.W. Becker, J.B. Kellogg, G. Ekstrom, and R.J. O'Connell. Comparison of azimuthal seismic anisotropy from surface waves and finite strain from global mantle-circulation models. *Geophys. J. Int.*, 155(2):696–714, 2003.
8. D. Bercovici. Generation of plate tectonics from lithosphere-mantle flow and void-volatile self-lubrication. *Earth Planet. Sci. Lett.*, 154:139–151, 1998.
9. D. Bercovici and Y. Ricard. Tectonic plate generation and two-phase damage: Void growth versus grain size reduction. *J. Geophys. Res.*, 110:B03401, 2005.
10. M.I. Billen. Modeling the dynamics of subducting slabs. *Annu. Rev. Earth Planet. Sci.*, 36(1):325–356, 2008.
11. P. Bird. Testing hypotheses on plate-driving mechanisms with global lithosphere models including topography, thermal structure, and faults. *J. Geophys. Res.*, 103:10115–10130, 1998.
12. R.C. Bostrom. Global tectonics under g having a minute westward tilt. *Terra Nova*, 18:55–62(8), 2006.
13. H.-P. Bunge, C.R. Hagelberg, and B.J. Travis. Mantle circulation models with variational data assimilation: Inferring past mantle flow and structure from plate motion histories and seismic tomography. *Geophys. J. Int.*, 152(2):280–301, 2003.
14. H.-P. Bunge, M.A. Richards, and J.R. Baumgardner. A sensitivity study of three-dimensional spherical mantle convection at 10^8 Rayleigh number: Effects of depth-dependent viscosity, heating mode and an endothermic phase change. *J. Geophys. Res.*, 102:11991–12007, 1997.
15. H.-P. Bunge, M.A. Richards, and J.R. Baumgardner. Mantle-circulation models with sequential data assimilation: Inferring present-day mantle structure from plate-motion histories. *Phil. Trans. Royal Soc. A: Math. Phys. Engng. Sci.*, 360:2545–2567, 2002.
16. H.-P. Bunge, M.A. Richards, C. Lithgow-Bertelloni, J.R. Baumgardner, S.P. Grand, and B.A. Romanowicz. Time scales and heterogeneous structure in geodynamic Earth models. *Science*, 280:91–95, 1998.
17. E. Burov and G. Toussaint. Surface processes and tectonics: Forcing of continental subduction and deep processes. *Global and Planetary Change*, 58:141–164, 2007.
18. F.A. Capitanio, G. Morra, and S. Goes. Dynamic models of downgoing plate-buoyancy driven subduction: Subduction motions and energy dissipation. *Earth Planet. Sci. Lett.*, 262:284–297, 2007.
19. U.R. Christensen. Heat transport by variable viscosity convection II: Pressure influence, non-Newtonian rheology and decaying heat sources. *Phys. Earth Planet. Int.*, 37:183–205, 1985.
20. P.A. Cundall. Numerical experiments on localization in frictional materials. *Archive of Applied Mechanics (Ingenieur-Archiv)*, 59:148–159, 1989.
21. J.P. Davidson and R.J. Arculus. The significance of Phanerozoic arc magmatism in generating continental crust. In M. Brown and T. Rushmer, editors, *Evolution and Differentiation of the Continental Crust*, pages 135–172. Cambridge Univ. Press, Cambridge, UK, 2006.
22. M. Faccenda, T.V. Gerya, and S. Chakraborty. Styles of post-subduction collisional orogeny: Influence of convergence velocity, crustal rheology and radiogenic heat production. *Lithos*, In Press, 2008.
23. M. Gerbault, J. Martinod, and G. Hérail. Possible orogeny-parallel lower crustal flow and thickening in the Central Andes. *Tectonophysics*, 399:59–72, 2005.

24. T.V. Gerya, J.A. Connolly, D.A. Yuen, W. Gorczyk, and A.M. Capel. Seismic implications of mantle wedge plumes. *Phys. Earth Planet. Int.*, 156:59–74, 2006.
25. T.V. Gerya, J.A.D. Connolly, and D.A. Yuen. Why is terrestrial subduction one-sided? *Geology*, 36:43–46, 2008.
26. T.V. Gerya, D.A. Yuen, and W.V. Maresch. Thermomechanical modelling of slab detachment. *Earth Planet. Sci. Lett.*, 226:101–116, 2004.
27. D.H. Green and T.J. Falloon. Pyrolite: a Ringwood concept and its current expression. In I. Jackson, editor, *The Earth's Mantle. Composition, Structure and Evolution*, pages 311–378. Cambridge Univ. Press, Cambridge, UK, 1998.
28. U. Hansen and D.A. Yuen. High Rayleigh number regime of temperature-dependent viscosity convection and the Earth's early thermal history. *Geophys. Res. Lett.*, 20:2191–2194, 1993.
29. O. Heidbach, G. Iaffaldano, and H.-P. Bunge. Topography growth drives stress rotations in the central andes: Observations and models. *Geophys. Res. Lett.*, 35:L08301, 2008.
30. A. Heuret, F. Funiciello, C. Faccenna, and S. Lallemand. Plate kinematics, slab shape and back-arc stress: A comparison between laboratory models and current subduction zones. *Earth Planet. Sci. Lett.*, 256:473–483, 2007.
31. D. Hindle and J. Kley. Displacements, strains and rotations in the Central Andean plate boundary zone. In S. Stein and J. Freymuller, editors, *Plate boundary zones*, Geodynamics Series 30, pages 135–144. AGU, Washington, D.C., 2002.
32. A.W. Hofmann. Sampling mantle heterogeneity through oceanic basalts: Isotopes and trace elements. In RW. Carlson, editor, *Treatise on Geochemistry, Vol.2: The Mantle and the Core*, pages 61–101. Elsevier, Amsterdam, 2003.
33. S. Honda and Y. Iwase. Comparison of the dynamic and parameterized models of mantle convection including core cooling. *Earth Planet. Sci. Lett.*, 139:133–145, 1996.
34. R.D. Hyndman, C.A. Currie, and S.P. Mazzotti. Subduction zone backarcs, mobile belts, and orogenic heat. *GSA Today*, 15:4–10, 2005.
35. G. Iaffaldano, H.-P. Bunge, and T. Dixon. Feedback between mountain belt growth and plate convergence. *Geology*, 34:893–896, 2006.
36. B. Isacks. Uplift of the central Andean plateau and bending of the Bolivian orocline. *J. Geophys. Res.*, 93(B4):3211–3231, 1988.
37. S. Kay and J. Abbruzzi. Magmatic evidence for Neogene lithospheric evolution of the central Andean "flat-slab" between 30S and 32S. *Tectonophysics*, 259(1-3):15–28, 1996.
38. H. Keppler and J.R. Smyth. *Water in Nominally Anhydrous melts*, volume 62 of *Rev. Min. Geochem.* Mineral. Soc. of Am., Washington, D.C., 2006.
39. J. Kley. Transition from basement-involved to thin-skinned thrusting in the Cordillera Oriental of southern Bolivia. *Tectonics*, 15(4):763–775, 1996.
40. J. Kley and C. Monaldi. Tectonic inversion in the Santa Barbara System of the central Andean foreland thrust belt, northwestern Argentina. *Tectonics*, 21(6):1061, 2002.
41. J. Kley, C. Monaldi, and J. Salfity. Along-strike segmentation of the Andean foreland: causes and consequences. *Tectonophysics*, 301:75–94, 1999.
42. J. Kley and T. Vietor. Subduction and mountain building in the central Andes. In O. Oncken et al., editors, *The Andes. Active subduction orogeny*, Frontiers in Earth Sciences, pages 624–659. Springer, Berlin, 2006.
43. C. Köstler. *Iterative solvers for modeling mantle convection with strongly varying viscosity.* PhD thesis, Friedrich-Schiller Univ. Jena, 2009.

44. M. Larin and A. Reusken. A comparative study of efficient iterative solvers for generalized Stokes equations. *Numer. Linear Algebra Appl.*, 15(1):13–34, 2008.
45. C. Lithgow-Bertelloni and M.A. Richards. The dynamics of Cenozoic and Mesozoic plate motions. *Rev. Geophys.*, 36:27–78, 1998.
46. M.T. McCulloch and V.C. Bennett. Progressive growth of the Earth's continental crust and depleted mantle: Geochemical constraints. *Geochim. Cosmochim. Acta*, 58:4717–4738, 1994.
47. S. Medvedev, Y. Podladchikov, M.R. Handy, and E.E. Scheuber. Controls on the deformation of the central and southern Andes (10-35S): Insight from thin-sheet numerical modeling. In O. Oncken et al., editors, *The Andes. Active subduction orogeny*, Frontiers in Earth Sciences, pages 475–494. Springer, Berlin, 2006.
48. S.E. Medvedev and Y.Y. Podladchikov. New extended thin-sheet approximation for geodynamic applications-I. Model formulation. *Geophys. J. Int.*, 136(3):567–585, 1999.
49. K. Mierdel, H. Keppler, J.R. Smyth, and F. Langenhorst. Water solubility in aluminous orthopyroxene and the origin of the Earth's asthenosphere. *Science*, 315:364–368, 2007.
50. M. Müller. *Toward a robust Terra code.* PhD thesis, Friedrich-Schiller Univ. Jena, 2008.
51. R. Müller, W. Roest, J.-Y. Royer, L. Gahagan, and J. Sclater. Digital isochrons of the world's ocean floor. *J. Geophys. Res.*, B 102:3211–3214, 1997.
52. M. Ogawa. Plate-like regime of a numerically modeled thermal convection in a fluid with temperature-, pressure-, and stress-history-dependent viscosity. *J. Geophys. Res.*, 108(B2):2067, 2003.
53. O. Oncken, D. Hindle, J. Kley, K. Elger, P. Victor, and K. Schemmann. Deformation of the central Andean upper plate system - facts, fiction, and constraints for plateau models. In O. Oncken et al., editors, *The Andes. Active subduction orogeny*, Frontiers in Earth Sciences, pages 3–27. Springer, Berlin, 2006.
54. A.N.B. Poliakov, Y. Podladchikov, and C. Talbot. Initiation of salt diapirs with frictional overburdens: Numerical experiments. *Tectonophysics*, 228:199–210, 1993.
55. K. Regenauer-Lieb and D.A. Yuen. Modeling shear zones in geological and planetary sciences: Solid- and fluid-thermal-mechanical approaches. *Earth Science Reviews*, 63:295–349, 2003.
56. R.L. Rudnick, M. Barth, I. Horn, and W.F. McDonough. Rutile-bearing refractory eclogites: Missing link between continents and depleted mantle. *Science*, 287:278–281, 2000.
57. W.P. Schellart, J. Freeman, D.R. Stegman, L. Moresi, and D. May. Evolution and diversity of subduction zones controlled by slab width. *Nature*, 446:308–311, 2007.
58. H. Schmeling, R. Monz, and D.C. Rubie. The influence of olivine metastability on the dynamics of subduction. *Earth Planet. Sci. Lett.*, 165:55–66, 1999.
59. G. Schubert, D.L. Turcotte, and T.R. Olson. *Mantle Convection in the Earth and Planets.* Cambridge Univ. Press, Cambridge, UK, 2001.
60. B. Scoppola, D. Boccaletti, M. Bevis, E. Carminati, and C. Doglioni. The westward drift of the lithosphere: A rotational drag? *Geol Soc Am Bull*, 118(1-2):199–209, 2006.
61. M. Sébrier and P. Soler. Tectonics and magmatism in the peruvian andes from late oligocene time to the present. In RS. Harmon and CW. Rapela, editors,

Andean Magmatism and its tectonic setting., volume 265 of *Geol. Soc. Am. Spec. Paper*, pages 259–278. Geol. Soc. of America, Boulder, Col., 1991.
62. P. Silver, R. Russo, and C. Lithgow-Bertelloni. Coupling of south American and African plate motion and plate deformation. *Science*, 279:60–63, 1998.
63. S. Sobolev, A. Babeyko, I. Koulakov, and O. Oncken. Numerical study of weakening processes in the central Andean back-arc. In O. Oncken et al., editors, *Mechanism of the Andean Orogeny: Insight from Numerical Modeling*, Frontiers in Earth Sciences, pages 513–535. Springer, 2006.
64. S.V. Sobolev and A.Y. Babeyko. What drives orogeny in the Andes? *Geology*, 33(8):617–620, 2005.
65. V. Steinbach and D.A. Yuen. Effects of depth-dependent properties on thermal anomalies produced in flush instabilities from phase transitions. *Phys. Earth Planet. Int.*, 86:165–183, 1994.
66. P.J. Tackley. Self-consistent generation of tectonic plates in time-dependent, three-dimensional mantle convection simulations. Part 1. Pseudoplastic yielding. *Geochem. Geophys. Geosys.*, 1:2000GC000036, 2000.
67. P.J. Tackley. Self-consistent generation of tectonic plates in time-dependent, three-dimensional mantle convection simulations. Part2. Strain weakening and asthenosphere. *Geochem. Geophys. Geosys.*, 1:2000GC000043, 2000.
68. E. Tan, E. Choi, P. Thoutireddy, M. Gurnis, and M. Aivazis. Geoframework: Coupling multiple models of mantle convection within a computational framework. *Geochem. Geophys. Geosys.*, 7:Q06001, 2006.
69. Y. Torii and S. Yoshioka. Physical conditions producing slab stagnation: Constraints of the Clapeyron slope, mantle viscosity, trench retreat, and dip angles. *Tectonophysics*, 445:200–209, 2007.
70. R. Trompert and U. Hansen. Mantle convection simulations with rheologies that generate plate-like behavior. *Nature*, 395:686–689, 1998.
71. R. Trumbull, U. Riller, O. Oncken, E. Scheuber, K. Munier, and F. Hongn. The time-space distribution of cenozoic volcanism in the south-central andes: a new data compilation and some tectonic implications. In O. Oncken et al., editors, *The Andes. Active subduction orogeny*, Frontiers in Earth Sciences, pages 29–43. Springer, 2006.
72. T. Vietor. Numerical modeling of plateau growth including some similarities to the central andes. In *2001 Margins meeting*, Schriftenreihe dt. geol. Ges., 14, page 223, Kiel, Germany, 2001.
73. U. Walzer and R. Hendel. Mantle convection and evolution with growing continents. *J. Geophys. Res.*, 111:B09405, doi:10.1029/2007JB005459, 2008.
74. U. Walzer and R. Hendel. Predictability of Rayleigh-number and continental-growth evolution of a dynamic model of the Earth's mantle. In *High Perf. Comp. Sci. Engng. '08*, accepted. Springer, Berlin, 2008.
75. U. Walzer, R. Hendel, and J. Baumgardner. The effects of a variation of the radial viscosity profile on mantle evolution. *Tectonophysics*, 384:55–90, 2004.
76. U. Walzer, R. Hendel, and J. Baumgardner. Whole-mantle convection, continent generation, and preservation of geochemical heterogeneity. In W.E. Nagel, W. Jäger, and M. Resch, editors, *High Perf. Comp. Sci. Engng. '07*, pages 603–645. Springer, Berlin, 2008.
77. W.-S. Yang. *Variable viscosity thermal convection at infinite Prandtl number in a thick spherical shell.* PhD thesis, University of Illinois, Urbana-Champaign, 1997.

78. W.-S. Yang and J.R. Baumgardner. A matrix-dependent transfer multigrid method for strongly variable viscosity infinite Prandtl number thermal convection. *Geophys. Astrophys. Fluid Dyn.*, 92:151–195, 2000.
79. X. Yuan, S.V. Sobolev, R. Kind, O. Oncken, G. Bock, G. Asch, B. Schurr, F. Graeber, A. Rudloff, W. Hanka, K. Wylegalla, R. Tibi, C. Haberland, A. Rietbrock, P. Giese, P. Wigger, P. Rower, G. Zandt, S. Beck, T. Wallace, M. Pardo, and D. Comte. Subduction and collision processes in the central Andes constrained by converted seismic phases. *Nature*, 408:958–961, 2000.
80. G. Zandt, A. Velasco, and S. Beck. Composition and thickness of the southern Altiplano crust, Bolivia. *Geology*, 22:1003–1006, 1994.
81. S. Zhong, N. Zhang, Z.-X. Li, and J.H. Roberts. Supercontinent cycles, true polar wander, and very long-wavelength mantle convection. *Earth Planet. Sci. Lett.*, 261:551–564, 2007.

Hybrid Code Development for the Numerical Simulation of Instationary Magnetoplasmadynamic Thrusters

M. Fertig[1], D. Petkow[1], T. Stindl[1], M. Auweter-Kurtz[2], M. Quandt[3], C.-D. Munz[3], J. Neudorfer[4], S. Roller[4], D. D'Andrea[5], and R. Schneider[5]

[1] Institut für Raumfahrtsysteme, Abt. Raumtransporttechnologie, Universität Stuttgart, Germany fertig@irs.uni-stuttgart.de
[2] Universität Hamburg, Germany Praesidentin@HVN.uni-hamburg.de
[3] Institut für Aerodynamik und Gasdynamik, Universität Stuttgart, Germany munz@iag.uni-stuttgart.de
[4] Höchstleistungsrechenzentrum Stuttgart, Stuttgart, Germany roller@hlrs.de
[5] Institut für Hochleistungsimpuls- und Mikrowellentechnik, Forschungszentrum Karlsruhe, Germany rudolf.schneider@ihm.fzk.de

Summary. This paper describes the numerical modeling of rarefied plasma flows under conditions where continuum assumptions fail. We numerically solve the Boltzmann equation for rarefied, non-continuum plasma flows, making use of well known approaches as PIC (Particle in Cell) and as DSMC (Direct Simulation Monte Carlo). The mathematical and numerical modeling is explained in some detail and the required computational resources are investigated.

1 Introduction

Within the small satellite program of the Institute of Space Systems at University of Stuttgart a lunar satellite is under development. The main propulsion system will consist of a cluster of instationary magnetoplasmadynamic (IMPD) thruster, also known as pulsed plasma thruster. The duration of a single pulse of the IMPD thruster is of the order of 8 μs. The current of about 30 kA allows an acceleration of the propellant mass bit leading to exhaust velocities of about 12 km/s, i.e. a specific impulse of approximately 1200 s [24]. Due to the instationary operation and the degree of rarefaction, discontinuities in the space distribution of the propelling plasma are to be expected. In order to significantly enhance the understanding of such an electric space propulsion system a realistic simulation of the observed rarefied plasma flow is needed. In this paper we describe the development of such a plasma flow simulation tool for complex three-dimensional geometries. The circumstance of rarefied plasma flow requires a kinetic description: The interaction of charged particles with electromagnetic fields requires the solution of the time-dependent full Maxwell-Vlasov system in three space dimensions for complex geometries. An attractive numerical method for this system is the Particle-in-Cell (PIC) approach [4, 14]. Unfortunately, this method does not take into account elastic and inelastic scattering between particles. However, since these interactions can play an important role for the thrust of electric propulsion systems, it is necessary to include them. In our approach the exchange of momentum and energy as well as chemical reactions is calculated via a Direct Simulation Monte Carlo (DSMC) method [3, 23]. A new developed Fokker-Planck solver [7], also using PIC techniques in velocity space, numerically models electron-electron and electron-ion Coulomb collisions in a self-consistent way. It is expected that the coupling of these models, i.e. PIC-DSMC-FP, allow for an accurate prediction of the thrust of electric space propulsion systems operating far from continuum.

In general, the coupling of different particle approaches for the numerical investigations of rarefied flows in the field of electric propulsion is a well known proceeding. In case of Hall thrusters it is common to treat the electrons as a fluid and the atoms and ions as particles, see e.g. [18, 6]. A rise in computational power allowed to apply fully kinetic (PIC-DSMC) approaches [17]. Most of the available Hall thruster studies are based on additional simplifying assumption concerning the neutral particle distribution, see e.g. [11].

In the field of pulsed plasma thrusters (PPT), much less particle based studies are available, see e.g. [12, 5, 15]. However, these models cannot resolve the question about how strong the deviation of the Maxwellian electron distribution is due to high gradients, unknown collisionality and electron transport.

2 Mathematical-Physical Model

2.1 Boltzmann Equation

According to the situation of rarefied, non-continuum plasma flow in the pulsed thruster, we have to start from the Boltzmann equation, given by

$$\frac{\partial f_i}{\partial t} + \mathbf{c}_i \cdot \nabla_x f_i + \frac{\mathbf{F}_i}{m_i} \cdot \nabla_c f_i = \left(\frac{\delta f_i}{\delta t}\right)_{Col} . \tag{1}$$

It describes the change of the velocity distribution function f_i for species i in time and phase space as a result of self and external forces and particle collisions. The right-hand side represents the Boltzmann collision integral [8]

$$\left(\frac{\delta f_i}{\delta t}\right)_{Col} = \sum_j n_j \int g(\mathbf{c}_i, \mathbf{c}_j)\, \sigma_{ij}(g, \mathbf{g}{\cdot}\mathbf{g}') \left[f_i'(\mathbf{c}\,_i')\, f_j'(\mathbf{c}\,_j') - f_i(\mathbf{c}_i)\, f_j(\mathbf{c}_j)\right] d\Omega\, d\mathbf{c}_j \tag{2}$$

which reflects the rate of change of f_i over time due to collisions. Here, the index j stands for all "scattering" populations, $n_j = n_j(\mathbf{x}, t)$ is the density of species j, $\mathbf{g} = \mathbf{c}_i - \mathbf{c}_j$ represents the relative velocity, $\sigma_{ij}(g)$ is the differential scattering cross section between the particles of the species i and j, and the differential solid angle $d\Omega$ is given by $d\Omega = \sin\theta \, d\theta \, d\phi$. The primed quantities refer to values after a collision and the unprimed ones denote the pre–collisional values. From the mathematical point of view, the Boltzmann equation is a complex integro-differential equation for the velocity distribution function. Up to now, a general solution of the Boltzmann equation to describe macroscopic problems is not possible. Hence, the different contributions to the Boltzmann equation have to be decoupled and treated separately according to the physical situation on hand.

2.2 Maxwell-Vlasov Equations

The Lorentz force $\mathbf{F}_i$ to particles with charge q_i in (1) is given by

$$\mathbf{F}_i = q_i \left[\mathbf{E}(\mathbf{x}, t) + \mathbf{c}_i \times \mathbf{B}(\mathbf{x}, t) \right] , \tag{3}$$

which depends on the velocity $\mathbf{c}_i$, the electric field $\mathbf{E}$, and the magnetic induction $\mathbf{B}$. Furthermore, in the case of highly ionized rarefied plasma flows the collision term can be neglected and the Boltzmann equation reduces to the Vlasov equation. The general solution of the Vlasov equation is given by its characteristics

$$\frac{d\mathbf{p}_i(t)}{dt} = \mathbf{F}_i(\mathbf{c_i}, \mathbf{x}, t) \quad \text{and} \quad \frac{d\mathbf{x}_i(t)}{dt} = \mathbf{c}_i(t), \tag{4}$$

that are called Lorentz equations in the following. Here, the relativistic momentum is given by $\mathbf{p}_i = m_i \gamma \mathbf{c}$ with the Lorentz factor $\gamma^2 = 1 + (\mathbf{p}_i)^2/(m_i c)^2$, where c denotes the speed of light. The difficulties in solving the Lorentz equations arise from the fact that $\mathbf{E}$ and $\mathbf{B}$ are not given explicitly. In fact, they have to be calculated at each time step in a self-consistent manner [22] from the Maxwell equations which consist of the two hyperbolic evolution equations

$$\frac{\partial \mathbf{E}}{\partial t} - c^2 \nabla \times \mathbf{B} = -\frac{\mathbf{j}}{\epsilon_0} , \quad \frac{\partial \mathbf{B}}{\partial t} + \quad \nabla \times \mathbf{E} = 0 \tag{5}$$

(Ampère's and Faraday's law) and the elliptic parts

$$\nabla \cdot \mathbf{E} = \frac{\rho}{\epsilon_0} , \quad \nabla \cdot \mathbf{B} = 0 \tag{6}$$

(Gauss' law and absence of magnetic monopoles), where the electric permittivity ϵ_0 and magnetic permeability μ_0 are related to the speed of light c according to $\epsilon_0 \mu_0 c^{\ 2} = 1$. For given charge and current densities ρ and $\mathbf{j}$, the Maxwell equations describe the temporal and spatial evolution of the electric field and the magnetic induction. With an integration over the entire momentum space, the charge ρ and current density $\mathbf{j}$ are obtained from

$$\rho = \sum_i q_i \int f_i(\mathbf{x}, \mathbf{p}, t) d^3p \quad \text{and} \quad \mathbf{j} = \sum_i q_i \int \mathbf{c}_i(\mathbf{p}) f_i(\mathbf{x}, \mathbf{p}, t) d^3p \,. \tag{7}$$

Up to this point, no numerical approximations were made and the description is exact. For the numerical realization of the Maxwell-Lorentz system, the PIC method is applied [13], where f_i is expressed by a weak approximation [22], yielding the following expressions for charge and current density

$$\rho^* = \sum_i q_i \sum_{k=1}^{N_i} \delta \left[\mathbf{x} - \mathbf{x}_i^{(k)}(t) \right] \,, \quad \mathbf{j}^* = \sum_i q_i \sum_{k=1}^{N_i} \mathbf{c}_i^{(k)} \delta \left[\mathbf{x} - \mathbf{x}_i^{(k)}(t) \right] \,, \tag{8}$$

where the superscript $^{(k)}$ denotes the kth particle of species i, N_i is the total number of particles within this group and δ represents the Dirac function. In order to determine the contribution of all charged particles, shape-functions are used to calculate ρ and $\mathbf{j}$ at the grid nodes. With these charge and current densities the new electromagnetic fields are computed at the nodes and afterwards interpolated to the local particle postions. This procedure has to be repeated at each time step.

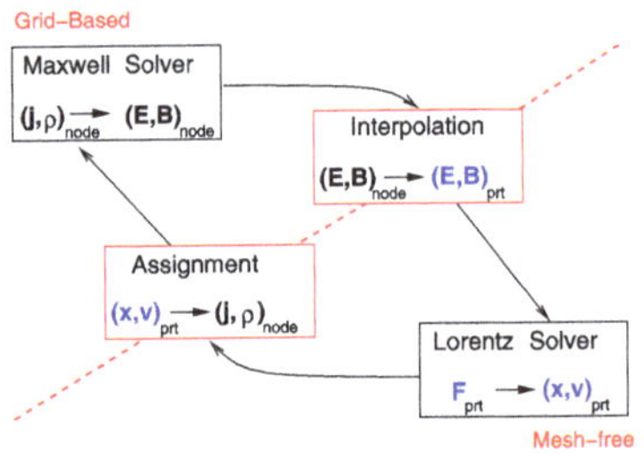

Fig. 1. Standard PIC concept

The numerical methods used to solve the Maxwell equations is based on explicit finite volume techniques combined with a hyperbolic divergence correction [21] to ensure the divergence constraints (6), yielding a very efficient and highly flexible Maxwell solver module for PIC applications on unstructured grids and for parallel computing.

2.3 Modeling of Long Range Charged Particle Collisions

We consider a plasma where electron-electron (e,e) collisions play an important role. In such plasmas the shape of the electron energy distribution function (EEDF) is mainly determined by the (e,e)-interactions which themselves try to drive the EEDF towards a Maxwellian distribution. However, in this situation a competition with inelastic electron-neutral reactions occurs, which depletes the high-energy tail. Energetic considerations indicate that the high-energy tail controls reactions like atomic excitation and ionization, and to some extent the plasma chemistry. Obviously, since the EEDF determines many properties of the plasma, it is essential to model (e,e)-collisions as realistic as possible. In the following, we describe the formulation that allows to include (e,e)-collisions into a PIC framework in a natural way. A well-established and for our purposes suitable mathematical model for (e,e)-collisions is given by the Fokker-Planck (FP) equation

$$\left(\frac{\delta f}{\delta t}\right)_{Coll} = -\frac{\partial}{\partial c_j}\left[\mathcal{F}_j\, f\right] + \frac{1}{2}\frac{\partial^2}{\partial c_j \partial c_k}\left[\mathcal{D}_{jk}\, f\right] , \tag{9}$$

where Einstein's summation convention has been adopted. This model describes the evolution of the electron distribution function $f = f_e(\mathbf{x}, \mathbf{c}, t)$ as a result of small-angle scattering of Coulomb point particles. The key quantities to determine the coefficients of dynamical friction $\mathcal{F}_j(\mathbf{x}, \mathbf{c}, t) \sim \partial\mathcal{H}/\partial c_j$ and diffusion $\mathcal{D}_{jk}(\mathbf{x}, \mathbf{c}, t) \sim \partial^2\mathcal{G}/\partial c_j \partial c_k$ are the Rosenbluth potentials [20]

$$\mathcal{H}(\mathbf{x}, \mathbf{c}, t) = 2\int\limits_{-\infty}^{\infty} \frac{f(\mathbf{x}, \mathbf{w}, t)}{|\mathbf{g}|}\, d^3w \; , \quad \mathcal{G}(\mathbf{x}, \mathbf{c}, t) = \int\limits_{-\infty}^{\infty} |\mathbf{g}|\, f(\mathbf{x}, \mathbf{w}, t)\, d^3w \; , \tag{10}$$

where $\mathbf{g} = \mathbf{c} - \mathbf{w}$ is the difference between the velocity of the scattered-off electrons and the velocity of the electrons that serve as scatterer. Clearly, the friction force $\boldsymbol{\mathcal{F}}$ and the diffusion tensor $\mathbb{D}$ themselves depend on the velocity $\mathbf{c}$ and, hence, the FP model generally is a complex non-linear problem, that has to be solved numerically in a self-consistent manner. The FP equation (9) for the evolution of f is equivalent to the stochastic differential equation (SDE) in the Itô sense [10, 16]

$$d\mathbf{C}(t) = \boldsymbol{\mathcal{F}}(\mathbf{C}, t)\, dt + \mathbb{S}(\mathbf{C}, t)\, d\mathbf{W}(t) \; , \tag{11}$$

where $\mathbf{W}(t)$ represents the three-dimensional Wiener process and the matrix $\mathbb{S}$ is related to the diffusion matrix according to $\mathbb{D} = \mathbb{S}\,\mathbb{S}^T$. As indicated, both quantities $\boldsymbol{\mathcal{F}}$ and $\mathbb{S}$ now depend on the stochastic variable $\mathbf{C} = \mathbf{C}(t)$, which will be identified later as the velocity of a single electron. Hence, the use of the SDE (11) fits in a remarkable way into the standard PIC approach, which is one basic concept of the present code development.

The assumption of an isotropic but non-Maxwellian velocity distribution of the scatterer implies an enormous reduction of the problem [20]. In order to be free of any model or assumption, we start from the Rosenbluth potentials (10) and apply Fourier transformation techniques to compute the integrals in velocity space which results in first principle, fully self-consistent determination of the deterministic friction and stochastic diffusion [7].

2.4 Inelastic Particle Collisions: The DSMC Approach

The DSMC technique is an attractive and well-established approach for the simulation of rarefied gas flows [3]. Developed during the 1970s by Bird [2], the method has been continuously improved and used, see e.g. [1, 9, 19].

The main underlying assumption within DSMC is that particle interactions take place instantaneously, thus enabling a separate treatment of particle interaction and movement by choosing a time step size smaller than the mean time between collisions and a cell size smaller than the mean free path. This is applicable especially to short-range interactions between neutral particles or between charged and neutral particles such as electrons and atoms. Consequently, each cell can be treated separately and particles inside one cell are not affected by anything that happens outside. Particle interaction is computed by use of the the Variable Hard Sphere model (VHS) [2] which is based on an energy dependant particle diameter. Species data is obtained by fitting numerical viscosity data to experimental values.

Recently, the VHS model was extended by replacing the diameter approach by a cross section approach [25]. Technically, this yielded the simple hard sphere model but enabled in general the treatment of a huge variety of interaction types as long as the cross section date is available. The model was applied to a simple Argon plasma allowing to perform verification simulations.

3 Program Structure and Code Coupling

Originally, the PIC concept (see Figure 1) was developed to tackle the numerical solution of the Maxwell-Vlasov equations. For that, the plasma is represented by a sample of different species of charged particles. In each time step the electromagnetic fields obtained by the numerical solution of the Maxwell equations are interpolated to the actual locations of these particles. According to the Lorentz force (3) the charges are redistributed and the new phase-space coordinates are determined by solving the usual laws of dynamics. To close the chain of self-consistent interplay, the particles have to be located with respect to the computational grid in order to assign the contribution of each charged particle to the changed charge and current density at the nodes which are the sources for the Maxwell equations in the subsequent time step. Such a basic PIC scheme is the starting point to build up the necessary program structure of the coupled code (see Figure 2), which allows for flexible combination of different modules and extensions. Realistic simulation of the rarefied plasma flow in the pulsed thruster requires the inclusion of different classes of particle interaction. For that, first the PIC cycle is extended by two additional modules: The Fokker-Planck solver treats the long range character of the Coulomb force to model collision relaxation of electrons and the DSMC module is able to handle short range interactions between like and unlike particles and the complex plasma chemistry. In contrast to the standard PIC Lorentz solver (cf. Figure 1) we now take into account all velocity changes, those due to the Lorentz force as well as short and long range interactions respectively. This results in the separation of the particle push routine from the Lorentz solver.

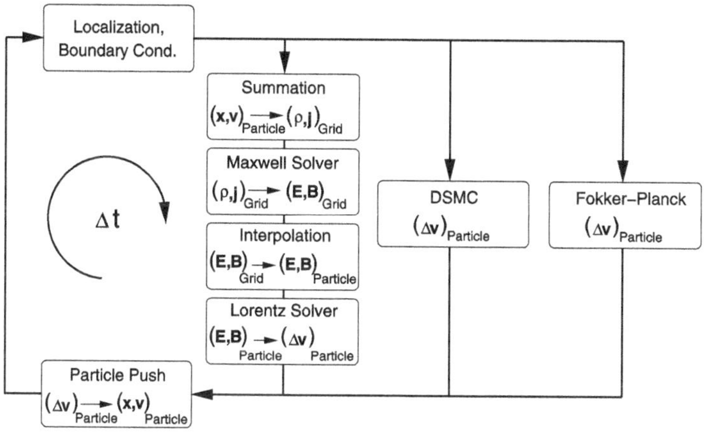

Fig. 2. Coupling concept

4 Numerical Results

For a successive build up of the program, each part of the code was separately tested and verified. In order to assess the interplay of the different building blocks of the Maxwell-Lorentz solver we applied the testcase of crossed particle beams between two condensator plates with constant external E- and B-field. Numerical results obtained with the 2D (left) and 3D (right) Maxwell-Lorentz module are depicted in Figure 3. Up to 300 million particles were used.

The quality and reliability of the numerical approximation methods used in the 3D Fokker-Planck solver for self-consistent intra-species collisional relaxation is demonstrated by the results presented in the Figures 4 and 5 for a typical code assessment experiment.

Here, the numerical experiment is initialized by Gaussian-shaped velocity distributions in v_x, v_y and v_z direction with mean $\mu = 0$ and three different variances ($\sigma_x^2 = 1$ (open squares), $\sigma_y^2 = 2.25$ (open triangles) and $\sigma_z^2 = 4$ (open circles)) as seen in Figure 4 (left). Additionally, the final states of the simulation (after $3 \cdot 10^3$ temporal iterations with $\Delta t = 0.05$) are plotted in this figure (red full line) which are an isotropic distribution function with $\sigma_f^2 \approx 2.4$. Essentially the same results are once again seen in Figure 4 (right plot), where now the velocity distributions are computed as functions of the modulus of the velocity. Obviously, the numerically obtained Maxwellian-shaped distribution function (blue open circles) is very close to the exact result (red full line). In Figure 5 the temporal evolution of the variance of the different initial distribution functions are plotted (lines with open symbols). At time $t \approx 110$, all variances have the same value, according to the equipartition principle. However, we observe that the "final" variance increases systematically for longer times due to different approximation errors. This numerical heating

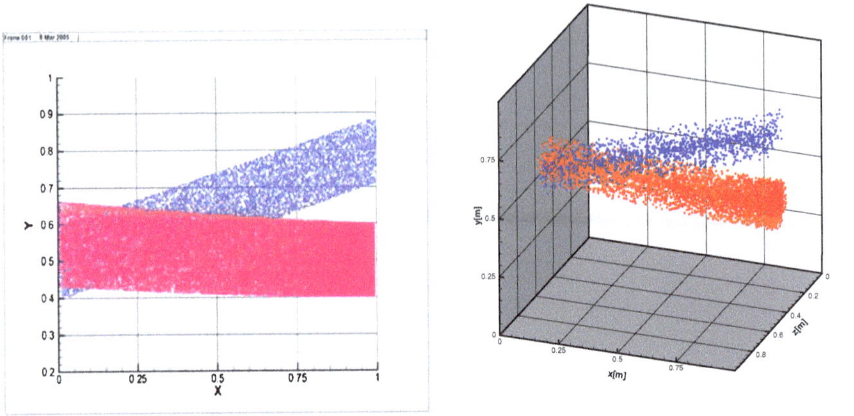

Fig. 3. Example of crossed particle beam (blue electron emitted from the left and red protons from the right) in two and three dimension between condensator plates with constant E- and B-field

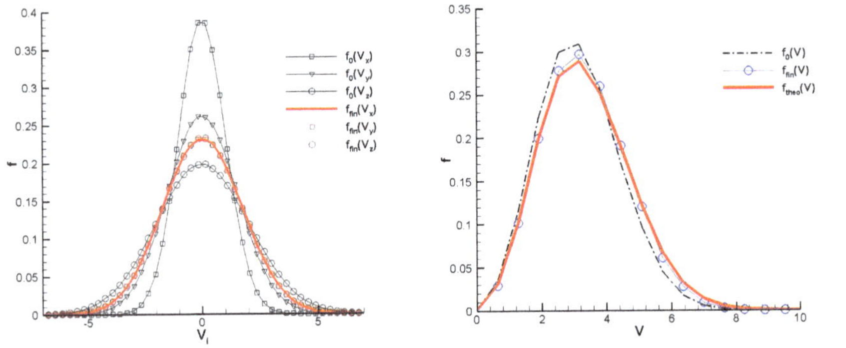

Fig. 4. Left plot: Initial (lines with open symbols) and final (full lines and symbols on it) velocity distribution functions. Right Plot: Initial (dashed-dotted line), final (open symbols) and exact equilibrium (full line) distributions functions over the modulus of the velocity

can be circumvented by applying an advanced renormalization technique (see, [7]). It is obvious from Figure 5 (full, dashed and dashed-dotted line) that this extended approach suppresses the numerical heating successfully and the equilibrium variance stays constant in the long-time regime.

Two 3D particle beam testcases (Figures 6 and 7) are presented in order to demonstrate the DSMC based particle interaction as well as the coupling to the underlying PIC framework, see Figure 2. Neither PIC nor Fokker–Planck do contribute to the simulation except with the time step size of 10^{-13} s which is predetermined by PIC. The green beam consists of C and F atoms. The blue one is an electron beam crossing the atom beam orthogonally. Each

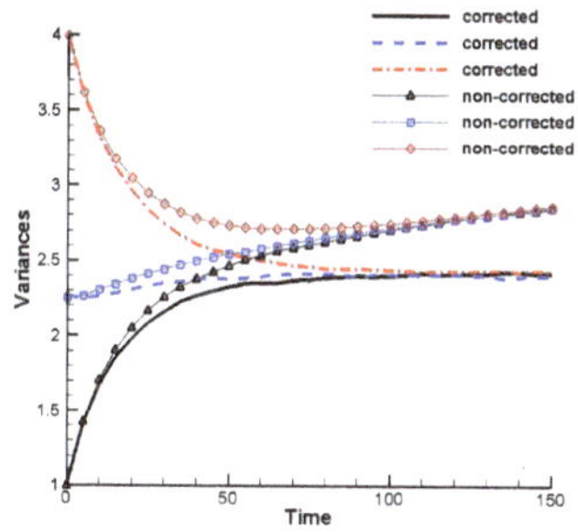

Fig. 5. Temporal evolution of the different initial variances. lines with open symbols: no correction; full, dashed, and dashed-dotted line: correction with renormalization technique

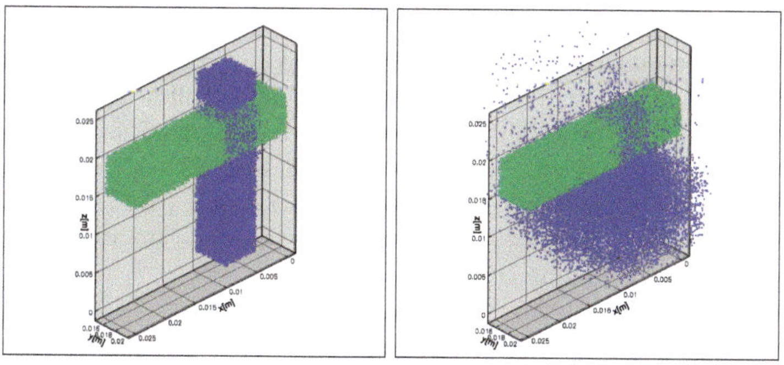

Fig. 6. 3D PIC-DSMC testcase showing elastic electron (blue) scattering on C and F atoms (green) after 1000 (left) and 10000 (right) iteration steps

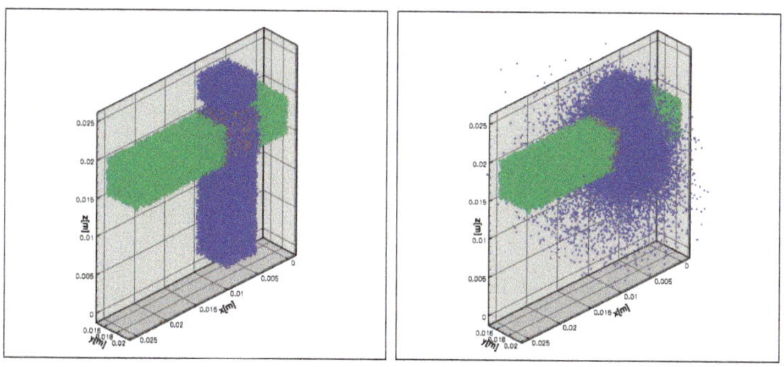

Fig. 7. The same testcase as in Figure 6 showing electron impact ionization (red) of ground–state C and F atoms after 1000 (left) and 10000 (right) iteration steps

simulated particle represents 10^{14} real particles. The reaction evaluation is based on the scheme discussed in [25]. All particles are initiated with $T = 300\ K$ with an additional one–dimensional macroscopic velocity of 1000 m/s

for the atoms and $5 \cdot 10^6\ m/s$ for the electrons. The kinetic electron velocity corresponds to approximately 70 eV which is far above the ionization energy of $E_{i,C} = 11.3\ eV$ and $E_{i,F} = 17.4\ eV$. In the first case, ionization is switched off leading to elastic scattering behavior. Figure 6 depicts an early and a late phase of the simulation. In the second case, electron impact ionization yields a high degree of ionization in the crossing area. This results in an unresisted transition of the subsequent electrons, see Figure 7.

4.1 Parallelization Concept

In order to tackle the challenging task of rarefied plasma simulation, strategies for efficient parallelization and load balancing of dynamically moving particles and their neighborhood relationships have to be developed.

The optimal parallelization strategy for the mesh-based Maxwell solver is the standard domain decomposition technique. Concurrently, the optimal parallelization strategy for the individual movement of charged particles would be a mesh-independent distribution of the particles. The following re-localization of the particles in the mesh cells again needs mesh-information. When particles leave the domain/processor they are located in, also non-local information is needed. Thus, the basic concept for parallelization is a particle-weighted domain decomposition, guaranteeing that all particles belonging to the same domain remain on the same computational node.

Besides reduction of computing time, the other major goal of parallelizing is to increase the usable memory, thus to allow for larger number of particles.

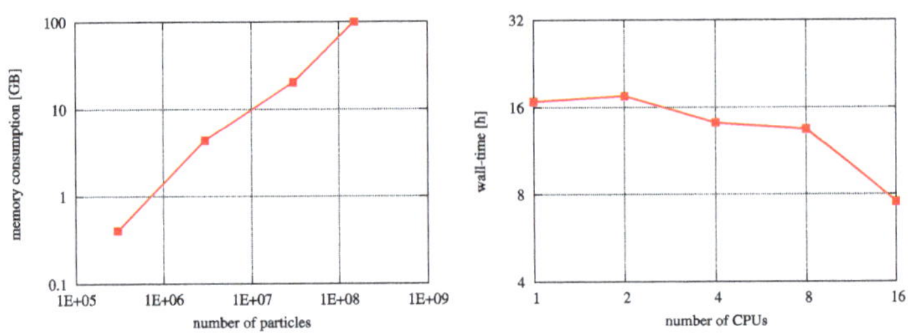

Fig. 8. Memory consumption for a given number of particles (left). The mostly poor speedup in the current parallelization reflects poor load balancing (right)

As figure 8 shows, the memory consumption for the crossed particle beams testcase increases strongly with the number of (macro-)particles. Therefore, an MPI-parallelization as described above was implemented, enabling the simulation of an increased number of particles distributed over the memory of several nodes. Several parallel computations have been performed, the largest of which was distributed over 64 CPUs, allocating a total of 846 GB of memory for a maximum number of computed particles of $64 \cdot 2 \times 10^7 = 1.28 \times 10^9$.

However, figure 8 (right) also demonstrates the limits of the currently implemented parallelization: Due to a poor load balancing of the particle algorithm, the speedup is poor. This is caused by the domain decomposition. Since the domain decomposition is currently based solely on the number of cells, not on the particles contained within, the load balancing inevitably suffers in the case of an inhomogeneous distribution of the particles, such as in the current testcase. For the case of a decomposition into eight domains, the disregard of the particle density results in more than half the number of particles being located within the domain of a single process and five domains without a single particle inside. The next step, therefore, will be the improvement of the load balancing through an improved domain decomposition, taking into account the number of particles in a domain as weights in the ParMetis call.

5 Summary and Conclusions

In order to model the gas flow in electric space propulsion systems under conditions where continuum assumptions fail a statistically based numerical scheme is developed. The scheme combines well known approaches from Particle-in-Cell (PIC) and Direct Simulation Monte Carlo (DSMC) with a new solver based on the Fokker Planck (FP) equation in order to account for long range particle interactions of charged species. While the simulation of unsteady devices like pulsed plasma thrusters mainly requires for a large number of simulated particles the description of tethered propulsion like CETEP is associated with a vast computational area as well. Therefore, big memory requirements combined with extreme demands concerning computational work do arise. In order to allow for an efficient and accurate solving the scheme is being parallelized and optimized. The combination of the PIC, DSMC and FP modules which constitutes a combination of grid based and particle based operations requires a balance between different optimisation and parallelization techniques.

So far, mainly the separate building blocks have been tested separately. In future, focus will be increased on coupled examples.

Acknowledgements

We gratefully acknowledge the Landesstiftung Baden-Württemberg who funded the development within the program "Modeling and Simulation on High Performance Computers" from 2003–2005 and the Deutsche Forschungsgemeinschaft (DFG) for funding within the project "Numerische Modellierung und Simulation hochverdünnter Plasmaströmungen". T. Stindl wishes to thank the State of Baden-Württemberg and the Erich-Becker-Stiftung, Germany, for their financial support. Computational resources have been provided by the Bundes-Höchstleistungsrechenzentrum Stuttgart (HLRS).

References

1. J. Balakrishnan, I.D. Boyd, and D.G. Braun. Monte Carlo simulation of vapor transport in physical vapor deposition of titanium. *J. Vac. Sci. Technol. A*, 18(3):907–916, May/Jun 2000.
2. G.A. Bird. Monte-Carlo simulation in an engineering context. Progress in Astronautics and Aeronautics;Rarefied Gas Dynamics, Vol. 74, Part 1, edited by Sam S. Fisher, 1981, pp. 239-255, 1981.
3. G.A. Bird. *Molecular Gas Dynamics and the Direct Simulation of Gas Flows.* Clarendon Press, Oxford, 1994.
4. C.K. Birdsall and A.B. Langdon. *Plasma Physics via Computer Simulation.* Adam Hilger, Bristol, Philadelphia, New York, 1991.
5. I.D. Boyd, M. Keidar, and W. McKeon. Modeling of a pulsed plasma thruster from plasma generation to plume far field. *Journal of Propulsion and Power*, 37(3):399–407, 2000.
6. S. Cheng, M. Santi, M. Celik, M. Martinez-Sanchez, and J. Peraire. Hybrid pic-dsmc simulation of a hall thruster plume on unstructured grids. *Computer Physics Communications*, 164:73–79, 2004.
7. D. D'Andrea, C.-D. Munz, and R. Schneider. Modeling of electron-electron collisions for particle-in-cell simulations. *FZKA 7218 Research Report, Forschungszentrum Karlsruhe – in der Helmholtz-Gemeinschaft*, 2006.
8. D. Diver. *A Plasma Formulary for Physics, Technology, and Astrophysics.* Wiley-VCH Verlag, Berlin, 2001.
9. D.J. Economou and T.J. Bartel. Direct Simulation Monte Carlo (DSMC) of Rarefied Gas Flow During Etching of Large Diameter (300-mm) Wafers. *IEEE Trans. Pl. Sci.*, 24, No. 1:131–132, Feb 1996.
10. C.W. Gardiner. *Handbook of Stochastic Methods.* Springer Verlag, Berlin, Heidelberg, New York, 1985.
11. L. Garrigues, A. Heron, J.C. Adam, and J.P. Boeuf. Hybrid and particle-in-cell models of a stationary plasma thrusters. *Plasma Sources Sci. Technol.*, 9:219–226, 2000.
12. N.A. Gatsonis and X. Yin. Hybrid (particle-fluid) modeling of pulsed plasma thruster plumes. *Journal of Propulsion and Power, Vol. 17, No. 5, September-October 2001, 945–958*, 2001.
13. R. Hockney and J. Eastwood. *Computer Simulation using Particles.* McGraw-Hill, New York, 1981.
14. G.B. Jacobs and J.S. Hesthaven. High-order nodal discontinuous Galerkin particle-in-cell method on unstructured grids. *J. Comput. Phys.*, 214:96–121, 2006.
15. M. Keidar, I.D. Boyd, E. Antonsen, and G.G. Spanjers. Electromagnetic effects in the near-field plume exhaust of a micro-pulsed-plasma thruster. *Journal of Propulsion and Power*, 20(6):961–969, November 2004.
16. P.E. Kloeden and E. Platen. *Numerical Solution of Stochastic Differential Equations.* Springer-Verlag, Berlin, Heidelberg, New York, 1999.
17. K. Komurasaki, S. Yokota, S. Yasui, and Y. Arakawa. Particle simulation of plasma dynamics inside an anode-layer hall thruster. 40th AIAA/ASME/SAE/ASEE Joint Propulsion Conference and Exhibit, 11-14 July, Fort Lauderdale, Florida, USA, 2004.
18. J.W. Koo and I.D. Boyd. Computational model of a hall thruster. *Computer Physics Communications*, 164:442–447, 2007.

19. M. Laux. *Direkte Simulation verdünnter, reagierender Strömungen.* PhD thesis, Institut für Raumfahrtsysteme, Universität Stuttgart, Germany, 1996.
20. M. Mitchner and C. Kruger. *Partially Ionized Gases.* Wiley, New York, 1973.
21. C.-D. Munz, P. Omnes, and R. Schneider. A three-dimensional finite-volume solver for the Maxwell equations with divergence cleaning on unstructered meshes. *Computer Physics Communications*, 130:83–117, 2000.
22. C.-D. Munz, R. Schneider, E. Sonnendrücker, E. Stein, U. Voß, and T. Westermann. A finite-volume particle-in-cell method for the numerical treatment of the Maxwell-Lorentz equations on boundary-fitted meshes. *Int. J. Numer. Meth. Engng.*, 44:461–487, 1999.
23. K. Nanbu, T. Morimoto, and M. Suetani. Direct simulation monte carlo analysis of flows and etch rate in an inductively coupled plasma reactor. *IEEE Trans. Pl. Sc.*, 27(5):1379–1388, 1999.
24. A. Nawaz, M. Auweter-Kurtz, G. Herdrich, and H. Kurtz. Investigation and optimization of an instationary MPD thruster at IRS. International Electric Propulsion Conference, Princeton, USA, 2005.
25. D. Petkow, M. Fertig, G. Herdrich, and M. Auweter-Kurtz. Ionization Model within a 3D PIC-DSMC-FP Code. AIAA-2007-4261, 39th AIAA Thermophysics Conference, Miami, FL, USA, 2007.

Doing IO with MPI and Benchmarking It with SKaMPI-5

Joachim Mathes, Argirios Perogiannakis, and Thomas Worsch

IAKS, Universität Karlsruhe, Germany, worsch@ira.uka.de

Summary. SKaMPI-5 is a micro-benchmark for MPI implementations, designed to be easily extensible.

Besides a "global" parallel file system at least some parallel machines offer (many) hard disks which are local to (processors or) computing nodes. While the MPI2 standard includes functions for doing disk IO, MPI is unable to use this resource because by (our) definition the local disks are not visible to remote processors/nodes. But it is possible to write thin layers of code which allow this kind of access. If used with many hard disks, theoretically very high disk IO bandwidths are possible. In the present paper we have a closer look at one such approach, and give an overview of the extension of SKaMPI-5 for the evaluation of this. We also present a few selected results of benchmark runs for MPI-IO, giving an impression of what is possible with the MPI-IO measurements available in SKaMPI-5. In addition we point out several problems one is facing.

1 Introduction

"The MPI standard defines a set of powerful collective operations useful for coordination and communication between many processes. Knowing the quality of the implementations of collective operations is of great interest for application programmers. [...]"

The above text (Worsch et al., 2003) describes one of the major motivations for the development of SKaMPI, the Special Karlsruher MPI-Benchmark. In this text "collective communication" can of course replaced with other parts of MPI-1 or MPI-2. As a consequence, for example benchmarks for one-sided communication have been integrated into SKaMPI-5 (Augustin et al., 2005).

In the present paper we report on some experiences with and results obtained with SKaMPI benchmarks for MPI-IO (Perogiannakis, 2006). These come in two flavors: On one hand we give some numbers in order to demonstrate the possibilities offered. On the other hand we describe a small set of functions which give users relatively convenient possibilities to use local hard disks in a "parallel" way (Mathes, 2008). These functions were benchmarked with SKaMPI-5, too, and a comparison with the capabilities of global file systems is worthwhile.

SKaMPI (`http://liinwww.ira.uka.de/~skampi/`) measures the performance of an MPI implementation on a specific underlying hardware. By providing not simply one number, but detailed data about the performance of single MPI operations under specific circumstances, e.g. the presence or absence of specific file hints, one can judge the consequences of design decisions regarding the performance of the system to be built.

The rest of this paper is organized as follows: In Section 2 we give a very short overview over MPI-IO and benchmarks available for it. Section 3 describes on a not too detailed level the infrastructure offered by SKaMPI-5 for MPI-IO. Section 4 is devoted to first results obtained with it. Section 5 describes a first attempt at making use of many hard disks, which on some machines are available locally to processors. We conclude this paper in Section 6. The present paper is based on the diploma theses of the first author (Mathes, 2008) and the second author (Perogiannakis, 2006).

2 Overview

SKaMPI-5 offers many different functions for investigating different performance aspects of MPI-IO functions in an MPI-2 library. Before giving a rough impression of what is possible in the following section, we review some related work concerning MPI-IO benchmarks. At the same time we will already point out a few advantages and deficiencies of the current implementation in SKaMPI-5.

2.1 Related Other MPI Benchmarks

In this subsection we concentrate on the two benchmarks which seem to be the most relevant concerning IO.

The b_eff_io *Benchmark*

The b_eff_io benchmark has been developed at the High Performance Computing Center at Stuttgart (HLRS) in cooperation with Lawrence Livermore National Laboratory (LLNL). Its aim is the measurement of the effective parallel bandwidth of IO (Rabenseifner et al., 2004). The goal is to get

- "a characteristic average number for the I/O bandwidth achievable with parallel MPI-I/O applications",[1] as well as
- "detailed information about several access patterns and buffer lengths".[1]

[1] Citations from the manpage of b_eff_io.

To this end different standard IO patterns, namely scattering, strided collective, non-collective separated and (non-)collective segmented accesses, are benchmarked. For each pattern first write, rewrite and read accesses are benchmarked for chunk sizes (on disks) some of which are "wellformed" i.e. some powers of 2, while others are "non-wellformed", i.e. larger than a power of 2.

It is one of the advantages of b_eff_io, that it distinguishes between the "first write" and "rewrite" cases. Currently corresponding results can be obtained with SKaMPI-5 only in a somewhat inconvenient way. One has to run each measurement twice; first on a newly created file, which is written only once, and then a on a newly created file, which is written repeatedly.

An important aspect of b_eff_io is the fact, that it is designed to finish within approximately 30 to 60 minutes (at least on machines that are able to write their total memory to disk in about 10 minutes). After our first experiences with the MPI-IO part of SKaMPI-5 we consider this a very useful aspect. Of course, this comes with the price of getting only a restricted view on the system; if the time limit for the measurement of one parameter set is exceeded, the measurement is aborted and the next parameter set is considered. In the current version 2.1 of b_eff_io the elementary datatype ("`etype`") for all operations is `MPI_DOUBLE`. Furthermore, in order to be able to stay within some time limit, b_eff_io uses only some access patterns, and restricts itself to some fixed filetypes. Nevertheless one does get a good overall impression of the system.

Intel MPI Benchmark

The Intel MPI Benchmark (IMB) is currently available as version 2.3. It can be used to do MPI-IO benchmarks. To this end the times needed by single calls to MPI routines measured. There are several categories:

- Single transfer benchmarks: In this so-called "local mode" only one process is active (and only non-collective operations are benchmarked).
- Parallel transfer benchmarks: In this "global mode" several processes are writing to or reading from disjoint parts of a file or different files. Several processes are writing to or reading from disjoint segments of a common file or separate files. Again, only functions using no shared file pointer are considered.
- Collective transfer benchmarks: These considering collective file accesses; all processes collectively call the same I/O-routine.

IMB does not do any measurements for standard access patterns. Data are transferred in one chunk between memories and files. Chunk sizes are always powers of 2, as are the sizes of communicators taken into account.

IMB exclusively uses `MPI_Byte` as `etype` and for the representation of data in memory. The MPI datatype for the filetype is either the `MPI_Byte` or a derived datatype consisting of as many bytes as the message is long. Datatypes with holes are not considered.

Others

Of course, there are some further MPI-IO benchmarks, for example IOR_MPIIO (Loewe et al., 2007), the Tile Reader benchmark (Ching et al., 2003), the MPI-IO test from Los Alamos National Lab (LANL) (Nunez, 2006), the NPBIO2.4-MPI benchmark, or PRIOmark, which was developed at the University of Rostock as part of the IPACS project and allows benchmarks of file systems and hard disks. But all of these systems (as well as others) usually cover only few of the possibilities offered by MPI-IO.

2.2 Related Papers

Among the papers on benchmarking parallel IO we want to mention the recent work by (Saini et al., 2007) where by using different benchmarks (most of them mentioned above) for the assessment of parallel I/O on two machines, as a matter of fact benchmarks are compared. It can be seen that there is another aspect which sounds (and is) obvious but should not be forgotten: It does make a difference which program is used to make some measurements.

In the future the easy extensibility of SKaMPI-5 might be useful to provide different approaches to benchmarking e.g. MPI-IO within one framework. This would make comparison like the one just mentioned easier. And it would be easier to try out new alternatives.

2.3 Problems

Whenever a benchmark produces some output numbers, it only means that it is possible to meet circumstances when and where it was possible to obtain these numbers.

E.g., when benchmarking collective communication operations, what one really benchmarks are latency and bandwidth realizable by the MPI implementation on the inter-processor network *at the time of the benchmark*. If one does not have exclusive access to the network, it may happen that another communication intensive application causes is significant degradation of the network throughput.

When benchmarking file IO, the situation becomes even worse. Not only is the network connecting the compute processors with the storage system a shared resource; the (on large machines: parallel) file system "itself", i.e. the processors and hard disks realizing it, are such a shared resource, too. It may happen, and it does happen that an IO-intensive job severely influences the benchmark results.

In our experience this problem cannot be taken too seriously. For a typical example of what can happen have a look at Figure 1. The diagrams are small, but it is not important to be able to read off some numbers. The important point is, that exactly the same measurement (of `MPI_File_close`) with identical parameters has been executed on different days on possibly different node groups of a parallel machine. In the left diagram the time axis goes up to 100 milliseconds, in the right up to 80 milliseconds. Furthermore, the "quality" of of the curves is clearly different. This example clearly shows how difficult it is to get reproducible benchmark numbers. Therefore we want to emphasize that the surprising results reported in later sections have been reproduced by different benchmark runs on different days.

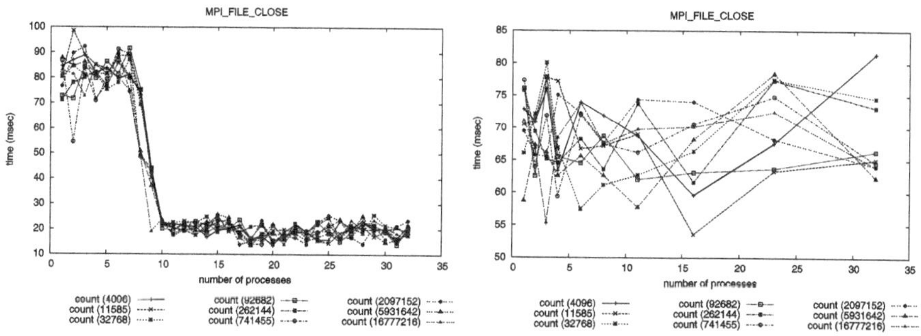

Fig. 1. Time needed by `MPI_File_close` on a parallel machine as measured by two separate benchmark runs on different days; on the x-axis the number of processes is plotted; different lines are for different sizes of the file to be closed

3 What SKaMPI-5 Offers for Benchmarking MPI-IO

The important part of SKaMPI-5's MPI-IO benchmarks are a number of pairs of SKaMPI measurement functions, one for writing and one for reading. Besides some varying parameters like `memsize`, `blocklen`, `stride`, etc., (which specify the lengths of some data segments in memory or in a file and which we will not discuss in detail here) all of these functions have four parameters in common:

- `MPI_Datatype etype`
- `MPI_Datatype filetype`
- `MPI_Datatype memtype`
- `int daccess`

The first two directly correspond to the parameters of the same name in the official documentation of `MPI_File_set_view`. The third parameter is the datatype used by MPI_File IO operations. The fourth parameter is SKaMPI specific: This integer is interpreted as a vector of bits, which are used to select whether

- positioning is via explicit offsets, individual or a shared file pointers,
- blocking or non-blocking MPI operations are to be used, and

- collective or non-collective MPI operations are to be used.

The pairs of write/read functions are:

- `MPI_IO_write_strided()` / `MPI_IO_read_strided()`
  These correspond to measurements of pattern type 0 or 1 in b_eff_io.
- `MPI_IO_write_separate()` / `MPI_IO_read_separate()`
  These correspond to measurements of pattern type 2 in b_eff_io.
- `MPI_IO_write_segmented()` / `MPI_IO_read_segmented()`
  These correspond to measurements of pattern type 3 or 4 in b_eff_io.
- `MPI_IO_write_noncontiguous()` / `MPI_IO_read_noncontiguous()`
  These can be used for benchmarks with non-contiguous filetypes.
- `MPI_IO_write_random()` / `MPI_IO_read_random()`
  These are a generalization of the strided functions.

In addition there are further functions for other purposes.

SKaMPI-5 tries to give its users much flexibility, for example for choosing the positioning method, the coordination (non-/collective), and the synchronism (non-/blocking) in many cases. Furthermore for many of the above functions it makes sense to use non-trivial datatypes. Unlike all benchmarks mentioned in Section 2.1, SKaMPI-5 allows to use such datatypes.

Since the performance of MPI-IO implementations (may) depend on the values of certain additional informations, in particular file hints, SKaMPI-5 also offers some helper functions for taking care of that. Among them are the following:

- `set_file_info()` / `reset_file_info()` / `print_file_info()`
- `set_io_working_dir()` / `get_io_working_dir()`
- `set_io_datarep()` / `get_io_datarep()`
- `set_io_atomicity()` / `get_io_atomicity()`
- `set_io_unique_open()` / `get_io_unique_open()`
- `set_io_preallocation()` / `get_io_preallocation()`
- `set_io_file_sync()` / `get_io_file_sync()`

File hints can be used to analyze the cooperation of the MPI library and the file system used in even more detail. See Subsection 4.4 for an example of a surprising result.

Furthermore SKaMPI-5 offers functions to benchmark the time needed by calls to some other MPI-IO related functions. The names below clearly indicate which calls measured:

- `MPI_IO_preallocate()`
- `MPI_IO_file_seek_shared()`
- `MPI_IO_delete_on_close()`
- `MPI_IO_close()`
- `MPI_IO_file_seek()`
- `MPI_IO_delete()`
- `MPI_IO_open()`
- `MPI_IO_open_close()`

A very much simplified excerpt of the example SKaMPI configuration file for benchmarking MPI-IO coming with the SKaMPI-5 sources looks like this:

```
FALSE = 0
TRUE  = 1

IO_EXPLICIT_OFFSETS          = 0x001
IO_INDIVIDUAL_FILE_POINTERS = 0x002
IO_SHARED_FILE_POINTER       = 0x004
IO_BLOCKING                  = 0x010
IO_NONBLOCKING_SPLIT         = 0x020
IO_NONCOLLECTIVE             = 0x100
IO_COLLECTIVE                = 0x200

set_io_datarep("native")
set_io_atomicity(TRUE)
set_io_preallocation(FALSE)
set_io_unique_open(TRUE)
set_io_file_sync(TRUE)

set_file_info ("romio_cb_read", "disable")
set_file_info ("romio_ds_read", "disable")

minlen = 4096
maxlen = 4096*4096
np = get_comm_size(MPI_COMM_WORLD)

daccess = IO_EXPLICIT_OFFSETS | IO_BLOCKING | IO_COLLECTIVE

begin measurement "MPI-IO read - segmented"
  for groupsize = 1 to np step *sqrt(2) do
    for count = minlen to maxlen step *sqrt(8) do
      measure comm(groupsize) :
        MPI_IO_read_segmented(count, count, count,
                        MPI_BYTE, MPI_BYTE, MPI_BYTE, daccess)
    od
  od
end measurement
```

Obviously it would be very tedious if one would have to write configurations like the above repeatedly for different specifications of daccess. To alleviate the task, it is also possible to program loops over different values for daccess in a simply way.

4 Some Results

There are many possible combinations of parameters which can be used for benchmarks of MPI-IO. The three possibilities for positioning, combined with blocking versus non-blocking and collective versus non-collective result in 12 different settings. In addition one can use file hints to adjust the behavior to different situations. There are binary switches like for example disabling or enabling data sieving and collective buffering, as well as the possibilities to set certain buffer sizes. In the present section we just show two examples of benchmark results which show some surprises and in our opinion support the point of view that one should whenever possible not base important implementation decisions on only a few benchmark data.

The results presented in this and the following section have been obtained on an HP XC6000 (Itanium 2 processors with a Quadrics interconnect, at Univ. of Karlsruhe), an HP XC4000 (dual core Opteron processors with Infiniband interconnect, at Univ. of Karlsruhe) both which have a Lustre parallel file system, and a NEC SX-8 (at HLR Stuttgart). We will refer to the machines as “XC6000”, “XC4000” and “SX8”.

4.1 Powers of Two, Once Again

It is not unusual to observe differences in running times depending on whether the size of messages is a power of two or not. This shows up again and again, for example when benchmarking collective communication. It also shows up in MPI-IO. A still rather moderate case has been observed on the SX8. Figure 2 shows the times needed by a single process for reading a whole file as one

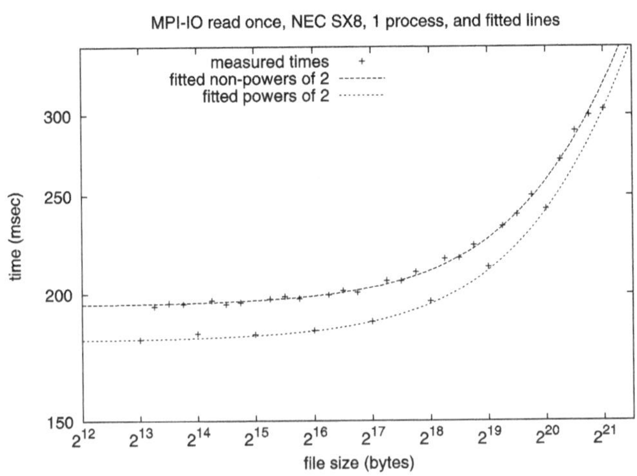

Fig. 2. SX8, a single process reading a single file in one chunk

chunk. In addition two lines lines are shown; one was fitted manually for the cases where the chunk size was a power of 2 and one for the case of non-power of 2 sizes. Figure 3 shows a more extreme case, a measurements on the XC4000 with SKaMPI's `MPI_IO_File_write_strided` function. Each process wrote a total of 4 MiB to disk in chunks of varying size (plotted on the x axis). Different lines in the diagram are for different numbers of processes. The stride was identical to the chunk length, hence there were no gaps between the chunks from different processes. One can see the generally not unusual, but in this case significant preference of data segments whose length is a power of 2, if the number of processes is larger than 4, i.e. larger than the size of one computation node.

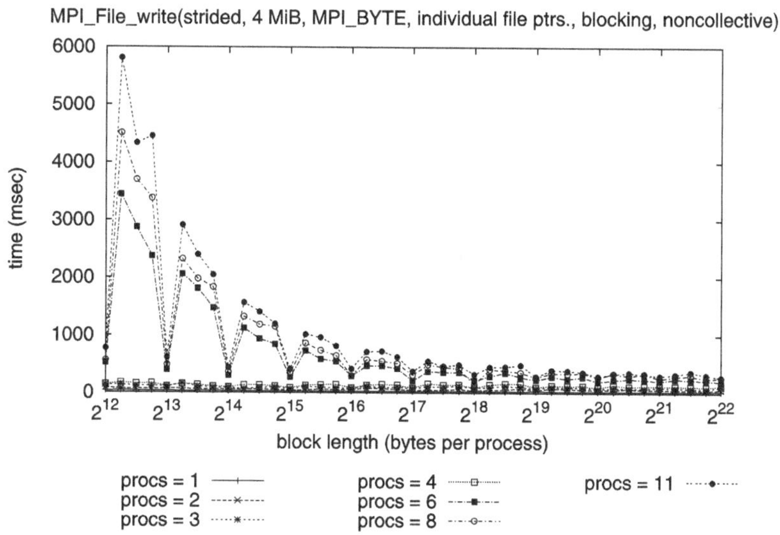

Fig. 3. XC4000 MPI File write strided

4.2 An Example Where Collective Versus Non-Collective IO Does (Not) Make a Difference

Another interesting aspect was found when comparing the results for `MPI_IO_File_write_once` and `MPI_IO_File_read_once` with individual file pointers and blocking synchronization on the XC4000 and the XC 6000. All processes access their own private separate files. In the case of *writing* to the file system there is no essential difference between the collective case (using `MPI_File_write_all`) and the non-collective case (`MPI_File_write`). In the case of *reading* from the file system one gets the diagrams in Figure 4. This time on the XC6000 there is a notable difference between the non-collective (left) and collective (right) version (but not on the XC4000). The diagrams are too small to read absolute numbers. But the important point is the qualitative behavior which should be obvious.

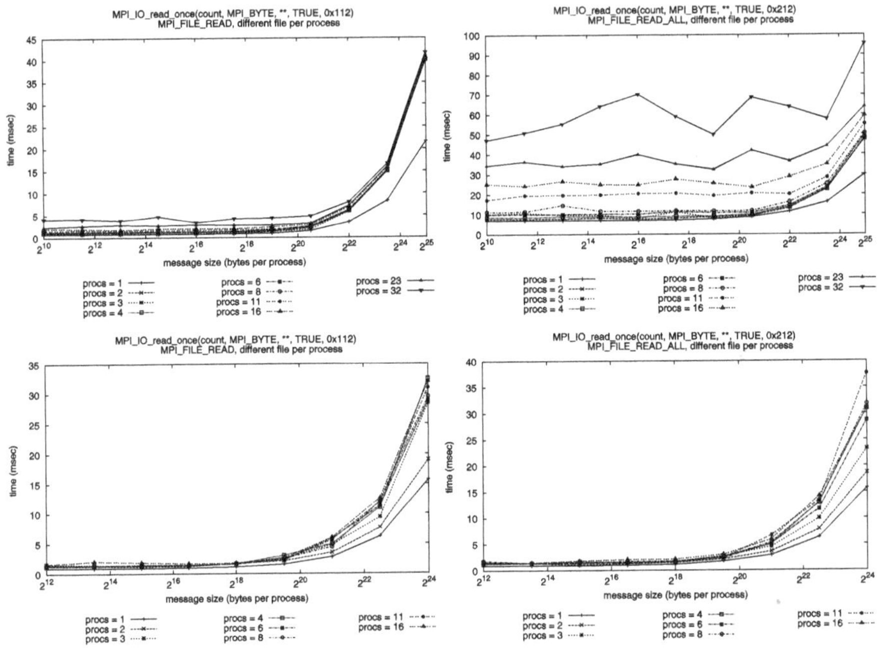

Fig. 4. `MPI_IO_File_read_once`: time needed on XC6000 (top) and XC4000 (bottom) for non-collective (left) and collective (right) version

4.3 An Example Where Caching Makes a Difference

It is almost a triviality to state that one can observe a difference in running times depending on whether the size of a message is smaller or larger than the size of a cache. Yet, the extent to which this is observable sometimes is unexpected.

Figures 5 and 6 show results for measurements where again only 1 process is running and it is reading one file of constant size; but it is read in contiguous chunks of different sizes. Below we have plotted the bandwidth (in MB/s) depending on the chunk size. In Figure 5 the file resided on a "local" file system, and in Figure 6 on a global parallel file system.

4.4 An Example Where File Hints Make a Difference

In (Gropp et al., 1999) the authors present four "levels" of how several processes can access data in a file in such a way that the optimization techniques

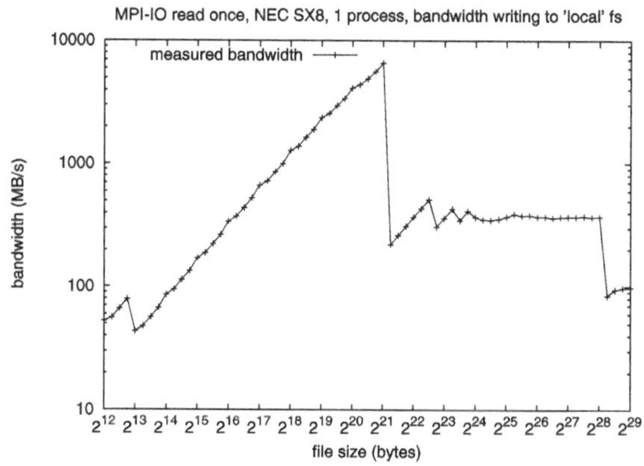

Fig. 5. SX8, a single process reading a file from a "local" file system in chunks of different sizes

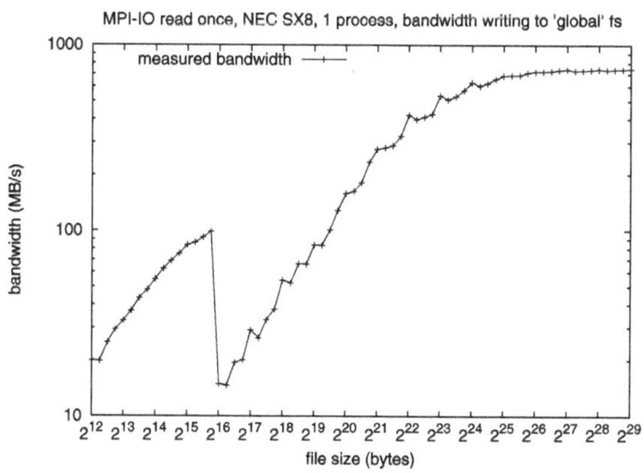

Fig. 6. SX8, a single process reading a file from a "global" file system in chunks of different sizes

data-sieving and *two-phase IO* may help to speed up the file access. Level 0 is the most primitive approach. Level 2 may profit from data sieving and level 3 may profit from two-phase IO.

ROMIO, which is the basis of MPI-IO in many MPI implementations, accepts the file hints `ds_read` and `ds_write` (or something similar named) for data sieving and `cb_read` and `cb_write` for two-phase IO.

Figure 7 shows a comparison of numbers obtained on the XC6000 for some benchmark. Surprisingly, for many chunk sizes, the fastest method is to use level 2 *without* data sieving, and level 3 access *without* collective buffering is better than with it.

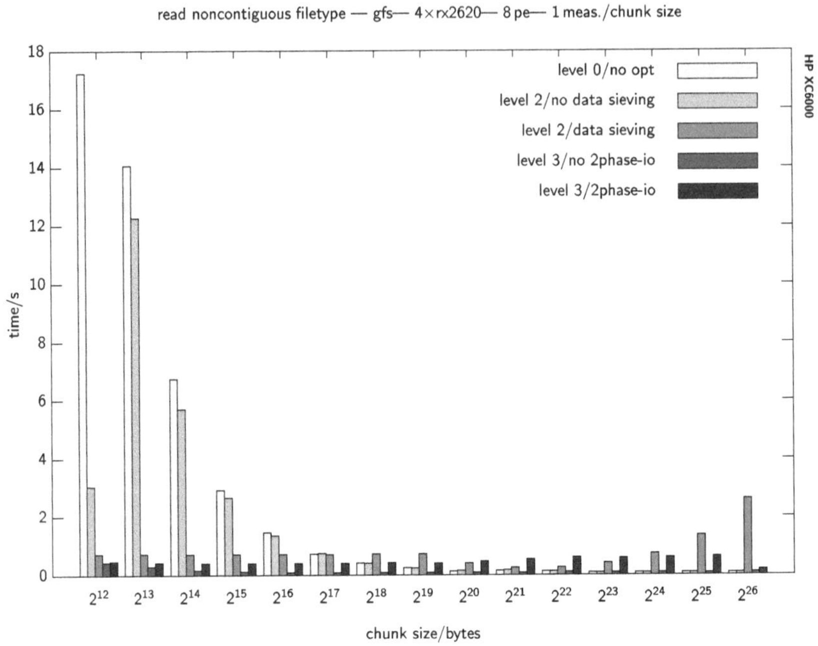

Fig. 7. XC6000, comparing different access strategies with different file hints

5 RFA: Remote Access to Local Hard Disks

The abbreviation RFA is a combination of RMA, "remote memory access" as in MPI's one-sided communication and MMAP, a concept of mapping files to memory which is available e.g. on Linux (but not on the SX8). (Furthermore, if really one-sided communication is required, it will not work on machines which require the use of `MPI_Alloc_mem`.)

This is a small library offering

- a concept called *global file* which is a file that is stored distributed on the local hard disks of (possibly many) computation nodes,
- two functions `gf_open` and `gf_close`, and
- two functions `gf_read` and `gf_write`.

The code is based on the Global Arrays library, which is provided as an example for Remote Memory Access in the book (Gropp et al., 1999, Sec. 6.4).

At the moment this is a proof of concept implementation which assumes that on each local hard disk exactly the same number of data elements is stored. In order to explain the important aspects, we first have a look at the signature of `gf_open`:

```
int gf_open(MPI_Comm ⟨comm⟩, const char* ⟨file_name⟩,
            gf_amode ⟨access_mode⟩, MPI_Datatype ⟨data_type⟩,
            GF* ⟨gf⟩, gf_cmode ⟨comm_mode⟩)
```

The ⟨*access_mode*⟩ can be `GFO_RDONLY`, `GFO_WRONLY` or `GFO_RDWR` as usual. The ⟨*comm_mode*⟩ for exchanging data between processes can be `GFO_ONESIDED` or `GFO_TWOSIDED`.

`gf_open` is a collective operation. All processes participating in ⟨*comm*⟩ first call the POSIX `open()` to open the file locally called ⟨*file_name*⟩ and then `mmap` to map the file contents to memory.

If the ⟨*comm_mode*⟩ for exchanging data between processes is `GFO_ONESIDED`, MPI one-sided communication is used for `gf_read` and `gf_write`. If the ⟨*comm_mode*⟩ for exchanging data between processes is `GFO_TWOSIDED`, then calls to `gf_read` and `gf_write` are

- collective if RFA has been compiled with a flag indicating that `MPI_Alltoall` should be used;
- non-collective if RFA has been compiled without this flag, in which case `MPI_Isends` etc. are used for communication.
  (One may experiment with both alternatives.)

The parameter ⟨*gf*⟩ is used to return something to be use as a file handle for the rfa library. For example the signature of `gf_read` looks like this:

`int gf_open(GF` ⟨*gf*⟩`, void*` ⟨*buffer*⟩`, MPI_Offset` ⟨*offset*⟩`, size_t` ⟨*count*⟩`)`

From the "global file" specified by ⟨*gf*⟩ it will read into ⟨*buffer*⟩ ⟨*count*⟩ many elements, starting at element number ⟨*offset*⟩. For the counting one has to imagine that the elements of all local files are concatenated (in the order of the ranks of the processes belonging to the communicator ⟨*comm*⟩ of the `gf_open` call.

Figure 8 shows results of one of the first benchmarks. Each of 4 processes was reading data belonging to the local file of the process with the "next" rank in ⟨*comm*⟩ using onesided communication. The diagram plots the bandwidth achieved by each process. The two processes which could communicate with another process on *the same node* (using shared memory) are using a much higher band-with than those processes which have to read data through the network.

6 Conclusion

MPI-IO is a part of MPI-2 which has been available at least in some MPI implementations before all the rest of MPI-2 had been implemented. While this gave application developers more freedom to implement their I/O as it would seem appropriate (and "look fast") for quite some time, it also makes it considerably harder to aim for peek performance.

While for example for collective communication there are reasonable performance models, this is much less so for I/O, because there are many more aspects and influences (Rabenseifner et al., 2004). Therefore the results of a

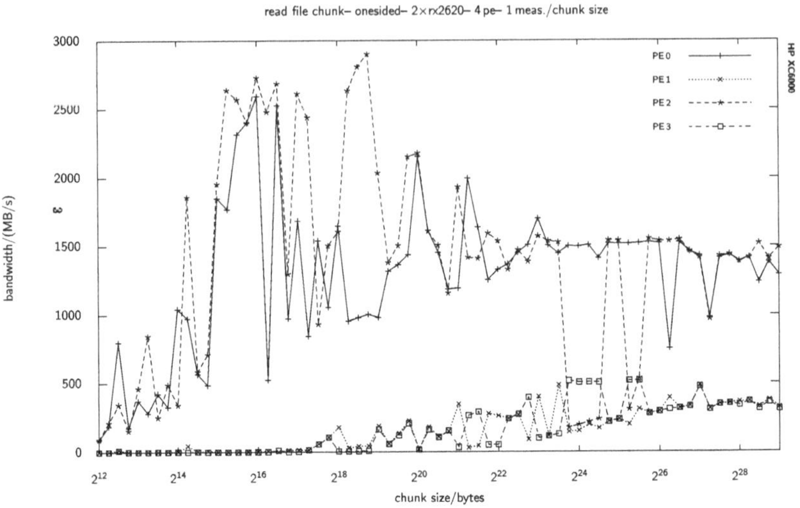

Fig. 8. XC6000, using RFA on 2 nodes with 2 processes each

tool like SKaMPI-5 are essentially for learning what's going on. In addition to the many already provided measurement methods it is very easy to write some new ones, which are specifically tailored and focused to the individual problems an application developer might face.

6.1 Use of Parallel Machines of HWW

The use of the different machines of the HWW is an inevitable and invaluable help for the development of portable software. The widespread use of SKaMPI all over the world can be attributed to the fact that it compiles and runs everywhere where MPI is available.

When carrying out the first measurements, it turned out that SKaMPI-5 by default is not very nice to parallel file systems. Even worse, we had to realize that it is a highly non-trivial task for MPI-IO implementors to bring together the semantics of parallel file systems and the semantics specified by the MPI2 standard for MPI-IO. SKaMPI-5 triggered subtle bugs on several machines. We would like thank all the members at the Computing Center of the University of Karlsruhe for their help, assistance and understanding.

6.2 Outlook

Some directions for further work have already been pointed out in the previous sections. As mentioned in Section 2.2 one could use the easy extensibility of SKaMPI-5 to try to provide functionality offered by other benchmarks. But at least in some interesting cases, like b_eff_io's differentiation between first write and rewrite, this will not be completely straightforward. But we expect the MPI-IO part of SKaMPI-5 to become as useful as its other parts. This includes the solution to problems mentioned in Section 2.3.

The current RFA implementation and variations therefore will be investigated in more detail.

All parts of SKaMPI are available under the GPL and further contributions to SKaMPI-5 are of course welcome.

References

W. Augustin, M.-O. Straub, and T. Worsch. Benchmarking mpi one-sided communication with SKaMPI-5. In W.E. Nagel, W. Jäger, and M. Resch, editors, *High Performance Computing in Science and Engineering '05*, pages 329–340. Springer-Verlag, 2005. ISBN 978-3540283775.

A. Ching, A. Choudhary, K. Coloma, W. Liao, R. Ross, and W. Gropp. Noncontiguous access through MPI-IO. In *Proc. of the IEEE/ACM Int. Symp. on Cluster Computing and the Grid*, pages 104–111, 2003.

W. Gropp, E. Lusk, and R. Thakur. *Using MPI-2: Advanced Featuers of the Message-Passing Interface.* MIT Press, 1999.

W. Loewe, T. McLarty, and C. Morrone. IOR Benchmark. ftp://ftp.llnl.gov/pub/siop/ior/, 2007.

J. Mathes. Unterstützung paralleler Datei-Ein-/Ausgabe unter MPI. Diploma thesis, Fak. f. Informatik, Univ. Karlsruhe, 2008.

J. Nunez. Los Alamos National Lab MPI-IO Test, User's Guide, version 1.0, 2006.

A. Perogiannakis. Leistungsmessung paralleler Ein-/Ausgabe in MPI-Bibliotheken (MOI-IO). Diploma thesis, Fak. f. Informatik, Univ. Karlsruhe, 2006.

R. Rabenseifner, A.E. Koniges, J.-P. Prost, and R. Hedges. The parallel effective I/O bandwidth benchmark: b_eff_io. In Christophe Cerin and Hai Jin, editors, *Parallel I/O for Cluster Computing*, chapter 4, pages 107–132. Kogan Page Ltd, 2004. ISBN 1-903996-50-3.

S. Saini, D. Talcott, R. Thakur, P. Adamidis, R. Rabenseifner, and R. Ciotti. Parallel I/O performance characterization of Columbia and NEC SX-8 superclusters. In *Proc. IPDPS 2007*, 2007.

T. Worsch, R. Reussner, and W. Augustin. Benchmarking collective operations with SKaMPI. In E. Krause and W. Jäger, editors, *High Performance Computing in Science and Engineering '02*, pages 491–502. Springer-Verlag, 2003. ISBN 3-540-43860-2.

Kolibri 360 GmbH
Ferdinand-Jühlke-Straße 7
99095 Erfurt, Deutschland
produktsicherheit@kolibri360.de